Handbook of Ecological Restoration

The two volumes of this handbook provide a comprehensive account of the rapidly emerging and vibrant science of the ecological restoration of both habitats and species.

Habitat restoration aims to achieve complete structural and functional, self-maintaining biological integrity following disturbance. In practice, any theoretical model is modified by a number of economic, social and ecological constraints. Consequently, material that might be considered as rehabilitation, enhancement, reconstruction, or recreation is also included. Re-establishment and maintenance of viable, self-sustaining wild populations are the aims of species-centred restoration.

Principles of Restoration defines the underlying principles of restoration ecology, in relation to manipulations and management of the biological, geophysical and chemical framework. The accompanying volume, *Restoration in Practice*, details state-of-the-science restoration practice in a range of biomes within terrestrial and aquatic (marine, coastal and freshwater) ecosystems. Policy and legislative issues on all continents are also outlined and discussed.

The Handbook of Ecological Restoration will be an invaluable resource to anyone concerned with the restoration, rehabilitation, enhancement or creation of habitats in aquatic or terrestrial systems, throughout the world.

MARTIN PERROW is an ecological consultant at ECON, an organisation which he founded in 1990 to bridge the gap between consultancy and research. He specialises in the restoration and rehabilitation of aquatic habitats, and is a leading exponent of biomanipulation. In addition to presenting his work to the scientific community, Martin has endeavoured to communicate his findings to the general public, through appearances on television and radio and contributions to newspapers and magazines. Martin has travelled extensively, leading eco-tours, and is an award-winning natural history photographer.

ANTHONY (TONY) DAVY is Head of Population and Conservation Biology in the School of Biological Sciences at the University of East Anglia, where he has taught and researched a wide variety of topics in ecology and plant biology. His research interests include genetic variation and the evolutionary and physiological responses of plants to their environments.

Tony is the Executive Editor of the *Journal of Ecology*, Associate Editor of the *Biological Flora of the British Isles*, and was the Honorary Meetings Secretary of the British Ecological Society for a number of years.

Handbook of Ecological Restoration

Volume 1
Principles of Restoration

Edited by
Martin R. Perrow
ECON
University of East Anglia
and
Anthony J. Davy
University of East Anglia

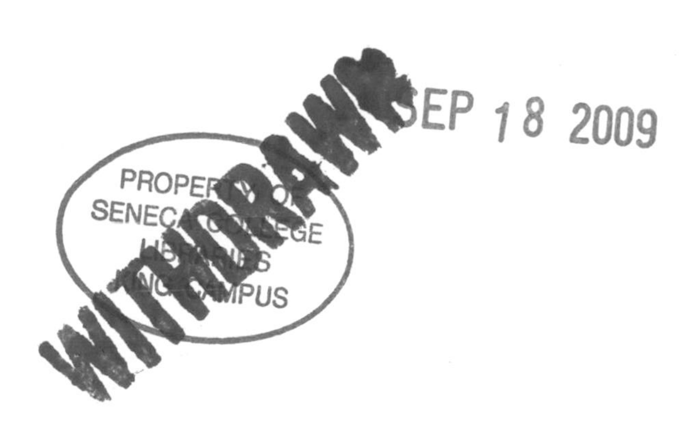

CAMBRIDGE UNIVERSITY PRESS
Cambridge, New York, Melbourne, Madrid, Cape Town, Singapore, São Paulo

Cambridge University Press
The Edinburgh Building, Cambridge CB2 8RU, UK

Published in the United States of America by Cambridge University Press, New York

www.cambridge.org
Information on this title: www.cambridge.org/9780521791281

First published 2002
Reprinted 2003
This digitally printed version 2008

A catalogue record for this publication is available from the British Library

ISBN 978-0-521-79128-1 hardback
ISBN 978-0-521-04983-2 paperback

This volume is dedicated to the late Sidney and Alice Abbs, Linda Davy, Audrey and the late Thomas Davy, and Maureen and Bill Perrow for love and steadfast support.

Contents

Contributors

Michael F. Allen
Center for Conservation Biology
University of California
Riverside CA 92521, USA

Lubomir Bisevac
Department of Environmental Biology
Curtin University of Technology
PO Box U 1987
Perth
WA 6845, Australia

Anthony D. Bradshaw
School of Biological Sciences
University of Liverpool
Liverpool L69 3BX, UK

Karl E. C. Brennan
Department of Environmental Biology
Curtin University of Technology
PO Box U 1987
Perth
WA 6845, Australia

Peter Brimblecombe
School of Environmental Sciences
University of East Anglia
Norwich NR4 7TJ, UK

David P. Butcher
Department of Land-Based Studies
Nottingham Trent University
Brackenhurst
Southwell NG25 0QF, UK

John Cairns Jr
Department of Biology
Virginia Polytechnic Institute and
State University
Blacksburg VA 24061, USA

Anthony J. Davy
Centre for Ecology, Evolution and Conservation
School of Biological Sciences
University of East Anglia
Norwich NR4 7TJ, UK

Jody W. Enck
Department of Natural Resources
Fernow Hall
Cornell University
Ithaca NY 14853, USA

Alan J. Gray
Institute of Terrestrial Ecology
Furzebrook Research Station
Wareham BH20 5AS, UK

Richard J. Hobbs
School of Environmental Science
Murdoch University
Murdoch
WA 6150, Australia

Karen D. Holl
Department of Environmental Studies
University of California
Santa Cruz CA 95064, USA

Michael J. Jackson
School of Biological Sciences
University of East Anglia
Norwich NR4 7TJ, UK
and
Department of Forest Sciences
University of British Columbia
Vancouver
British Columbia V6T 1Z4, Canada

David A. Jasper
Soil Science and Plant Nutrition
School of Agriculture
University of Western Australia
Nedlands
WA 6009, Australia

Carl G. Jones
Mauritian Wildlife Foundation
Black River
Mauritius
Indian Ocean

Jillian C. Labadz
Department of Land-Based Studies
Nottingham Trent University
Brackenhurst
Southwell NG25 0QF, UK

David W. Macdonald
Wildlife Conservation Research Unit
Department of Zoology
University of Oxford
Oxford OX1 3PS, UK

Jonathan D. Majer
Department of Environmental Biology
Curtin University of Technology
PO Box U 1987
Perth
WA 6845, Australia

Robert H. Marrs
School of Biological Sciences
Derby Building
University of Liverpool
Liverpool L69 3BX, UK

Thomas P. Moorhouse
Wildlife Conservation Research Unit
Department of Zoology
University of Oxford
Oxford OX1 3PS, UK

Malcolm D. Newson
Department of Geography
University of Newcastle upon Tyne
Newcastle upon Tyne NE1 7RU, UK

Martin R. Perrow
ECON, Ecological Consultancy
School of Biological Sciences
University of East Anglia
Norwich NR4 7TJ, UK

John Pitlick
Department of Geography
University of Colorado–Boulder
260 UCB
Boulder CO 80309, USA

John S. Richardson
Department of Forest Sciences
University of British Columbia
Vancouver
British Columbia V6T 1Z4, Canada

Wilhelm Ripl
Department of Limnology
Technische Universität Berlin
HS1
D-14195 Berlin, Germany

David A. Sear
Department of Geography
University of Southampton
Highfield
Southampton SO17 1JB, UK

José Maria Cardoso da Silva
Universidade Federal de Pernambuco
Centro de Ciências Biológicas
Departamento de Zoologia
Recife PE 50670–901, Brazil

Dennis Sinnott
Department of Environmental Management
University of Central Lancashire
Preston PR1 2HE, UK

Martin Søndergaard
Department of Lake and Estuarine Ecology
National Environmental Research Institute
PO Box 314
DK-8600 Silkeborg, Denmark

John A. Strand
Rural Economy and Agricultural Society
Lilla Böslid
S-310 31 Eldsberga, Sweden

Fran H. Tattersall
Wildlife Conservation Research Unit
Department of Zoology
University of Oxford
Oxford OX1 3PS, UK

Mark L. Tomlinson
ECON, Ecological Consultancy
School of Biological Sciences
University of East Anglia
Norwich NR4 7TJ, UK

Peter D. Vickery
Department of Wildlife Conservation
University of Massachusetts
Amherst MA 01773, USA

Stefan E. B. Weisner
Department of Limnology
Lund University
Ecology Building
S-223 62 Lund, Sweden

Steve G. Whisenant
Department of Rangeland Ecology and Management
Texas A&M University
College Station TX 77843, USA

Klaus-Dieter Wolter
Department of Limnology
Technische Universität Berlin
HS1
D-14195 Berlin, Germany

John C. Zak
Ecology Program
Department of Biological Sciences
Texas Technical University
Lubbock TX 79409, USA

Luis Zambrano
Colección Nacional de Peces
Instituto de Biología, UNAM
Adpo Post. 70-153
México DF

Foreword

Ecological restoration is necessary because the relationship between human society and natural systems is not as mutualistic as it should be. Although natural systems constitute the biological life support system of the planet, societal practices do not acknowledge human society's dependence on these ecosystems. The human aspiration to live sustainably on the planet must recognise that the elimination of many of the planet's species and habitats is not a sustainable practice. A necessary first step to correcting this situation is to achieve a balance between the rates of damage and the restoration of ecosystems. In other words, the anthropogenic forces destroying natural systems must be restrained and the restorative practices increased. Arguably, the most persuasive argument supporting ecological restoration is the ethical responsibility for the well-being of other life forms. Even in the unlikely event that humans could survive without undomesticated species, would they really care to do so? Additionally, humans should be making decisions not only for themselves but to benefit their descendants and to enable them to lead a quality life.

Effective ecological restoration must involve all major levels of ecological organisation from component species to entire systems. Ideally, the science should be robust at all levels; however, in a newly emerging field, the evidence base will obviously benefit from additional research. Volume 1 of the *Handbook* discusses ecological principles from the systems level of biological organisation to the species components. Furthermore, the dynamic aspects, including evolution, are given the attention they deserve.

Most restoration projects require manipulation of the physical and chemical environment, as well as the biota, and all are likely to require sound professional judgement on a continuing basis. Temporal scales are important in restoration projects, and both temporal and spatial scales are almost certain to increase dramatically in the future. Attention is also given in this volume to both structural and functional attributes of both species and ecosystems. All of these aspects of restoration are followed by a discussion on monitoring, a procedure to ensure that previously established quality-control conditions have been met.

Ecological restoration is a multidimensional activity. Volume 1 provides a sound overview of the ecological principles in each of the major dimensions over a broad range, including genetic variation, island biogeography, corridors between habitat 'islands', colonisation dynamics and effects of anthropogenic stresses. Almost all restoration projects involve some manipulation of the physical environment; in some cases, this will be substantial. Both natural processes and artificial means are discussed; artificial means may include removal of sediments from lakes or manipulation of water tables for terrestrial systems. Manipulation of the chemical environment may include reduction of toxicity, exchanges between sediment and water column, and manipulation of air quality.

The rapidly emerging field of restoration ecology is firmly rooted in science. Professor Bradshaw remarked many years ago that restoration is the 'acid test' of ecological science. However, implementing ecological restoration will be the 'acid test' of human society's relationship with other species and the interdependent web of life that they collectively represent. The practices of human society often do not proclaim that humans are 'a part' of the web, but 'apart' from it. The interdependent web of life existed for billions of years before humans appeared on the planet. Arguably, humans are the first species to influence so markedly the fate of so many other species, often in deleterious ways. Moreover, humans have a choice either to restore damaged ecosystems and their integrity or to affect other species adversely and have adverse effects upon ecosystem integrity.

The *Handbook of Ecological Restoration* provides both the science and the methods and procedures for implementing restoration. Ecological restoration provides a means for partially offsetting the environmental surprises of human society's vast uncontrolled experiment with the planet's biosphere. However, the rate and extent of the healing processes are far behind the rate and extent of the ecologically damaging processes. One hopes the *Handbook* will play a major role in addressing this unfortunate situation and will be a major step toward a mutualistic relationship between human society and natural systems, which are so important to the quality of human life.

John Cairns Jr.

Preface

Restoration ecology is a discipline whose time has come. In recent years it has advanced almost explosively on a range of fronts, with attempts to restore habitats, species and human cultural values. We view it as timely to try and provide researchers and practitioners with a comprehensive review – the 'state of the science', shortly after the turn of the century. The increasing need for ecological restoration is an inevitable consequence of the relentless growth of the global human population, its increasing cultural and technological sophistication and the concomitant consumption of resources. Our central aspiration to understand ecosystems sufficiently well to be able to restore or replace them may be regarded with dismay by those who believe that it will serve to encourage destruction in the first place. However, any knowledge potentially can be abused and the benefits of restoration surely greatly outweigh any abuse. Clearly, at any site, conservation of the existing organisms in their undamaged environment is unequivocally preferable to subsequent restoration *in situ*, or reconstruction of an equivalent system elsewhere by way of mitigation. Unfortunately, conservation cannot always be wholly successful and so long as its effectiveness is less than perfect, the world's biodiversity is on an inexorable 'ratchet' to extinction; the main uncertainty is the rate at which this will happen, and restoration will be the only means to counteract the decline.

Although ecology is a relatively young science, ecological restoration has a history that is nearly as old as ecology itself, even though its distinctiveness has become widely appreciated only in the last 20 years. For many years the prime focus of ecology involved mainly analytical and descriptive approaches. As has now been documented many times, Aldo Leopold's pioneering reconstruction of tallgrass prairie at the University of Wisconsin Arboretum, which started in 1935 in the aftermath of economic depression and the development the Dust Bowl in the American Midwest, signalled the advent of a truly synthetic approach to ecology. The current output of the journal *Restoration Ecology* alone is ample testimony to Leopold's vision and innovation. Ecological restoration is inevitably a broader church than its scientific core, because of the economic, cultural, landscape, aesthetic and political dimensions to the ownership and use of land or water; ecologists pursuing the science, ourselves included, have often found it convenient to disregard influences of the dominant, human species. This is something that we cannot sensibly continue to do. Having said that, we must also emphasise that science is a central, essential prerequisite for restoration. Unless we understand the way in which ecosystems work and how their organisms interact with their environments, there can be no restoration worthy of the name. Equally, however, we hope that this volume illuminates graphically the important contribution that the restoration approach has to offer towards the advancement of the science of ecology. This contribution partly arises from the power of the inherently synthetic, holistic approach to test our understanding of ecosystem structure and function (as exemplified by Leopold's 'law of intelligent tinkering' and Bradshaw's 'acid test'). Of no less importance from our perspective is the strong tradition of experimentation, particularly in the field, that is the key to unravelling the complex problems often posed by restoration. We hope that the full potential of the restoration approach will be exploited increasingly by ecologists.

This volume is the first of the two that make up the *Handbook*. It is concerned with the essential principles of ecological restoration; in other words, we aim to provide a coherent account of the science that underpins and informs restoration. This volume is also designed as a logical precursor to its complement, which covers the more practical aspects of restoring many particular types of ecosystem in considerable detail.

The following account of 'principles' comprises 21 chapters organised into five parts, dealing sequentially with the overall background to restoration, the manipulation of the physical environment, the manipulation of the chemical environment, the manipulation of the biota, and the monitoring and appraisal of restored systems. We recognise that the allocation of complex issues and topics to any such logical framework involves a certain degree of arbitrariness; we have striven to offset this in the detailed coverage and coherence of chapters, or alternatively by cross-referencing between them. Part 1 ('The background') consists of five chapters that discuss, respectively, the philosophy of restoration, the rationale for restoration, the ecological context from a landscape perspective, the ecological context from a species population perspective and the evolutionary context from a species perspective. In subsequent parts terrestrial and aquatic systems and their respective biota tend to be dealt with in separate chapters, because of the necessarily different problems and approaches encompassed. Hence Part 2 ('Manipulation of the physical environment') has three chapters that deal specifically with terrestrial systems, wetlands and still waters, and the fluvial geomorphology of running water. Part 3 ('Manipulation of the chemical environment') likewise comprises three chapters that examine the chemical environment of the soil, the chemical treatment of water and sediments and atmospheric chemistry. Part 4 ('Manipulation of the biota') is the largest element of the book, with nine chapters covering the particular issues relating to different types of organism and their various roles in restoration: plant populations and communities in terrestrial systems; plants in aquatic systems; micro-organisms; terrestrial invertebrates; aquatic invertebrates; fish; reptiles and amphibians; birds; and mammals. The volume concludes with Part 5, comprising a single chapter addressing the all-important topic of monitoring and appraisal.

Our contributors provide any strengths the book may have. The weaknesses in its structure and scope are ours. We are deeply indebted to all those who have contributed their time and efforts in what has proved to be a massive undertaking. It has been an honour to have Tony Bradshaw, John Cairns Jr. and Erik Jeppesen on the editorial board and we are grateful for their unfaltering support throughout the project's long history. We are grateful to CUP for taking up the challenge when we needed a new publisher, and particularly to Alan Crowden, who made the transition seamless. Shana Coates provided editorial assistance from the press. We are also grateful to the rest of the CUP team and for the essential support and hard work of Mark Tomlinson of ECON in the preparation of manuscripts. We thank the reviewers, many of whom were also authors, for generously sharing their insights and suggesting improvements to chapters.

Tony Davy and Martin Perrow

Part I • The background

1 • Introduction and philosophy

ANTHONY D. BRADSHAW

DEGRADATION AS A UNIVERSAL CONCOMITANT OF HUMAN SOCIETIES

The place of human beings in nature has always been ambivalent. At the present time it is easy to see that we need to cherish nature because, for so many reasons, it supports us. The physiognomy and well-being of our planet depends on its living skin, without which the land would become unstable and, more importantly, its atmosphere would lose its crucial oxygen content, and life as we know it would perish.

Yet at the same time human beings have never been able to live in any sort of stability and comfort without subduing nature to some extent. The early hunter–gatherer activity caused a small amount of damage to existing ecosystems but this was well within the recuperative powers of nature. But, as a result of increasing human populations, inevitably requiring more resources, and the development of techniques of exploitation, such a balanced situation was not to last. The domestication of animals allowed the stocking of selected areas, the domestication of plants allowed them to be grown to order, both permitting higher densities of human populations, with the formation of coherent societies, but with the concomitant destruction of original ecosystems as an incidental necessity.

It is from these considerations that the Judaeo-Christian justification of a domineering approach to nature stems – that man should 'subdue [the earth]; and have dominion over the fish of the sea, and over the fowl of the air, and over every living thing that moveth upon the earth.' (Genesis 1:28) – no sign of any kindness or respect. This attitude remains a fundamental part of the exploitive capitalist cultures which dominate the world, which Aldo Leopold eloquently deplored (Leopold, 1949). Despite Leopold's wish for an environmental ethic, human existence is not possible without damage – every individual needs space to move and to live with protection.

The situation has been exacerbated by demands for greater comfort, affecting the size and luxury of our living space. How many people now would accept a housing standard of 3.5 square metres of space per person not long ago applicable in China, or accept a building without an elaborate electricity network, of copper wires, running through it? For all this, quite apart from increasing amounts of simple building materials such as clay and cement, special raw materials such as iron, copper, chromium and nickel have had to be found. And the major source of energy for achieving this construction as well as for the comfort of warmth has been coal. All these resources lie in the ground, and can only be got at by major land disturbance.

This population has, of course, developed other demands, such as for increasing mobility and therefore for more roads, for increasing recreation and therefore for increased recreation space. Winding rural roads have turned into motorways, and narrow paths through sand dunes into eroded blowouts. Damage at first insignificant and unnoticed has had an uncomfortable way of becoming catastrophic.

To support this burgeoning population arable agriculture has spread to almost every part of the globe. The land surface has not been completely destroyed. But the original vegetation has been removed, the land surface cultivated and crops grown. At the same time there has been the widespread grazing of animals, and the felling of trees for timber. These may not initially have disturbed the soil, but have grossly altered the vegetation and allowed widespread erosion of the soil and the land surface, and extended to the disturbance and degradation of adjacent water (Jacks & Whyte, 1939; Lowdermilk, 1953).

TOTAL LOSSES OF LANDSCAPES, ECOSYSTEMS AND SPECIES: OF USEFUL LAND AND COMMUNITIES

It was the most severe types of degradation from industrial operations which particularly caught our attention, because their effects were so radical, in areas often close to where people live – the people who had once gained their livelihood from what had been there. They have been a particular challenge, because the original ecosystems have been totally destroyed. However now, because these areas are treated and our ecological sensitivity is sharper, we are turning our attention to less degraded situations, where perhaps only a few species have been lost. They are a more subtle challenge.

Each country discovered the problem growing within it. So there is no one global account. But some indications of how the worst problems were realised and began to be faced up to, is to be found, for Britain, in Whyte & Sisam (1949) and Senior (1964) and for North America in Caudill (1976) and Gunn (1995).

If nothing is done, these different types of degradation accumulate, with obvious consequences. But the background is not one of a static world population, but one that continues to grow substantially. So the need for restoration as an integral part of the philosophy and activities of all human societies is crucial (Cairns, 1995; this volume).

AN ECOLOGICAL VIEW OF THE DAMAGE AND ITS RESTORATION: DEFINITIONS

On every area of the land surface of the world there is normally a cover of vegetation. It is made up of a characteristic spectrum of species. That vegetation roots into a mixture of weathered rock and organic matter derived from the plants, the soil. The soil and vegetation support a characteristic population of animals and micro-organisms. The whole is both supported and challenged by the local climate. Aquatic systems have an analogous structure.

All this constitutes an ecosystem (Tansley, 1935). There are no bounds to an ecosystem, because the earth's living cover is effectively continuous. But we arbitrarily recognise different ecosystems based on location or on species. The crucial characteristic of an ecosystem is that its components, which could be a major element such as soil or an individual species, interact physically, so that a change in one component can lead to corresponding changes in others. At the same time the components share, and circulate between them, materials. Many important materials, such as phosphorus, cycle almost completely within an ecosystem and do not move out; others such as nitrogen are more labile.

An ecosystem therefore has two major attributes, structure and function, each made up of different elements. They can be used to define and illustrate the damage that ecosystems can suffer (Magnuson *et al.*, 1980; Bradshaw, 1987*a*) (Fig. 1.1). An original ecosystem will typically (although not always) have high values for both. Degradation drives one or both attributes downwards, often to nearly nothing. If the area is left to its own devices, the natural processes of primary succession will restore the ecosystem to its starting-point (Miles & Walton, 1993). It is these processes that, unaided, originally produced the variety of natural ecosystems we see today.

In aquatic systems if the disturbing factor is removed natural recovery can be quick, but on land natural succession is a rather slow process and may be very slow where the degradation has left an unnatural inhospitable substrate. It is this that prompts us to undertake restoration. We expect a short timescale. It is hardly acceptable that children living beside a colliery spoil heap should spend the whole of their childhood with this as their environment. Half would be more than enough. If a childhood is from two to 12 years, this means that the restoration should take no more than five years.

But what should constitute 'restoration'? In ecological restoration four words are in common use – *restoration, rehabilitation, remediation, reclamation* – although there are others (Bradshaw, 1997*a*). Perhaps the most complete guidance is from the *Oxford English Dictionary* (1971).

The relevant definition of *restoration* is: 'the act of restoring to a former state or position . . . or to an

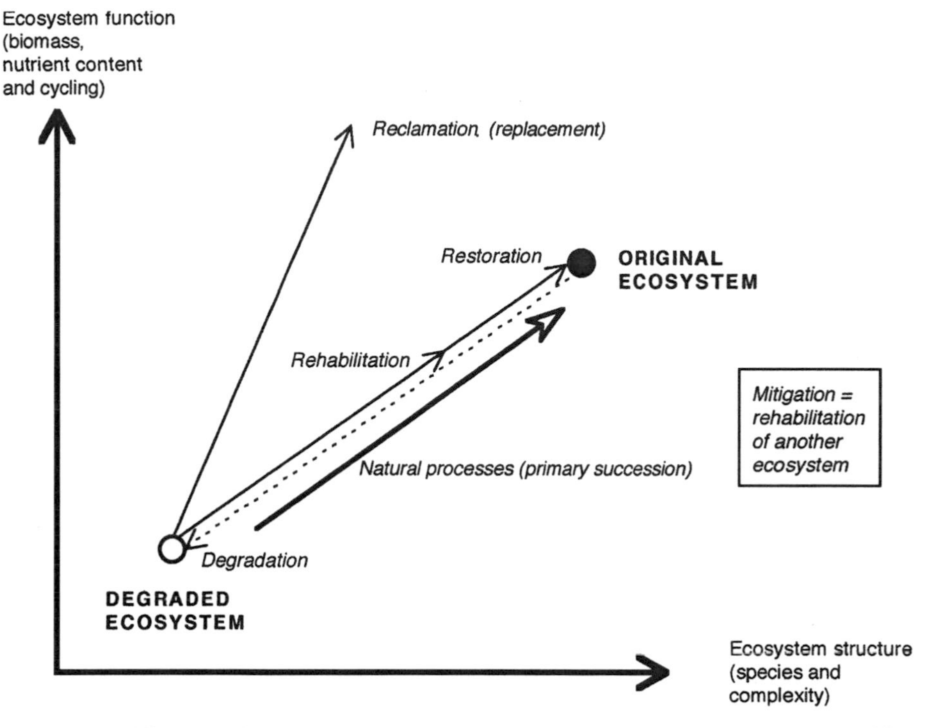

Fig. 1.1. The different options for the improvement of a degraded ecosystem can be expressed in terms of the two major characteristics of structure and function. When degradation occurs both characteristics are usually reduced, although not necessarily equally. Used in its narrow sense, *restoration* implies bringing back the ecosystem to its original or previous state in terms of both structure and function. There are then a number of other alternatives, including *rehabilitation* in which this is not totally achieved, and *replacement* of the original by something different. All of these alternatives are covered, by many people, by the general term *reclamation*. *Mitigation* is a different consideration. (See Bradshaw, 1987*a*.)

unimpaired or perfect condition'. To *restore* is: 'to bring back to the original state . . . or to a healthy or vigorous state'. There is the implication of returning to an original state, and to a state that is perfect and healthy. This seems to be the way in which we continue to use the word on both sides of the Atlantic (Box, 1978), even although it does have perfectionist implications (Francis *et al.*, 1979). In fact the term has been taken to have substantial implications:

Restoration is defined as the return of an ecosystem to a close approximation of its condition prior to disturbance. In restoration, ecological damage to the resource is repaired. Both the structure and the functions of the ecosystem are recreated. Merely recreating the form without the functions, or the functions in an artificial configuration bearing little resemblance to a natural resource, does not constitute restoration. The goal is to emulate a natural, functioning, self-regulating system that is integrated with the ecological landscape in which it occurs. (National Research Council, 1992.)

This has been broadened, perhaps excessively, by the recently formed Society for Ecological Restoration to 'ecological restoration is the process of assisting the recovery and management of ecological integrity. Ecological integrity includes a critical range of variability in biodiversity, ecological processes and structures, regional and historical context, and sustainable cultural practices.' (Society for Ecological Restoration, 1996.)

Rehabilitation is defined in the *Oxford English Dictionary* as: 'the action of restoring a thing to a previous condition or status'. This appears rather similar to restoration, but there is little or no implication of perfection. In common usage, something that is rehabilitated is not expected to be in as original or as healthy a state as if it had been restored (Francis *et al.*, 1979). For this reason the word can be used to indicate any act of improvement from a degraded state (Box, 1978; Wali, 1992).

Remediation is: 'the act of remedying'. To *remedy* is: 'to rectify, to make good'. Here the emphasis is on the process rather than on the end point reached.

Reclamation is a term used by many practitioners, especially in Britain but also in North America. It is defined as: 'the making of land fit for cultivation'. But to *reclaim* is given as: 'to bring back to a proper state'. There is no implication of returning to an original state but rather to a useful one. *Replacement* may therefore be involved. To *replace* is: 'to provide or procure a substitute or equivalent in place of' (although an alternative meaning is to 'to restore').

Enhancement is sometimes used in the USA to indicate the establishment of an alternative ecosystem (Pratt & Stevens, 1992). This seems an unsatisfactory use of the word which is defined in the *Oxford English Dictionary* as: 'to raise in degree, heighten, intensify; or to increase in value, importance, attractiveness'. This is in fact the use suggested by Francis *et al.* (1979). There is no implication of making something bad better, but of making something already good better.

Mitigation is often used when restoration is considered. But it is nothing directly to do with restoration. To *mitigate* is to 'appease . . . or to moderate the heinousness of something'. So although mitigation can be an outcome of restoration (or rehabilitation or reclamation) it is a separate consideration. It may well involve the improvement of quite another ecosystem. All of these options are represented in Fig. 1.1.

HUMAN PERCEPTIONS OF WHAT IS NEEDED: PHILOSOPHICAL PROBLEMS

It is easy to believe that restoration should only have one aim, that of restoration in a narrow sense – to put back faithfully what was there before. But this has led to endless problems, particularly in North America. Should it be what was there just before the area was damaged – which may be a cultural landscape heavily influenced by human beings, or should it be what was there before human beings started to modify it? In the light of Fig. 1.1 it would seem that, whether the term restoration or the more general term reclamation is used, all types of restoration should be acceptable, and what is actually carried out should be based on pragmatic considerations.

There may be important reasons, such as maintenance of biodiversity, for putting back precisely the original native ecosystem, as in the Australian mineral sand industry. But there may equally be good reasons for putting back the man-made ecosystem which existed previously, as arable agriculture after surface mining of coal in Illinois. But it may be imperative to establish a vegetation cover that will rapidly provide stability and prevent surface erosion, as in the establishment of grassland after the Aberfan mining disaster in South Wales. Equally it may be entirely justifiable to allow natural successional processes to go where they will, which may not be to replace what was there before, as in many gravel pits and hard rock quarries, thereby contributing to biodiversity (Bradshaw & Chadwick, 1980).

Context should influence our understanding of what restoration is. As Parker & Pickett (1997) argue, restoration is more to be considered as a process, with the degree of active intervention being determined by contextual circumstances. This is the attitude adopted in the recent overview (Fox *et al.*, 1998), significantly entitled 'Land reclamation: achieving sustainable benefits'.

This pragmatic approach is, however, only a small step away from serious philosophical problems. Is it possible, anyway, to achieve restoration in the narrow sense without an inordinate passage of time? If it is not, are restoration and restoration ecology misleading terms? This is the view taken by some philosophers, notably Katz (1996) and Elliot (1997) – restoration is faking nature. This attitude seems to miss the essential qualities of restoration, that whatever is carried out involves nature and natural processes, in the achievement of a functioning ecosystem. Restoration would only be faking nature if the ground was being covered with synthetic plastic turf.

The trouble is that common usage has accepted restoration as the term to describe a variety of different operations/objectives. At one stage the normal word to cover these was 'reclamation', seen from the titles of seminal books such as Oxenham (1966) *Reclaiming Derelict Land*; Hutnik & Davies (1973) *Ecology and Reclamation of Devastated Land*; Schaller & Sutton (1978) *Reclamation of Drastically Disturbed Lands*. For the UK Government 'reclamation' has always been the standard term. If, now, we have to live with 'restoration' it will be a reasonable usage if it is applied not just to putting back what was there before but 'as a blanket term to describe all those activities which seek to upgrade damaged land or to recreate land that has been destroyed and to bring it back into beneficial use, in a form in which the biological potential is restored.' (Bradshaw & Chadwick, 1980, p. 2). This will involve attending to all the component ecosystem characteristics, well discussed by Ehrenfeld & Toth (1997). With the overlay of 'ecological', restoration can (and should) therefore be applied to the individual components of an ecosystem, as illustrated in Fig. 1.1, rather than to the ecosystem in its entirety (Ehrenfeld, 2000). The value of using the word restoration in this way is that it encourages us to think of all the fundamental processes by which an ecosystem works (Cairns, 1988), and the importance of natural processes in restoration, especially those involved in succession (Bradshaw, 1997*b*; Parker, 1997).

It also means that people from many different disciplines are likely to be involved in restoration, from the ecologist to the engineer, from the landscape architect to the community worker, and indeed from the politician to the ordinary person. Successful restoration can only be achieved if many different disciplines are involved in the process (Higgs, 1997).

RESTORATION AS AN ACID TEST FOR ECOLOGY

When can restoration be deemed to have been achieved, assuming that the word restoration is being used in the narrow sense? The only answer is that this will be a matter of arbitrary definition, because the end point ecosystem is not a fixed entity. But it will require careful experimental design (Michener, 1997). If restoration is being used in a wide sense, then there is more flexibility, because achievement can be in relation to a single character or process. In either case the purist may decide that perfect restoration is an unattainable end point and that what we should expect and settle for is rehabilitation.

Nevertheless the level of achievement in restoration is something about which we should be concerned, not only to fight back accumulating damage, but because it is a fundamental test of our ecological understanding. Restoration is not only a problem-solving matter; it is a tool for ecological research (Jordan *et al.*, 1987). It is not difficult to take a piece of machinery to pieces without understanding it properly. But putting it together again and making it work is a test of our real understanding. There are innumerable papers in which mechanisms and qualities purporting to be important in ecosystems have been described. To show whether they are or not is another matter, and there is always the uncertainty as to whether some other undescribed character is not much more important. Restoration is a crucial test – if we put an ecosystem together and it does not work, or does not work properly, our knowledge is faulty (Bradshaw, 1987*b*), a concept now widely accepted (e.g. Niering, 1997). There is only one problem, arising from the self-restoring properties of ecosystems. We might put it together wrongly and yet it could still begin to work. Under such circumstances it is necessary to include a rate of achievement as a criterion. In the absence of any treatment, self-restoration of almost any ecosystem, at least on land, is relatively slow. So our test should look for success within a relatively short period of time.

Because of its significance, restoration ecology should be an expanding subject within ecology. In many ways it is, with two dedicated journals, *Restoration Ecology* and *Ecological Restoration*, each with increasing circulation, and a continuous flow of papers into other journals and edited volumes, and a focused professional society. Thirty years ago the flow was minuscule, although a few overviews, on reclamation, were beginning to appear, already mentioned. Yet despite its clear heuristic value, because of its origins it remains a discipline rooted in practice. But readers can judge whether a 'quite new ecological science is emerging to give us the necessary knowledge to take control of our flora and fauna' . . . to become . . . 'not just tinkerers but craftsmen and engineers' (Harper, 1987).

REFERENCES

Box, T.W. (1978). The significance and responsibility of rehabilitating drastically disturbed land. In *Reclamation of Drastically Disturbed Lands*, eds. F.W. Schaller & P. Sutton, pp. 1–10. Madison, WI: American Society of Agronomy.

Bradshaw, A.D. (1987*a*). The reclamation of derelict land and the ecology of ecosystems. In *Restoration Ecology*, eds. W.R. Jordan, M.E. Gilpin & J.D. Aber, pp. 53–74. Cambridge: Cambridge University Press.

Bradshaw, A.D. (1987*b*). Restoration: an acid test for ecology. In *Restoration Ecology*, eds. W.R. Jordan, M.E. Gilpin & J.D. Aber, pp. 23–29. Cambridge: Cambridge University Press.

Bradshaw, A.D. (1997*a*). What do we mean by restoration? In *Restoration Ecology and Sustainable Development*, eds. K. M. Urbanska, N.R. Webb & P.J. Edwards, pp. 8–16. Cambridge: Cambridge University Press.

Bradshaw, A.D. (1997*b*). Restoration of mined lands: using natural processes. *Ecological Engineering*, **8**, 225–269.

Bradshaw, A.D. & Chadwick, M.J. (1980). *The Restoration of Land*. Oxford: Blackwell Scientific Publications.

Cairns, J. (1988). Increasing diversity by restoring damaged ecosystems. In *Biodiversity*, ed. E.O. Wilson, pp. 333–343. Washington, DC: National Academy Press.

Cairns, J. (ed.) (1995). Restoration ecology: protecting our national and global life support systems. In *Rehabilitating Damaged Ecosystems*, ed. J. Cairns, pp. 1–12. Boca Raton, FL: Lewis Publishers.

Caudill, H.M. (1976). *The Watches of the Night*. Boston, MA: Little, Brown.

Ehrenfeld, J.G. (2000). Defining the limits of restoration: the need for realistic goals. *Restoration Ecology*, **8**, 2–9.

Ehrenfeld, J.G. & Toth, L.A. (1997). Restoration ecology and the ecosystem perspective. *Restoration Ecology*, **5**, 307–317.

Elliot, R. (1997). *Faking Nature*. London: Routledge.

Fox, H.R., Moore, H.M. & McIntosh, A.D. (eds.) (1998). *Land Reclamation: Achieving Sustainable Benefits*. Rotterdam: Balkema.

Francis, G.R., Magnuson, J.R., Regier, H.A. & Talhelm, D.R. (1979). *Rehabilitating Great Lakes Ecosystems*. Ann Arbor, MI: Great Lakes Fishery Commission.

Gunn, J.M. (ed.) (1995). *Restoration and Recovery of an Industrial Region: Progress in Restoring the Smelter-Damaged Landscape near Sudbury, Canada*. New York: Springer-Verlag.

Harper, J.L. (1987). The heuristic value of ecological restoration. In *Restoration Ecology*, eds. W.R. Jordan, M.E. Gilpin & J.D. Aber, pp. 35–45. Cambridge: Cambridge University Press.

Higgs, E. (1997). What is good ecological restoration? *Conservation Biology*, **11**, 338–348.

Hutnik, R.J. & Davies, G. (eds.) (1973). *Ecology and Reclamation of Devastated Land*. New York: Gordon & Breach.

Jacks, G.V. & Whyte, R.O. (1939). *The Rape of the Earth*. London: Faber & Faber.

Jordan, W.R., Gilpin, M.E. & Aber, J.D. (1987). Restoration ecology: ecological restoration as a technique for basic research. *In Restoration Ecology*, eds. W.R. Jordan, M.E. Gilpin & J.D. Aber, pp. 3–21. Cambridge: Cambridge University Press.

Katz, E. (1996). The problem of ecological restoration. *Environmental Ethics*, **18**, 222–224.

Leopold, A. (1949). *A Sand County Almanac*. New York: Oxford University Press.

Lowdermilk, W.C. (1953). *Conquest of the Land over Seven Thousand Years*. US Department of Agriculture Bulletin 99. Washington, DC: US Government Printing Office.

Magnuson, J.J., Regier, H.A., Christien, W.J. & Sonzogi, W.C. (1980). To rehabilitate and restore Great Lakes ecosystems. In *The Recovery Process in Damaged Ecosystems*, ed. J. Cairns, pp. 95–112. Ann Arbor, MI: Ann Arbor Science.

Michener, W.K. (1997). Quantitatively evaluating restoration experiments: research design, statistical design, statistical analysis, and data management considerations. *Restoration Ecology*, **5**, 324–338.

Miles, J. & Walton, D.W.H. (1993). *Primary Succession on Land*. Oxford: Blackwell.

National Research Council (1992). *Restoration of Aquatic Ecosystems: Science, Technology and Public Policy*. Washington, DC: National Academy Press.

Niering, W.A. (1997). Editorial: Human-dominated ecosystems and the role of restoration ecology. *Restoration Ecology*, **5**, 273–274.

Oxenham, J.R. (1966). *Reclaiming Derelict Land*. London: Faber & Faber.

Oxford English Dictionary (1971). *The Oxford English Dictionary.* Oxford: Oxford University Press.

Parker, V.T. (1997). Scale of successional models and restoration objectives. *Restoration Ecology*, **5**, 301–306.

Parker, V.T. & Pickett, S.T.A. (1997). Restoration as an ecosystem process: implications of the modern ecological paradigm. In *Restoration Ecology and Sustainable Development*, eds. K.M. Urbanska, N.R. Webb & P.J. Edwards, pp. 17–32. Cambridge: Cambridge University Press.

Pratt, J.R. & Stevens, J. (1992). Restoration ecology: repaying the national debt. In *Proceedings of the High Altitude Revegetation Workshop no.10*, eds. W.G. Hassell, S.K. Nordstrom, W.R. Keammerer & W.J. Todd, pp. 40–49. Fort Collins, CO: Colorado State University.

Schaller, F.W. & Sutton, P. (eds.) (1978). *Reclamation of Drastically Disturbed Lands.* Madison, WI: American Society of Agronomy.

Senior, D. (1964). *Derelict Land: A Study of Industrial Dereliction and How It May Be Redeemed.* London: Civic Trust.

Society for Ecological Restoration (1996). *Ecological Restoration: Definition.* http://www.ser.org.

Tansley, A.G. (1935). The use and abuse of vegetational concepts and terms. *Ecology*, **16**, 284–307.

Wali, M. (1992). Ecology of the rehabilitation process. In *Ecosystem Rehabilitation*, ed. M. Wali, vol. 1, pp. 3–26. The Hague: SPB Academic Publishing.

Whyte, R.O. & Sisam, J.W.B. (1949). *The Establishment of Vegetation on Industrial Waste Land*, Publication no. 14. Aberystwyth: Commonwealth Bureau of Pastures and Field Crops.

2 • Rationale for restoration

JOHN CAIRNS JR

INTRODUCTION

While the world's economic systems have enjoyed unprecedented expansion, ecological systems have been degraded and diminished at an appalling rate. For example, Dahl (1990) notes that approximately 117 million acres of wetlands have been lost in the United States since the 1780s. Excluding Alaska, this wetland loss is approximately 53% (Dahl, 1990). The National Research Council (1992) estimates 4.3 million acres of degraded lakes in the United States and 3.2 million miles of rivers and streams that would benefit from restoration. The degradation of aquatic ecosystems is usually not uniformly distributed throughout most political units, as the loss of wetlands in the United States illustrates (Fig. 2.1). This complicates the funding of and responsibility for aquatic ecosystem restoration. Public awareness that human society's practices are unsustainable was raised significantly at a global scale by the report of the World Commission on Environment and Development (1987). However, awareness must be accompanied by changed behaviour, and there is scant evidence that substantive changes have occurred.

Adults who spent their childhoods in dysfunctional families report that they were unaware of behaviours other than the ones they experienced. Many of these individuals have been fortunate enough to find a model of behaviour other than the dysfunctional one and have incorporated the changed behaviours into their adult lives. First, however, they had to become aware of a different model before they learned of a viable, superior and alternative paradigm. Similarly, a dysfunctional relationship between humans and the environment is the norm in today's society, and ecological restoration and rehabilitation are the exceptions. Clearly, dysfunctional relationships are not sustainable. Human society, thus, appears to prefer one enormous risk of producing an unsustainable planet to the series of continuing behavioural adjustments that would be necessary in the transitional period before sustainability can be achieved. Even then, constant adjustment would be necessary since nature is a pulsating system. Adjustment to such a system requires constant change based on a knowledge of the amplitude, duration and quality of the changes; this knowledge is lacking for most of the world's ecosystems. This argument might suggest to some that no restoration be carried out until the knowledge base improves. However, numerous case histories (see examples in this volume and National Research Council, 1992) provide persuasive evidence that ecological restoration of damaged ecosystems results in markedly improved ecological attributes with the present state of knowledge. Of course, methodology and models can be improved, but the present loss of species and habitat requires immediate remedial measures.

The biosphere supports human society, and human society should both protect the environment and restore it in return. This relationship is healthy and sustainable. A primary reason for ecological restoration is to provide models of alternative, less destructive relationships between humans and natural systems.

WHY SHOULD WE CARE?

Compelling reasons for carrying out ecological restoration

As Bradshaw (1983) notes, 'The acid test of our understanding is not whether we can take ecosystems

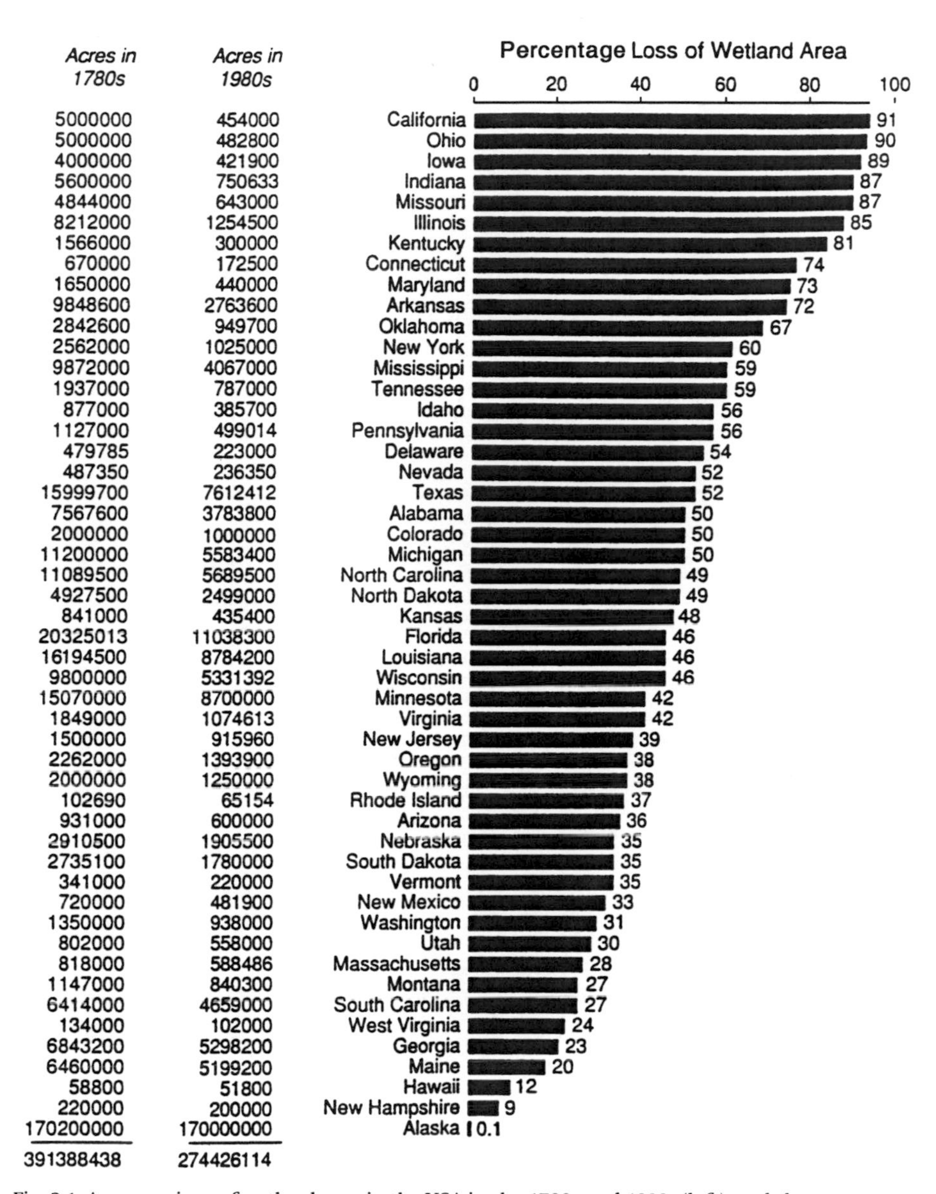

Fig. 2.1. A comparison of wetland area in the USA in the 1780s and 1980s (left), and the percentage of wetland area lost by the 1980s (right). All data courtesy of the US Fish and Wildlife Service.

to bits on pieces of paper, however scientifically, but whether we can put them together in practice and make them work.' If the basic planning in a restoration project is done correctly, ten years or less may often be enough for nature to take over. If the ecological stressors are removed, nature usually does quite well with recovery from damage. Cairns (1998*a*) believes that human society as currently known will not survive if ecological restoration and preservation are not widely practised. There are also ethical responsibilities for both future generations and the other species with which humans share the planet (e.g. Cairns, 1998*b*). Hawken *et al.* (1999) believe that natural capital is the basis for all other forms of capital and should be accumulated. Of course, political instability could result from continued bad water management (e.g. Postel, 1999). Cairns (1994) believes that ecological restoration proclaims a new, improved relationship of human society to natural systems – humans are both the guardians of natural systems and their dependants.

There are some very compelling reasons for carrying out ecological restoration.

1. Human society must protect and enhance the delivery of ecosystem services (e.g. Daily, 1995, 1997). From a political standpoint, this reason may be the most persuasive since many people grasp the value of life-support systems. The danger of emphasising this reason is that ecosystems might be viewed as commodities rather than living systems that deserve compassionate treatment regardless of the services they provide. 'Short-termism' may override long-term planning as it often does in overharvesting fish stocks.
2. Human society's practices are the best indication of its ethos or set of guiding beliefs. Ecological restoration is a positive statement of co-operation with natural systems. Preserving those systems still undamaged and protecting those restored would be an even more positive statement, especially if accompanied by major restorative efforts for presently damaged systems.
3. The ability to estimate the cost of restoration will be markedly improved as the number of projects increases. When the full cost of ecological restoration is better documented, it may well act as a deterrent to further damage because the dollar costs can be incorporated into comparisons of alternative actions. These full cost numbers will also enable the amount of money in restoration bonds to be determined with more precision.
4. Having restoration projects in each ecoregion, and preferably in each major area of the country, will provide demonstrations for local citizens, and this visibility will vastly increase environmental literacy.

Linkages between human and environmental health

As was the case for human health, publications on environmental health have been initially preoccupied with symptoms resulting from toxicity, stress (e.g. thermal) or physical conditions (e.g. increased suspended solids). Just as the medical focus for humans has gone beyond mere absence of symptoms to consideration of the attributes of health and well-being, environmental health is experiencing the same paradigm shift. Not surprisingly, evidence for numerous, significant linkages is becoming more and more apparent.

Kaplan (1983) remarks that nature is important to people and an important ingredient in a quality life. Wilson (1993) believes that humans still rely on nature's rules, even in the minority of peoples who have existed for more than one or two generations in wholly urban environments. Earlier, Wilson (1984) described *biophilia* as 'the innately emotional affiliation of human beings to other living organisms'. Wilson (1993) believes that biophilia is not a single instinct, but a complex set of learning rules. Ulrich (1993) goes further to propose that evolution selects those individuals who both learn and retain associations and practices related to natural systems that enhance the prospects of survival. However, Ulrich (1993) is also careful to analyse biophobic responses, which are reminders of the severe penalties resulting from dangerous interactions with various species or habitats. Human ancestors must have been fairly adept at risk analysis.

DESTRUCTIVE ACTIONS AND PHILOSOPHIES

Preference for enormous risk compared to a series of behavioural adjustments

Human society's preference for one enormous risk compared to a series of behavioural adjustments suggests that such actions occur because human society does not have enough information. However, Orr & Ehrenfeld (1995) believe that human society has enough information but is in denial. Numerous examples exist, such as cigarette smoking, unhealthy diets, sexually transmitted diseases, and various forms of addictions. Hardin (1999) discusses this myopia with regard to human population size and affluence. Ehrlich & Ehrlich (1996) believe that misinformation is a major factor in the anti-science backlash. Publications espousing protection of the biosphere have been generally available for decades (e.g. Caldwell, 1972) and have influenced many people (but not enough) to preserve the biosphere. Clearly, something more than sound reasoning is needed because much damage to biospheric integrity has occurred in the nearly three decades since Caldwell's book was published. Environmental historian McNeill (2000) notes that, for the first time in human history, human society has altered ecosystems with increased intensity, on large scales, and with much speed. This larger problem is the one that those interested in restoring ecosystems face at the outset of the twenty-first century.

Disassociation from, and reassociation with, nature

The rate of ecological damage and degradation dramatically exceeds the rate of ecological restoration and rehabilitation, clearly indicating a problem in human society's relationship with natural systems. Cairns (1994) views ecological restoration as a re-establishment of a harmonious relationship between human society and natural systems. The desire for a deep connection with nature still exists, as evidenced by the popularity of ecotourism, nature channels on television, bird-and whale-watching, and the like. Why does human society abuse the natural systems it still loves? The classic paper of White (1967) describes the alienation in terms of human cultural history. Nash (1982) adds to the discussion that this separation from nature resulted from human cultural development. Hamilton & Cairns (1961) also discuss the diminished relationship, although in a somewhat less ecological context. More recently, Abram (1996) traces the damaged relationship with natural systems to the development of written language. If this hypothesis is valid, then personal computers and the internet may have deepened the separation.

In Canada, the United States, Australia, and some other parts of the 'developed world', people still glorify the pioneer spirit, which includes the conquest of nature, but often at the expense of indigenous peoples and species. Poet Walt Whitman (1967) celebrates the felling of primeval forests and the damming of rivers. However, John Muir and David Thoreau were both advocating a spiritual relationship with nature long before urban sprawl encroached upon the semi-wild places, and people (who had never worked for a living with their hands) emerged from the urban areas in huge 'suburban utility vehicles' to play pioneer.

These degrees of 'apartness' appear minor when compared to the exemptionalist position of some economists (e.g. Simon, 1983): technology, ingenuity and creativity free human society from the iron laws of nature that restrict other species. In Genesis, God instructs humans to be fruitful and multiply; fill the land and conquer it; dominate the fish of the sea, the birds of the sky, and every beast that walks the land. However, the human relationship with other creatures should not be based on a position of human dominance as some Christians interpret this injunction (see Bradshaw, this volume).

Janzen's (1988) efforts to restore the Guanacaste dry forest in Costa Rica (see also Janzen, volume 2) is a splendid example of a reassociation with nature following a disassociation from nature. Arguably, the most important feature is that the restoration occurred in a country with a modest per capita income and was supported by its president and the local citizenry. This effort puts inhabitants of wealthier countries to shame when they claim they cannot afford ecological restoration. This restoration was also robust science, and Janzen received a number of awards (e.g. Craaford Prize, Blue Planet Prize) for what he terms biocultural restoration.

Diamond (1999) relates the advancement of various human societies to the biota available for domestication. These ties to domesticated animals outside of agriculture are remarkably strong. For example, Beck & Myers (1996) report that companion animals (pets) reside in 56% of the households in the United States. Wilson (1984) notes that more people visit zoos each year than attend all professional sporting events. Anderson *et al.* (1992) provide persuasive evidence that pet ownership significantly lowers systolic blood pressure in both men and women.

Association with plants is also viewed as beneficial, even in the workplace (e.g. Randall *et al.*, 1992). Lewis (1996) describes the therapeutic benefits of gardening as a means of improving mental health. Not surprisingly, humans experience benefits from landscapes (e.g. Korpela & Hartig, 1996) and have preferences for various types of landscapes (e.g. Purcell *et al.*, 1994). An important finding for restoration ecologists is that benefits appear to cover quite a range of cultures (e.g. Hull & Revell, 1989).

This admittedly cursory survey of the literature shows that ecological restoration would satisfy human psychological needs and provide valuable health benefits. Why, then, is it not more common?

War, terrorism, peace and restoration

War damages ecosystems in a variety of ways. Younquist (1997) describes the environmental warfare unleashed by Saddam Hussein during the so-called Gulf War in the early 1990s, and Earle (1992) recounts the damage a year later. Myers & Kent (1999) have an excellent discussion of environmental degradation caused by refugees from military conflict. War also diverts resources from such activities as environmental protection and restoration. Cairns (2000) states that sustainability, as now envisioned, cannot be achieved in the absence of peace. Even terrorism might well divert resources that otherwise would be used for ecological restoration. The Tofflers' (1993) disquieting book persuasively argues that society is prepared for neither twenty-first century war or peace, but they do see possibilities for the latter.

A grim picture indeed! But, as Hippocrates (*c.* 400-377 BC) noted: 'He will manage the cure best who foresees what is to happen from the present condition of the patient.' Restoration ecologists cannot avoid the obstacles hampering their activities nor be paralysed by their magnitude. If human society is to have a quality future, ecological destroyers must be restrained and ecological healers encouraged! As Nobel Laureate Joshua Lederberg (quoted in Garrett, 1994) stated: 'Nature is not benign . . . The survival of the human species is *not* a preordained evolutionary program.' Ecological healers have a major role to play in the continued survival of the human species. However, hope should include more than mere survival! A quality life should be the objective, and in this context, ecological restoration will be enormously useful if a major effort is launched in time!

Restoring society's economic perspective

Most of human society has become obsessed with growth – more shopping malls, highways and other anthropogenic artifacts. Arguably, the most outspoken advocate of growth was economist Julian Simon:

> We have in our hands now – actually in our libraries – the technology to feed, clothe, and supply energy to an ever-growing population for the next 7 billion years . . . Even if no new knowledge were ever gained . . . we would be able to go on increasing our population forever. (Quoted in Bartlett, 1998.)

Since one might well question the soundness of the growth described by Simon over such a long period on a finite planet, attempts have been made to dull the misgivings of those who question the growth policy by using the terms 'smart growth' or 'managed growth'. For example, the President's Council on Sustainable Development (1996) states (p. 92): 'While some growth is necessary, it is the nature of the growth that makes the difference.' Growth is clothed in a battle of words, but the basic problem is that growth displaces other species, fragments habitats, and generally damages ecological health and integrity.

The well-being of cultures and societies is shaped by short-term, competitive 'needs'. Ironically, much environmental damage is financed by government subsidies (e.g. Myers & Kent, 1999). Merely reducing or eliminating these subsidies would markedly reduce the pressures on restoration ecologists and simultaneously leave more habitat to use as a source of recolonising species for damaged ecosystems. Arguably, the integrity of ecosystems and the human communities and societies associated with them are closely linked. Economist Gordon Tullock (1994) has studied the economic systems of non-human societies, which have been in place for a longer time period than humans have been on the planet. The striking feature of these societies to an environmental biologist is that no money is used, natural capital is protected, and almost no material is produced that cannot be reincorporated into natural systems. In fact, most 'wastes' are eagerly sought by one or more other species. Moreover, the practices of these non-human societies are sustainable, as demonstrated by their long history of success. Each of these species would probably have continued for countless more generations were it not for human interference. Whatever ecological damage these non-human societies do is easily repaired by the natural resiliency of natural systems. Assisted recovery (i.e. ecological restoration) is generally unnecessary.

Selfish, short-term view

Although most people can see the value of an ecological life-support system, it appears that a selfish, short-term view is responsible for continuing ecological damage and failure to repair much of the damage. It is highly probable that most readers of this volume will be able to find a damaged ecosystem near their residence, but only a fortunate few will have comparable access to a sizable, robust ecosystem with a high degree of ecological integrity. This volume offers abundant answers to 'How should we restore damaged ecosystems?' and much evidence that is persuasive to professionals in the field of ecological restoration in answer to 'Why should we . . . ?' Other contributors have attempted to answer 'Why don't we . . . ?' I can only speculate on why the general public has not embraced ecological restoration. Ecological restoration is basically the healing of unhealthy, wounded natural ecosystems. However, awareness of any but the grossest damage requires a fairly high level of ecological literacy that is not common in society as a whole. Two imperatives are most likely to cause a major paradigm shift that will lead to widespread restoration of damaged ecosystems: (1) a substantial increase in ecological literacy in both the general public and its representatives and (2) ecological damage so severe that even the ecologically illiterate are aware of it. Homer (*The Iliad*) was well aware of this: 'Once damage has occurred, even a fool can understand it.' Havel, a poet in Czechoslovakia who later became President of that country, attributed society's inaction to other paradigms (Havel, 1992):

> What a paradox for our generation that man, the greater collector of information, has all the knowledge about the dangers facing the species, but is almost incapable of acting, because the knowledge is either too poorly organized or because it is rendered useless due to the paradigms that govern our lives.

It seems obligatory for restoration professionals to give at least 10% of their working time to increase societal literacy as rapidly as possible. However, the resistance to major paradigm shifts is fierce, so professionals must be prepared to cope with widespread major ecological disequilibrium with far fewer recolonising species than are now available. A disquieting prospect at best, but not improbable. Lest this seem unduly pessimistic, it is noteworthy that environmental emergencies are the recurring themes in interviews with world-renowned scientists (Newbold, 2000). Jane Lubchenco's presidential address in 1997 to the American Association for the Advancement of Science calls for 'a more effective, interdisciplinary . . . effort on the environment.' Ehrlich (1997) strongly endorses working with a variety of other disciplines, including economics, in order to understand what is driving human decisions on resource use and even further '*how to change some aspects of human behaviour*' (emphasis his). Arguably, if resource use does not change, ecological restoration does not have a bright future.

In order to influence resource use, one must be able to convince such administrators as the County Planning Officer of Lancashire, UK, or the Town Manager of Blacksburg, VA, USA, that repairing damaged ecosystems is a sensible, civic responsibility. Working with local officials might be termed the 'bottom – up' approach. However, bioregional planning that is essential to effective ecological restoration also requires a 'top – down' approach to ensure integrated environmental management. Clearly, no single approach will produce the necessary imperatives. These imperatives, working through people's lives and livelihoods, must be continually addressed. A number of books on these seemingly obvious points were written in the last half of the century, but have not yet produced a paradigm shift that results in an ethos (a guiding set of values) on societal responsibility for repairing damaged ecosystems.

WHAT CAN BE ACCOMPLISHED?

Fortunately, abundant case histories of ecological restoration and rehabilitation exist. One interesting example from a historic perspective is the restoration and recovery of the Thames Estuary (Gameson & Wheeler, 1977). In the mid nineteenth century, the condition of the Thames had become so foul that sheets soaked in disinfectant were hung in the Houses of Parliament in an attempt to counteract the stench. By 1858, the smell at Westminster had become so overpowering that its control became a matter of strong personal interest to Members of Parliament, and interceptor sewers were constructed. In addition to this historic example, the National Research Council (1992) volume cites some interesting case histories of aquatic ecosystem restoration. More recent restoration examples are given in this volume and in many of the journals cited in the literature sections of the individual chapters.

The National Research Council (1992, pp. 354–355) recommends that inland and coastal wetlands in the United States be restored at a rate that offsets any further loss of wetlands and contributes to an overall gain of 10 million wetland acres by the year 2010. This number represents less than 10% of the total number of acres of wetlands in the United States lost in the last 200 years. The recommendation for restoration of rivers and streams in the United States is 400 000 miles within the next 20 years (from 1992). The recommended magnitude of this restoration represents approximately 12% of the 3.2 million miles of streams and rivers in the United States and was recommended because it is comparable to the miles of streams and river affected by point source and urban runoff. Excluding the North American Great Lakes, as well as flood control and water supply reservoirs, the report recommends that 1 million acres of lakes be restored, in association with wetland and river restoration, by the year 2000, increasing to 2 million acres in the long term. These figures and dates illustrate the magnitude of the undertaking. Data on progress toward these goals are not easily gathered nor has enough time elapsed except on phase 1 of the lake restoration; however, this limited information does depict the magnitude of the problem for just the United States.

CONSTRAINTS ON RESTORATION

Ethical problems in ecological restoration

Does restoration ecology represent a new trend in human society's relationship with natural systems, enhancing a benign co-evolution? Or, are restoration ecologists merely running a group of environmental 'body shops' that repair damaged ecosystems without appreciable effect on either rates of ecological destruction or on human society's set of guiding beliefs? At its worst, ecological restoration could be used as another justification for continued damage to natural systems (e.g. as the equivalent of a repair shop for automotive damage). Furthermore, the rate of ecological destruction on a global level is so enormous that the comparatively few attempts to repair ecological damage are dwarfed by comparison. Indeed, there are some ethical problems associated with ecological restoration.

1. Most ecological restoration is carried out (through management) to repair damage caused by human mismanagement. If management is the disease, how can it be the cure? Noss (1985) states: 'This is the irony of our age: "hands-on" management is needed to restore "hands-off" wilderness character'.
2. Some mitigative restoration is carried out on relatively undamaged habitats of a different kind. For example, created wetlands may replace an upland forest, or an upland forest may be destroyed to create a 'replica' of the savanna that once occupied a particular area. Logically, this secondarily damaged habitat should be replaced by yet another mitigative action. Sacrificing a relatively undamaged habitat to provide mitigative habitat of another kind deserves more caution than it has been given.
3. At the current state of knowledge, restoration projects are likely to have unforeseen outcomes. Ecological restoration carried out by the most skilled professionals will occasionally, perhaps frequently, omit some very important variables. Or, episodic

events may occur at inconvenient times. Some of the unforeseen results may offset any ecological benefits likely to result from a particular restoration project.

4. Well-meaning restoration efforts may displace the species best able to tolerate anthropogenic stress. By attempting to return an ecosystem to its predisturbance condition, the evolution of a species capable of co-existing with human society may be hampered. Attempts to manipulate the environment in such a way as to promote the success of one or two species may impede both the natural successional process and also exclude other species that would otherwise be there.
5. Similarly, if ecological restoration is carried out on an extremely large scale, human-dominated successional processes could become the 'norm'.
6. Finding sources of recolonising species for damaged ecosystems is increasingly difficult. Should one remove them from quality ecosystems and risk damaging that ecosystem, or use pioneer species, or, worse yet, exotics with the hope that the more desirable species will eventually colonise naturally?

Global climate change

Damaged ecosystems are likely to recover or be restored most quickly when climatic conditions suit the species that once inhabited them. Conditions unsatisfactory for recolonisation by once resident species provide an excellent opportunity for exotic, invasive species. Significant climate change is likely to destabilise already damaged ecosystems further and to lessen the recolonisation by appropriate species from undamaged sites. In fact, significant climate change will almost certainly damage healthy ecosystems. Ecological restoration may be nearly impossible in a period of appreciable climate change.

In this instance, policy change could make a substantial difference. As Retallack (1999) notes, the United States is responsible for nearly 25% of total greenhouse gas emissions, despite having only 4% of the world's population. The United States emits five times the global average – more than any other country in the world (Retallack & Bunyard, 1999). A poll released in October 1998 by the World Wildlife Fund notes that 57% of Americans believe climate change is already happening, 79% support the Kyoto Agreement to reduce greenhouse gases, and over 66% think that the United States should act now, unilaterally, to reduce carbon dioxide emissions. Results of this poll are remarkable support for the position of mainstream scientists worldwide. Delay in fulfilling the voluntary commitment the United States made at the 1992 Rio Conference (to reduce its greenhouse gas emissions to 1990 levels by the year 2000) shows that this obviously was an empty promise. Precautionary action is now essential. As the Romans maintained: 'Bis dat qui cito dat' – one gives doubly who gives quickly. Climate change could quickly spiral out of control (Bunyard, 1999). A destabilised climate would undoubtedly cause a refugee crisis (e.g. Foley, 1999) that would worsen the environmental crisis and adversely affect ecological restoration.

Hostile lawsuits

Space does not permit extensive coverage of lawsuits, vandalism, and other attempts to block ecological restoration. In the aggregate, they deserve serious attention. One example of a lawsuit concerns a ranching group's attempt to block the reintroduction of the Mexican grey wolf into the US Southwest (Taugher, 1999*a*). The New Mexico Cattle Growers Association sued the US Fish and Wildlife Service to block the reintroduction, claiming the captive-bred wolves contained the genes of dogs or coyotes and were not deserving of a reintroduction effort. Biologists noted that the wolves were living only in zoos and sanctuaries pending release and that new DNA testing techniques established no evidence for the ranchers' claim. Even with this ruling, at least five wolves were shot to death after the reintroduction. Taugher (1999*b*) also notes that the Rio Grande silvery minnow, listed as an endangered species in the United States in 1993, has suffered serious habitat reduction since then. The problem of restoring a declining river ecosystem and meeting the water demand of irrigation for farmers and a burgeoning urban population has led to a call by US Senator Pete Domenici to restore upstream habitat now unsuitable for the minnow to justify rendering presently suitable downstream habitat unsuitable (for details, see Editorial, 2000; Soussan 2000). To further complicate this situation, the US Bureau of Reclamation asserts that there have been misrepresentations of facts on water supply issues by the Middle Rio Grande Conservancy District senior staff (Gabaldon & Leutheuser, 2000). This brief summary depicts an illustrative and complicated legal issue in which ecological restoration may often be invoked – in this case, justifying damage to presently suitable habitat for an endangered species on the assumption that ecological restoration will make a presently unsuitable habitat suitable and that the endangered species will successfully recolonise the restored area.

The global commons and environmental quality control of private property

The National Research Council (1992) recommends ecological restoration at the landscape level whenever possible. However, restoration projects are often approached piecemeal. Part of this problem is due to fragmentation through assignment of government agencies to a portion of environmental management. As Leopold (1990) notes: 'Each agency acts as if it is the only flower facing the sun.' Each government agency and most of human society is looking at ecosystems in terms of the uses that might be made of them, including alternative uses that might threaten the ecosystem's existence. No organisation is responsible for maintaining the integrated processes and relationships that collectively make an ecosystem what it is and make sustainable use possible.

Restored ecosystems are more likely to be self-maintaining if restoration is carried out at the landscape level (e.g., National Research Council, 1992). Landscape-level restoration will almost always involve public property (especially where hydrologic systems are concerned) and a mixture of organisational and personal private property. Consequently, another formidable barrier to a landscape approach is the inevitable conflicts between environmental protection and property rights. The individual property owner with a small wetland is likely to be irate when told that filling, draining or altering the wetland in major ways is illegal. This property, the owner sometimes says, is private 'and I will do as I wish with my property'. However, private property rights are not sacred, even in societies with strong views on this subject. Each person lives not only on private property, but in a larger ecological landscape shared with others. So, a key question (at which environmental literacy, ethics, and human institutions such as law and economics interact) is: To what extent should individual, organisational or national behaviour and attitudes be modified for the betterment of others of the human species and for other species as well?

Hardin's (1968) classic paper and numerous subsequent publications describe the inequities that occur in the use and abuse of the commons. Hardin suggests that private property ownership is one way of avoiding the tragedy, since it gets rid of the destructive competition between common-ground users. On the other hand, the insistence of many property owners that they have the right to do whatever they choose with their private property often damages adjacent or even distant ecosystems. Much more attention should be given to the many different systems in which use and maintenance of commons were ensured up to quite recent times. That these problems have existed for at least 4000 years is evidenced by the inclusion of several laws in the famous Code of Hammurabi that dealt with the use of irrigation water (Postel, 1999). Why restore any damaged ecosystem if misuse of the commons and/or private property can, in a short period of time, erase the labour of years or decades? However, these problems are not insurmountable, as documented by Janzen's (1988) Guanacaste Forest restoration and a substantial number of other projects (e.g. the Adobe Creek restoration of a salmon run by a group of high-school students [described in Cairns, 1998*a*]).

All zoning ordinances restrict property rights, and examples can be found in the news of cases where the proper balance between property rights and common good is delimited. Ecological restoration should not be impaired by practices on private land. Plans for constructing an incinerator for hazardous waste next to an elementary school predictably are met with fierce resistance. They are both in the same airshed, and unrestricted exercise of property rights on one property may unilaterally devalue the property rights of others in the same airshed. Helicopter flights in the Grand Canyon create noise that historically has never existed in the Grand Canyon. If, hypothetically, a proposal were made to permit the construction of a fast-food restaurant opposite a war memorial, a public outcry would be quite predictable.

OVERCOMING THE CONSTRAINTS AND SETTING CRITERIA FOR SUCCESS

Responsibility for restoration

Clearly, ecosystem restoration without concomitant protection of unimpaired ecosystems would be senseless (e.g. Woodwell, 1994). However, for this discussion, the key issue becomes: how does one prevent restored ecosystems from being re-damaged? The only possible means is continual, direct surveillance and monitoring of the health of the ecosystem while maintaining equal vigilance on proposed or actual activities that threaten ecosystem health. This design requires a system of ecosystem guardians for each restored or undamaged ecosystem.

Ecosystem protection and restoration will require the collaboration of ordinary citizens who can be especially attentive to the actions and proposed actions of individuals and organisations that might threaten the ecosystem. In addition, skilled professionals who can gather the hard evidence necessary for policy and regulatory decisions are needed. Paul R. Ehrlich (pers. comm.) has recommended that professional biologists tithe their time on projects beneficial to the general well-being of ecosystems and, consequently, human society. In the case of restoration and the maintenance of ecosystem health, this donation of expertise could well be extended to engineers, chemists, economists, sociologists, and people from almost any other discipline.

A number of scientific measurements, some well within the capabilities of highly motivated but relatively untrained individuals, can furnish very useful information. The Save Our Streams program, administered by Trout Unlimited, provides one example. Regrettably, most of these measurements are at the population or community level, and there are relatively few, generally accepted measurements of integrity or condition at the ecosystem or landscape level. Measurements that do exist tend to be experimental, require skilled professionals for reliable measurement and analysis, or are generally quite expensive.

In examining various measurements of ecosystem integrity, primarily for large river systems of the world, Cairns (1997) suggests that examination of a selected list of practices and guiding beliefs of human society might accurately predict the general health and condition of the ecosystems in which these societies live. For example, if economic development is the highest value of a particular society, one would not expect ecosystems to fare well. If human society is not willing to modify its present behaviour (e.g. living on floodplains and expecting engineering solutions for protection, and wishing to move water where people are rather than people to where water is), it is quite likely that society will live in managed, rather than natural, ecosystems. This drawback is not intended to denigrate scientific measurements of ecological integrity or to hamper further development in this area. It is, rather, to suggest that, since the fate of natural systems is in the hands of human society, the practices and guiding beliefs of this society must be examined as a useful means of predicting the condition of ecosystems associated with that particular society.

Responsibility after restoration

Any ecosystem, including restored ecosystems, can be damaged at any time. Ideally, early-warning monitoring systems (see Holl & Cairns, this volume) will detect a deviation from established quality control criteria in time to take remedial action before serious damage is done. Ecological restoration is an investment in the future and deserves the same level of stewardship that a financial investment receives.

Paying for restoration

Holl & Howarth (2000) give a detailed analysis of how to pay for ecological restoration. However, if the owner of the damaged ecosystem has disappeared or no owner exists because the damage occurred on common ground, the government must intervene and furnish money. All natural resources in any state require maintenance, so financial support must always be available. The ecological infrastructure (i.e. natural resources) has just not fared well in the cash flow. Since the methods and procedures for ecological restoration are now reliable and sound, the big challenge is to devise a sustainable ecological infrastructure maintenance system!

How to restore

Reflections on methods and successes

The Kissimmee River restoration and other case histories serving as illustrations in the National Research Council (1992) report will encourage other regions to engage in ecological restoration by providing evidence of capabilities for restoration. Whether the cost and effort involved will dissuade citizens from doing so remains to be seen. Evidence in publications by Caldwell (1972), the National Research Council (1992, 1996), the National Academy of Engineering (1994, 1996, 1997), Odum (1989), Passmore (1974) and Hardin (1993) show heightened concern.

The first step in re-evaluating human society's relationship with natural systems will be to quantify rates of ecological damage and repair. The establishment of bioregions in which such evidence is gathered would facilitate this process and simultaneously furnish local citizens with evidence that they can confirm personally. Additionally, since degrees of ecological damage or repair require considerable professional judgement, it will be necessary to establish a qualified 'blue-ribbon committee' of ecologists knowledgeable in the determination of both ecosystem health and degree of ecological restoration. It would be advisable for these groups to use criteria and standards that are as homogeneous as the differences between and among bioregions permit. A national blue-ribbon committee could furnish both information and judgement to the regional committees. This group should be responsible for continually modifying and revising criteria and standards when enough new knowledge is available to justify revision from a scientific perspective.

BEGIN CHANGES NOW

Logic suggests that the present rate of population growth and concomitant ecological destruction cannot continue indefinitely without severe effects on human quality of life. Either human society will re-examine its relationship with natural systems and alter society's impact on them, or, eventually, natural processes will regulate human society's numbers and level of affluence. The first goal should almost certainly be to ensure that the rate of ecological damage does not exceed the rate of ecological repair or restoration. However, achieving a balance between destruction and repair merely increases the probability that situations will get no worse unless the population continues to grow. Ensuring a net gain in quality ecosystems is a more desirable goal, especially if the human population size is stabilised or even decreased over the long term. These actions would enhance the accumulation of ecological capital, such as old-growth forests, topsoil and species, through habitat improvement. The longer human society waits to discuss and examine its relationship with natural systems, the less likely it will be that quality ecosystems will be available as models or that species will be available to recolonise damaged ecosystems. The rate of ecosystem damage should be exceeded by the rate of ecosystem restoration and an overall net gain should be made annually in ecological capital and robust ecosystems with an exemplary level of ecological integrity.

The global ecological blight directly affects people, their homes, their surroundings, the attitudes of others in their areas, and their whole livelihoods. The mission of restoration ecologists is to diminish, perhaps even eliminate, this blight. This mission goes well beyond an interesting academic problem – it addresses the core of the human condition. Planning officers and policy officials are derelict in their responsibilities if they ignore the deteriorating condition of the planet's ecological life-support system. These imperatives were sufficiently unobvious to citizens and their legislative representatives that numerous books were written to draw attention to them. In 1992, the Union of Concerned Scientists (1992) issued the *World scientists' warning to humanity* which is signed by a huge number of world renowned scientists, including many Nobel Laureates. There is much to do and the time to do it is now!

ACKNOWLEDGMENTS

I am indebted to Amy Ostroth for transferring the handwritten original to the word processor. Darla Donald has provided her usual skilled editorial assistance. E. Scott Geller and B.R. Niederlehner provided some useful references. E. Scott Geller, Alan Heath, Karen Holl and John Heckman provided useful comments on the first draft and A.D. Bradshaw made some helpful suggestions on subsequent drafts. The Cairns Foundation supplied funds for the processing of this manuscript.

REFERENCES

Abram, D. (1996). *The Spell of the Sensuous: Perception and Language in a More-Than-Human World.* New York: Vintage Books.

Anderson, W., Reid, C. & Jennings, G. (1992). Pet ownership and risk factors for cardiovascular disease. *Medical Journal of Australia*, **157**, 298–301.

Bartlett, A.A. (1998). Malthus marginalized. *The Social Contact*, **8**(3), 239–251.

Beck, A.M. & Myers, N.M. (1996). Health enhancement and companion animal ownership. *American Review of Public Health*, **17**, 247–257.

Bradshaw, A.D. (1983). The reconstruction of ecosystems. *Journal of Applied Ecology*, **20**, 1–17.

Bunyard, P. (1999). How climate change could spiral out of control. *Ecologist*, **29**, 68–74.

Cairns, J., Jr (1994). *Ecological Restoration: Re-examining Human Society's Relationship with Natural Systems*, The Abel Wolman Distinguished Lecture. Washington, DC: National Research Council.

Cairns, J., Jr (1997). Eco-societal restoration: creating a harmonious future between human society and natural systems. In *Watershed Restoration: Principles and Practices*, eds. J.E. Williams, C.A. Wood & M.P. Dombeck, pp. 487–499. Bethesda, MD: American Fisheries Society.

Cairns, J., Jr (1998*a*). Can human society exist without ecological restoration. *Annals of Earth*, **16**, 21–24.

Cairns, J., Jr (1998*b*). Replacing targeted compassion with multidimensional compassion: an essential paradigm shift to achieve sustainability. *Speculations in Science and Technology*, **21**, 45–51.

Cairns, J., Jr (2000). World peace and global sustainability. *International Journal of Sustainable Development and World Ecology*, **7**, 1–11.

Caldwell, L.K. (1972). *In Defense of Earth: International Protection of the Biosphere*. Bloomington, IN: Indiana University Press.

Dahl, T.E. (1990). *Wetland Losses in the United States: 1780s to 1980s*. Washington, DC: US Department of the Interior, US Fish and Wildlife Service.

Daily, G. (1995). Restoring value to the world's degraded lands. *Science*, **269**, 350–355.

Daily, G. (ed.) (1997). *Nature's Services: Societal Dependence on Natural Ecosystems*. Washington, DC: Island Press.

Diamond, J. (1999). *Guns, Germs, and Steel*. New York: W.W. Norton.

Earle, S.A. (1992). Persian Gulf pollution: assessing the damage one year later. *National Geographic Magazine*, **181**, 122–134.

Editorial (2000). Domenici idea keeps minnow in limelight. *Albuquerque Journal*, 28 August, p. A8.

Ehrlich, P.R. (1997). *A World of Wounds: Ecologists and the Human Dilemma*. Oldendorf/Luhe, Germany: Ecology Institute.

Ehrlich, P.R. & Ehrlich, A.H. (1996). *Betrayal of Science and Reason: How Environmental Anti-Science Threatens Our Future*. Washington, DC: Island Press.

Foley, G. (1999). The looming refugee crisis. *Ecologist*, **29**, 96–97.

Gabaldon, M. & Leutheuser, R. (2000). Reclamation active cog in water-supply process. *Albuquerque Journal*, 25 August, p. A17.

Gameson, A.L. H. & Wheeler, A. (1977). Restoration and the recovery of the Thames Estuary. In *Recovery and Restoration in Damaged Ecosystems*, eds. J. Cairns, Jr, K.L. Dickson & E.E. Herricks, pp. 72–101. Charlottesville, VA: University Press of Virginia.

Garrett, L. (1994). *The Coming Plague: New Emerging Diseases in a World Out of Balance*. New York: Farrar, Straus & Giroux.

Hamilton, E. & Cairns, H. (eds.) (1961). *Plato: The Collected Dialogues*. Princeton, NJ: Princeton University Press.

Hardin, G. (1968). The tragedy of the commons. *Science*, **162**, 34–35.

Hardin, G. (1993). *Living within Limits: Ecology, Economics, and Population Taboos*. Oxford: Oxford University Press.

Hardin, G. (1999). *The Ostrich Factor*. Oxford: Oxford University Press.

Havel, V. (1992). Considering the carrying capacity of the Earth: finite or infinite? As quoted by B. Rosborough in American Academy of Arts and Sciences, Cambridge, MA, 5 March.

Hawken, P., Lovins, A. & Lovins, H. (1999). *Natural Capitalism*. New York: Little, Brown.

Holl, K.D. & Howarth, R.B. (2000). Paying for restoration. *Restoration Ecology*, **8**, 260–267.

Hull, R.B. & Revell, G.R.B. (1989). Cross-cultural comparison on landscape and scenic beauty evaluations: a case study in Bali. *Journal of Environmental Psychology*, **9**, 177–191.

Janzen, D.H. (1988). Guanacaste National Park: tropical ecological and biocultural restoration. In *Rehabilitating Damaged Ecosystems*, vol. 2, ed. J. Cairns, Jr, pp. 143–192. Boca Raton, FL: CRC Press.

Kaplan, R. (1983). The role of nature in the urban context. In *Behaviour and the Natural Environment*, eds. I. Altman & J. Wohlwill. Cambridge, MA: Perseus Publishing.

Korpela, K. & Hartig, T. (1996). Restorative qualities of favorite places. *Journal of Environmental Psychology*, **16**, 221–233.

Leopold, L. (1990). Ethos, equity and the water resource. *Environment*, **32**, 16–20, 37–42.

Lewis, C.A. (1996). *Green Nature/Human Nature: The Meaning of Plants in Our Lives*. Urbana, IL: University of Illinois Press.

McNeill, J.R. (2000). *Something New under the Sun: An Environmental History of the Twentieth-Century World*. New York: W. W. Norton.

Myers, N. & Kent, J.V. (1999). *Perverse Subsidies: Tax $s Undercutting Our Economies and Environments Alike*. Winnipeg, Manitoba: International Institute for Sustainable Development.

Nash, R. (1982). *Wilderness and the American Mind*. New Haven, CT: Yale University Press.

National Academy of Engineering (1994). *The Greening of Industrial Systems*. Washington, DC: National Academy Press.

National Academy of Engineering (1996). *Engineering within Ecological Constraints*. Washington, DC: National Academy Press.

National Academy of Engineering (1997). *Technological Trajectories and the Human Environment*. Washington, DC: National Academy Press.

National Research Council (1992). *Restoration of Aquatic Ecosystems: Science, Technology, and Public Policy*. Washington, DC: National Academy Press.

National Research Council (1996). *Linking Science and Technology to Society's Environmental Goals*. Washington, DC: National Academy Press.

Newbold, H. (ed.) (2000). *Life Stories: World-Renowned Scientists Reflect on Their Lives and the Future of Life on Earth*. Berkeley, CA: University of California Press.

Noss, R.T. (1985). Wilderness recovery and ecological restoration: an example for Florida. *Earth First*, **5**, 18–19.

Odum, E.P. (1989). *Ecology and Our Endangered Life-Support System*. Sunderland, MD: Sinauer Associates.

Orr, D.W. & Ehrenfeld, D. (1995). None so blind: the problem of ecological denial. *Conservation Biology*, **9**, 985–987.

Passmore, J. (1974). *Man's Responsibility for Nature: Ecological Problems and Western Tradition*. New York: Charles Scribner's Sons.

Postel, S. (1999). *Pillar of Sand: Can the Irrigation Miracle Last?* New York: W.W. Norton.

President's Council on Sustainable Development (1996). *Sustainable America: A New Consensus*. Washington, DC: US Government Printing Office.

Purcell, A.T., Lamb, R.J., Peron, E.M. & Falchero, S. (1994). Preference or preferences for landscape? *Journal of Environmental Psychology*, **14**, 195–209.

Randall, K., Shoemaker, C.A., Relf, D. & Geller, E.S. (1992). Effects of plantscapes in an office environment on worker satisfaction. In *The Role of Horticulture in Human Well-Being and Social Development*, ed. D. Relf, pp. 106–109. Portland, OR: Timber Press.

Retallack, S. (1999). How US politics is letting the world down. *Lapis*, **9**, 11–17.

Retallack, S. & Bunyard, P. (1999). We're changing our climate! Who can doubt it? *Ecologist*, **29**, 60–63.

Simon, J. (1983). *The Ultimate Resource*. Princeton, NJ: Princeton University Press.

Sousssan, T. (2000). Sen. suggests moving minnow. *Albuquerque Journal*, 24 August, pp. A1, A2.

Taugher, M. (1999*a*). Anti-wolf suit called "frivolous." *Albuquerque Journal*, 21 September, p. C3.

Taugher, M. (1999*b*). Survey suggests minnow habitat shrinking. *Albuquerque Journal*, 22 September, p. B3.

Toffler, A. & Toffler, H. (1993.) *War and Anti-War: Survival at the Dawn of the 21st Century*. New York: Little, Brown.

Tullock, G. (1994). *The Economics of Non-Human Societies*. Tucson, AZ: Pallas Press.

Ulrich, R.S. (1993). Biophilia, biophobia, and natural landscapes. In *The Biophilia Hypothesis*, eds. S.R. Kellert & E.O. Wilson, pp. 73–137. Washington, DC: Island Press.

Union of Concerned Scientists (1992). *World Scientists' Warning to Humanity*, Cambridge, MA: Union of Concerned Scientists .

White, L., Jr (1967). The historic roots of our ecological crisis. *Science*, **155**, 1203–1207.

Whitman, W. (1967). 'Pioneers! O Pioneers!' In *Leaves of Grass*. Seacaucus, NJ: Longriver Press.

Wilson, E.O. (1984). *Biophilia: The Human Bond with Other Species*. Cambridge, MA: Harvard University Press.

Wilson, E.O. (1993). Biophilia and the conservation ethic. In *The Biophilia Hypothesis*, eds. S. R. Kellert & E. O. Wilson, pp. 31–41. Washington, DC: Island Press.

Woodwell, G.M. (1994). Ecology: the restoration. *Restoration Ecology*, **2**, 1–3.

World Commission on Environment and Development (1987). *Our Common Future*. Oxford: Oxford University Press.

Younquist, W. (1997). *Geodestinies*. Portland, OR: National Book Company.

3 • The ecological context: a landscape perspective

RICHARD J. HOBBS

INTRODUCTION

Restoration ecology has developed from, and has been practised primarily on, a site-based approach. The restoration of a well-defined area such as a minesite, wetland or a degraded ecosystem of some description is generally attempted. However, it is clear that relatively large areas of the earth are in need of some form of restoration, following degradation through overuse or inappropriate management, which has impaired the functioning or altered the structure of the landscape as a whole (MacMahon, 1998). Thus, there is a need to expand the scope of restoration ecology to embrace broader scales and tackle landscape-scale problems. While this is increasingly recognised, the science of landscape-scale restoration is still in a formative phase (Bell *et al.*, 1997). In this chapter I present a summary of landscape structure and function, discuss the impacts of human modification of landscapes, and present a series of options for developing guidelines for landscape restoration. Some of the material in this chapter is modified from work presented elsewhere (Hobbs, 1995, 1999; Hobbs & Harris, 2001; Hobbs & Lambeck, in press).

LANDSCAPES: STRUCTURE AND FUNCTION

A landscape is defined as an area of land, at the scale of hectares to square kilometres, which consists of a collection of different, but interacting patches (also called landscape elements). Patchiness focuses on the spatial matrix of ecological processes, and emphasises the fluxes of materials and organisms within and between parts of the landscape. It is a form of spatial heterogeneity in which boundaries are discernible, and in which units appear as contrasting, discrete states of physical or ecological phenomena (Ostfeld *et al.*, 1997). Patches may comprise different ecosystems (e.g. lakes, rivers, forest, grassland), different land uses (e.g. urban, agricultural, nature reserve), or different community types, successional stages or alternative states within a particular ecosystem (e.g. post-fire, pole-stage and old-growth forest stands).

Landscape ecology considers three main aspects of landscapes: structure (or pattern), function (or process) and change (Forman & Godron, 1986; Turner, 1989; Forman, 1995; Pickett & Cadenasso, 1995; Turner *et al.*, 1995). The characteristics of individual patches and their spatial relationship with other patches determine landscape structure, while landscape function is determined by physical, chemical and biotic transfers between patches. Landscape change results from either changes in individual patches or changes in patch configurations and interrelations. The three aspects of landscapes are closely interlinked, since structure strongly influences function, which can feed back into structure, and landscape change can affect both structure and function.

Structure

Landscape mosaics are commonly complex entities consisting of numerous patches of varying types in a variety of configurations. Our understanding of landscape patterns has increased greatly with the advent of remote sensing and geographic information system (GIS) technologies. Remote sensing, using airborne or satellite sensors, offers the possibility to acquire large amounts of data on the characteristics of the earth's surface in a relatively easy, repeatable and analysable way. Geographic information systems allow us to develop spatially explicit databases on a wide range of landscape features, which again allows quantitative analysis of pattern (Haines-Young *et al.*, 1993).

An assessment of landscape pattern requires that the scale of investigation is defined, and this scale will be determined by the types of question being asked and the types of organism or process being studied. Many problems, misunderstandings, conflicting results and misapplication of research findings arise from failures to determine the relevant scale of study or to ensure that studies are conducted at similar scales. While the landscape level is the primary focus here, it is important to recognise that individual landscapes occur in a regional setting. Indeed, boundaries between landscapes are frequently little more than convenient lines drawn by humans. A frequently used natural landscape unit is the catchment or watershed (the area of land within which water drains into one watercourse), since this has natural topographically determined boundaries.

Also of importance is the patch level – i.e. the level of the units which make up landscapes. Individual patches have a set of characteristics (e.g. size, composition, age) which can influence processes both within the patch and at the landscape level. The recognition that patterns and processes at one level can be influenced by patterns and processes at other (higher and lower) levels is essential if we are to come to grips with the complexity of large-scale ecological patterns and processes.

Pattern in the landscape is the result of the interaction of many influences. At a broad scale, patterns of vegetation composition and structure can be related to regional gradients in climatic variables such as temperature and rainfall, and to changes in soil and landform type and topography. Within any given region, climate, soil type and landform generally determine the broad vegetation patterning. Within that broad patterning, however, numerous smaller scales of pattern may be present. These may be determined by finer-scale variations in soil characteristics or microclimate, but may also be the result of other factors. Species turnover between different parts of a landscape and patchy distribution of populations of individual species can add to the complexity of vegetation patterning. This patchiness can be as a result of chance dispersal events, response to localised disturbances or response to micro-environmental variation.

The history of disturbances determines the distribution of patches of different age and successional stage within a landscape, and may, in some cases, be responsible for the development of mosaics of patches in alternative semi-permanent vegetation states (Turner, 1987; Baker, 1992; Hobbs, 1994). The components of the disturbance regime will determine the scale and pattern of variation observed, and each disturbance type may produce different vegetation responses. Individual disturbance types can also produce different responses depending on factors such as environmental variations within the disturbed areas, weather characteristics following the disturbance, and interactions with other disturbances (e.g. Suffling, 1993). Important landscape-scale disturbances include fire, drought, infrequent frosts or periods of higher than normal temperatures, severe storms, localised soil disturbance by animals, tree falls and insect outbreaks. In addition to the natural disturbance regime, human disturbance is an important component of many ecosystems. In regions with long histories of human habitation, human-induced disturbance has frequently been an important influence in shaping and maintaining ecosystems and landscapes (e.g. le Houérou, 1981; Thirgood, 1981). In many parts of the world, aboriginal inhabitants exerted a powerful influence on the landscape, but this was usually significantly modified following invasion by European settlers (e.g. Flannery, 1994). Current human activities such as vegetation removal, timber or soil extraction and introduction of non-native plants and animals continue to produce a further overlay of variation on the landscape, as well as altering the natural or aboriginal regime.

Heterogeneity

Landscape heterogeneity is a complex multi-scale phenomenon, involving the size, shape and composition of different landscape units and the spatial (and temporal) relations between them. Adequate methods for measuring and assessing its significance have yet to be developed. Landscape heterogeneity is frequently examined at two levels, between landscapes (regional scale) and within landscapes (landscape scale). Various indices have been developed for use at both these scales, and include various combinations of estimates of numbers of different landscape units present, areas occupied by different landscape units and lengths of landscape unit perimeters (Turner & Gardner, 1991; Turner *et al.*, 1991; Hargis *et al.*, 1998). Other approaches consider the underlying patterns of plant species richness and variations in evenness (Scheiner, 1992). To date, most studies using these indices have investigated their ability to quantify landscape pattern in space and with time. In general, the indices have been considered useful if they have indicated differences between landscapes already known to differ substantially, but there are few instances of indices providing new insights, allowing extrapolation to other situations. Similarly, the relationship between the index and functional aspects of heterogeneity have rarely been explored (Cale & Hobbs, 1994).

Changing the scale of measurement affects the results obtained. Changing the grain (spatial resolution) and extent (total area of study) affects measures of heterogeneity because they are sensitive to the number of patches detected, which changes with changes in scale (Turner *et al.*, 1991). The spatial configuration of patches influences the rate of change in their number with changing scale.

An adequate description of landscape heterogeneity must include not only a description of the number, sizes and configurations of patches, but also some characterisation of the structure and composition within them. Study of landscape heterogeneity has tended to focus on one of these scales (i.e. configuration of landscape units within the landscape or configuration within landscape units), but seldom both together.

The interpretation of differences in heterogeneity between landscapes is difficult. If one landscape has a higher index than another, this may or may not have any significance when particular processes or functions are considered. The significance of the measured difference in heterogeneity depends on how well measured heterogeneity corresponds to functional heterogeneity. Functional heterogeneity is process-dependent and hence can only be defined in the context of a particular process (e.g. animal movement, water flow) (Cale & Hobbs, 1994). The measures of landscape heterogeneity currently available are frequently difficult to interpret and should be used with caution in both research and management.

Ecotones and edges

A component of landscape structure which has received increasing attention recently is the ecotone, or edge between adjacent patches. Ecotones can be considered at a variety of scales, ranging from the biome down to the individual patch. Ecotones are considered important because they represent the boundary between different patches through which various landscape flows pass (Holland *et al.*, 1991; Hansen & di Castri, 1992; Gosz, 1993). One component of this is that the ecotone between different biome or vegetation types could be expected to be the place where the first indications of responses to global climatic changes would be detected. Patch size and shape strongly affect the amount of edge that is present. A set of phenomena known as 'edge effects' is associated with edges. These result from physical, chemical and biotic transfer into patches from adjacent patches (see below).

Landscape fluxes

Connectivity and movement of biota

Connectivity refers to the degree to which flows (for example, of animals or materials) are possible between different patches. For biota, populations in one area may be linked to other neighbouring populations through dispersal or source – sink relations. More generally, biota often need to move across the landscape for a variety of reasons, including dispersal and resource acquisition, and that movement is required to counter the potential effects of fragmenting populations into small, isolated units. Movement can be thought of as minimising the impacts of demographic stochasticity and inbreeding depression.

A habitat network can be defined as an interconnected set of habitat elements which together allow

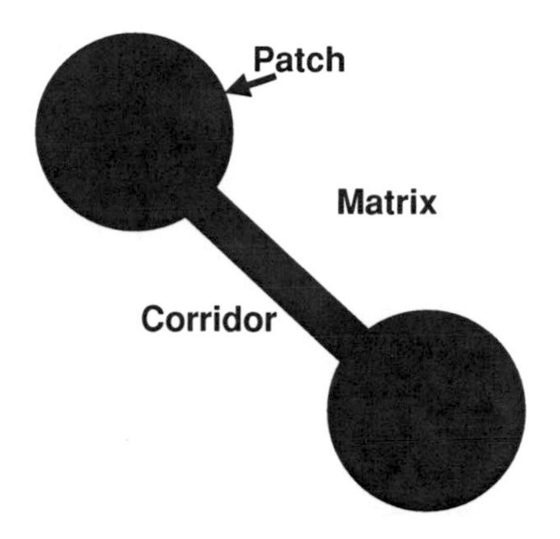

Fig. 3.1. Simple patch–corridor–matrix model of landscapes, which dichotomises landscape elements into habitat and non-habitat (matrix), with corridors linking habitat patches. Derived from Forman (1995).

for movement of biota and enhance population survival probabilities. The basis of this concept is the assumption that patches of habitat are generally embedded in a matrix of 'non-habitat'. A common perception among conservation biologists is that the matrix is hostile to the organisms within the relatively small fragments. This concept has been refined to more generally describe landscapes in terms of patch, corridor and matrix (Forman, 1995), with corridors representing narrow strips of habitat or, if not habitat, at least vegetation that allows biotic movement between patches (Fig. 3.1). The retention or provision of corridors between fragments is often seen as an important element of landscape management and conservation (Harris, 1984; Hudson, 1991). Some argue, however, that the requirement for faunal movement may have been overstated, and that corridors may not be required to foster it when it is necessary (Simberloff & Cox, 1987; Simberloff *et al.*, 1992). While movement along corridors is frequently assumed to occur, there have been relatively few studies that have shown that corridors are actually required for movement (Hobbs, 1992). Studies that have been frequently cited as illustrating corridor use for faunal movement, do not, in fact, provide clear evidence. The types of study required to establish unequivocally that corridors are important for faunal movement are difficult and costly to design and implement and require intensive, long-term observations. Few studies provide good data on animal movement, although recent studies of marked or radio-tagged animals have indicated that some species do use corridors for movement, in preference to moving across open ground. Other studies have shown that corridors can act in complex ways, enhancing movement of some species in some cases, while inhibiting that of others (e.g. Hill, 1995).

The universality of the patch–corridor–matrix model has been questioned recently, particularly in relation to the influence of the matrix, which may not always be completely hostile to all elements of the biota (McIntyre & Hobbs, 1999). Although corridors have received widespread attention and are frequently part of conservation plans and activities, their utility is often debated, and they are in reality only one part of a broader picture (Hobbs, 1992; Dawson, 1994; Wilson & Lindenmayer, 1996; Hobbs & Wilson, 1998; Bennett, 1999). Probably a more constructive approach is to consider overall landscape connectivity, the extent to which different elements of the landscape are functionally connected from the viewpoint of particular biotic elements. Connectivity in a landscape depends on the relative isolation of habitat elements from one another and the extent to which the matrix represents a barrier to movement of organisms. There have been attempts to derive some generalities concerning connectivity from modelling simple geometric relationships arising from the distribution of habitat patches in landscapes with different proportions of habitat and non-habitat (With & Crist, 1995; Pearson *et al.*, 1996; Wiens, 1997; With, 1997). These studies have indicated that there may be thresholds where small changes in the proportion of habitat present result in large changes in connectivity. However, the problem remains that different species will perceive the landscape differently, and landscape connectivity will depend on the mobility and habitat specificity of the species involved (Cale & Hobbs, 1994; Pearson *et al.*, 1996; Kolasa & Waltho, 1998). Also, most modelling efforts consider straightforward habitat versus non-habitat dichotomies; they therefore do not deal with the possibility that the matrix may be more or less permeable, or conversely, resistant to movement. Hence the role of all landscape elements in facilitating or inhibiting movement needs to be considered (Taylor *et al.*, 1993; Wiens, 1997; McIntyre & Hobbs, 1999).

An important function of animal movement is the recolonisation of patches that have suffered local extinction. The concept of metapopulations has received increasing attention recently, especially in the context of fragmented landscapes. This concept assumes that species' populations can exist as a series of interlinked sub- or metapopulations which persist in separate habitat patches, but may be linked by interpatch movement (McCulloch, 1996; Hanski & Gilpin, 1997). More explicitly, the metapopulation approach assumes that 'populations are spatially structured into assemblages of breeding populations and that migration among the local populations has some effect on local dynamics, including the possibility of population re-establishment following extinction' (Hanski & Simberloff, 1997). Subpopulations characteristically undergo periodic extinctions and re-establish following recolonisation by dispersers from other elements of the metapopulation. Determining whether metapopulation dynamics actually occur in fragmented landscapes is, however, relatively difficult, and there is still some question as to the general validity of the concept.

Metapopulation theory deals with the factors that determine the likelihood of local extinctions and subsequent recolonisation. In general, subpopulations in small isolated patches are considered more likely to go extinct while large well-connected patches are more likely to be recolonised, resulting in different probabilities of species occurring on patches of different sizes and connectivity. Clearly, the validity of these relationships depends on the extent to which our perceptions of the influence of size and connectivity match the actual influence of these parameters on the species involved. It is also possible that metapopulations exhibit non-linear dynamics and may collapse unexpectedly to extinction (Hanski *et al.*, 1995), making it difficult to predict the occurence or dynamics of populations in fragmented landscapes.

Water, nutrients and material

Fluxes of water, nutrients and material are often important determinants of landscape patterning, and can also be strongly affected by that patterning and its dynamics (Hornung & Reynolds, 1995). Surface and subsurface hydrology is determined by geology, landform and surface characteristics. Patch characteristics and configurations can influence interception, infiltration, evapotranspiration and runoff patterns. Thus, for instance, perennial vegetation in semi-arid areas can intercept and utilise more water than annual cropland, with the result that runoff and water input to groundwater will be greater under an annual cropping regime. Distribution of water across the landscape clearly influences the distribution of patch types – for instance, riparian forests and swamps can develop only where there are large amounts of water available close to the surface.

Nutrient and material fluxes across the landscape result primarily from the processes of erosion, leaching and transport by wind and water. Redistribution of nutrients over long time periods has resulted in areas of the landscape accumulating nutrients which have been eroded from other areas. This in turn exerts a strong influence on the types of patch present in each area. Resource-rich patches may be particularly important from many points of view, both in terms of vegetation composition and faunal assemblages, and also in terms of potential human utilisation. Landscape patterning can influence erosional processes, with some patch types acting as interceptors of eroded material. For instance, in managed landscapes windbreaks act both to reduce the degree of wind erosion and to intercept eroded material. Riparian strips also intercept eroded material and nutrients transported in runoff, and hence affect nutrient and material loadings in waterways. Riparian strips are frequently viewed as 'buffer zones', and this concept can be extended to any type of patch which protects adjacent patches from nutrient or material inputs.

Nutrients can be transferred across the landscape in a variety of other ways. Particularly important vectors are animals, which may feed in one patch and defaecate in another, resulting in a transfer of nutrients between patches. In extreme cases, this can significantly increase the nutrient input into recipient patches, for instance in the case of seabird colonies. Fire also redistributes nutrients in smoke and ash, although it is difficult to quantify this effect.

Movement of pollutants and excess nutrients across the landscape is important in determining the nutrient and pollution loading in any given location. It is becoming increasingly recognised that landscape and regional processes have to be considered when dealing with pollution problems in particular ecosystems. For instance, deposition of nitrogen or sulphur arising from regional atmospheric pollution can have important local impacts (Appleton, 1995), and applications of phosphorus fertiliser to agricultural land can run off and cause pollution problems in adjacent lakes or estuaries (e.g. Boggess *et al.*, 1995; Flaig & Reddy, 1995). For large river systems, pollution problems may have to be tackled not just at the landscape scale, but also at the national or continental level (Malle, 1996).

LANDSCAPE CHANGE

Natural change

Natural landscapes are in a constant state of flux. Patch composition and configuration changes in response to a variety of processes, particularly climatic changes and disturbance regimes. Climate varies greatly over a number of different time-scales, ranging from the long-term changes which take place over thousands of years in response to glacial/interglacial cycles to relatively short-term phenomena such as drought cycles over periods of decades. A considerable body of evidence is available which shows that natural vegetation responds to long-term climatic changes by migrating across the landscape (Delcourt & Delcourt, 1991). However, individual species migrate at different rates and, at any given time, the composition of the landscape depends on which species are already present, which species are migrating in, and how these two sets of species interact. Past landscapes in any particular area are liable to have been quite different to those present today.

Infrequent episodic climatic events are particularly important agents of landscape change. Events such as droughts, exceptionally high rainfalls or windstorms are capable of switching patches from one type to another very quickly (Hobbs, 1994). For instance, a grassland patch can become a shrubland patch following a year of exceptionally high rainfall. In this respect, episodic events can be considered as particular types of disturbance. Any form of disturbance, such as fire, flood, storm or landslide can also act in this way. As discussed earlier, the distribution of patches within a landscape is determined by the overall disturbance regime.

Landscape change may also be biotically forced. Species migrations, as discussed above, are important here, but the effects of animals and pathogens should also be noted. Changes in abundances of herbivores can lead to changes in patch types and configurations, and pathogen spread can significantly alter landscape composition and structure (Knight, 1987).

There has been considerable debate over the idea that, while individual landscape patches are in a constant state of flux, the landscape as a whole is in a state of equilibrium. In other words, over the entire landscape, the distribution of patches of different types or ages should remain constant through time. For this to be the case a certain minimum area is required, known as the 'minimum dynamic area' (Pickett & Thompson, 1978; Baker, 1992). The validity of the assumption of equilibrium landscapes depends greatly on the scale considered (Turner *et al.*, 1993), and recent writers have questioned its generality. Examination of areas where large tracts of natural ecosystems persist indicates a distribution of patch types and ages that is far from equilibrium (Sprugel, 1991).

Human-induced change: modification and fragmentation

Fragmentation of natural ecosystems has been called one of the most pervasive changes in terrestrial ecosystems across the earth. It occurs wherever land transformation results in the removal of the pre-existing land cover and its replacement with other cover types, be it urban, agriculture, production forestry or other anthropogenic land uses. Such activities may remove most of the pre-existing cover types, transforming landscapes and regions in relatively short periods. Examples include deforestation in tropical areas such as Amazonia, and clearance for agriculture in the Midwest of the United States and in Australia (Hobbs & Saunders, 1993; Laurance & Bierregaard, 1997; Schwartz, 1997). In these and many other cases, the result of such activities is that the pre-existing cover types are left as patches in a modified matrix of production land. These patches may be of varying sizes and may be isolated from other such patches to varying degrees.

Impact of fragmentation on biota

Considerations of fragmented landscapes have focused on factors that influence the likelihood of persistence of the biota retained in the remaining fragments of habitat. Obviously, the major impact of transforming one ecosystem type to another is the dramatic reduction in the amount of habitat left for those species dependent on the pre-existing ecosystem type. The actual area of many ecosystem types is reduced to a few percent of their original extents, and this not only affects their representation in the landscape and in any reserve system that is developed, but also affects the ability of individual species to survive. Species vary greatly in their habitat requirements, and it is likely that in fragmented systems, populations of some species are limited simply by the amount of habitat available.

In addition to the simple amount of habitat remaining, the spatial arrangement of that habitat may also be important. Considerable debate centred for many years on whether fragmented systems would perform better in terms of retaining species if they consisted of a few large fragments versus lots of small patches (Simberloff, 1982; Willis, 1984; Lomolino, 1994; Boecklen, 1997). The so-called SLOSS (single large or several small) debate stemmed from ideas generated from the theory of island biogeography, which related species numbers on islands to the island size and isolation. There is no real resolution to this debate, since different types of organism will probably require different amounts and distributions of habitat. For instance, large mammalian predators probably require large uninterrupted tracts of habitat, whereas insects and plants can probably survive well in much smaller areas.

Other factors may also limit populations, however. For instance, some species may be limited, not by the areal extent of habitat, but by the availability of a particular resource on which they depend. An example may be nectarivorous birds which depend on a continuous supply of nectar throughout the year. If the fragmentation process removes a particular plant community or species which provided nectar resources at a critical time of year, nectarivorous birds could no longer survive in the area.

Connectivity and corridors

As noted above, a further factor influencing population viability may be the ability of species to move around the landscape. Species need to move for numerous reasons, such as to fulfil resource requirements, to disperse to new territories, or to migrate. Some species may move only short distances and hence be able to exist entirely within an individual habitat patch, but others may need to move either between different habitat types or across the landscape between different fragments. Others may need to move across entire continents as they follow migration routes.

In an unfragmented landscape, the ability of species to move depended entirely on the 'permeability' of the different ecosystem types present, i.e. the extent to which different patch types either aided or hindered movement. The permeability of different landscape elements is likely also to vary from species to species. Some species will only move in thickly vegetated cover types, whereas other species favour more open country. In a fragmented landscape, dense wooded cover is often replaced with low, open crop or pasture. In this situation we might expect that the permeability of the landscape would decrease for species which require wooded cover. The exact impact of the fragmentation depends both on its extent and pattern and also on how particular species view the landscape. Using simple geometric models, it can be shown that the extent to which the various habitat patches in the landscape remain 'connected' declines as the extent of habitat removal increases. However, there is also evidence that the relationship is not linear, and that there is a threshold in 'connectivity' at about 60% habitat removal, at which the level of connectivity declines sharply.

Edge effects and processes

In addition to the actual removal of the pre-existing ecosystem type and the creation of the patch/matrix-type landscape discussed above, ecosystem modification can also occur, whereby the pre-existing ecosystem type is not removed entirely, but is altered by human use, such as by grazing, timber harvesting and so on. The two processes of fragmentation and modification can occur simultaneously, and often fragments remaining in a modified matrix are also severely impacted by modifying influences (Hobbs, 1993*b*; McIntyre & Hobbs, 1999). Thus, often it is important to consider not only the amount of a particular ecosystem which is left, but also the condition of the remaining patches.

Factors which can influence the condition of patches include both factors operating within individual patches and factors arising from the surrounding landscape. An important feature of fragmented ecosystems is that the pre-existing disturbance regime can no longer operate (Hobbs, 1987). For instance, fire frequency and intensity may increase or decrease in isolated fragments, and this in turn affects the survival and regeneration of the plant communities in the fragments. Also, if movement of faunal species around the landscape becomes restricted, seed dispersal, and hence gene flow, of plant species may be reduced or even cease. As species are lost from fragmented systems, so too are any functions that these species performed. For instance, if burrowing mammals disappear from fragmented systems, soil disturbance and turnover is reduced, and hence nutrient cycling and water infiltration to the soil may be affected.

The isolation of fragments in a modified matrix opens the fragments up to a range of microclimatic, hydrological and biotic changes (Saunders *et al.*, 1991). A primary result of this isolation is the creation of an edge between the pre-existing ecosystem and the modified matrix. At this edge, the conditions prevalent in the matrix impinge into the fragment, and a series of 'edge effects' can occur. The edges of isolated fragments experience different microclimatic conditions, receive more nutrients transferred from adjacent patches and may have a higher incidence of weed invasion or predation than the interior of the fragment (Kapos, 1989; Matlack, 1993; Murcia, 1995). Such changes can result in changes in vegetation structure and floristic and faunal species compositions. The distance to which edge effects permeate a patch varies with the type of patch and the feature being considered: for instance microclimatic changes may occur only over tens of metres, while changes in predation rates or weed invasion may extend much further. For instance, edge-related impacts can result in considerable mortality of the species up to 100 m from the fragment edge (Laurance *et al.*, 1997), and this impact can be particularly severe in small fragments where virtually the entire fragment becomes 'edge' rather than 'interior' habitat.

In addition to the localised edge effects, the surrounding landscape can influence fragments more generally. For instance, significant changes in hydrology can result from the process of replacing one cover type with another. These large-scale changes influence not only the production matrix but also the ecosystems remaining in fragments (George *et al.*, 1995). Also, animals and plants prevalent in the management matrix may exert profound influences on the biota in fragments. In particular, domestic stock can dramatically alter the structure and composition of fragmented systems by grazing and trampling. Similarly, feral predators can be responsible for the decline of faunal species which would otherwise survive in the fragmented system. Finally, invasive plant species may disperse into fragmented systems from the surrounding matrix, and again these have the potential to significantly alter the structure and dynamics of the fragmented system. Hence fragmentation results not only in the formation of small, isolated fragments, but also in landscape-scale system changes.

THE BASIS OF LANDSCAPE-SCALE RESTORATION: UNDERSTANDING THRESHOLDS AND SETTING GOALS

Most of the information and methodologies on ecological restoration centre on individual sites, and ultimately restoration activities have to be conducted in particular sites. However, site-based restoration has to be placed in a broader context, and is often insufficient on its own to deal with large-scale restoration problems. Landscape- or regional-scale processes are often either responsible for ecosystem degradation at particular sites, or alternatively have to be restored to achieve restoration goals. Hence restoration is often needed both within particular sites and at a broader landscape scale.

How are we then to go about restoration at a landscape scale? What are the relevant aims? What landscape characteristics can we modify to reach these aims, and do we know enough to be able to confidently make recommendations on priorities and techniques?

There are a number of steps in the development of a programme of landscape-scale restoration, which can be outlined as follows:

1. Assess whether there is a problem which requires attention, for instance:
   (a) changes in biotic assemblages (e.g. species loss or decline, invasion)
   (b) changes in landscape flows (e.g. species movement, water and/or nutrient fluxes)
   (c) changes in aesthetic or amenity value (e.g. decline in favoured landscape types).
2. Determine the causes of the problem, for instance:
   (a) removal and fragmentation of native vegetation
   (b) changes in pattern and abundance of vegetation/landscape types
   (c) cessation of historic management regimes.
3. Determine realistic goals for restoration, for instance:
   (a) retention of existing biota and prevention of further loss
   (b) slowing or reversal of land or water degradation processes
   (c) maintenance or improvement of potential for biological production
   (d) integration of approaches to tackle multiple goals.
4. Develop cost-effective planning and management tools for achieving agreed goals:
   (a) determining priorities for action in different landscape types and conditions
   (b) spatially explicit solutions
   (c) acceptance and 'ownership' by managers and landholders
   (d) an adaptive approach which allows course corrections when necessary.

This short list hides a wealth of detail, uncertainty and science yet to be done. For instance, the initial assessment of whether there is a problem or not requires the availability of a set of readily measurable indicators of landscape 'condition' or 'health'. This ties in with recent attempts to use the concept of ecosystem health as an effective means of discussing the state of ecosystems (Costanza *et al.*, 1992; Cairns *et al.*, 1993; Shrader-Frechette, 1994). Central elements of ecosystem health are the system's vigour (e.g. production), organisation (or the diversity and number of interactions between system components) and resilience (the system's capacity to maintain structure and function in the presence of stress (Rapport *et al.*, 1998). Attempts have also been made to produce readily measurable indices of ecosystem health for a number of different ecosystems, although there is still debate over whether these are useful or not. In the same way, there have been recent attempts to develop a set of measures of landscape condition.

Aronson and Le Floc'h (1996*b*) present three groups of what they term 'vital landscape attributes' which aim to encapsulate (1) landscape structure and biotic composition, (2) functional interactions among ecosystems and (3) degree, type and causes of landscape fragmentation and degradation. Their list of 16 attributes provides a useful start for thinking about these issues, but many of the attributes are either difficult to measure or hard to interpret, or both. It thus remains difficult to conduct a practical assessment of whether a particular landscape is in need of restoration, and if so, what actions need to be taken. Steps towards this are being developed, at least for landscape flows, in the Landscape Function Analysis approach developed for Australian rangelands (see Ludwig *et al.*, 1997; Tongway & Ludwig, volume 2), which aims to provide easily measurable and interpretable indices of landscape function. These authors indicate that the interpretation of measures of landscape condition must be based on the goals set for that particular landscape. A particular condition may indicate that the landscape is quite acceptable for one type of use, but in need of restoration to allow the continuation or adoption of another land use.

Once a problem has been perceived, the correct diagnosis of its cause and prescription of an effective treatment is by no means simple. The assumption underlying landscape ecology is that landscape processes are in some way related to landscape patterns. Hence, by determining the relationship between pattern and process, better prediction of what will happen to particular processes (biotic movement, metapopulation dynamics, system flows, etc.) if the pattern of the landscape is altered in particular ways, can occur. Thus, we are becoming increasingly confident that we can, for instance, predict the degree of connectivity in a landscape from the proportion of the landscape in different cover types. Similarly, as landscapes become more fragmented, a greater proportion of the biota drops out, and again there may be thresholds or breakpoints where relatively large numbers of species are lost.

As part of this process, we need to develop ecological response models which capture the essence of the landscape and its dynamics. These models can be simple or complex, quantitative or conceptual, and there needs to be consideration of both general characteristics of landscapes and more specific elements relating to specific cases. General features of many systems seem to be the potential for the system to exist in a number of different states, and the likelihood that restoration thresholds exist, which prevent the system from returning to a less-degraded state without the input of management effort (Aronson *et al.*, 1993; Scheffer *et al.*, 1993; Milton *et al.*, 1994; Hobbs & Norton, 1996). Whisenant (1999) has recently suggested that two main types of such threshold are likely, one which is caused by biotic interactions, and the other caused by abiotic limitations. The type of restoration response needed will depend on which, if any, thresholds have been crossed (Fig. 3.2a). If the system has degraded mainly due to biotic changes (such as grazing-induced changes in vegetation composition), restoration efforts need to focus on biotic manipulations which remove the degrading factor (e.g. the grazing animal) and adjust the biotic composition (e.g. replant desired species). If, on the other hand, the system has degraded due to changes in abiotic features (such as through soil erosion or contamination), restoration efforts need to focus first on removing the degrading factor and repairing the physical and/or chemical environment.

In the latter case, there is little point in focusing on biotic manipulation without first tackling the abiotic problems. In other words, it is important that system functioning is corrected or maintained before biotic composition and structure are considered. Considering system function provides a useful framework for the initial assessment of the state of the system and the subsequent selection of repair measures (Tongway & Ludwig, 1996; Ludwig *et al.*, 1997). Where function is not impaired, goals for restoration can legitimately focus on composition and structure.

The same scheme can be considered at a landscape scale. At broad scales it becomes even more difficult to decide what should be restored, where and how. It can be hypothesised that restoration thresholds might exist at the landscape scale as are apparent in particular ecosystems or sites (Fig. 3.2b). One type of threshold relates to the loss of biotic connectivity as habitat becomes increasingly fragmented and modified, while another relates to whether landscape modification has resulted in broad-scale changes in landscape physical processes, such as hydrology. If the landscape has crossed a biotic threshold, restoration needs to aim at restoring connectivity. If, on the other hand, a physical threshold has been crossed, this needs to be treated as a priority. Hence, for instance, in a fragmented forested landscape, the primary goal may be the provision of additional habitat or re-establishing connectivity for particular target species, whereas in a modified river or wetland system, the primary need may be to re-establish water flows (Middleton, 1999). Restoration activities required to overcome particular physical changes may also act to overcome biotic thresholds. An example of this would be where extensive revegetation is required to counteract hydrological imbalances, and at the same time can have a positive impact on biotic connectivity (Hobbs, 1993*a*; Hobbs *et al.*, 1993).

If we accept that different types of thresholds in landscape function are possible, a number of important questions have to be asked in terms of restoration. First, does the threshold work the same way on the way up as it did on the way down, or is there a hysteresis effect? In other words, in a landscape where habitat area is being increased, will species return to the system at the same rate as they dropped out when habitat was being lost? Second, what happens when pattern and process are not tightly linked? For instance, studies in central Europe have illustrated the important role of

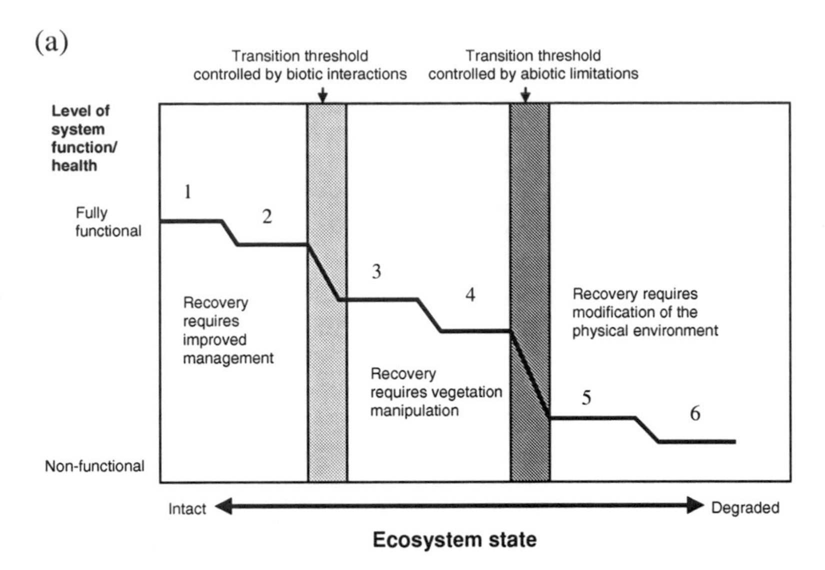

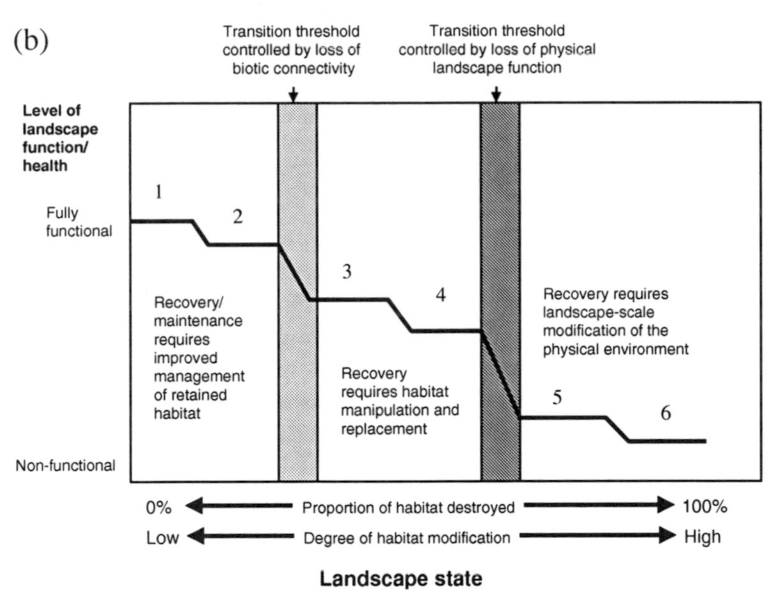

Fig. 3.2. (a) Conceptual model of system transitions between states of varying levels of function, illustrating the presence of two types of restoration threshold, one controlled by biotic interactions and one controlled by abiotic limitations. Adapted from Whisenant (1999). (b) A similar model applied to landscapes, indicating transition thresholds controlled by loss of biotic connectivity and loss of physical landscape function. From Hobbs & Harris (2001).

traditional management involving seasonal movement of sheep between pastures in dispersing seeds around the landscape (Bakker *et al.*, 1996; Fischer *et al.*, 1996; Poschlod *et al.*, 1996). The long-term viability of some plant species may be threatened by the cessation of this process, and restoration in this case will not involve any modification of the existing landscape pattern – rather it will entail the reinstatement of a management-mediated process of sheep movement. Hence, correct assessment of the problem and its cause and remedy require careful examination of the system and its components rather than generalised statements of prevailing dogma.

DEVELOPING GUIDELINES FOR LANDSCAPE-SCALE RESTORATION

Several approaches are possible to developing guidelines and options for landscape restoration. A hierarchy of guidelines can be considered, depending on the scope and goals of the restoration and management, and the level and detail of information available for making decisions. This can be set out as follows:

1. General guidelines which are applicable in most types of landscape (Whisenant, 1999).
2. Guidelines which are related to particular broad categories of landscape (McIntyre & Hobbs, 1999, in press).
3. Guidelines which are derived from consideration of specific species or suites of species in a given landscape (Lambeck, 1997).
4. Spatially specific options which are developed for particular landscapes in relation to existing landscape structure, target species and specified local management goals.

General guidelines

How do you go about determining how to conduct restoration at a landscape scale? What sort of landscape-level management and restoration is appropriate for different landscapes? A set of general principles, derived from island biogeography theory, suggest that bigger patches are better than small patches, connected patches are better than unconnected, and so on. For fragmented landscapes, such principles can be translated into the need to retain existing habitat patches, especially large ones, and existing connections, and to revegetate in such a way as to provide larger patches and more connections (Hobbs, 1993*a*). Whisenant (1999) provided a set of guidelines for landscape repair, which included the following:

1. Treat causes rather than symptoms.
2. Emphasise process repair over structural replacement.
3. Design repair activities at the proper scale.
4. Design landscapes to increase retention of limiting resources.
5. Design spatial variation into landscapes.
6. Design landscapes to maintain the integrity of primary processes.
7. Design linkages into landscapes.
8. Design propagule donor patches into landscapes.
9. Design landscapes to encourage animal dispersal of desired seed.
10. Design landscapes to encourage wind dispersal of desired seed.
11. Design landscapes to encourage positive animal interactions.

While these guidelines were primarily directed at semi-arid areas, they may have broader application. Whisenant (1999) suggests that the guidelines provide a framework for considering landscape interactions, but recognises that they do not produce specific quantitative designs. It is important that we attempt to move from these generalised guidelines to more specific recommendations for particular types of landscape, and finally to spatially explicit options for individual landscapes. Hence, I now consider attempts to categorise broad landscape types and identify the priority activities in each, and then examine options for developing spatially explicit landscape restoration plans.

Guidelines for broad landscape types

If we consider fragmented landscapes as an example, priority actions are likely to involve: (1) the protection of existing habitat patches; (2) their effective management; and (3) restoration both within patches and at a broader landscape scale. But where do we go from there? Which are the priority areas to retain? Should we concentrate on retaining the existing fragments or on restoration, and relatively how many resources (financial, manpower, etc.) should go into each? How much restored habitat is required, and in what configuration? When should we concentrate on providing corridors versus additional habitat? If we are to make a significant impact in terms of conserving remaining fragments and associated fauna, these questions need to be addressed in a strategic way.

McIntyre & Hobbs (1999) have examined these questions in terms of the range of human impacts on landscapes. They identified four broad types of landscapes (Table 3.1, Fig. 3.3), with intact and relictual landscapes at the extremes, and two intermediate states, variegated and fragmented. In variegated landscapes, the habitat still forms the matrix, whereas in fragmented landscapes, the matrix comprises 'destroyed habitat'. Each of the four levels described in Table 3.1 is associated with a particular degree of habitat destruction, and the categories are not entirely arbitrary. For instance, the distinction between variegated and fragmented landscapes reflects suggestions discussed earlier that landscapes in which habitats persist over more than 60% of the area are operationally not fragmented, since they consist of a continuous cluster of habitat. This broad division can be regarded as a 'first cut', and the provision of names for each category is for convenience rather than to set up a rigid classification. Further investigation is required to test these categories and to examine the need for further subcategories. For instance, functionally different types of 'fragmented' landscapes could be recognised.

Habitat modification alters the condition of the remaining habitat, and can occur in any of the situations illustrated in Table 3.1. Modification acts to create a layer of variation in the landscape over and above the straightforward spatial patterning caused by vegetation destruction. There is a tendency for habitats to become progressively more modified with increasing levels of destruction, owing to the progressively greater proportion of edge in remaining habitats.

The framework in Table 3.1 can assist in deciding where on the landscape to allocate greater and lesser efforts towards different management actions (McIntyre & Hobbs, in press). Three types of action could be applied to habitats for their conservation management:

1. *Maintain* the existing condition of habitats by removing and controlling threatening processes. It is generally much easier to avoid the effects of degradation than it is to reverse them.
2. *Improve* the condition of habitats by reducing or removing threatening processes. More active management may be needed to initiate a reversal of

Table 3.1. *Four landscape states defined by the degree of habitat destruction*

Landscape type	Degree of destruction of habitat (% remaining)	Connectivity[a] of remaining habitat	Degree of modification[b] of remaining habitat	Pattern of modification[b] of remaining habitat
Intact	Little or none (>90%)	High	Generally low	Mosaic with gradients
Variegated	Moderate (60–90%)	Generally high but lower for species sensitive to habitat modification	Low to high	Mosaic which may have both gradients and abrupt boundaries
Fragmented	High (10–60%)	Generally low but varies with mobility of species and arrangement on landscape	Low to high	Gradients within fragments less evident
Relictual	Extreme (<10%)	None	Generally highly modified	Generally uniform

[a]From Pearson *et al.* (1996).
[b]Modified from McIntyre & Hobbs (in press).

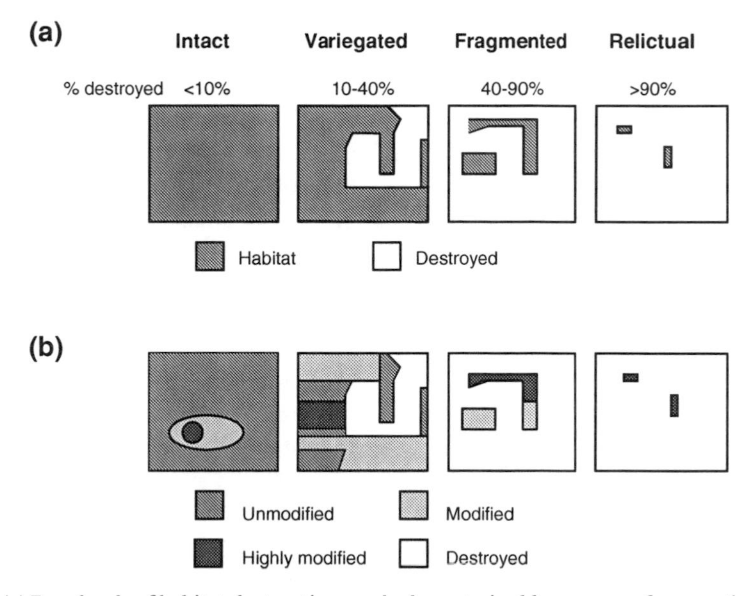

Fig. 3.3. (a) Four levels of habitat destruction, each characterised by a range of proportions of habitat destroyed. (b) Pattern of habitat modification overlying landscape patterns of habitat destruction depicted in (a). Although any combination of destruction and modification levels is theoretically possible, those considered to be typical of different destruction levels are illustrated. Modified from McIntyre and Hobbs (1999, in press).

condition (e.g. removal of exotic species, reintroduction of native species) in highly modified habitats.

3. *Reconstruct* habitats where their total extent has been reduced below viable size using replanting and reintroduction techniques. Here reconstruction implies the return of key habitat elements to areas where habitat has previously been completely removed. As this is so difficult and expensive, it is a last resort action that is most relevant to fragmented and relictual landscapes. We have to recognise that restoration often does not come close to restoring habitats to their unmodified state, and this reinforces the wisdom of maintaining existing ecosystems as a priority.

The next stage is to link these activities to specific landscape components (matrix, connecting areas, buffer areas, fragments) in which they would be most effective, and to determine priorities for management action in different landscape types. A general approach might be to build on strengths of the remaining habitat by filling in gaps and increasing landscape connectivity, increasing the availability of resources by rehabilitating degraded areas, and expanding habitat by revegetating to create larger blocks and restore poorly represented habitats. The first priority is the maintenance of elements which are currently in good condition. This will be predominantly the vegetated matrix in intact and variegated landscapes and the remnants which remain in good condition in fragmented landscapes. There may well be no remnants left in good condition in relict landscapes. Maintenance will involve ensuring the continuation of population, community and ecosystem processes which result in the persistence of the species and communities present in the landscape. Maintaining fragments in good condition in a fragmented system may also require activities in the matrix to control landscape processes, such as hydrology. For instance, numerous examples have been documented where maintenance of habitat patches depends on the effective management of surface and subsurface water movement in the surrounding landscape (Rowell, 1986; Barendregt *et al.*, 1995; George *et al.*, 1995).

The second priority is the improvement of elements that have been modified in some way. In variegated landscapes, buffer areas and corridors may be a priority, while in fragmented systems, improving the surrounding matrix to reduce threatening processes will be a priority, as indicated above. In relict landscapes, improving the condition of fragments will be essential for their continued persistence. Improvement may involve simply dealing with threatening processes such as stock grazing or feral predators, or may involve active management to restore ecosystem processes, improve soil structure, encourage regeneration of plant species, or reintroduce flora or fauna species formerly present there (Hobbs & Yates, 1997).

Reconstruction is likely to be necessary only in fragmented and relict areas. Primary goals of reconstruction will be to provide buffer areas around fragments, to increase connectivity with corridors, and to provide additional habitat (Hobbs, 1993*a*). While some basic principles of habitat reconstruction have been put forward, the benefits of such activities have rarely been quantified. Questions remain about which characteristics of 'natural' habitat are the most important to try to incorporate into reconstruction, and what landscape configurations are likely to be most effective.

In order to answer such questions, it becomes very important to specify clearly what the conservation goals are for the area. Lambeck (1997) has recently contended that more efficient solutions to conservation problems can be developed if we take a strategic approach rather than a generalised one. This involves developing a clear set of conservation objectives rather than relying on vague statements of intent. One set of objectives relates to the achievement of a comprehensive, adequate and representative set of reserves or protected area networks. Another, complementary set of objectives relate to the adequacy of the existing remnant vegetation (not only reserves). The process of setting conservation objectives in any given area can be simplified by identifying a set of key or 'focal' species which are most at risk from the main threats identified in the area, in essence a multi-species indicator/umbrella species approach.

To identify focal species, Lambeck (1997) recognised three distinct sets of species, each of which were likely to be limited or threatened by particular characteristics of the landscape (Fig. 3.4). These were:

1. Area or habitat limited species, i.e. species whose numbers are limited by the availability of large enough patches of suitable habitat.

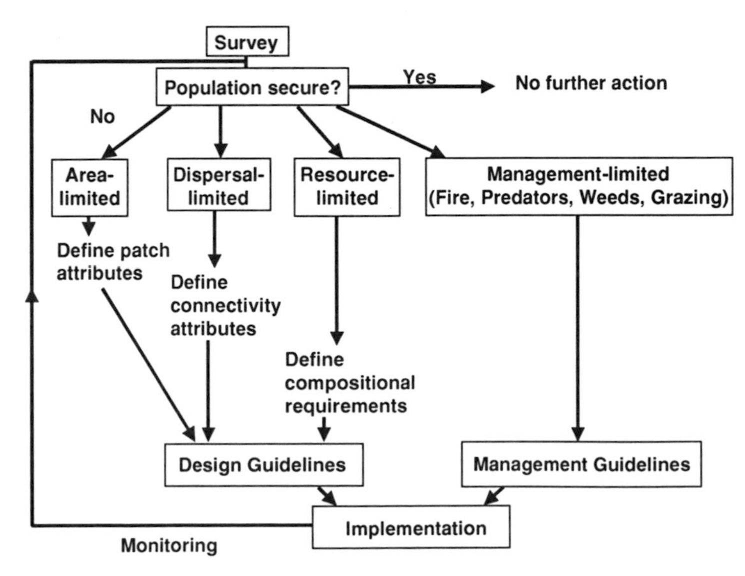

Fig. 3.4. A process for selecting focal species in any given landscape, based on whether species are limited by area, movement or management, and resulting in landscape design guidelines and management guidelines. Modified from Lambeck (1997).

2. Movement limited species, i.e. species whose numbers are limited by the degree to which they can move between habitat patches.
3. Management limited species, i.e. species whose numbers are limited by processes such as predation, disturbance, fire and the like, which can be manipulated within particular sites.

Design of landscape reconstructions is based on the requirements of the most sensitive species in each of these categories. For instance, if you can identify which species have the requirement for the largest areas of habitat, you can start assessing the adequacy of the current landscape for that species, and hence all other species with less-demanding habitat requirements, and can also start making recommendations on where and how much habitat reconstruction needs to be undertaken.

Spatially explicit options for specific landscapes

The ability to make spatially explicit recommendations as to where restoration activities should occur is essential if real solutions are to be developed and implemented. Moving from generalised guidelines to specific options for a particular area requires that we are able to translate the generalised guidelines into concrete recommendations of what to do where. There are relatively few examples where this translation has been accomplished successfully. Part of the problem lies in the fact that every landscape is unique in terms of its biophysical characteristics and its pattern of human alteration. Translating a general guideline thus has to take account of the unique characteristics of the area being managed/restored and refer directly to the spatial realities of each situation. Outlined in Box 3.1 is one attempt to do so in the Western Australian wheat belt.

While this approach enables us to identify the minimum patch sizes required for species to have a reasonable probability of occurrence, and to identify patches that were too isolated to be occupied, the results do not ensure that populations will persist in the long term. To assess this, further more detailed research and monitoring will be required. In the meantime, land managers are being advised to ensure that their landscape has patches of habitat equal to or exceeding the specified size, and that these patches are separated by distances less than those specified. These patches need to be distributed from one boundary of the management area to the other and connected by high-quality strips of linear habitat. Such a design aims to ensure that populations of all species are linked across the area being managed.

This method clearly requires a minimum of spatial landscape data and observations of the biota, and further refinement is possible with more detailed information. Even with the minimal data requirements, the approach is still very time-consuming, and requires that each case be analysed independently. For the more general application of the approach, it may be possible to work back from the particular to a more generalisable approach, if landscapes with similar characteristics (biophysical and anthropogenic) can be identified and grouped. In that case, the approach adopted in one landscape could be applied to other landscapes with similar characteristics.

CONCLUDING REMARKS

In this chapter I have summarised what we know about how landscapes work, and how this might be utilised for the planning and implementation of landscape-scale restoration. Our level of understanding of landscape patterns and processes is increasing, and our ability to work effectively at the landscape scale is being facilitated by improved tools such as remote sensing and GIS.

Landscape-scale restoration is still largely in its infancy. The recognition of the importance of embarking on restoration at landscape and regional scales is increasing, and examples of landscape-scale projects are beginning to accumulate. However, these are often still in their early stages, and it will take time to assess their efficacy. Because of the temporal and spatial scales involved, novel and integrated approaches need to be taken to the problem. Often the only way to assess different scenarios is by computer modelling, and it is usually impossible to design replicated landscape-scale experiments. Hence, there is a need to learn as we

Box 3.1 The Western Australian wheat belt

The wheat belt of Western Australia has been extensively cleared for agriculture, leaving native vegetation as mostly small isolated fragments (Hobbs & Saunders, 1993; Saunders *et al.*, 1993). While the lower parts of the landscape are under threat from salinisation (George *et al.*, 1995), in the remainder of the landscape the primary threats are habitat loss and isolation. Restoration options are currently being sought to deal with threats to the conservation of the region's highly diverse biota.

Using the scheme devised by Lambeck (1997), species that are considered to be threatened by each of the threatening processes are grouped and ranked in terms of their sensitivity. Presence/absence surveys of vegetation remnants indicate the distribution of the various species whose populations are limited by the amount of habitat available or by the degree of isolation of habitat patches. Analysis of the spatial attributes of the vegetation remnants in the landscape undertaken using GIS routines then determines the characteristics of habitat patches where species do and don't occur. This enables the specification of the minimum fragment size and the maximum interpatch distance that is required for these species to have a specified probability of occurring. It is then possible to identify all remnants that do not meet these criteria and specify the amount of habitat reconstruction required to produce fragments of adequate size. Similarly, it is possible to identify all remnants that are too isolated for the most dispersal-limited species and identify the need for the construction of intermediate habitat patches or corridors. Maps are then produced which indicate the extent to which each patch needs to be expanded or connected in order to have an equivalent probability of being occupied by the species which had the greatest demand for that patch type.

This approach has been used to determine the requirements for bird habitat in four watersheds in the wheat belt of Western Australia, each covering an area of approximately 20–30 000 ha (see Wallace, 1998; Lambeck, 1999). An example is given in Fig. B3.1.

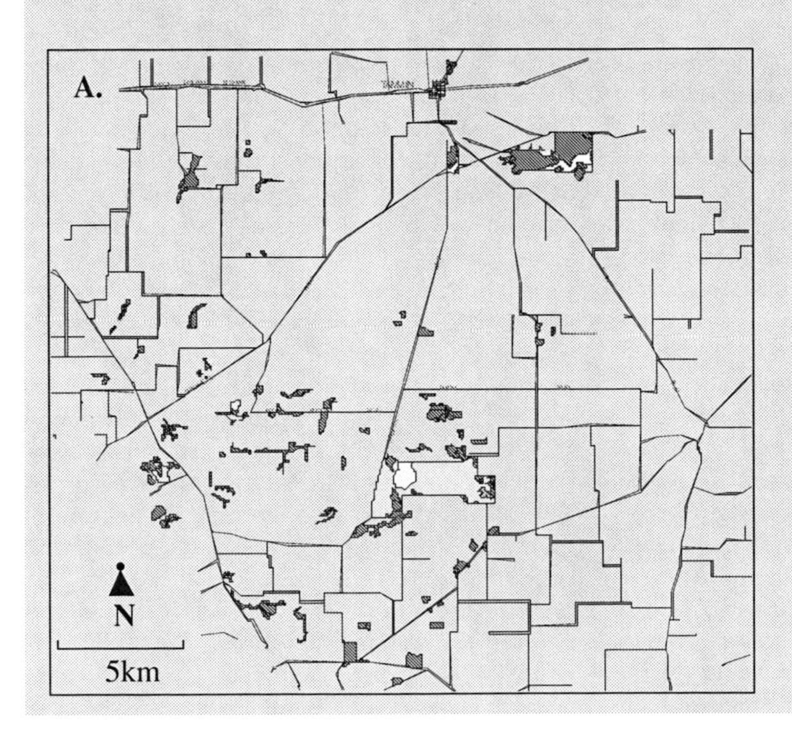

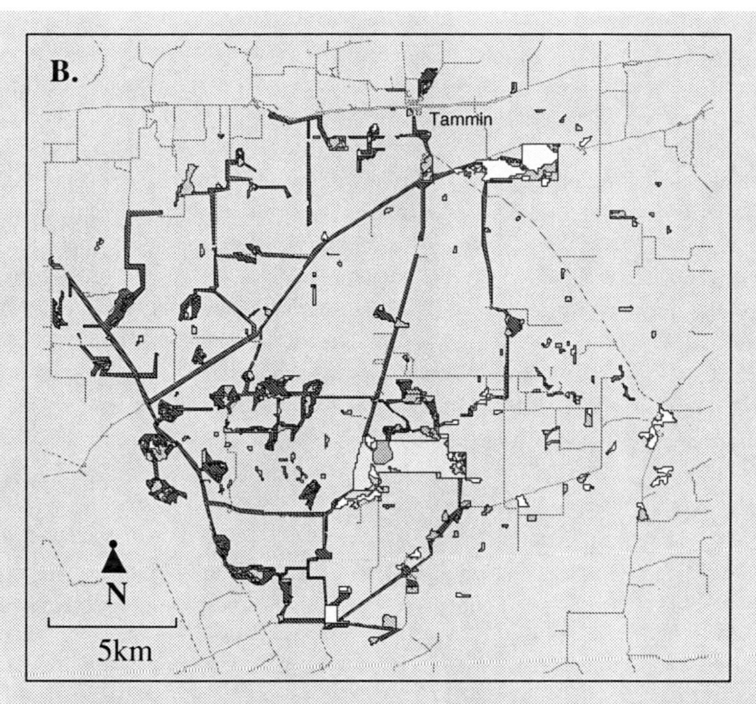

Fig. B3.1. An example of the development of management options for a highly fragmented landscape, based on the estimated habitat and connectivity requirements of focal bird species. (a) The landscape as it currently exists in the South Tammin subcatchment (watershed) in southwest Western Australia. Shaded areas are native vegetation of different types. (b) Recommendations for revegetation to enlarge existing vegetation patches and connect isolated patches. From Hobbs & Lambeck (in press).

go, and to build a degree of adaptability into the restoration designs.

Once the options for restoration have been derived from guidelines or more detailed ecological response models, these then have to be considered in the broader context of individual and societal goals (Hobbs & Lambeck, in press). Restoration is not an alternative to conservation or sound management, but is often a necessary part of these activities. Restoration at the landscape scale often also involves tackling multiple issues at once, and balancing conservation and production. The idea of 'reintegrating' landscapes embodies the idea that successful restoration must encompass the biophysical, social and economic realities of the situation (Aronson & Le Floc'h, 1996*a*, *b*; Hobbs & Saunders, 1993).

To succeed, restoration activities need not only to be based on sound ecological principles and information, but also to be economically possible and practically achievable. Almost anything is possible if there are enough human and financial resources available, but usually the cost and social desirability of foregoing other options limits the potential extent of restoration activities. Selling the benefits of restoration is often difficult due to the relatively long timeframes involved in achieving anything. Discount rates often mean that short-term costs greatly outweigh any long-term benefits. Often another primary driver in deciding which options will be pursued is the prevailing political climate, which drives government support and funding for restoration activities.

Broad-scale restoration is going to become increasingly necessary as humans continue to modify and use the earth and its resources, and we need to ensure that we continue to improve our ability to provide restoration options at these broad scales.

REFERENCES

Appleton, E.L. (1995). A cross-media approach to saving the Chesapeake Bay. *Environmental Science and Technology*, **29**, 550A–555A.

Aronson, J. & Le Floc'h, E. (1996*a*). Hierarchies and landscape history: dialoging with Hobbs and Norton. *Restoration Ecology*, **4**, 327–333.

Aronson, J. & Le Floc'h, E. (1996*b*). Vital landscape attributes: missing tools for restoration ecology. *Restoration Ecology*, **4**, 377–387.

Aronson, J., Floret, C., Le Floc'h, E., Ovalle, C. & Pontanier, R. (1993). Restoration and rehabilitation of degraded ecosystems in arid and semiarid regions. 1: A view from the South. *Restoration Ecology*, **1**, 8–17.

Baker, W.L. (1992). The landscape ecology of large disturbances in the design and management of nature reserves. *Landscape Ecology*, **7**, 181–194.

Bakker, J.P., Poschlod, P., Strykstra, R.J., Bekker, R.M. & Thompson, K. (1996). Seed banks and seed dispersal: important topics in restoration ecology. *Acta Botanica Neerlandica*, **45**, 461–490.

Barendregt, A., Wassen, M.J. & Schot, P.P. (1995). Hydrological systems beyond a nature reserve, the major problem in wetland conservation of Naardermeer (The Netherlands). *Biological Conservation*, **72**, 393–405.

Bell, S.S., Fonseca, M.S. & Motten, L.B. (1997). Linking restoration and landscape ecology. *Restoration Ecology*, **5**, 318–323.

Bennett, A.F. (1999). *Linkages in the Landscape: The Role of Corridors and Connectivity in Wildlife Conservation.* Gland, Switzerland: IUCN.

Boecklen, W.J. (1997). Nestedness, biogeographic theory, and the design of nature reserves. *Oecologia*, **112**, 123–142.

Boggess, C.F., Flaig, E.G. & Fluck, R.C. (1995). Phosphorus budget–basin relationships for Lake Okeechobee tributary basins. *Ecological Engineering*, **5**, 143–162.

Cairns, J.J., McCormick, P.V. & Niederlehner, B.R. (1993). A proposed framework for developing indicators of ecosystem health. *Hydrobiologia*, **263**, 1–44.

Cale, P. & Hobbs, R.J. (1994). Landscape heterogeneity indices: problems of scale and applicability, with particular reference to animal habitat description. *Pacific Conservation Biology*, **1**, 183–193.

Costanza, R., Norton, B.G. & Haskell, B.D. (eds.) (1992). *Ecosystem Health: New Goals for Environmental Management.* Washington, DC: Island Press.

Dawson, D. (1994). *Are Habitat Corridors Conduits for Animals and Plants in a Fragmented Landscape? A Review of the Scientific Evidence.* London: English Nature.

Delcourt, H.R. & Delcourt, P.A. (1991). *Quaternary Ecology: A Paleoecological Perspective.* New York: Chapman & Hall.

Fischer, S.F., Poschlod, P. & Beinlich, B. (1996). Experimental studies on the dispersal of plants and animals on sheep in calcareous grasslands. *Journal of Applied Ecology*, **33**, 1206–1222.

Flaig, E.G. & Reddy, K.R. (1995). Fate of phosphorus in the Lake Okeechobee watershed, Florida, USA: overview and recommendations. *Ecological Engineering*, **5**, 127–142.

Flannery, T. (1994). *The Future Eaters: An Ecological History of the Australasian Lands and People.* Port Melbourne, Vic: Reed Books.

Forman, R.T.T. (1995). *Land Mosaics: The Ecology of Landscapes and Regions.* Cambridge: Cambridge University Press.

Forman, R.T.T. & Godron, M. (1986). *Landscape Ecology.* New York: John Wiley.

George, R.J., McFarlane, D.J. & Speed, R.J. (1995). The consequences of a changing hydrologic environment for native vegetation in south Western Australia. In *Nature Conservation*, vol. 4, *The Role of Networks*, eds. D.A. Saunders, J.L. Craig & E.M. Mattiske, pp. 9–22. Chipping Norton, NSW: Surrey Beatty.

Gosz, J.R. (1993). Ecotone hierarchies. *Ecological Applications*, **3**, 369–376.

Haines-Young, R., Green, D.R. & Cousins, S. (eds.) (1993). *Landscape Ecology and GIS.* London: Taylor & Francis.

Hansen, A.J. & di Castri, F.D. (eds.) (1992). *Landscape Boundaries: Consequences for Biotic Diversity and Ecological Flows.* New York: Springer-Verlag.

Hanski, I.A. & Gilpin, M.E. (eds.) (1997). *Metapopulation Biology: Ecology, Genetics, and Evolution.* New York: Academic Press.

Hanski, I. & Simberloff, D. (1997). The metapopulation approach, its history, conceptual domain, and application to conservation. In *Metapopulation Biology: Ecology, Genetics, and Evolution*, eds. I.A. Hanski & M.E. Gilpin, pp. 5–26. New York: Academic Press.

Hanski, I., Pöyry, J. Pakkala, T. & Kuussaari, M. (1995). Multiple equilibria in metapopulation dynamics. *Nature*, **377**, 618–621.

Hargis, C.D., Bissonette, J.A. & David, J.L. (1998). The behaviour of landscape metrics commonly used in the study of habitat fragmentation. *Landscape Ecology*, **13**, 167–186.

Harris, L.D. (1984). *The Fragmented Forest: Island Biogeographic Theory and the Preservation of Biotic Diversity.* Chicago, IL: University of Chigago Press.

Hill, C.J. (1995). Linear strips of rain forest vegetation as potential dispersal corridors for rain forest insects. *Conservation Biology*, **9**, 1559–1566.

Hobbs, R.J. (1987). Disturbance regimes in remnants of natural vegetation. In *Nature Conservation: The Role of Remnants of Native Vegetation*, eds. D.A. Saunders, G.W. Arnold, A.A. Burbridge & A.J.M. Hopkin, pp. 233–240. Chipping Norton, NSW: Surrey Beatty.

Hobbs, R.J. (1992). Corridors for conservation: solution or bandwagon? *Trends in Ecology and Evolution*, **7**, 389–392.

Hobbs, R.J. (1993*a*). Effects of landscape fragmentation on ecosystem processes in the Western Australian wheatbelt. *Biological Conservation*, **64**, 193–201.

Hobbs, R.J. (1993*b*). Can revegetation assist in the conservation of biodiversity in agricultural areas? *Pacific Conservation Biology*, **1**, 29–38.

Hobbs, R.J. (1994). Dynamics of vegetation mosaics: can we predict responses to global change? *Ecoscience*, **1**, 346–356.

Hobbs, R.J. (1995). Landscape ecology. In *Encyclopedia of Environmental Science*, vol. 2, ed. W.A. Nierenberg, pp. 417–428. San Diego, CA: Academic Press.

Hobbs, R.J. (1999). Restoration ecology and landscape ecology. In *Issues in Landscape Ecology*, eds. J.A. Wiens & M.R. Moss, pp. 70–77. Guelph, Ontario: International Association of Landscape Ecology.

Hobbs, R.J. & Harris, J.A. (2001). Restoration Ecology: repairing the Earth's ecosystems in the new millenium. *Restoration Ecology*, **9**, 239–246.

Hobbs, R.J. & Lambeck, R.L. (in press). Landscape science and management in Western Australia. In *Integrating Landscape Ecology into Natural Resource Management*, eds. J. Liu & W.W. Taylor. Cambridge: Cambridge University Press.

Hobbs, R.J. & Norton, D.A. (1996). Towards a conceptual framework for restoration ecology. *Restoration Ecology*, **4**, 93–110.

Hobbs, R.J. & Saunders, D.A. (eds.) (1993). *Reintegrating Fragmented Landscapes: Towards Sustainable Production and Conservation.* New York: Springer-Verlag.

Hobbs, R.J. & Wilson, A.-M. (1998). Corridors: theory, practice and the achievement of conservation objectives. In *Key Concepts in Landscape Ecology*, eds. J.W. Dover & R.G.H. Bunce, pp. 265–279. Preston, UK: International Association for Landscape Ecology.

Hobbs, R.J. & Yates, C.J. (1997). Moving from the general to the specific: remnant management in rural Australia. In *Frontiers in Ecology: Building the Links*, eds. N. Klomp & I. Lunt, pp. 131–142. London: Elsevier Science.

Hobbs, R.J., Saunders, D.A. & Arnold, G.W. (1993). Integrated landscape ecology: a Western Australian perspective. *Biological Conservation*, **64**, 231–238.

Holland, M.M., Risser, P.G. & Naiman, R.J. (eds.) (1991). *Ecotones: The Role of Landscape Boundaries in the Management and Restoration of Changing Environments.* New York: Chapman & Hall.

Hornung, M. & Reynolds, B. (1995). The effects of natural and anthropogenic environmental changes on ecosystem processes at the catchment scale. *Trends in Ecology and Evolution*, **10**, 443–449.

Hudson, W.E. (ed.) (1991). *Landscape Linkages and Biodiversity.* Washington, DC: Island Press.

Kapos, V. (1989). Effects of isolation on the water status of forest patches in the Brazilian Amazon. *Journal of Tropical Ecology*, **5**, 173–85.

Knight, D.H. (1987). Parasites, lightning, and the vegetation mosaic in wilderness landscapes. In *Landscape Heterogeneity and Disturbance*, ed. M.G. Turner, pp. 59–83. New York: Springer-Verlag.

Kolasa, J. & Waltho, N. (1998). A hierarchical view of habitat and its relationship to species abundance. In *Ecological Scale: Theory and Applications*, eds. D. Peterson & V.T. Parker, pp. 55–76. New York: Columbia University Press.

Lambeck, R.J. (1997). Focal species: a multi-species umbrella for nature conservation. *Conservation Biology*, **11**, 849–856.

Lambeck, R.J. (1999). *Landscape Planning for Biodiversity Conservation in Agricultural Regions*, Biodiversity Technical Paper no. 2. Canberra: Department of the Environment and Heritage.

Laurance, W.F. & Bierregaard, R.O. (eds.) (1997). *Tropical Forest Remnants: Ecology, Conservation and Management of Fragmented Communities.* Chicago, IL: University of Chicago Press.

Laurance, W.F., Laurance, S.G., Ferreira, L.V., Rankin-de Merona, J.M., Gascon, C. & Lovejoy, T.E. (1997). Biomass collapse in Amazonian forest fragments. *Science*, **278**, 1117–1118.

le Houérou, H.N. (1981). Impact of man and his animals on Mediterranean vegetation. In *Mediterranean-Type Shrublands*, eds. F. di Castri, D.W. Goodall & R.L. Specht, pp. 479–521. Amsterdam: Elsevier.

Lomolino, M.V. (1994). An evaluation of alternative strategies for building networks of nature reserves. *Biological Conservation*, **69**, 243–249.

Ludwig, J., Tongway, D., Freudenberger, D., Noble, J. & Hodgkinson, K. (eds.) (1997). *Landscape Ecology, Function and Management: Principles from Australia's Rangelands.* Melbourne, Vic: CSIRO Publishing.

MacMahon, J.A. (1998). Empirical and theoretical ecology as a basis for restoration: an ecological success story. In *Successes, Limitations, and Frontiers in Ecosystem Science*, eds. M.L. Pace & P.M. Groffman. New York: Springer-Verlag.

Malle, K.-G. (1996). Cleaning up the River Rhine. *Scientific American*, **274**, 70–73.

Matlack, G.R. (1993). Microenvironment variation within and among forest edge sites in the eastern United States. *Biological Conservation*, **66**, 185–194.

McCulloch, D.R. (ed.) (1996). *Metapopulations and Wildlife Conservation.* Washington, DC: Island Press.

McIntyre, S. & Hobbs, R.J. (1999). A framework for conceptualizing human impacts on landscapes and its relevance to management and research. *Conservation Biology*, **13**, 1282–1292.

McIntyre, S. & Hobbs, R.J. (in press). Human impacts on landscapes: matrix condition and management priorities. In *Nature Conservation*, vol. 5, *Nature Conservation in Production Environments*, eds. J. Craig, D.A. Saunders & N. Mitchell. Chipping Norton, NSW: Surrey Beatty.

Middleton, B. (1999). *Wetland Restoration: Flood Pulsing and Disturbance Dynamics.* New York: John Wiley.

Milton, S.J., Dean, W.R.J., du Plessis, M.A. & Siegfried, W.R. (1994). A conceptual model of arid rangeland degradation: the escalating cost of declining productivity. *BioScience*, **44**, 70–76.

Murcia, C. (1995). Edge effects in fragmented forests: implications for conservation. *Trends in Ecology and Evolution*, **10**, 58–62.

Ostfeld, R.S., Pickett, S.T.A., Shachak, M. & Likens, G.E. (1997). Defining the scientific issues. In *The Ecological Basis of Conservation: Heterogeneity, Ecosystems and Biodiversity*, eds. S.T.A. Pickett, R.S. Ostfeld, M. Shachak & G.E. Likens, pp. 3–10. New York: Chapman & Hall.

Pearson, S.M., Turner, M.G., Gardner, R.H. & O'Neill, R.V. (1996). An organism-based perspective of habitat fragmentation. In *Biodiversity in Managed Landscapes: Theory and Practice*, eds. R.C. Szaro & D.W. Johnston, pp. 77–95. New York: Oxford University Press.

Pickett, S.T.A. & Cadenasso, M.L. (1995). Landscape ecology: spatial heterogeneity in ecological systems. *Science*, **269**, 331–334.

Pickett, S.T.A. & Thompson, J.N. (1978). Patch dynamics and the design of nature reserves. *Biological Conservation*, **13**, 27–37.

Poschlod, P., Bakker, J., Bonn, S. & Fischer, S. (1996). Dispersal of plants in fragmented landscapes. In *Species Survival in Fragmented Landscapes*, vol. 35, eds. J. Settele, C. Margules, P. Poschlod & K. Henle, pp. 123–127. Dordrecht: Kluwer.

Rapport, D.J., Costanza, R. & McMichael, A.J. (1998). Assessing ecosystem health. *Trends in Ecology and Evolution*, **13**, 397–402.

Rowell, T.A. (1986). The history of drainage at Wicken Fen, Cambridgeshire, England, and its relevance to conservation. *Biological Conservation*, **35**, 111–142.

Saunders, D.A., Hobbs, R.J. & Margules, C.R. (1991). Biological consequences of ecosystem fragmentation: a review. *Conservation Biology*, **5**, 18–32.

Saunders, D.A., Hobbs, R.J. & Arnold, G.W. (1993). The Kellerberrin project on fragmented landscapes: a review of current information. *Biological Conservation*, **64**, 185–192.

Scheffer, M., Hosper, S.H., Meijer, M.-L., Moss, B. & Jeppesen, E. (1993). Alternative equilibria in shallow lakes. *Trends in Ecology and Evolution*, **8**, 275–279.

Scheiner, S.M. (1992). Measuring pattern diversity. *Ecology*, **73**, 1860–1867.

Schwartz, M.W. (ed.) (1997). *Conservation in Highly Fragmented Landscapes.* New York: Chapman & Hall.

Shrader-Frechette, K.S. (1994). Ecosystem health: a new paradigm for ecological assessment. *Trends in Ecology and Evolution*, **9**, 456–457.

Simberloff, D.S. (1982). Island biogoegraphic theory and the design of wildlife refuges. *Ékologiya*, **4**, 3–13.

Simberloff, D.S. & Cox, J. (1987). Consequences and costs of conservation corridors. *Conservation Biology*, **1**, 63–71.

Simberloff, D.S., Farr, J.A., Cox, J. & Mehlman, D.W. (1992). Movement corridors: conservation bargains or poor investments. *Conservation Biology*, **6**, 493–504.

Sprugel, D.G. (1991). Disturbance, equilibrium, and environmental variability: what is 'natural' vegetation in a changing environment? *Biological Conservation*, **58**, 1–18.

Suffling, R. (1993). Induction of vertical zones in subalpine valley forests by avalanche-formed fuel breaks. *Landscape Ecology*, **8**, 127–138.

Taylor, P.D., Fahrig, L., Henein, K. & Merriam, G. (1993). Connectivity as a vital element of landscape structure. *Oikos*, **68**, 571–573.

Thirgood, J.V. (1981). *Man and the Mediterranean Forest: A History of Resource Depletion*. London: Academic Press.

Tongway, D.J. & Ludwig, J.A. (1996). Rehabilitation of semi-arid landscapes in Australia. 1: Restoring productive soil patches. *Restoration Ecology*, **4**, 388–397.

Turner, M.G. (1987). *Landscape Heterogeneity and Disturbance*. New York: Springer-Verlag.

Turner, M.G. (1989). Landscape ecology: the effect of pattern on process. *Annual Review of Ecology and Systematics*, **20**, 171–197.

Turner, M.G. & Gardner, R.H. (eds.) (1991). *Quantitative Methods in Landscape Ecology: The Analysis and Interpretation of Landscape Heterogeneity*, vol. 82. New York: Springer-Verlag.

Turner, M.G., Romme, W.H., Gardner, R.H., O'Neill, R.V. & Kratz, T.K. (1993). A revised concept of landscape equilibrium: disturbance and stability on scaled landscapes. *Landscape Ecology*, **8**, 213–227.

Turner, M.G., Gardner, R.H. & O'Neill, R.V. (1995). Ecological dynamics at broad scales. Ecosystems and landscapes. *BioScience*, Supplement 1995, S29–S35.

Turner, S.J., O'Neill, R.V., Conley, W., Conley, M.R. & Humphries, H.C. (1991). Pattern and scale: statistics for landscape ecology. In *Quantitative Methods in Landscape Ecology: The Analysis and Interpretation of Landscape Heterogeneity*, vol. 82, eds. M.G. Turner & R.H. Gardner, pp. 17–49. New York: Springer-Verlag.

Wallace, K.J. (ed.) (1998). *Dongolocking Pilot Planning Project for Remnant Vegetation, Final Report (Phase 1)*. Perth, WA: Department of Conservation and Land Management.

Whisenant, S.G. (1999). *Repairing Damaged Wildlands: A Process-Orientated, Landscape-Scale Approach*. Cambridge: Cambridge University Press.

Wiens, J.A. (1997). Metapopulation dynamics and landscape ecology. In *Metapopulation Biology: Ecology, Genetics, and Evolution*, eds. I. A. Hanski & M. E. Gilpin, pp. 43–62. New York: Academic Press.

Willis, E.O. (1984). Conservation, subdivision of reserves, and the anti-dismemberment hypothesis. *Oikos*, **42**, 396–401.

Wilson, A.-M. & Lindenmayer, D. (1996). *Wildlife Corridors and the Conservation of Biodiversity: A Review*. Canberra: Greening Australia Ltd.

With, K.A. (1997). The application of neutral landscape models in conservation biology. *Conservation Biology*, **11**, 1069–1080.

With, K.A. & Crist, T.O. (1995). Critical thresholds in species' responses to landscape structure. *Ecology*, **76**, 2446–2459.

4 • The ecological context: a species population perspective

DAVID W. MACDONALD, TOM P. MOORHOUSE AND JODY W. ENCK

INTRODUCTION

Restoration in context

Wheeler (1995) offers the generalisation, derived from consideration of plant community restoration, that a suitable end point of restoration is that restoration 'must result in the development of a self-sustaining semi-natural habitat (under a given management regime)'. This is also applicable to restoring animal communities, a criterion for success being a viable, self-sustaining population in the wild (Griffith *et al.*, 1989; Ebenhard, 1995; Saltz & Rubenstein, 1995; Nolet & Baveco, 1996; Sarrazin & Barbault, 1996; IUCN 1998). The requirements of restoration are very much the same for both plant and animal communities; the end product must be a self-sustaining population or community.

Restoration may be appropriate as a conservation action, either to reverse a process of deterioration and loss, or to reinstate a species which has been lost. In either case, the process of restoration may involve diverse activities. Obviously, it is restoration to repair, or even recreate, the habitat that a particular species or community require. Less obviously, restoration may involve not only the fostering of beneficial factors, but the removal of inimical ones. Thus, in the case of the water vole (*Arvicola terrestris*), now a threatened species in the UK, the creation of riparian habitat and the extermination of invasive American mink (*Mustela vison*) are equal components of restoration (Macdonald & Strachan, 1999). Similarly, there is some evidence that the restoration of rare or threatened plant species in some ecosystems can be facilitated by limiting the impact of phytophagous insects (Louda, 1994). For many plant communities, restoration will require lowering the nutrient status of the substrate, especially in cases where the land has been reclaimed following agricultural intensification (Bekker *et al.*, 1997; Bakker & Berendse, 1999). In such cases, high nutrient levels can be seen as inimical to the restoration of a species-rich sward, and restoration will require management either to reduce the nutrient levels in the soil via cropping (Marrs, 1985; Grootjans & van Digglen, 1995) or to prevent dominance of particular species over others under the prevalent nutrient status by grazing (Gibson *et al.*, 1987) or mowing (see Marrs, this volume).

There are many situations where the restoration of habitats necessitates the removal of invasive mammals (a notable example being the need to remove goats, cats and rats from islands such as the Galapagos archipelago). Similarly, habitat restoration, for mammalian species, may include increasing, even reintroducing, prey numbers for predators. For example, the restoration of wolf (*Canis lupus*) populations in Italy required the reintroduction of red deer (*Cervus elephas*). It may also involve reintroducing the species itself, such as beavers (*Castor liber*), to the UK (see Macdonald *et al.*, this volume).

The principle of restoration is so broad that it scarcely differs from conservation, the distinction being that the goal of conservation is to prevent deterioration, while that of restoration is to reverse it. Since most ecosystems are deteriorated, most conservation is restoration. Whatever the restoration measures may be, they need to have clearly defined objectives, the attainment of which allow the restoration to be considered a 'success'. This will generally involve an end point which incorporates the particular species or community to be restored within a fully functioning ecosystem. In practice, however, the potency of charismatic, emblematic species (usually mammals or birds) can provide a flagship which facilitates the conservation of a community of plants and animal species and their habitat (Mallinson, 1995).

In devising performance indicators by which to judge a restoration project, the link between the flagship and the rest of the 'flotilla' is an important consideration. For example, restoring open woodland/pine–grassland communities in the southeast United States for the benefit of red-cockaded woodpeckers (*Picoides borealis*) had the additional effect of increasing the richness and abundance of small mammal species in the area (Masters *et al.*, 1998). Conservation is, however, essentially consumer-driven, and the desire for one species in an area may be inimical to the presence of another. Restoration ecologists may thus face awkward choices.

Decision-making context within which choices are made

Understanding how best to restore populations requires recognition of the decision-making context within which scientific findings can be brought to bear on the problem. The prevailing decision-making context is that various legal and scientific conventions support society's interest in ensuring that species do not become extinct.

In the UK and much of Europe, and within the global influence of the IUCN/SSC's guidelines, the EC Habitats and Species directives aim to ensure rare species are protected and extirpated species are restored. A national endangered species act in the USA and various state/provincial laws serve the same purposes in the USA and Canada. Usually, the legal responsibility for restoration under these mandates falls on governmental agencies, often with support from non-governmental conservation organisations.

Rare and declining species are, however, not the only candidates for restoration. Hundreds of examples of restoration attempts exist for species that do not exist locally, but may be common elsewhere (Griffith *et al.*, 1989; Hunter & Hutchinson, 1994). Given the combination of legal mandates and increasing public interest in restoration from ecosystem and biodiversity perspectives (Kellert, 1986), conservation agencies are faced with the daunting task of choosing from among a myriad of potential candidate species.

The question therefore arises of what problem restoration is supposed to solve. The answer is that restoration – that is actively recreating conditions advantageous to a particular species or community – may remedy problems arising from any or all of direct exploitation, introduction of predators, competitors or disease, and habitat loss or fragmentation (Diamond, 1989). Restoration may take all or some of various forms, including reintroduction of the species, removal of the predators, competitors or disease, and restoration of the habitat. Each of these raises both practical and scientific issues, but they are united by one feature that pervades the full suite of proximate causes of the problem: namely, all are ultimately associated with human activities. For example, direct exploitation of a species occurs to meet human needs for food, hides, recreation or relief from damage; introductions may result from an anticipated financial return or other perceived advantage or, at worst, negligence. Habitat fragmentation and loss both are the consequences of land use decisions. So, the proximate causes of many of the ecological problems have an ultimate human dimension (Table 4.1). Further, because of a lack of large-scale publicly owned land on which restoration of most species can be attempted, government agencies responsible for restoration necessarily must work with private landowners and others who will be affected in important ways by restoration decisions. That is why, in this chapter, and using mammalian examples, we seek to present an interdisciplinary view that binds biological and socio-economic perspectives.

Biological considerations

Species cannot be restored unless the correct habitat, in the correct quantities, is available for their persistence (Gilpin & Soulé, 1986). Griffith *et al.* (1989) state that 'without high habitat quality, translocations [of organisms to a given location for restoration purposes] have low chances of success regardless of how many organisms are released or how well they are prepared for the release', and Balmford *et al.* (1996) state that

Table 4.1. *Several challenges exist for those who must make decisions about restorations including the type(s) of candidate species being considered, proximate causes of extirpation that need to be overcome and ultimate causes of extirpation that provide part of the context within which restoration decisions are made*

Categories of candidate species	Ultimate causes of extirpation	Proximate causes of extirpation
Rare, declining locally	Direct exploitation	Problems associated with small populations
Rare, declining globally	Introduction of predators, competitors or disease	
Extirpated locally, common elsewhere	Changes in land use	Problems associated with habitat loss and fragmentation
Extirpated locally, rare elsewhere		

'by far the strongest determinant of re-introduction success is the availability of suitable habitat'.

From an individual species restoration perspective, therefore, it will often be necessary to have first addressed any requirement for restoration of the appropriate habitat before concentrating upon the restoration of the species in question. This is an example of the wider benefits that the restoration of a 'flagship' species (see above) can bring for the ecological communities present.

Wheeler (1995) provides a framework for site-based restoration considerations, albeit from a bottom–up perspective; i.e. habitat restoration for the creation of a specific habitat type, with any benefits to wildlife being included in the decision analysis, as opposed to habitat restoration specifically for a single species, with any benefits to the plant community or other wildlife being included in the decision analysis:

1. The feasibility of an objective. Namely, is the objective possible or technically reasonable? This will depend largely upon the existing environmental conditions at a site.
2. The former character of the site. That is: restoration should perhaps concentrate upon recreating the site in its last known state, even if this means damaging the natural state.
3. The rarity of the species which would benefit from the restoration. This is easily defended as species with small populations have a greater potential for extinction and it is also an objective measure. However several rare species from different taxa may have differing habitat requirements, and the debate over whether to restore an area for a charismatic bird as opposed to a dowdy insect is likely to be swayed by popular appeal.
4. The scarcity of opportunity for achieving a particular end point. For example, species-rich fen is associated with uncommon, low-productivity habitat, irrigated by nutrient-poor water. To create a reedbed when rich fen vegetation could be created could be seen as a wasted opportunity.

These four points capture most of the biological considerations, notwithstanding that Green (1995) and Stewart & Hutchings (1996) stress that a prerequisite for successful restoration is likely to be understanding the reasons for the original decline. However, as we have alluded above and will show in more detail below, biological considerations cannot generally be taken in isolation from the human dimension.

Sociological considerations

Restoration of habitats and the species within them raises questions of which habitats or which species should be restored, and where? What conditions should be met before a project is approved? What protocols facilitate quality control at each stage of the operation? What education and communication strategy will be integrated with the biological work, and how will responsibilities be allocated and discharged? Finally, how will success be defined, and against what performance indicators will it be measured? Notwithstanding general principles, the answers to these questions may require radically different approaches when addressed to different habitats and species.

The 'human dimension' to the question of whether a species should be restored encompasses the potential costs and benefits to people living in the area if restoration proceeded. For example, restoration of a top-level predator could raise the need to prevent or mitigate predation on pets, livestock or desirable game species. Restoration of a large ungulate, e.g. elk (*Cervus elaphus canadensis*) or moose (*Alces alces*), could raise the need to manage herbivory on agricultural, ornamental or other plant species. Restoration of particular plant species may, for example, require the restoration of appropriate disturbance regimen; indeed, for many plant species disturbance regimes are one of the most important considerations for *in situ* conservation (Pavlovic, 1994). In many cases, managing such disturbance will require management of levels of anthropogenic activities, with resultant impacts upon the lives of the people in the locality. The appropriateness of this action will depend upon the extent of the management concerned and the reaction of the local inhabitants to this. Restoration could also bring new visitors who want to view or listen to the restored species. To residents of local municipalities, visitors have the potential to be either a new source of revenue or an added burden (e.g. in terms of needed services, crowding, and soil and vegetation trampling).

Enck *et al.* (1998) suggest that negative impacts from restoration could be experienced by almost all municipalities. Clearly, social feasibility for restoration will be highest for those municipalities with the greatest capacity to identify and prevent/mitigate negative impacts, and to identify and realise positive impacts. This capacity has been referred to variously as community resiliency (Harris *et al.*, 1996), vitality (McNamara & Deaton, 1996), or well-being (Eberts & Khawaga, 1988). Regardless of the name, community capacity is an index to the degree to which a community can anticipate and deal with impacts related to whatever changes it faces (Swanson, 1996). This capacity is affected by: quality of local leadership, the degree to which a community has a proactive planning mechanism for achieving community goals, economic and social capital that can be brought to bear to achieve those goals, and high-quality social services that meet the needs of local residents (Enck *et al.*, 1998).

One of the most important challenges in considering the human dimensions of restoration is in predicting the likelihood of municipalities experiencing the costs or benefits. These are likely to differ at national versus local levels (Glass *et al.*, 1990). Further, various municipalities at the local level have different capacities to respond to, or to assist with, restoration because they have varying capacities to accrue costs and benefits associated with restoration (Enck *et al.*, 1998; Enck & Brown, 2000). Municipalities that take their capacities into consideration have a much better basis for: (1) providing input into restoration decisions; (2) helping to make decisions in the context of community-based management; and (3) making changes at the local level that are necessary to ensure the successful implementation of restoration actions (Enck *et al.*, 1998).

For example, a concern about wolf restoration in the northeastern USA has been the possibility of predation on livestock (Responsive Management, 1996) while a countervailing benefit has been the possibility of increased ecotourism (Heberlein, 1976). However, Enck & Brown (2000) found that some municipalities in northern New York State were more likely than others to be able to develop effective depredation mitigation plans and to benefit economically from ecotourism. Indeed, some communities whose residents were supportive of restoration for its ecotourism potential had little prospect of realising that potential because they lacked both the necessary physical infrastructure and the social and economic capital to build that infrastructure. Sillero-Zubiri & Laurenson (2001) discuss the generality of restoring and conserving carnivores alongside people. Thus, the 'community capacity approach' for assessing social feasibility can be used by conservationists, in conjunction with ecological information, to identify geographic areas in which to pursue restoration.

Against this background, our aim in this chapter is to describe the general ecological processes and considerations relevant to species restoration.

THE FUNDAMENTALS

Habitat loss and fragmentation

An analogy is frequently drawn between nature reserves and land-bridge islands – islands that were formerly connected to the mainland and were created by a rise in the level of the ocean (Newmark, 1987). According to island biogeography theory (MacArthur & Wilson, 1967), the number of species occurring on an island (or isolated habitat patch) tends towards an equilibrium level between the rate of colonisation by new species and the rate of extinction of resident species, colonisation being determined by the ability of species to reach the island, which in turn is a function of the degree of isolation from a source of dispersers. Such a land-bridge island may be considered supersaturated with species at the outset, in that the ratio of island-to-mainland species numbers is higher than would be predicted from the area of the island. Newmark (1987) examined a selection of extant parks and found that the three predictions made by the land-bridge hypothesis were upheld in each park. These were that: (1) the total number of extinctions would exceed the number of colonisations within the reserve; (2) the number of extinctions should be inversely related to reserve size; and (3) the number of extinctions should be directly related to reserve age.

The land-bridge island perspective of nature reserves is a special case of habitat fragmentation in which the remaining habitat patches are relatively large. Habitat loss and fragmentation due to habitat destruction result in reduced population sizes which increase the probability of extinction from demographic and environmental stochasticity (Burkey, 1995; see below). Habitat destruction is considered the critical initial threat to plant and animal populations (Terborgh & Winter, 1980) to the extent where it has been described as the major cause of species extinctions (Tilman *et al.*, 1994). Hayes (1991) listed habitat destruction, dependence on habitats that are vulnerable to destruction and dependence on habitats that are naturally uncommon as the most common factors causing extinction vulnerability in mammal species.

Habitat destruction incorporates two components: namely habitat loss and habitat fragmentation. Although the effects of fragmentation on populations differ from those of habitat loss *per se*, they are often considered together for the reason that they typically co-occur (Fahrig, 1997). Fahrig (1997) separates loss from fragmentation in that if habitat loss results in a constant number of smaller patches, then patch size effects are due to habitat loss alone: only when the number of patches increases by the breaking apart of habitats are both habitat loss and fragmentation involved in the decreasing size and increasing isolation of habitat patches. However, a decrease in patch size over a constant number of patches will necessarily increase the amount of matrix habitat between the patches and therefore patch isolation, thus contributing to fragmentation as defined by Bright (1993), the 'process leading to increasing isolation and decreased area of habitat patches'.

The distinction between habitat loss and fragmentation is important in the sense that separating the effects of the area of each individual habitat patch and of their spatial arrangement (their number and degree of isolation) may have implications for the conservation of individual species' populations. For the purposes of this chapter, therefore, it is convenient to refer to habitat fragmentation as resulting from two attributes of patchy habitats, the size of each separate patch and the isolation (in terms of distance) of each from the others.

Explaining the effects of habitat loss and fragmentation

Extinctions are predicted to be more common on small islands, or small patches of the environment, which support smaller areas of habitat and so smaller populations (Simberloff, 1998). One set of considerations concerns the consequences of small populations, and another concerns the linked consequences of patch size, configuration and dispersion. Both are embraced by Caughly's (1994) two paradigms for species conservation biology: (1) the small population paradigm, which focuses on the extinction risk inherent in small populations, and (2) the declining population paradigm, which seeks to reveal the 'tangible cause' which reduces the population to the small size in the first place.

The essentials of restoration theory fall within the ambit of both these non-exclusive paradigms. For example, habitat fragmentation, as a cause of extinction, is representative of the declining population paradigm, whereas metapopulation dynamics come under the small population paradigm in explaining the effects of small habitat patches upon subdivided populations. We will summarise the essentials of both families of concepts below.

For the purposes of habitat restoration, an examination of the factors which relate to habitat patch sizes and spatial organisation and how these affect population viability (small population paradigm) should underpin any attempt at habitat restoration or recreation. How much habitat to restore, and in what arrangement to best benefit the population under consideration, are important considerations in determining the requirements of meeting the endpoint of a self-sustaining species. These considerations should be complemented by the consideration of in what geographic location should the habitat be restored. This latter point encompasses both biological and human dimensions and is addressed later in this chapter.

Problems with small populations

Small populations are vulnerable to extinction through demographic stochasticity, environmental stochasticity, loss of heterozygosity, genetic drift and inbreeding depression (for a review, see Caughly, 1994). Some species, because of their complex societies, may also be prone to an Allee effect. For example, packs of African wild dogs (*Lycaon pictus*), when below a critical threshold, are disadvantaged by lost opportunities for co-operation (Courchamp & Macdonald, in press).

Demographic stochasticity

Demographic stochasticity results from random variations in birth and death rates among individuals in a population. The smaller the population, the greater the impact on the average rate due to variability in each individual's contribution. The analogy presented by Belovsky *et al.* (1994) is of coin tossing, in which each toss is an individual in a population, and a head represents not producing offspring in a particular year. The probability of tossing all heads in five tosses is 0.5^5, whereas the probability in two tosses is much greater, 0.5^2. Small populations may go extinct literally through increased susceptibility to a run of bad luck.

Environmental stochasticity

Environmental stochasticity is the effect of environmental fluctuations upon a population's demographic parameters (Caughly, 1994). The environmental impact may be direct (via catastrophes such as flooding or fire) or indirect (e.g. annual variations in weather affecting primary production). The persistence of the population of a given size is affected by the severity of the environmental fluctuations (V_e), that is the amount of the variation in r, the intrinsic rate of increase of a population, which can be explained by environmental fluctuations. When $\bar{r} > V_e$, persistence time will curve upward with population size. However, if $V_e > \bar{r}$, then the increase in persistence time for a given population size will tend to taper off (Lande, 1994, cited in Caughly, 1994). Environmental stochasticity therefore increases the probability of extinction by adding to the background variation in r that is an inescapable consequence of demographic stochasticity.

Loss of heterozygosity

Mean heterozygosity, H, is the proportion of loci that are heterozygous in the average individual in a population. Rate of decline of heterozygosity due to genetic drift is a function of N, the population size, such that heterozygosity in the population is reduced by $1/(2N)$ each generation (Wright, 1931). This loss is counterbalanced by the mutation rate such that $H^* = 2Nm$, where H^* is the equilibrium between m, the mutational input to genetic variance, and loss from drift. Thus, the higher N, the larger the value to which H converges. Individuals from populations with low heterozygosity may have reduced fitness, due to selection acting upon semi-lethal recessive characteristics that would be less likely to be expressed in a population of the same species with high heterozygosity (Caughly, 1994; Stewart & Hutchings, 1996; Gray, this volume).

Inbreeding depression

Mating between full sibs born of unrelated parents results in offspring which, on average, exhibit 75% of the heterozygosity of individuals taken at random from the population as a whole: a 25% decrease (Caughly, 1994). If a population remains small over several generations, matings between close relatives may lead to an 'extinction vortex' (Stewart & Hutchings, 1996) whereby reduced heterozygosity exposes the young to the effects of semi-lethal recessive alleles, reducing fecundity and increasing mortality. This leads ultimately to the population becoming even smaller and continuing the downward spiral. However, for a population which survives this purging effect, inbreeding may cease to be a problem, and fitness may even increase as the result of the semi-lethal alleles being removed from the gene pool (Caughly, 1994).

Problems with fragmented habitats

It is not always correct to treat a population as a single unit. If conditions in different localities are sufficiently different so as to produce a degree of independence in the dynamics of different local populations, and if their spatial arrangement is such that separation or isolation affects the dispersal and colonisation rates between them, then population dynamics in the whole landscape cannot be understood by averaging dynamics at a local scale (Ritchie, 1997). These spatially explicit population dynamics can be categorised as either metapopulations or source–sinks.

Metapopulation dynamics occur when local populations are isolated sufficiently to produce low colonisation rates over a large network of habitat patches. Each population has a substantial chance of stochastic local extinction, but this is balanced by recolonisation of empty but suitable habitat and immigration of breeding individuals into populations which would otherwise go extinct (Hanski, 1999). This latter is termed the 'rescue effect' (Brown & Kodric-Brown, 1977).

Metapopulations grade into source–sink dynamics when 'certain local populations can consistently sustain net positive growth rates, have little risk of extinction and produce colonists to other localities with net negative growth rates and/or high risks of random extinction' (Ritchie, 1997). Local populations are linked by dispersal such that individuals from the source patches make up the complement in the breeding population of the sink patches, rescuing them from extinction (e.g. Pulliam & Danielson, 1991).

The continued existence of a source–sink or metapopulation depends on: (1) a threshold level of habitat availability, below which the metapopulation will descend to extinction because colonisation is too infrequent to overcome local extinction (Kareiva & Wennergren, 1995; Hanski *et al.*, 1996) and (2) the ability of dispersers to reach unoccupied, suitable habitat or to 'rescue' local populations that are close to extinction.

Metapopulation dynamics

The 'classical' metapopulation model (Levins, 1969, 1970) gives the instantaneous rate of chance of the fraction of currently occupied patches p as:

$$dp/dt = mp(1-p)-ep$$

where m and e are parameters for the rates of colonisation and extinction respectively.

During habitat destruction, a fraction, $1-h$, of habitat patches may be permanently destroyed, thus leaving a density of h patches suitable for colonisation. The above equation then becomes:

$$dp/dt = mp(h-p)-ep$$

with the equilibrium position of the fraction of empty patches being given as $h-p^* = e/m$ (Hanski *et al.*, 1996), where p^* is the fraction of occupied patches at steady state.

It would be practically useful if this model could be extended to provide an estimate of the minimum area of suitable habitat (MASH) necessary to ensure metapopulation survival. Such an estimate seems to be at hand insofar as when $h < e/m$, the metapopulation will become extinct, leading to the rule of thumb that a 'necessary and sufficient condition for metapopulation survival is that the remaining number of habitat patches following a reduction in patch number exceeds the number of empty but suitable patches prior to patch destruction' (Hanski *et al.*, 1996). However, Hanski *et al.* (1996) demonstrate that this relationship is likely to provide an underestimate of the MASH necessary to ensure metapopulation survival, due to three processes:

1. Where immigration into populations occurs, thereby decreasing the risk of stochastic extinction due to the bolstering of numbers within the population (the 'rescue effect'), the Levins rule can be rewritten as the discrete time model:

$$h - p^* = 1 - F/F(1/G - h)$$

   where F is the probability of a population surviving one time interval and G is the probability of recolonisation given that the fraction of occupied patches at a given time = 1. In this equation, the fraction of empty patches is no longer independent of h but grows with increasing h. The metapopulation goes extinct if $h < (1-F)/G$, but the 'fraction of empty patches can no longer be used to estimate the amount of habitat necessary for metapopulation survival'.
2. In a small number of local patches/populations, stochasticity in colonisation and extinction may lead to the extinction of populations predicted to persist by the Levins equation, a source of variability analogous to demographic stochasticity (above).
3. The current snapshot of the metapopulation may not be at an equilibrium state, but *en route* to a new state after a bout of habitat destruction. This is termed the extinction debt by Tilman *et al.* (1994) and has three attributes which complicate the estimation of the MASH. These are: that long time lags are predicted before extinction of a metapopulation following even moderate habitat destruction; the amount of extinction caused by habitat destruction is an accelerating curve; and the species most prone to extinction may often be abundant, superior competitors, which may have traded greater competitive ability for lower recolonisation rates. It has yet to be seen whether there is a widespread negative correlation between dispersal ability and competitive ability (Kareiva & Wennergren, 1995), but the possibility is sufficiently plausible for its implications for conservation to have been modelled (Dytham, 1995; see below). This may mean that the dominant species could potentially decline more rapidly than would be predicted from its initial abundance.

Individual species responses to fragmentation

Tilman *et al.* (1994) raise the point that species interactions and properties of individual species can complicate predictions from a metapopulation model. Dytham (1995) presents a model of two competing species, a superior (poor disperser) and an inferior (good disperser) competitor, based on a grid of habitable and uninhabitable cells. The model suggests that the dispersal ability of species in such a pairing determines their resistance to extinction. The better disperser in competition with a dominant may actually reach a peak of numbers with a moderate degree of habitat loss. Indeed, the pattern of habitat destruction has bearing upon persistence of both species. Random habitat loss has relatively severe effects upon population persistence. However, if there is a gradient of destruction across the grid, the outcome includes some areas where patch density is much higher than the mean for the grid and this allows local persistence and a postponement of extinction.

Clearly, such generalisations must be tuned to species' ecology. Plant species dispersal rates have been shown to be rather slow for some species. Seed rain data from chalk grassland indicates that chalk grassland species would be comparatively slow invaders at 0.3–3.5 m per annum (Hutchings & Booth, 1996; Hutchings & Stewart, volume 2). However, seed banks of viable propagules are often present in the soil, partly removing the necessity for recolonisation to rely upon dispersal (Hodgeson & Grime, 1990). For animal species, predators are likely to need more space than do prey, and populations in the tropics may be more dense than those in temperate environments, changing the calculation of reserve area required to sustain a given population size (Belovsky, 1987). These considerations are especially important for relatively large, wide-ranging organisms, and are illustrated for carnivorous mammals by Woodroffe (2001) amongst others in Gittleman *et al.* (2001).

Extinction rates are higher for habitat specialists or long-lived species exhibiting extreme life-history traits (Sarrazin & Barbault, 1996). Each species is a special case, and thus each must be considered separately when it comes to restoration. This raises difficulties because, often, species-specific parameters are unknown. This is especially so regarding dispersal. Metapopulations rely upon dispersal in order to exist: the effects of spatial isolation upon species in fragmented habitats are entirely due to restriction of dispersal, which limits capacity to supplement declining populations, to recolonise habitats where extinctions have occurred or to colonise newly suitable habitats (Bennett, 1999). Dispersal is, however, extraordinarily difficult to study, making the transfer of theory into practice an imprecise alchemy (Kareiva & Wennergren, 1995; Macdonald & Johnson, 2001). Thus while the metapopulation is a central concept in restoration thinking, the theory frustratingly fails to provide a rule of thumb for operational decisions regarding the MASH, when seeking to assess population viability.

Where to restore

Peripheral versus central populations

There is a qualitative distinction between populations of a species in the central parts of its distribution and those at the periphery (Lesica & Allendorf, 1995; Lomolino & Channell, 1995). Many species at the edge of their range occur in unusual or atypical environments and are more likely to be isolated from one another. Therefore natural selection and genetic drift are expected to promote divergence of peripheral populations (Lesica & Allendorf, 1995). Peripheral populations tend to be less dense and more unstable, and thus are assumed to have higher extinction probabilities. Consequently, a declining species may collapse inwards, from the outside to the centre of its geographic range (Lomolino & Channell, 1995). However, in reality, of 31 species of non-volant, terrestrial mammals in which geographical range collapse had been recorded, 74% collapsed towards the periphery of their range, rather than the reverse. This led Lomolino & Channell (1995) to speculate that since contemporary extinctions result from anthropogenic factors which 'spread like a contagion across the landscape', peripheral populations may be isolated from these events by their very nature and so have heightened persistence. Interestingly, and counter-intuitively, Macdonald *et al.* (in press) found that water voles survived better in small, isolated fragments of habitat because these were less likely to be devastated by an invasive predator, the American mink.

Given their genetic and morphological divergence, peripheral populations may be seen as disproportionately important for protecting genetic diversity relative to their size and frequency. This adds a further conundrum to those already faced by the restoration biologist striving to decide where to focus activity. There may be different advantages to focusing on optimal habitat in the central areas of the historical distribution and to atypical habitat at the periphery.

Landscape connectivity

There are four obvious options regarding land-use planning in order to benefit animal species: (1) expanding the area of protected habitats; (2) maximising the quality of existing habitats; (3) minimising the impacts from surrounding habitats; and (4) promoting connectivity of natural habitats to counter the effects of isolation (Bennett, 1999).

Any feature of a fragmented landscape which can facilitate dispersal between habitat patches may have benefits to species persistence (Hess, 1996). The creation in the landscape of stepping-stones or corridors of suitable habitat may therefore be a goal of restoration (see Hobbs, this volume).

Expanding the area of protected natural habitats is the essential basis for nature conservation, in that large tracts of habitat are more likely than small areas to support self-sustaining populations of animals and plants. Increasing the habitat size also enhances the capacity for an area to retain a greater richness of species. Maximising the quality of existing habitats involves active management in order to enhance resources essential for the native fauna.

Minimising the impacts from surrounding land uses is necessary in fragmented landscapes, where processes arising from outside of fragments may have major effects upon populations and communities within fragments. Mitigating tactics such as zoning of land uses around protected areas, use of buffer zones around conservation areas and management programmes to control the numbers and impacts of pest species of plants and animals may all be useful.

Management to promote connectivity of natural habitats depends on the specificity of requirements of the species in question. Those with more adaptable requirements may be able to disperse through various habitats in the landscape mosaic, whereas those with highly specific requirements may require a corridor of a specific habitat.

Some species can only disperse along particular corridors. Anderson (1995) noted that the distance between habitat patches and the presence of linking habitats such as ditches and hedges may be critical in providing connectivity for animals such as amphibians and voles. Fitzgibbon (1997) found that the use of farm woodlands by wood mice (*Apodemus sylvaticus*) and bank voles (*Clethrionomys glareolus*) was strongly influenced by the landscape around the woods as well as the habitat within them. In this latter case, in the spring, wood mice were less likely to be found in woods with relatively fewer hedges nearby, and in autumn, both species were more abundant in woods with many adjoining hedges.

The efficacy of conservation corridors is a contentious issue (Hobbs 1992; Hess, 1996; Bennett, 1999; Hobbs, this volume). Conceivable disbenefits include the spread of disease or unwanted species, exposure to competitors, parasites, predation, hunting and poaching by humans, and genetic effects such as outbreeding depression and hybridisation between previously disjunct taxonomic forms.

ANALYTICAL TOOLS FOR ASSESSMENT

Assessment of both the ecological and human dimensions are required for restoration to be successfully implemented. The two major techniques which have a role in ecological assessment are geographic information systems (GIS) and population viability analysis (PVA). GIS provide technologies for the storage and analysis of cartographic data and have been used extensively in assessing the suitability of landscapes for different species (Aspinall & Veitch, 1993).

Population viability analysis

Definitions

Population viability analysis is an attempt to incorporate the factors of demographic stochasticity, genetic variation and environmental stochasticity into a formal assessment of extinction risk and to incorporate these into plans to manage a population (e.g. Macdonald *et al.*, 1998). This may involve, for instance, stochastic, individual-based population models, in which a starting population is specified with an age or stage structure, and then a sequence of events characterising the annual life history is imposed upon it.

Genetic factors may be included by allocating allelic variability to individuals in the population. The model is then run for a large number of iterations, usually in the order of 1000 runs in order to account for the stochastic variation between runs. GIS have made it possible to incorporate PVA models into spatially explicit virtual landscapes.

These 'process-based models' rely on the premise that the distribution of a species in the landscape arises from interactions between individual behavioural processes such as home ranges, territoriality and dispersal, and life-history processes of births and deaths. In these models the habitat data act as templates on which the population processes occur and the distribution of organisms amongst habitats emerges as the model is run.

These approaches are more complex than straightforward habitat availability estimates, but have been used to predict the distributions of other species with complex rapidly changing distribution patterns (e.g. Rushton *et al.*, 1997). Nonetheless, models in wildlife management should not be taken as representations of the truth, but should be used more as problem-solving tools (Starfield, 1997).

A common usage of PVA in biological feasibility studies is to estimate a minimum viable population (MVP) size, that is, the minimum number of individuals in a population that has a good chance of surviving for some relatively long period of time (Hanski *et al.*, 1996). This can then be used to estimate the amount of habitat required for effective conservation of a species – arguably the single most important datum needed to determine whether the objectives of the restoration are biologically feasible. For example, a feasibility study of reintroducing wild boar (*Sus scrofa*) to Scotland estimated that the minimum viable population size required for a >95% probability of survival for over 50 years would be at least 300 animals initially. From the perspective of biological feasibility, no area in Scotland is currently large enough to support such a population (Howells & Edwards-Jones, 1997). From the perspective of social feasibility, conservationists might go on to ask the questions 'Why is no area currently large enough?' and 'How would human attitudes and beliefs influence the likely success of additional habitat restoration schemes?'.

Model limitations

Models may be at their most useful in identifying and ranking causes of species decline, setting research priorities and focusing conservation (Haight, 1996). Modelling is often useful in the evaluation of conservation options for species which it is impossible, or improper, to manipulate in reality, and where decisions cannot wait for protracted investigation, i.e. it is better to make mistakes in a virtual world than the real one. However, models can be no better than their assumptions or the data with which they are parameterised. Their elegance gives them an apparent credibility and accuracy which may be spurious: for example, in a comparison of several different PVA models seeking to identify the extinction risks posed by different factors affecting British mammals, Macdonald *et al.* (1998) found wide variation in the specific answers, but general agreement as to the ranking of threats.

Performance indicators for assessments of human dimensions

One aspect of the human dimension is understanding public attitudes towards restoration. Measurement of public attitudes has been used by governmental decision-makers to assess the degree to which wildlife restoration is socially feasible (e.g. Bath, 1989; Kellert, 1991). In the case of carnivore restorations, these human dimensions have proved pivotal (e.g. Breitenmoser *et al.*, 2001; Clark *et al.*, 2001).

Public perceptions, whether well founded or misinformed, have an enormous impact on the feasibility of conservation actions, as illustrated by Macdonald & Johnson's (2000, in press) exploration of farmers' attitudes to conservation on their land. Particularly in the context of mammalian predators, the complex interactions of biology and socio-economics in forming public attitude to wildlife stretch the current capabilities of conservation science (as illustrated by the case of hunting mammals in Britain: Macdonald *et al.*, 2000). Despite the obvious vagaries of public attitudes, at some stage in the restoration process, the level of public support or opposition will be used as an index of a scheme's feasibility. For example, if there is a policy mandate to restore an endangered species, and biological considerations point to a particular region where this is most feasible, it will be vital to ascertain the local stakeholders' attitudes (Griffith *et al.*, 1989). It will be equally vital to disentangle fact from fancy as a first step to planning communication and education programmes to provide reassurance, and to focus attention of management required to mitigate those risks that are substantial (e.g. Kellert, 1991; Clark *et al.*, 2000). This raises a series of questions about how opinion should be canvassed, on what spatial (political scale) should it be sampled, and by what process should decisions then be made.

The most useful answers are those which reveal the perceptions, and their veracity, that underpin attitude. In this context, asking whether citizens support or oppose a particular project is likely to prove less informative than asking them about the likelihood of various impacts occurring and whether those impacts would be good or bad (Bright & Manfredo, 1996; Pate *et al.*, 1996). These more evaluative questions force people towards a risk analysis rather than a snap judgement. Risk analysis opens the door to assessment of the realism of the perceived risks. Attitude is labile, and reflects education. A single attitude survey is a very blunt instrument, but a series of them can be a revealing barometer of the success of an education programme or the resilience of the science on which it is based.

At what scale is it sensible to sample stakeholder opinions? The impacts of a restoration project are likely to vary with local circumstances. For example, restoring a predator in a stock-farming area is a different proposition to doing so in wilderness. It is therefore important that schemes are evaluated in the context of the relevant human community, each of which has its own leadership, budget, vision for the future, and social characteristics. Communities, that is non-overlapping geographic units encompassing a mosaic of private and public land, are the appropriate scale at which to consider the human dimensions of wildlife restoration (Enck *et al.*, 1998).

The extent to which public attitudes indicate the social feasibility of a restoration project, and how they should influence the decision-making process is a politically complex issue. Deciding what percentage of support constitutes a mandate for action is difficult, especially so when attitudes may be based on faulty information and may be very labile (e.g. Heberlein, 1976; Responsive Management, 1996; Enck & Brown, 2000). There may be a low probability that wolves will prey on livestock (Thompson, 1993) or that moose will cause vehicle accidents (Hicks & McGowan, 1992), but level of concern about those issues may be quite high (e.g. Bath, 1989; Lauber & Knuth, 1998).

Hitherto, the incorporation of the human dimension into the restoration process has often been *ad hoc*. Some projects have involved protracted and thoughtful consultation but others have not. For example, Enck *et al.* (1998) reviewed literature pertaining to restoration attempts of large herbivores in Canada and the USA. Although biological feasibility was always assessed in these studies, social feasibility was assessed in only about half of them and none determined community capacity.

This absence of community capacity assessment greatly restricted the involvement of local residents in decision-making or implementation. Consequently, wildlife agencies generally took the responsibility of interpreting whether the level of acceptance expressed through surveys or public meetings was sufficient to proceed with restoration. However, such attitudes are a questionable basis for decision-making. Stable attitudes generally stem from personal experience (Heberlein, 1976), which is unavailable to stakeholders in an area from which a species has been extirpated when they are asked about its restoration (see Enck & Decker, 1997).

Enck *et al.* (1998) found that ignoring information on community capacity when evaluating potential costs and benefits of a proposed restoration nearly always resulted in polarisation of attitudes (e.g. Anonymous, 1992; van Deelen *et al.*, 1997). Without an opportunity to consider why restoration may or may not be in the best interest of a municipality, some factions supported restoration while others opposed it, often with minimal evidence of whether their hopes or concerns were well founded (Lohr *et al.*, 1996). Indeed, Enck's review revealed that when the barometer of social feasibility was merely a snapshot of public attitude, the likely result was that local residents scrambled to avoid being selected for the 'next government experiment in restoration'. Under these circumstances, social feasibility was a minimalist concept which meant little more than that local people were prepared to accept the (unknown) consequences of a restoration project. In the absence of any thoughtful exploration of the real costs and benefits to each stakeholder group, there was little local ownership of the scheme, and thus little stake in its success.

Public support is integral to the success of any restoration programme (Kellert *et al.*, 1996). However, it should be obvious, but seemingly has not been, that asking people to voice approval when they are inadequately informed is an abdication of the wildlife biologist's responsibility (Enck & Decker, 1997). A municipality might appropriately be asked to decide whether to permit a wildlife agency to restore a particular species based on their capacity to identify and achieve restoration-related opportunities and to mitigate any negative consequences of restoration. Decisions at the community level then can be used to inform agency decisions about whether and where to restore a wildlife species.

This brings the restoration decisions into a context of reality based upon existing municipality characteristics. Further, the roles of the wildlife agency and municipality with respect to 'seller' and 'buyer' of the idea of restoration can be reversed (Enck *et al.*, 1998). Criteria for deciding whether to advance with a restoration project may thereby include: (1) biological information; (2) requests for restoration from local municipalities; and (3) evidence that requesting municipalities have conducted an assessment of their capacity.

CRITICAL ASPECTS OF REINTRODUCTION

Definitions

Where a species has been extirpated, or its numbers have fallen below the level from which they can recover, restoration options are reintroduction or supplementation. Following the IUCN (1998), *reintroduction* is an attempt to establish a species in an area which was once part of its range, but from which it has been extirpated or become extinct. Animals for this purpose may be captive-bred stock, or be *translocated*. Translocation is the deliberate and mediated movement of wild populations from one part of their range to another.

Attempts to establish a species outside of its recorded distribution but within an appropriate habitat and geographical area is termed an *introduction*. The likelihood is that, in the face of widespread environmental degradation, over-zealous exploitation or persecution, and shifts in biogeographic zones with climate change, these techniques will be relied upon increasingly (Griffith *et al.*, 1989). It is inappropriate, however, to consider reintroduction/translocation as conservation tools until the causes of the original decline have been both understood and remedied (Sarrazin & Barbault, 1996). In the vast majority of cases, this will trace back ultimately to human activities.

Almost inevitably, reintroduced/translocated populations start small (Nolet & Baveco, 1996; Sarrazin & Barbault, 1996) and therefore newly restored populations are likely to be subject to the same threats that jeopardise all small populations. The founding population should therefore be as large as possible and have a high rate of increase (Griffith *et al.*, 1989; Dunham, 1997). Species with high fecundity (and therefore population growth rate) will tend to retain more of the original genetic diversity of the founders than will a species with low fecundity (Lesica & Allendorf, 1995), although there may be exceptions (Leberg, 1993). Furthermore, the level of genetic variation in a population could influence its ability to adapt to new environmental conditions (Leberg, 1993).

Captive breeding

Reintroductions using translocated individuals are more likely to be successful than those using captive-bred animals (Griffith *et al.*, 1989; Mallinson, 1995). However, finding sufficiently large extant populations which can act as a source of propagules may be problematic; therefore captive breeding from declining populations may be justified to fuel a reintroduction or supplementation programme.

It is important to maximise genetic variation in captive-bred and reintroduced populations, and this may be achieved by selecting individuals with known pedigree, individuals with high allozyme heterozygosity, or individuals coming from geographically separated populations (Sarrazin & Barbault, 1996). Preserving the genetic variation within a captive population is of equal importance to both animal and plant populations, arguments of avoiding inbreeding – and outbreeding – depression and selecting a source gene pool with high levels of diversity being relevant to both. In addition, for plant species, the restoration and maintenance strategies must be dictated in part by the range of different mating systems which different plant species employ: upwards of 20% of plant species are primarily selfing and the rest exhibit varying degrees of outcrossing (Fenster & Dudash, 1994).

Captive breeding is a costly process, and these costs increase with the biomass of individuals within the species, especially within animal taxa. Since the majority of captive breeding is undertaken in zoos, however, efforts are usually targeted towards zoo preferences, and hence towards larger-bodied animals, even amongst mammalian taxa. A difficult question is which animals to select for the ark. The current predilections of zoos are such that, by taxonomic order, the percentage of threatened species and subspecies covered by existing programmes increases with body size (Balmford *et al.*, 1996).

Captive breeding may serve various purposes, including education and provision of a sort of museum. However, few would dissent from the view that its primary justification should be to fuel reintroduction. Insofar as this is rather rarely achieved, doubts have been raised over the legitimacy of many captive-breeding programmes. Beck *et al.* (1994) concluded that only 16 (11%) of 145 reintroduction projects resulted in successful reintroduction. One cause of failure can be behavioural deficiencies in released animals (Snyder *et al.*, 1996). In the words of Shepherdson (1994), 'unlike genes, behaviour cannot be frozen in a test tube'. Attempting to ensure that captive animals develop behavioural competence requires ingenuity and can raise ethical concerns (e.g. in predators acquiring practice with live prey).

Inbreeding

Despite its theoretical dangers, some populations appear unaffected by severe inbreeding (Ebenhard, 1995). A small number of immigrants may significantly reduce inbreeding depression, making introduction of fresh individuals an important management option, while bearing in mind the opposite risk of outbreeding depression (Leberg, 1993; Ebenhard, 1995). Both translocation and captive breeding are a balancing act between achieving a well-adapted, genetically variable propagule, and causing outbreeding depression.

Disease

Endangered species may tend toward enhanced susceptibility to disease because of reduced genetic diversity resulting from reduced population size. Certainly, disease problems have been common in captive populations of endangered species (Snyder *et al.*, 1996). Wildlife reintroduction may represent a significant disease risk (Viggers *et al.*, 1993). For this reason, veterinary involvement is important, to advise on quarantine and screening (Woodford & Rossiter, 1994).

Population modelling

Population modelling offers a route to predicting the likely consequences of reintroduction schemes. Obviously, even the best model is only an approximation of reality and must be treated with appropriate caution.

Performance indicators

The use of performance indicators is important. Success is not a single end point, but a series of steps towards a predefined state with vigilant surveillance thereafter. Careful thought to indicators against which these milestones will be judged should be given before the restoration begins.

What, however, constitutes a reintroduction success? Beck *et al.* (1994) define a success as attaining a population size of greater than 500 individuals, which are free of provisioning or human support, or for which population viability analysis predicts that the population will be self-sustaining. Considering variation in life-history traits, habitat quality or the eventual metapopulation structure, the figure of 500 is arbitrary. Sarrazin & Barbault (1996) prefer, as criteria for success, extinction probability estimates that combine population size, growth rate and growth rate variance.

CONCLUDING REMARKS

The vast majority of cases where restoration is required have anthropogenic disturbance as a causal factor. Each species, however, differs in its susceptibility to this factor and, indeed, many species thrive in habitats created by anthropogenic disruption, whilst others have been pushed to, and past, the brink of extinction by the self-same phenomenon. A prerequisite for a restoration attempt of a given species must be an in-depth understanding of the peculiar ecology of not only the species in question but of the community of which it is a part. Failure to appreciate a critical aspect of a species' life history may result in failure of the restoration project and, apart from anything else, this can damage the public credibility of conservationists. Perhaps the only universal truth is that for almost any species the more habitat available the more likely the species is to recover and persist in the absence of any other limiting factors. However, how much habitat is enough? Restoration is limited both by space and finance. In such cases, knowledge of the reproductive ecology, dispersal rates, metapopulation dynamics, predator–prey interactions, likely conflicts with human populations and a plethora of other considerations become necessary. Many of these considerations may be addressed via population modelling based on the best knowledge of the species' background biology, but it is necessary for such models to be thoroughly ground-truthed to allow the predictions to be based in ecological reality as well as theoretical plausibility.

ACKNOWLEDGMENTS

We gratefully acknowledge discussion with our colleagues in the WildCRU.

REFERENCES

Anderson, P. (1995). Ecological restoration and creation: a review. *Biological Journal of the Linnean Sociey*, **56**, 187–211.

Anonymous (1992). *Environmental Assessment for Permitting an Experimental Elk Reintroduction Research Study.* Chequamegon National Forest, WI: *US Forest Service*.

Aspinall R. & Veitch, N. (1993). Habitat mapping from satellite imagery and wildlife survey data using a bayesian modeling procedure in a GIS. *Photogrammetric Engineering and Remote Sensing*, **59**, 537–543.

Bakker, J.P. & Berendse, F. (1999). Constraints on the restoration of ecological diversity in grassland and heathland communities. *Trends in Ecology and Evolution*, **14**, 63–68.

Balmford, A., Mace, G.M. & Leader-Williams, N. (1996). Designing the Ark: setting priorities for captive breeding. *Conservation Biology*, **10**, 719–727.

Bath, A.J. (1989). The public and wolf re-introduction in Yellowstone National Park. *Society and Natural Resources*, **2**, 297–306.

Beck, B.B., Rapaport, L.G., Stanley-Price, M.R. & Wilson, A.C. (1994). Reintroduction of captive-born animals. In *Creative Conservation: Interactive Management of Wild and Captive Animals*, eds. P.J.S Olney, G.M. Mace & A.T.C Feistner, pp. 265–286. London: Chapman & Hall.

Bekker, R.M., Verweij, G.L., Smith, R.E.N., Reine, R., Bakker, J.P. & Schnieder, S. (1997). Soil seed banks in European grasslands: does land use affect regeneration perspectives? *Journal of Applied Ecology*, **34**, 1293–1310.

Belovsky, G.E. (1987). Extinction models and mammalian persistence. In *Viable Populations for Conservation*, ed. M.E. Soulé, pp. 35–57. Cambridge: Cambridge University Press.

Belovsky, G.E., Bissonette, J.A., Dueser, R.D., Edwards, T.C., Jr, Lueke, C.M., Richie, M.E., Slade, J.B. & Wagner, F.H. (1994). Management of small populations: concepts affecting the recovery of endangered species. *Wildlife Society Bulletin*, **22**, 307–316.

Bennett, A.F. (1999) *Linkages in the Landscape: The Role of Corridors and Connectivity in Wildlife Conservation*. Gland, Switzerland: IUCN.

Breitenmoser, U., Breitenmoser-Wursten, C., Carbyn, L.N. & Funk, S.N. (2001). Assessment of carnivore reintroductions. In *Carnivore Conservation*, eds. J.L. Gittleman, S.M. Funk, D.W. Macdonald & R.K. Wayne, pp. 241–281. Cambridge: Cambridge University Press.

Bright, A.D. & Manfredo, M.J. (1996). A conceptual model of attitudes towards natural resource issues: a case study of wolf re-introduction. *Human Dimensions of Wildlife*, **1**, 1–21.

Bright, P.W. (1993). Habitat fragmentation: problems and predictions for British mammals. *Mammal Review*, **23**, 101–111.

Brown, J.H. & Kodric-Brown, A. (1977). Turnover rates in insular biogeography: effects of immigration on extinction. *Ecology*, **58**, 445–449.

Burkey, T.V. (1995). Extinction rates in archipelagos: implications for populations in fragmented habitats. *Conservation Biology*, **9**, 527–541.

Caughly, G. (1994). Directions in conservation biology. *Journal of Animal Ecology*, **63**, 215–244.

Clark, T.W., Mazur, N., Cork, S.J., Dovers, S. & Harding, R. (2000). Koala conservation policy process: appraisal and recommendations. *Conservation Biology*, **14**, 681–690.

Clark, T.W., Mattson, D., Reading, R.P. & Miller, B. (2001). Interdisciplinary problem solving in carnivore conservation: an introduction. In *Carnivore Conservation*, eds. J.L. Gittleman, S.M. Funk, D.W. Macdonald & R.K. Wayne, pp. 223–240. Cambridge: Cambridge University Press.

Courchamp, F. & Macdonald, D.W. (in press). Crucial importance of pack size in the African wild dog Lycaon pictus. *Animal Conservation*, **4**, 169–174.

Diamond, J.M. (1989). The present, past and future of human-caused extinctions. *Philosophical Transactions of the Royal Society B*, **325**, 469–477.

Dunham, K.M. (1997). Population growth of mountain gazelles *Gazella gazella* reintroduced to Central Asia. *Biological Conservation*, **81**, 205–214.

Dytham, C. (1995). The effect of habitat destruction on species persistence: a cellular model. *Oikos*, **74**, 340–344.

Ebenhard, T. (1995). Conservation breeding as a tool for saving animal species from extinction. *Trends in Ecology and Evolution*, **10**, 438–443.

Eberts, P.R. & Khawaga, M. (1988). *Changing Socioeconomic Conditions in Rural Localities in the 1980s: Experiences in New York State*, Rural Sociology Bulletin no. 152. Ithaca, NY: Cornell University.

Enck, J.W. & Brown, T.L. (2000). *Preliminary Assessment of Social Feasibility for Reintroducing Gray wolves to the Adirondack Park in Northern New York.* Human Dimensions Research Unit Publication no. 00-3. Ithaca, NY: New York State College of Agriculture and Life Sciences, Cornell University.

Enck, J.W. & Decker, D.J. (1997). Examining assumptions in wildlife management: a contribution of human dimensions inquiry. *Human Dimensions of Wildlife*, **2**, 56–72.

Enck, J.W., Porter, W.F., Didier, K.A. & Decker, D.J. (1998). *The Feasibility of Restoring Elk to New York: A Final Report to the Rocky Mountain Elk Foundation.* Ithaca, NY: State University of New York, College of Agriculture and Life Sciences at Cornell University, and Syracuse, NY: College of Environmental Science and Forestry.

Fahrig, L. (1997). Relative effects of habitat loss and fragmentation on population extinction. *Journal of Wildlife Management*, **61**, 603–610.

Fenster, C.B. & Dudash, M.R. (1994). Genetic considerations for plant population restoration and conservation. In *Restoration of Endangered Species: Conceptual Issues, Planning and Implementation*, eds. M. Bowles & C. Whelan, pp. 34–62. Cambridge: Cambridge University Press.

Fitzgibbon, S.D. (1997). Small mammals in farm woodlands: the effect of habitat, isolation and surrounding land use patterns. *Journal of Applied Ecology*, **34**, 530–539.

Gibson, C.W.D., Watt, T.A. & Brown, V.K. (1987). The use of sheep grazing to recreate species-rich grassland from abandoned arable land. *Biological Conservation*, **42**, 165–193.

Gilpin, M.E. & Soulé, M.E. (1986). Minimum viable populations: processes of species extinction. In *Conservation Biology*, ed. M.E. Soulé, pp. 19–35. Sunderland, MA: Sinauer Associates.

Gittleman, J.L., Funk, S.M., Macdonald, D.W. & Wayne, R.K. eds. (2001). *Carnivore Conservation.* Cambridge: Cambridge University Press.

Glass, R.J., More, T.A. & Stevens, T.H. (1990). Public attitudes, politics, and extramarket values for reintroduced wildlife: examples from New England. *Transactions of the North American Wildlife and Natural Resources Conference*, **55**, 548–557.

Green, R.E. (1995). Diagnosing causes of bird population declines. *Ibis*, **137** (Suppl. 1 JAN), S47–S55.

Griffith, B., Johnston, C.A., Scott, J.M., Carpenter, J.W. & Reed, C. (1989). Translocation as a species conservation tool: status and strategy. *Science*, **245**, 477–480.

Grootjans, A. & van Digglen, R. (1995). Assessing the restoration prospects of degraded fens. In *Restoration of Temperate Wetlands*, eds. B.D. Wheeler, S.C. Shaw, W. Fojt & R.A. Robertson, pp. 73–90. New York: John Wiley.

Haight, R.G. (1996). Wildlife models: predicting the effects of habitat restoration. *Journal of Forestry*, 4–6.

Hanski, I. (1999). Habitat connectivity, habitat continuity and metapopulations, in dynamic landscapes. *Oikos*, **87**, 209–219.

Hanski, I., Moitanen, A. & Gyllenberg, M. (1996). Minimum viable metapopulation size. *American Naturalist*, **147**, 527–541.

Harris, C., Brown, G. & McLaughlin, W. (1996). *Rural Communities in the Inland Northwest: Final Report, Parts 1 and 2, Interior Columbia River Basin Ecosystem Management Project.* Walla Walla, WA: US Forest Service and Bureau of Land Management.

Hayes, J.P. (1991). How mammals become endangered. *Wildlife Society Bulletin*, **19**, 210–215.

Heberlein, T.A. (1976). Some observations on alternative mechanisms for public involvement: the hearing, the public opinion poll, the workshop, and the quasi-experiment. *Natural Resources Journal*, **16**, 197–212.

Hess, G. (1996). Disease in metapopulation models: implications for conservation. *Ecology*, **77**, 1617–1632.

Hicks, A.C. & McGowan, E.M. (1992). Proposed moose translocation to northern New York. *Alces*, **28**, 243–248.

Hobbs, R.J. (1992). The role of corridors in conservation: solution or bandwagon? *Trends in Ecology and Evolution*, **7**, 389–392.

Hodgeson, J.C. & Grime, J.P. (1990). The role of dispersal mechanisms, regenerative strategies and seed banks in the vegetation dynamics of the British Landscape. In *Species Dispersal in Agricultural Habitats*, eds. R.G.H. Bunce & D.C. Howard, pp. 65–81. London: Belhaven Press.

Howells, O. & Edwards-Jones, G. (1997). A feasibility study of reintroducing wild boar *Sus scrofa* to Scotland: are existing woodlands large enough to support minimum viable populations? *Biological Conservation*, **81**, 77–89.

Hunter, M.L. & A. Hutchinson, A. (1994). The virtues and shortcomings of parochialism: conserving species that are locally rare but globally common. *Conservation Biology*, **8**, 1163–1165.

Hutchings, M.J. & Booth, K.D. (1996). Studies on the feasibility of recreating chalk grassland on ex-arable land. 1: The potential roles of seed bank and the seed rain. *Journal of Applied Ecology*, **33**, 171–181.

IUCN (1998). *IUCN/SSC Guidelines for Reintroductions.* Gland, Switzerland: IUCN/SSC Reintroduction Specialist Group, IUCN.

Kareiva, P. & Wennergren, U. (1995). Connecting landscape patterns to ecosystem and population processes. *Nature*, **373**, 299–302.

Kellert, S.R. (1986). Social and perceptual factors in the preservation of animal species. In *The Preservation of Species: The Value of Biological Diversity*, ed. B. Norton, pp. 50–73. Princeton, NJ: Princeton University Press.

Kellert, S.R. (1991). Public views of wolf restoration in Michigan. *Transactions of the North American Wildlife and Natural Resources Conference*, **56**, 152–161.

Kellert, S.R., Black, M., Rush, C.R. & Bath, A.J. (1996) Human culture and large carnivore conservation in North America. *Conservation Biology*, **10**, 977–990.

Lande, R. (1994). Risks of population extinction from demographic and environmental stochasticity and random catastrophes. *American Naturalist*, **142**, 911–927.

Lauber, T.B. & Knuth, B.A. (1998). Refining our vision of citizen participation: lessons from a moose reintroduction proposal. *Society and Natural Resources*, **11**, 411–424.

Leberg, P.L. (1993). Strategies for population reintroductions: effects of genetic variability on population growth and size. *Conservation Biology*, **7**, 194–199.

Lesica, P. & Allendorf, F.W. (1995). When are peripheral populations valuable for conservation? *Conservation Biology*, **9**, 753–760.

Levins, R. (1969). Some demographic consequences of environmental heterogeneity for biological control. *Bulletin of the Entomological Society of America*, **15**, 237–240.

Levins, R. (1970). Extinction. In *Some Mathematical Problems in Biology*, ed. M. Gesternhaber, pp. 77–107. Providence, RI: American Mathematical Society.

Lohr, C., Ballard, W.B. & Bath, A. (1996). Attitudes toward gray wolf reintroduction to New Brunswick. *Wildlife Society Bulletin*, **24**, 414–420.

Lomolino, M. & Channell, R. (1995). Splendid isolation: patterns of geographic range collapse in endangered mammals. *Journal of Mammalogy*, **76**, 335–347.

Louda, S.M. (1994). Experimental evidence for insect impact on populations of short-lived, perennial plants, and its application in restoration ecology. In *Restoration of Endangered Species: Conceptual Issues, Planning and Implementation*, eds. M. Bowles & C. Whelan, pp. 118–138. Cambridge: Cambridge University Press.

MacArthur, R.H. & Wilson, E.O. (1967). *The Theory of Island Biogeography.* Princeton, NJ: Princeton University Press.

Macdonald, D.W. & Johnson, D.D.P. (2001). Dispersal in theory and practice: consequences for conservation biology. In *Dispersal*, eds. J. Clobert, E. Danchin, A.A. Dhondt & J.D. Nichols, pp. 358–372. Oxford: Oxford University Press.

Macdonald, D.W. & Johnson, P.J. (2000). Farmers and the custody of the countryside: trends in loss and conservation of non-productive habitat 1981–1998. *Journal of Biological Conservation*, **94**, 221–234.

Macdonald, D.W. & Johnson P.J. (in press). Farmers as custodians of the countryside: links between perception and practice. In *Farming and Mammals*, Occasional Publication of the Linnean Society, eds. F. H. Tattersall & W. Manley.

Macdonald, D. & Strachan, R. (1999). *The Mink and the Water Vole: Analyses for Conservation.* Wildlife Conservation Research Unit and Environment Agency.

Macdonald, D.W., Mace, G. & Rushton, S. (1998). *Proposals for Future Monitoring of British mammals.* London: Department of the Environment, Transport and the Regions.

Macdonald, D.W., Tattersall, F.T., Johnson, P.J., Carbone, C., Reynolds, J., Rushton, S., Earley, M. & Langbein, J. (2000). *Managing British Mammals: Case Studies From the Hunting Debate.* Wildlife Conservation Research Unit.

Macdonald, D.W., Sidorovich, V.E., Anisomova, E.I., Sidorovich, N.V. & Johnson, P.J. (in press). The impact of American mink *Mustela vison* and European mink *Mustela lutreola* on water voles *Arvicola terrestris* in Belarus. *Ecography.*

Mallinson, J.J.C. (1995). Conservation breeding programs: an important ingredient for species survival. *Biodiversity and Conservation*, **4**, 615–635.

Marrs, R.H. (1985). Techniques for reducing soil fertility for nature conservation purposes: a review in relation to research at Roper's Heath, Suffolk, England (UK). *Biological Conservation*, **34**, 307–332.

Masters, R.E., Lochmiller, R.L., McMurry, S.T. & Bukenhofer, G.A. (1998). Small mammal responses to pine–grassland restoration for red-cockaded woodpeckers. *Wildlife Society Bulletin*, **26**, 148–158.

McNamara, K.T. & Deaton, B.J. (1996). Education and rural development. In *Rural Development Research*, eds. T.D. Rowley, D.W. Sears, G.L. Nelson, J.N. Reid & M.J. Yetley. Westport, CT: Greenwood Press.

Newmark, W.D. (1987). A land-bridge island perspective on mammalian extinctions in western North American parks. *Nature*, **325**, 430–432.

Nolet, B.A. & Baveco, J.M. (1996). Development and viability of a translocated beaver *Castor fiber* population in the Netherlands. *Biological Conservation*, **75**, 125–137.

Pate, J., Manfredo, M. Bright, A. & Tishbein, G. (1996). Coloradoan's attitudes toward reintroducing the gray wolf in Colorado. *Wildlife Society Bulletin*, **24**, 421–428.

Pavlovic, N.B. (1994). Disturbance-dependent persistence of rare plants: anthropogenic impacts and restoration implications. In *Restoration of Endangered Species: Conceptual Issues, Planning and Implementation*, eds. M. Bowles & C. Whelan, pp. 159–193. Cambridge: Cambridge University Press.

Pulliam, H.R. & Danielson, B.J. (1991). Sources, sinks and habitat selection: a landscape perspective on population dynamics. *American Naturalist*, **137** (Suppl. Jun. 1991), S50–S66.

Responsive Management (1996). *Public Opinion on and Attitudes Toward Reintroduction of the Eastern Timber Wolf to Adirondack Park.* Harrisonburg, VA.

Ritchie, M.E. (1997). Populations in a landscape context: sources, sinks and metapopulations. In *Wildlife and Landscape Ecology: Effects of Pattern and Scale*, ed. J.A. Bisonette, pp. 160–184. New York: Springer-Verlag.

Rushton, S.P., Lurz, P.W.W., Fuller, R. & Garson, P.J. (1997). Modelling the distribution of the red and grey squirrel at the landscape scale: a combined GIS and population dynamics approach. *Journal of Applied Ecology*, **34**, 1137–1154.

Saltz, D. & Rubenstein, D.I. (1995). Population dynamics of a reintroduced Asiatic wild ass (*Equus hemionus*) herd. *Ecological Applications*, **5**, 327–335.

Sarrazin, F. & Barbault, R. (1996). Reintroductions: challenges and lessons for basic ecology. *Trends in Ecology and Evolution*, **11**, 474–478.

Shepherdson, D. (1994). The role of environmental enrichment in the captive breeding and reintroduction of endangered species. In *Creative Conservation: Interactive Management of Wild and Captive Animals*, eds. P.J.S. Olney, G.M. Mace & A.T.C. Feistner, pp. 167–177. London: Chapman & Hall.

Sillero-Zubiri, C. & Laurenson, K. (2001). Interactions between carnivores and local communities: conflict or co-existence. In *Carnivore Conservation*, eds. J. L. Gittleman, S.M. Funk, D.W. Macdonald & R.K. Wayne, pp. 282–312. Cambridge: Cambridge University Press.

Simberloff, D. (1998). Small and declining populations. In *Conservation Science and Action*, ed. W.J. Sutherland, pp. 315–337. Oxford: Blackwell Science.

Snyder, N.F.R., Derrickson, S.R., Beissinger, S.R., Wiley, J.W., Smith, T.B., Toome, W.D. & Miller, B. (1996). Limitations of captive breeding in endangered species recovery. *Conservation Biology*, **10**, 338–348.

Starfield, A.M. (1997). A pragmatic approach to modeling for wildlife management. *Journal of Wildlife Management*, **61**, 261–269.

Stewart, A.J.A. & Hutchings, M.J. (1996). Conservation of populations. In *Conservation Biology*, ed. I.F. Spellerberg, pp. 122–140. Harlow, UK: Longman.

Swanson, L.E. (1996). Social infrastructure and economic development. In *Rural Development Research*, eds. T.D. Rowley, D.W. Sears, G.L. Nelson, J.N. Reid & M.J. Yetley, pp. 103–119. Westport, CT: Greenwood Press.

Terborgh, J. & Winter, B. (1980). Some causes of extinction. In *Conservation Biology: An Evolutionary Ecological Perspective*, eds. M.E. Soulé & B.A. Wilcox, pp. 119–133. Sunderland, MA: Sinauer Associates.

Thompson, J.G. (1993). Addressing the human dimensions of wolf reintroduction: an example using estimates of livestock depredation and costs of compensation. *Society and Natural Resources*, **6**, 165–179.

Tilman, D., May, R.M., Lehman, C.L. & Newark, M.A. (1994). Habitat destruction and the extinction debt. *Nature*, **371**, 65–66.

van Deelen, T.R., McKinney, L.B., Joselyn, M.G. & Buhnerkempe, J.E. (1997). Can we restore elk to southern Illinois? The use of existing digital land-cover data to evaluate potential habitat. *Wildlife Society Bulletin*, **25**, 886–894.

Viggers, K.L., Lindermayer, D.B. & Spratt, D.M. (1993). The importance of disease in reintroduction programmes. *Wildlife Research*, **20**, 687–698.

Wheeler, B.D. (1995). Introduction: restoration and wetlands. In *Restoration of Temperate Wetlands*, eds. B.D. Wheeler, S.C. Shaw, W. Fojt & R.A. Robertson, pp. 1–18. New York: John Wiley.

Woodford, M.H. & Rossiter, P.B. (1994). Disease risks associated with wildlife translocation projects. In *Creative Conservation: Interactive Management of Wild and Captive Animals*, eds. J.P.S. Olney, G.M. Mace & A.T.C. Feistner, pp. 176–200. London: Chapman & Hall.

Woodroffe, R.B. (2001). Strategies for carnivore conservation: lessons from contemporary extinctions. In *Carnivore Conservation*, eds. J.L. Gittleman, S.M. Funk, D.M. Macdonald & R.K.Wayne, pp. 61–92. Cambridge: Cambridge University Press.

Wright, S. (1931). Evolution in Mendelian populations. *Genetics*, **16**, 97–159.

5 • The evolutionary context: a species perspective

ALAN J. GRAY

INTRODUCTION

The restoration of a process

The emerging science of restoration ecology has long been influenced by genetic and evolutionary ideas. The seminal work of A. D. (Tony) Bradshaw and his colleagues in the 1970s on the reclamation of industrial wasteland was fundamentally based on studies of the evolution and selection of metal-tolerant plants (Smith & Bradshaw, 1979; Bradshaw & McNeilly, 1981). In tracing the transition in the UK Nature Conservation movement from 'nature preservation' to 'creative conservation' Sheail *et al.* (1997) note that the first attempts to create species-rich grassland communities on road verges and amenity areas emphasised the importance of using local cultivars and plants from native sources (Wells *et al.*, 1982). At the same time, largely in North America, the foundations of conservation genetics were being laid by Michael Soulé and others in a series of texts which emphasised the genetic aspects of small population size and species extinctions (e.g. Soulé & Wilcox, 1980; Frankel & Soulé, 1981). As we shall see, the evolutionary implications of small population size and scarcity are recurring themes in restoration ecology.

Accompanied by developments in the underpinning science of ecology, these strands have brought us to a contemporary view of restoration science which emphasises process and function. The modern paradigm is the restoration not of a species assemblage or community but of a functioning ecosystem, which includes the process of genetic and evolutionary change. Ironically this could also involve speciation and even extinction!

The challenge, therefore, is not simply to select genotypes for restoration schemes which are adapted to survive in the restored environments or communities (although this is an essential first hurdle). We must also consider the long-term perspective of the evolutionary time-scale and ask whether the genetic diversity of the species we have introduced will enable it to escape extinction and to respond to future change. This is not a trivial challenge. Despite the wealth of general theory, the relationship between genetic diversity and species persistence is not well understood, but is clearly not straightforward. There may also be a conflict between short-term genetic adaptedness and long-term evolutionary adaptability. However, we will return to these issues later.

Patterns of genetic diversity

If a goal of restoration is the restoration of the evolutionary process, where might we begin in understanding what that process is? The first, and best, clue to the outcome of the evolutionary process – and one which provides a guide to action – is the genetic structure of contemporary species' populations. In other words the spatial pattern of genetic diversity (since we rarely have information on temporal changes) frequently displays indirect evidence of the forces that have shaped it. Of course, natural selection is only one of these forces – others include various forms of genetic drift and migration, but extensive studies of genetic variation in natural populations (especially of plants) have indicated that selection is a powerful and ubiquitous one (Gray, 1996*a*).

These studies began in the 1920s with the pioneering work of the Swedish botanist Gote Turesson (who coined the word 'ecotype') and included the extensive transplant and common garden experiments of the Carnegie group (Jens Clausen, David Keck and

William Hiesey) in North America in the 1930s and 1940s. More than 50 years ago such studies had already demonstrated, first, that patterns of genetic variation among populations of plant species are often correlated with environmental variation, and, second, that plants in populations from similar habitats often display similar traits.

An enormous body of published work since then has confirmed the almost universal occurrence of what Turesson (1922) called a 'genotypical response to habitat'. This response occurs at a great range of spatial scales and is elicited by a variety of environmental factors. Thus population differentiation (i.e. significant heritable differences among populations of the same species) has been demonstrated in relation to climate, altitude, soil type, soil moisture, salinity, pathogen distribution, herbivory, plant competition and density, herbicide application, fertiliser treatment, and many other environmental variables (for a fuller review with some examples see Gray, 1996*a*). Genetically differentiated populations may be separated by many kilometres or, in cases of high selection pressure such as metal-contaminated soils, by a few centimetres. Population genetics textbooks are rich in similar examples of genetic differentiation among populations of animals, especially small and relatively sessile ones such as molluscs and butterflies. More recently, thanks to the development of molecular techniques and especially assays of variation in mitochondrial DNA, greater population genetic substructuring has been revealed among a wide range of animal species, including several with relatively large and overlapping ranges (see e.g. Avise, 1994; Avise & Hamrick, 1996).

The obvious implication of such patterns of genetic diversity is that the choice of genotype for specific restoration schemes should be guided in the first instance by the principle that local individuals from similar habitats are likely to be the best adapted. The chances of successful introduction will be optimised by transferring individuals from nearby similar habitats.

That is not to say that all species display clear genotype–environmental correlations or do not have populations from non-local sources which are as well (or even better) adapted to particular habitats. In fact the plant literature includes one or two examples of apparent non-adaptation (Rapson & Wilson, 1988) and even maladaptation (Rice & Mack, 1991). Nevertheless local populations have been exposed to local selection pressures and, to a greater or lesser degree, have demonstrated their fitness in that particular local environment. Thus, where they are available, and in the absence of other information (discussed later), local genotypes are to be preferred. Of course, local genotypes may not always be available, or indeed the most genetically appropriate for transfer to a restoration site. These issues are taken up in the next section.

THE CHOICE OF GENOTYPES

Native versus non-native: is local best?

The expansion of creative conservation in Britain in the 1980s, and particularly the use of wild flower mixtures to create species-rich grasslands, generated a demand for seed and a rapid growth in the seed supply industry to meet that demand. It soon became clear that non-native cultivars made up a significant proportion of the seed being sown in restoration schemes – a situation which caused dismay among botanists (Akeroyd, 1994) and even spawned a pressure group, *Flora Locale*, dedicated to ensuring that only native plants of British, and preferably local, origin are introduced. Concerns about introducing non-native, usually agricultural, varieties of species such as kidney vetch (*Anthyllis vulneraria*), bird's-foot trefoil (*Lotus corniculatus*) and salad burnet (*Sanquisorba minor*) centred not on the fact that they might be maladapted, and hence not survive, but on their potentially harmful impact on native plants. Not only might their introduction confuse natural species distribution patterns but it could lead to them becoming weeds, outcompeting natives, and, of greatest concern and relevance to this chapter, hybridising with native genotypes.

This viewpoint sees the prime reason for using local species to be the conservation of local genetic diversity. Non-natives are a threat to the genetic integrity of local populations and to patterns of native diversity. The fear is of genetic 'swamping' of local genotypes, and the literature espousing these concerns abounds with terms such as genetic 'purity', genetic 'contamination' and genetic 'pollution' (e.g. Millar & Libby, 1991; Akeroyd, 1994; Butler, 1994; Jones & Hayes, 1999). Local genetic diversity is important as part of the total diversity within a species, and local populations have heritage value containing evidence of their history in the same way as historic buildings or ancient monuments (Daniels & Sheail, 1999). Loss of identity by hybridisation and introgression, and even the extinction of small populations, or endemic taxa, may be more widespread than is suggested by the relatively few well-known examples (such as the threat to narrow-ranging and endemic duck species from the ubiquitous mallard [*Anas platyrhynchos*] or the North American ruddy duck [*Oxyura jamaicensis*]), particularly among flowering plants and among fishes. Rhymer & Simberloff (1996) provide a comprehensive review of extinction by hybridisation.

An opposing view is that, far from being a threat, non-natives can bring novel genetic variation to potentially depauperate populations. Hybridisation and introgression are seen as creative forces empowering continuing evolutionary change and increasing fitness. This view sees local genotypes as products of past selection (and other forces such as genetic drift and gene flow) which may or may not be well adapted to local conditions. If they are well adapted, local selection will continue to ensure that genes of adaptive significance survive any bouts of introgression with non-natives. Emphasis is placed on the dynamic, evolving, nature of populations in which genetic changes are continually occurring by selection, drift, gene flow and mutation. A more likely threat than the production of maladapted individuals by hybridisation and introgression is the threat from reduced variation and fitness caused by mating among relatives in small populations. In other words outbreeding depression is not as frequent, or important, a phenomenon as inbreeding depression (see below).

Which of these views should guide the choice of genotypes in a restoration scheme? The answer, of course, is that both views have some validity, and that the relevance of one or the other to a particular restoration scheme depends entirely on the species and the objectives of the scheme. Thus one might tend towards the conservatism of the first view when introducing a new population of a narrow-ranging endemic, and towards the free-for-all of the second view when dealing with a widespread and less obviously differentiated species. Scale is of critical importance, as is the grain of the environment as perceived by the species concerned. There is no generic answer to the question 'How local is local?' Patterns of genetic variation generated by spatial differences may be small or large scale, clinal or abrupt, static or changing. The extent to which a species will respond to such differences varies enormously between species depending on their size, range, population structure, breeding system, history and behaviour.

Nevertheless, in the absence of appropriate data for each species, the approach of restorationists has been to take a local or, perhaps more accurately, regional perspective when choosing sources of germplasm. There are at least three good reasons for this.

First, the chances of successful establishment and persistence are increased by using local genotypes. Millar & Libby (1989) give examples of tree species in the USA where failure to use local stock for replanting or reseeding has resulted either in failure to establish (e.g. seedlings of ponderosa pine [*Pinus ponderosa*] from California and Oregon being killed by frost when planted in Colorado) or failure to survive longer-term periodic events such as drought or the outbreak of disease (e.g. Douglas-fir [*Pseudotsuga menziesii*] from inland and coastal populations). Indeed, long experience of replanting in North American forestry has led to the development of guidelines based on 'seed zones' in which only seed or plants from the same seed zone are used for restoration (Kitzmiller, 1990) (see below). Jones *et al.* (in press) report a similar pattern of more successful establishment in a local provenance of hawthorn (*Crataegus monogyna*). A local ecotype and several commercially available provenances from Britain and continental Europe were planted in experimental hedges in Wales where, over three years, the locally obtained plants had not only the highest survivorship, but also significantly later bud-burst, were most thorny and were less diseased than imported provenances. These traits are considered to be important in producing effective stock-proof hedges with potential wildlife benefits, and the results have important implications for the widespread use of commercial hawthorn provenances in hedge restoration.

A second reason for choosing local germplasm, which we have already touched on, is the maintenance of the genetic integrity of local populations. In addition to the arguably purist view that local gene pools should be retained as examples of the genetic diversity within a species, there is, where restoration involves adding new individuals to an existing, often depleted, population, the further possibility that hybridisation with local genotypes will have a significant effect on fitness. Where this effect is positive it may lead to the loss of the local gene pool through introgressive hybridisation; where it is negative it may lead to the decline, or even loss, of the local population through the production of maladapted genotypes (outbreeding depression). There are well-documented examples of both these phenomena – particularly where the newly interacting populations have been recognised as distinct taxa (frequently as separate subspecies but also as species). The gene pool of the native red deer (*Cervus elaphus*) in Scotland is threatened by hybridisation with introduced sika deer (*C. nippon*), as is that of the wildcat (*Felis silvestris*) by feral housecats (*F. catus*) and the polecat (*Mustela putorius*) by escaped domestic ferrets (*M. furo*). There are many similar examples, particularly among birds, fish and flowering plants (see Rhymer & Simberloff, 1996). Cases where outbreeding depression is known to have caused serious demographic decline are rarer, but include the elimination of the Tatra mountain ibex (*Capra ibex*) from Czechoslovakia by interbreeding with introduced ibex subspecies. Recently Keller *et al.* (2000) demonstrated negative outbreeding effects in hybrids between local Swiss plants of three arable weed species and plants obtained from commercial wildflower seed mixtures. Using shoot biomass as a measure of fitness, F_1 plants of corncockle (*Agrostemma githago*), common poppy (*Papaver rhoeas*) and white campion (*Silene latifolia* ssp. *alba*) all demonstrated hybrid vigour (heterosis), but the F_2 generations had significantly reduced biomass compared to the local parents.

The difficulty, of course, is predicting in advance what will happen when an introduced population meets and exchanges genes with an existing local population – a difficulty which has been highlighted in recent years by the introduction of genetically modified crops and the need to assess the risks from gene flow to their wild relatives. There are no absolute rules, although general evolutionary and genetic principles may help. A highly successful widespread genotype must be regarded as a potential threat to a narrow range, isolated endemic (except where the latter occupies an obviously extreme environment such as a saltmarsh or a metal-contaminated soil). Where the two populations have been isolated for a long time, or by considerable distance, the results of genetic mixing are likely to be more dramatic – in either direction, producing severe outbreeding depression or extensive hybrid vigour. Where populations share many genes by descent the effects of gene flow on fitness may be less extreme. Fenster & Dudash (1994) provide a detailed discussion of the issues of combining gene pools by mixing plants from different sources in restoration schemes.

In fact loss of unique local genetic identity and increase in fitness due to hybridisation are really different sides of the same coin. Occasionally the introduction of foreign genes may be a means of rescuing unfit or endangered populations. Classical examples include the dusky seaside sparrow (*Ammodramus maritimus nigrescens*) in Florida where the last few wild birds were caught and mated in captivity with other coastal subspecies (but see Avise [1996] for the outcome and full story) and the lakeside daisy (*Hymenoxys acaulis* var. *glabra*) in Illinois where the few remaining wild plants shared the same self-incompatibility alleles (DeMauro, 1994). Fenster & Dudash (1994) consider genetic manipulation to restore local species vigour, perhaps following a period of inbreeding depression, to be a legitimate goal of conservation genetics. Similarly the reintroduction of a population which has been extirpated in nature is seen as an opportunity to use genetic principles to restore short- or long-term vigour.

A third reason for choosing local genotypes is to avoid the possibility that individuals introduced from further afield may be invasive and lead to changes in ecosystem function and diversity. Several studies have reported more vigorous growth in introduced plants than plants of the same species in their native range (Crawley *et al.*, 1996; Rees & Paynter, 1997) – examples include St John's wort (*Hypericum perforatum*) from Australia which grew taller than native European plants (Pritchard, 1960) and purple loosestrife (*Lythrum salicaria*) plants from America, where it is introduced, being more vigorous than plants from its native European range (Blossey & Kamil, 1996). Whilst this may be due to more favourable conditions in the non-native environment, including escape from natural enemies and pathogens, Blossey & Nötzold (1995) have proposed that evolutionary changes in that environment lead to increased competitive ability (as genotypes which grow rapidly but are poorly defended against herbivores are favoured by selection). This hypothesis (evolution of increased competitive ability) has relevance for the introduction of genotypes in restoration schemes. It suggests that we may sometimes anticipate ecological release (an increase in population growth rates where factors affecting population regulation have been removed) in species populations that have evolved in different environments.

Finally, in this section on choosing local or non-local genotypes, it is important to emphasise that the power and ubiquity of natural selection suggest that, where species occur over a range of habitats, the matching of the donor and recipient habitats is at least as important as, and probably overrides, considerations of geographical distance. Organisms which have evolved in a similar habitat elsewhere are more likely to be pre-adapted to conditions at the new site. This is spectacularly true of those selected by extreme environments – e.g. a plant tolerant of, say, soils contaminated by copper is more likely to survive and establish on a distant copper-contaminated site than a local plant of the same species from ordinary soils – and is probably true over a broad range of conditions. Barratt *et al.* (1999) report significant differences in reproductive effort and biomass allocation between intraspecific variants of bird's-foot trefoil (*Lotus corniculatus*) and common knapweed (*Centaurea nigra*) in response to a range of experimentally manipulated water-table depths, and emphasise the importance of selecting the appropriate variant for wet grassland restoration schemes. Jones & Hayes's (1999) study of the effect of provenance on the establishment of several wildflower species in established swards demonstrates the importance of matching provenance to site conditions (and especially sward management). Albeit in a short-term study, local provenances were not always superior to non-locals for seedling establishment and survivorship. Based on the patterns of quantitative trait variation in Californian populations of purple needlegrass (*Nassella pulchra*), Knapp & Rice (1998) conclude that the greatest risk of transplant failure would be in moving seed between interior and coastal sites of the same region rather than between similar sites in northern and southern California (although the data from isozyme variation gave a different picture; see below).

These three recent examples from the restoration ecology literature provide empirical confirmation that the principle of assuming a 'genotypical response to habitat', evident from the work of the early genecologists described above, has considerable utility. It enables the restorationist to exploit past evolution.

Problems of small populations: how many founders?

A key question in restoration ecology is 'How large should reintroduced populations be in order to avoid potential genetic and evolutionary problems?'. Even if we can select appropriate genotypes (as we discussed above, and return to later) how many individuals are needed to ensure that the species survives and responds to future evolutionary change?

It was, of course, concerns about the consequences of small population size which preoccupied the founders of conservation genetics (Soulé & Wilcox, 1980; Frankel & Soulé, 1981). They recognised that lowered fitness may combine with demographic processes to drive small populations to local extinction – the 'extinction vortex' of Gilpin & Soulé (1986). A decrease in fitness was seen as a more or less inevitable outcome of the genetic consequences of small population size. These are: (1) an increase in inbreeding (more mating between relatives) and (2) genetic drift (more chance that alleles will be fixed or lost by random fluctuations between generations). Inbreeding increases homozygosity, and may lead to the reduced vigour described as 'inbreeding depression'. This may happen by exposing to selection deleterious recessive alleles in the homozygous condition (the dominance model of inbreeding) or by reducing overall heterozygosity (the over-dominance model, the heterozygote being superior to either homozygote), or both together. Genetic drift in small populations reduces within-population genetic diversity, and increases among-population differentiation, by reducing the number of alleles and the levels of heterozygosity – hence, it is argued, reducing the ability of populations to adapt to future change.

In considering how large populations may need to be in order to avoid deleterious effects from inbreeding and genetic drift, one must also take account of the fact that the size of the breeding population is unlikely to be the same as the actual observed population. This is because, for many reasons, individuals do not equally contribute genes to the next generation. The reasons include non-random mating (as in social grouping of animals or limited gene flow in plants), differences in fertility, unequal sex ratios, variation in age structure and fluctuations in the breeding population. Whatever the cause, the genetically effective population size (N_e) – which is the size of an idealised population in which all individuals contribute genes equally to the gamete pool and which has the same inbreeding rate and allele frequency variation as the observed population – is likely to be smaller, and may be very much smaller, than the actual number of individuals (N). Crawford's (1984) review of empirical estimates of effective population sizes based on allozyme data in animals shows they can range from 1% to 95% of the census number.

Despite these difficulties Frankel & Soulé (1981), based on the earlier work of Franklin (1980) and Soulé (1980), developed the '50/500 rule of thumb'. This states that a genetically effective population size (N_e) of 50 is required for a tolerable level of inbreeding and to avoid serious reduction of fitness, and that 500 is the minimum genetically effective size needed to prevent the erosion of genetic variation and empower future adaptive and evolutionary change. Although the 50/500 rule has been widely criticised, especially with regard to its universal applicability, it embraces an idea, mentioned at the beginning of this chapter, which is germane to considerations of ecological restoration. This is that there are two distinct dimensions to the problem – short-term fitness and long-term evolutionary potential. The numbers of individuals required to achieve these two goals may be an order of magnitude different.

Even more important, the strategy required to sample the species' gene pool may also be different, depending on the specific goal. Maintenance of the evolutionary potential of a species by including as much genetic diversity in a restored population as possible may require the inclusion of individuals from many populations or lineages. By contrast, restoration of a population adapted to the local habitat may best be achieved, as we have seen, by using only local genotypes. In fact, from a genetic and evolutionary perspective, the key question is not 'How many individuals should be introduced?' (although this remains a key population biology issue) but 'How representative of the species genetic diversity should the introduced population be in order to achieve the goals of restoration?' The answer depends not only on the specific goals of the restoration but on the distribution of genetic diversity within the species. This in turn is dependent on a range of the species' intrinsic biological properties, such as its mating system, life history and social structure, and extrinsic dynamic processes such as population distribution, history of selection, past fluctuations in population size and so on.

It is unlikely in most cases that we will know anything in advance about the genetic diversity or the population genetic structure of the species we are introducing. For a widespread species, especially if there is evidence that it may be highly differentiated into subpopulations, breeding groups or habitat-specialists (e.g. ecotypes or local demes), the sensible approach to optimise short-term fitness would be to utilise a local population from a similar habitat. Sampling such a population is discussed below. Longer-term evolutionary flexibility could be built into such a system by ensuring that the population within the restoration remains connected genetically, i.e. there is gene flow, with other nearby populations. Population genetics theory suggests that the rate of gene flow need only be very low (averaging one migrant per generation) to ensure such connectivity.

Where the newly introduced population is isolated from other populations and a goal of restoration is to maximise genetic diversity to facilitate a future evolutionary response, as in establishing populations of scarce or threatened species, then quite different rules apply. In this case it may be necessary to construct an 'artificial' population with samples from the range of the species' natural populations. The guidelines to be used for sampling to create such a population are essentially those we might use for the *ex situ* conservation of genetic diversity in gene banks or captive populations of animals. The theory, and some advice, has been covered in a number of texts (see for example chapters by Barrett & Khon, Brown & Briggs, Falk, Millar & Libby and Templeton in Falk & Holsinger [1991]; and chapters by Snaydon and Schaal & Leverich in Kapoor-Vijay & White [1992]). The way in which sampling should be stratified across populations, i.e. how to decide the number of populations to sample per species and the number of individuals to sample per population, depends critically on what can be gleaned in advance about the particular species' mating system and population biology. In general terms where the objective is to capture as much as possible of the genetic diversity within a species, the number of populations sampled should be increased (at the expense of the number of individuals per population) where there is evidence of between-population differentiation, where populations are isolated or fragmented with no gene flow, and where inbreeding is suspected (e.g. self-fertilising plants). Similar principles apply to sampling single populations to maximise genetic diversity; where substructuring into genetic neighbourhoods or inbreeding are suspected, more individuals should be sampled by reducing the number of offspring or seed per individual.

Ideally a survey of the species' genetic diversity should be carried out in advance of any restoration or reintroduction. Unless we know how genetic diversity is distributed within and among populations we cannot design a sampling scheme which will capture most of that diversity. Blind sampling could give very different results for different species. Templeton (1991) gives an example of his work on the collared lizard (*Crotaphytus collaris*) in Missouri, which genetic surveys revealed to have highly subdivided populations, lizards from individual rocky glades within a forest being almost genetically identical and those from different glades having many genetic differences. As Templeton points out, a sample of lizards taken from a single glade would miss almost all the genetic diversity present in the species, however large the sample size. The contrasting population genetic structure of two grass species in southwest Britain provides a further example. All 30 populations of bristle-bent grass (*Agrostis curtisii*) contained almost all the alleles found in an isoenzyme survey of the species (and most plants were heterozygous) whereas populations of nit grass (*Gastridium ventricosum*) tended to be fixed for a different allele at each locus (and plants were mostly homozygous) (Gray, 1996*b*). To capture all the allelic diversity in the latter most, if not all, populations would have to be sampled whilst sampling only a single population of the former would be sufficient.

If it is not possible to survey genetic diversity in advance are there any generalisations to guide the sampling design? Kay & John (1996) believe not and advocate genetic surveys of each and every species (in this case of rare and declining plants). Gray (1996*a*, *b*) is more optimistic, pointing out that, within phylogenetic groups, useful patterns may emerge from an understanding of the species biology. For example, within the grass family annuals are more likely to be diploid, inbreeding and to occupy open ephemeral habitats – features which tend to restrict genetic recombination and lead to individual homozygosity and high between-population differentiation – whereas perennial species tend to be polyploid, outbreeding and occur in mainly closed swards – encouraging heterozygosity and low between-population differentiation. The patterns described above in *Agrostis curtisii*, which is perennial and self-incompatible, and *Gastridium ventricosum*, an annual inbreeder, provide an extreme example of this contrast. Although there are certain to be exceptions, a good general rule would be to sample from more populations of an annual grass than a perennial grass, at least where the objective is to include as much genetic diversity as possible.

So far I have avoided giving actual numbers, emphasising instead the importance of taking into account both the aim of the restoration and the way in which genetic diversity is distributed among species populations. Recommended numbers in the literature on seed collections for genetic conservation in plants range from one seedhead from 200–300 plants from 500 populations in winter wild oat (*Avena sterilis*) (Allard, 1970), through five seeds from 10 plants per population for outbreeding plants and one seed from 200 individuals for inbreeding plants (Snaydon, 1992), to a (varying) number of seeds from 10–15 individuals from each of up to five populations (Centre for Plant Conservation, in Falk & Holsinger, 1991). The considerably smaller numbers in these last two cases reflect an appreciation of the strong law of diminishing returns in sampling allelic diversity, the first ten organisms randomly sampled from a population being equally or more important than a further 90 (Brown & Briggs, 1991). Lawrence *et al.* (1995) show that rather few plants (about 172) sampled at random from a population are needed to conserve nearly all the polymorphic genes segregating in that population, irrespective of the breeding system (in fact this number can be divided across populations and will still capture all but the rarest alleles, i.e. those at a frequency below 0.05). Such numbers, as Guerrant (1992) and Guerrant & Pavlick (1998) suggest, are surprisingly low but reflect the fact that, on average, almost 80% of the allozyme diversity at polymorphic loci in plants occurs within populations (Hamrick *et al.*, 1991). Similarly, a relatively small founding population of animals can contain a sizeable proportion of the species' genetic diversity (Templeton, 1991). In contrast, based on a consideration of quantitative genetics, Lynch (1996) argues that the current population sizes (of several hundred individuals or fewer) at which conservationists recognise a potential threat to the genetic integrity of species are in fact two or three orders of magnitude below the stage at which such a risk occurs. He suggests that effective population sizes of at least 1000 and 10 000 reproductive adults respectively are required to avoid genetic degradation and allow future adaptive change.

Thus, there is no generic answer to the question 'How many individuals, or genotypes, should be introduced to a specific restoration?' There is no universally applicable number guaranteed to prevent a future reduction in diversity or to enable a response to future evolutionary change. We must take account of differences between species in how genetic variation is partitioned among populations and passed between generations. We must also account for scale, the heterogeneity of the environment and the past, and likely future, connectedness between populations.

Restoration strategies: mix or match?

Discussing the sort of issues described in the two preceding sections, Lesica & Allendorf (1999) recently expressed the problem (at least for plants) as a question of whether to 'mix or match'. Should one choose local plants, or plants from habitats that 'match' the site to be restored, or should one use 'mixtures' of genotypes from different sources in an attempt to capture more diversity? We will return to their insights later. First we should briefly consider how different methods of measuring genetic diversity may inform that choice.

Most of what is known about genetic diversity in plant and animal populations has been learned from studies of variation in soluble enzymes. Where different forms of an enzyme differ in shape or electrical charge, electrophoresis can be used to detect them as bands on a starch or acrylamide gel – representing the alternative products, or alleles, of a single gene (allozymes) or products of different genes (isozymes). Despite disadvantages such as the fact that only a proportion of the underlying variation is detected and the relationship between isoenzyme and other genetic variation is unclear, enzyme electrophoresis remains the ideal technique for rapid survey of genetic variation. It is relatively cheap and easy to use, almost any tissue can be analysed, and genes coding for allozyme variation usually display co-dominance. Its widespread use has enabled genetic parameters to be calculated for a wide range of organisms, comparisons to be made between species, and insights gained into the genetic structure of individual species populations. Using the models and assumptions of population genetics theory it has been possible to estimate rates of inbreeding, levels of gene flow between subdivided populations and the relatedness of various breeding groups. The advent of several techniques enabling analysis of variation at the

DNA level (see Gray, 1996*c*) has refined and revised many of these estimates, and although they are currently generally more expensive and difficult to use than enzyme electrophoresis, such 'molecular tools' (Avise, 1994) have provided powerful insights into genetic diversity, not least because they have made it possible to measure variation in nuclear and cytoplasmic genes, in non-coding DNA and in mitochondrial (mt DNA) and chloroplast DNA (cp DNA). Access to diversity in DNA evolving at different rates and with different modes of transmission has opened up new vistas in evolutionary biology. In particular in animal species the study of mt DNA variation (which is maternally inherited) has enabled the detection of discrete evolutionary lineages. In plants, cp DNA, also mostly maternally inherited, has been found to be useful in revealing phylogenetic patterns, although not usually at the intraspecific level (exceptions include a study of different lineages in populations of monkey flower [*Mimulus guttatus*]: Fenster & Ritland, 1992). Where independent and possibly ancient evolutionary lineages can be detected, the option of deliberately including different lineages in a reintroduction scheme may be considered as a way of maintaining the species' evolutionary potential.

Finally, there is the possibility of measuring variation in morphological and other traits, including those which are likely to affect Darwinian fitness. Usually such traits are under the control of many genes, vary in a continuous rather than a discrete way, and are often affected by variation in the environment (particularly in plants). Thus the analysis of the genetical control of such traits is not straightforward and requires, as a minimum, that groups of related individuals are cultured in a uniform environment. This usually means that information about these quantitative traits is not available prior to any restoration (with the exception of a few well-studied rare or endangered species). It is worth remembering that most of the variation highlighted by the genecological studies referred to earlier has been in traits under polygenic control. Such variation usually correlated with environmental variation and inevitably of adaptive significance, was confirmed, or revealed, by growth in a uniform environment – the common garden experiment. Such experiments can provide powerful insight into the distribution of genetic variation among plant species populations and, where there is time and the opportunity to do so, a carefully designed common garden experiment in advance of reintroduction has much to recommend it. The sampling design will depend on the objectives of the reintroduction and what is known in advance about the species' breeding system. As long as strict rules of experimental design are followed the patterns of variation in traits such as height, leaf shape and number, flowering time, seed number, seed size and so on, plus an indication of their heritability, can be estimated from such experiments (Gray & Scott, 1980; Lawrence, 1984). A particularly useful approach is to sample natural progenies by sampling seed from several individual plants which, providing all plants are grown in a single completely randomised block, allows variation to be partitioned within and between families (see Lawrence [1984] for background, analysis and some other designs).

Thus different methods of measuring genetic diversity will give different information. Those based on variation in enzymes or DNA, because they are assumed to reveal variation which is selectively neutral, provide information on genetic structure, phylogenetic relationships, breeding systems and gene flow. (We should note that the assumption of neutrality may be violated in the case of isozymes: Riddock, 1993; Prentice *et al.*, 1995, 2000). Such variation may not bear any relationship to morphological variation or to variation in genes presumed to be under selection. It would certainly be foolish to extrapolate from isoenzyme or molecular data to quantitative traits, something of a problem for conservation genetics in general which has largely had to rely on the former (Lynch, 1996). In their comparison of isozymes and quantitative trait variation in Californian populations of *Nasella pulchra* Knapp & Rice (1998) discuss the value to a restoration programme of the highly dissimilar picture which emerged from the two methods (and also usefully review earlier studies that have compared isozyme and morphological variation). They suggest that neutral markers are less useful than quantitative trait variation for evaluating material for restoration or reintroduction schemes. Because the acquisition of data

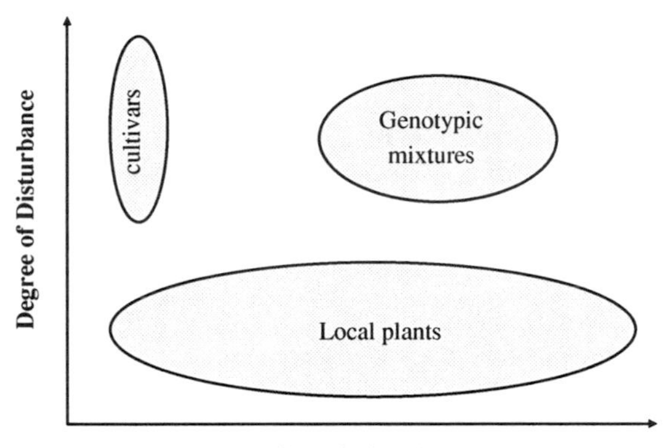

Fig. 5.1. A general relationship between the degree and size of disturbance and the possible sources of plants which might be used for restoration. After Lesica & Allendorf (1999).

on quantitative trait variation is time-consuming and labour-intensive, major environmental variation patterns such as regional variations in climate may be a better guide than marker genes to the appropriate scales for translocation.

This last point brings us back to the central dilemma, framed at the beginning of this section as a question of 'mix or match?'. Lesica & Allendorf's (1999) solution to this question recognises three broad categories of genotypes which might be used in restoration projects. These are 'local plants' or plants from habitats which closely match the habitat to be restored, 'genotypic mixtures' which are mixtures of genotypes from ecologically distinct populations, and 'cultivars' or genotypes selected for their ability to colonise stressed or highly disturbed sites. The choice of which to use should be guided by the size and the degree of disturbance of the area being restored (Fig. 5.1). Local habitat-matched genotypes are preferred when disturbance is low and the environment is similar to what it was before the disturbance, their use where large areas are to be restored reducing the possible impact on local gene pools. Genotypic mixtures or their hybrids are advocated in large, highly disturbed sites where the abiotic environment is very different from that prevailing prior to the disturbance (e.g. roadsides, quarry sites). Cultivars adapted to short-term survival in highly disturbed sites are preferred where the disturbance is small in area, allowing rapid establishment and reducing potential problems from gene flow to surrounding indigenous genotypes.

The framework provided by Lesica & Allendorf (1999) is not intended to be too prescriptive, but to guide the restoration ecologist in thinking about which genotypes to include in a restoration scheme – specifically when to use local genotypes, high-performance genotypes or a mixture of genotypes which maximise the genetic diversity of the introduction.

Rather more prescriptive recommendations have been offered by Millar & Libby (1989) based on their experience of replanting conifer forests, and by Knapp & Rice (1994) who look at the genetic issues involved in using native grasses for restoration. Both have considerable generality. Box 5.1 summarises seven specific recommendations made by Millar & Libby (1989) which incorporate thinking about genetic principles and could apply to a wide range of plants (and even be relevant for the introduction of certain animals). Knapp & Rice (1994), working with grasses, concentrate on the strategies to adopt when collecting, and bulking-up, seed for restoration schemes. Again they emphasise matching the collecting scheme to the scale of the genetic variation in the species and to the heterogeneity of the environment. Equalising the contribution of each parent, avoiding bias by sampling well-spaced plants at random irrespective of phenotype, and sampling over time are all suggested as ways of maximising the genetic diversity of the sample (and hence increasing the chances of future establishment and persistence). In discussing the bulking-up of seed for restoration by cultivation of selected genotypes, Knapp & Rice highlight the need to avoid potentially deleterious genetic shifts (alterations in the genetic diversity of the sample such that it is not representative of the original collection). Growing plants in a matching, non-agricultural, environment for as few generations as possible, with equal sampling, multiple harvest times and isolated from gene flow are among the suggestions for avoiding genetic shifts. Similar problems are faced in managing the genetic diversity in captive animals where the goal of captive breeding is to provide a population for restoration to the wild. Lacy (1987, 1994)

Box 5.1 General recommendations for restoration plantings

1. Do not buy seed or plants unless absolutely necessary, but collect from, or near, the restoration site sampling the range of variation in elevation, slope, aspect, drainage, soils, etc.
2. If collecting off site, collect from adjacent matching habitats-again sampling variation in environment.
3. If forced to collect from more distant sites try to obtain information about genetic variation in the species. If none available, match sites and see if data on breeding system, life history and ecology is available to guide site matching and collecting strategy.
4. Collect from many rather than a few parents, sampling each equally.
5. If you have to buy from a nursery use one that records origins of seed/stock or consider buying from several nurseries.
6. Consider giving nature a chance first. Natural regeneration may do the job for you.
7. If native or matched populations are unavailable build a new 'landrace' from a wide range of donor populations, allowing *in situ* selection to produce adapted individuals.

Source: Summarised and adapted from Millar & Libby (1989)

provides a comprehensive review of possible genetic consequences of captive breeding and of ways to minimise genetic change prior to reintroduction.

The strategies derived from reafforestation and grassland restoration clearly have general application, but the details will vary from species to species and depend on the objective of the restoration. For example the 'seed zones' referred to earlier which guide the replanting of conifer forests by the US Forest Service (Kitzmiller, 1990) are delineated on the basis of broad latitudinal, geographic and altitudinal divisions and are based on the empirical realities of higher survival and performance of within-zone translocations. Such zones may include subzones based on ecological criteria, reflecting a finer scale of genetic differentiation within the broader regional and climatic divisions. It is to be expected that other species (plants and animals) will perceive their environment at a range of spatial scales, the extent to which they have become genetically differentiated varying with factors such as their mating system, dispersal biology, plasticity, average population size, and so on. This differentiation will be species-specific. However broad regional divisions, equivalent to seed zones, may be discernible from climatic and environmental data, e.g. the bioclimatic zones of Great Britain recognised by Preston & Hill (1997), the biogeographical zones of Scotland defined by Carey *et al.* (1995) or the phytogeographic regions of Switzerland (Landolt, 1991). This last classification was suggested as a basis for regulating the use of wildflower mixtures in the study by Keller *et al.* (2000) mentioned earlier. With more detailed understanding of the population genetic structure of individual species it would be possible to identify within such broad zones finer scales of differentiation and management, perhaps equivalent to the genetic resource management units (GMRUs) prescribed for forests (Millar & Libby, 1991).

RESTORING THE EVOLUTIONARY PROCESS

Maintaining higher-order interactions

In considering genetic and evolutionary aspects of restoration ecology this chapter has dealt almost exclusively with single populations of single species – largely using case studies from populations of plants, of which there is most experience. Of course restoration schemes invariably involve reintroducing many different species. Exceptions include the reintroduction of endangered species or species with highly specific functions such as forest trees or metal-tolerant grasses. Even where single species are introduced, it is unlikely that future genetic change will be unaffected by the numbers and types of the associated species they encounter. In extreme cases an introduction may fail completely unless the appropriate mutualists (e.g. mycorrhizae, pollinators) or prey species are also introduced (a classic example being the need to create conditions suitable for the specific red ant host species [*Myrmica sabuleti*] before the large blue butterfly [*Maculinea arion*] could be reintroduced to sites in the UK [Thomas, 1994]). Unless the restored site exactly mimics the numbers, proportion and impact of the associated interacting species in the donor site, which seems unlikely, genetic changes in the founder population wrought by novel species interactions appear inevitable. They may involve small changes in gene frequencies or major genetic shifts, of the type which are postulated for example by the 'evolution of increased competitive ability' theory mentioned earlier, a change specifically driven by decreased herbivory. Any differences in management between a donor and restored site can be expected to cause genetic shifts; for example changes in grazing regime will alter the proportion of grazed and ungrazed ecotypes. Of course such genetic change under shifting selection pressures is precisely what evolution is about, and is entirely consistent with the modern stated objective of restoration – to conserve the range of processes that allow evolutionary change to occur rather than to fix levels of genetic diversity. Nonetheless it is important to remind ourselves that the evolutionary context of restoration ecology has a community, as well as a species, perspective.

It is similarly unrealistic to treat single populations as if they will be isolated from other populations or to ignore the effects of population subdivision or fragmentation. The dynamics which determine future genetic change may be those that operate at the metapopulation level. The effect of metapopulation structure on genetic variation and evolution is an area of great interest to population geneticists and ecologists, and one rich in theory built largely on Wright's (1931, 1951) island model of population structure. The relative size, isolation, persistence of, and rate of gene flow between the populations comprising a metapopulation will fundamentally affect how genetic variation is distributed among them, and how the species as a whole responds to selection. Such theory has guided decisions about whether or not to invigorate small and endangered populations by introducing new genotypes (Ellstrand & Elam, 1993; Pegtel, 1998; van Groenendael *et al.*, 1998; Keller *et al.*, 2000), whether to subdivide populations into small inbred demes (Maruyama, 1970; Templeton *et al.*, 1987), and what regional conservation policies will optimise the maintenance of gene flow between otherwise isolated populations (Avise & Hamrick, 1996). Nevertheless there remains a huge gap between theory and empirical testing in natural populations. Designing the size and spacing of species populations within a restoration (or the arrangement of introduced populations in relation to existing ones) should take account of what is known about existing rates of gene flow in the species and their effect on genetic diversity. There is a clear trade-off between, at one extreme, outbreeding effects such as genetic 'swamping' of one population by another and, at the other, the effects of genetic drift and inbreeding in small isolated populations. Where a particular species should be placed along the continuum between these two extremes will, as we have seen, vary according to its past history, mobility, dispersal, breeding system and life history.

CONCLUDING REMARKS

Restorationists share with conservation biologists the need to act now. Frankel & Soulé (1981) famously described 'the luxury and excitement of adversary science' which may take 'years or decades . . . before a clear resolution is reached', as being something conservationists cannot afford. Likewise the restoration of ecosystems cannot await a full understanding of the implications for genetic and evolutionary change in the component species. The many inherent conflicts – between short-term adaptation and long-term evolutionary flexibility, between the homogenising effect of extensive gene flow and selection of locally adapted genotypes, between the probability and likely effects of outbreeding depression versus those of the more widely observed inbreeding depression – are unlikely to be resolved for more than a handful of well-studied species. Even predicting what the effects of inbreeding may be in a particular population or species requires an understanding of a range of factors including the genetic basis of inbreeding effects (see above).

Despite these uncertainties, and the difference in philosophy which underlies the *Flora Locale* and the 'evolutionary free-for-all' viewpoints described earlier, useful generalisations and some practical advice have emerged from a consideration of the genetic issues.

Whether preserving genetic integrity of native populations is scientifically questionable (or even ethically dubious: Peretti, 1998) or not, the practical value of using local germplasm (to increase chances of establishment and survival, to avoid problems which may stem from hybridisation and to avoid invasive genotypes) makes it a sensible option in most cases. Similarly, the utility of taking a regional perspective, of matching genotype to environment, of sampling, and cultivating off-site, in relation to existing patterns of genetic differentiation and of designing mixtures from a range of donor populations and lineages, are reinforced by considering the underlying genetic diversity.

Finally, we must acknowledge that the uncertain relationship between genetic diversity and species persistence highlighted at the outset demands that the evolutionary perspective is placed in context with other factors important in restoration ecology. The assumption that high genetic diversity is necessary for long-term survival is challenged by the occurrence of many successful but genetically depauperate species and by the failure to attribute any particular extinction to low genetic diversity alone (Gray, 1996*c*). Genetic principles can guide restoration programmes but their success may depend on a deeper understanding of species' population biology and natural history.

REFERENCES

Akeroyd, J. (1994). *Seeds of Destruction? Non-Native Wildflower Seed and British Floral Biodiversity*. London: Plantlife.

Allard, R.W. (1970). Population structure and sampling methods. In *Genetic Resources in Plants: Their Exploration and Conservation*, eds. O.H. Frankel & E. Bennett, pp. 97–107. Oxford: Blackwell.

Avise, J.C. (1994). *Molecular Markers, Natural History and Evolution*. New York: Chapman & Hall.

Avise, J.C. (1996). Toward a regional conservation genetics perspective: phylogeography of faunas in the south eastern United States. In *Conservation Genetics: Case Histories from Nature*, eds. J.C. Avise & J.L. Hamrick, pp. 431–470. New York: Chapman & Hall.

Avise, J.C. & Hamrick, J.L. (eds.) (1996). *Conservation Genetics: Case Histories from Nature*. New York: Chapman & Hall.

Barratt, D.R. Walker, K.J., Pywell, R.F., Mountford, J.O. & Sparks, T.H. (1999). Variation in the responses of intraspecific variants of wet grassland species to manipulated water levels. *Watsonia*, **22**, 317–328.

Blossey, B. & Kamil, J. (1996). What determines the increased competitive ability of nonindigenous plants? In *Proceedings of the 9th International Symposium on Biological Control of Weeds*, pp. 3–9. Cape Town: University of Cape Town.

Blossey, B. & Nötzold, R. (1995). Evolution of increased competitive ability in invasive nonindigenous plants: a hypothesis. *Journal of Ecology*, **83**, 887–889.

Bradshaw, A.D. & McNeilly, T. (1981). *Evolution and Pollution*. London: Edward Arnold.

Brown, A.H.D. & Briggs, J.D. (1991). Sampling strategies for genetic variation in *ex situ* collections of endangered species. In *Genetics and Conservation of Rare Plants*, eds. D.A. Falk & K.E. Holsinger, pp. 99–119. New York: Oxford University Press.

Butler, D. (1994). Bid to protect wolves from genetic pollution. *Nature*, **370**, 497.

Carey, P.D., Preston, C.D., Hill, M.O., Usher, M.B. & Wright, S.M. (1995). An environmentally defined biogeographical zonation of Scotland designed to reflect species distributions. *Journal of Ecology*, **83**, 833–845.

Crawford, T.J. (1984). What is a population? In *Evolutionary Ecology*, ed. B. Shorrocks, pp. 135–174. Oxford: Blackwell.

Crawley, M.J., Harvey, P.H. & Purvis, A. (1996). Comparative ecology of the native and alien floras of the British Isles. *Philosophical Transactions of the Royal Society of London B*, **351**, 1251–1259.

Daniels, R.E. & Sheail, J. (1999). Genetic pollution: concepts, concerns and transgenic crops. In *Gene Flow and Agriculture: Relevance for Transgenic Crops*, BCPC Symposium no. 72, ed. P.J. W. Lutman, pp. 65–72. Farnham, UK: British Crop Protection Council.

DeMauro, M.M. (1994). Development and implementation of a recovery program for the federal threatened Lakeside Daisy (*Hymenoxys acaulis* var. *glabra*) In *Restoration of Endangered Species*, eds. M.L. Bowles & C.J. Whelan, pp. 298–321. Cambridge: Cambridge University Press.

Ellstrand, N.C. & Elam, D.R. (1993). Population genetic consequences of small population size implications for plant conservation. *Annual Review of Ecology and Systematics*, **24**, 217–242.

Falk, D.A. & Holsinger, K.E. (eds.) (1991). *Genetics and Conservation of Rare Plants*. Oxford: Oxford University Press.

Fenster, C.B. & Dudash, M.R. (1994). Genetic considerations for plant population restoration and conservation. In *Restoration of Endangered Species*, eds. M.L. Bowles &

C.J. Whelan, pp. 34–62. Cambridge: Cambridge University Press.

Fenster, C.B. & Ritland, K. (1992). Chloroplast DNA and isozyme diversity in two *Mimulus* species (Scrophulariaceae) with contrasting mating systems. *American Journal of Botany*, **79**, 1440–1447.

Frankel, O.H. & Soulé, M.E. (1981). *Conservation and Evolution*. Cambridge: Cambridge University Press.

Franklin, I.A. (1980). Evolutionary change in small populations. In *Conservation Biology: An Evolutionary–ecological Perspective*, eds. M.E. Soulé & B.A. Wilcox, pp. 135–50. Sunderland, MA: Sinauer Associates.

Gilpin, M.E. & Soulé, M.E. (1986). Minimum viable populations: the process of species extinctions. In *Conservation Biology: The Science of Scarcity and Diversity*, ed. M.E. Soulé, pp. 13–43. Sunderland, MA: Sinauer Associates.

Gray, A.J. (1996*a*). Genetic diversity and its conservation in natural populations of plants. *Biodiversity Letters*, **3**, 71–80.

Gray, A.J. (1996*b*). Genecology, the genetic system and the conservation genetics of uncommon British grasses. In *The Role of Genetics in Conserving Small Populations*, ed. T.E. Tew, T.J. Crawford, J.W. Spencer, D.P. Stevens, M.B. Usher & J. Warren, pp. 56–64. Peterborough, UK: Joint Nature Conservation Committee.

Gray, A.J. (1996*c*). The genetic basis of conservation biology. In *Conservation Biology*, ed. I.F. Spellerberg, pp. 107–122. Harlow, UK: Longman.

Gray, A.J. & Scott, R. (1980). A genecological study of *Puccinellia maritima* Huds. (Parl.). 1: Variation estimated from single-plant samples from British populations. *New Phytologist*, **85**, 89–107.

Guerrant, E.O. (1992) Genetic and demographic considerations in the sampling and reintroduction of rare plants. In *Conservation Biology: The Theory and Practice of Nature Conservation and Management*, eds. P.L. Fielder & S.K. Jain, pp. 321–346. New York: Chapman & Hall.

Guerrant, E.O. & Pavlick, B.M. (1998). Reintroduction of rare plants: genetics, demography and the role of *ex situ* conservation methods. In *Conservation Biology for the Coming Decade*, 2nd edn, eds. P.L. Fielder & P.M. Kareiva, pp. 80–108. New York: Chapman & Hall.

Hamrick, J.L., Godt, M.J.W., Murawski, D.A. & Loveless, M.D. (1991). Correlations between species traits and allozyme diversity: implications for conservation biology. In *Genetics and Conservation of Rare Plants*, eds. D.A. Falk & K.E. Holsinger, pp. 75–86. Oxford: Oxford University Press.

Jones, A.T. & Hayes, M.J. (1999). Increasing floristic diversity in grassland: the effects of management regime and provenance on species introduction. *Biological Conservation*, **87**, 381–390.

Jones, A.T., Hayes, M.J. & Sackville Hamilton, N.R. (in press). The effect of provenance on the performance of *Crataegus monogyna* in hedges. *Journal of Applied Ecology*.

Kapoor-Vijay, P. & White, J. (eds.) (1992). *Conservation Biology: A Training Manual for Biological Diversity and Genetic Resources*. London: Commonwealth Secretariat.

Kay, Q.O.N. & John, R. (1996). Patterns of variation in relation to the conservation of rare and declining plant species. In *The Role of Genetics in Conserving Small Populations*, eds. T.E. Tew, T.J. Crawford, J.W. Spencer, D.P. Stevens, M.B. Usher & J. Warren, pp. 41–55. Peterborough, UK: Joint Nature Conservation Committee.

Keller, M., Kollmann, J. & Edwards, P. J. (2000). Genetic introgression from distant provenances reduces fitness in local weed populations. *Journal of Applied Ecology*, **37**, 647–659.

Kitzmiller, J.H. (1990). Managing genetic diversity in a tree improvement program. *Forest Ecology and Management*, **35**, 131–149.

Knapp, E.E. & Rice, K.J. (1994). Starting from seed: genetic issues in using native grasses for restoration. *Restoration and Management Notes*, **12**, 40–45.

Knapp, E.E. & Rice, K.J. (1998). Comparison of isozymes and quantitative traits for evaluation patterns of genetic variation in purple needlegrass (*Nassella pulchra*). *Conservation Biology*, **12**, 1031–1041.

Lacy, R.C. (1987). Loss of genetic diversity from managed populations: interacting effects of drift, mutation, immigration, selection and population subdivision. *Conservation Biology*, **1**, 143–158.

Lacy, R.C. (1994). Managing genetic diversity in captive populations of animals. In *Restoration of Endangered Species*, eds. M.L. Bowles & C.J. Whelan, pp. 63–89. Cambridge: Cambridge University Press.

Landolt, E. (1991). *Gefährdung der Farn und Blütenpflanzen in der Schweiz mit gesamtschweizerischen und regionalen roten Listen*. Bern: Bundesamt für Umwelt, Wald und Landschaft.

Lawrence, M.J. (1984). The genetical analysis of ecological traits. In *Evolutionary Ecology*, ed. B. Shorrocks, pp. 27–63. Oxford: Blackwell.

Lawrence, M.J., Marshall, D.F. & Davies, P. (1995). Genetics of genetic conservation. 1: Sample size when collecting germplasm. *Euphytica*, **84**, 89–99.

Lesica, P. & Allendorf, F.W. (1999). Ecological genetics and the restoration of plant communities: mix or match? *Restoration Ecology*, **7**, 42–50.

Lynch, M. (1996). A quantitative-genetic perspective on conservation issues. In *Conservation Genetics: Case Histories from Nature*, eds. J.C. Avise & J.L. Hamrick, pp. 471–501. New York: Chapman & Hall.

Maruyama, T. (1970). Rate of decrease of genetic variation in a subdivided population. *Biometrika*, **57**, 299–312.

Millar, C.I. & Libby, W.J. (1989). Disneyland or native ecosystem: genetics and the restorationist. *Restoration and Management Notes*, **7**, 18–24.

Millar, C.I. & Libby, W.J. (1991). Strategies for conserving clinal, ecotypic, and disjunct population diversity in widespread species. In *Genetics and Conservation of Rare Plants*, eds. D.A. Falk & K.E. Holsinger, pp. 149–70. New York: Oxford University Press.

Pegtel, D.M. (1998), Rare vascular plant species at risk: recovery by seeding? *Applied Vegetation Science*, **1**, 67–74.

Peretti, J.H. (1998). Nativism and nature: rethinking biological invasion. *Environmental Values*, **7**, 183–92.

Prentice, H.C., Lönn M., Lefkovitch, L.P. & Runyeon, H. (1995). Associations between allele frequencies in *Festuca ovina* and habitat variation in the alvar grass-lands on the Baltic island of Öland. *Journal of Ecology*, **83**, 391–401.

Prentice, H.C., Lönn, M., Lager, H., Rosén, E. & Van der Maarel, E. (2000) Changes in allozyme frequencies in *Festuca ovina* populations after a 9-year nutrient/water experiment. *Journal of Ecology*, **88**, 331–347.

Preston, C.D. & Hill, M.O. (1997). The geographical relationships of British and Irish vascular plants. *Botanical Journal of the Linnean Society*, **124**, 1–120.

Pritchard, T. (1960). Race formation in weedy species with special reference to *Euphorbia cyparissias* L and *Hypericum perforatum* L. In *The Biology of Weeds*, ed. J. Harper, pp. 61–66. Oxford: Blackwell.

Rapson, G.L. & Wilson, J.B. (1988). Non-adaptation in *Agrostis capillaris* L. (Poaceae). *Functional Ecology*, **2**, 479–490.

Rees, M. & Paynter, Q. (1997). Biological control of scotch broom: modelling the determinants of abundance and the potential impact of introduced insect herbivores. *Journal of Applied Ecology*, **34**, 1203–1221.

Rhymer, J.M. & Simberloff, D. (1996). Extinction by hybridisation and introgression. *Annual Review of Ecology and Systematics*, **27**, 83–109.

Rice, K.J. & Mack, R.N. (1991). Ecological genetics of *Bromus tectorum*. 3: The demography of reciprocally sown populations. *Oecologia*, **88**, 91–101.

Riddock, R.J. (1993). The adaptive significance of electrophoretic mobility in phosphoglucose isomerase (PGI). *Biological Journal of the Linnean Society*, **50**, 1–17.

Sheail, J., Treweek, J.R. & Mountford, J.O. (1997). The UK transition from nature preservation to 'creative conservation'. *Environmental Conservation*, **24**, 224–235.

Smith, R.A.H. & Bradshaw, A.D. (1979). The use of metal-tolerant plant populations for the reclamation of metalliferous wastes. *Journal of Applied Ecology*, **16**, 595–612.

Snaydon, R.W. (1992). Sampling plant populations for genetic conservation. In *Conservation Biology: A Training Manual for Biological Diversity and Genetic Resources*, eds. P. Kapoor-Vijay & J. White, pp. 87–92. London: Commonwealth Secretariat.

Soulé, M.E. (1980) Thresholds for survival: maintaining fitness and evolutionary potential. In *Conservation Biology: An Evolutionary–Ecological Perspective*, eds. M.E. Soulé & B.A. Wilcox, pp. 111–24. Sunderland, MA: Sinauer Associates.

Soulé, M.E. & Wilcox, B.A. (1980) *Conservation Biology: An Evolutionary–Ecological Perspective*, Sunderland, MA: Sinauer Associates.

Templeton, A.R. (1991). Off-site breeding of animals and implications for plant conservation strategies. In *Genetics and Conservation of Rare Plants*, eds. D.A. Falk & K.E. Holsinger, pp. 182–194. New York: Oxford University Press.

Templeton, A.R., Davis, S.K. & Read, B. (1987). Genetic variability in a captive herd of Speke's gazelle (*Gazella spekei*). *Zoo Biology*, **6**, 305–313.

Thomas, J.A. (1994). The ecology and conservation of *Maculinea arion* and other European species of large blue. In *Ecology and Conservation of Butterflies*, ed. A. Pullin, pp. 180–196. London: Chapman & Hall.

Turesson, G. (1922). The genotypical response of the plant species to the habitat. *Hereditas*, **3**, 211–350.

van Groenendael, J.M., Ouborg, N.J. & Hendriks, R.J.J. (1998). Criteria for the introduction of plant species. *Acta Botanica Neerlandica*, **47**, 1–3.

Wells, T.C.E., Bell, S. & Frost, A. (1982). *Creating Attractive Grasslands Using Native Plant Species*. Peterborough, UK: Nature Conservancy Council.

Wright, S. (1931). Evolution in Mendelian populations. *Genetics*, **16**, 97–159.

Wright, S. (1951). The genetic structure of populations. *Annals of Eugenics*, **15**, 323–354.

Part 2 • Manipulation of the physical environment

6 • Terrestrial systems

STEVE G. WHISENANT

INTRODUCTION

Mining, overgrazing, deforestation, cultivation and soil compaction dramatically alter the physical environment of terrestrial ecosystems. Among the more serious changes are damaged hydrologic processes (infiltration and runoff), accelerated erosion (fluvial and aeolian) and unfavourable micro-environmental conditions (wind, temperature and relative humidity). These changes inhibit both natural recovery processes and our ability to direct successional development with ecological restoration.

Properly functioning ecosystems have natural recovery processes that maintain sustainable flows of soil, nutrients, water and organic materials. During degradation, positive feedback mechanisms reinforce and accelerate damaging processes (Fig. 6.1), leading to irreversible vegetation change once a site's capacity for self-repair has been exceeded (Rietkerk & van de Koppel, 1997). Contemporary succession theory describes this catastrophic change as having crossed a transition threshold that inhibits natural recovery (Friedel, 1991; Laycock, 1991; Walker, 1993; Rietkerk & van de Koppel, 1997). Designing restoration strategies that overcome threshold barriers to natural recovery processes is one of the more important challenges for ecological restoration. That requires an understanding of treatment strategies that reduce threshold barrier effects.

Two types of threshold barriers limit the natural recovery of damaged ecosystems (Whisenant, 1999). It is important to distinguish between the two, because they require different restoration approaches. The first is controlled by interference from other organisms (biotic interactions), usually invasive weeds or other plants that prevent natural recovery (Fig. 6.2). Reducing problematic species (e.g. selective plant removal with herbicides, fire, mechanical, or hand treatments) and/or adding appropriate species are the most effective strategies for these circumstances. The second barrier operates when dysfunctional hydrologic processes or harsh micro-environments create abiotic limitations to recovery (Fig. 6.2). These seriously degraded areas usually require improvements to their physical environment.

After identifying limiting features of the physical environment, we can design restoration strategies that 'jump-start' the ecosystem's self-repairing mechanisms (Whisenant, 1999). Thus, understanding and directing the role of the physical environment is an essential component of ecological restoration. Two aspects of the physical environment are most relevant to ecological restoration: (1) physical controls over resource fluxes, and (2) physical controls over micro-environmental conditions. Recognising and restoring ecosystem processes that influence the physical environment is the focus of this chapter.

Physical controls over resources fluxes

Healthy ecosystems capture and retain limiting resources through a combination of biotic and abiotic controls over the resources flowing through the landscape (areas A and B in Fig. 6.2). Resource losses will increase as the dominant mechanisms of resource control shift from biotic to abiotic processes (Davenport *et al.*, 1998). Although each ecosystem has a unique combination of processes contributing to proper functioning, healthy ecosystems have sustainable resource fluxes. Resource losses are offset by resource gains to the system. In contrast, the recovery of damaged ecosystems, with depleted resource pools, requires additional resources (soil, water, nutrients,

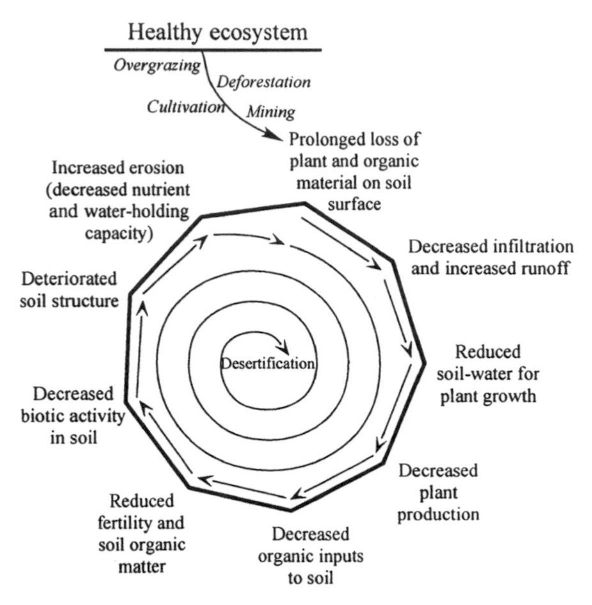

Fig. 6.1. Soil degradation cycle illustrating the importance of the soil surface in the continuing downward spiral of soil condition. Although other degradation pathways are possible, this is most common. While soil surface condition is not the causal factor in all soil degradation processes, it is the most widespread. Adapted, with changes, from Whisenant (1999).

basic cations or organic materials), that must first be captured and then retained. Consequently, it is useful to understand the processes that control the flows of water, soil, nutrients and organic material through the landscape. Comparing the retention of resources in healthy versus damaged ecosystems provides insights with clear implications for ecological restoration.

Both living and dead organic materials contribute to biotic control over resource flows. Severely damaged ecosystems, with fewer plants and organic materials, exert little biotic influence over limiting resources (area C in Fig. 6.2). Therefore, most resource retention occurs within certain landform and microtopographic features of those landscapes. These naturally occurring resource-retaining features suggest management opportunities for initiating autogenic restoration that will improve biotic control mechanisms (Whisenant *et al.*, 1995). On the most damaged sites, initial plant establishment may not occur until physical manipulations to the soil surface increase resource availability. That allows plants to establish and begin to exert biotic control over limiting resources.

Resource flows from severely damaged landscapes, with few biotic controls, usually occur through fluvial or aeolian transport of soil, water, nutrients, cations and organic materials. At larger scales, geomorphic processes and landforms influence resource flows by creating zones of resource depletion or resource deposition (Ludwig *et al.*, 1994; Ludwig & Tongway, 1995; Tongway & Ludwig, 1997*a*). Landforms provide visual clues to fluvial and aeolian processes that influence resource retention. The relative topographic position of a site suggests the magnitude of fluvial processes operating across that landscape (Fig. 6.3). Relative position within the landform affects erosion potential, runoff rate and the water-retention potential of closed basins. Concave landforms, of any size, have a relatively high potential to capture resources. For example, wetlands are concave landforms that capture a high percentage of nutrients and organic materials flowing through those landscapes. In contrast, without strong biotic influences, convex sites and steep slopes exert little control over resource flows (Whisenant, 1999).

At smaller scales, resource-conserving processes operate at the scale of microtopographic depressions in the soil surface (micro-gilgai, hoofprints, root channels, animal-created holes, cracks in vertisols or microcatchments) and aboveground obstructions (rocks, logs or vegetation). Damaged soil surfaces trigger positive feedback degradation systems (Rietkerk & van de Koppel, 1997; van de Koppel *et al.*, 1997) (illustrated in Fig. 6.1) leading to irreversible vegetation changes (Fig. 6.4). Surface condition assessments provide useful information about stability (ability to withstand erosive forces), hydrologic processes (infiltration, runoff, deep drainage and depth to the water table) and nutrient cycling (Aronson *et al.*, 1993; National Research Council, 1994; Tongway & Ludwig, 1997*b*).

Physical controls over micro-environmental conditions

Degraded ecosystems often have greater temperature extremes and higher wind speeds. These changes reduce soil-water levels, create difficult

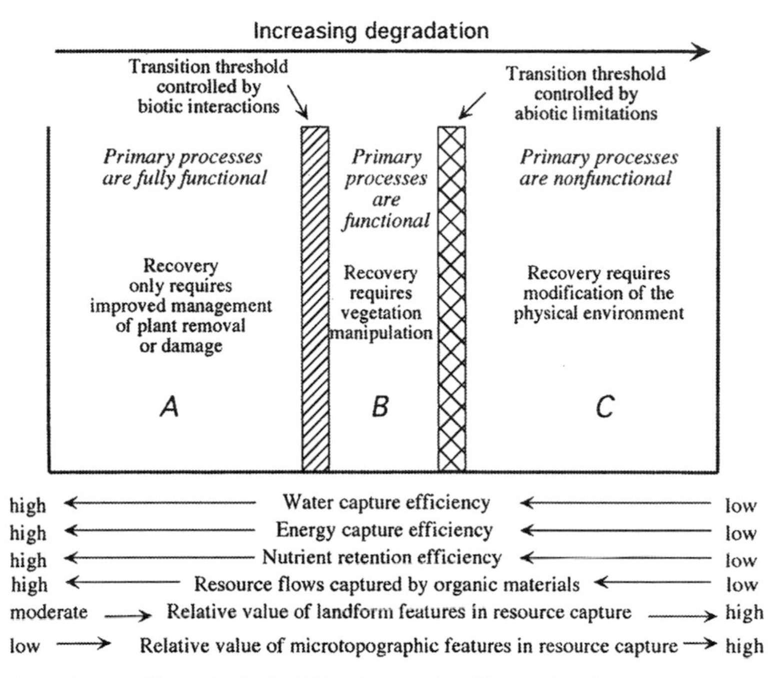

Fig. 6.2. Degradation of hypothetical wildland vegetation illustrating the two common transition thresholds that separate the three vegetative groups (A, B and C) of functional significance. These groups are defined by their functional integrity, rather than species composition.

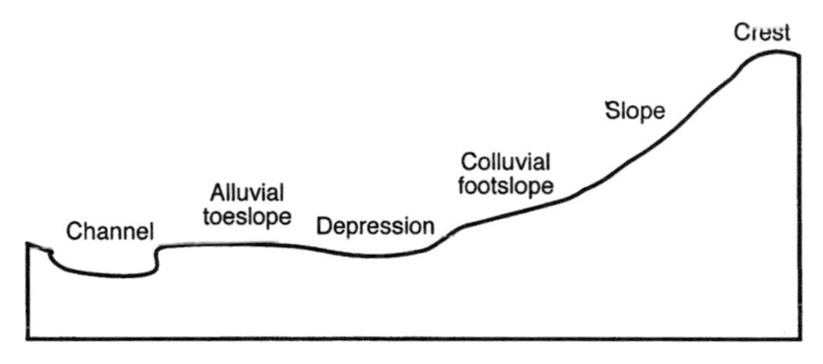

Fig. 6.3. Hypothetical arrangement of landforms illustrating the relationship of landform types with their relative position among other landform types. Pedogenic processes associated with vertical subsurface soil water movement dominate crest sites. Slopes transport material by mass movement (flow, slide, slump, creep), surface and subsurface water action. Colluvial footslopes are dominated by the redeposition of material by mass movement, surface wash, fan formation, creep and subsurface water action. Depressions may be of any size and occur in several different areas of a landscape. Wherever located, depressions hold water and catch materials from surface wash. Alluvial deposition and processes associated with subsurface water movements are most active on alluvial toeslopes. Slumping and falling materials shape channel walls, while the channel bed is shaped by the downslope surface water transport of materials.

abiotic environments and reduce plant growth (Lugo, 1992; Vetaas, 1992; Brown & Lugo, 1994; Guariguata *et al.*, 1995; Fimbel & Fimbel, 1996; Ashton *et al.*, 1997). Protection from wind can increase productivity through indirect and micro-environmental effects (Cleugh, 1998) and protection from the direct force of the wind (Cleugh *et al.*, 1998). Direct mechanical effects of wind include plant

Fig. 6.4. Bare, sealed soil surfaces, like this lateritic soil in southwest Niger, greatly reduce the infiltration of water into the soil. Most of the water reaching this soil surface will continue as runoff. Consequently, fewer plants are produced, soil structure deteriorates and the site becomes progressively less favourable for plant growth.

motions due to turbulence and the drag force of the mean wind; uprooting of plants when the wind drag force exceeds the stem or root/soil strength. Plant leaves are also subject to wind damage caused by tearing, stripping and abrasion. Sandblasting damage to leaves includes both abrasion and tearing.

Woody plants have strong ameliorating effects on the micro-environment. The importance of this relationship is a central theme of both ecological restoration and agro-forestry. Although woody plants compete with understorey plants for light, the benefits of habitat amelioration often outweigh any negative effects (Holmgren *et al.*, 1997). As an example, agronomic crops sheltered by wind barriers often grow taller, contain more dry matter, have a larger leaf area index and produce higher yields (Vandermeer, 1989).

MANIPULATING THE PHYSICAL ENVIRONMENT ON RESTORATION SITES

Manipulating the physical environment is a critical aspect that becomes increasingly more important as ecosystem damage increases. Altering the physical environment allows us to tackle fundamental problems of resource fluxes (both surface and subsurface flows) and adverse micro-environmental conditions. Surface flows are addressed by roughening the soil surface or adding aboveground obstructions (Whisenant, 1999). In extreme situations, such as mining, the entire landform must be reshaped. The position of the landform type, relative to adjacent areas, affects both surface and subsurface flows. Groundwater levels can be managed by manipulating inputs (subsurface inflows and surface drainage) and outputs (transpiration and subsurface outflows) of water through the site.

Trees, herbaceous vegetation, litter or duff can significantly change the micro-environment. Herbaceous vegetation can also be used to alter micro-environmental conditions during the critical, but brief, seedling establishment phase. Woody vegetation is a more effective option when those changes must last for a longer period. Fire is often an effective and practical tool for manipulating micro-environmental conditions. Fire is most commonly used to expose the soil surface or lower vegetation strata to direct sunlight, wind and less humid conditions.

Manipulating surface flows

Fluvial flows are often managed with engineered structures that concentrate and accelerate runoff. Although common, this approach raises streamflow energy and increases downstream erosion hazards. Most terrestrial ecosystems benefit from strategies that promote the retention and use of water where it falls, rather than encouraging runoff. Additional water is lost to above- or belowground flows, but the site makes full use of the available water. When successful, this approach produces vegetation that does not allow water to develop enough velocity to cause erosion.

Surface flows of limiting resources move through fluvial or aeolian transport systems. These resources can be captured with treatments that focus on soil surface conditions (Whisenant, 1999). On bare soil surfaces, the initial treatment usually involves one or two general approaches. The first increases soil surface roughness with pits, contour furrows, basins, ripping or chiselling. The second approach is to add aboveground obstructions such as logs, rocks, woody debris, herbaceous litter or man-made erosion control products. Although temporary, these treatments can provide a window of opportunity for establishing the desired plants. Once established, those plants provide self-sustaining erosion control and site development. These techniques do not attempt rapid improvement of the soil structure. Rather, they increase infiltration by ponding water on the surface. This stabilises soil resources, increases infiltration and captures more soil, nutrients, seed and organic materials from fluvial flows.

Following extreme damage the landform of large areas must be changed. Land recontouring typically follows mining or other activities that cause massive alterations in the original landform structure. The objectives of land recontouring are greater slope stability and reconstruction of the original drainage basin topography. These changes reduce soil erosion and create suitable surfaces for establishing plants that provide continuing stabilisation.

Roughen the soil surface

Hydrologic processes are of critical importance and usually deserve early consideration when planning an ecological restoration programme. Both natural and man-made features may be used to increase water availability on relatively barren sites. Natural depressions in the soil accumulate water, nutrients and organic materials that increase plant establishment in arid and semi-arid regions (Ahmed, 1986; Kennenni & Maarel, 1990). Man-made depressions concentrate limiting resources and initiate soil-repairing processes (Whisenant *et al.*, 1995). For example, surface modifications on slowly permeable soils increased precipitation use efficiency (PUE) more than 100% (Wight & Siddoway, 1972). PUE is the annual, aboveground primary production/annual precipitation. The same treatments increased PUE about 20% on sites with high infiltration capacities. These changes facilitated vegetative development that captured an increasing percentage of the organic matter and nutrients moving across the landscape in wind and water. Therefore, surface soil treatments often focus on reducing runoff and erosion by retaining precipitation where it falls. These treatments should be designed to increase surface roughness, reduce the length of unobstructed soil surface, and ultimately to increase vegetative cover.

Initial restoration efforts should focus on establishing species that not only grow under existing conditions; they should initiate autogenic processes that improve ecosystem functioning. Treatments that focus on roughening the soil surface or creating aboveground obstructions provide temporary, but often essential benefits (Whisenant, 1999). They are implemented to halt degradation (e.g. erosion and runoff) and improve conditions for plant establishment and growth (Whisenant, 1995).

Water limitations are most common in arid and semi-arid ecosystems, but also occur on severely damaged sites in humid environments and saline soils. Where precipitation falls in widely spaced, intense events, much of the water is lost from the site. Seedlings experience water limitations, even in relatively high-precipitation regions, if soil surface conditions are poor. Specialised surface soil modifications that collect runoff water require additional investments, but are the most reliable establishment technique in many areas (Weber, 1986). Some arid-land farming systems harvest water from areas treated with latex, asphalt or wax to improve runoff efficiency (Ffolliott *et al.*, 1994), but the most common strategies seek to harvest or concentrate runoff water or to trap wind-blown snow.

The direct benefits of water-harvesting strategies such as pitting and contour furrowing are generally short lived. These soil modifications have a finite life that is determined by erosion rate, depth and precipitation events. However, even during a short period, they may establish plants with a lasting, self-perpetuating impact. Water-harvesting techniques that establish shrubs to change micro-environmental conditions and harvest wind-blown soil, nutrients and propagules may have long-term benefits in arid and semi-arid ecosystems (Whisenant, 1995; Whisenant & Tongway, 1995; Whisenant *et al.*, 1995).

Creating depressions, such as microcatchments, in the soil surface can concentrate water and improve seedling survival (Fig. 6.5). They are most appropriate in arid regions with high runoff coefficients, with the basin-to-catchment ratio (ratio of water holding area to water harvest area) being determined by slope, rainfall characteristics,

Fig. 6.5. Recently created microcatchments used to establish woody plants on severely crusted soils in Niger. Where previously little water moved into these soils, these microcatchments held enough water to establish large woody plants that began to improve soil and micro-environmental conditions.

runoff rate and the requirements of planted species. Water-harvesting methods do not guarantee success. Seedlings can fail, even with water harvesting, during very dry years. Water harvesting is often unnecessary during wet years. However, water harvesting increases seedling establishment and plant production during the years that are neither too dry nor too wet. Like other risk-reduction strategies, water harvesting increases the probability of success, but does not eliminate failure.

Controlling wind erosion involves the application of two basic principles: (1) reduce the wind velocity near the soil surface, and (2) increase the resistance of the soil surface to wind drag (Lal, 1990). Management practices that improve soil structure and conserve soil moisture also increase the soil's resistance to wind erosion. Tillage and soil management practices, prior to planting, can reduce erosion and improve success. Surface ridges produced by tillage will reduce wind erosion. The height and lateral frequency of ridges, furrow shape, orientation relative to wind direction and the proportion of erodible to non-erodible grains determine the impact of surface ridges (Middleton, 1990). Tillage ridges are more effective when oriented at right angles to erosive winds. Furrows that are parallel to the wind may accelerate soil loss by increasing the scouring influence through the furrows. Tillage practices reduce wind erosion by slowing saltation (soil particles bouncing along the soil surface) and surface creep (rolling or sliding of particles along the ground) if they use crop residue or produce a very rough cloddy seedbed with furrows perpendicular to the prevailing wind direction (Lal, 1990; Potter *et al.*, 1990). Practices that improve soil aggregation by adding soil organic matter, mulch or soil conditioners also reduce wind erosion. Since moist soil is less susceptible to wind erosion, cultural practices that conserve water are particularly helpful. These practices include mulches, preparatory crops, cover crops or even irrigation where possible.

Add aboveground obstructions

Aboveground obstructions reduce the flow rate of wind and water across the soil surface. Effective obstructions capture water, nutrients and organic materials and increase the infiltration rate. In Niger, rock barriers laid on the contour of gentle slopes reduced water movement, increased infiltration and reduced erosion (Fig. 6.6). Rocks and logs can provide effective barriers that enable plants to establish on previously barren sites. Biotic obstructions include vegetation (density, cover, height and stiffness) and non-living organic obstructions such as woody debris (Fig. 6.7) and herbaceous litter on the soil surface (Whisenant, 1999). Organic mulches confer greater resistance to raindrop impact, while reducing wind and water flow velocities.

Steep or otherwise erodible slopes are susceptible to excessive erosion losses before a protective plant cover is established. Without adequate protection, slopes can suffer severe soil erosion, making it even more difficult to establish the desired plant cover. Geotextiles are permeable textiles used with soil, rock, earth or other materials. They provide immediate protection against soil loss and improve seedling establishment. Geotextiles used for erosion control are three-dimensional meshes, mats, blankets or honeycomb-shaped webs (geocells) (Rickson, 1995). They can be organic or of synthetic materials and designed for surface or buried applications. Some geotextiles are considered temporary, lasting only long enough for the protective plant cover to establish. Other geotextiles provide a permanent protection and are designed to function in combination with established plants. The selection of an appropriate geotextile material should be based on the expected flow velocities (Rickson, 1995). Some materials tolerate very little water flow, while others are specifically designed to accommodate very high velocities. Reputable geotextile dealers provide good advice on selecting the appropriate geotextile materials for a specific application.

Ultimately, plants are the preferred obstructions for fluvial flows. They are self-sustaining, reduce flow velocities, improve soil stability and increase the amount of water moving into the soil. To be effective obstacles to fluvial flows, plants must be rigid enough to remain erect against small to moderate water flows. Consider the importance of flexual rigidity by contrasting a turf-forming grass with a stiff, erect bunchgrass. Bermuda grass

Fig. 6.6. Rocks placed along the slope contour to serve as aboveground obstacles in Niger. The rocks trapped soil, water, nutrients, organic materials and naturally dispersed seeds. Although the area was not seeded, herbaceous plants established and continued to improve soil and hydrologic conditions.

(*Cynodon dactylon*) protects the soil surface from raindrop impact and erosive water flows, but does not reduce water velocity. Nor does it trap much of the soil, nutrients or organic materials in the water, because the flexible stems lie over under relatively low water flows. In contrast, the stiff stems of switchgrass (*Panicum virgatum*) remain erect and reduce water velocity (Kemper *et al.*, 1992), causing suspended materials (i.e. soil and organic materials) to fall from the water stream.

Lowering the wind velocity near the soil surface and increasing the soil surface's resistance to wind drag reduces wind erosion (Lal, 1990). Wind velocity near the soil surface is reduced with afforestation, temporary cover crops, plant residue (Fig. 6.8), rocks, gravel or logs on the surface. Plants affect wind erosion in at least five ways (Morgan, 1995). First, plant foliage reduces wind speed by exerting a drag on airflows. Plant biomass, projected foliage area facing the wind, leaf area density, leaf orientation and leaf shape all influence a plant's ability to reduce wind erosion. Vegetation effects on wind speed vary seasonally, particularly with deciduous species. Second, the foliage traps moving sediments. This not only reduces erosion, it has important implications for nutrient dynamics on the site since aeolian dust has substantially more nutrients than degraded soils (Drees *et al.*, 1993). Third, vegetative cover protects the soil surface. Fourth, plant root systems increase the resistance of the soil to erosional processes. Fifth, vegetation influences soil moisture through uptake, transpiration and micro-environmental modifications.

Plants reduce airflows by transferring the momentum from the air to the vegetation. The vegetation serves as a momentum sink that reduces wind

Fig. 6.7. The vegetation on this previously forested slope in Chipinque Ecological Park near Monterrey, Mexico was removed by a wildfire. Woody debris was laid along the slopes, as aboveground obstructions, to reduce erosion and facilitate the establishment of new trees.

velocity. Since canopy density and height are the main factors determining canopy roughness, sparse, tall canopies will create more drag on airflows. Fluttering leaves reduce wind velocities more than rigid leaves (Vogel, 1984, 1989), but are more susceptible to leaf stripping and tearing damage (Cleugh *et al.*, 1998).

Our understanding of how plants alter wind erosion is insufficient to simulate their effects reliably. However, we can effectively design restoration programmes that reduce wind erosion by increasing vegetative cover, plant height and soil surface roughness while reducing the length of unprotected soil. Recent studies indicated plant area index and canopy cover are highly correlated with the transport capacity of wind and provide reasonably safe indicators of soil protection (Armbrust & Bilbro, 1997). In general, taller, finer-leafed plants with large surface areas are more effective in reducing wind erosion (Middleton, 1990).

Snow fences, in Kansas, increased water availability by capturing and storing snow until it melted. For example, eastern redcedar (*Juniperus virginiana*) seedling survival increased from 70% to 90% and Scots pine (*Pinus sylvestris*) seedling survival increased from 0 to 90% when planted within 7.6 m of 1.2-m tall snow fences (60% open) (Dickerson *et al.*, 1976). *Juniperus* seedlings planted near the snow fences were 33% taller than other seedlings. Shelter-belts and individual woody plants can provide similar snow-trapping benefits without the maintenance requirements of fences.

Landform manipulations

Recontouring requires the transport of large amounts of soil or overburden material. Consequently, it is

Fig. 6.8. Aboveground obstacles used to reduce sand deposition on the highway through the Taklimakan Desert in western China. The sand fence and the 1 m squares were created from crop stubble and hand thatched into the sand. These structures are often used as a temporary measure to facilitate the establishment of seeded species. However, here they are maintained permanently, because no vegetation grows in this portion of the Taklimakan Desert.

very expensive and is only considered for small areas, following severe disturbances (such as surface mining), or high-priority locations. Although thorough planning and co-ordinated soil or overburden handling can reduce those costs, recontouring easily exceeds the combined costs of seedbed preparation, seed, soil amendments, planting and seedlings.

Hillslope geomorphology is destroyed by disturbances such as mining. Post-disturbance landforms that are too steep or do not conform to predisturbance drainage patterns are unstable and unlikely to support sustainable vegetation. Slopes are considered stable when they are not eroding at an accelerated rate. In geomorphic terms, this is a temporary equilibrium resulting from the interaction of climate, slope geomorphology, soil and the vegetation growing on the slope. Climatic factors that influence slope stability and the ability to grow vegetation are season, amount, intensity and type (rain or snow) of precipitation, temperature regime, wind patterns (speed), and relative humidity. Slope geomorphology includes the shape, length and gradient of the landform.

Soil properties, such as porosity and bulk density, are particularly important because of their influence on hydrologic processes. Mining, construction or other activities that reduce porosity and increase bulk density can accelerate both surface erosion and mass failure. They also reduce vegetation growth, thus limiting the recovery potential. Recontoured spoils should be covered with an adequate growth medium. Toxic, or potentially toxic, materials should be buried under non-toxic layers. Toxic spoils should be placed deeper than 120 cm and probably as deep as 200 to 250 cm to ensure an adequate soil depth for healthy root systems (Weaver, 1920; Coupland & Johnson, 1964; Foxx *et al.*, 1984).

Recontoured slopes are subject to shear stresses that act to displace the slope mass through mass failure. The slope material, biomass and water comprise most of the shear stress forces, but shear stresses are increased by earthquakes, equipment movements and other physical forces. Shear strength or resistance includes the forces acting to resist movement of the slope mass and is a function of the frictional properties of the soil and rock along its weakest plane (for non-vegetated, cohesion-less slope materials). The slope angle at which shear stress is equal to shear resistance in cohesion-less material is known as the natural angle of repose (also known as the angle of internal friction) (Satterlund & Adams, 1992). Slope failure occurs when shear stress exceeds shear strength or resistance.

The vegetation on recontoured slopes ultimately provides most of the slope's stability. While undisturbed slopes near the natural angle of repose may appear erosion free, stability is the result of slope lithology, shape, length, gradient and surface vegetation. Well-established vegetation greatly increases slope shear strength, which has a strong slope-stabilising influence. The soil is held tightly to the near-surface material by a maze of plant roots intertwined with uneven surface and broken segments of the parent rock. Mature grasses, forbs, shrubs and trees send their roots deep into the hillslope and near-surface soil. Stems and leaves break the force of raindrop impact on the soil surface. Vegetation and the coarse texture of the root zone organic materials create preferential flow paths for water to move into the soil. Each of these factors exerts strong slope-stabilising influences. Plant growth and survival is reduced on hillslopes because of greater water loss. The water budget on a slope is determined by infiltration rate, water-holding capacity of the soil and runoff.

Mining or construction activities, such as blasting or relocating, destroy the stabilising influence of plant roots. Thus, the developmental processes that created stable plant and soil communities must be repeated before vegetation can exert the same stabilising influence on the new slope. The goal of the new landform should be to duplicate pre-disturbance exposure and slope parameters. Recontouring landforms to the approximate original contour (AOC) improves success. Although widely used in open-pit coal-mining, AOC is not easily achieved in metal mines. They tend to be underground, in mountainous terrain with very steep slopes, or in large open pits that do not lend themselves to backfilling and grading during or after mining. Excess spoil can be properly placed in heads of hollows or as valley fills. A lack of spoil also occurs in some situations. However, even in these types of mines, the return to AOC slopes less steep than the natural angle of repose reduces the environmental impact.

The stable landforms of undisturbed landscapes are usually the result of thousands of years of natural weathering, erosion, deposition and plant growth. These same processes also create an approximate equilibrium on entire watersheds. Massive disturbances destroy this equilibrium. Hillslopes and vegetation must evolve toward a new equilibrium with the climate of the area. Achieving the same level of soil stability found on well-vegetated hillslopes may take centuries. Since geomorphic conditions of the new site will not be in equilibrium with the climate, vegetation and hydrologic forces of the new drainage, loss of lithologic or geologic control must be compensated. This compensation may take the form of reduced slope gradient, slope length or rapid vegetative growth on the disturbance (Munshower, 1994).

Gentler slope gradients reduce erosion and increase the probability of successful plant establishment. Successful, long-term revegetation and stabilisation of slopes at or near the natural angle of repose is very rare. Plant performance is seldom acceptable on disturbed slopes greater that 2:1 (Munshower, 1994). Depending on the disturbance size, it may be possible to reduce slope length. Erosion increases rapidly with increasing slope gradient (Meyer *et al.*, 1975). Short, steep slopes are less likely to exhibit severe erosion than are long, somewhat less-steep slopes. The recontoured slopes may be concave, convex or straight. Concave landforms are more erosion resistant and stable than straight or convex slopes.

New slopes may be constructed with a single type of surface feature (simple) or grouped into complex surfaces. Complex slopes are more stable than concave, convex or straight slopes. Restored hillslopes that are concave or complex in profile with low gradients and short lengths will be most stable. This configuration will facilitate infiltration and support more vigorous vegetation growth.

Recontoured slopes should flow smoothly into the undisturbed landscape and the new drainage systems should form a smooth transition from restored to untouched stream channels. These reconstructed landforms should have a smooth, continuous sequence of hillslopes and valley floors that blend into the surrounding landscape. Subsequent vegetation establishment patterns should continue to reduce the differences between undamaged and recontoured landforms.

Naturally eroded channels usually produce less sediment than man-made channels. Thus, it has been suggested that first-order streams (the smallest channels in the watershed) should not be constructed, but should be allowed to develop naturally (Munshower, 1994). The amount and velocity of water that can flow through a drainage channel is influenced by the channel's cross-sectional area, rate of fall, length and vegetation. For any given stretch of valley floor, longer channels (i.e. greater sinuosity) reduce water velocity and increase the amount of water the channel can safely transport. Herbaceous plants, shrubs and trees along the drainage channel filter sediment, slow overland flows and increase sediment deposition, thus reducing the erosive force of major runoff events. This reduces large-scale changes in the channel structure and encourages additional plant growth and changes in basin structure.

Even the most carefully recontoured landscapes require time for the geomorphic and vegetative components to achieve a new equilibrium with their climate. Thus, in the near term, they are susceptible to erosion, mass movement and even catastrophic failure. Consequently, prudence suggests we overdesign drainage basins to provide additional near-term stability. Hillslopes, valley floors and stream channels should be designed to achieve greater stability than existed before the disturbance. This is particularly important in arid and semi-arid ecosystems, where vegetation establishment is slow and high-intensity rainfall events are common (Munshower, 1994). The accelerated runoff that typifies disturbed slopes easily exceeds the stabilising capacity of newly seeded slopes. Reducing the gradient and length of hillslopes and increasing channel sinuosity may provide enough additional stability to allow mature vegetation to establish.

Manipulating subsurface flows

Damaged surface soils reduce the amount of water entering the subsurface flows. Thus, improving surface conditions increases water flow into the soil. In addition, compacted subsurface soil layers severely limit both vertical and horizontal flows of water through the soil. After removing the limitations to surface water the possibility of subsurface problems should be considered. Loosening compacted soils, soil amendments or using plants to transpire more or less groundwater are all effective strategies for managing subsurface flows of water and associated resources.

Loosen compacted soils

Compacted soil surfaces reduce the amount of water that moves into soils. Compaction problems are common on old mine sites (Davies *et al.*, 1992; Ashby, 1997), abandoned roads (Cotts *et al.*, 1991; Luce, 1997), following timber harvest (Guariguata & Dupuy, 1997; Whitman *et al.*, 1997), old oil-field sites (Chambers, 1989) and following cultivation or hay production (Lal, 1996; Bell *et al.*, 1997). Compacted soils damage hydrologic processes by reducing infiltration and increasing runoff. Water movement through compacted soil layers is greatly reduced. The compacted layer may occur on the surface and/or in subsurface layers of the soil. Compacted soils have poor aeration and restrict the physical movement of larger soil organisms. This creates soils with less available water, oxygen limitations and disrupted nutrient cycling processes. These problems not only occur where mechanical equipment has operated, but are also caused by livestock (Stephenson & Veigel, 1987; Lal, 1996). Although

Fig. 6.9. Deep ripping (0.4–0.6 m) this highly compacted substrate in west Texas was necessary before water would move into the soil. Ripping or deep ploughing was necessary to get any establishment of planted species in this seedbed.

some compacted soils improve slowly through the actions of freezing, thawing, root penetration and shrink–swell actions, they usually require deep ploughing or deep ripping before plants can establish (Berry, 1985; Ashby, 1997; Bell *et al.*, 1997; Luce, 1997) (Fig. 6.9).

Soil amendments

Soil amendments can immediately alter the rate of subsurface flows and may be used to begin improving soil structure. Many soil physical problems are reduced with surface or subsurface additions of organic materials (Fig. 6.10). Although the benefits of organic matter are well known, large-scale applications are impractical. Subsurface flow modifications usually require that the materials be incorporated into the soil matrix. This adds substantially to the cost of the project. Thus, incorporating organic materials is usually restricted to relatively small, high-priority sites. There are many potential organic amendments with different benefits, but selecting a specific organic amendment depends on its local availability, transportation costs, application costs and local regulations. Inexpensive organic materials provide an important opportunity that may greatly facilitate restoration efforts.

Surface organic materials confer greater resistance to raindrop impact, while reducing wind and water velocities. In Niger, mulching bare, crusted soils increased soil moisture, seed capture, development of ground cover and the germination of woody species after a single rainy season (Chase & Boudouresque, 1987). Incorporated organic materials improve drainage and aeration. For example, adding 1% to 6% organic matter (w/w) in the early stages of decomposition (less than one year)

Fig. 6.10. Mouldboard plough loosening compacted seedbed and incorporating leaf litter.

increased the size of mineral soil aggregates and decreased erodibility (Chepil, 1955). However, as those initial organic materials broke down over the next four years, they lost their cementing properties. Maintaining stable aggregates requires continuing organic inputs, since aggregate stability declines in the absence of continuing organic inputs.

Manipulating groundwater

Understanding how vegetation affects hydrologic processes provides insights with powerful restoration implications. Water yields (both surface and subsurface) often increase when: (1) trees are removed or thinned; (2) vegetation is converted from deep-rooted species to shallow-rooted species; and (3) plant cover changes from species with high interception capacities to species with low interception capacities (Brooks *et al.*, 1991). While absolute water yield changes are smaller in semi-arid regions, their ecological importance is often greater.

Disrupting landscape-scale hydrologic functioning can trigger unexpected consequences on other parts of the landscape. This occurs because shrubs and trees on one part of a landscape can regulate the hydrology for the remainder of that landscape (Burel *et al.*, 1993; Hobbs, 1993; Ryszkowski, 1995). Removing woody vegetation on upper reaches of a watershed can significantly increase groundwater levels in lower landforms. This may occur when woody vegetation is cleared and replaced with shallow-rooted, annual crop species. The diminished water-use of these crops increases groundwater recharge and raises water tables. This additional groundwater acquires additional salts and may seep out at lower parts of the landscape. Dryland salinity occurs when this water evaporates and leaves salts behind on the soil surface. In Western Australia, converting native forests to herbaceous pasture raised the water table from 18 m to 3 m within 12 years (Ruprecht & Schofield, 1991). Since this water was salty, the conversion seriously damaged agricultural productivity and indigenous biodiversity.

Planting trees and shrubs into affected watersheds is a common strategy for reducing dryland salinity problems. Lowering the water table requires that annual evapotranspiration plus streamflow losses equal or exceed the rainfall and inflow of water from other sources. In central Queensland, Australia, the water table under a single *Casuarina glauca* tree (four years old and 5 m tall) was depressed by 130 mm compared to the water table 10 m from the tree (Walsh *et al.*, 1995). Reforesting 5–10% of a landscape affected by secondary salinisation (high saline water table) reduced the groundwater level 100–200% (Schofield, 1992) whereas reforesting 25% of the area lowered the water table by approximately 800%. Groundwater simulations of a southern Australia landscape suggested afforestation could significantly alter groundwater recharge over substantial areas (Pavelic *et al.*, 1997). However, small-scale efforts (<100 ha) are not expected to affect groundwater levels.

The location of the transpiring trees, within the watershed, may also be critical to long-term success. In a study from Western Australia, groundwater salinity increased under trees after five years, particularly where tree densities were highest (Stolte *et al.*, 1997). Since the continuing flow of saline groundwater to the trees was believed to increase groundwater salinity, tree plantations on discharge zones may not be viable long-term strategies. Consequently, a more sustainable long-term strategy for the control of secondary salinisation, caused by raising water tables, is to establish species on the recharge zones that use all the water that falls there (Stolte *et al.*, 1997).

Since wetlands slow the flow of water, increase organic matter accumulation, and have high evapotranspiration rates, they play a major role in the hydrologic conditions of the landscape. Damaged and destroyed wetlands reduce the retention time of water within that landscape, which increases downstream flooding potential. Large wetland areas, in Byelorussia, were drained and cultivated during the 1960s. The resulting hydrologic changes accelerated organic matter loss, increased plant pest problems and significantly reduced crop yields (Susheya & Parfenov, 1982). Problems caused by the widespread loss of wetlands led to the creation of new wetland restoration programmes, designed to return numerous bogs to the large, cultivated areas. These restored bogs were credited with repairing enough of the damaged ecosystem processes to reduce the damage done by the drainage operations.

Manipulating micro-environmental conditions

Manipulating micro-environmental conditions can be done rapidly by removing living or dead vegetation or more slowly through vegetation development. Micro-environmental conditions are made more arid, sunny and windy by removing plants or ground litter. Overstorey trees are removed or killed with logging or by chemical or mechanical treatments. In some situations, fire may be used to remove either living vegetation, litter or duff on the soil surface. These changes stimulate the natural regeneration of species that require direct sunlight and/or bare mineral soil seedbeds.

Environmental modification by the vegetation has important implications for ecological restoration. Since vegetation responds to and strongly modifies its immediate environment, we have the opportunity to direct those changes toward restoration objectives. Plants modify their environment through both passive and active mechanisms. They passively affect their immediate environment with their physical structure by shading the soil and altering wind movements. This reduces wind speed, lowers the extremes of air and soil temperatures and increases relative humidity. Plant structures trap wind-blown soil, nutrients and the propagules of micro-organisms and other plants. Metabolic processes actively change the environment by altering temperature, humidity and the physical and chemical processes of soils. Plants gradually increase soil organic carbon and improve the water and nutrient holding capacities of the soil. Consequently, the combination of passive and active mechanisms means that the capacity of plants to modify their environments is roughly proportional to their vegetation biomass, stature and rate of metabolic activity (Roberts, 1987). Thus, sparse desert vegetation is less able to alter its environment than forest vegetation. However, the lesser plant-induced environmental alterations in arid ecosystems may still have significant biological impacts.

Mulches to improve seedbed micro-environments

Seedbed mulches reduce soil erosion (Siddoway & Ford, 1971), lessen temperature extremes, conserve soil moisture, increase seed germination and increase seedling growth (Zak & Wagner, 1967; Eck *et al.*, 1968; Singh & Prasad, 1993). Because mulches change the nature of the seedbed, the type and amount of mulch affects the species that establish (Luken, 1990; Munshower, 1994). Mulches that improve seedbed conditions include straw, hay, wood chips, shredded bark, peat moss, corncobs, sewage sludge, sugar-cane trash, manure, plastic, and synthetic petroleum products (Luken, 1990; Singh & Prasad, 1993). The benefits of mulches are probably greatest in arid environments (Winkel *et al.*, 1991; Singh & Prasad, 1993; Roundy *et al.*, 1997) and where weed competition is a serious obstacle.

Gravel, stones, rocks, and even oil are useful for certain applications. Gravel mulches increase germination under water-limiting conditions, unless they are too deep (Winkel *et al.*, 1991, 1993). In arid regions of India, gravel mulches reduced moisture losses from planting sites and were more stable during high winds (Mertia, 1993). Gravel, stone and rock mulches increased seedling establishment during natural or artificial recovery of disturbed sites in arctic Alaska (Bishop & Chapin, 1989*a*, *b*). Where readily available, rock mulches shield woody plant seedlings from temperature extremes and provide some protection from herbivory. Three or more 10- to 20-cm diameter rocks arranged around seedlings have enough thermal mass to provide thermal buffering and reduce evaporation (Bainbridge *et al.*, 1995).

Seedbed improvement with temporary herbaceous crops

Biological seedbed preparation includes the use of nurse crops (also called companion crops) preparatory crops, and woody plants to ameliorate harsh soil and micro-environmental conditions. Although each of these three methods requires two separate plantings, the timing of those plantings is different. Nurse crops and woody plants are usually grown simultaneously with the desired species, but preparatory crops are grown and harvested (or ploughed under) prior to planting the final species. Effective restoration strategies not only address initial establishment concerns, they initiate autogenic processes that continue to improve seedbeds, and facilitate the long-term recruitment of additional plants (Danin, 1991; Jones *et al.*, 1994; Whisenant, 1995; Whisenant *et al.*, 1995).

Nurse crops grow at the same time as the desired final species. Thus, they are most appropriate where water will not limit establishment. Under these conditions, planting nurse crops at or near the time when the perennial species are planted has several advantages, including: (1) reduced wind and water erosion; (2) less weed competition; (3) shelter for the seedlings from wind and severe temperature; and (4) forage provided by the nurse crop before the perennial species are fully developed. On damaged acidic soils, switchgrass was found to facilitate the natural recruitment of *Populus* spp. (aspens), *Salix* spp. (willows) and *Betula* spp. (birches) by physically capturing wind-blown seed and acting as a nurse crop (Choi & Wali, 1995).

Oats (*Avena fatua*) and barley (*Hordeum vulgare*) are common nurse crops for establishing perennial plants. However, common rye (*Secale cereale*) is too competitive to be a good nurse crop, and wheat (*Triticum aestivum*) is somewhat intermediate. Competition from nurse crops is decreased by: (1) reducing the seeding rate of oats or barley to between 7 and 11 kg ha^{-1}; (2) drilling nurse crops and perennial species at 90° angles or in alternate rows; and (3) harvesting the nurse crop early (Vallentine, 1989). The competition from nurse crops must be controlled (partitioned in time or space) to increase perennial species establishment. Nurse crops are less frequently used on water-limited ecosystems, or where soil fertility is limited. Annual legumes are often used as nurse crops, but they greatly reduced the establishment of *Artemisia californica* in the southern California coastal sage scrub (Marquez & Allen, 1996).

The preparatory crop method involves planting annual, residue-producing crops during the growing season prior to directly seeding a perennial species into the residue. Grain sorghum (*Sorghum* spp.) is the most common preparatory crop in sandy, semi-arid west Texas. Grain sorghum is planted and the grain is harvested and sold within normal farming practices. Selling the grain partially offsets restoration costs. The sorghum stalks remain standing to reduce wind erosion, trap snow and retain soil water. The following spring, a no-till drill is used to plant perennial grasses into the standing sorghum stubble. The stalks reduce erosion before and after planting, and increase perennial plant establishment. Erect residues of the preparatory crop are more effective than horizontal residues because vertical residues absorb more wind energy (Siddoway *et al.*, 1965). The height, diameter and number of stalks determine the effectiveness of standing residue, because they determine the silhouette area through which winds pass (Bilbro & Fryear, 1994). In north central Texas, seeding into preparatory crop litter was 88% successful, while seeding into clean, tilled seedbeds was 67% successful (Great Plains Agricultural Council, 1966).

Strip cropping is a variation of preparatory cropping that has been used in the semi-arid portions of the North American Great Plains where wind erosion is a serious hazard. During strip cropping, mechanically fallowed strips (each 10 m wide) are alternately seeded to perennial grasses (Bement *et al.*, 1965). Grass strips alternate with similar-sized strips planted to annual crops, such as cotton (*Gossypium hirsutum*), wheat or grain sorghum. Then, after the grass strips are established, the previously cropped strips are fallowed one year and planted to grasses the next year. This strategy reduces wind erosion during the entire establishment process compared to planting perennial grasses alone.

Shelter-belts and interim tree crops

Afforestation, with adapted species, reduces erosion by decreasing wind speed, protecting the soil surface and increasing litter. Shelter-belts are most effective and most practical when the trees provide additional benefits to the ecosystem and/or the local economy (e.g. amelioration of micro-environment, nitrogen fixation, or provision of fuel-wood, fodder or wildlife habitat). Shelter-belts are highly effective and widely used to reduce wind speed, erosion and evapotranspiration in areas susceptible to wind erosion. They should be planted perpendicular to prevailing winds. Taller trees provide protection for greater distances than shorter trees, but may also require rows of shorter trees or shrubs to fill in the lower-level gaps. In semi-arid parts of Australia, planting 5% of the land area to shelter-belts, timber-belts or tree blocks reduced wind speed 30% to 50% and reduced soil loss up to 80% (Bird *et al.*, 1992). Shelter-belts should be planted with several rows that are perpendicular to prevailing winds. The most effective arrangement puts tall trees in the center, with many rows of dense shrubs or shorter trees on the outside. Shelter-belts with trees and dense shrubs are relatively impermeable to wind and have few gaps (Mohammed *et al.*, 1996). The best shelter-belt species grow rapidly, have long life spans, tolerate existing stresses, and provide valuable products for local inhabitants.

Studies from many environments have shown that wind-breaks can increase plant productivity (Lugo, 1992; Vetaas, 1992; Brown & Lugo, 1994; Guariguata *et al.*, 1995; Fimbel & Fimbel, 1996; Ashton *et al.*, 1997; Cleugh *et al.*, 1998). However, despite numerous reports of positive benefits, the impact of wind-breaks on plant growth is complex. Shelter may either increase or decrease evaporation rates, depending on water status and prevailing weather conditions (Cleugh, 1998). In dry environments, the micro-environment in the quiet zone (where wind speeds are reduced) is more conducive to plant growth – with less wind, less evaporation, and higher humidity. In the wake zone (turbulent area downwind of the quiet zone), plant water use is often greater than in similar non-sheltered areas (Cleugh, 1998). Sheltering plants with abundant water supplies can stimulate photosynthetic activity by increasing humidity within the canopy. In the quiet zone, sheltered well-watered plants do not use less water, but photosynthetic activity and water-use efficiency are likely to be increased (Cleugh, 1998). In contrast, where water availability is limited, photosynthetic activity and assimilation are often reduced, regardless of the sheltering effect. Additional benefits of wind-breaks are due to protection from damage rather than micro-environmental amelioration (Cleugh, 1998; Cleugh *et al.*, 1998). Reduced leaf abrasion, tearing and stripping can be important reasons for the greater productivity of sheltered plants. A reduced incidence of sand-blasting at the seedling stage is important in certain ecosystems (Cleugh *et al.*, 1998).

Once converted to pastureland, tropical rainforest recovery is inhibited by seed shortage, seed predation, and a harsh micro-environment that kills developing seedlings (Uhl, 1988). Restoration of these tropical forests requires strategies that: (1) increase natural seed dispersal; (2) reduce the impact of seed predators; and (3) ameliorate harsh micro-environmental conditions. Overcoming these limitations is very costly unless natural processes are stimulated to achieve those objectives. Because interior forest species cannot tolerate existing environmental conditions, it may not be possible to begin with native rainforest species. Fortunately, several studies have demonstrated the feasibility of planting adapted tree species (either native or exotic) as a first step toward developing a native forest (Uhl, 1988; Lugo, 1992; Brown & Lugo, 1994; Guariguata *et al.*, 1995; Fimbel & Fimbel, 1996; Ashton *et al.*, 1997). In Puerto Rico, tree plantations modify soil and micro-environmental conditions enough to facilitate the natural immigration of native species (Lugo, 1992). As the plantation trees grew, they made micro-environmental conditions more suitable for native forest species. The native species accumulated litter on the forest floor, which increased nutrient retention and reduced erosion. The plantation trees also accelerated the return of native plant species by attracting animals that imported native seed.

Exotic tree plantations in the moist and wet tropics do not remain monocultures (Lugo, 1992), because native trees invade the understorey and penetrate the canopy of the plantation species. In the absence of extreme site damage, native forests often replace exotic plantations. Where the damage is more extreme, the resulting community may develop into a combination of native species and plantation trees (Lugo, 1992). Similar procedures were effective in Sri Lanka (Ashton *et al.*, 1997) and Uganda (Fimbel & Fimbel, 1996) where Caribbean pine (*Pinus caribaea*) was used as a nurse plant to improve micro-environmental conditions enough that native tree species became established.

Despite the obvious and important advantages of using exotic species to modify environmental conditions for subsequent establishment of native species, caution is advisable. To avoid creating major problems, the behaviour of introduced species in their new environment should be well understood. Each tree species creates a unique environment that facilitates the development of a different native flora. The establishment and growth of native tree species may vary between plantation species, because each species creates a unique light environment (Guariguata *et al.*, 1995). For example, fast-growing fleshy-fruited trees were found to create new habitat-forming islands in abandoned tropical pastures because they attract fruigivores that brought additional species (Nepstad *et al.*, 1991). Using tree plantations to restore tree richness is effective when managers match species to particular site conditions and overcome limiting factors that prevent the regeneration of native species (Lugo, 1997). After a forest canopy is returned, micro-environmental conditions change and animals that bring seed are attracted. However, some plantation species inhibit native species (Murcia, 1997) and animal transport of seed is not always fully effective. Although tree plantations can effectively recreate favourable environments for native tree species, animals should not be expected to return all species to the area (Parrotta *et al.*, 1997).

CONCLUDING REMARKS

Healthy ecosystems have natural recovery processes that maintain sustainable flows of soil, nutrients, water and organic materials. One of the most important opportunities for ecological restoration is the development of strategies that return the capacity for self-repair to severely damaged sites. Contemporary succession theory describes this catastrophic change as having crossed a transition threshold that inhibits natural recovery. Understanding and overcoming threshold barriers to natural recovery is one of the more important challenges for ecological restoration. Two types of threshold barriers limit the natural recovery. The first is controlled by interference from other organisms (biotic interactions), usually invasive weeds or other plants that prevent natural recovery. Reducing problematic species and/or adding appropriate species are the most effective strategies for these circumstances. The second barrier operates when dysfunctional hydrologic processes or harsh micro-environments create abiotic limitations to recovery.

After identifying limiting features of the physical environment, we should design restoration strategies that 'jump-start' self-repairing mechanisms of the damaged ecosystem. Thus, understanding and directing the role of the physical environment is an essential component of ecological restoration. Two aspects of the physical environment are most relevant to ecological restoration: (1) physical controls over resource fluxes, and (2) physical controls over micro-environmental conditions. Recognising and restoring ecosystem processes that influence the physical environment is critical and is an area where much progress can be made in the future.

Manipulating the physical environment of damaged ecosystems becomes increasingly more important as damage increases. Altering the physical environment allows us to address the problems of resource fluxes (both surface and subsurface flows) and adverse micro-environmental conditions. We address surface flows by roughening the soil surface or adding aboveground obstructions. In extreme situations, such as mining, the entire landform must be reshaped. The position of the landform type, relative to adjacent areas, affects both surface and subsurface flows. We can manage groundwater levels by manipulating inputs (subsurface inflows and surface drainage) and outputs (transpiration and subsurface outflows) of water through the site.

Mechanical, chemical and fire-based treatments can create rapid changes in the micro-environment of restoration sites. By opening sites to more sunlight and wind, they favour certain species over others. However, ameliorating harsh micro-environmental conditions is more effectively done by vegetation. Environmental modification by plants has important implications for ecological restoration. Since vegetation responds to and strongly modifies its immediate environment, it provides a mechanism to direct those changes toward restoration objectives. Plants modify their environment through both passive and active mechanisms. They passively affect their immediate environment with their physical structure by shading the soil and altering wind movements. This reduces wind speed, lowers the extremes of air and soil temperatures, and increases relative humidity. Plant structures trap wind-blown soil, nutrients and the propagules of micro-organisms and other plants. Metabolic processes actively change the environment by altering temperature, humidity and the physical and chemical processes of soils. Plants gradually increase soil organic carbon and improve the water and nutrient holding capacities of the soil. Consequently, the combination of passive and active mechanisms means that the capacity of plants to modify their environments is roughly proportional to their vegetation biomass, stature and rate of metabolic activity. Trees, herbaceous vegetation, litter or duff can significantly change local micro-environments. Herbaceous vegetation can alter micro-environmental conditions during the critical, but brief seedling establishment phase, but woody vegetation is a more effective option when those changes must last for a longer period.

REFERENCES

Ahmed, H.A. (1986). Some aspects of dry land afforestation in the Sudan with special reference to *Acacia tortilis* (Forsk.) Hayne, *Acacia senegal* Wild. and *Prosopis chilensis* (Molina) Stutz. *Forest Ecology and Management*, **16**, 209–221.

Armbrust, D.V. & Bilbro, J.D. (1997). Relating plant canopy characteristics to soil transport capacity by wind. *Agronomy Journal*, **89**, 157–162.

Aronson, J., Floret, C., Le Floc'h, E., Ovalle, C. & Pontanier, R. (1993). Restoration and rehabilitation of degraded ecosystems in arid and semi-arid lands. 1: A view from the south. *Restoration Ecology*, **1**, 8–17.

Ashby, W.C. (1997). Soil ripping and herbicides enhance tree and shrub restoration on stripmines. *Restoration Ecology*, **5**, 169–177.

Ashton, P.M.S., Gamage, S., Gunatilleke, I.A.U.N. & Gunatilleke, C.V.S. (1997). Restoration of a Sri Lankan rainforest – using Caribbean pine *Pinus caribaea* as a nurse for establishing late-successional tree species. *Journal of Applied Ecology*, **34**, 915–925.

Bainbridge, D.A., Fidelibus, M. & MacAller, R. (1995). Techniques for plant establishment in arid ecosystems. *Restoration and Management Notes*, **13**, 190–197.

Bell, M.J., Bridge, B.J., Harch, G.R. & Orange, D.N. (1997). Physical rehabilitation of degraded krasnozems using ley pastures. *Australian Journal of Soil Research*, **35**, 1093–1113.

Bement, R.E., Barmington, R.D., Everson, A.C., Jr & Remenga, E.E. (1965). Seeding of abandoned croplands in the central Great Plains. *Journal of Range Management*, **18**, 53–59.

Berry, C.R. (1985). Subsoiling and sewage sludge aid loblolly pine establishment on adverse sites. *Reclamation and Revegetation Research*, **3**, 301–311.

Bilbro, J.D. & Fryear, D.W. (1994). Wind erosion losses as related to plant silhouette and soil cover. *Agronomy Journal*, **86**, 550–553.

Bird, P.R., Bicknell, D., Bulman, P.A., Burke, S.J.A., Leys, J.F., Parker, J.N., Sommen, F.J.v.d. & Volker, P. (1992). The role of shelter in Australia for protecting soils, plants and livestock. *Agroforestry Systems*, **20**, 59–86.

Bishop, S.C. & Chapin, F.S. (1989*a*). Establishment of *Salix alaxensis* on a gravel pad in arctic Alaska. *Journal of Applied Ecology*, **26**, 575–583.

Bishop, S.C. & Chapin, F.S. (1989*b*). Patterns of natural revegetation on abandoned gravel pads in arctic Alaska. *Journal of Applied Ecology*, **26**, 1073–1081.

Brooks, K., Ffolliott, P.F., Gregersen, H.M. & Thames, J.L. (1991). *Hydrology and the Management of Watersheds*. Ames, IA: Iowa State University Press.

Brown, S. & Lugo, A.E. (1994). Rehabilitation of tropical lands: a key to sustaining development. *Restoration Ecology*, **2**, 97–111.

Burel, F., Baudry, J. & Lefeuvre, J. (1993). Landscape structure and the control of water runoff. In *Landscape Ecology and Agroecosystems*, eds. R.G.H. Bunce, L. Ryszkowski & M.G. Paoletti, pp. 41–47. Boca Raton, FL: Lewis Publishers.

Chambers, J.C. (1989). *Native Species Establishment on an Oil Drill Pad Site in the Unitah Mountains, Utah: Effects of Introduced Grass Density and Fertilizer*, Report no. INT-402. Ogden, UT: US Department of Agriculture Forest Service, Intermountain Research Station.

Chase, R. & Boudouresque, E. (1987). Methods to stimulate plant regrowth on bare Sahelian forest soils in the region of Niamey, Niger. *Agriculture, Ecosystems and Environment*, **18**, 211–221.

Chepil, W.S. (1955). Factors that influence clod structure and erodibility of soil by wind. 5: Organic matter at various stages of decomposition. *Soil Science*, **80**, 413–421.

Choi, Y.D. & Wali, M.K. (1995). The role of *Panicum virgatum* (switchgrass) in the revegetation of iron-mine tailings in northern New York. *Restoration Ecology*, **3**, 123–132.

Cleugh, H.A. (1998). Effects of windbreaks on airflow, microclimates and crop yields. *Agroforestry Systems*, **41**, 55–84.

Cleugh, H.A., Miller, J.M. & Böhm, M. (1998). Direct mechanical effects of wind on crops. *Agroforestry Systems*, **41**, 85–112.

Cotts, N.R., Redente, E.F. & Schiller, R. (1991). Restoration methods for abandoned roads at lower elevations in Grand Teton National Park, Wyoming. *Arid Soil Research and Rehabilitation*, **5**, 235–249.

Coupland, R.T. & Johnson, R.E. (1964). Rooting characteristics of native grassland species in Saskatchewan. *Journal of Ecology*, **53**, 475–507.

Danin, A. (1991). Plant adaptations in desert dunes. *Journal of Arid Environments*, **21**, 193–212.

Davenport, D.W., Breshears, D.D., Wilcox, B.P. & Allen, C.G. (1998). Viewpoint: sustainability of piñion–juniper ecosystems–a unifying perspective of soil erosion thresholds. *Journal of Range Management*, **51**, 231–240.

Davies, R., Younger, A. & Chapman, R. (1992). Water availability in a restored soil. *Soil Use and Management*, **8**, 67–73.

Dickerson, J.D., Woodruff, N.P. & Banbury, E.E. (1976). Techniques for improving tree survival and growth in semi-arid areas. *Journal of Soil and Water Conservation*, **31**, 63–66.

Drees, L.R., Manu, A. & Wilding, L.P. (1993) Characteristics of aeolian dusts in Niger, West Africa. *Geoderma*, **59**, 213–233.

Eck, H.V., Dudley, R.F., Ford, R.H. & Gantt, C.W., Jr (1968). Sand dune stabilization along streams in the southern Great Plains. *Journal of Soil and Water Conservation*, **23**, 131–134.

Ffolliott, P.F., Brooks, K.N., Gregersen, H.M. & Lundgren, A.L. (1994). *Dryland Forestry: Planning and Management*. New York: John Wiley.

Fimbel, R.A. & Fimbel, C.C. (1996). The role of exotic conifer plantations in rehabilitating degraded tropical forest lands: a case study from the Kibale Forest in Uganda. *Forest Ecology and Management*, **81**, 215–226.

Foxx, T.S., Tierney, G.D. & Williams, J.M. (1984). *Rooting Depths of Plants relative to Biological and Environmental Factors*, Report no. LA-10254-MS. Los Alamos, NM: US Department of Energy, Los Alamos National Laboratory.

Friedel, M.H. (1991) Variability in space and time and the nature of vegetation change in arid rangelands.

Proceedings of the 4th International Rangeland Congress, eds. A. Gaston, M. Keinick & H.-N.L. Houérou, pp. 114–118. Montpellier, France: Association Française de Pastoralisme.

Great Plains Agricultural Council (1966). *A Stand Establishment Survey of Grass Plantings in the Great Plains*, Report no. 23. Lincoln, NE: Nebraska Agriculture Experiment Station.

Guariguata, M.R. & Dupuy, J.M. (1997). Forest regeneration in abandoned logging roads in lowland Costa Rica. *Biotropica*, **29**, 15–28.

Guariguata, M.R., Rheingans, R. & Montagnini, F. (1995). Early woody invasions under tree plantations in Costa Rica: implications for forest restoration. *Restoration Ecology*, **3**, 252–260.

Hobbs, R.J. (1993). Effects of landscape fragmentation on ecosystem processes in the western Australian wheatbelt. *Biological Conservation*, **64**, 193–201.

Holmgren, M., Scheffer, M. & Huston, M.A. (1997). The interplay of facilitation and competition in plant communities. *Ecology*, **78**, 1966–1975.

Jones, C.G., Lawton, J.H. & Shachek, M. (1994). Organisms as ecosystem engineers. *Oikos*, **69**, 373–386.

Kemper, D., Dabney, S., Kramer, L., Dominick, D. & Keep, T. (1992). Hedging against erosion. *Journal of Soil and Water Conservation*, **47**, 284–288.

Kennenni, L. & Maarel, E.V.D. (1990). Population ecology of *Acacia tortilis* in the semi-arid region of Sudan. *Journal of Vegetation Science*, **1**, 419–424.

Lal, R. (1990). *Soil Erosion in the Tropics: Principles and Management*. New York: McGraw-Hill.

Lal, R. (1996). Deforestation and land-use effects on soil degradation and rehabilitation in Western Nigeria. 1: Soil physical and hydrological properties. *Land Degradation and Development*, **7**, 19–45.

Laycock, W.A. (1991). Stable states and thresholds of range conditions on North American rangelands: a viewpoint. *Journal of Range Management*, **44**, 427–433.

Luce, C.H. (1997). Effectiveness of road ripping in restoring infiltration capacity of forest roads. *Restoration Ecology*, **5**, 265–270.

Ludwig, J.A. & Tongway, D.J. (1995). Spatial organisation of landscapes and its function in semi-arid woodlands, Australia. *Landscape Ecology*, **10**, 51–63.

Ludwig, J.A., Tongway, D.J. & Marsden, S.G. (1994). A flow-filter model for simulating the conservation of limited resources in spatially heterogeneous, semi-arid landscapes. *Pacific Conservation Biology*, **1**, 209–213.

Lugo, A.E. (1992). Tree plantations for rehabilitating damaged forest lands in the tropics. In *Ecosystem Rehabilitation*, 2nd edn, vol. 2, *Ecosystem Analysis and Synthesis*, ed. M.K. Wali, pp. 247–255. The Hague: SPB Academic Publishing.

Lugo, A.E. (1997). The apparent paradox of reestablishing species richness on degraded lands with tree monocultures. *Forest Ecology and Management*, **99**, 9–19.

Luken, J.O. (1990). *Directing Ecological Succession*. New York: Chapman & Hall.

Marquez, V.J. & Allen, E.B. (1996). Ineffectiveness of two annual legumes as nurse plants for establishment of *Artemisia californica* in coastal sage scrub. *Restoration Ecology*, **4**, 42–50.

Mertia, R.S. (1993). Role of management techniques for afforestation in arid regions. In *Afforestation of Arid Lands*, eds. A.P. Dwivedi & G.N. Gupta, pp. 73–77. Jodhpur, India: Scientific Publishers.

Meyer, L.D., Foster, G.R., & Romkens, M.J. (1975). *Sources of Soil Eroded by Water from Upland Slopes, in Present and Prospective Technology for Prediction of Sediment Yields and Sources*, Report no. ARS-S-40. Washington, DC: US Department of Agriculture, Agriculture Research Service.

Middleton, N.J. (1990). Wind erosion and dust-storm control. In *Techniques for Desert Reclamation*, ed. A.S. Goudie, pp. 87–108. New York: John Wiley.

Mohammed, A.E., Stigter, C.J. & Adam, H.S. (1996). On shelter-belt design for combating sand invasion. *Agriculture, Ecosystems and Environment*, **57**, 81–90.

Morgan, R.P.C. (1995). Wind erosion control. In *Slope Stabilization and Runoff Control: A Bioengineering Approach*, eds. R.P.C. Morgan & R.J. Rickson, pp. 191–220. New York: E. & F.N. Spon.

Munshower, F.F. (1994). *Practical Handbook of Disturbed Land Revegetation*. Boca Raton, FL: Lewis Publishers.

Murcia, C. (1997). Evaluation of Andean alder as a catalyst for the recovery of tropical cloud forests in Colombia. *Forest Ecology and Management*, **99**, 163–170.

Nepstad, D.C., Uhl, C. & Serro, E.A.S. (1991). Recuperation of a degraded Amazonian landscape: forest recovery and agricultural restoration. *Ambio*, **20**, 248–255.

National Research Council (1994). *Rangeland Health: New Methods to Classify, Inventory, and Monitor Rangelands*. Washington, DC: National Academy Press.

Parrotta, J.A., Knowles, O.H. & Wunderle, J.M. (1997). Development of floristic diversity in 10-year-old restoration forests on a bauxite mined site in Amazonia. *Forest Ecology and Management*, **99**, 21–42.

Pavelic, P., Narayan, K.A. & Dillon, P.J. (1997). Groundwater flow modelling to assist dryland salinity management of a coastal plain of southern Australia. *Australian Journal of Soil Research*, **35**, 669–686.

Potter, K.N., Zobeck, T.M. & Hagan, L.J. (1990). A microrelief index to estimate soil erodibility by wind. *Transactions of the American Society of Agricultural Engineers*, **33**, 151–155.

Rickson, R.J. (1995). Simulated vegetation and geotextiles. In *Slope Stabilization and Runoff Control: A Bioengineering Approach*, eds. R.P.C. Morgan & R.J. Rickson, pp. 95–131. New York: E. & F.N. Spon.

Rietkerk, M. & van de Koppel, J. (1997). Alternate stable states and threshold effects in semi-arid grazing systems. *Oikos*, **79**, 69–76.

Roberts, D.W. (1987). A dynamical system perspective on vegetation theory. *Vegetatio*, **69**, 27–33.

Roundy, B.A., Abbott, L.B. & Livingston, M. (1997). Surface soil water loss after summer rainfall in a semi-desert grassland. *Arid Soil Research and Rehabilitation*, **11**, 49–62.

Ruprecht, J.K. & Schofield, N.J. (1991). Effects of partial deforestation on hydrology and salinity in high salt storage landscapes. 1: Extensive block clearing. *Journal of Hydrology*, **129**, 19–38.

Ryszkowski, L. (1995). Managing ecosystem services in agriculural landscapes. *Nature and Resources*, **31**, 27–36.

Satterlund, D.R. & Adams, P.W. (1992). *Wildland Watershed Management*. New York: John Wiley.

Schofield, N.J. (1992). Tree planting for dryland salinity control in Australia. *Agroforestry Systems*, **20**, 1–23.

Siddoway, F.H. & Ford, R.H. (1971). Seedbed preparation and seeding methods to establish grassed waterways. *Journal of Soil and Water Conservation*, **26**, 73–76.

Siddoway, F.H., Chepil, W.S. & Armbrust, D.V. (1965). Effect of kind, amount, and placement of residue on wind erosion. *Transactions of the American Society of Agricultural Engineers*, **8**, 327–331.

Singh, S.B. & Prasad, K.G. (1993). Use of mulches in dry land afforestation programme. In *Afforestation of Arid Lands*, eds. A.P. Dwivedi & G.N. Gupta, pp. 181–190. Jodhpur, India: Scientific Publishers.

Stephenson, G.R. & Veigel, A. (1987). Recovery of compacted soil on pastures used for winter cattle feeding. *Journal of Range Management*, **40**, 46–48.

Stolte, W.J., Mcfarlane, D.J. & George, R.J. (1997). Flow systems, tree plantations, and salinisation in a western Australian catchment. *Australian Journal of Soil Research*, **35**, 1213–1229.

Susheya, L.M. & Parfenov, V.I. (1982). The impact of drainage and reclamation on the vegetation and animal kingdoms on Byelo-Russian bogs. In *Proceedings of International Scientific Workshop on Ecosystem Dynamics in Freshwater Wetlands and Shallow Water Bodies*, vol. 1, pp. 218–226. Moscow: UNEP and SCOPE.

Tongway, D.J. & Ludwig, J.A. (1997*a*). The nature of landscape dysfunction in rangelands. In *Landscape Ecology Function and Management: Principles for Australia's Rangelands*, eds. J. Ludwig, D. Tongway, D. Freudenberger, J. Noble & K. Hodgkinson, pp. 49–62. Collingwood, Vic: CSIRO Publishing.

Tongway, D.J. & Ludwig, J.A. (1997*b*). The conservation of water and nutrients within landscapes. In *Landscape Ecology Function and Management: Principles for Australia's Rangelands*, eds. J. Ludwig, D. Tongway, D. Freudenberger, J. Noble & K. Hodgkinson, pp. 13–22. Collingwood, Vic: CSIRO Publishing.

Uhl, C. (1988). Restoration of degraded lands in the Amazonian Basin. In *Biodiversity*, eds. E.O. Wilson & F.M. Peter, pp. 326–332. Washington, DC: National Academy Press.

Vallentine, J.F. (1989). *Range Developments and Improvements*, 3rd edn. New York: Academic Press.

van de Koppel, J., Rietkerk, M. & Weissing, F.J. (1997). Catastrophic vegetation shifts and soil degradation in terrestrial grazing systems. *Trends in Ecology and Evolution*, **12**, 352–356.

Vandermeer, J. (1989). *The Ecology of Intercropping*. New York: Cambridge University Press.

Vetaas, O.R. (1992). Micro-site effects of trees and shrubs in dry savannas. *Journal of Vegetation Science*, **3**, 337–344.

Vogel, S. (1984). Drag and flexibility in sessile organisms. *American Zoologist*, **24**, 37–44.

Vogel, S. (1989). Drag and configuration of broad leaves in high winds. *Journal of Experimental Botany*, **40**, 941–948.

Walker, B.H. (1993). Rangeland ecology: understanding and managing change. *Ambio*, **22**, 80–87.

Walsh, K.B., Gale, M.J. & Hoy, N.T. (1995). Revegetation of a scalded saline discharge zone in Central Queensland. 2: Water use by vegetation and watertable drawdown. *Australian Journal of Experimental Agriculture*, **35**, 1131–1139.

Weaver, J.E. (1920). *Root Development in Grassland Formation*, Report no. 292. Washington, DC: Carnegie Institute.

Weber, F.R. (1986). *Reforestation in Arid Lands*. Arlington, VA: Volunteers in Technical Assistance.

Whisenant, S.G. (1995). Initiating autogenic restoration on degraded arid lands. In *Proceedings of the 5th International Rangeland Congress*, vol. 1, ed. N.E. West, pp. 597–598. Salt Lake City, UT: Society for Range Management.

Whisenant, S.G. (1999). *Repairing Damaged Wildlands: A Process-Oriented, Landscape-Scale Approach.* Cambridge: Cambridge University Press.

Whisenant, S.G. & Tongway, D. (1995). Repairing mesoscale processes during restoration. In *Proceedings of the 5th International Rangeland Congress*, vol. 2, ed. N.E. West, pp. 62–64. Salt Lake City, UT: Society for Range Management.

Whisenant, S.G., Thurow, T.L. & Maranz, S.J. (1995). Initiating autogenic restoration on shallow semi-arid sites. *Restoration Ecology*, **3**, 61–67.

Whitman, A.A., Brokaw, N.V.L. & Hagan, J.M. (1997). Forest damage caused by selection logging of mahogany (*Swietenia macrophylla*) in Northern Belize. *Forest Ecology and Management*, **92**, 87–96.

Wight, J.R. & Siddoway, F.H. (1972). Improving precipitation-use efficiency on rangeland by surface modification. *Journal of Soil and Water Conservation*, **27**, 170–174.

Winkel, V.K., Roundy, B.A. & Cox, J.R. (1991). Influence of seedbed microsite characteristics on grass seedling emergence. *Journal of Range Management*, **44**, 210–214.

Winkel, V.K., Medrano, J.C., Stanley, C. & Walo, M.D. (1993). Effects of gravel mulch on emergence of galleta grass seedlings. In *Wildland Shrub and Arid Land Restoration Symposium*, vol. INT-GTR-315, eds. B. Roundy, E.D. McArthur, J.S. Haley & D.K. Mann, pp. 130–134. Las Vegas, NV: US Department of Agriculture Forest Service.

Zak, J.M. & Wagner, J. (1967). Oil-base mulches and terraces as aids to tree and shrub establishment on coastal sand dunes. *Journal of Soil and Water Conservation*, **22**, 198–201.

7 • Wetlands and still waters

JILLIAN C. LABADZ, DAVID P. BUTCHER AND DENNIS SINNOTT

INTRODUCTION

Wetland and still water habitats are elements of a continuum that depend upon water storage and supply for their continued existence and ecological stability. If inputs of water are significantly reduced, or outputs are increased, the water balance will change and wetland ecology is likely to suffer. Many types of wetland have frequently been damaged by artificial drainage, in what has been termed 'reclamation' for economic exploitation such as agriculture, forestry or peat production (see, for example, Gilman, 1994). Also significant in many situations may be a change in water-storage capacity, either by manipulation of the ground surface (peat extraction) or by increased sedimentation caused by accelerated soil erosion as a result of human land-use changes. Cooke *et al.* (1993) indicated that 25% of lakes assessed in the USA could be described as 'impaired' and 20% as 'threatened' in some way. Often this was by nutrient enrichment, but other stresses included siltation and inappropriate water levels.

Allen & Feddema (1996) suggest that wetland restoration involves actions aimed at improving damaged or deteriorating areas that were once healthy wetlands. In many countries, these natural wetlands are now only a fraction of their historical extent and have been identified as priority habitats for ecological restoration (e.g. Cooke *et al.*, 1993; Allen & Feddema, 1996; Tallis, 1998).

This chapter relates to the fundamental principles that control the physical environment of wetlands and still waters. It concentrates upon fresh waters rather than estuarine or coastal wetlands. Mitsch & Gosselink (1993) stated that 'hydrology is probably the single most important determinant of the establishment and maintenance of specific types of wetlands and wetland processes.' This in turn influences the biochemical conditions and processes operating, which are also of vital importance for ecological restoration but are outside the immediate scope of this chapter (see Søndergaard *et al.*, Weisner & Strand, this volume).

It must be acknowledged that hydrology alone cannot provide a prescription for successful ecological restoration. There must also be setting of clear ecological objectives, consideration of the institutional and economic situation and careful land management (Armstrong *et al.*, 1995). Ward *et al.* (1995) discuss planning of restoration of wetlands for birds, and state that suitable water regime and appropriate vegetation types are prerequisites for success. For peatlands, quality is usually associated with the capacity to form peat. Thus conservation priorities have often focused on sites that can be categorised as active bog within the EC Habitats Directive, with the presence of *Sphagnum* species being generally considered to be the most reliable indication of such activity (Tallis, 1998). However, as Heathwaite (1995*b*, p. 416) reiterates: 'An understanding of the key hydrological processes in disturbed mires is important because strategies may then be developed for greater control of the water balance, thus ensuring greater success to ecological strategies for mire conservation.'

This chapter focuses upon this need to understand hydrological processes and the water balance in any attempt to restore the ecology of a mire, wetland or lake. The principles of water movement and components of the water balance and typical water relations in a variety of wetlands and lakes are discussed. Two of the types of physical damage most commonly affecting wetlands and still waters (accelerated removal of surface water and excessive accumulation of sediments) are outlined and the techniques most commonly used to effect ecological restoration in each case are considered.

Further discussion of the practice of ecological restoration for lakes (Jeppesen & Sammalkorpi) and Wetlands (Wheeler *et al.*) can be found in volume 2.

A HYDROLOGICAL CLASSIFICATION OF WETLANDS

The nature of a wetland is fundamentally controlled by hydrological processes. The very existence of wetlands depends upon a positive water balance, resulting in waterlogging, but the characteristics of an individual wetland are the product of the origin and volume of water, the chemical quality of that water and the variability of water supply. Researchers have classified wetlands using various criteria. For example, Hughes & Heathwaite (1995) have identified the source of the water as a fundamental control and have classified wetlands accordingly (Table 7.1).

Other workers have included the hydrotopography of the wetlands as a characteristic feature, important in their classification (Wheeler, 1995). Von Post & Granlund (1926), for example, subdivided mires into three main types:

1. Ombrogenous mires – those that developed under the exclusive influence of precipitation.
2. Topogenous mires – controlled by horizontal flows of 'mineral soil-water'.
3. Soligenous mires – developed on sloping sites where laterally mobile 'mineral soil-water' maintained wet conditions.

All of these attempts at classification emphasise the origin and flow pathway of the water supply in controlling the nature of the wetland, but fewer workers have noted the variability of supply as being important. The vegetation that characterises a wetland is dependent not only on the source, but also the variability of supply, and it is this latter factor which is most affected where wetlands are damaged. Relationships between water supply and plants are discussed in texts such as Baird & Wilby (1999) and particular examples are given by Budelsky & Galatowitsch (2000), Mountford & Chapman (1993), Wheeler & Shaw (1995*b*) and Grosvernier *et al.* (1997), but detailed information is not yet available for many wetland species.

PRINCIPLES AND PROCESSES OF WATER MOVEMENT

Any attempt to restore a lake or wetland to a sustainable system depends upon an understanding of function. The natural circulation of water near the surface of the earth, driven by the radiant energy received from the sun, is commonly termed the 'hydrological cycle'. It is estimated that, at any one time, less than 0.025% of global water is held as soil moisture or in rivers and lakes (Shaw, 1994).

'Direct precipitation' describes rain falling directly upon a water body, such as a lake. Depending upon the particular conditions occurring in the catchment, this may represent a major or minor component of the water inputs. A water body or wetland must therefore be considered in the context of its catchment, which contributes water by the processes described below. In some cases the majority of the

Table 7.1. *A hydrological classification of wetlands*

	Extent		
Source of water	Small (<50 ha)	Medium (50–1000 ha)	Large (>1000 ha)
Rainfall	Part of some basin mires	Raised bogs	Blanket bogs
Springs	Flushes: acid valley and basin mires	Fen basins: acid valley and basin mires	Fen massif
Floods	Narrow floodplains	Valley floodplains	Floodplain massif

Source: From Hughes & Heathwaite (1995).

water will be derived from subsurface flow through the soil matrix or the underlying rock, but in other cases surface flows will form the major input.

The concept of the 'water balance' embodies the fundamental principle of the conservation of mass (Gilman, 1994). The water balance of any storage body over a predefined period of time can be described using the following equation (Eq.7.1). All terms must be expressed in identical units, preferably megalitres (1×10^6 litres, equal to $1000\ m^3$).

$$DP + SFI_c + SFI_d + GWI_s + GWI_d = ET + SFO_c + SFO_d + GWO \pm \Delta STORAGE \quad (7.1)$$

where

DP = direct precipitation
ET = evapotranspiration
GWI_s = groundwater spring inflow
GWI_d = groundwater diffuse (seepage) in flow
GWO = groundwater seepage outflow
SFI_c = surface channel inflow
SFO_c = surface channel outflow
SFI_d = surface diffuse inflow
SFO_d = surface diffuse outflow
$\Delta STORAGE$ = change in storage of water

The assessment of the various components of the water balance is discussed in detail in a number of standard hydrological texts such as Ward & Robinson (1990) and Shaw (1994). A brief outline is given below. It is important to remember that the flow processes have an element of temporal variability, so water balances calculated over a year are likely to be more 'accurate' than those derived for shorter periods. However, the short-term variation in hydrological conditions may be critical for ecological behaviour. Spatial variability may also be important, even at relatively small scales. The water balance is therefore essential in developing an understanding of the ecology of a water body. Moreover, although simple in concept, the water balance is difficult to measure in practice (Hillel, 1971).

Precipitation

Precipitation is the basic input driving the hydrological cycle. Rain and snow are the most frequent inputs, although fog and dew ('occult precipitation') may be important in some locations (see Shaw, 1994, for further information). Precipitation may be expressed in terms of its total over a specified time period (mm depth per unit area), its duration, or its intensity over very short time periods (for example, an intensity of $24\ mm\ h^{-1}$ for 30 minutes). All of these, together with antecedent wetness conditions, will contribute to determining the fate of the water as it moves through the hydrological cycle.

Box 7.1 Assessment of precipitation

Whilst the measurement of precipitation may appear to be a simple matter, it can be fraught with difficulties. There may be large fluctuations within and between years (e.g. Marsh & Bryant, 1991), and areal precipitation over a catchment is unlikely to equal that measured at any one point (Shaw, 1994). Recent development in radar technology have provided areal data in some countries, but even here ground-truthing is required (Harrison *et al.*, 1997).

Precipitation records are usually expressed in mm, as a depth over a surface. In order to calculate the volume of water involved it is therefore necessary to know the surface area of the water body, either from an original survey or estimated from published maps. Automatic weather stations which will record precipitation at intervals of 0.1 mm are now reasonably freely available to researchers in developed countries. Managers and researchers in less well-resourced situations may need to rely upon manual storage gauges or secondary data. In Britain, for example, daily precipitation records can be obtained from the Meteorological Office or, in certain circumstances, the Environment Agency. For the USA many rainfall records are now freely available on the World Wide Web. At worst, it may be necessary to obtain general indications of typical rainfall for a region and use these to estimate the likely direct precipitation.

Box 7.2 Assessment of evapotranspiration

There are various methods of assessing evaporation directly, but all have attendant limitations. For a free water surface, evaporating pans are cheap and easy to maintain, but correction factors are needed to relate results to a larger area and they do not represent the effect of vegetation. Lysimeters usually involve a large block of undisturbed soil isolated from their surroundings in a watertight container in the ground (such as a galvanised dustbin). These have the advantage that they can be vegetated, but they are much more complex and costly to install and operate.

In many cases, evapotranspiration is estimated using an equation rather than measured directly. This may be via an energy budget approach, applied at a particular site using the Bowen ratio (which is accurate but expensive in terms of instruments and labour), or at a regional scale using remotely sensed data (e.g. Farah & Bastiaanssen, [2001] for the catchment of Lake Naivasha, Kenya). Alternatively, evapotranspiration can be estimated by means of a simple empirical equation such as Thornthwaite's (which uses mean monthly temperatures) or a more complicated equation such as the Penman Monteith. Details for calculation by the various methods are given in the standard texts such as Shaw (1994) and Ward & Robinson (1990).

Examples of assessment of various methods of evaporation calculation are given by Winter (1995) for Williams Lake, Minnesota, and by Fermor *et al.* (2001) for reedbeds with *Phragmites australis.*

Most water bodies will receive at least some direct precipitation, but the significance of this input will range enormously. It may be minimal in small topogenous or soligenous groundwater fed lakes, marshes or fens, but may represent the only source of water for many ombrogenous mires (raised bogs and blanket bogs) which lie above the regional water table. Most lakes receive a mixture of surface and groundwater inputs with a smaller amount of direct precipitation but Wetzel (1999) cites, as an extreme example, Lake Victoria in east Africa receiving more than 70% of its water from precipitation directly onto the surface.

Average annual precipitation totals range considerably from less than 100 mm to more than 1500 mm in the USA, and from less than 600 mm to more than 2400 mm in the British Isles (see Ward & Robinson, 1990).

Evaporation, transpiration and interception storage

Evaporation is the conversion to water vapour and return to the atmosphere of precipitation which has reached the earth's surface. Total evaporation (consumptive use) includes transpiration (water vapour escaping from within plants). It requires input of energy and a removal process (turbulence, etc.). Evaporation from a free water surface is influenced by meteorological factors (temperature gradient, humidity gradient, wind speed) and physical characteristics of the water (Ward & Robinson, 1990; Shaw, 1994).

Thornthwaite (1944) defined potential evaporation (PE) as 'the water loss which will occur if at no time there is a deficiency of water in the soil for the use of vegetation' whilst Penman (1956) said PE is 'evaporation from an extended surface of short green crop, actively growing, completely shading the ground, of uniform height and not short of water'. Some authors also discuss 'wet surface evaporation'.

For other land surfaces, actual evaporation (AE) will be dependent upon temperature, wind speed, amount of incoming radiation, relative humidity, soil moisture conditions and land use/vegetation. The latter is important because of the ability of vegetation to increase the water losses by means of transpiration. Evaporation and transpiration are commonly treated as a single 'loss process'. Vegetation type and condition will influence the amount of water lost to evapotranspiration (see Bosch & Hewlett, 1982; Ingram, 1983; Newson, 1997; Wetzel, 1999).

Infiltration and infiltration-excess overland flow

If water does not immediately evaporate, it may pass through the surface of the soil via pores or small

Box 7.3 Assessment of infiltration

Burt *et al.* (1990) discuss infiltration rates and infiltration-excess overland flow in peatlands. Studies using ring infiltrometers suggested that infiltration capacities on cotton grass (*Eriophorum* spp.) moorland were as low as 2 mm h^{-1}, whereas on crowberry (*Empetrum nigrum*) the mean infiltration capacity was just less than 30 mm h^{-1}. However, experiments with sprinkling infiltrometers suggested rather higher infiltration rates, at least into the top few centimetres of decaying vegetation and peat. Holden *et al.* (2001) used tension infiltrometers and produced an average figure of 12.4 mm h^{-1} for the surface layers of blanket peat, with *Sphagnum*-covered peat and bare peat giving slightly higher values than that covered by *Calluna* or *Eriophorum*.

openings, into the soil mass. Any given soil will have an infiltration capacity, a maximum rate at which the soil can absorb falling rain when it is in a specified condition. As soil becomes saturated, the infiltration capacity falls asymptotically to a final, minimum, constant level. Infiltration capacity varies with soil texture, such that sandy soils have much higher final infiltration capacity than fine clay soils or peats (see Shaw, 1994).

Where precipitation is of greater intensity than the infiltration capacity of the soil, diffuse 'infiltration-excess overland flow' will occur. Where rainfall is particularly intense or soil is particularly fine-textured, this situation can occur long before the soil is actually saturated with water.

Saturation-excess overland flow

When the soil profile is completely saturated, excess water (added from above by precipitation, or laterally by overland or subsurface flow) cannot be accommodated. Hewlett & Hibbert (1967) first put forward the hypothesis that in many catchments precipitation will enter the soil and, by a combination of infiltration and lateral flow through the soil, the areas immediately around the stream channels become saturated. With time the lower valley slopes may also become saturated as the water table rises. Burt *et al.* (1990) described the frequent occurrence of 'saturation-excess overland flow' in peatlands, explaining how expansion of the saturated zone during precipitation leads in turn to expansion of the area experiencing overland flow. Holden *et al.* (2001) confirmed the importance of saturation-excess overland flow in blanket peats in northern England.

Pipe and macropore flow

Almost as rapid as surface flow may be macropore flow, in pores above capillary size which allow water to move by gravity, bypassing the soil matrix and providing rapid recharge for the water table. Baird (1997) suggests that many authors take macropores to exert 3 cm suction, which means that they are more than 1 mm in diameter. He found that they accounted for over 50% of flow even in a highly humified fen peat. Holden *et al.* (2001) investigated macropore flow in blanket peat in the northern Pennines, England, and found that macropores accounted for an average 36% of flow through the peat, with 51% in *Sphagnum*-covered peat. They also report a change with depth in bare, *Calluna*- and *Eriophorum*-covered peats, such that macropores were most significant in the 5-cm zone, accounting for 47–53% of flow, and had reached a minimum (13–22%) by 20 cm below the surface. A number of studies have also been made of rapid subsurface flow which occurs in even larger features known as 'pipes' in peat-covered catchments. Jones & Crane (1984) found that almost half the water to enter streams during flow events ('stormflow') originated in discrete pipes. The precise role of pipes as sites for gully extension and stream channel initiation remains unclear, but they can play a significant role in the hydrology of some peat-covered catchments.

Subsurface flow through rock or soil matrix

Most soil or rock matrices will be permeable to water to some extent, giving rise to both lateral and vertical flows within the ground. The rate of flow in any rigid medium is controlled by three factors

Box 7.4 Assessment of surface flows

The first step in identifying surface inflows is to delimit the surface catchment of the water body of interest. In the case of an ombrogenous mire, this may be limited to the peat expanse and there may be no need to consider any further inflows. In most other cases, however, there is a strong likelihood of inputs from a wider catchment area. For example, Aqualate Mere is a National Nature Reserve in Staffordshire, UK, which has a designated area of only 241 ha but has been identified as having a surface catchment extending to 5924 ha (almost 60 km^2) (Environment Consultancy of the University of Sheffield, 2001).

Catchment boundaries can be determined initially by inspection of topographic contour maps (e.g. Ordnance Survey sheets at 1:10 000) and drawn according to the likely divergence and convergence of any surface drainage water which may occur. Contour maps also provide some indication as to the direction of flow of surface drainage infrastructures. However, the results are extremely dependent upon the scale and age of the map employed. Fine details of gradient are not always discernible and it is possible for agricultural drainage ditches to have been deepened to such an extent that they run counter to the general topography. The use of available stereo air photographs and a suitable mirror stereoscope viewer should be attempted as this can contribute to the desktop identification of the relative topography and watershed features. In general, it is recommended that this method be used in support of the inspection of contour maps. Photogrammetry is especially useful where slopes are discernible but becomes more difficult where gradients are very slight. In addition, the photographs are often dated or are not available as stereo pairs for the entire catchment. In some countries, catchment boundaries may be obtained digitally (e.g. in Britain from the Centre for Ecology and Hydrology, being derived from a digital terrain model at a scale of 1:50 000). However, these will not be as accurate as is possible using detailed examination of the larger-scale maps. It is always desirable, and is arguably essential, to visit the sites for confirmation of details not apparent from the maps. Site surveys could range from a simple walk over survey for site familiarisation to a detailed hydrotopographic survey (using conventional levelling techniques, a total station or a differential global positioning system, as in Sinnott *et al.*, in press) to determine slopes of ditch bases.

If time and resources permit, surface channel inflows identified in the process described above should be continuously gauged to assess their relative importance to each other, to other sources and temporal variability. In some cases, a government body (e.g. the Environment Agency in England and Wales) may have a gauging station within the catchment. Such data should be used, preferably at a daily resolution, so that fluctuations can be assessed. In most cases, however, such records will not exist. If there is particular concern of fluctuating water levels or drought-induced stress, then it may be appropriate to consider installing a water level recorder in a rated section or at a weir or flume. Such instruments operate using either a float system or a pressure transducer (see Shaw [1994] for further detail).

Where continuous recording of surface channel inflows is not possible because of time or resource limitations, a broad indication of the relative importance of different inflows may be obtained by instantaneous 'spot' gauging. Most frequently this is done by accurate measurement of the cross-sectional area of the flow in the channel, combined with a measure of flow velocity at various points in the cross-section. The estimate of velocity is best made using a calibrated impeller flow meter (e.g. manufactured by Ott or Valeport) or, especially useful in small and slow flows, an electromagnetic current meter (e.g. supplied by Aqua Data Systems Ltd). Alternatively, in small channels where there is limited macrophyte growth and ecological conditions are not especially sensitive, a technique called dilution gauging may be used. This involves injecting a known quantity of a tracer such as sodium chloride or fluorescent dye upstream and recording the passage of the tracer downstream (see Shaw [1994] for further detail).

An additional technique of some limited value in assessing the water budget may be the use of the channel morphology to estimate 'bankful discharge', the maximum rate that can be contained before overbank flooding occurs. Here velocity is most frequently estimated by means of Manning's formula, which requires estimation of a roughness coefficient dependent upon the nature of the channel bed and banks (see Shaw, 1994).

If no direct measurements of channel runoff are available then it may be necessary to resort to secondary data from similar catchments. Ward & Robinson (1990, p. 250) present a very generalised, global map of mean annual runoff, ranging from less than 20 mm to more than 1000 mm of equivalent depth, whilst Rochelle *et al.* (1989) discuss the use of runoff contour maps developed from existing data for broad regions.

In Britain, the Centre for Ecology and Hydrology have used existing flow data to produce the *Flood Estimation Handbook* (Centre for Ecology and Hydrology, 1999) and the *Hydrology of Soil Types* (HOST) soil classification (Boorman *et al.*, 1995), use of which allows some estimation of flow in ungauged catchments. Soils have a major influence on hydrological processes, since their physical properties govern the storage and transmission of water, whilst at the same time they influence water quality by acting as chemical buffers and biological filters. The HOST classification (Boorman *et al.*,1995) is based on conceptual models and incorporates consideration of the permeability of the substrate and the depth to water table, in addition to soil characteristics and wetness regimes. Twenty-nine HOST classes are defined. They may be used to estimate low flows at ungauged sites as well as standard percentage runoff (SPR), which is the percentage of rainfall which causes a short-term increase in flow at the catchment outlet as opposed to contributing to evapotranspiration or longer-term storage in the ground. The dominant HOST classes for each soil association (map unit) are given in Appendix B of the report, or they are available in digital format at a 1-km grid definition.

according to Darcy's law:

$$Q = KIA \tag{7.2}$$

where

Q = groundwater discharge ($m^3\ s^{-1}$)
K = hydraulic conductivity ($m\ s^{-1}$)
I = hydraulic gradient (dimensionless, also known as 'head' per unit distance)
A = cross-sectional area across which flow occurs (m^2)

K, the hydraulic conductivity of a material, reaches a maximum when all available pore spaces are already filled with water. This is known as the saturated hydraulic conductivity, sometimes abbreviated to Ksat. As a result, groundwater movements will be greatest when the wetland or lake is in hydraulic continuity with the local or regional groundwater.

Baird (1997) has pointed out, however, that drained peatlands commonly have water tables more than 70 cm below the surface for much of the year, so consideration of hydraulic conductivity in the unsaturated zone is important in such cases.

WATER RELATIONS IN WETLANDS AND LAKES

It is not feasible to describe water relations in the wide variety of wetland types, in detail, here. Instead, examples will be given from the experience of the authors, which relates to blanket bogs, raised bogs and lakes. For information on fens and marshes the reader is referred to Wheeler *et al.* (volume 2) and to texts such as Grootjans & van Diggelen (1995), Lloyd & Tellam (1995) and Armstrong *et al.* (1995).

Water relations and runoff generation in blanket bogs

Studies of the hydrology and fluvial geomorphology of blanket peat moorland remain relatively scarce despite a history of scientific investigation stretching back almost 40 years. Of the early studies, Conway & Millar's (1960) is perhaps the most notable. Water balance calculations showed that a relatively uneroded *Sphagnum*-covered basin retained significantly more water than another basin which had been both drained and burnt over. Paradoxically, this result may have revived the traditional view that peat-covered catchments act

Box 7.5 Assessment of groundwater flows

The definition of subsurface (groundwater) catchments is even more problematic than for surface water catchments. A likely consequence of the topography, connectivity and permeability of underlying rocks and sediments is that some of the hydrological inputs to the surface catchment may not be transferred to the core wetland, whilst some areas not contributing to surface flow will be intimately connected with the wetland by subsurface flow.

Strictly speaking, the subsurface catchment of any wetland can be identified by reference to geological maps and borehole information. By reference to the extent, juxtaposition and lithology of the various strata it is possible to identify the maximum extent of the area for which continuity of groundwater may apply. Once the extent of the potential area has been identified it is necessary to consider the slope of the regional water table and the presence of any local water table as a result of an impeding stratum. It is also important to note that water tables vary temporally, so care must be taken in interpreting a slope from test water levels obtained at different times. This whole process may also be more complicated in areas of significant drift cover, because boreholes will frequently not have penetrated the drift and the nature of underlying strata is assumed rather than mapped from observation.

It is also important to consider the effects of hydraulic conductivity (permeability of the strata) upon the time taken for groundwater to travel a given distance. Shaw (1994) lists typical values for (permeability) of aquifers and suggests that 10 m per day would be possible in sand, whereas 3 m per day is a more typical velocity in sandstone, 0.1 m per day in semi-porous silt and 0.01 to 0.0001 m per day common in clays. Baird (1997) describes use of a porous disc tension infiltrometer to measure hydraulic conductivity in the unsaturated zone at various pressure heads and stresses the importance of this flow contribution in drained peats.

Most countries will have a source of published geological information. Information on geology and groundwater can also frequently be found in existing management plans and reports produced by previous workers. In the UK, local authority departments of engineers or surveyors may also hold some borehole data, and the Environment Agency may hold groundwater records for a particular area and it is their practice to identify 'groundwater units' which warrant consideration as a body if any action has implications for the hydrology of the area.

Measurement or estimation of the groundwater flow is not a simple matter as it requires installation and monitoring of nests of piezometers to establish the hydraulic gradients operating, execution of pump tests to establish hydraulic conductivities and geological investigations to establish the boundaries for flow if Darcy's law (see above) is to be applied (see Shaw [1994] and Price [1996] for further detail). Within lakes, Cooke *et al.* (1993) describe the use of submersed seepage meters (steel barrel halves with collecting bags attached, which are inserted into the lake floor by scuba divers) to assess spatial and temporal variation in groundwater inputs. These may particularly useful if it is suspected that there are discrete springs contributing water to the lake body.

Winter (1995) describes various methods for measuring and modelling groundwater flux in lake systems. He concludes that the low cost and simplicity of direct measurement methods such as seepage meters and potentiomanometers are advantageous. Disadvantages are that measurements are often not continuous, and a large number of time-consuming (and therefore expensive in labour) measurements are needed to characterise heterogeneous geology. An approach of 'groundwater flow systems analysis' with a continuously monitored network of piezometers and dipwells is able to give much greater understanding, but is expensive to construct and model.

After a period of data gathering a detailed analysis of the spatial variation of water table level and flow across the site can be carried out, normally followed by a model construction process in which a detailed mathematical model of the site would be built (e.g. MODFLOW, used by Bromley & Robinson [1995] and Reeve *et al.* [2000]). The model would then be calibrated and a number of rainfall scenarios run. The purpose underlying such a strategy is to evaluate those times when a site is under hydrological stress and the places where that stress is most apparent. The final aspect of this hydrological evaluation is to consider the type of vegetation across those areas deemed to be under hydrological stress. The whole approach is clearly expensive and too time-consuming to be practicable for many managing agencies, and results are not necessarily transferable to other sites.

Box 7.6 Assessment of change in storage of water

Most limnological investigations for purposes of ecological restoration will require a knowledge of changing water storage. Many attempts at restoration for conservation purposes have been hampered by a lack of knowledge of what would be considered 'normal' water level fluctuations for a particular lake.

The simplest technological approach to monitoring water levels in lakes and open water bodies is to use fixed stage boards, which are essentially rulers marked at centimetre intervals providing a means of assessing water level relative to an arbitrary datum. These can be read by observers at intervals and have the advantage of low cost and low maintenance requirements. The disadvantage is that time intervals between measurements are likely to exceed those of the natural fluctuations of level in many lakes. Where lake levels are known to be variable, installation of a continuous stage recorder offers much greater detail and insight into the behaviour of the water body.

The next step is to relate fluctuations in water level to changes in water storage in the lake. This requires an understanding of lake morphometry and production of a bathymetric map, showing depth contours at regular intervals. This can be used to calculate a depth–capacity table for the lake, allowing assessment of change in volume for any given change in water surface level. Increasing accuracy and availability of digital echo sounders attached to differential global positioning systems (DGPS) has meant that detailed lake surveys can be produced much more effectively. Results can be stored digitally and analysed in a GIS or surveying CAD package to produce a depth–capacity relationship. Examples include Sullivan (1996) for Possum Kingdom Lake in Texas and Halcrow (2000) for three small reservoirs in England.

Where accessibility is limited or very large water bodies are being investigated, the use of remote sensing or photogrammetry may offer some indication of fluctuating water levels over time, but will not allow conversion to changes in water storage.

Assessment of changes in water storage in wetlands other than open water bodies is a rather more complex procedure. Storage of water in the soil is usually monitored by measurement of the elevation of the water table, the level at which water pressure is equal to atmospheric pressure and so water will stand in a well hydraulically connected to the groundwater body.

In order to calculate changes in water storage it is necessary to determine:

- The morphology of the water table
- The volume of the strata available for water storage
- The relationship between change in volume of water stored and the observed change in water table (known as the 'specific yield').

These are simple in principle, if time-consuming and therefore costly, to monitor. Topography can be monitored using standard surveying techniques or, more recently, GPS technology (Sinnott *et al.*, in press). The morphology of the water table can then be determined by use of dipwells, simple tubes with perforations to allow water ingress, which allow measurement of the water table relative to the ground surface at a point. The volume of peat available for storage can be determined by simple rodding of peat depths or by use of ground-penetrating radar. Specific yield can be estimated with a lysimeter, or from daily observations of precipitation inputs and water table fluctuations. Gilman (1994) has described the monitoring of the storage in wetlands in more detail than is possible here, and has stressed the importance of accurate survey to the long-term value of the records.

Complications in assessment of changing water storage arise, however, because there will also be changes above the water table, as evidenced by changes in moisture content of the soil. Also in soils such as peat, where volumetric moisture contents are extremely high, the peat surface itself will fluctuate with the change in water storage over time (a phenomenon known as 'mooratmung': see Ingram, 1983). Gilman (1994) suggests that this is most significant in floating mires ('Schwingmoor').

like aquifers, storing rainfall up and releasing it gradually during dry periods. However, as will be demonstrated below, nothing could be further from the truth as even intact blanket peat is highly productive of storm runoff and, by contrast, generates little baseflow.

Most blanket peat bogs within the United Kingdom are found in the upland areas of the north and west, where rainfall occurs on two out of three days each year. Annual rainfall totals show little temporal variability (compared to the lowlands) and are usually above 1000 mm. There have, however, been few direct studies of evaporation from blanket peat moorland (Burt, 1994).

In undrained peat the water table is at or near the ground surface for most of the year. Figure 7.1 demonstrates this for the UK's Environmental Change Network (ECN) site at Moor House in the northern Pennines, indicating the fall in the water table during the drought of 1995 and the more typical response (reaching only 20 cm below the surface) during the summer of 1996. This has two important implications for runoff generation: the widespread development of saturation-excess overland flow and the generation of shallow subsurface stormflow in the acrotelm. Burt *et al.* (1990) noted that eroded or drained blanket peat areas will tend to have lower water tables with a consequent decrease in the incidence of saturation-excess overland flow, whereas uneroded blanket bogs may experience frequent near-surface flow over large areas. Labadz *et al.* (1991) recorded an average of 35% of incident rainfall appearing as storm runoff on a small but heavily eroded headwater catchment.

The runoff regime of blanket peat catchments can best be described as 'flashy'. Discharge in streams on blanket peat responds quickly to precipitation events, typically peaking within 1 hour of maximum precipitation and subsiding to baseflow within 7 to 12 hours of the cessation of rainfall (Crisp & Robson, 1979; Robinson & Newson, 1986; Labadz *et al.*, 1991). The rapid rise in discharge in response to a rainfall event is associated with a fall in solute concentrations (as evidenced by specific electrical conductivity) but an immediate increase in suspended sediment concentration, particularly where there is an actively eroding gully network. These relationships are both explained by the increase in surface and near-surface flow, as the water has a very limited residence time in contact with the peat and it gains

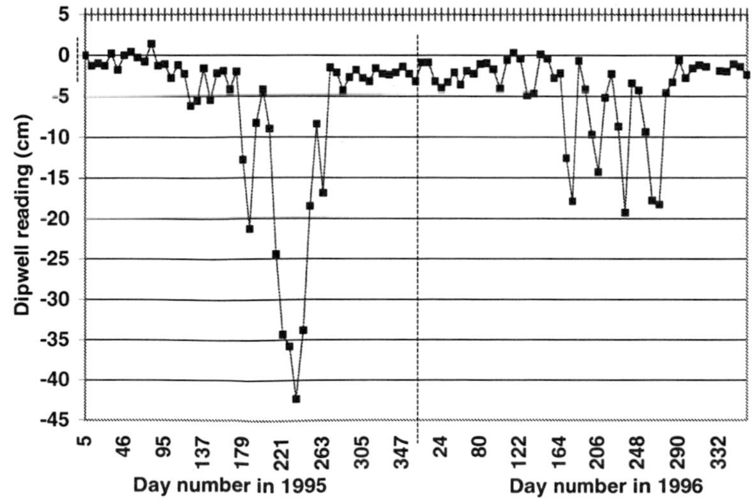

Fig. 7.1. Variations in water table in blanket peat over two years. From John Adamson, Environmental Change Network, in Burt *et al.* (1997).

little soluble load. In contrast, rapid surface runoff favours acquisition of a large sediment load. At the site studied by Labadz *et al.* (1991), there was some variation in the relative timing of discharge and suspended sediment peaks, but positive hysteresis was evident in about half the storms monitored over a two-year period. This suggests a need for sediment to be made available by mechanisms such as desiccation or frost action, even in eroded areas with much bare peat.

Blanket bogs remain wet because their rate of drainage is so slow. Typical saturated hydraulic conductivities for peat in the surface layers (acrotelm) of an undisturbed mire are likely to be high (around 2 m per day) but in the more dense and humified catotelm peat this decreases to 0.1 m per day or lower. Burt (1995) summarised studies of the effects of drainage upon the water balance of peat mires and indicated that the exact impact of drainage upon the hydrological regime is dependent not only upon peat properties and the density and location of the drains, but also upon antecedent moisture conditions. It is possible for flows to be enhanced in some circumstances and reduced in others at the same site.

Drainage will, however, tend to be associated with increased storm runoff where it is also coincident with another factor serving to reduce infiltration rates, such as burning which may induce hydrophobicity in the peat, or removal of vegetation and - erosion of the surface layers to expose the less permeable peat of the catotelm. In this case, runoff percentages may be in the order of 60% (Labadz *et al.*, 1991).

Damage to the peat mass in a blanket bog, such as desiccation or burning, will often form a 'crust' which may serve to reduce infiltration capacity and increase the proportion of surface flow (Tallis, 1998) leading to the suggestion that rainfall intensities as low as 3 mm h^{-1} may be sufficient to generate surface runoff on sloping ground.

Anderson *et al.* (1995) investigated the effects of afforestation on blanket peat but concentrated upon water tables rather than measuring flows. They found that shallow ploughing significantly lowered the water table compared with the control, followed by subsidence of the ground surface by a few centimetres as a result of consolidation of the peat at all depths.

Water relations and runoff generation in raised bogs

Over the last 60 years a new perspective on the conservation value of bogs has emerged with increased environmental and ecological awareness. It is a stated objective of nature conservation agencies to provide for the future sustainability of raised mire habitat (Joint Nature Conservation Committee, 1994) as 'active raised bogs' and 'degraded raised bogs capable of regeneration' are now listed under the EC Habitats and Species Directive (1992) as a priority habitat. It is now widely acknowledged that the lowland raised bogs of the UK are threatened landscapes that have been greatly reduced in area and number (Wheeler & Shaw, 1995*a*). Some form of human intervention has damaged most, if not all, existing raised bog sites. This is primarily a result of the continuing threats of drainage for agriculture, forestation and peat extraction. For example, the estimate in the National Peatland Resource Inventory (quoted in Joint Nature Conservation Committee, 1994) is that of 37,413 ha of land in England with raised bog soils, only 493 ha (1% of the area, in 15 sites) retain near-natural bog vegetation.

The general quality of a bog is assessed by the degree to which it has remained capable of active peat growth (Joint Nature Conservation Committee, 1994). This requires the continued existence of sufficient hydrological integrity of the mire complex: 'Only mires which are sufficiently hydrologically intact can form peat' (Immirzi *et al.*, 1992, p. 28). Thus an essential element of any approach to wetland restoration is the assessment of damage, or threat of damage, to hydrological conditions, together with consideration of appropriate options for remediation. In summary, Maltby (1997) has emphasised that peatland ecosystems are not very resilient to stress in terms of water relations: 'Such is the subtlety of the relationship between micro-habitat conditions, especially water table, and species distribution, that the biodiversity assemblage is highly vulnerable to perturbation.' (p. 127.)

Water relations in raised bogs have been widely discussed in publications such as Gilman (1994, 1997), Heathwaite & Gottlich (1993) and Hughes & Heathwaite (1995). The classic peat bog is often described as being diplotelmic, with an active acrotelm above a lower catotelm (Ivanov, 1981). The acrotelm is the layer in which concentration by evaporation and dilution by rain are most immediately and directly felt; it is also the rooting zone for the plant cover and one in which the chemical processes are far from fully understood (Proctor, 1995). The hydraulic conductivity of the acrotelm is very much higher than that of the catotelm, and its relative thinness results in a limited storage capacity for water. Material is added annually to the catotelm by decomposition from the acrotelm, so that it becomes deeper and denser over time, reducing the hydraulic conductivity of the mire and enabling the water-retentive bog vegetation to maintain a high water table despite a continually deepening catotelm. Unlike the acrotelm, the catotelm remains permanently saturated but rates of water movement are very low indeed. It is this relationship which ensures that, despite storing large amounts of water, peat-covered catchments are poor suppliers of baseflow (Burt *et al.*, 1990). Low hydraulic conductivities within the catotelm help to maintain a water table close to the ground surface, a condition which is essential to the continuing functioning of surface vegetation, and any disturbance of the catotelm and acrotelm hydrology has a consequent impact on surface vegetation.

It is helpful, when considering water exchanges between raised bogs and their surrounding environment, to examine broad water transfer pathways of the water balance process. There is still debate as to the relative importance of different pathways, with precipitation, evapotranspiration, surface runoff, seepage, pipe flow and channel flow all of potential importance (Ingram, 1983). In an intact peat mass, the prevailing high water table means that there is little capacity to retain water supplied as precipitation. Most of the water not lost to evapotranspiration will discharge at the peat margins through the acrotelm, to pass ultimately into the drainage channels surrounding. So long as the infiltration capacity is not exceeded, most runoff will initially be as lateral seepage or pipe flow within the peat mass. During heavy rain, however, infiltration-excess overland flow may occur and in prolonged rainfall events there is also the likelihood of saturation-excess and return flow (Burt & Gardiner, 1984; Burt *et al.*, 1990). Recent work by MacAllister & Parkin (1999) suggests the most significant hydrological pathway transferring water away from raised mires is surface runoff, a pathway largely ignored in most models and studies.

Intact raised mires typically have an internal drainage system which transfers rainfall outwards from the centre to the peripheral lagg zone; it has often been assumed that this occurs through the peat along centrifugal flow lines and that discharge may be measured using Darcy's law (Ingram, 1982; Heathwaite, 1993). This and other flow equations are described in texts such as Hillel (1971).

Raised mires are characteristically domed, rising up to 10 m above the surrounding land and often being 0.5 km or more in diameter (Bromley & Robinson, 1995). There is some consensus amongst researchers that the principal controls on the formation and condition of raised mire systems are hydrological. It is now generally accepted that the dome shape is sustained by the low hydraulic conductivity of the mass of the raised bog which restricts the outflow of water.

The 'groundwater mound' theory (Ingram, 1982; see also Bragg, 1995) indicates that the profile and dimensions of the catotelm are a function of: (1) the shape and size of the basal plan of the bog; (2) the net recharge to the catotelm; and (3) the overall permeability of the accumulating peat. Ingram's theory involves a water mound sustained by dynamic equilibrium between recharge and seepage together with groundwater table geometry (Childs, 1969) to derive the relationship between a mire's shape and dimensions, net recharge and hydraulic conductivity of the peat. It does not take into account surface runoff, since it assumes that the mound is able to absorb all received rainfall. This omission may explain why natural raised mires are rarely as high as Ingram's theory would suggest. Other workers (e.g. Kneale, 1987; Armstrong, 1995) have attempted to apply the groundwater mound to particular situations and have found it difficult to achieve realistic results in terms of the height of the resultant dome. In practice even damaged mires generate substantial volumes of saturation-excess overland flow.

Water and sediment movements in lakes

Most natural lakes are small in area, shallow in depth and occur in terrain of gentle slopes, with wetland–littoral components dominating in the productivity and synthesis of organic matter (Wetzel, 1999). Artificially constructed reservoirs, on average, are larger and deeper. Cooke *et al.* (1993) note that natural lakes and reservoirs have many biotic and abiotic processes in common. They may have identical habitats (pelagic, benthic, profundal and littoral zones) and processes such as dissolved oxygen depletions or internal nutrient releases are commonly found in both. There are, however, some significant differences, which must be understood if successful management and restoration is to occur. Natural lakes often tend to have relatively small catchments, with water entering at least partly by seepage from a local or regional groundwater table, and surface inputs often involve small streams, which traverse wetland or littoral areas. Reservoirs, by contrast, are often built on major rivers with much larger catchments in relation to the volume of the water body.

This section can provide only a very brief review of the processes occurring within lakes. For more detail on particular aspects, readers are referred to texts such as Wetzel (1983), Lerman *et al.* (1995), Hakanson & Jansson (1983) and Morris & Fan (1998). Successful lake management or restoration depends upon a holistic view of lake and watershed processes. In any consideration of techniques for restoration, it is vital to consider the lake or reservoir in its catchment context.

Lakes interact with surrounding surface and groundwater systems which can be highly dynamic. Wetzel (1999) suggests that the evapotranspiration capacities of emergent and floating-leaved aquatic plants can be so effective that the rates of movement of groundwater in surficial sediments can be altered appreciably, and presents a summary table of evapotranspiration rates for various species from the literature. Changes in water table can occur both seasonally (with a cone of depression causing seepage into the lake during the growing season) and diurnally.

'Retention time' is a characteristic value for a lake calculated from the total inflow minus outflow divided by the total volume of the lake. Groundwater-dominated lakes, supported by a regional water table with no discrete outflow channel, will typically have a very long retention time, sometimes of the order of several years. For surface-water dominated lakes fed by a large catchment, the retention time may be only a matter of weeks. However, Wetzel (1999) notes that theoretical retention time is realised only approximately in most lakes. Seasonal rates of inflow and stratification changes as a result of temperature or density variations may mean that river water entering a lake is at certain times channelled over, through or beneath the main water body. Wetzel (1983) suggested that the primary determinant of the physical, chemical and biological interactions of a lake is the phenomenon of thermal stratification which occurs in temperate and many subtropical lakes during summer months. The warmer epilimnion at the surface is separated from the colder hypolimnion by the thermocline, a layer where there is a rapid decrease in temperature with depth.

Most of the sediment accumulating in most reservoirs is allochthonous, sourced from outside the water body. Some may be derived from direct atmospheric inputs but, in most cases, by far the greatest proportion is a product of catchment erosion, delivered to the reservoir via the natural fluvial system and its anthropogenic extensions. This will include both mineral and organic components. Small amounts of the sediment may be generated within the reservoir basin by wave-induced landslips or bank erosion, but only in exceptional cases does this form a major contribution to the total sediment deposited in the reservoir basin. The other source of sediment which is significant in most natural lakes and may be important in some reservoirs is autochthonous biological production by littoral and wetland plants (Wetzel, 1999), which again will include both organic and mineralised fractions. Cooke *et al.* (1993) suggest that many natural lakes have high productivity because they tend to be small in area and shallow, allowing a larger area of warmer epilimnetic sediments with rooted plants and algae associated with leaf and sediment surfaces.

Reservoirs and lakes tend to trap both coarse and fine sediment particles as a result of the decrease in velocity and shear stress experienced as water flows from upstream. In consideration of reservoir sedimentation, 'sediment yield' is defined as the sediment transport comprising both a bedload component (typically sands and gravels transported along the bed of a river) and a suspended sediment component typically comprising finer particles of silt and clay. The mechanics of how sediment becomes deposited in reservoirs is covered in the technical literature (Bruk, 1985; Mahmood. 1987).

PRINCIPLES OF ECOLOGICAL RESTORATION

Reversing the effects of surface drainage or abstraction

One of the major types of damage to wetlands and lakes is desiccation caused either by surface drainage or direct abstraction of surface water. Groundwater abstraction may also cause problems in some situations, but space limitations preclude its discussion here.

Mire ecosystems differ greatly from most land-based ecosystems in that their characteristic hydrological regime arrests the decay process, leading to diagenesis of organic matter and the autochthonous production of peat. This propensity actively to produce peat is the criterion by which the quality of a mire is judged and its understanding is fundamental to the implementation of any rehabilitation strategy. The hydrological condition of a raised mire system is largely a product of the balance between two factors: the effective rainfall input into the system and the losses of water through evaporation, surface and subsurface runoff. In practice managers are clearly not able to control the rainfall input but it is important to stress that the degree of rainfall will control the sensitivity of the mire to any damage. Those mires, such as Thorne and Hatfield Moors in South Yorkshire, UK, which are close to the threshold of rainfall required for *Sphagnum* growth will be more sensitive to drainage since there is less replenishment of the system. The impact of drainage on the mire is to increase surface loss, to cause increased hydraulic gradients within the peat body and thus cause greater subsurface flow into the ditch systems. Any wetland restoration designed to enable provision of appropriate and desired site conditions will usually require restoration of hydrological conditions as the precursor to development of vegetation and habitat (see, for example, Heathwaite & Gottlich [1993] and Gilman [1994]). This is likely to apply even if hydrological restoration in itself is not sufficient to enable subsequent recolonisation.

MacAllister & Parkin (1999) have suggested that the most significant hydrological pathway transferring water away from raised mires may be surface runoff, a pathway largely ignored in most models and studies. However the volume of surface runoff is not, in itself, a satisfactory indicator of good condition since in a healthy raised mire there would be substantial volumes of overland flow. A raised mire in good hydrological condition will generate substantial saturation overland flow whenever there is an excess of rainfall. In a damaged raised mire system the increased surface flow through drainage ditches will cause localised reductions in water table level. The significance of that reduction will be related to the hydraulic conductivity of the peat, the drainage density and the surface topography.

It is of note that much of the existing management strategy with regard to restoration of lowland raised mires is based on the hydrology of mires in their undisturbed state and associated with groundwater mound theory (Ingram, 1982). The relationship between hydrological conditions in an undisturbed mire and those within a cut-over mire, however, exhibit significant differences, as discussed by Eggelsmann *et al.* (1993).

1. The fragmentary nature of residual peat structures in a cut-over mire does not allow the creation of a groundwater mound in any recognisable form.
2. The rapid transfer of water through the ditch systems to the edge of the mire acts as a significant control on general water table levels within the mire.
3. As a result of the increased area in which rapid drainage is taking place, hydraulic gradients in the

peat are likely to be significantly greater than in an undisturbed system.

4. The drying of peat over time is likely to have increased the hydraulic conductivities to be found in the peat. Furthermore desiccation cracks within the peat will allow a far higher hydraulic conductivity than would normally be the case in an undisturbed mire. The increased heterogeneity in the hydraulic conductivity across the mire is of great significance where flow predictions are made, particularly if a distributed model is to be used.

The overall effect of these factors is to produce an increase in the heterogeneity of the hydrology of the mire. The mire is unlikely to act as a uniform hydrological unit; many distinct units may be discerned often with significant threshold changes or high gradients between them. The creation of a groundwater mound is therefore considered an unobtainable goal for most cut-over mires and furthermore, may not be appropriate to the satisfactory regeneration of the site.

The primary aim of the hydrological management of damaged and fragmentary mires is normally to minimise water loss through a strategy of ditch blockage or through some attempt at sealing the boundary of the mire to prevent the loss of water. Most attempts at restoration to date have concentrated their efforts within the boundary of the peatland area and often within the boundary designated for nature conservation, which may be considerably smaller than the original peat extent. Only in recent years have workers considered approaches using buffer zones outside the area of peat.

Techniques have been applied at a wide variety of scale and cost, with or without detailed monitoring to assess the effectiveness of the works. English Nature's Lowland Peatland Programme (LPP) funded the preparation of rehabilitation management plans of key sites and many County Wildlife Trusts and other bodies have also attempted such restoration, often on an *ad hoc* basis. The majority of attempts at hydrological management and restoration of raised bog mesotopes have focused upon reducing water loss from those outputs that are situated within the designated site boundary. This usually involves a programme of ditch blocking (Fojt, 1995), often in the absence of any associated hydrological monitoring (e.g. Sinnott *et al.*, 1999).

Ditches have a dual effect on peatland areas, lowering water tables though subsurface flow and increasing the total surface outflow from the site by shortening overland flow paths to the channel network. Boelter (1972) suggested that open ditches have little influence on the water table of moderately decomposed peat but in less decomposed peats the influence may extend to 50 metres from the ditch. In raised bogs the 50-m zone of influence has been confirmed by the authors at a raised mire site in Yorkshire, UK (Harding *et al.*, 1997).

If ditches are not maintained, changes over time will occur as they fill in with vegetation and sediment, losing their effectiveness in water removal (e.g. Fisher *et al.*, 1996). Indeed, this 'benign neglect' of ditches may be one of the simplest management strategies proposed to return raised mires towards favourable condition. Van Strien *et al.* (1991) even suggest that reduction in the frequency of ditch cleansing will have a beneficial effect upon species richness. However, Van Seters & Price (2001), working on a naturally regenerated cut-over bog in Quebec, found that *Sphagnum* had not re-established even after 25 years from abandonment of peat working. They concluded that, without suitable management such as ditch blocking, *Sphagnum* regeneration would be very slow to occur.

Many restoration projects have therefore concerned the reclamation of cut-over sites by means of deliberate ditch blockage (e.g. Mawby, 1995). At Wedholme Flow, Cumbria, UK, a strategy of small ditch blockages using either peat plugs with a polythene membrane or tin sheets was employed. These blockages were installed at a high density. Results suggested that variations in water table reflected changes in rainfall, with cut-over peat experiencing fluctuations of at least 500 mm on an annual basis and in some cases 700 mm. In the first year following damming, the peat was shown to become saturated during the winter months, but water levels still dropped considerably during the summer. Monitoring of peat anchors was used to show that the peat surface had risen following damming, but the cut-over areas at Wedholme continued to show a drop in

Table 7.2. *Mean water table levels before and after restoration, Transect 1, Wedholme Flow.*

	Pre-restoration mean water table level (m below ground surface)			Post-restoration mean water table level (m below ground surface)		
	North	South	All	North	South	All
Mean	−0.103	−0.295	−0.196	−0.056	−0.063	−0.063
Median	−0.069	−0.298	−0.147	−0.046	−0.0455	−0.047
Minimum	−0.627	−0.65	−0.65	−0.36	−0.582	−0.448
Maximum	0.09	0.023	0.09	0.109	0.217	0.217
Standard deviation	0.118	0.171	0.174	0.065	0.137	0.107

Source: Data of F. J. Mawby, after White & Butcher (1994).

the drier summer months whereas undisturbed bog showed only very slight fluctuations. More detailed analysis of Mawby's hydrological data was undertaken for English Nature by White & Butcher (1994). Transect 1 monitored an area of primary active bog and an area of abandoned cut-over bog which had been dammed. The southern half of the transect underwent damming from January 1992, and thus 'before and after' comparisons by White & Butcher used 31 December 1991 as a cut-off date.

The behaviour of the northern and southern ends of Transect 1 can be regarded in terms of an absolute difference between restored and primary bog, and in terms of the temporal context given by the four-year record. Summary data for before and after restoration conditions can be found in Table 7.2.

In absolute terms, relative to the ground surface at each dipwell, Fig. 7.2 illustrates the difference along the transect between water table depths before and after damming. It is clear from these data that the impact on water table depths on the southern side of Transect 1 has been spectacular. Mean water table depth has increased in all wells, though the rise in mean water table depth becomes less dramatic in the dipwells nearer the perimeter drain. It is clear that the restoration programme has had an impact on water table levels in this part of the mire.

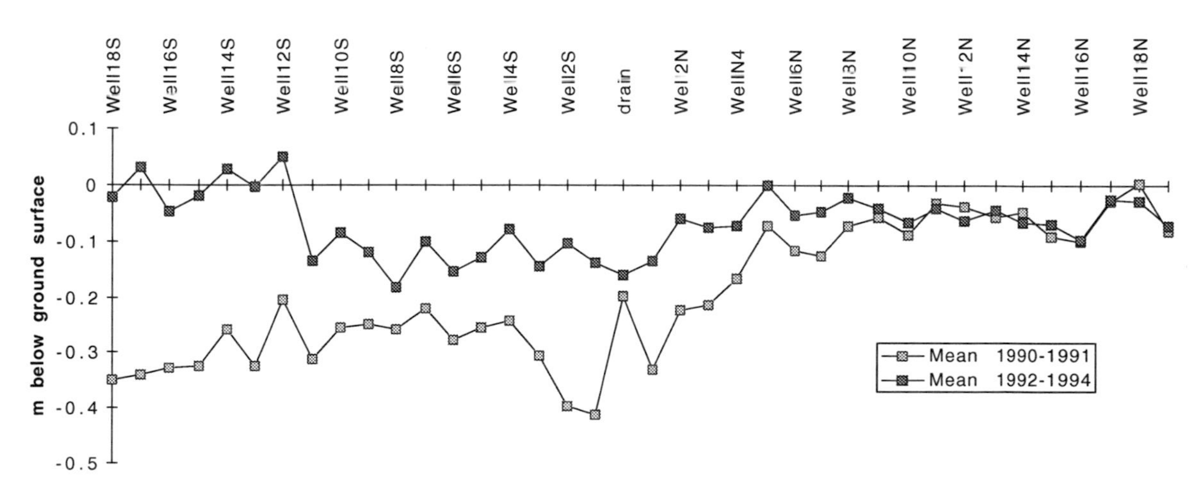

Fig. 7.2. Wedholme Flow, Transect 1, mean water table depth before and after damming. Data of F. J. Mawby, after White & Butcher (1994).

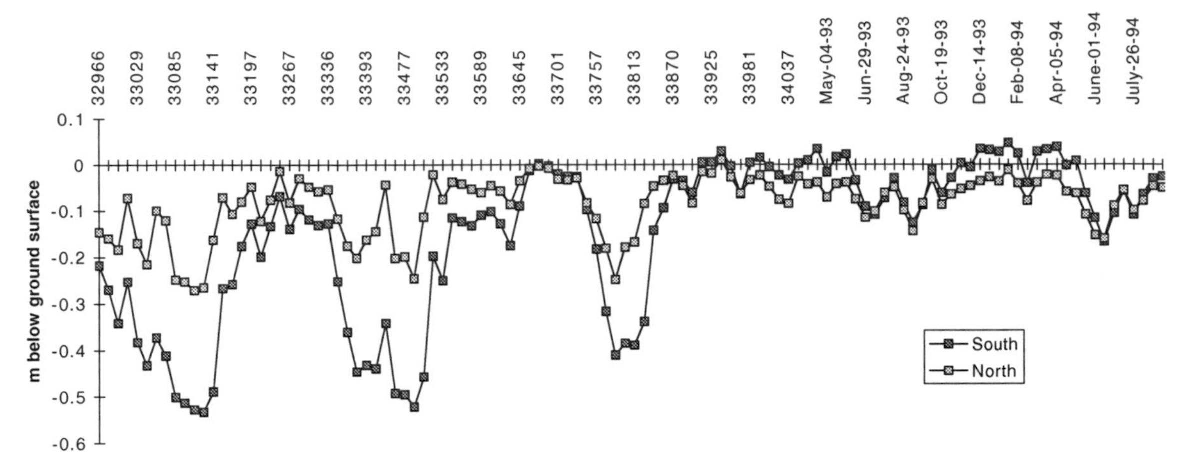

Fig. 7.3. Wedholme Flow, Transect 1, mean water table relative to ground surface, north and south, 1990–4. Data of F. J. Mawby, after White & Butcher (1994).

The influence of the restoration programme becomes much clearer when viewed in a temporal context. Figure 7.3 illustrates the behaviour of the northern and southern parts of Transect 1 over the whole sampling period. Both northern and southern dipwells experienced a cyclical fluctuation in water table depth, with maximum depths experienced in summer during relatively dry conditions, and minimum water table depth in the winter. Although both northern and southern dipwells show a similar pattern, the amplitude of variation in water table depth is much smaller in the undisturbed bog (north) than in the cut-over bog (south).

For the period before the commencement of damming, both sections of the transect appear to exhibit a slow decline in water tables from March to August/September, followed by a relatively faster rise to a stable winter level. The programme of restoration is given as commencing in January 1992, and the graph shows an almost immediate response with a high degree of correspondence between winter and early spring data for both sites. For the first summer after restoration commenced, as noted above, this correspondence broke down and water levels on the southern transect still experienced a much steeper decline than those of the northern transect. Despite this decline, southern water table levels did not fall back to minimum levels experienced in previous years (a minimum mean value of −0.412 m compared with −0.533 m and −0.523 m for 1990 and 1991 respectively), whereas water levels in northern dipwells fell to a level very close to those of the previous two years (a mean minimum of −0.236 m compared with −0.244 m and −0.231 m).

A more extreme hydrological intervention on more damaged sites where there has been extensive peat stripping may involve the creation of open water reservoirs to increase local water storage in a peatland area. LaRose *et al.* (1997) described the excavation of such reservoirs in a Canadian peatland (Lac-St-Jean, Quebec) and concluded that such reservoirs increased the local water storage, raised the water table to a higher and more stable level, and resulted in *Sphagnum* establishment on the experimental baulks.

Other intensive hydrological approaches to raise water table levels in cut-over mires have included the use of pump recharge systems, as described by Heathwaite (1995*a*) at Thorne Moors, Yorkshire, UK. A ditch recharge system was used in an attempt to control the hydraulic gradient from the periphery of the nature reserve to the active cutting area. This active approach, unlike ditch blockages, requires constant management and is not self-sustaining. Furthermore, Heathwaite records the failure of this system through seepage losses.

An additional method for reducing water losses from a peatland area which has been previously afforested is the removal of trees in an effort to reduce the levels of evapotranspiration. Brooks & Stoneman (1997) report such a restoration project at Langlands Moss, Scotland, where trees were removed by helicopter to leave a largely brash-free surface. A total of 150 dams were then installed, using a mix of plyboard for smaller dams and PVC piling for larger ones. This project was expensive (£3200 per hectare after timber sales) but was reported to be effective, with a notable increase in *Sphagnum* spp. cover. Other workers have carried out similar projects to reduce evapotranspiration on a more modest scale, by felling invasive tree cover such as birch.

The majority of attempts at hydrological management and restoration of peatlands described above focused upon reducing water loss from those outputs that are situated within the designated site boundary. Only relatively recently has there been awareness that it may also be necessary to look beyond the designated conservation boundary and to consider hydrological management of at least some parts of the surrounding land. Charman (1997) says that philosophy is changing, and that the theoretical basis for inclusion of such buffer zones is relatively well understood, but still difficult to apply in practice. The Joint Nature Conservation Council stated:

> 'Bogs must be protected at their margins from potentially damaging activities, especially those activities likely to cause hydrological disturbance by maintaining or increasing water run-off by artificial drainage' (Joint Nature Conservation Council, p. 16).

More specifically they state:

> 'Site boundaries must be chosen to include all land judged necessary to provide and maintain the hydrological functions needed to conserve the special features of the site' (Joint Nature Conservation Council, p. 17).

Schouwenaars (1995) discusses the selection of internal and external (buffer zone) water management options. He suggests that in hydrologically isolated bog remnants or cut-over bogs, efforts should focus on the selection of sites with good prospects for rewetting by internal measures. In areas with greater lateral and downward water losses, however different water management options should be analysed and evaluated prior to implementation, using hydrological models.

Loss of surface water by abstraction may also be an issue requiring restoration in natural lakes. Llyn Cwellyn (an oligotrophic mountain lake in Snowdonia National Park, North Wales, UK) has been proposed as a Special Area of Conservation (SAC) under the European directive on the conservation of natural habitats and of wild flora and fauna. Confirmation as a SAC would bring an obligation to maintain or restore habitat examples to 'favourable conservation status'. Llyn Cwellyn has maximum depth 36 m but the outflow has been regulated for water supply for the Caernarfon area – the top 2 m can be drawn off by gravity. Duigan *et al.* (1998) described concern that fluctuating water levels might have an adverse effect upon populations of littoral flora and fauna because they do not have time to adapt or migrate as environmental conditions alter, and that amplitude and timing of fluctuations may affect populations of fish such as trout (*Salmo trutta*) and particularly Arctic charr (*Salvelinus alpinus*). Consultation has therefore taken place between the water company and the statutory conservation body in relation to the use of the lake as a reservoir, and Countryside Council for Wales is consulted on applications for water abstraction.

Reversing the effects of sediment accumulation in lakes and reservoirs

A second common type of damage requiring ecological restoration is the accumulation of excessive amounts of sediment in a lake or reservoir. Labadz *et al.* (in press) have summarised issues regarding reservoir sedimentation.

Cooke *et al.* (1993) state that the first and most obvious step towards protection and restoration of a lake or reservoir is to limit, divert, or treat excessive nutrient, organic and silt loads. Nutrient loads are not dealt with here (see Søndergaard *et al.*, this volume) but silt loads, more correctly known as sediment loads, are a serious issue in the physical environment of many lakes and reservoirs. Some sedimentation is inevitable and lakes are famously known as 'ephemeral features in the landscape', with sedimentation providing an opportunity for ecological study (e.g. Oldfield, 1977) but many lakes and reservoirs globally are suffering ecological degradation as a result of excessive sedimentation. It might be argued that, in the strictest sense, reservoirs cannot be restored because they are by definition a relatively recent anthropogenic disturbance of the river system. However, some reservoirs certainly do experience ecological degradation, and may be restored to an earlier and more desirable condition. Oglethorpe & Miliadou (2000), for example, describe Lake Kerkini, a Ramsar site in Greece with rich and diverse flora and fauna. This flood protection and irrigation reservoir was constructed 1932, but in 1982 the operators built a larger dam because of excessive sedimentation. Over the last 15 years at least six of the commercial fish species have disappeared and several other important bird and mammal populations have declined.

Oglethorpe & Miliadou (2000) say that lakes require sustainable management, taking into account both economic interests and conservation management. It is suggested that this may be achieved by trapping sediment upstream before it can enter the water body, not just raising of lake water level with a bigger dam and attendant ecosystem losses each time.

Basson & Rooseboom (1999) discuss various options for reservoir sedimentation control and divide these into options which: (1) minimise sediment loads entering reservoirs, (2) minimise deposition of sediment within the reservoir basin, (3) remove previously accumulated sediment, and (4) replace lost reservoir capacity. Most of these techniques could equally be applied to natural lakes. The first obvious step to minimising sediment loads entering a reservoir would appear to be soil conservation techniques in the catchment. Workers such as Morgan (1995) have described approaches to soil conservation based upon agricultural practices and mechanical methods of reducing soil transport. It seems self-evident that the nature of the vegetation and land use in a catchment will have some impact on the sediment yield of rivers draining it, although this will not necessarily be demonstrable in a simple statistical analysis of the data because of the many other influencing factors (White *et al.*, 1996).

In Britain, Newson (1988) described some of the work of the Institute of Hydrology and discussed implications for upland land-use planning and land management. He included the two scenarios of land *allocation* (on the basis of sensitivity or capability maps) and *accommodation*, where land is managed by a combination of 'free' market forces and technical dialogue. He suggested that in practice a middle way was likely, with 'keep off' attitudes only prevailing for very sensitive sites or for persistent and deliberate contributions to the deterioration of upland water quality.

Many individual studies on British lakes and reservoirs have reported the effects of land-use change upon sedimentation rates. For example Rowan *et al.* (1995) provisionally linked a period of increased sedimentation in Abbeystead reservoir, Lancashire, UK, from 1930 to 1948 with agricultural land improvement in the drive to increase food production associated with the Second World War. Sedimentation in this period peaked at 373 t km^{-2} yr^{-1}, compared to only 78 t km^{-2} yr^{-1} for the previous 55 years. Heathwaite (1993) measured sedimentation in Slapton Ley, a natural lake in southwest England, and identified an increase in erosion since 1945 which was possibly associated with a post-war increase in the area of arable and temporary grassland. A major peak in sediment influx in 1987 (to over 1 g cm^{-2} yr^{-1}) was tentatively linked to recent conversion of permanent to temporary grassland, including ploughing of riparian areas.

Dearing *et al.* (1981) investigated sedimentation in Llyn Peris, North Wales, and linked a much increased rate since 1965 (approximately 41 t km^{-2} yr^{-1}, compared to less than 15 t km^{-2} yr^{-1} for most of the rest of the twentieth century) with dramatically increased sheep population in Snowdonia. It was suggested that overgrazing leads to a decrease in tree cover and an increase in peat erosion and stream channel erosion. The possible influences of tourism (trampling) and dust from quarrying were also mentioned. Van der Post *et al.* (1997) also discussed the influence of grazing pressure in a study of sedimentation at Blelham Tarn in the English Lake District. They plotted the annual sediment accumulation against the number of sheep on agricultural census returns for the two parishes adjacent to the lake and found an extremely close correspondence.

Several studies have considered the impact of afforestation upon sedimentation. Battarbee *et al.* (1985) compared non-afforested with recently afforested sites in Scotland and suggested that a 20-fold increase in sedimentation rates resulted, although they did acknowledge that this effect may only last for approximately ten years, until the forest canopy closes and drainage channels stabilise. Burt *et al.* (1984) and Francis & Taylor (1989) reported large increases in short-term stream sediment loads (2.5 to 4.8 times the previous values) on catchments ploughed prior to afforestation. Stott (1997) provides a helpful summary table of other studies in the literature, which again suggests that the short-term increase in suspended sediment yield may be associated with ploughing or felling rather than with established forest. Longer-term effects were also considered by Dearing (1992), who investigated sedimentation in Lyn Geirionydd, North Wales, UK, and concluded that afforestation in the twentieth century did not appear to have significantly increased sediment yields, whereas unvegetated spoil heaps from nearby mining did act as significant point sources. Foster *et al.* (1987) summarised data from 20 studies and indicated that recent maximum sediment yields under cultivation and moorland were noticeably higher than those found under forest, given the same catchment:lake volume ratio. More recently Foster & Lees (1999*a*, *b*) have undertaken studies of nine lakes and reservoirs in northern England as part of the Land–Ocean Interaction Study (LOIS) project and suggested that average post-1953 sediment yields from pasture, arable, moorland and forested catchments were respectively 13, 31, 29 and 13 t km^{-2} yr^{-1}.

Severe erosion of peat hags has created problems in areas where reservoirs have been built immediately downstream of the eroding blanket peat moorlands. The low density of peat means that there is relatively little in-channel storage (Labadz *et al.*, 1991) until flow velocities are dramatically reduced on entering a lentic water body. As a result of their high trap efficiencies, reservoirs provide an opportunity to assess longer-term average rates of erosion from the catchment. Labadz *et al.* (1995) summarised the results of a study of over 120 reservoirs in the southern Pennines, UK, many with blanket peat in at least part of the catchment area. Mean annual area-specific sediment yield was 124.5 t km^{-2} yr^{-1}, but the median value was 77 t km^{-2} yr^{-1}. Although these rates of erosion are low compared with those reported globally, they are relatively high for the UK (Newson, 1986). Moreover, as discussed above, peat has a low bulk density so that a small mass of erosion represents a large loss of material and, equally, a large volume of infill in the reservoir. The southern Pennine data base indicates a loss by infilling of 7.5% of the total original quoted capacity. Techniques for restoring eroded peat moorlands were discussed above. However, some recent workers such as Tallis (1998) have indicated that some gully erosion may be the natural end point to blanket peat accumulation and that, as such, it would not be appropriate to seek to restore all eroded areas.

If it is not practical or sufficient to reduce catchment erosion, the next logical step is to reduce the amount of sediment entering reservoirs by trapping it immediately upstream. Liu Chuin Ming (1985) described how sediment yields in Taiwan have been drastically reduced by the use of check dams. These have been found to fill completely with coarse material within two to three years, and thus sediment downstream is relatively fine. Similarly, Gong (1987) reported that sediment delivery in the Yellow River had been reduced by 48.7% as a result of a comprehensive programme of silt dam construction, compared with 8.7% reduction caused by detention in reservoirs. The beneficial impact of such measures as bywash channels and residuum lodges in the southern Pennines, UK, was demonstrated by White *et al.* (1996).

Dredging of reservoirs in the UK has usually only been undertaken where necessitated by engineering works (e.g. Appleton, 1976), but in many cases overseas it has been seen as a possible means to sustain reservoir capacity and thus prolong life (see Morris and Fan, 1998; Basson & Rooseboom, 1999). An unpublished report by Mott MacDonald, commissioned by a water company in northern England in 1995 and kindly made available to the authors, investigated the feasibility of options for desilting reservoirs. This concluded that desilting a reservoir full of water would produce dredgings with a very high moisture content, making disposal both difficult and expensive. Desilting of reservoirs whilst the water level is drawn down is certainly cheaper, although moisture content may still be a problem. It was recommended that pilot trials be considered to establish the feasibility of transporting the silt and if necessary to evaluate the relative benefits of possible dewatering techniques. These might include harrowing or heaping to encourage natural drainage, use of a mobile filter-belt plant (with or without chemical treatment) and use of a mobile centrifuge plant (with polyelectrolyte dosing). The costs of excavation, transport and disposal of sediment depend upon a large number of factors, many of which are site-specific. One major problem may be the identification of a local site suitable for disposal. If it is possible to use the material for capping and landscaping an existing landfill site, or to use it as an agricultural soil conditioner, then the costs of disposal will be much less than if the material is put into landfill and taxed as such. Transport of large volumes of material on minor roads may cause physical damage, noise and air pollution and may not be deemed acceptable in environmentally sensitive areas, National Parks, etc. The economics of desilting also depends upon the equivalent cost of alternative resource options. The report suggested that in 1995 the typical cost might be around £10 per m^3 plus potential landfill tax (at that stage likely to be another £2 per m^3, but now much higher). On the basis of preliminary calculations based upon a 50% reduction in sedimentation, specific recommendations were:

1. Clean out those existing silt traps (residuum lodges) which are currently full of sediment.
2. Consider the construction of new silt traps where the economic case is feasible.
3. Consider erosion control measures (gabion weirs, contour channels, tree planting) in rapidly eroding catchments where the water company owns the land or can reach agreement with the landowners.

One possibility not fully investigated in that report was the possible economic value of the sediment from the reservoir or from silt traps upstream. This was explored in Spain by Fonseca *et al.* (1998) who suggested that there was scope for reservoir sediments to be used as mineral fertilisers for agricultural soils. There is little published information available on such use in Britain to date, but a company in West Yorkshire (West Riding Organics) has for some years been removing sediment from reservoirs or from the sediment lodges just upstream for incorporation into potting compost marketed as 'Moorland Gold'. Trials reported by Griffiths (1993) of Harlow Carr Botanical Gardens showed promising results, with Moorland Gold Compost giving the largest final size plants in geranium, impatiens, nicotiana and petunia trials when compared to various coir-and peat-based products. Given the increasing pressure to conserve Britain's lowland raised bogs, the possibility of a useful alternative to peat composts which has the advantages of economic revenue and helping to maintain reservoir capacity must be worthy of further consideration.

Bruk (1985) summarised the views of an international panel of expert contributors and concluded that, in the long run, watershed management is the best way to reduce the yield of sediment and its entry into the reservoir. For large basins, however, this may be a slow and prohibitively expensive process. The construction of auxiliary check dams may have a quicker effect but these will in turn fill with sediment and so may not last long unless actively managed, which again increases the costs. Bypassing of sediment laden flows is another effective method recommended for consideration in the design stages of any project. Flood flushing and venting of turbid currents may prove effective means of reducing deposition in reservoirs but these depend on availability of suitable bottom outlets and an excess of water, and the environmental consequences downstream must be considered and may be prohibitive. Dredging of sediment deposits is a costly operation, which may be justified in certain circumstances by the economic value of the water and the impossibility of replacing lost reservoir capacity. Disposal of the excavated silt may also cause difficulties unless it can be used for the improvement of surrounding agricultural land.

What is clear is that changing sedimentation rates cannot (as has sometimes been the case) be considered only as evidence of changing sediment inputs, but that changing retention and management must also be evaluated (Labadz *et al.*, 1995). Catchment management is the only permanent solution to excessive lake or reservoir sedimentation. However, it may be a slow and expensive or even politically unacceptable option in some circumstances. The river 'transfer system' may also be used to minimise delivery, and operation of a reservoir itself may be seen as the final stage in the reduction of sedimentation. Pattinson *et al.* (1994) outlined the use of a holistic 'staged catchment management' approach whereby the catchment, the transfer network and the reservoir itself are seen as components in a system to minimise water treatment costs. It would seem appropriate that a similar framework be adopted for the management of lake and reservoir sedimentation, in such a way as to maximise water yield and minimise undesirable effects.

CONCLUDING REMARKS

Wetlands are complex systems where multiple processes operate in combination. A significant amount of work towards ecological restoration has taken place in wetlands and still waters, but a great deal of this work has been carried out on a pragmatic or even an *ad hoc* basis. This reflects the urgency of the requirement to protect important sites and the frequent shortfalls in available funding. Whilst there is a body of knowledge relating to the hydrological processes of still waters and wetlands, too often managers, through time and resource constraints, have been required to act with only a limited understanding of the functioning of their particular site.

Often, when ecological restoration is attempted, several interventions are employed at the same time. In addition, a particular restoration activity may have several disparate aims, as when sediment removal from a lake is intended to reduce phosphorus release and lower trophic status as well as to improve the physical environment of open water. Restoration work has often been completed with limited prior monitoring, and it has therefore been difficult to sustain scientific assessments for a sufficient time period in order to evaluate success (Carpenter & Lathrop, 1999) or to disentangle the precise effects of particular interventions. Burt (1994) stresses the importance of long-term observation of the natural environment as a basis for environmental policies.

Zedler & Callaway (1999) found little congruence between the predicted trajectories of recovery and those available detailed and sustained records of habitat development at one wetland restoration site. Many laudable results have been achieved by the hard work and detailed on-the-ground knowledge of managers such as Mawby (1995), but there remain many sites where restoration has been a hit-and-miss affair, where time and money has been wasted because the hydrological functioning of the system has been poorly understood.

ACKNOWLEDGMENTS

The opinions expressed are those of the authors and not of any other body. Support for research on peatlands, lakes and reservoirs has been received from NERC, the DETR (now DEFRA), English Nature, the Environment Agency, the Wildlife Trusts, Yorkshire Water Services and North West Water. Much of the initial work was carried out whilst JCL was at the University of Huddersfield. The work of Rachel Johnson, Alan Potter, Tim Burt, Paul White, Colin Knightley, Vicky Pattinson, Marcus Beasant, Julie Walker, Steve Pratt and others is gratefully acknowledged.

REFERENCES

Allen, A.O. & Feddema, J.J. (1996). Wetland loss and substitution by the Section 404 Permit Program in S California, USA. *Environmental Management*, **20**, 263–274.

Anderson, A.R., Pyatt, D.G. & White, I.M.S. (1995). Impacts of conifer plantations on blanket bogs and prospects of restoration. In *Restoration of Temperate Wetlands*, eds. B.D. Wheeler, S.C. Shaw, W.J. Fojt & R.A. Robertson, pp. 533–548. Chichester, UK: John Wiley.

Appleton, B. (1976). Silt-sucker cuts a dash 250m above o.d. *New Civil Engineer*, 2 December, 30–31.

Armstrong, A.C. (1995). Hydrological model of peatmound form with vertically varying hydraulic conductivity. *Earth Surface Processes and Landforms*, **20**, 473–477.

Armstrong, A.C., Caldow, R., Hodge, I.D. & Treweek, J. (1995). Re-creating wetlands in Britain: the hydrological, ecological and socio-economic dimensions. In *Hydrology and Hydrochemistry of British Wetlands*, eds. J. Hughes & A.L. Heathwaite, pp. 445–466. Chichester, UK: John Wiley.

Baird, A.J. (1997). Field estimation of macropore functioning and surface hydraulic conductivity in a fen peat. *Hydrological Processes*, **11**, 287–295.

Baird, A.J. & Wilby, R.L. (eds.) (1999). *Eco-Hydrology Plants and Water in Aquatic and Terrestrial Ecosystems*. London: Routledge.

Basson, G.R. & Rooseboom, A. (1999). *Dealing with Reservoir Sedimentation: Guidelines and Case Studies*, ICOLD Bulletin no. 115. Paris: International Commission on Large Dams.

Battarbee, R.W., Appleby, P.G., Odell, K. & Flower, R.J. (1985). ^{210}Pb dating of Scottish lake sediments, afforestation and accelerated soil erosion. *Earth Surface Processes and Landforms*, **10**, 137–142.

Boelter, D.H. (1972). Water table drawdown around an open ditch in organic soils. *Journal of Hydrology*, **15**, 329–340.

Boorman, D.B., Hollis, J.M. & Lilly, A. (1995). *Hydrology of Soil Types: Hydrologically Based Classification of the Soils of the United Kingdom*, Institute of Hydrology Report no. 126. Wallingford, UK: Institute of Hydrology.

Bosch, J.M. & Hewlett, J.D. (1982). A review of catchment experiments to determine the effect of vegetation changes on water yield and evaporation. *Journal of Hydrology*, **55**, 3–23.

Bragg, O.M. (1995). Towards an ecohydrological basis for raised mire restoration. In *Restoration of Temperate Wetlands*, eds. B.D. Wheeler, S.C. Shaw, W.J. Fojt & R.A. Robertson, pp. 305–314. Chichester, UK: John Wiley.

Bromley, J. & Robinson, M. (1995). Groundwater in raised mire systems: models mounds and myths. In *Hydrology and Hydrochemistry of British Wetlands*, eds. J.M.R. Hughes & A.L. Heathwaite, pp. 95–110. Chichester, UK: John Wiley.

Brooks, S. & Stoneman, R. (1997). Tree removal at Langlands Moss. In *Conserving Peatlands*, eds. L. Parkyn, R.E. Stoneman & H.A.P. Ingram, pp. 315–322. Wallingford, UK: CAB International.

Bruk, S. (rapporteur) (1985). *Methods of Computing Sedimentation in Lakes and Reservoirs*, Technical Report under Project IHP- II Project A 2.6.1. Paris: UNESCO.

Budelsky, R.A. & Galatowitsch, S.M. (2000). Effects of water regime and competition on the establishment of a native sedge in restored wetlands. *Journal of Applied Ecology*, **37**, 971–985.

Burt, T.P. (1994). Long-term study of the natural environment: perceptive science or mindless monitoring? *Progress in Physical Geography*, **18**, 475–496.

Burt, T.P. (1995). The role of wetlands in runoff generation from headwater catchments. In *Hydrology and Hydrochemistry of British Wetlands*, eds. J. Hughes & A.L. Heathwaite, pp. 21–38. Chichester, UK: John Wiley.

Burt, T.P. & Gardiner, A.T. (1984). Runoff and sediment production in a small peat-covered catchment: some preliminary results. In *Catchment Experiments in Fluvial Geomorphology*, eds. T.P. Burt & D.E. Walling, pp. 133–151. Norwich, UK: Geobooks.

Burt, T.P., Donohoe, M.A. & Vann, A.R. (1984). A comparison of suspended sediment yields from two small upland catchments following open ditching for forestry drainage. *Zeitschrift fur Geomorphologie*, **51**, (Suppl.), 51–62.

Burt, T.P., Heathwaite, A.L. & Labadz, J.C. (1990). Runoff production in peat-covered catchments. In *Process Studies in Hillslope Hydrology*, eds. M.G. Anderson & T.P. Burt, pp. 463–500. Chichester, UK: John Wiley.

Burt, T.P., Labadz, J.C. & Butcher, D.P. (1997). The hydrology and geomorphology of blanket peat: implications for integrated catchment management. In *Blanket Mire Degradation: Causes, Consequences & Challenges*, proceedings of a conference of the Mires Research Group of the British Ecological Society, University of Manchester, April 1997, eds. J.H. Tallis, R. Meade & P.D. Hulme, pp. 121–127. Aberdeen, UK: Macaulay Land Use Research Institute on behalf of Mires Research Group.

Butcher, D.P., Claydon, J., Labadz, J.C., Potter, A.W.R. & White, P. (1992). Reservoir sedimentation and colour problems in southern Pennine reservoirs. *Journal of the Institute of Water and Environmental Management*, **6**, 418–431.

Carpenter, S.R. & Lathrop, R.C. (1999). Lake restoration: capabilities and needs. *Hydrobiologia*, **395/396**, 19–28.

Centre for Ecology and Hydrology (1999). *Flood Estimation Handbook*. Wallingford, UK: Centre for Ecology and Hydrology (previously Institute of Hydrology).

Charman, D.J. (1997). Peatland palaeoecology and conservation. In *Conserving Peatlands*, eds. L. Parkyn, R.E. Stoneman & H.A.P. Ingram, pp. 65–73. Wallingford, UK: CAB International.

Childs, E.C. (1969). *An Introduction to the Physical Principles of Soil Phenomena*. Chichester, UK: John Wiley.

Conway, V.M. & Millar, A. (1960). The hydrology of some small peat-covered catchments in the North Pennines. *Journal of the Institute of Water Engineers*, **14**, 415–424.

Cooke, G.D., Welch, E.B., Peterson, S.A. & Newroth, P.R. (1993). *Restoration and Management of Lakes and Reservoirs*. Boca Raton, FL: Lewis Publishers.

Crisp, D.T. & Robson, S. (1979). Some effects of discharge upon the transport of animals and peat in a north Pennine headstream. *Journal of Applied Ecology*, **16**, 721–736.

Davis, M.B. & Ford, M.S. (1982). Sediment focusing in Mirror Lake, New Hampshire. *Limnology and Oceanography*, **27**, 137–150.

Dearing, J. (1992). Sediment yields and sources during the past 800 years. *Earth Surface Processes and Landforms*, **17**, 1–22.

Dearing, J., Elner, J.K. & Happey-Wood, C.M. (1981). Recent sediment flux and erosional processes in a Welsh upland catchment based on magnetic susceptibility measurements. *Quarternary Research*, **16**, 356–372.

Duigan, C.A., Allott, T.E.H., Monteith, D.T., Patrick, S.T., Lancaster, J. & Seda, J.M. (1998). The ecology and conservation of Llyn Idwal and Llyn Cwellyn (Snowdonia National Park, north Wales, UK)–two lakes proposed as Special Areas of Conservation in Europe. *Aquatic Conservation: Marine and Freshwater Ecosystems*, **8**, 325–360.

Egglesmann, R., Heathwaite, A.L., Grosse-Brauckmann, G., Küster, E., Naucke, W., Schuch, M. & Schweickle, V. (1993). Physical processes and properties of mires. In *Mires: Process, Exploitation and Conservation*, eds. A.L. Heathwaite & K.H. Gottlich, pp. 171–262. Chichester UK: John Wiley.

Environmental Consultancy of the University of Sheffield (2001). *West Midlands Meres and Mosses Conservation Plans*. Reports to English Nature and Environment Agency by ECUS (Environmental Consultancy of the University of Sheffield) in collaboration with Nottingham Trent University (J.C. Labadz) and University of East Anglia (M.R. Perrow). Sheffield, UK: Environmental Consultancy of the University of Sheffield.

Farah, H.O. & Bastiaanssen, W.G.M. (2001). Impact of spatial variations of land surface parameters on regional evaporation: a case study with remote sensing data. *Hydrological Processes*, **15**, 1585–1607.

Fermor, P.M., Hedges, P.D., Gilbert, J.C. & Gowing, D.J.G. (2001). Reedbed evapotranspiration rates in England. *Hydrological Processes*, **15**, 621–631.

Fisher, A.S., Podniesinski, G.S. & Leopold, D.J. (1996). Effects of drainage ditches on vegetation patterns in abandoned agricultural peatlands in central New York. *Wetlands*, **16**, 397–409.

Fojt, W.J. (1995). The nature conservation importance of fens and bogs and the role of restoration. In *Restoration of Temperate Wetlands*, eds. B.D. Wheeler, S.C. Shaw, W.J. Fojt & R.A. Robertson, pp. 33–48. Chichester, UK: John Wiley.

Fonseca, R., Barriga, F. & Fyfe, W. S. (1998). Reversing desertification by using dam reservoir sediments as agricultural soils. *Episodes*, **21**, 218–224.

Foster, I.D.L. & Lees, J.A. (1999*a*). Changing headwater suspended sediment yields in the LOIS catchments over the last century: a palaeolimnological approach. *Hydrological Processes*, **13**, 1137–1153.

Foster, I.D.L. & Lees, J.A. (1999*b*). Changes in the physical and geochemical properties of suspended sediment delivered to the headwaters of LOIS river basins over the last 100 years: a preliminary analysis of lake and reservoir bottom sediments. *Hydrological Processes*, **13**, 1067–1086.

Foster, I.D.L., Dearing, J.A., Grew, R. & Orend, K. (1987). The sedimentary database: an appraisal of lake and reservoir sediment based studies of sediment yield. In *Erosion Transport and Deposition Processes*, IAHS Publication no. 189, pp. 19–43: Wallingford, UK: International Association of Hydrological Sciences.

Francis, I.S. & Taylor, J.A. (1989). The effect of forestry drainage operations on upland sediment yields: a study of two peat-covered catchments. *Earth Surface Processes and Landforms*, **14**, 73–83.

Gilman, K. (1994). *Hydrology and Wetland Conservation*. Chichester, UK: John Wiley.

Gilman, K. (1997). Bog hydrology: the water budget of the mire surface. In *Conserving Peatlands*, eds. L. Parkyn, R.E. Stoneman & H.A.P. Ingram, pp. 132–138. Wallingford, UK: CAB International.

Gong, S. (1987). The role of reservoirs and silt-trap dams in reducing sediment delivery into the Yellow River. *Geografisker Annaler*, **69**, 173–179.

Griffiths, P. (1993). *Peat-Free Multipurpose Compost Trials.* Harrogate, UK: Harlow Carr Botanical Gardens.

Grootjans, A. & van Diggelen, R. (1995). Assessing the restoration prospects of degraded fens. In *Restoration of Temperate Wetlands*, eds. B.D. Wheeler, S.C. Shaw, W.J. Fojt & R.A. Robertson, pp. 73–90. Chichester, UK: John Wiley.

Grosvernier, P., Matthey, Y. & Buttler, A. (1997). Growth potential of three *Sphagnum* species in relation to water table level and peat properties with implications for their restoration in cutover bogs. *Journal of Applied Ecology*, **34**, 471–483.

Hakanson, L. & Jansson, M. (1983). *Principles of Lake Sedimentology.* Berlin: Springer-Verlag.

Halcrow, W. (2000). *Sedimentation in Storage Reservoirs*, Final Report to Department of the Environment, Transport and the Regions, February 2001. http://www.defra.gov.uk/environment/water/rs/index.htm

Harding, R.J., Labadz, J.C. & Butcher, D.P. (1997). *Initial Report on Water Table Data from Swarth Moor, July 1996 to February 1997*, Report to English Nature. Huddersfield, UK: University of Huddersfield.

Harrison, D.L., Driscoll, S.J. & Hughes, P.A. (1997). Benefits of quality control and correction of radar estimates of precipitation. In *Proceedings of the 6th National Hydrology Symposium*, University of Salford, September 1997, pp. 6.13–6.20. Wallingford, UK: Institute of Hydrology.

Heathwaite, L. (1993) Catchment controls on the recent sediment history of Slapton Ley, South-West England. In *Landscape Sensitivity*, eds. D.S.G. Thomas & R.J. Allison, pp. 241–259. Chichester, UK: John Wiley.

Heathwaite, L. (1995*a*). Problems in the hydrological management of cut-over raised mires, with especial reference to Thorne Moors, South Yorkshire. In *Restoration of Temperate Wetlands*, eds. B.D. Wheeler, S.C. Shaw, W.J. Fojt & A. Robertson, pp. 315–330. Chichester, UK: John Wiley.

Heathwaite, L. (1995*b*). The impact of disturbance on mire hydrology. In *Hydrology and Hydrochemistry of British Wetlands*, eds. J. Hughes, & L. Heathwaite, pp. 401–417. Chichester, UK: John Wiley.

Heathwaite, L. & Gottlich, K. (eds). (1993). *Mires: Process, Exploitation and Conservation.* Chichester, UK: John Wiley.

Hewlett, J.D. & Hibbert, A.R. (1967). Factors affecting the response of small watersheds to precipitation in humid areas. In *Forest Hydrology*, eds. W.E. Sopper & H.W. Lull, pp. 275–290. Oxford: Pergamon Press.

Hillel, D. (1971). *Soil and Water: Physical Principles and Processes.* New York: Academic Press.

Holden, J., Burt, T.P. & Cox, N.J. (2001). Macroporosity and infiltration in blanket peat: the implications of tension disc infiltrometer measurements. *Hydrological Processes*, **15**, 289–303.

Hughes, J. & Heathwaite, L. (1995) *Hydrology and Hydrochemistry of British Wetlands.* Chichester, UK: John Wiley.

Immirzi, C.P. & Maltby, E. with Clymo, R.S. (1992). *The Global Status of Peatlands and their Role in Carbon Cycling*, A Report for Friends of the Earth by the Wetland Ecosystems Research Group, Department of Geography, University of Exeter. London: Friends of the Earth.

Ingram, H.A.P. (1982). Size and shape in raised mire ecosystems: a geophysical model. *Nature*, **297**, 300–303.

Ingram, H.A.P. (1983). Hydrology. In *Ecosystems of the World*, Vol. 4A, *Mires Swamp, Bog, Fen and Moor, ED.* A.J.P. Gore, pp. 67–158. Amsterdam: Elsevier. Irish Peatland Conservation Council (1998). http://www.aoife.indigo. ie/~ipcc/index.html

Ivanov, K.E. (1981). *Water Movement in Mirelands.* London: Academic Press.

Joint Nature Conservation Committee (1994). *Guidelines for Selection of Biological SSSIs: Bogs.* Peterborough, UK: English Nature.

Jones, A.A. & Crane, F.G. (1984). Pipeflow and pipe erosion in the Maesnant experimental catchment. In *Catchment Experiments in Fluvial Geomorphology*, eds. T.P. Burt & D.E. Walling. pp. 55–72. Norwich, UK: Geobooks.

Kirkham, D. & van Bavel, C. H. M. (1948). Theory of seepage into auger holes. *Proceedings of the Soil Science Society of America*, **13**, 75–81.

Kneale, P. (1987). Sensitivity of the groundwater mound model for predicting mound topography. *Nordic Hydrology*, **18**, 193–202.

Labadz, J.C., Burt, T.P. & Potter, A.W.R. (1991). Sediment yield and delivery in the blanket peat moorlands of the Southern Pennines. *Earth Surface Processes and Landforms*, **16**, 255–271.

Labadz, J.C, Butcher, D.P., Potter, A.W.R. & White, P. (1995). The delivery of sediment in upland reservoir systems. *Physics and Chemistry of the Earth*, **20**, 191–197.

Labadz, J.C., Butcher, D., White, P. & Green, A.E. (in press). Siltation in surface storage reservoirs: issues, measurement and management. In *Siltation*, eds. I. Jefferson & M. Rosenbaum. London: Thomas Telford.

LaRose, S., Price, J. & Rochefort, L. (1997). Rewetting of a cut-over peatland: hydrologic assessment. *Wetlands*, **17**, 416–423.

Lerman, A., Imboden, D.M. & Gat, J.R. (1995) *Physics and Chemistry of Lakes*. Berlin: Springer- Verlag.

Liu Chuin Ming (1985) Impact of check dams on steep mountain channels in north-eastern Taiwan. In *Soil Erosion and Conservation*, eds. S.A. El-Swaify, W.C. Moldenhauer & A. Lo, pp. 540–548. Ames, IA: Soil Conservation Society of America.

Lloyd, J.W. & Tellam, J.H. (1995). Groundwater-fed wetlands in the UK. In *Hydrology and Hydrochemistry of British Wetlands*, eds. J. Hughes & L. Heathwaite, pp. 39–61. Chichester, UK: John Wiley.

MacAllister, C. & Parkin, G. (1999). Towards a whole-system model for the hydrology of peat mires. In *Patterned Mires and Mire Pools*, eds. V. Standen, J.H. Tallis & R. Meade, pp. 116–126. London: Mires Research Group of British Ecological Society.

Mahmood, K. (1987). *Reservoir Sedimentation: Impact, Extent and Mitigation*, World Bank Technical Paper no. 71. Washington, DC: World Bank.

Maltby, E. (1997). Peatlands: the science case for conservation and sound management. In *Conserving Peatlands*, eds. L. Parkyn, R.E. Stoneman & H.A.P. Ingram, pp. 121–131. Wallingford, UK: CAB International.

Marsh, T. & Bryant, S. (1991). 1988–1990: a ride on a hydrological rollercoaster. *Geographical Review*, **14**, 30–37.

Mawby, F.J. (1995). Effects of damming peat cuttings on Glasson Moss and Wedholme Flow, two lowland raised bogs in North-West England. In *Restoration of Temperate Wetlands*, eds. B.D. Wheeler, S.C. Shaw, W.J. Fojt & R.A. Robertson, pp. 349–358. Chichester, UK: John Wiley.

Mitsch, W.J. & Gosselink, J.G. (1993). *Wetlands*. New York: Van Nostrand Reinhold.

Morgan, R.P.C. (1995). *Soil Erosion and Conservation*, 2nd ed. Harlow, UK: Longman.

Morris, G.L. & Fan, J. (1998). *Reservoir Sedimentation Handbook*. McGraw-Hill: New York.

Mountford, J.O. & Chapman, J.M. (1993). Water regime requirements of British wetland vegetation: using the moisture classifications of Ellenberg and Londo. *Journal of Environmental Management*, **38**, 275–288.

Newson, M.D. (1986). River basin engineering: fluvial geomorphology. *Journal of the Institute of Water Engineers and Scientists*, **40**, 307–324.

Newson, M.D. (1988). Upland land use and land management. In: *Geomorphology and Environmental Planning*, ed. J.M. Hooke, pp. 19–32. Chichester, UK: John Wiley.

Newson, M.D. (1997). Plantation forestry: a sustainable resource. In *Proceedings of a Discussion Meeting, Institution of Chartered Foresters*, University of York, April 1996, ed. H.G. Miller, pp. 63–77. Edinburgh: Institution of Chartered Foresters.

Oglethorpe, D.R. & Miliadou, D. (2000). Economic valuation of the non-use attributes of a wetland: a case-study for Lake Kerkini. *Journal of Environmental Planning and Management*, **43**, 755–767.

Oldfield, F. (1977) Lakes and their drainage basins as units of sediment-based ecological study. *Progress in Physical Geography*, **1**, 460–504.

Pattinson, V.A., Butcher, D.P. & Labadz, J.C. (1994). The management of water colour in peatland catchments. *Journal of the Institute of Water and Environmental Management*, **8**, 298–307.

Penman, H.L. (1956). Evaporation: an introductory survey. *Netherlands Journal of Agricultural Science*, **4**, 9–29.

Price, M. (1996). *Introducing Groundwater*. London: Chapman & Hall.

Proctor, M.C.F. (1995). The ombrogenous bog environment. In *Restoration of Temperate Wetlands*, eds. B.D. Wheeler, S.C. Shaw, W.J. Fojt & R.A. Robertson, pp. 287–304. Chichester, UK: John Wiley.

Reeve, A.S., Siegel, D.I. & Glaser, P.H. (2000). Simulating vertical flow in large peatlands. *Journal of Hydrology*, **227**, 207–217.

Robinson, M. & Newson, M.D. (1986). Comparison of forest and moorland hydrology in an area with peat soils. *International Peat Journal*, **1**, 49–68.

Rochelle, B.P., Stevens, D.L. & Church, M.R. (1989). Uncertainty analysis of runoff estimates from a runoff contour maps. *Water Research Bulletin*, **25**, 491–498.

Rowan, J.S., Goodwill, P. & Greco, M. (1995). Temporal variability in catchment sediment yield determined from repeated bathymetric surveys: Abbeystead Reservoir, UK. *Physics and Chemistry of the Earth*, **20**, 199–206.

Schouwenaars, J.M. (1995). The selection of internal and external water management options for bog restoration. In *Restoration of Temperate Wetlands*, eds. B.D. Wheeler, S.C. Shaw, W.J. Fojt, & R.A. Robertson, pp. 331–346. Chichester, UK: John Wiley.

Shaw, E.M, (1994). *Hydrology in Practice*. 3rd edn. London: Chapman & Hall.

Sinnott, D., Butcher, D.P. & Gateley, P.S. (1999). *Hydrological Evaluation of Three Raised Mires*, Research Contract no. BAT/97/98/70, Report to Scottish Natural Heritage. Preston, UK: University of Central Lancashire.

Sinnott, D., Labadz, J.C. & Butcher, D.P. (in press). Assessment of the hydrological condition of raised mire systems using differential GPS to acquire high quality topographic data. *Earth Surface Processes and Landforms*, **27**.

Stott, T. (1997). A comparison of stream bank erosion processes on forested and moorland streams in the Balquhidder catchments, central Scotland. *Earth Surface Processes and Landforms*, **22**, 383–399.

Sullivan, S.A. (1996). DGPS and GIS improve lake sedimentation survey procedures. In *Proceedings of the International Conference on Reservoir Sedimentation*, Fort Collins, September 1996, eds M. L. Albertson, A. Molinas, & R. Hotchkiss, pp. 255–262. Fort Collins, CO: Colorado State University.

Tallis, J.H. (1998). Growth and degradation of British and Irish blanket mires. *Environmental Review*, **6**, 81–122.

Thornthwaite, C.W. (1944). A contribution to the report of the committee on transpiration and evaporation, 1943-1944. *Transactions of the American Geophysical Union*, **25**, 686–693.

van der Post, K.D., Oldfield, F., Haworth,E.Y., Crooks, P.R.J. & Appleby, P.G. (1997). A record of accelerated erosion in the recent sediments of Blelham Tarn in the English Lake District. *Journal of Palaeolimnology*, **18**, 103–120.

van Seters, T.E. & Price, J.S. (2001). The impact of peat harvesting and natural regeneration on the water balance of an abandoned, cut-over bog, Quebec. *Hydrological Processes*, **15**, 233–248.

van Strien, A.J., van der Berg, T., Rip, W.J. & Strucker, R.C.W. (1991). Effects of mechanical ditch management on the vegetation of ditch banks in Dutch peat areas. *Journal of Applied Ecology*, **28**, 501–513.

von Post, L. & Granlund, E. (1926). *Södra Sveriges Tortillangar I.* °Arsb. **19**, Series C, no. 335. Stockholm: Sveriges Geologiska Undersokning.

Ward, D.E., Hirons, G.J. & Self, M.J. (1995). Planning for the restoration of peat wetlands for birds. In *Restoration of Temperate Wetlands*, eds. B.D. Wheeler, S.C. Shaw, W.J. Fojt & R.A. Robertson, pp. 207–222. Chichester, UK: John Wiley.

Ward, R.C. & Robinson, M. (1990). *Principles of Hydrology*, 3rd edn. New York: McGraw-Hill.

Wetzel, R.G. (1983). *Limnology*. 2nd edn. Philadelphia, PA: W.B. Saunders.

Wetzel, R.G.(1999). Plants and water in and adjacent to lakes. In *Eco-Hydrology: Plants and Water in Terrestrial and Aquatic Environments*, eds. A.J. Baird & R.L. Wilby, pp. 269–299. London: Routledge.

Wheeler, B.D. (1995). Introduction: restoration and wetlands. In *Restoration of Temperate Wetlands*, eds. B.D. Wheeler, S.C. Shaw, W.J. Fojt & R.A. Robertson, pp. 1–18. Chichester, UK: John Wiley.

Wheeler, B.D. & Shaw, S.C. (1995*a*). *Restoration of Damaged Peatlands*. London: HMSO.

Wheeler, B.D. & Shaw, S.C. (1995*b*). Plants as hydrologists? An assessment of the value of plants as indicators of water conditions in fens. In *Hydrology and Hydrochemistry of British Wetlands*, eds. J. Hughes, & L. Heathwaite, pp. 63–82. Chichester, Uk: John Wiley.

White, P. & Butcher, D.P. (1994). *Hydrological Monitoring of Rehabilitation Work on Lowland Peatland NNRs: An Evaluation*, English Nature Research Report no 72. Peterborough, UK: English Nature.

White, P., Labadz, J.C. & Butcher, D.P. (1996). The management of sediment in reservoired catchments. *Journal of the Chartered Institute of Water and Environmental Management*, **10**, 183–189.

Winter, T.C. (1995). Hydrological processes and the water budget of lakes. In *Physics and Chemistry of Lakes*, eds. A. Lerman, D.M. Imboden & J.R. Gat, pp. 37–62. Berlin: Springer-Verlag.

Zedler, J.B. & Callaway, J.C. (1999). Tracking wetland restoration: do mitigation sites follow desired trajectories? *Restoration Ecology*, **7**, 69–73.

8 • Running water: fluvial geomorphology and river restoration

MALCOLM D. NEWSON, JOHN PITLICK AND DAVID A. SEAR

INTRODUCTION

Perhaps the overwhelming perception from fluvial geomorphologists in much of the developed world is that human modification to the basin sediment system now constitutes a major control on river morphology (Gregory, 1995). The term 'river training', once used by civil engineers when required by the modernist agenda to minimise the risk of flooding and erosion, sums up the morphological strictures placed on many (if not most) channels in Europe and North America. In the humid temperate parts of the developed world there has been no part of the channel network safe from 'channelisation' to protect human communities and the restoration effort is mainly characterised, as in the UK, by efforts to restore the 'natural' morphology directly, via a new kind of engineering works. However, in the semi-arid lands of the developed world, the engineering emphasis has been upon drought protection by dam construction to achieve 'river regulation' and it is via the profound changes to the flow regime that 'damage' to morphology – and hence physical habitat – has occurred. Logically, it is under an agenda of restoring an environmentally beneficial flow regime (see Downs *et al.*, volume 2) that fisheries managers and ecologists seek geomorphological guidance.

This chapter brings together experiences of restoration via direct morphological intervention (mainly in the UK) and via flow management (mainly in the semi-arid realm of the USA). In both climatic zones a third vital element of the in-stream physical habitat must be considered: the sediment supply from the catchment. There are many ways in which development processes have changed this supply and channels may be expected to react as a result. Furthermore, sediment-supply 'drivers' of channel response may be relatively localised in space and time, making transient behaviour more likely (in both space and time) and, therefore, proper assessment of the words 'natural' and 'damaged' more difficult. Nevertheless, empirical assessment of the balance between naturally driven and anthropogenically driven river behaviour (and restoration of the former where appropriate) is a major element of the international research effort in geomorphology.

The role of geomorphological concepts and information

Arguably, the science of river channel process and form – fluvial geomorphology – should be at the core of all proposals for stream restoration (see also Downs *et al.*, volume 2). However, there are obvious objections to this hegemony: within science, the aims of restoration are clearly ecological and the necessary precautions hydraulic, offering at least two other disciplines (biology and engineering) a key role. In practical terms, too, restoration involves community vision, political expediency, financial assessment and engineering management (Kondolf & Downs, 1996), none of which relates directly to the often academic skills of the fluvial geomorphologist! As a result, interdisciplinary teams are at the core of most current and completed schemes of river restoration; the degree of geomorphological input may depend on the personality mix in such teams, or the extent to which a 'purist' view, e.g. of the restored river length nested within a dynamic context of flows and sediment fluxes, can be condoned or exploited for the site in question.

The geomorphological concept which most clearly embraces the needs and options for restoration is the basin sediment system, in which sources, transfers and sinks for sediments (Schumm, 1977) combine to produce characteristic forms which, in turn, become the physical basis of habitat: both when covered, to various depths, by flow and also when exposed as bars. The basic premise of fluvial geomorphology is that rivers adjust their morphology (width, depth, slope and planform) to carry the water and sediment supplied from the drainage basin; during the passage downstream the calibre of the bed material declines as channel gradient decreases. Collectively, these changes allow the river to maintain a mass balance of energy and material. The concept of a continuum in fluvial processes is the basis of traditional models in both geomorphology (Leopold & Maddock, 1953) and aquatic ecology (Vannote *et al.*, 1980).

However, in the field, the pattern of source–transfer–sink is only partly mappable on to the gradually declining gradient and bed material calibre from mountains to ocean (Sear, 1996). Instead, a combination of historical legacies (e.g. glaciation, changing sea-level) and extreme flow variability produces patchy or repeated morphological sequences in all but low energy reaches. Thus, in recent years fluvial geomorphologists have tended to move away from simplified statements of channel properties evolving downstream (such as stream order and 'hydraulic geometry': Park, 1977) to develop a process-based understanding of river dynamics within stream segments and reaches (see Frissell *et al.* [1986] for terminology). Simultaneously, however, there has been considerable effort in the field of morphological channel classification by linking and merging the information gathered at individual sites and sections (e.g. National Rivers Authority, 1996; Rosgen, 1996). There is now healthy rivalry between geomorphological restoration designs based upon (apparently) universal river channel classifications (i.e. by restoring the 'appropriate' channel type for a location), and those based upon process measurements or the application of physically based theories or empirical equations (Kondolf & Downs, 1996).

Unlike its neighbouring sciences in the modern interdisciplinary teams that manage river restoration, fluvial geomorphology lacks major national and international schemes of data gathering. Practical problems of measurement (e.g. of bed material loads and of morphological inventories at an appropriate, normally small, scale) make primary survey an almost universal requisite to provide sufficient geomorphological information. The latter situation is, however, improving as techniques involving remote sensing and terrain modelling become more sophisticated and as empirical surveys become financed on a broader scale (for recent developments in the United Kingdom, see Newson *et al.*, 2001).

Pattern, scale and 'damage': a geomorphological view

A major difficulty in prescribing geomophological inputs to restoration schemes is the relative rarity of collaborative research between geomorphologists and freshwater ecologists. Whilst certain geomorphological principles and parameters (e.g. stream order) enter the major system-wide models in freshwater ecology, it is only with the recent rise of 'ecohydraulics' that any concerted effort to reconcile concepts and scales of investigation have been made. In the face of dire river ecology management dilemmas, simplistic hydraulic models of physical habitat have emerged. In principle, their use is an apparently successful strategy, but they have been heavily criticised by ecologists and geomorphologists alike. However, interdisciplinary schemes of basic physical habitat assessment have been slow to develop and it has, as yet, proved virtually impossible to exploit the potential of the hierarchical model of habitats (Frissell *et al.*, 1986) in practical schemes of restoration (or flow management).

Geomorphologists, in 'purist' mode, tend to demarcate very carefully the semantics of the broad 'restoration' agenda. The large historical component of geomorphological assessment, ranging over time-scales from centuries to millennia, demands an analysis of the word 'restoration' (Brookes & Shields, 1996; and see below). Geomorphologists begin their assessments of restoration potential from three standpoints:

1. The concept of 'damage' to river channels and floodplains must be assessed against an analysis of the flow regime and sediment source–transfer–sink system of the particular basin. In other words, they must be empirically assessed against a theoretical

background but also utilising techniques of environmental reconstruction of past channels and floodplains to help formulate a 'vision' of restoration.
2. The context of restoration, normally 'at-a-site' (in the tradition of civil engineering responding to community desires) must be basin-scale and must extend laterally from the channel to include the floodplain and valley floor. It must also anticipate future channel dynamics in the light of developments in catchment land use and water management, and in the context of climate change. Both spatial and historical analyses are essential (Sear, 1994; Kondolf & Larson, 1995).
3. The restored morphology for a reach (whether achieved by flow or form modifications) must be expected to be dynamic and to respond to both intrinsic and extrinsic changes; fluvial morphology is often transient in nature as it responds to, perhaps, distant and long-term signals of this sort.

In detail, the concept of 'damage', essential to restoration strategies and designs, gains expression in fluvial geomorphology in a variety of ways:

- From flow manipulations which distort the spatial or temporal regime of water level variation in relation to key form elements
- Flow manipulations which distort the broad spatial or temporal workings of the sediment system, both in-channel and in relation to the floodplain, particularly through lateral and vertical channel change (depending on local dynamics)
- Flow manipulations which impact on the detail of river bedforms such as the sorting of sediment sizes, both laterally and vertically
- Direct 'river training' to create artificial planforms, sections and dimensions which relate to society's conventional development needs of the river (e.g. flood protection)
- Sediment-related 'maintenance' which tends to distort channel dimensions and reduces the diversity of sediment sizes and forms at all scales
- The sediment impacts of catchment and river management, particularly of dam construction and sediment trapping
- A variety of secondary impacts from changes to the vital ecotones between channel and floodplain, notably the riparian vegetation zone.

Clearly each type and source of damage has very specific impacts on different members of the biotic community at different life stages but ecologists can often make broad assessments in relation to a geomorphological assessment of change in physical habitat. It is also important to stress that the impact of many forms of geomorphological damage is temporary, that is, recovery may occur over a variety of time-scales, particularly in channels with sufficient flow energy and substrate material to reactivate basic geomorphological processes (Brookes & Sear, 1996). It is the authors' view, therefore, that restoration schemes which focus on assisted natural recovery are likely to be most cost-beneficial and sustainable.

RELATIONSHIPS BETWEEN FLOW AND SEDIMENT TRANSPORT: GEOMORPHOLOGICAL TOOLS FOR RESTORATION

It is inappropriate here to present a lengthy treatise on sediment transport. However, particularly in major rivers impacted by reservoir flow regulation (such as the Colorado River in the western USA, Box 8.1), an understanding of the relationship between channel discharge and the mobility of channel sediments is fundamental to proposals for environmentally beneficial flow release strategies (Downs *et al.*, volume 2) from dams. Such an understanding is also highly relevant to direct restoration of the 'natural' morphology of smaller, unregulated channels (e.g. in the UK) because reconciliation of the desired morphology with the operative in-channel processes secures a sustainable design; failure to do so results in wash-out, or sedimentation and failure (Seal *et al.*, 1998).

Basic processes of sediment transport: habitat implications for flow-based restoration

The following discussion of sediment transport processes is arranged in a hierarchical order that roughly mimics the scale of ecosystem processes, from local-level to reach-scale concerns. Among the local-level (grain-scale) processes we emphasise the need for accurate information on flow hydraulics and sediment properties, both of which affect estimates of streambed mobility and sediment transport rates. Among the reach-scale processes we emphasise the importance of maintaining a mass balance in the total sediment load, which, if modified, may lead to widespread changes in macro-scale morphology and complexity of a river channel.

Techniques for estimating rates and modes of sediment transport continue to evolve as new information on fluid-sediment interactions becomes available. However, a number of basic equations are applicable, provided allowances are made for boundary conditions in the field, e.g. local controls on sediment supply. Perhaps the most useful and fundamental sediment transport relation is the Shields' parameter:

$$\tau^* = \frac{\tau}{(\rho_s - \rho)\, g\, D} \qquad (8.1)$$

where τ^* is the dimensionless shear stress, t is the near-bed shear stress, ρ_s and ρ are the densities of sediment and water, respectively, g is the gravitational acceleration, and D is the grain size. Equation 8.1, when applied locally at the grain scale, represents a force balance between the fluid drag and lift acting on a particle versus the weight force or frictional resistance provided by the particle and its surrounding grains. When applied at the reach scale, Eq. 8.1 may be considered as an index of sediment-transport intensity, rather than a true force balance between the flow and the grains on the bed. Application of Eq. 8.1 at either of these scales requires certain information and assumptions, and appropriate formulation of the individual terms, especially τ and D.

The Shields' parameter (Eq. 8.1) has a distinct advantage over other measures of flow strength, such as mean flow velocity or unit stream power, in that it explicitly accounts for the effect of bed material sediment size on substrate mobility. Typically, portions of the streambed that experience high velocity and shear stress – riffles, for example – will contain the coarsest particles, whereas areas of low velocity and shear stress will be floored by finer particles. The arrangement of fine and coarse particles is not often perfect, however, and thus there is a tendency for bed mobility to vary both spatially and temporally. This is illustrated in Fig. 8.1, which presents field data showing how the active portion of the bed surface of gravel rivers changes with increasing Shields stress. Research on the ecological effects of substrate mobility indicates that some aquatic communities benefit from low to moderate levels of disturbance, whereas high levels of disturbance and instability can reduce the density of benthic insects, and crush or sweep away incubating salmonid embryos. A key goal in the collaborative efforts between geomorphologists and ecologists is, therefore, to characterise the effects of spatial

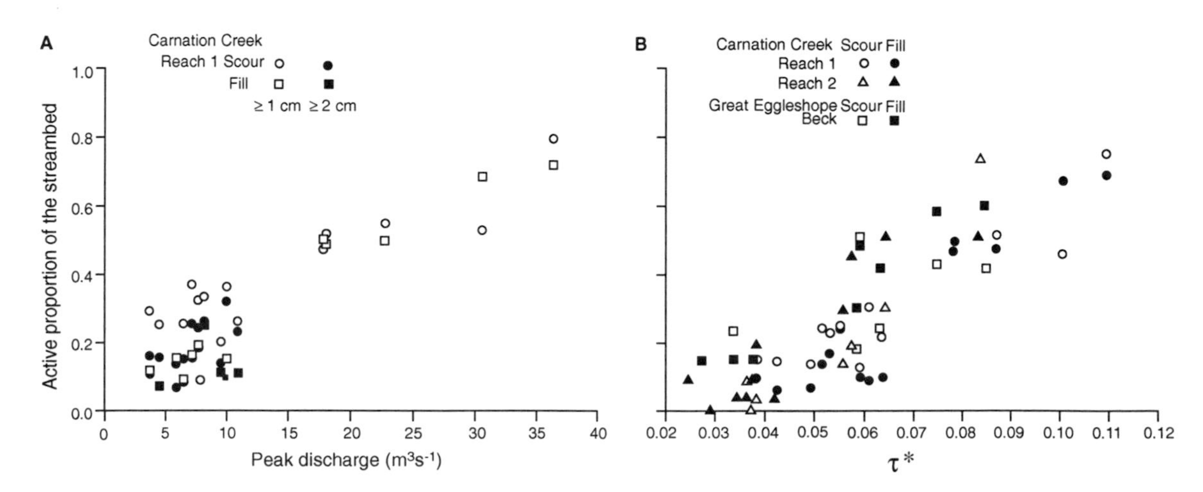

Fig. 8.1. Change in the active proportion of the bed surface with (A) increasing discharge and (B) dimensionless shear stress. From Haschenburger (1999).

differences in streambed mobility at scales that are ecologically meaningful, not just to one species or taxon, but to a full suite of aquatic organisms.

Sediment transport phases

The transport and deposition of sediment in rivers may affect habitats used by fish and benthic invertebrates in several ways, some desirable, others not. In gravel-bed rivers, periodic movement of the framework grains is necessary to winnow fine sediment from the bed to maintain spawning habitat or interstitial void space (Kondolf & Wilcock, 1996). In other rivers, the goal is to prevent fine sediment (silt and sand) from settling on the bed (Milhous, 1998), or to fully suspend the material and deposit it elsewhere. Whatever the objective, empirically derived relations for τ^* can be used to infer different phases of sediment transport, ranging from initial motion to significant motion to full suspension. The initial motion phase denotes the onset of bedload transport, also defined as the critical dimensionless shear stress, τ_c^*. In this phase, very few particles on the streambed are moving and bedload transport rates are very low (Wilcock & MacArdell, 1993; Andrews, 1994). This phase is nonetheless important for maintaining micro-habitats in gravel-bed rivers because it marks the point at which coarse particles first start moving. The lower limit for this transport phase is a matter of some debate. Values of τ_c^* derived from flume experiments and field observations suggest that the threshold ranges by a factor of more than 2, from $\tau_c^* = 0.030$ to 0.070 (Buffington & Montgomery, 1997). The wide range in reported values of τ_c^* is somewhat unsettling, but partly explained by differences in sediment properties (sorting), measurement techniques, and interpretation. Practitioners faced with the problem of estimating a transport threshold in the field must, nonetheless, choose a value of τ_c^*. To avoid an arbitrary choice, the threshold for motion can be estimated empirically by tracking the displacement of tagged particles or by sampling the material in motion with a bedload sampler. Neither of these techniques is foolproof, but field observations of τ_c^* can provide some basis for comparison with laboratory-derived values.

The second transport phase, significant motion, or complete mobilisation, is characterised by near-continuous movement of most particles on the bed. A lower limit for this phase is not very well established. Pitlick & Van Steeter (1998) reasoned that significant motion should occur at flows approaching the bankfull discharge, since these flows shape the channel and therefore rework most of the bed (Box 8.1). Using data from gravel-bed reaches of the Colorado River near Grand Junction, Colorado, they found that the bankfull discharge produced an average τ^* of 0.047, which is about 1.5 times the commonly cited threshold value of $\tau_c^* = 0.030$.

The third transport phase involving full suspension occurs at Shields stresses much higher than those noted above. Such conditions may occur often in rivers that carry silt and very fine sand, but they are infrequent in rivers that carry very coarse sand; sand in the intermediate size ranges may move either as bedload or suspended load, thus for a broad range of rivers it becomes important to distinguish the transport mode. As discharge and flow strength increase, particles are suspended higher and higher in the flow, resulting in a more uniform distribution of sediment concentration. The latter process makes it possible to move fine sediment on to higher elevation surfaces, such as bar tops or floodplains, that might serve as habitat for semi-aquatic species.

Box 8.1 Restoration of channel habitats damaged by regulated flows: the Colorado River

BACKGROUND INFORMATION

The interior region of the western United States is characterised by a semi-arid to arid climate, with annual runoff averaging only 10–20% of precipitation (Lins *et al.*, 1990). The region is drained by relatively few large rivers, most of which have been dammed and diverted to satisfy the water resource needs of an ever-expanding population. The Colorado River is among the largest and most regulated of these rivers. Dam construction in the Colorado River basin took place over a period of about 40 years, beginning in the mid-1930s; the total reservoir storage capacity in the basin now amounts to about three years of annual flow. Dams on the main stem of the Colorado River and its principal tributaries (the Green River and San Juan River) have significantly altered the flow of water and sediment through the basin, and it has become increasingly apparent that dam operations (storage and hydro-power generation) are seriously affecting the ecological integrity of the entire Colorado River system. Among the many different impacts, much of the present concern focuses on populations trends of four endangered fish: the Colorado pikeminnow (*Ptychocheilus lucius*), razorback sucker (*Xyrauchen texanus*), humpback chub (*Gila cypha*) and bonytail (*Gila elegans*). Historic accounts indicate that these species were once plentiful in the Colorado River basin; however, in recent years their populations have declined to very low and potentially non-sustainable levels. Water-management operations have contributed to the decline of native fishes by altering the timing and magnitude of peak flows, and by reducing the amount and quality of in-channel habitat (van Steeter & Pitlick, 1998).

The downstream effects of reservoir regulation in the Colorado River basin vary widely depending on the geographic location of the dam (or series of dams) with respect to major water and sediment sources. Runoff carried by the Colorado River is derived mostly from snowmelt in forested high-elevation basins of the central and southern Rocky Mountains. The river then flows long distances across the Colorado Plateau, through sparsely vegetated terrain that yields little additional runoff, but large amounts of sediment (Iorns *et al.*, 1965; Andrews, 1986; van Steeter & Pitlick, 1998). The available suspended sediment data indicate that concentrations of suspended sediment at main-stem gauging stations have not changed appreciably in the last 50 years (van Steeter & Pitlick, 1998), suggesting that unregulated tributaries are supplying sediment more or less as they have in the past. However, since flows are now regulated, both rivers have lost some of their capacity to carry sediment, resulting in 30–50% decreases in annual suspended sediment loads (Andrews, 1986; van Steeter & Pitlick, 1998). Related studies of channel geomorphology document 20–30 m decreases in main-channel width in response to these decreases in sediment load (Andrews, 1986; van Steeter & Pitlick, 1998; Allred & Schmidt, 1999).

The nature of the channel changed following the period of reservoir construction from the late 1930s through the 1960s (Fig. B8.1). Pre- and post-reservoir channel patterns were reconstructed by comparing different sets of black-and-white aerial photographs covering about 150 km of the river centred on the area around Grand Junction, Colorado. The outlines of visible features such as channel banks, islands, secondary channels and backwaters were digitised and adjusted to a common scale to calculate changes in surface area. These data were then transferred to a geographic information system (GIS) for final analysis. The analysis indicates that, since the late 1930s, the average width of the main channel of the Colorado River has decreased by about 10%, while the surface area of side channels and backwaters has decreased by 25–30% (van Steeter & Pitlick, 1998). The changes in channel morphology coincide with a distinct period from the late 1950s to the late 1970s when annual suspended sediment loads decreased by 40–60% because of reservoir construction and operation (van Steeter &

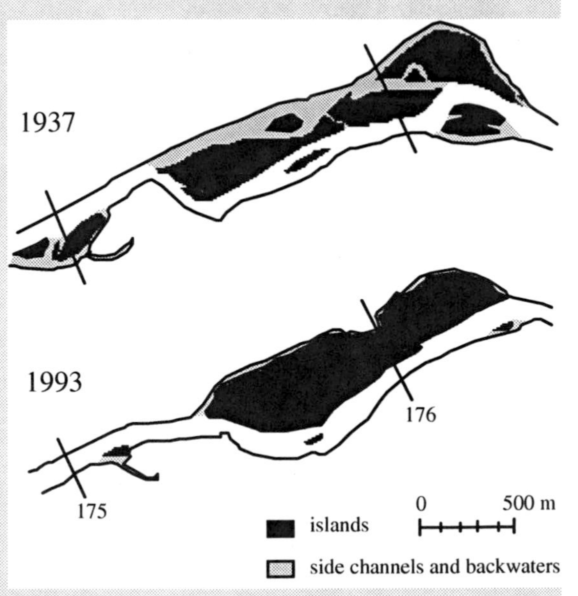

Fig. B8.1. Digitised channel maps of a section of the Colorado River showing historical changes in channel complexity. Markers indicate mileage upstream from the Green River confluence.

Pitlick, 1998). Clearly, significant amounts of sediment would have been deposited in the river during this time, with a disproportionate amount finding its way into backwaters and side channels where flow depth and velocity are typically low. Loss of these important habitats, when combined with changes in water quality and the introduction of non-native fishes, has potentially had long-lasting impacts on the native fish community of the Colorado River.

RESTORATION STRATEGY

The present situation requires a restoration strategy that extends well beyond the single species, single life-stage approach. The Colorado pikeminnow is the top predator in the system, but these fish require different resources and habitats throughout their life history. For example, adult pikeminnow spawn by broadcasting their eggs over riffles and submerged gravel bars, and spawning success is apparently enhanced where the substrate consists of loose, open-framework particles with deep interstitial voids. Thus, one of the restoration objectives is to specify 'flushing' flows (Kondolf & Wilcock, 1996) that will mobilise the gravel substrate and prevent silt and fine sand from accumulating on the bed. Presumably these flushing flows will benefit other members of the food chain, including smaller forage fish, and benthic invertebrates. Another set of restoration objectives is geared toward maintaining channel complexity and habitat heterogeneity. Adult pikeminnow show a preference for morphologically complex river reaches where flooding and lateral migration maintain a mosaic of features, including mid-channel bars, islands, secondary channels and backwaters. Habitat heterogeneity is thought to be essential for maintaining diversity in aquatic communities; however, as noted above, the historic trend in the Colorado River is towards a more simple channel. Possible solutions for reversing this trend are discussed below, but the problem of channel narrowing and simplification, so common to regulated rivers in the western USA, clearly arises as a consequence of changes in the mass balance of sediment, caused here by reductions in peak flows. Maintaining the mass balance of sediment is fundamentally important, because sediment that is not transported through a given reach of river will be deposited somewhere, resulting in further channel simplification and lower habitat heterogeneity.

APPLICATION OF SEDIMENT TRANSPORT THEORY TO RESTORATION

Common applications of the Shields and associated equations include evaluation of the transport thresholds discussed above, and calculation of sediment transport rates for particular flows. For these purposes, it is often necessary to form a continuous relation between discharge, Q, and τ^*. Figure B8.2 shows an example of such a relation, derived from a series of field measurements in a 500-m gravel-bed reach of the Colorado River, near Grand Junction, Colorado. These data show that the relation between Q and τ^* is non-linear, with the Shields stress increasing rapidly in the range from low to intermediate discharges, and more slowly thereafter. Based on this relation, and field evidence of gravel transport, we estimate that a flow of 500 $m^3\ s^{-1}$ – about half the bankfull discharge – is required to reach the threshold for initial motion ($\tau^* = 0.030$). Hydrologic data from a streamflow gauge at this site indicate that this discharge is exceeded about 30 days per year, on average. The bankfull discharge of 1050 $m^3\ s^{-1}$ produces a Shields stress of 0.047 (about 1.5 times the threshold value), and is exceeded approximately 5 days per year. Additional calculations of sediment load indicate that roughly 50% of the annual sediment load is carried by flows in this range. On the basis of these results, and similar results from other

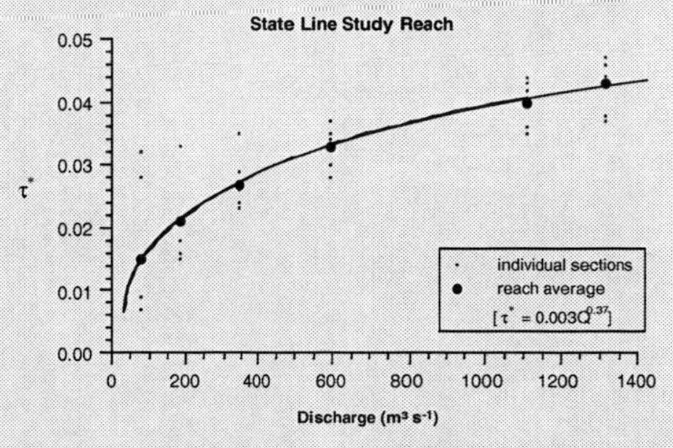

Fig. B8.2. Relation between discharge and dimensionless shear stress for a reach of the Colorado River near Grand Junction, Colorado.

reaches, Pitlick & van Steeter (1998) provided a series of recommendations to the US Fish and Wildlife Service for flows that perform important geomorphological and ecological functions, including flushing of interstitial fines and maintaining the mass balance of sediment.

Our approach for solving the more vexing problem of channel complexity derives from recent theoretical work on the formation of stable alluvial channels (Parker, 1978*a*,*b*, 1979; Ikeda *et al.*, 1988; Cao & Knight, 1998). This work incorporates some of the same sediment transport concepts noted earlier in predicting the stable width and depth of a channel that carries a finite sediment load. However, rather than a stable channel, we seek one that is somewhat unstable and laterally active. To increase lateral activity and channel complexity, we proposed that the river must be allowed to widen, which will, in turn, create the space for new bars and islands to form. A straightforward method for predicting conditions required to initiate channel widening is provided in a series of papers by Parker (1978*a*, *b*, 1979). Parker's analysis yields a result that the bankfull width and depth of a gravel channel are adjusted to a shear stress that is high enough to transport the load, but not so high as to erode the banks; theoretically that the bankfull Shields stress, τ_b^*, should be about 20% higher than the τ_c^*.

Pitlick & van Steeter (1998) extended Parker's theoretical reasoning to the case of the Colorado River by taking evenly spaced measurements of bankfull width and depth along a length of 90 km. These data were used with corresponding measurements of the reach-averaged slope and surface grain size to derive individual values of the τ_b^* for 58 cross-sections. The results, shown in Fig. B8.3, indicate that τ_b^* varies from 0.027 to 0.072 (less than a factor of 3), and averages 0.047 for the reach as a whole. The consistency in τ_b^* reflects the combined effect of a downstream decrease in slope and grain size, coupled with a proportional increase in bankfull depth (Pitlick & van Steeter, 1998). These data show rather clearly that gravel-bed reaches of the Colorado River are adjusted to a bankfull Shields stress that is, on average, about 1.5 times the critical Shields stress. This result implies that, if the slope and grain size of a gravel-bed river are known or specified in advance, the equilibrium channel geometry can be predicted well from Parker's theory. Perhaps more importantly, the consistent trend in τ_b^* provides broad-based support for site-specific flow recommendations, giving ecologists and water managers some assurance that prescribed flows derived from detailed analyses such as those discussed will achieve the desired effects on a widespread basis.

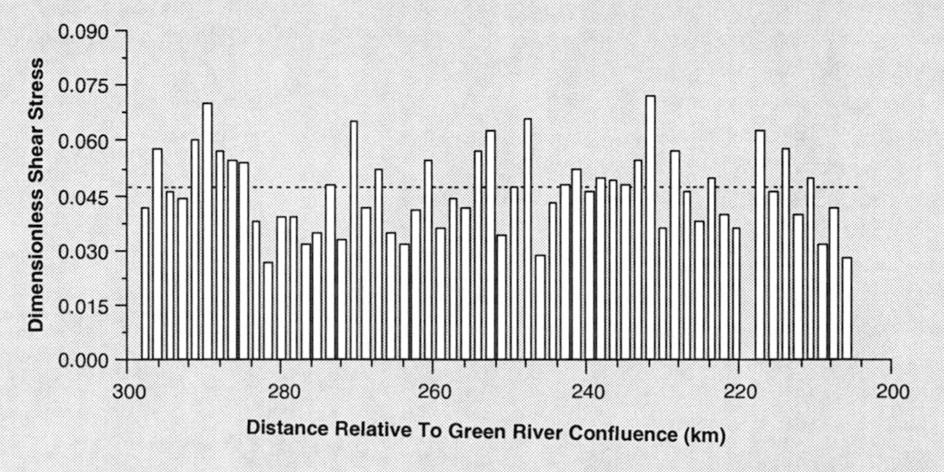

Fig. B8.3. Downstream trends in the bankfull dimensionless shear stress of the Colorado River. These data are derived from alluvial reaches that contain critical habitats for the endangered Colorado pikeminnow. From Pitlick & van Steeter (1998).

RESTORING CHANNEL DIMENSIONS AND FEATURES: THE UK EXPERIENCE

The relatively small size of most elements of the river channel network in the UK ('creeks' on the scale of the Colorado) and the fact that settlement and infrastructure reach far into rural catchments have resulted in both extensive and profound modifications to the 'natural' in geomorphological terms. Analysis of data from the recent River Habitat Surveys (RHS) indicates that 65% of a randomly selected survey net of 6000 sites showed extensive modification via dredging, revetment and other interventions to reduce out-of-bank flooding (National Rivers Authority, 1996; Raven *et al.*, 1996, 1998). Erosion control structures are less common: outside the uplands, low rates of sediment supply in relation to available stream power have normally not created 'wild' rivers in terms of morphological change. However, management for flood control, together with impacts from river regulation (similar to, but smaller in scale than, those in Colorado) have created an expensive problem of routine sediment management along 35 000 km of channels in England and Wales alone (Brookes *et al.*, 1983). The geomorphological input to both strategy and projects, therefore, attempts to raise awareness of the precise impacts of management ('What's been lost?' questions), the design of replacement or mimic fluvial features ('Where? How large?') and the sustainable long-term maintenance of restored or rehabilitated channels, given that most will also be expected to carry floods of greater than bankfull return periods ('What else needs doing, or undoing?').

Contexts for UK river restoration

The river restoration situation in the UK can, to date, be considered something of a chaotic concept (despite the successes of the River Restoration Project, now the River Restoration Centre; see below). Financial opportunism for schemes with a conservation or fisheries (often fishing) aim, together with the severe habitat damage perpetrated during the 'high period' of interventionist engineering, have meant that a very wide range of strategies and operations has qualified as 'restoration' and projects have been rapidly completed at a number of sites, mainly in the lowlands. Fig. 8.2 illustrates the 'spotty' nature of the movement as a whole and excludes a myriad of localised, informal schemes.

For more than 15 years geomorphologists in the UK have been benefiting from collaboration with the formal conservation movement in its broad desire to reduce the loss of in-channel and corridor habitats during traditional river management. For example, in the production of the *Rivers and Wildlife Handbook* (Lewis & Williams, 1984) a chapter was included on 'River processes and form' (Newson, 1984); the successor volume (Ward *et al.*, 1994) raised the sophistication of the geomorphological input with a chapter on 'River morphology and fluvial processes' (Newson & Brookes, 1994).

The river restoration movement, working through the River Restoration Project (RRP: Perrow & Wightman, 1993; Holmes & Nielsen, 1998; Vivash *et al.*, 1998) has rapidly achieved a very influential position in river management policies, partly by carrying out two prestige schemes (on the Rivers Cole and Skerne) and partly by providing guidance on such central issues as environmentally acceptable ways of controlling bank erosion (River Restoration Centre, 1999).

The RRP achieved a major impact on two lowland channel lengths (not reaches, *sensu stricto*), both 2 km long, in the catchments of the Skerne (250 km^2) and the Cole (129 km^2) (Kronvang *et al.*, 1998). Neither length has a high stream power and it might be claimed that the longer-term 'stability' of restored channel designs has not yet been tested in risky conditions (see Brookes [1990] and Brookes & Sear [1996] for discussions of stream power approaches to channel adjustment). However, partly as the result of weak/variable bed and bank materials, adjustments in the restored channel of the River Cole (and downstream impacts) have both been significant (Sear *et al.*, 1998).

The RRP pioneered the use of two forms of geomorphological survey (and indeed of post-project appraisal); these formalised procedures resulted from a research and development programme sponsored by the National Rivers Authority and

Fig. 8.2. Map showing river restoration and rehabilitation sites in the different Environment Agency regions of England and Wales, and Northern Ireland. From Sear et al. (2000).

subsequently the Environment Agency in England and Wales. Both the Skerne and the Cole were given Catchment Baseline Surveys and Fluvial Audits (described in Table 8.1) before works began (Kronvang *et al.*, 1998), but channel dynamics and structures on both streams remained within the engineering field. Thorough geomorphological survey (catchment, corridor and channel scales) and hydraulic treatment of rehabilitation schemes, such as those on the River Waveney (Newson *et al.*, 1999) (Box 8.2) and on the River Idle (Downs & Thorne, 1998) may now, however, become a norm, especially following recent fatal and damaging flood events in the English lowlands.

Table 8.1. *Geomorphological procedures recommended by the Environment Agency, 1998*

Stage	Planning	Project		
Procedure	Catchment Baseline Study	Geomorphological Fluvial Audit	Geomorphological Dynamic Assessment	Channel Design
Aims	Overview of the basin sediment system and morphology	To suggest sustainable, geo-morphologically based options for problem sites	To relate reach processes and rates to morphology, allowing management of either	To design channels within context of the basin system and local processes
Scale	Often a gauged catchment (modal size 100–300 km^2)	Those reaches identified in Catchment Base-line Study or by managers	Problem or project reach	Project length
Methods	Data collation, including River Habitat Surveys; consultation; reconnaissance fieldwork at key points throughout catchment	Detailed field studies of sediment sources, sinks, transport processes, floods and land-use impacts	Field survey of channel form and flows; hydrological and hydraulic data	Quantitative description of dimensions and location of features, substrates, revetments
Core information	Characterisation of river lengths and sediment management problems	Identifies range of options and 'potentially destabilising phenomena'	Sediment transport rates and morphological stability/trends; 'regime' approach where appropriate	The 'appropriate' features and their dimensions within a functionally designed channel
Outputs	5–10 page report; maps or GIS; field forms and photos	Maps at 1:10000; time chart; report; recommendations	Quantitative guidance to intervention (or not) and predicted impacts on reach and beyond	Plans, drawings, tables and report suitable as input to quantity surveying and engineering costings
Destination	Local Environment Agency Action Plans (LEAPs); Feasibility studies for rehab/restoration	Investment/ management staff or policy forums	Engineering managers and project steering groups	Funded projects of flood defence, erosion protection, rehabilitation or restoration
Follow-up	Geomorphological Post-Project Appraisal			

Selecting and designing morphologies: to classify or compute?

The smaller scale of UK channels and the relatively smaller impact of regulation (compared with semi-arid lands) has promoted restoration of the morphological elements of the channel and riparian ecotones. Early policy guidance to improve degraded channels came from landscape architecture, implying that 'gardening' channels, their margins and structures was required (Water Space Amenity Commission, 1980, 1983). This phase predates the formal movement to

Box 8.2 Geomorphological inputs to a lowland rehabilitation scheme: the River Waveney, East Anglia, UK

A number of channels in the Anglian Region (Eastern England) of the Environment Agency have undergone rehabilitation schemes, e.g. the River Wensum (Newson *et al.*, 1997), Harper's Brook (Ebrahimnezhad & Harper, 1997) and the River Deben (Kemp *et al.*, 1999).

The River Waveney drains a catchment of 889 km^2, 670 km^2 of which is non-tidal. The catchment is of low relief and everywhere is below 100 m AOD. The main channel profile is generally virtually flat with an average non-tidal gradient of 1:2250; engineering has reduced this further in many places (e.g. between Billingford and Earsham the gradient drops to 1:5500). Many of the tributary channels are, in contrast, significantly steeper, lack control structures and actively transport fine sediment produced on the surrounding catchment (e.g. from roads and intensive farming). Catchment-scale issues are very important in assessing the sustainability of UK schemes of restoration and rehabilitation. Amongst the terms of reference for the Waveney geomorphological surveys, including Catchment Baseline, Fluvial Audit and Dynamic Assessment, were to describe and map:

- Features considered as typical of this type of river in this part of Britain
- Location of the segments/reaches suitable for works
- Potential threats posed by current sediment dynamics and channel/catchment management
- Design specification of the features selected
- Stability of the features once emplaced
- Influence of the features on physical habitat (flow types/biotopes).

The Catchment Baseline Study (CBS) identified more than 20 'lengths' based on channel character in the field and heavily influenced by the backwater conditions created by the many mill structures in the channel. The basic channel character of the Waveney is the result of the amount of available gradient (and therefore flow types/morphological features), local sediment sources and riparian tree cover (which in turn controls instream macrophyte growth). The more active tributaries were, however, divided into reaches (in the geomorphological sense) and the Fluvial Audit became an essential basis for a precautionary approach to catchment management after rehabilitation (see below).

INSTALLATION OF 'RIFFLES'

After consultation with angling interests the Fisheries function of Environment Agency Anglian Region decided that 'riffles' were appropriate and feasible target features for rehabilitation of the Waveney channel. It was not clear at this stage what physical aspect of in-channel habitat was to be recreated by 'riffles' (substrate, flow field, spawning, aeration), but the target fish species were dace (*Leuciscus leuciscus*) and chub (*L. cephalus*). The reason for using the term 'riffle' in quotation marks reflects the feeling amongst geomorphologists that it is only with more discipline in the use of terminology that the specific aims, objectives and measures of success for river 'restoration' actions will emerge more clearly for debate and decision-making. True riffles are major components of an active bed material transport process and their hydraulics reflect this; what was required on the Waveney (under the term rehabilitation, rather than restoration) was a series of mimic features based upon natural riffles.

To assist in the design of riffle spacing, the literature was reviewed and a new empirical equation derived from a data set of 85 separate streams covering the following range in variables: riffle spacing (17.1–1200 m), river bed slope (0.00093–0.0215), bankfull width (5.2–76.6 m). Figure B8.4 uses this data set to illustrate how riffle spacing increases as bed slope declines and how spacing increases with bankfull width. As channel gradients increase beyond the values covered in this data set, spacing reduces still further and riffles become replaced by steps irrespective of channel width. Riffle spacing may initially be predicted as follows:

$$\square_r = 7.36\, w^{0.896}\, S^{-0.03} \qquad r^2 = 0.67,\ p > 0.001 \qquad (8.2)$$

Where $\square_r$ is riffle spacing in metres, w is bankfull width in metres and S is the channel bed slope through the pool–riffle sequence.

a

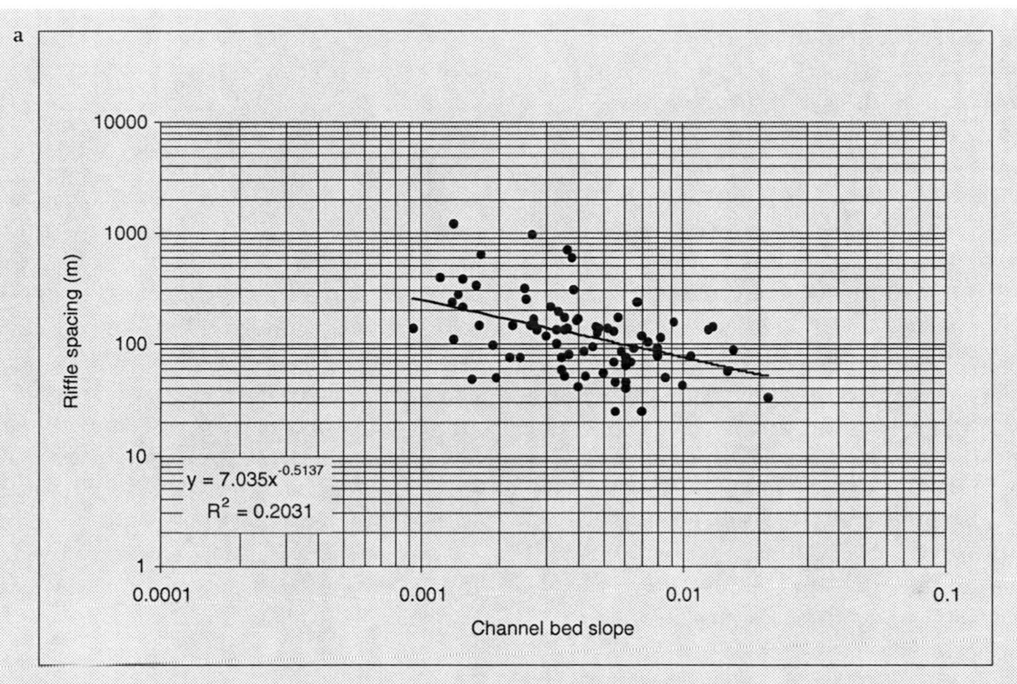

b

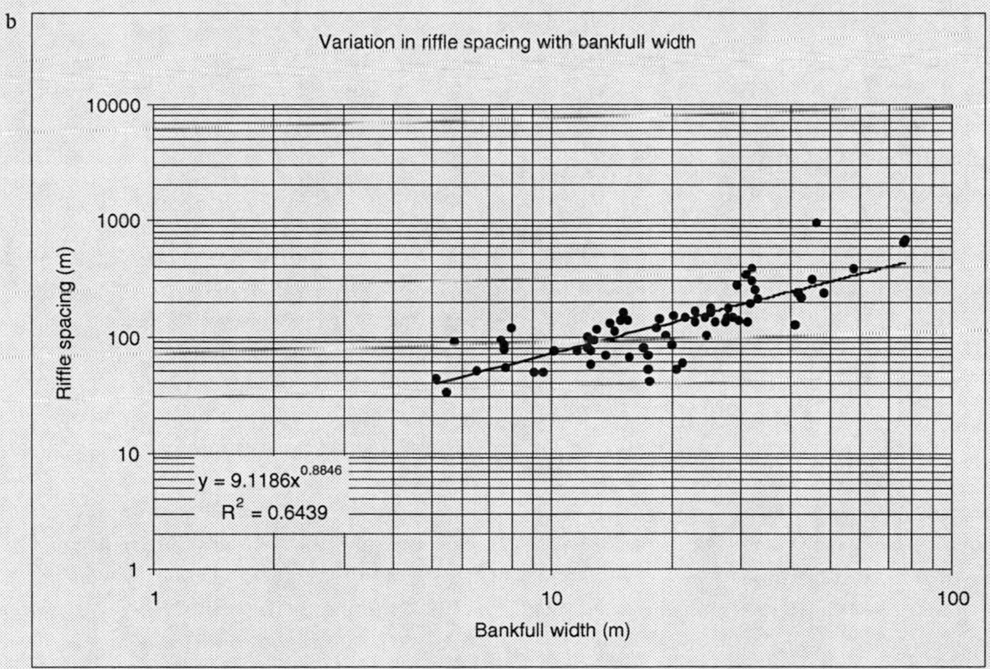

Fig. B8.4. Design guidance to riffle spacing from the geomorphological literature, based upon (a) channel bed slope and (b) bankfull channel width. After Newson *et al.* (1999).

Values of riffle amplitude are time-dependent as pools tend to fill with sediments and riffles tend to scour during floods, whilst both may fill when sediment transfer through a reach is increased. The scientific literature suggests that riffle bed-widths should be 7–16% wider on average than pools. Given the low gradients and the effect that this might have on conveyance, it was recommended that banks should be reprofiled at the riffle crests to provide a maximum crest width 15% greater than the reach average.

Bankfull stream power assessment provides guidance on the likelihood of erosional or depositional adjustments at each reach, based on proximity to an empirically derived threshold of 35 W m^{-2} (Brookes, 1990; Brookes & Sear, 1996; Sear, 1996). Above this threshold, sites may be expected to experience erosional adjustment (depending on boundary materials), below 10 W m^{-2} then depositional adjustment may be expected. Shear stress calculations indicate that gravels of intermediate diameters up to 47 mm may be transported under bankfull flow conditions; however generally material above 5–10 mm would be stable. This being so, it was recommended that the gravel be composed principally of material of the order of 10–20 mm with smaller proportions of larger material. However, a compromise was needed between this guidance, the local availability of materials and the needs, for spawning, of chub and dace, for which there is little specific guidance. Cowx & Welcomme (1999) recommend, for dace, an optimum spawning substrate size of 30–350 mm, but neither dace nor chub create true redds for spawning (like salmonids: see Perrow *et al.*, this volume) and the cleanliness of the gravels may be paramount, rather than size. Both fish species breed best in flow velocities of 20–50 cm s^{-1}. Because fines are constantly entering the Waveney main channel at points close to some of the rehabilitation sites (and because large concentrations are available nearby) a trapping action by the new features is inevitable. Excess fine sediment accumulations may need to be removed by regular maintenance if silt or sand begins to cloak the 'riffles'. Cementation of a 'riffle' may be important from a fisheries point of view but should not affect hydraulic performance provided that sands do not completely cover the gravels and reduce the grain resistance of the surface at low flows.

IMPACT OF REHABILITATION ON FLOOD CONVEYANCE

To assess the influence of riffle rehabilitation on overall water surface elevations at low flows a further one-dimensional hydraulic modelling exercise was conducted using HECRAS (modification of HEC-2 US Army Corps of Engineers step backwater model) for four potential rehabilitation sites. HECRAS also indicates the effect on velocities and so can indicate the potential change in physical habitat conditions resulting from the rehabilitation proposals. HECRAS is a program formulated to determine longitudinal water surface profiles, based on solution of the one-dimensional energy equation with energy loss due to friction over a fixed-bed calculated using the Manning equation:

$$U = R^{2/3}S^{1/2}n^{-1} \tag{8.3}$$

where U is the section averaged velocity, S is the energy slope and, for sufficiently wide reaches, the hydraulic radius (R) is equal to average depth (d). In the absence of sudden and major changes in channel width, energy losses are accounted for by channel bed and bank roughness defined by Manning's n. The model is known to over-predict water surface elevation at low discharges and therefore water surface elevations are expressed as percentage increases on the modelled water surface elevations for existing conditions. The scale of the relative increase is therefore given for each 'riffle' rehabilitation option. Model runs were conducted for a range of scenarios using the survey and discharge measured in the field.

Conclusions from the HECRAS simulations were that much of the hydraulic adjustment resulting from bed elevation changes is taken up via velocity changes, a desirable outcome for rehabilitation. At low flows the minor increases in water surface elevation predicted by the model result from the accommodation of discharge by increased bed width, and by increased flow velocities. Given a functional objective of flow aeration generated by rough turbulent flow over the 'riffles' the model results seem encouraging.

However, at high flows, aeration is unlikely to be effective once the features are drowned out. A full hydraulic monitoring programme is in progress at one of the sites where the installation of 'riffles' has recently gone ahead.

CONCLUSIONS

One of the clearest geomorphological conclusions from the work carried out on this project is that there are relatively few active natural sources of sediment within the main stem of the Waveney channel: bank erosion and bed scour are highly localised and transport distances limited by the low stream power developed by the river in flood. At the same time, however, sediment transport (notably of sands and finer materials) is active in a number of tributaries. We also include under catchment management any alterations in routine channel maintenance protocols to maintain or protect the emplaced rehabilitation features. These will include:

- Desilting at some of the rehabilitation sites to permit a firm footing for the gravels
- Desilting a length of channel upstream of the features to delay the onset of infilling and cementation
- 'Ploughing' (or equivalent) of cemented gravels if this becomes a problem.

incorporate ecologically relevant surveys into river conservation management policy via River Corridor Surveys, River Habitat Surveys (Raven *et al.*, 1998) and System for Evaluating Rivers for Conservation (SERCON) (Boon *et al.*, 1998).

The River Habitat Surveys have provided, for the first time, a morphological database for UK rivers. At first it was thought possible to construct an objective channel classification which could serve to indicate deviations from 'natural' in geomorphological terms and also to indicate 'appropriate' restoration agendas. However, the conventional classification agenda proved frustrating (Newson *et al.*, 1998), with the result that the database is now considered to be more useful in providing guidance on the 'natural' features of UK channels in statistical space than in Cartesian space (Jeffers, 1998). The abandonment of a strategy based solely upon stream classifications of the type designed by Rosgen (1994, 1996) is perhaps appropriate at this time, given the warnings of Kondolf (1995) about the importance of local processes in geomorphological restoration. Work in Australia on 'river styles' (Brierley & Fryirs, 2000) is a robust response to this criticism and is purposefully aimed at restoration strategies, but it is likely that every physiographic context will produce a different response and reaction to river typologies.

Whilst a fuller measure of ecological integrity is not yet available (Harper *et al.*, 2000), the RHS database is also now proving valuable as the basis for simple scoring procedures which will indicate numerically the degree of damage to each surveyed site and thereby help to drive a national strategic approach to restoration targets. Strategic use of RHS data to guide a catchment approach to restoration has already been achieved for the River Sankey in northwest England (Environment Agency, 1999).

Lessons from UK river restoration to date: a geomorphological perspective

Undoubtedly, there are two major messages emerging from the UK experiences of channel restoration which are critical for geomorphological inputs to river management:

1. In many cases it will be impossible to return to a fully functioning sediment transfer system, even at the reach scale, and those features installed to improve habitat or amenity are likely, therefore, to need maintenance. In this way they are unsustainable.
2. Apparent costs of 'full' restoration (e.g. £100 000 per kilometre: Vivash et al., 1998) means that the evaluation of rivers for channel habitat quality and the active conservation of important fluvial geomorphological features in rivers are vital in future to ensure that such costs are avoided. Another implication is that the River Restoration Project's high costs of evaluation and monitoring must be reduced in future by use of simplified and widely approved science-based protocols.

Most purists (not only geomorphologists, but other natural scientists) would claim that the deficiencies of current public policy in relation to 'restoration' include:

- Insufficient attention to prior appraisal of sites and options
- Unwillingness to finance or to follow geomorphological protocols
- Domination of 'vision' in the qualitative sense, rather than 'design' in the technical sense
- Use of 'mimic' channel features, rather than the restoration of a functioning sediment transfer system
- Selection of demonstration reaches within catchments whose sediment and habitat dynamics remain unrestored
- Costs of the operations in relation to benefits and sustainability
- Lack of post-project monitoring to enable proper appraisal and to refine methodologies.

These words partly echo those of Kondolf (1998) in passing on the Californian experience with channel restoration. From the geomorphological research perspective, post-project monitoring is needed on the widest possible scale given that Sear *et al.* (1998) suggest that adjustments of both restored channels and downstream 'impact reaches' can be profound and unexpected by the general community; they point up an almost complete lack of adjustment models to cope with applied needs in this respect.

CONCLUDING REMARKS

From the above it is clear that the geomorphologist participating in an interdisciplinary river restoration project needs to place the following 'purist' concerns in the context of the project's aims and objectives (assuming that improved or conserved biodiversity lies within the aims and objectives):

1. That the scheme will utilise, where possible, the principles of 'assisted natural recovery' through a proper evaluation of 'damage' and appropriate location of the works.
2. That flow regimes through restored reaches will be appropriate to habitat requirements.
3. That the scheme will involve appropriate channel dimensions and profiles.
4. That it will encourage or conserve an appropriate suite of morphological elements/forms.
5. That it will promote the operation of a natural set of dynamic processes sufficient, if appropriate, to promote sediment erosion, transport and deposition (and therefore channel change).
6. That proposals for post-project maintenance involve proper understanding of the restored channel dynamics, the impacts of post-project adjustments (both in the restored sections and upstream/downstream) and the nature of stress to biota caused by regular channel works.
7. That the catchment context of the scheme will be assessed and catchment planning devices used to protect the investment (e.g. against siltation from upstream sources).

At this, strategic, point it is appropriate to briefly repeat the warnings of others about the culture and semantics (Brookes & Shields, 1996) of restoration schemes. Whilst the 'purist' geomorphological approach to restoration has the concerns listed above, geomorphological inputs are also relevant to less comprehensive ambitions such as 'rehabilitation' and 'enhancement'. The accompanying chapter by Downs *et al.* (volume 2) addresses these thematic problems.

We have polarised the contrast between flow-based restoration schemes in large US dryland rivers and form-based restoration in the smaller streams of the UK. It is, however, the interaction of flow and form and the resulting spatially and temporarily varied physical habitat which must be understood. In the absence of such understanding our case studies appear to reveal a managerial approach—tackling the worst excesses of previous flow modification or channel engineering first 'because they are there'. Nevertheless, the managerial agenda has limits: dams on the Colorado cannot produce a completely natural regime (unless they are demolished!) and in the flood-prone UK, most channels must contain flows of a higher return period than is natural. Thus, the science base needs to prepare for the

Table 8.2. *Pre-restoration and post-restoration distribution of physical habitats (as percentages of surface channel cover inferred from flow types) in the Rivers Cole and Skerne*

	Cole channel		Skerne channel	
Biotope	Pre-restoration	Post-restoration	Pre-restoration	Post-restoration
Glide	91	65	90.0	88
Deadwater	0.2	7	0	9
Pool	4.5	0	–	–
Run	3.7	12	6	3
Riffle	0	12	3	0
Mixed: boil/rapid	0.6	4	0.1	0

Source: After Kronvang *et al.* (1998).

inevitable interactive agenda—channels and flows managed together to optimise ecosystem benefits. As revealed in depth by Downs *et al.* (volume 2) collaborative approaches between the relevant sciences remain elusive, despite the arrival of terms such as 'ecohydraulics' and 'hydro-ecology'. The use of flow types in the UK's River Habitat Surveys represents the rapid incorporation of this hydraulic indicator resulting from the research efforts of fieldworkers such as Jowett (1993), Wadeson (1994) and Padmore (1998). Kronvang *et al.* (1998) illustrate the comparative performance of the two RRP schemes in diversifying the channel hydraulics in terms of physical biotopes indicated by flow type (Table 8.2). Such meso-scale approaches may, eventually, hold the key to the hierarchical scale switching which will be essential in reconciling site restoration with catchment management (Harper *et al.*, 1999; Newson & Newson, 2000).

The contemporary contribution from fluvial geomorphology to an improved understanding of river habitats is therefore that of an empirical understanding of a suite of erosional forms, a suite of depositional forms (both of particular relevance to freshwater ecology) and the broad way in which these may be typical of particular locations over evolutionary time-scales. Of equal importance is an understanding of degree and frequency to which these forms are covered by water, hydraulically 'drowned' and interact via backwater effects. The study of physical biotopes in New Zealand, South Africa and the UK is illuminating these relationships; currently proposed research will investigate the linkages between flow variability (measured from gauging station records at 15-minute intervals: Archer, 2000) and the spatial extent of biotopes (mapped in the field). The contribution of hydraulic habitat models such as the foremost physical habitat simulation model (PHABSIM) adds the critical assessment of the 'usable area' of channel open to biota in different life stages. It is also of note that the empirical understanding of the role of flow patterns (notably secondary flow cells which diverge from the main downstream movement of water) has become both of vital importance to geomorphological process studies and of relevance to the true nature of lotic and benthic environments. Co-ordination of such research in the next decade, matched with the increasing contribution from new technologies of measurement and monitoring, will greatly refine the restoration agenda without, hopefully, making it completely rational and free from 'vision'.

REFERENCES

Allred, T.M. & Schmidt, J.C. (1999). Channel narrowing by vertical accretion along the Green River near Green River, Utah. *Geological Society of America Bulletin*, **111**, 1757–1772.

Andrews, E.D. (1986). Downstream effects of Flaming Gorge Reservoir on the Green River, Colorado and Utah. *Geological Society of America Bulletin*, **97**, 1012–1023.

Andrews, E.D. (1994). Marginal bed load transport in a gravel-bed stream, Sagehen Creek, California, *Water Resource Research*, **30**, 2241–2250.

Archer, D. (2000). Indices of flow variability and their use in identifying the impact of land use changes. *British Hydrological Society, 7th National Hydrology Symposium*, Newcastle, 2.67–2.73. London: Institution of Civil Engineers.

Boon, P.J., Wilkinson, J. & Martin, J. (1998). The application of SERCON (System for Evaluating Rivers for Conservation) to a selection of rivers in Britain. *Aquatic Conservation: Marine and Freshwater Ecosystems*, **8**, 597–616.

Brierley, G.J. & Fryirs, K. (2000). River styles: a geomorphic approach to catchment characterisation: implications for river rehabilitation in the Bega catchment, NSW, Australia. *Environmental Management*, **25**, 661–679.

Brookes, A. (1990). Restoration and enhancement of engineered river channels: some European experiences. *Regulated Rivers: Research and Management*, **5**, 45–56.

Brookes, A. & Sear, D.A. (1996). Geomorphological principles for restoring channels. In *River Channel Restoration: Guiding Principles for Sustainable Projects*, eds. A. Brookes & F.D. Shields, pp. 75–101. Chichester, UK: John Wiley.

Brookes, A. & Shields, F.D. (1996). Perspectives on river channel restoration. In *River Channel Restoration:Guiding Principles for Sustainable Projects*, eds. A. Brookes & F.D. Shields, pp. 1–19. Chichester, UK: John Wiley.

Brookes, A., Gregory, K.J. & Dawson, F.H. (1983). An assessment of river channelization in England and Wales. *Science of the Total Environment*, **27**, 97–112.

Buffington, J.M. & Montgomery, D.R. (1997). A systematic analysis of eight decades of incipient motion studies, with special reference to gravel-bedded rivers. *Water Resource Research*, **33**, 1993–2029.

Cao, S. & Knight, D.W. (1998). Design for hydraulic geometry of alluvial channels, *Journal of Hydraulic Engineering*, **124**, 484–492.

Cowx, T. & Welcomme, G.R. (1999). *Rehabilitation of Rivers for Fish*. London: UN/FAO and Fishing News Books.

Downs, P.W. & Thorne, C.R. (1998). Design principles and suitability testing for rehabilitation in a flood defence channel: the River Idle, Nottinghamshire. *Aquatic Conservation: Marine and Freshwater Ecosystems*, **8**, 17–38.

Ebrahimnezhad, M. & Harper, D.M. (1997). The biological effectiveness of artificial riffles in river rehabilitation. *Aquatic Conservation: Marine and Freshwater Ecosystems*, **7**, 187–197.

Environment Agency (1999). Sankey: river rehabilitation. In *River Habitat Survey: Applications*. Warrington, UK: Environment Agency.

Frissell, C.A., Liss, W.J., Warren, C.E. & Hurley, M.D. (1986). A hierarchical framework for stream habitat classification: viewing streams in a watershed context. *Environmental Management*, **10**, 199–214.

Gregory, K.J. (1995). Human activity and palaeohydrology. In *Global Continental Palaeohydrology*, eds. K.J. Gregory, L. Starkel & V.R. Baker, pp. 151–172. Chichester, UK: John Wiley.

Harper, D.M., Ebrahimnezhad, M., Taylor, E., Dickinson, S., Decamp, O., Verniers, G. & Balbi, T. (1999). A catchment-scale approach to the physical restoration of lowland UK rivers. *Aquatic Conservation: Marine and Freshwater Ecosystems*, **9**, 141–157.

Harper, D.M., Kemp, J.L., Vogel, B. & Newson, M.D. (2000). Towards the assessment of 'ecological integrity' on running waters of the UK. *Hydrobiologia*, **422**, 133–142.

Haschenburger, J.K. (1999). A probability model of scour and fill depths in gravel-bed channels. *Water Resources Research*, **35**, 2857–2869.

Holmes, N.T.H. & Nielsen, M.B. (1998). Restoration of the rivers Brede, Cole and Skerne: a joint Danish and British EU-LIFE demonstration project. 1: Setting up and delivery of the project. *Aquatic Conservation: Marine and Freshwater Ecosystems*, **8**, 185–196.

Ikeda, S., Parker, G. & Kimura, Y. (1988). Stable width and depth of straight gravel rivers with heterogeneous bed material, *Water Resources Research*, **24**, 713–722.

Iorns, W.V., Hembree, C.H. & Oakland, G.L. (1965). *Water Resources of the Upper Colorado River Basin*, Technical Report, US Geological Survey Professional Paper no. 441. Washington, DC: US Geological Survey.

Jeffers, J.N.R. (1998). Characterization of river habitats and prediction of habitat features using ordination techniques. *Aquatic Conservation: Marine and Freshwater Ecosystems*, **8**, 529–540.

Jowett, L.G. (1993). A method of objectively identifying pool, run and riffle habitats from physical measurements. *New Zealand Journal of Marine and Freshwater Research*, **27**, 241–248.

Kemp, J.L., Harper, D.M. & Crosa, G.A. (1999). Use of 'functional habitats' to link ecology with morphology and hydrology in river rehabilitation. *Aquatic Conservation: Marine and Freshwater Ecosystems*, **9**, 159–178.

Kondolf, G.M. (1995). Geomorphological stream classification in aquatic habitat restoration: uses and limitations. *Aquatic Conservation: Marine and Freshwater Ecosystems*, **5**, 127–141.

Kondolf, G.M. (1998). Lessons learned from river restoration projects in California. *Aquatic Conservation: Marine and Freshwater Ecosystems*, **8**, 39–52.

Kondolf, G.M. & Downs, P.W. (1996). Catchment approach to planning channel restoration. In *River Channel Restoration: Guiding Principles for Sustainable Projects*, eds. A. Brookes & F.D. Shields, pp. 129–148. Chichester, UK: John Wiley.

Kondolf, G.M. & Larson, M. (1995). Historic channel analysis and its application to riparian and aquatic habitat restoration. *Aquatic Conservation: Marine and Freshwater Ecosystems*, **5**, 109–126.

Kondolf, G.M. & Wilcock, P.R. (1996). The flushing flow problem: defining and evaluating objectives. *Water Resources Research*, **32**, 2589–2599.

Kronvang, B., Svendsen, L.M., Brookes, A., Fisher, K., Moller, B., Ottosen, O., Newson, M. & Sear, D. (1998). Restoration of the rivers Brede, Cole and Skerne: a joint Danish and British EU–LIFE demonstration project. 3: Channel morphology, hydrodynamics and transport of sediments and nutrients. *Aquatic Conservation: Marine and Freshwater Ecosystems*, **8**, 209–222.

Leopold, L.B. & Maddock, T. (1953). *The Hydraulic Geometry of Stream Channels and Some Physiographic Implications*, US Geological Survey Professional Paper no. 252. Washington, DC: US Geological Survey.

Lewis, G. & Williams, G. (1984). *Rivers and Wildlife Handbook: A Guide to Practices which Further the Conservation of Wildlife on Rivers*. Sandy, UK: RSPB/RSNC.

Lins, H.F., Hare, F.K. & Singh, K.P. (1990). Influence of the atmosphere. In *Surface Water Hydrology*, vol. 1, *Geology of North America*, eds. M.G. Wolman & H.C. Riggs, pp. 11–53. Boulder, CO: Geological Society of America.

Milhous, R. (1998). Modelling of instream flow needs: the link between sediment and aquatic habitat. *Regulated Rivers*, **14**, 79–94.

National Rivers Authority (1996). *River Habitats in England and Wales*, River Habitat Survey Report no. 1. Bristol, UK: National Rivers Authority.

Newson, M.D. (1984). Introduction 2: River processes and form. In *Rivers and Wildlife Handbook: A Guide to Practices which Further the Conservation of Wildlife on Rivers*, eds. G. Lewis & G. Williams, pp. 3–9. Sandy, UK: RSPB/RSNC.

Newson, M.D. & Brookes, A. (1994). River morphology and fluvial processes. In *The New Rivers and Wildlife Handbook*, eds. D. Ward, N. Holmes & P. Jose, pp. 19–30. Sandy, UK: RSPB.

Newson, M.D., Clark, M.J., Sear, D.A. & Brookes, A. (1998). The geomorphological basis for classifying rivers. *Aquatic Conservation: Marine and Freshwater Ecosystems*, **8**, 431–436.

Newson, M.D., Sear, D.A. & Heritage, G. (1999). *Rehabilitation of Selected Sub-Reaches of the River Waveney, Anglia Region, Environment Agency: A Geomorphological Assessment*. Ipswich, UK: Environment Agency.

Newson, M.D. & Newson, C.L. (2000). Geomorphology, ecology and river channel habitat: mesoscale approaches to basin-scale challenges. *Progress in Physical Geography*, **24**, 195–217.

Newson, M.D., Hey, R.D., Bathurst, J.C., Brookes, A., Carling, P.A., Petts, G.E. & Sear, D.A. (1997). Case studies in the application of geomorphology to river management. In *Applied Fluvial Geomorphology for River Engineering and Management*, eds. C.R. horne, R.D. Hey & M.D. Newson, pp. 311–363. Chichester, UK: John Wiley.

Newson, M.D., Thorne, C.R. & Brookes, A. (2001). The management of gravel-bed rivers in England and Wales: from geomorphological research to strategy and operations. In *Gravel-Bed Rivers V*, ed. M.P. Mosley, pp. 581–605. Wellington: New Zealand Hydrological Society.

Padmore, C.L. (1998). The role of physical biotopes in determining the conservation status and flow requirements of British rivers. *Aquatic Ecosystem Health and Management*, **1**, 25–35.

Park, C.C. (1977). World-wide variations in hydraulic geometry exponents of stream channels: an analysis and some observations. *Journal of Hydrology*, **35**, 133–146.

Parker, G. (1978*a*). Self-formed straight rivers with equilibrium banks and mobile bed. 1: The sand–silt river. *Journal of Fluid Mechanics*, **89**, 109–125.

Parker, G. (1978*b*). Self-formed straight rivers with equilibrium banks and mobile bed. 2: The gravel river. *Journal of Fluid Mechanics*, **89**, 127–146.

Parker, G. (1979). Hydraulic geometry of active gravel rivers. *Journal of Hydraulic Engineering*, **105**, 1185–1201.

Perrow, M.R. & Wightman, A.S. (1993). *The River Restoration Project, Phase 1: The Feasibility Study*. Oxford, UK: River Restoration Project.

Pitlick, J. & van Steeter, M.M. (1998). Geomorphology and endangered fish habitats of the Upper Colorado

River. 2: Linking sediment transport to habitat maintenance. *Water Resources Research*, **34**, 303–316.

Raven, P.J., Fox, P., Everard, M., Holmes, N.T.H. & Dawson, F.H. (1996). River habitat survey: a new system for classifying rivers according to their habitat quality. In *Freshwater Quality: Defining the Indefinable*, eds. P.J. Boon & D.L. Howell, pp. 215–234. Edinburgh: HMSO.

Raven, P.J., Holmes, N.T.H., Dawson, F.H., Everard, M., Fozzard, L. & Rouen, K.J. (1998). *River Habitat Quality: The Physical Character of Rivers and Streams in the United Kingdom and Isle of Man*, River Habitat Survey Report no. 2. Bristol, UK: Environment Agency.

River Restoration Centre (1999). *River Restoration Manual of Techniques: Restoring the River Cole and River Skerne, UK.* Silsoe, UK: River Restoration Centre.

Rosgen, D.L. (1994). A classification of natural rivers. *Catena*, **22**, 169–199.

Rosgen, D.L. (1996). *Applied River Morphology.* Pagosa Springs, CO: Wildland Hydrology.

Schumm, S.A. (1977). *The Fluvial System.* New York: John Wiley.

Seal, R., Stein, O.R. & Boelman, S.F. (1998). Performance of in-stream structures under flood conditions. In *Engineering Approaches to Ecosystem Restoration*, ed. D.F. Hayes, CD-ROM. Denver, CO: American Society of Civil Engineers.

Sear, D.A. (1994). River restoration and geomorphology. *Aquatic Conservation: Marine and Freshwater Ecosystems*, **4**, 169–177.

Sear, D.A. (1996). The sediment system and channel stability. In *River Channel Restoration: Guiding Principles for Sustainable Projects*, eds. A. Brookes & F.D. Shields, pp. 149–177. Chichester, UK: John Wiley.

Sear, D.A., Briggs, A. & Brookes, A. (1998). A preliminary analysis of the morphological adjustment within and downstream of a lowland river subject to river restoration. *Aquatic Conservation: Marine and Freshwater Ecosystems*, **8**, 167–183.

Sear, D.A., Wilcock, D., Robinson, M. & Fisher, K. (2000). River channel modification in the UK. In *The Hydrology of the UK: A Study of Change*, ed. M. Acreman, pp. 55–81. London: Routledge.

Vannote, R.L., Minshall, G.W., Cummins, K.W., Sedell, J.R. & Cushing, C.E. (1980). The river continuum concept. *Canadian Journal of Fish and Aquatic Sciences*, **37**, 130–137.

van Steeter, M.M. & Pitlick, J. (1998). Geomorphology and endangered fish habitats of the Upper Colorado River. 1: Historic changes in streamflow, sediment load and channel morphology. *Water Resources Research*, **34**, 287–302.

Vivash, R., Ottosen, O., Janes, M. & Sorensen, H.V. (1998). Restoration of the rivers Brede, Cole and Skerne: a joint Danish and British EU-LIFE demonstration project. 2: The river restoration works and other related practical aspects. *Aquatic Conservation: Marine and Freshwater Ecosystems*, **8**, 197–208.

Wadeson, R.A. (1994). A geomorphological approach to the identification and classification of instream environments. *South African Journal of Aquatic Sciences*, **20**, 1–24.

Ward, D., Holmes, N. & Jose, P. (eds.) (1994). *The New Rivers and Wildlife Handbook.* Sandy, UK: RSPB.

Water Space Amenity Commission (1980). *Conservation and Land Drainage Guidelines.* London: Water Space Amenity Commission.

Water Space Amenity Commission (1983). *Conservation and Land Drainage Guidelines*, 2nd edn. London: Water Space Amenity Commission.

Wilcock, P.R. & MacArdell, B.W. (1993). Surface-based frictional transport rates: mobilisation thresholds and partial transport of a sand-gravel sediment. *Water Resources Research*, **29**, 1297–1312.

Part 3 • Manipulation of the chemical environment

9 • Manipulating the chemical environment of the soil

ROBERT H. MARRS

INTRODUCTION

Soil is an important component of most terrestrial ecosystems. Soils, through their physical structure, physicochemical properties and biological activities, are driving variables controlling the structure and function of the ecosystem they support. The soil of course does not exist in isolation, it is part of the ecosystem, which includes soils, micro-organisms, plants, animals and dead material, which all interact. The entire system reflects the management that humans impose on it and, clearly, any management impact in any one part of the system may have serious implications for the structure of the system or its processes. As we more or less take this for granted in 'normal' ecosystems, it should be no surprise that we need to manipulate soil processes during restoration work.

There is, however, a fundamental difference in the way that soil is viewed by restoration ecologists compared with 'normal' ecosystems; the starting soil material is often inadequate to support the target ecosystem. The soil must, therefore be manipulated during the restoration process in order to rectify this problem, and in most instances an ecological approach can help match soil properties with an appropriate target ecosystem (Marrs, 1993; Marrs & Bradshaw, 1993). Often, soil manipulation can involve a large number of management techniques. Whereas some treatments affect the soil directly, indirect techniques that operate via plant growth or management (e.g. cropping, grazing or burning) are also often used.

The importance of soil: setting the scene for restoration

The importance of soils and the factors controlling their development was first described by Jenny (1941) and later extended to ecosystems (Jenny, 1980). He developed a short descriptive equation to describe the factors governing soil formation under 'normal conditions', which can be extended and used when soils are being created or modified. This well-known equation described soil formation (or ecosystem formation) in terms of five state factors (climate, biota, relief, parent material, time) and management by humans was added by Marrs & Bradshaw (1993) for land restoration.

Only some of the state factors can be modified during a restoration task. Clearly, it is impossible to control climate, but it is worth remembering that the climate will be an important overarching variable determining choice of species and ecosystem. The biota are one of the most important factors that are usually manipulated during land restoration; choice of species will have important consequences not only on the plant communities that develop (see Davy, this volume; Gray, this volume), but there may also be very important interactions with soil. For example Wedin & Tilman (1990) in a comparative study of grass species growing on the same soil showed that there were important differences in nitrogen availability between species after three years. Relief in most natural situations would be a fixed factor, but in land restoration schemes this can be altered as desired (see Whisenant, this volume). Similarly, parent material is usually considered to be a fixed factor, but in land restoration this can be seen almost as an experimental variable. The parent material can be buried with different materials, mixed with organic or inorganic soil-forming materials to improve permeability and/or chemical properties, or chemically modified by liming or acidification. Its compaction can be modified by compression or by ripping. Time is also an important factor as in many restoration schemes it is essential to plan for soil and ecosystem development; and it may take tens of years to obtain the desired ecosystem.

There is a huge armoury of techniques for manipulating soils and their properties towards different ecosystem end points. Not all will be suitable in any given situation but the land manager can use them in combination, and can perhaps be more creative in creating new ecosystems than in managing existing ones.

The importance of soil factors in structuring plant communities

The central importance of the ability of the soil to modify vegetation is well understood in agriculture and forestry. Indeed production systems are geared towards improving soil physicochemical properties. For non-agricultural, semi-natural, temperate ecosystems the relationship between species diversity (density, *y* axis) has been related to site/soil fertility (*x* axis), where fertility has been described using a surrogate measure (aboveground biomass), as a 'humpback' relationship (Grime, 1979) (Fig. 9.1a). Grime (1979) argued that species with different properties occupied different parts of the curve (Fig. 9.1a), and that modified relationships were found on different parent materials, with a greater number of species found in the 'hump' on neutral compared to acidic soils, where the available species pool was reduced (Fig. 9.1b). Tilman (1988) has argued from a theoretical perspective that the ratio between two key environmental factors which change during succession – light (decreases) and soil nitrogen (increases) – is the main factor controlling species sequences and co-existence at any given point. Irrespective of that, soil development is an important factor structuring plant communities and their productivity.

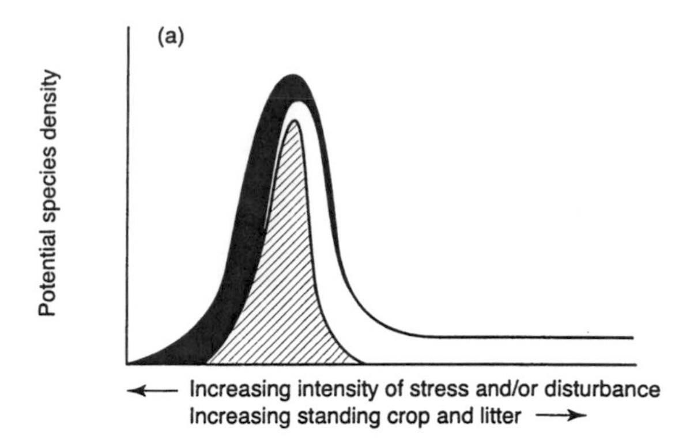

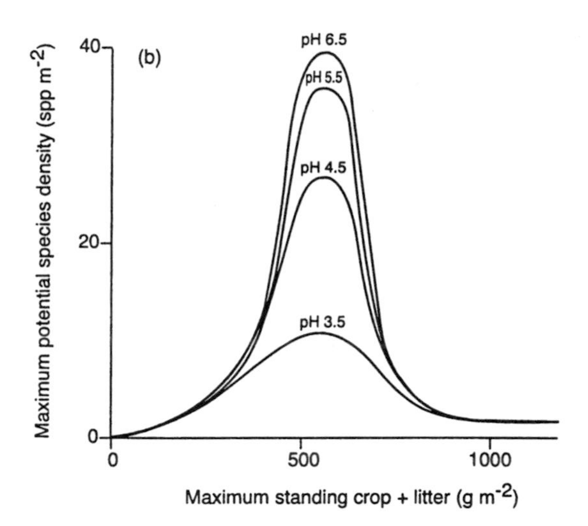

Fig. 9.1. Generalised humpback model proposed by Grime (1979) to illustrate the relationship between site fertility expressed as the mass of the vegetation plus litter and species diversity; (a) shows the parts of the proposed curve which contain potential dominants (unshaded), species or ecotypes adapted to the prevailing stress or disturbance (shaded), and species which are neither potential dominants nor are highly adapted (hatched); and (b) the effects of soil pH on this model relationship for herbaceous vegetation in the UK.

A generalised model for soils in land restoration

From a pragmatic viewpoint it is useful to consider ecological restoration using the generalised successional model of Odum (1971), who postulated, amongst other things, that as succession proceeded there was increased diversity, an increased niche specialisation, increased organic matter content, and increased proportion of the nutrients in organic forms and a tightening of nutrient cycles. This model can be visualised conceptually as an increase in complexity through time (Fig. 9.2). From a restoration ecologist's point of view there are two possible scenarios for soil manipulation as part of ecosystem development:

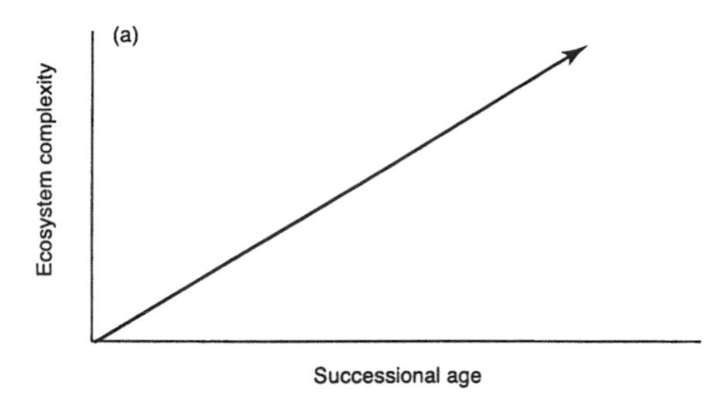

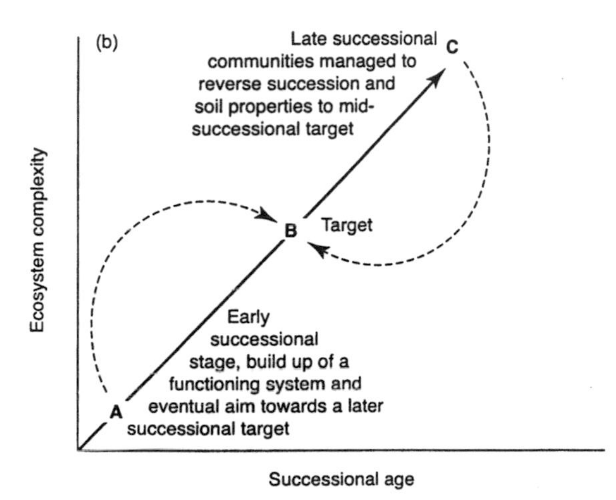

Fig. 9.2. An hypothetical model illustrating Odum's (1971) ideas that ecosystem complexity increases through successional time: (a) generalised relationship, and (b) how this model needs to be viewed for restoration ecology. Early-successional communities need to be developed towards a target (aggrading approach) and for late-successional communities processes need to be reversed (degrading approach).

Aggrading approach

Where the starting capital of soil nutrients is poor and, indeed, may be non-existent, there is a need to build a soil ecosystem from scratch. Here the starting position would be low (Point A, Fig. 9.2) and the aim would be to build soil structure and function towards an ecosystem near the middle of the graph (Point B, Fig. 9.2). Examples of this approach would be restoration of mining wastes, sands and other inert materials. The soil-forming material at the start has a low organic matter content, its water-holding capacity is poor and nutrient supplies may be impaired. In addition, toxicities may need to be tackled on some wastes. This aggrading approach process is akin to a primary succession *sensu stricto*, given that new ecosystems are often restored on raw mineral wastes where there is no existing biota (Marrs & Bradshaw, 1993).

Degrading approach

Where the starting soil is from a late-successional stage and the aim is to restore ecosystems from an earlier phase, the starting position would be high on the trajectory (Point C, Fig. 9.2) and the aim is to degrade the soil to one that would support a community near the middle of the graph (Point B, Fig. 9.2). The concept of deliberately degrading an ecosystem in successional terms may seem heretical to some land managers, but it is an essential process in the restoration of mid-successional communities, either after succession has occurred, or where semi-natural communities are to be recreated on abandoned agricultural land.

THE AGGRADING APPROACH: BUILDING ECOSYSTEMS FROM SCRATCH

Aggrading is typically used to establish a new ecosystem on pristine substrates, which are very low in available nutrients and contain almost no organic matter. Environmental factors identified as constraints for ecosystem development on man-made wastes include adverse physical properties, with problems relating to texture, stability, temperature, water retention and severe nutrient deficiency (Table 9.1). For convenience these waste materials can be subdivided into two main groups, based on the complexity of treatment needed:

1. Non-toxic mineral wastes, where there are no problems with toxicities and the only problem is soil and ecosystem development.
2. Mineral wastes similar to (1), but with the additional problems of some form of toxicity that must be overcome (acidity, toxic metals, salinity) before plants can grow well.

We will examine the principles of ecosystem development during natural successions on non-toxic

Table 9.1. *The physical and chemical problems to be overcome when restoring different types of derelict land materials*

	Physical properties				Chemical properties				
					Nutrients				
Materials	Texture/ structure	Stability	Water supply	Surface temperature	Macro	Micro	pH	Toxic materials	Salinity
Sand and gravel	○/o	o	o	o	○/o	o	○/o	o	o
Bauxite mining	○○/o	o	o	o	○○	o	o	o	o
Acid rocks	○○○	o	○○	o	○○	o	○	o	o
China clay wastes	○○○	○○	○○	o	○○○	o	○	o	o
Iron ore mining	○○○/o	○○/o	○/o	o	○○	o	o	o	o
Calcareous rocks	○○○	o	○○	o	○○○	o	●	o	o
Coastal sands	○○/o	○○○/o	○/o	o	○○○	o	o	o	o/●
Urban wastes	○○○/o	o	o	o	○○	o	o	o/●●	o
Roadsides	○○○/o	○○○	○○/o	○/o	○○	o	○/o	o	o/●●
Gold wastes	○○○	○○○	○	o	○○○	o	○○○	o	●●●
Land from sea	○○	o	o	o	○○	o	o/●	o	o/●●
Oil shale	○○	○○○/o	○○	o/●●	○○○	o	○○/o	o	o/●●
Strip mining	○○○/o	○○○/o	○○/o	o/●●●	○○○/o	o	○○○/o	o	o/●●
Colliery spoil	○○○	○○○/o	○/o	o/●●●	○○○	o	○○○/o	o	o/●●
Fly ash	○○/o	o	o	o	○○○	o	○/●●●	●●	o/●●
Heavy metal wastes	○○○	○○○/o	○○/o	o	○○○	o	○○○/●	○/●●●	o/●●●

Key: Deficiency — ○○○ Severe; ○○ Moderate; ○ Slight; o Adequate. Excess — ●●● Severe; ●● Moderate; ● Slight.

Source: Modified after Bradshaw & Chadwick (1980).

mineral wastes and consider the additional treatments that need to be used to restore them. We will then consider the additional treatments needed to restore materials when toxic substances are present.

Lessons from nature: ecosystem development on raw mineral wastes

Much of the theory on ecosystem development on non-toxic mineral wastes has been derived from comparing natural primary successions with their analogues on china clay wastes. The lessons learned from these successions provide a conceptual framework for deriving restoration techniques, where managers can use the general principles to accelerate succession.

China clay wastes are ideal for such general studies; they comprise gravel-sized particles, which are poorly compacted and have a high porosity. Chemically they are inert, have a low pH (4.0–5.0), low cation exchange capacity ($\leq$0.2 mmol kg^{-1}), are deficient in most plant nutrients except potassium and have no toxicity problems (Marrs *et al.*, 1983).

Bradshaw (1983) in a comparative study of natural and man-made primary successions divided the process of ecosystem development into three main stages: (1) an initial colonisation phase, where stochastic processes were important in structuring the vegetation that colonised; (2) an intermediate-development phase, in which nutrient accumulation is crucial; and (3) a late-development phase, including further colonisation and replacement.

A fundamental feature that emerged from this and similar studies was the prominent role that nitrogen plays in the first two phases (Robertson & Vitousek, 1981; Marrs & Bradshaw, 1993). This can be derived from first principles. First, nitrogen is required in larger amounts by plants than any other mineral element. Although the potassium concentration is of similar magnitude to that of nitrogen, plant tissues contain three to four times as many nitrogen atoms than potassium, and between eight and ten times the number of atoms of any other plant mineral nutrient element (Epstein, 1972). Second, nitrogen is held in most soils only in organic matter and is released mainly by decomposition processes (see Allen *et al.*, this volume). Since in temperate climates organic matter turnover is slow, there must be a large capital of nitrogen so that a sufficient supply is released to meet the annual vegetation requirements. In temperate systems litter decomposition constants are *c.* 0.0625 per year (k = 1/16) (*sensu* Jenny *et al.*, 1949; Olsen, 1963) and mineralisation rates are *c.* 0.02 (Reuss & Innis, 1977). In tropical climates mineralisation may be faster, although the rate will depend on rainfall (Singh *et al.*, 1989). As many ecosystems have an annual plant requirement of 100 kg N ha^{-1}, based on dry matter production of 5000 kg N ha^{-1} and a nitrogen concentration of 2% in vegetation, simple arithmetic shows that to provide this supply from either litter turnover or soil mineralisation the capital in the system must be between 1600 and 5000 kg N ha^{-1}.

As there is very little nitrogen present in most raw mineral wastes a nitrogen store must be built up during succession or restoration. In china clay wastes there is as little as 5–20 kg N ha^{-1} within the surface 21 cm. There is, therefore, an inescapable chronic shortage of nitrogen, especially when compared with developed soils, which usually contain 2000–10 000 kg N ha^{-1}, which encompasses the range predicted above.

If no nitrogen-fixing vascular species invade during the early stages of the primary succession, nitrogen takes a very long time to accumulate, because inputs must be derived from atmospheric sources. As nitrogen inputs in polluted rain can range from <10 kg N ha^{-1} yr^{-1} in relatively unpolluted areas to >40 kg N ha^{-1} yr^{-1} in more polluted areas (Brimblecombe & Stedman, 1982; Fowler, 1987), accumulation of the theoretical target capital of 1600 kg N ha^{-1} will take 32–160 years if there is a 100% efficiency of capture. As capture efficiency is likely to be well below 100%, especially during the early phases when vegetation is sparse, it might take several hundred years to reach this hypothetical target, especially in unpolluted areas.

Marrs & Bradshaw (1993) speculated that there were three possible strategies that would allow plants to colonise, accumulate nutrients and develop a functioning soil ecosystem.

1. Low relative growth rate (RGR). If colonising species had a relatively low RGR, their annual nutrient

requirement, given the same tissue concentration, will be correspondingly reduced. The observation in a comparative study that species from mine wastes have exceptionally low growth rates (Grime & Hunt, 1975) supports this hypothesis. An alternative hypothesis could be that initial colonists could have lower nitrogen concentrations, although the range of variation in this character appears to be much less than that of RGR (Chapin, 1980).

2. Scavenging nutrients. A species with a normal growth rate, but with a widely ramifying root system and distributed at low densities, may be able to scavenge enough nitrogen from a large area and concentrate it centrally. The growth per unit area exploited would be low, but on an individual plant basis could be normal. This type of species has been noted by Grubb (1986), and possible candidates include *Reynoutria japonica* (Hirose, 1986) the shrubs *Salix atrocinerea* and *S. caprea* and the grass *Holcus lanatus* (Marrs & Bradshaw, 1993).
3. Nitrogen fixation. This is the most obvious method of overcoming nitrogen shortage, with an unlimited atmospheric supply, providing the problems of energetics and relationships with nitrogen-fixing micro-organisms can be overcome (see Allen *et al.*, this volume). Generally, invasion by nitrogen-fixing higher plants is a feature of most primary successions. Studies of successions on natural materials, glacial moraines and sand dunes plus two on man-made materials – ironstone spoil (Leisman, 1957) and china clay waste (Roberts *et al.*, 1980) – confirm this view, with a wide range of nitrogen-fixing species involved (Table 9.2). Both leguminous and non-leguminous species such as *Dryas* spp. and *Alnus* spp. are common, and significant nitrogen-fixation rates have been detected (Lawrence *et al.*, 1967).

The nitrogen accumulation process in primary succession is perhaps the most often-quoted example of

Table 9.2. *Examples of nitrogen-fixing species which have been found on primary successions*

Substrate	Nitrogen-fixing species	Site location	Reference
Glacial moraines	*Alnus crispa* *Dryas drummondii*	Glacier Bay, Alaska, USA	Crocker & Major (1955)
Glacial moraines	*Astragalus alpinus* *Astragalus tananaica* *Astragalus nutzotinesis* *Dryas drummondii* *Dryas intergrifolia* *Sherperdia canadensis*	Muldrow Glacier, Alaska, USA	Viereck (1966)
Glacial moraines	*Lotus corniculatus* *Trifolium badium* *Trifolium thalli*	Hintereisferner and Aletsch Glacier, Europe	Friedel (1938*a*, *b*); Richard (1968)
Glacial moraines	*Coriaria* spp.	New Zealand	A.D. Bradshaw (unpubl. data)
China clay sand waste	*Lotus corniculatus* *Lupinus arboreus* *Sarothamnus scoparius* *Ulex europaeus* *Ulex gallii*	Cornwall, UK	Roberts *et al.* (1981)
Ironstone spoil	*Melilotus alba* *Trifolium repens*	Minnesota, USA	Leisman (1957)

Source: Marrs & Bradshaw (1993).

facilitation (*sensu* Connell & Slatyer, 1977), and we can guess empirically the times and nitrogen capitals required for sufficient facilitation by estimating when late-successional, non-nitrogen-fixing species invade. Generally the nitrogen capitals are in the same order of magnitude as our theoretical target (400–2200 kg N ha^{-1}) and the times taken to achieve these values were between 21 and >120 years (Marrs & Bradshaw, 1993). On china clay wastes 1000 kg N ha^{-1} (700 kg N ha^{-1} in the soil) and 1800 kg N ha^{-1} (1200 kg N ha^{-1} in the soil) were the levels at which non-nitrogen-fixing *Salix* scrub and *Betula–Quercus* woodland developed respectively, and it took about 70–100 years.

Given the obvious advantages of nitrogen fixation, and the preceding arguments about the crucial role of nitrogen accumulation in primary successions, we might expect a profusion of nitrogen-fixing species in the early stages. Whilst nitrogen-fixers are a feature of most primary successions, they do not necessarily predominate. The reason is that both the plant species and the endophyte must colonise the site and usually nitrogen-fixing species require a reasonable phosphorus supply. As there is no equivalent to nitrogen fixation for phosphorus, inputs must come slowly from atmospheric sources, usually <0–5 kg P ha^{-1} yr^{-1}, and gradual weathering of minerals. Scavenging for phosphorus may be enhanced by mycorrhizal associations, which are a common feature of plants growing in nutrient-poor soils. Mycorrhizae are often found in roots growing on man-made wastes during restoration (Daft & Hacskaylo, 1976; Stahl *et al.*, 1988). (Plants colonising early successional or severely damaged sites tend to lack mycorrhizae, certainly in comparison with later stages; see also Allen *et al.*, this volume, and Greipsson, volume 2).

It would be foolish to believe that nitrogen is the only factor controlling ecosystem development on man-made wastes. Indeed, we have already seen how nitrogen accumulation might be restricted by phosphorus deficiency. Depending on the starting waste material, other elements such as potassium, calcium, magnesium and especially phosphorus in its own right, may also play a role. Common experience suggests that potassium is not usually limiting, because of its widespread occurrence in clay minerals, and that calcium and magnesium are only limiting in certain materials, although often any positive effects are associated with their effects on pH. By contrast, phosphorus deficiencies are common. Certain clays used in brick-making provide an exception; the phosphorus levels on these wastes can be high and have a major impact on the rate of natural colonisation of these materials (Dutton & Bradshaw, 1982). On most wastes plant growth is severely limited unless phosphorus is given in combination with nitrogen (Bradshaw *et al.*, 1978; Smith & Bradshaw, 1979) (Fig. 9.3).

The general lesson learned is that ecosystem development on man-made wastes requires development of organic matter with a constituent nitrogen content. As this happens, there is also a need to ensure that decomposition processes are developed so that there is efficient nutrient. As the organic

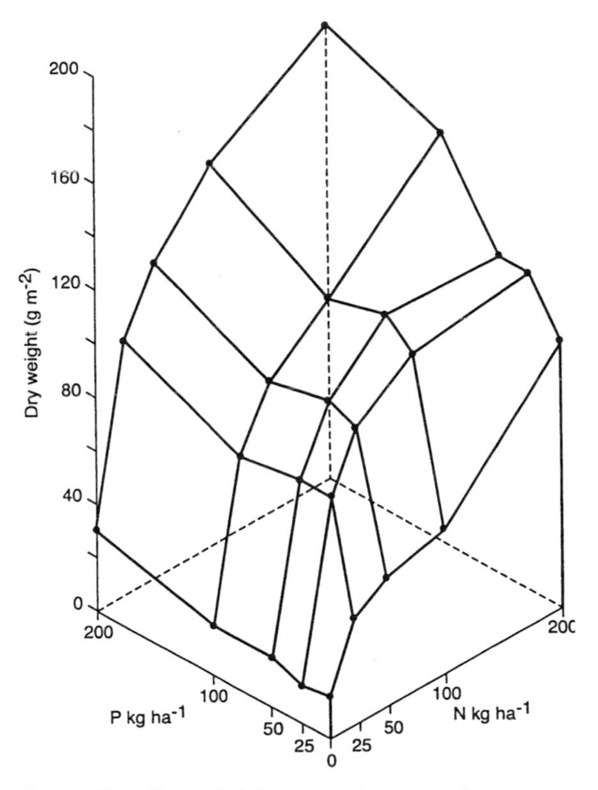

Fig. 9.3. The effects of adding both nitrogen and phosphorus fertilizers on the growth of *Festuca ovina* on limestone fines. From Bradshaw *et al.* (1978).

matter develops there will also be improved water-holding capacity and soil structure.

There are two major strategies that can be used to develop this pool of organic matter and nutrients within any raw material. The first is using a strategy based on man-modified successional processes (using time, natural accumulation of nutrients, nitrogen fixation, fertilisers, etc.), and the other is the importation of accumulated organic matter and nutrients from some other ecosystem (using topsoil, sewage sludge, green manures, etc.). Each has its place, but the restoration ecologist might view the latter option as cheating! Realistically, the strategy to be adopted will depend on (1) whether toxicity problems are present or not (see below), and (2) the desired productivity of the final ecosystem. Generally, if the aim is to produce a semi-natural ecosystem with low productivity then the first strategy will perhaps be suitable. However, if the aim is to restore an ecosystem with a modest or high productivity then it will be more appropriate to use an importation strategy for immediate increases in the capacity of the system to recycle sufficient nutrients to accommodate the required plant growth. Importation of organic matter is commonly used to aid ecosystem development on e.g. dunes and beaches, when 'topsoil' is not present (Walmsley, volume 2; Greipsson, volume 2).

The simplest scenario: building ecosystems on non-toxic mineral wastes

We have shown that nitrogen accumulation and its efficient recycling are important factors in ecosystem development. There are, however, only two ways that nitrogen accumulation can be accelerated, through the addition of fertilisers and through biological fixation. Biological fixation, if successful, should be cheaper and longer lasting. Many different species with symbiotic nitrogen-fixing micro-organisms (see Allen *et al.*, this volume) have been used to increase the nitrogen contents of soils in reclamation schemes. In temperate climates these include both agricultural and non-agricultural plant species. These include annuals (*Lupinus alba, L. angustifolius, L. luteus, Ornithopus sativus, Trifolium dubium, T. subterraneum, Vicia grandiflora, V. sativa*); biennials (*Medicago alba, M. lupulina* and *M. officinalis*) and perennials (*Alnus glutinosa, Anthyllis vulneraria, Coronilla varia, Cytisus scoparius, Lotus corniculatus, L. uliginosus, Lupinus arboreus, L. perennis, Medicago sativa, Robinia pseudoacacia, Trifolium hybridum, T. repens, T. pratense* and *Ulex europaeus*) (Jefferies *et al.*, 1981*a*). In subtropical and tropical climates the same principles apply with nitrogen-fixing agricultural legumes (for example, *Phaseolus* spp., *Stylosanthes* spp), but more usually tree species are used, for example *Acacia auriculiformis, Casuarina equisetifolia, Cassia siamea, Leucena* spp. and *Tamarina simica* (Marrs & Bradshaw, 1998).

This approach can fix substantial inputs of nitrogen. For example, inputs of 27, 49 and 72 kg N ha^{-1} yr^{-1} were measured for *Ulex europaeus, Trifolium repens* and *Lupinus arboreus*, respectively (Skeffington & Bradshaw, 1980), and net nitrogen accumulation rates with legumes can reach up to 295 kg N ha^{-1} yr^{-1} if they are managed in an appropriate manner (Jefferies *et al.*, 1981*a*). However, on poor substrates, legumes often do not achieve these levels, but their continuing contribution over many years, without significant repeated fertiliser costs, can be extremely important (Jefferies *et al.*, 1981*a, b*; Elias *et al.*, 1982). Significant contributions can also be obtained from non-leguminous nitrogen-fixing species such as *Alnus* spp., and nitrogen can be transferred to other non-nitrogen-fixing species (Fig. 9.4).

Clearly, to obtain a rapid accumulation of nitrogen cheaply it is sensible to use a combination of fertilisers and legumes. Where this is done it is recommended that fertiliser nitrogen levels be reduced to below 50 kg N ha^{-1} yr^{-1}, as inputs greater than this have adverse effects on nitrogen fixation (Skeffington & Bradshaw, 1980). Where agricultural forage legumes are used, inputs of lime and phosphorus may still be needed to maintain a large legume component and hence high nitrogen fixation rates (Marrs & Bradshaw, 1993). As both of these elements may be subject to large leaching losses (Dancer, 1975; Marrs & Bradshaw, 1980), additions of these elements may need to be continued for some time.

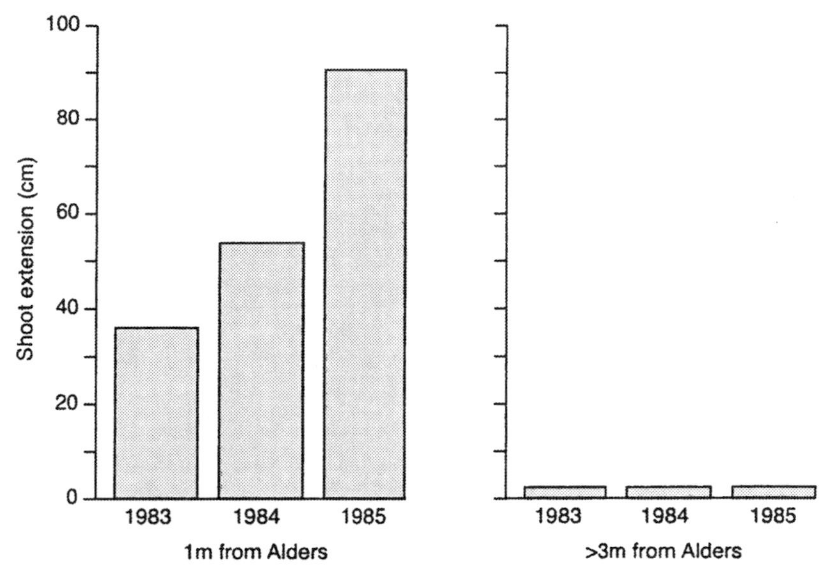

Fig. 9.4. The impact of nitrogen-fixing species on non-nitrogen-fixers growing nearby, illustrated by differential shoot extension rates of ten-year-old *Acer pseudoplatanus* growing in proximity to *Alnus* spp. growing on china clay wastes between 1983 and 1985. From Marrs & Bradshaw (1993).

Estimates of the times required to achieve the theoretical target derived from successions studies of the minimum theoretical target of 1600 kg N ha^{-1} yr^{-1} for china clay wastes show that if both legumes and fertilisers are used in combination, the time required to reach the target can be reduced from >130 to between seven and 20 years (Marrs & Bradshaw, 1993).

Once ecosystems have become established and the nitrogen capital is accumulating, successful restoration can be assisted by increasing the recycling of nitrogen by removing blockages. A major blockage is the low supply from soil mineralisation. Experience here suggests that it is not the mineralisation rate *per se* that is reduced but rather the soil nitrogen capital is too low. Once the capital builds up the supply increases from this source. A second blockage is that nitrogen is locked up in live and dead aboveground biomass. An easy way to remove this blockage is to increase the turnover by grazing or mowing. Grazing, for example, transfers nitrogen as faeces and urine to the soil pools, and thus provides a much more rapid cycling of the accumulated capital in vegetation and litter. On reclaimed china clay wastes, a minimum of 27% of the nitrogen taken up into the aboveground pool could be returned by a managed grazing regime (Marrs *et al.*, 1980). Presumably, the addition or encouragement of soil animals would also encourage this process, but little is known about their effects. Colonisation of restored colliery wastes by some faunal groups can be very rapid (less than two years), with the Collembola having an important influence on decomposition (Hutson, 1980*a*, *b*).

A more complex scenario: building ecosystems on toxic mineral wastes

In this part the additional problem of toxicity is considered. This occurs where some soil factor will prevent or impede the growth of plants and other organisms from growing as a result of direct toxic effects. Nevertheless, all of the principles derived above for non-toxic wastes also apply to wastes with toxicity problems. However, in order to deal with the additional complications of toxicity problems two major approaches have been used; the first is essentially a civil engineering approach where the soil is ameliorated or physically separated from the growing plants, and the second is a biological approach using tolerant species or, in recent times, hyperaccumulator species. Clearly, it is possible to use some of these approaches in combination. The relative merits of each approach for increasing severity of problems are shown in Table 9.3. It must also

Table 9.3. *Approaches to the revegetation of metalliferous soils of different metal content*

Waste characteristic	Reclamation technique	Problems encountered
Low toxicity Toxic metal content: <0.1% No major acidity or alkalinity problems	**Amelioration and direct seeding with agricultural or amenity grasses and legumes** Apply lime if pH <6. Add organic matter if physical amelioration required. Otherwise apply nutrients as granular compound fertilisers. Seed using traditional agricultural or specialised techniques.	Probable commitment to long-term maintenance. Grazing must be strictly monitored and excluded in some situations due to movement of toxic metals into vegetation.
Low toxicity plus climatic limitations Toxic metal content: <0.1% No major acidity or alkalinity problems Extremes of temperature, rainfall, etc.	**Amelioration and direct seeding with native species** Seed or transplant adapted native species using amelioration treatments (e.g. lime, fertiliser) where appropriate.	Irrigation is often necessary during establishment in arid climates. Expertise required on the selection of native flora.
High toxicity Toxic metal content: >0.1% High salinity	**Amelioration and direct seeding with tolerant ecotypes** Sow metal and/or salt-tolerant seed. Apply lime, fertiliser and organic matter, as necessary, before seeding.	Possible commitment to regular fertiliser applications. Relatively few species have evolved tolerant populations. Grazing inadvisable. Very few species a available commercially as tolerant varieties.
Extreme toxicity Very high toxic metal content Intense salinity or acidity	**Surface treatment and seeding with agricultural or amenity grasses and legumes** Amelioration with 10–50 cm of innocuous barrier material (e.g. overburden). Apply lime and fertiliser as necessary.	Regression will occur if shallow depths of amendments are applied or if upward movement of metal occurs. Availability and transport costs may be limiting.
	Barrier layer Surface treatment with 30–100 cm of innocuous barrier material (e.g. unmineralised rock) and surface covering with a suitable rooting medium (e.g. subsoil). Apply lime and fertiliser as necessary.	Susceptibility to drought according to the nature and depth of surface covering. High cost and potential limitation of availability of barrier material.
	Hyperaccumulators Novel approach for decontamination of toxic soils using species that can accumulate high concentrations of metals in their tissues.	Needs selection of appropriate species matched to metal problem.

Source: Modified after Williamson *et al.* (1982).

be acknowledged that some toxic factors may move through food chains and impinge on ecosystem processes such as decomposition.

Civil engineering approach

There are two ways that toxic wastes can be managed to minimise the effects of toxicities. The first is through an amelioration of the material by the addition of physical or organic amendments, and the second is through the physical separation of the growing plants from the toxic material.

Amelioration

The most common approach here is to add either topsoil (A horizon), subsoil (B and sometimes C horizons) or overburden (C horizon) which have already been stripped from the site. However, care must be taken to ensure that they do not contain toxic materials and that they are adequate for the planned restoration. The quality of topsoil, subsoil and overburden is extremely variable (Bloomfield *et al.*, 1981). Other inorganic materials may be available as a result of other industrial activities; for example limestone chipping, slate wastes, colliery wastes, pulverised fly ash (PFA) have all been used as potential amendments to contaminated land (Williamson *et al.*, 1982). Organic amendments can also be used; the most common include farmyard manures, poultry manures, sewage sludge, domestic refuse and a whole range of other waste products – straw, wood chippings, sawdust and bark, shoddy, soot, seaweed, spent hops, ground hoof and horn and paper wastes. Some types of sewage sludge may, however, add toxic chemicals in their own right, so care must be taken with their use (McGrath *et al.*, 1993).

The use of these materials can have many ameliorating effects on toxic starting materials including: (1) provision of good surface rooting conditions; (2) physical dilution of the concentration of the toxin concerned; and (3) chemical complexing so that the toxic element is locked up out of effective circulation, at least in the short term.

For certain severe pH problems, the most appropriate strategy may involve addition of chemical amendments. Severe alkalinity problems can be treated by the addition of elemental sulphur (Neilsen *et al.*, 1993), and acidification, a common feature on many pyrite-bearing coal and metal wastes, can be amended by the addition on limestone. As pyrite-containing spoils continue to produce sulphuric acid for many years, it is essential to be able to predict the amounts of limestone to be added. Huge amounts of $CaCO_3$, up to 400 t ha^{-1}, may need to be added, and these amounts need to be ripped into the surface layers to counteract not only the current low pH, but also the hidden acidity produced by continued pyrite oxidation (Costigan *et al.*, 1981). Jefferies (1981) suggested that coarse-grained limestone was more effective than fine-grained, and that 100 t ha^{-1} $CaCO_3$ was sufficient for the establishment of legume species. High salinity may also be encountered in many reclamation schemes. This is much more difficult to deal with and the most appropriate methods include the installation of good drainage to help remove the excess of salts by leaching, and additional irrigation with non-saline water to help this leaching process. If possible, topsoil or mulches should be added to the surface to help vegetation establish and reduce surface soil evapotranspiration. Almost certainly, salt-tolerant species appropriate to the climate will need to be selected. Such species might include *Atriplex nummularia*, *Choris gayana*, *Cistus salvifolius*, *Cynodon dactylon* and *Stipa tenacissima*.

Physical separation

This is to some extent an extension of the strategy of adding physical amendments, the difference here is that much greater depths tend to be used (>30 cm) to make sure that the roots of the plants do not grow into the toxic material and, in some cases, barrier layers (25–45 cm) of rock fill may also be put in place (Fig. 9.5). This approach is extremely expensive, especially if the barrier material and topsoil have to be imported onto the site. To put this in context a 75 cm deep layer of topsoil and rockfill is equivalent to 75 000 m^3 ha^{-1} of material that has to be obtained and moved into position by heavy machinery.

Biological approach

It is well known that plant species can evolve tolerance for adverse soil factors including heavy metals such as copper, zinc and lead (Smith & Bradshaw, 1979; Gray, this volume) but some species have

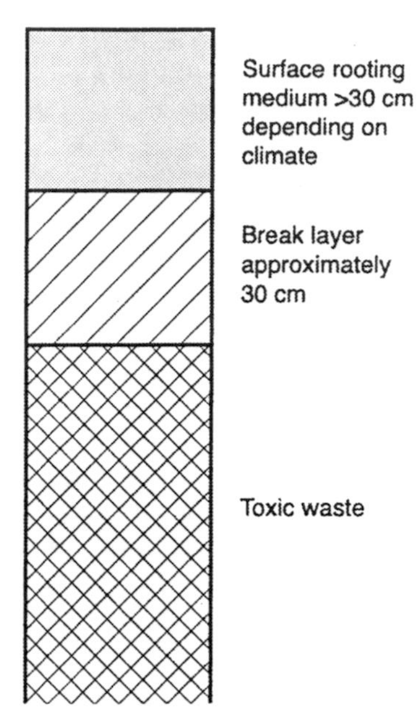

Fig. 9.5. Typical approach to the restoration of extremely toxic soils. The waste material is covered by a barrier or break layer of crushed rockfill (25–45 cm) plus a further covering of topsoil. After Williamson *et al.* (1982).

gone further and developed the ability to tolerate very large concentrations of certain elements. This latter group are termed hyperaccumulators, because they often contain very high concentrations of certain toxic elements. Clearly there is scope for using both groups of species in restoration schemes. At its simplest they can both be used to produce a vegetation cover on land that is contaminated with the element that they can tolerate, and it may be possible to go further and use hyperaccumulators to clean up the land by extracting the toxic element in question.

Use of tolerant materials

This approach was pioneered by Smith & Bradshaw (1979); they collected seed from metal-tolerant species, which had evolved tolerance naturally on metalliferous mining sites. They showed that these cultivars colonised faster, persisted longer (up to nine years), and produced a better stabilising cover than non-tolerant commercial varieties in contaminated sites. Three cultivars were produced commercially; *Agrostis capillaris* cv. Goginan, *Festuca rubra* cv. Merlin, *Agrostis capillaris* cv. Parys, tolerant to acidic lead/zinc-, calcareous lead/zinc- and copper-contaminated wastes respectively. This approach has not been adopted in many situations for metal mine revegetation, as amendment or barrier methods are preferred. However, where it has been used with locally collected tolerant material (Sudbury, Canada; Zimbabwe), it has been extremely successful (A.D. Bradshaw, pers. comm.).

Use of hyperaccumulators

Hyperaccumulator plant species are usually endemic to metalliferous soils, and are able to survive with much greater concentrations of toxic metals in their tissues than can be tolerated by most plant species. The threshold concentration used to define a species as a hyperaccumulator depends on the element in question, primarily because the background concentrations and relative toxicities vary between elements. Thus the threshold concentration on a dry weight basis is 100 $\mu g\ g^{-1}$ (0.01%) for cadmium, 1000 $\mu g\ g^{-1}$ (0.1%) for copper, cobalt and lead, but 10 000 $\mu g\ g^{-1}$ (1%) for manganese and zinc (Baker *et al.*, 2000; Whiting *et al.*, in press). Thresholds for other elements have not been defined.

The use of hyperaccumulator species is simply to grow the appropriate species on the contaminated soil, and use the growth of the harvested material as a metal-collecting device. The biomass is harvested, ashed to reduce its bulk, and the resultant ash can either be stored in a landfill site, or enter the metal ore market.

The efficacy of this approach depends on a number of factors. It is essential to have species that can accumulate large concentrations of the toxic element in a biomass that is worth harvesting. Thus, the ideal species would be high yielding and contain a large concentration of the element. However, many metal-tolerant species have evolved in infertile soils and hence have a low biomass; genetic engineering may be able to produce new breeds that can bring about the ideal clean-up species. Nevertheless, McGrath *et al.* (1993) managed to reduce the soil concentration of contaminated soil from 440 μg Zn g^{-1} to <300 μg Zn g^{-1}, i.e. below the threshold established by the Commission of the European Community (1986).

A range of hyperaccumulator species has been shown to have potential for phytoremediation, including: *Alyssum lesbiacum, A. tenium, A. murale, Brassica junceae, B. napa, B. rapa, Cardamine halleri, Cochlearia pyernica, Raphanus sativus, Reynoutria sachalinense, Thlaspi caerulescens, T. ochroleucum* among others (McGrath, 1995; Ebbs *et al.*, 1997; Robinson *et al.*, 1998). Baker *et al.* (2000) emphasise that hyperaccumulator plants are relatively rare, and are often in areas likely to be affected by mining activity. Hence there is an urgent need to document what is available as well as develop strategies for their use.

THE DEGRADING APPROACH: IMPOVERISHING THE SOIL NUTRIENT SUPPLIES

Reducing the soil nutrient supply is necessary where it has been increased to a level that can no longer support the target ecosystem. These high levels of nutrients can be brought about by three mechanisms: (1) natural succession, (2) pollution, and (3) fertilisation.

Natural successional processes

One of the original theories of succession 'autogenesis' (Clements, 1916), renamed 'relay floristics' and 'facilitation' (Egler, 1954; Connell & Slatyer, 1977), suggested that succession occurred via a series of steps and each community modified the site, and allowed invasion by species characteristic of the next state. Although not universal (Miles, 1979, 1987) it occurs in some successions, for example on the glacial moraines at Glacier Bay with *Dryas drummondii* as the pioneer nitrogen-fixer (Crocker & Major, 1955). Indeed, generalised models of succession (Odum, 1971; Gorham *et al.*, 1979) suggest an increase in total nitrogen during succession, and that there is an increased nutrient supply through cycling.

In some examples of succession, soil fertility does appear to increase. In the *Calluna* heathland to *Betula* woodland succession, Miles (1981) showed that soil structure changed from a podzol with very low rates of soil microbiological processes (nitrogen mineralisation was negative, indicating immobilisation of the soil inorganic nitrogen supply) to a brown-earth soil where decomposition and nutrient supply was much faster (Table 9.4). Moreover, when mid- to late-successional stages are removed and natural recolonisation allowed, a rapid invasion of high-yielding species often occurs. Where nitrogen-fixing *Ulex europaeus* was cut on sites which had formerly been chalk grassland, Green (1972, 1980) demonstrated 'soil seral eutrophication' with invasion by competitive nitrophiles (*Rubus idaeus, Chamerion angustifolium* and *Agrostis stolonifera*) rather than chalk grassland flora. Soil analysis indicated an increase in nitrogen of 770 kg ha^{-1}, which was similar to the 500 kg ha^{-1} found in the aboveground biomass of a 12-year-old stand of *U. europaeus*. Moreover, there was an increase of approximately 20 $\mu g\ g^{-1}$ in extractable nitrogen, an increase of almost 50% over grassland soils. This increased soil nitrogen supply under *Ulex* may be due to nitrogen fixation but, even where non-nitrogen-fixing species such as *Crataegus monogyna* are cleared from chalk grassland, and *Pteridium aquilinum* from heathlands, the same trend towards invasion by more productive species occurs (Grubb & Key, 1975; Marrs & Lowday, 1992). Grubb & Key (1975) suggested an invasion of *Galium aparine* and *Cirsium vulgare* after *Crataegus* removal was correlated with an increase in nitrogen mineralisation (from 7 to 34 $\mu g\ g^{-1}$) and extractable phosphorus (from 5 to 11 $\mu g\ g^{-1}$).

However, these increases are not universal. Robertson & Vitousek (1981) found a tenfold increase in nitrogen mineralisation in primary succession, but in a secondary succession the sequence was not consistent: the earliest stage dominated by annuals had the lowest rate (13 $\mu g\ N\ g^{-1}\ 30d^{-1}$) and the forest had the greatest (25 $\mu g\ N\ g^{-1}\ 30d^{-1}$), which were consistent with expectations; but the intermediate perennial-herb and shrub communities were transposed in sequence (17 and 13 $\mu g\ N\ g^{-1}\ 30d^{-1}$ respectively). Similar contradictory findings were found in a comparative study of two successional sequences on three types of parent material (sand, clay and limestone) in lowland England (Gough & Marrs, 1990*a*).

Thus, the relationship between soil change during succession is complex. In primary successions it is probable that nutrients and fertility increase (Marrs & Bradshaw, 1993). In secondary

Table 9.4. *Comparative data on a range of ecological processes across a chronosequence between* Calluna *and* Betula *woodland*

		Betula pendula				
	Calluna	18 years old	26 years old	38 years old	90 years old	LSD at 5% level
Mean number of earthworms per 1 m^2	1	5	27	127	78	26
Organic matter (g dm^{-3})	194	153	143	120	97	31
Cellulose decomposition						
At 0.4 cm	4.5	16	22	23	22	3.6
At 16–20 cm	3.2	6.0	9.0	8.9	17	4.2
Nitrogen mineralisation after 14 days incubation (mg dm^{-3} $week^{-1}$)	−1.3	25	41	45	40	10
pH	3.8	3.9	4.0	4.7	4.9	0.1
Exchangeable calcium (mg dm^{-3})	117	108	109	101	89	25
Total phosphorus (mg dm^{-3})	151	210	196	240	232	82
Carbon: nitrogen ratio	30	26	19	22	15	11
Carbon: phosphorus ratio	500	320	280	270	170	200
Carbon: potassium ratio	440	430	410	460	310	210
Bioassay (Radish) (mg)						
Shoots	87	18	43	59	66	18
Roots	3.3	16	39	74	77	30

Source: Miles (1981).

successions the results are less predictable, probably because of the differential positive and negative feedback effects on soil processes (Wedin & Tilman, 1990).

Pollution inputs

There have been there have been continual increases in the amounts of nitrogen added to soils in deposition over the last 100 years, both from rainfall and dry deposition. Estimates range from <4 to 12 kg N ha^{-1} yr^{-1} for rainfall inputs in Britain (Warren Springs Laboratory, 1987). However, values of 40–80 kg N ha^{-1} yr^{-1} near point sources have been recorded (Pitcairn *et al.*, 1991).

These increases have been implicated in changing ecosystem structure and function. In chalk grassland in the Netherlands, Bobbink & Willems (1987) documented an increase in the coarse grass *Brachypodium pinnatum* between 1956 and 1985, with a corresponding decrease in diversity. Moreover, During & Willems (1986) who repeated an earlier survey of bryophytes and lichens at four Dutch grasslands showed that after *c.* 30 years the lichens had almost completely disappeared, the acrocarpous and pleurocarpous mosses had been reduced, and species which tend to grow on litter (e.g. *Brachythecium rutabulum*) had increased. Similar studies on Dutch heaths have shown a long-term loss of *Calluna vulgaris*-dominated heathland, with change towards grasslands (Diemont & Heil, 1984). Heil & Diemont (1983) demonstrated that under a background rainfall input of 40 kg N ha^{-1} yr^{-1} and a small fertiliser input of 28 kg N ha^{-1} yr^{-1} (equivalent to the nitrogen transfers from *Calluna vulgaris* to the soil after defoliation from heather beetle attacks), *Deschampsia flexuosa* grassland developed. In this study phosphorus additions also produced a similar change towards grassland, although less than with nitrogen. Similar vegetation change has been found in Britain in areas and whilst high nitrogen deposition has been implicated (Lee *et al.*, 1987; Pitcairn *et al.*, 1991), complex interactions with other climatic and management factors are also implicated (Marrs, 1993; Britton, 1998).

Effects of residual fertilisers

It is well known that additions of fertilisers can reduce species diversity in hay meadows (Digby & Kempton, 1987; Smith, 1988), wet, mesotrophic grasslands (Mountford *et al.*, 1993), sand dune vegetation (Willis, 1963; Boorman & Fuller, 1982), chalk grassland (Smith *et al.*, 1971) and in old-field successions (Maly & Barrett, 1984; Hyder & Barrett, 1986). In all these examples fertilisers were added to relatively species-rich ecosystems. Where fertilisers and grazing have been used together in species-poor grasslands, species diversity increased (Jones, 1967; Harper, 1971). Here there were only one or two dominant species, and when fertiliser was applied, the composition of the grassland changed to a more mixed grassland community with more than 12 dicotyledons (essentially this is a demonstration of the effect described by Grime [1979] for soils of different pH; Fig. 9.1b). This change only occurred, however, when grazing was strictly controlled in a grazed–ungrazed cycle. Where grazing was unrestricted, little change occurred.

Fertiliser additions are never fully accounted for in increased plant growth, with some leaching out or being bound up in the soil as residues. These residues may remain in the soil for many years (Johnston & Poulton, 1977) and influence plant growth. This is a particular problem for restoration work on agricultural land that is being taken out of productive use with the aim of restoring communities of high conservation value, as is the case in many European countries where agricultural and environmental policies dictate reduced production and set increased biodiversity targets.

Irrespective of how the nutrient supply has been elevated, the first part of any restoration scheme is to define the problem accurately and determine how the problem has arisen: essentially setting the parameters on Fig. 9.2 in terms of soil chemical status and environmental drivers that may influence management towards the target. Where an aspect of soil fertility has been elevated beyond that needed to sustain the target community, there is a need to reduce those supplies. This can only be done in two ways: by increasing the losses relative to inputs, or by sequestration into an unavailable form. These translate into two practical strategies; directly, where the aim is to remove nutrient through cropping or offtake; or indirectly where plant-available nutrients are either accumulated in unavailable stores or lost through non-managed pathways. Whilst this simple classsification clarifies thinking, in reality, management strategies to reduce soil nutrients probably involve more than one method. In addition, in recent times, there has been a need to degrade soils by acidification, for example on high-pH, ex-arable soils so that they can support plant communities typical of much more acidic soils.

In any event, the treatment is best viewed from an ecosystem perspective (Marrs, 1993). The ultimate aim of any degrading strategy should be to identify all of the processes used in management so that: (1) nutrient offtake plus losses are greater than inputs; (2) the treatment effects a reduction in plant-available nutrient supplies; and (3) after treatment soil conditions approximate to the required target (Fig. 9.2).

Maximising offtake

Offtake can be manipulated by grazing, cropping, burning and total soil removal. These treatments represent an increasing scale of impact but decreasing scale of frequency. Grazing frequencies range from daily, through different seasonal patterns, to occasional treatments where animals graze tall vegetation down to a short sward or even bare ground. Cutting frequencies vary from several cuts per year to a cut in occasional years, burning is generally used on an infrequent basis perhaps, on a five- to ten-year cycle, and topsoil removal is a severe treatment that is usually only used once or on a very long-term rotation. Alternatively, offtake can occur through enforced leaching as a result of fallowing.

Grazing

It is often assumed that grazing removes a crop from the site and thus there must be a significant reduction in nutrients. However, when nutrient removal is measured actual losses may be small, and nutrient cycling may be enhanced, effectively negating the treatment.

A good example of this is the sheep grazing study (densities varied between 5–4 and 15–5 ewe units ha^{-1}) on *Agrostis–Festuca* grassland at Llyn Llydaw,

Wales, where some nutrients were removed as carcasses and wool. However, the amounts removed were low compared with measured nutrient inputs; the net removal relative to inputs were ×1.2 for phosphorus, ×3.6 for potassium, and there was no net removal of nitrogen, calcium or magnesium (Brasher & Perkins, 1978). Thus, losses of all elements were either marginal or compensated in rainfall. Moreover, rainfall inputs did not include occult deposition, an important source of nitrogen in the uplands (Fowler, 1987). Thus, the net offtake by the sheep may be less than estimated here. The amount of nutrients recycled by grazing was almost twice that removed from the site, probably because it removed the blockage to cycling in aboveground vegetation (Bülow-Olsen, 1980; Marrs *et al.*, 1980).

The impact of grazing could be enhanced by modifying grazing practice. Received wisdom from historical studies of grazing systems suggests that nutrient offtake can be increased if the animals are allowed to graze during the day, but removed from the site at night, where additional nutrients are lost through urine and manure deposition. This strategy is used to manage Dutch and German heaths (Gimingham & de Smidt, 1983). Although more nutrients should be taken off the site using this approach, this method has not been quantitatively tested.

An alternative may be to increase the harvest of natural grazing animals. It has often been suggested that removal of nutrients by European rabbits (*Oryctolagus cuniculus*) managed in warrens on many areas of semi-natural vegetation was one way that nutrients were lost in the past. In theory there should be a drain on nutrients because a crop is taken off the land with no nutrient imports as feedstuffs into the grazing system, as is the case with other forms of livestock husbandry. However, warreners often used to import new rabbit stock, and add feed, at critical times of the year (Crompton & Sheail, 1975). Using literature data Marrs (1993) showed that the nutrient offtake in warrened ecosystems was small (0.04 kg N ha^{-1} yr^{-1}; 0.031 kg P ha^{-1} yr^{-1}; 0.01 kg K ha^{-1} yr^{-1}), compared to sheep. Even so, these losses might have been of vital significance in maintaining ecosystems with very low productivity, especially when combined with sheep grazing.

Grazing is often combined with cropping for hay or silage. Whilst this combined treatment should increase annual offtake, as yet there have been few quantitative studies in restoration schemes.

Cropping vegetation

There are three ways of cropping vegetation to reduce soil fertility: arable cropping, hay/silage cropping and non-agricultural cropping.

Arable cropping

Taking an arable crop continuously without the addition of fertilisers will eventually reduce the nutrient capital and more importantly the nutrient supply. Much of our knowledge comes from a series of long-term continuous-cropping experiments at Rothamsted Experimental Station set up 100–150 years ago to assess the financial value of the residual manures which farmers had applied during their tenure of the land (Johnston, 1970). The long timescale of this work also allows us to determine how long it takes to reduce soil fertility in terms of both total and plant-available nutrients.

In 1903, nitrogen concentrations were elevated only in the plots to which farmyard manure had been applied (Johnston & Poulton, 1977), and inorganic nitrogen fertiliser additions had no effect on soil nitrogen concentration. Where a residue of nitrogen had accumulated in the manured soil (4850 kg ha^{-1}) the subsequent relationship between total soil nitrogen through time was modelled (Jenkinson & Johnston, 1977; Johnston & Poulton, 1977). They suggested that an equilibrium concentration of 2980 kg N ha^{-1} would eventually be achieved but there would be a depletion to 50% of the residue in 50 years.

For phosphorus and potassium the situation is more complex. Johnston & Poulton (1977) measured three parameters for soil P (total P, isotopically exchangeable P and bicarbonate-soluble P) and exchangeable K in stored soils, collected at intervals between 1903 and 1974 in ten different treatments. From these data, Marrs (1993) estimated the linear depletion rate for each parameter and the time needed to reach the 1903 untreated values. Depletion of plant-available phosphorus (bicarbonate-soluble and isotopically exchangeable forms) took

71 years, and total phosphorus took 90 years. For potassium, depletion rates showed much more variability. The quickest depletion was 10 years and the slowest was 386 years.

An implicit assumption in cropping is that because a crop is removed there will be an automatic removal of nutrients. However, in a study of cereal cropping designed to deplete nutrients on a nature reserve, Marrs (1985) showed that if grain alone was considered, there was no net removal of nitrogen, although there was a loss of both potassium (3 × inputs) and especially phosphorus (8 × inputs). If all of the aboveground material was removed, either as straw or by burning, a net loss of nitrogen (2 × inputs) occurred, and losses of potassium and phosphorus were greatly enhanced (Table 9.5). Depletion of phosphorus and potassium using cereal crops can be improved by adding inorganic nitrogen fertilisers. Although the strategy of adding fertilisers to reduce fertility may appear absurd, it is realistic because: (1) soil nitrogen does not build up under cereal cropping, and (2) plant growth and nutrient uptake, and hence offtake, are enhanced by nitrogen fertiliser addition. This strategy has been demonstrated to be effective in a long-term experiment at Broadbalk (Dyke *et al.*, 1983), where more nutrients were extracted from the soil of the nitrogen-treated plot than the untreated plot (Table 9.6) and the phosphorus store in the soil was almost completely exhausted.

Table 9.5. *Nutrient budgets of arable ecosystems: an estimate of nutrient inputs and outputs in a scheme to reduce soil fertility by cropping with cereal rye at Roper's Heath, Suffolk, UK*

	Element (kg ha^{-1})		
	Nitrogen	Phosphorus	Potassium
Inputs			
Rainfall	16.9	0.11	3.2
Cereal seed	2.1	0.6	1.6
Total	19.0	0.71	4.8
Removals			
Grain	18.0	5.0	14.0
Excess over inputs	0	×8.2	×2.9
Total aboveground standing crop	35.0	10.0	61.0
Excess over inputs	×1.9	×16.4	×12.7

Source: Marrs (1993).

Table 9.6. *Nutrient budgets of arable ecosystems: the effects of adding nitrogen fertiliser on the offtake of nutrients in the Broadbalk experiments (1970–5)*

	Yield	Nutrient content (kg ha^{-1})			
	(kg ha^{-1})	Phosphorus	Potassium	Calcium	Magnesium
No nitrogen	2.86	6.2	15	5	2.2
Nitrogen	5.38	9.0	25	10	3.3
Increased by nitrogen fertiliser	2.52	2.8	10	5	1.1

Source: After Marrs (1993).

Hay/silage cropping

In a Dutch nature reserve that had been cropped for hay for several decades with the liberal use of fertiliser until 1973, Petgel (1987) and Bakker (1989) showed that when the fertiliser applications were stopped, the crops were maintained at reasonably high levels for five years (6000–8000 kg ha^{-1}), but thereafter fell to below 4000 kg ha^{-1} in all years, except 1983 when the yield was >6000 kg ha^{-1} (Fig. 9.6). Soil changes were also documented: water-extractable phosphorus was reduced to an apparent equilibrium value within two years, but extractable potassium fell slowly over the nine-year period until 1982. In similar studies, Oomes (1990) found hay yields reduced from 10 200 kg ha^{-1} to 6500 kg ha^{-1} in four years and 4100 kg ha^{-1} in nine years on sandy soils. On clay there was an initial reduction in yield from 10 200 to 5000 kg ha^{-1}, but thereafter it increased again.

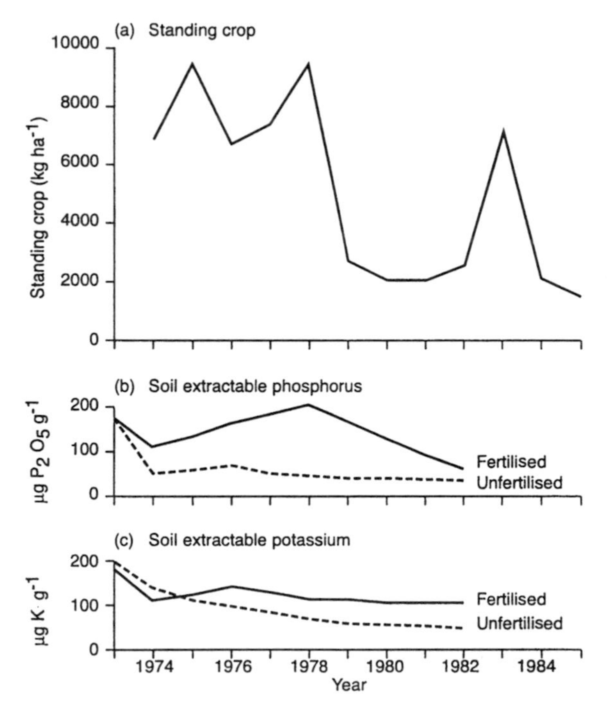

Fig. 9.6. The effects of continued cutting of grassland after fertiliser addition was stopped in 1973 on (a) standing crop, (b) soil-extractable phosphorus and (c) soil-exchangeable potassium. Soil data from fertilised (solid line) and unfertilised (dashed line) plots are shown. Data from Petgel (1987).

Hay cropping without fertilizers can, however, reduce the fertility of some soils relatively rapidly. Gough & Marrs (1990*a*, *b*) in chronosequence studies found depletion rates in a similar order of magnitude; 12 years on an arable sandstone site in Nottinghamshire, and four years on an arable clay site in Essex. In both these studies soil-extractable phosphorus concentrations had fallen to near semi-natural levels, and were supporting typical species-rich meadow vegetation. Hay cropping is also important in maintaining semi-natural vegetation; Wells (1980) showed that continuous removal of clippings for eight years reduced both the extractable phosphorus and exchangeable magnesium concentrations. Addition of nitrogen fertilisers to grassland to effect an increased removal of other elements has been shown to be effective (van der Woude *et al.*, 1994).

Non-agricultural crops

On many nature reserves cropping vegetation is often required to prevent succession. Examples include the removal of: (1) accumulated grass or *Calluna* in grassland and heathland management programmes; (2) late-successional species including bracken (*Pteridium*), invading shrubs and trees; and (3) accumulated litter (*Pteridium*, grass and *Calluna*). Few studies have documented the effects of these treatments on nutrient budgets, but presumably the impact will be similar, in terms of the amounts removed, to burning.

Prescribed burning

Burning vegetation removes nutrients that have accumulated in successional ecosystems in a similar way to cutting and is typically used as a management tool on heathlands and grasslands. A major advantage of burning is that, when done properly, it removes the majority of the aboveground biomass and litter. In grassland, all the accumulated thatch will be removed and, in heathlands, the old bushes and litter are removed with the bushes regenerating from the old stems.

Burning tends to remove a proportion of all nutrients in the smoke. Chapman (1967), for example, showed that controlled burning of lowland heath removed large amounts of the nutrients in the aboveground vegetation (95% of nitrogen, 26% of phosphorus, 21% of potassium), but that only nitrogen and phosphorus could not be replaced by rainfall inputs over a 12-year period. After burning, nutrients not lost in smoke are deposited on the soil as ash, and are either leached, fixed or taken up by developing vegetation. Kenworthy (1964) and Allen (1964) have shown that potassium was leached rapidly after burning but phosphorus was fixed, rendering it less available for plants in the short term. Chapman *et al.* (1989*a*, *b*) using a simulation model predicted that leaching losses of phosphorus were significant on the heaths studied, and that the phosphorus adsorption capacity of the soils was crucial in controlling the phosphorus dynamics in lowland heath ecosystems (Chapman *et al.*, 1989*b*). They also suggested that nitrogen losses after burning may limit growth of vegetation developing in the period immediately after burning.

Topsoil removal

Soil organic matter and nitrogen and phosphorus stores are typically concentrated in the surface layers in most ecosystems. This accumulated nutrient capital can be removed instantly by stripping, irrespective of whether it is in unavailable forms or not.

Topsoil removal has been used as part of the management of semi-natural vegetation for centuries. Cutting peat and surface mor from heathlands was often done as part of 'turbary rights' exercised on many commons in Britain, and a similar practice 'plaggen' was used on Dutch heathlands (Gimingham & de Smidt, 1983). Many of these practices are no longer exercised in the UK, with the consequence that the soil nutrient supplies are increasing. There have been several attempts to reintroduce plaggen in Holland (Bakker, 1989), and there have been some suggestions that similar conservation management should be attempted in Britain (Marrs, 1985; Chapman *et al.*, 1989*b*; Smith *et al.*, 1991; Traynor, 1995). This apparently drastic approach has several positive advantages: (1) the fertility can be reduced and (2) the topsoil can be sold. Moreover, soil from a nature reserve could be sold at a premium rate if it has a seed bank containing many species of high conservation interest (Putwain & Gilham, 1990). However, as the seeds are also distributed in the surface layers a balance must be struck between nutrient and seed bank removal.

Enforced leaching by fallowing

Fallowing can be used to enforce nutrient loss from topsoil through leaching. The aim is to prevent, by tillage or herbicide use, any plant growth and hence uptake of nutrients from the topsoil. Nutrients are still produced from soil microbial turnover and decomposition and these are susceptible to leaching, especially in moderate to high rainfall. This approach has been assessed experimentally at Rothamsted where the soil organic matter content of grassland was reduced more quickly under bare fallow than other cropping systems (Fig. 9.7a), but it still took 15 years to halve the organic carbon content. In simple lysimeter studies of abandoned arable soils where the intention was to restore chalk grassland, leaching losses were much greater under fallow than under grass (Fig. 9.7b). Losses of nitrogen were greater (up to 500 kg N ha^{-1}) than those of phosphorus (<10 kg P ha^{-1}) over an eight-month over-winter period (Marrs *et al.*, 1991).

Sequestration techniques to reduce nutrient availability

Nutrients can be sequestered within an ecosystem so that they no longer contribute to plant productivity. The total amount of the nutrient on an area basis remains the same but a fraction is locked out of circulation by sequestration in an unavailable pool. Clearly, the impact of sequestering nutrients in ecosystem pools can be enhanced if it is combined with an offtake strategy as described above.

Burial

As the nutrient content generally declines with depth, one simple way to reduce soil nutrient supply is to mix nutrient-rich surface layers with deeper nutrient-poor horizons. This treatment has two potential benefits. First, there is a dilution of the nutrients within a formerly nutrient-rich zone, and second, a proportion of the available nutrients is placed physically below the rooting zone of shallow-rooted plants and may be more susceptible to

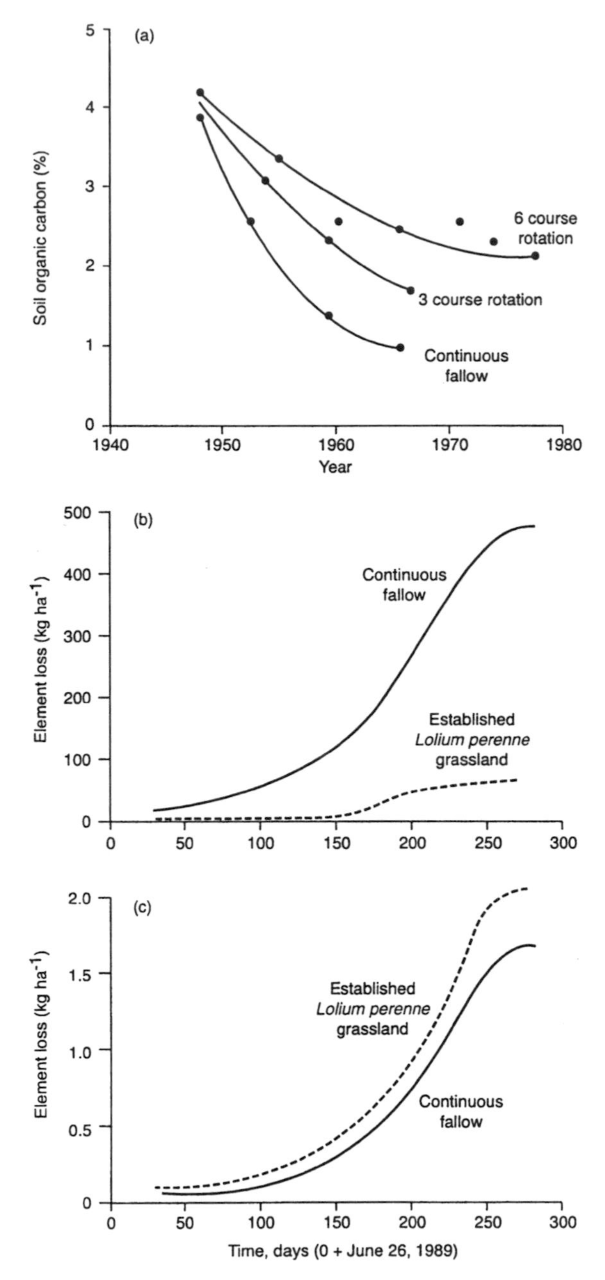

Fig. 9.7. The loss of nutrients through fallowing. (a) The effects of two farming systems and continuous fallow on soil organic matter (a surrogate measure of total soil nitrogen). From Jenkinson (1988). Leaching losses of (b) nitrogen and (c) phosphorus in a lysimeter study of chalk grassland soils. From Marrs *et al.* (1991).

leaching. Experimental studies of plant growth on mixtures of various parts of the soil profile in east-central Texas showed a 50% reduction in extractable phosphorus in a mixture of the 0–60 cm horizon, from 23 $\mu g\ g^{-1}$ in the 0–20 cm fraction to $<12\ \mu g\ g^{-1}$ (Chichester, 1983). Growth of test plants reflected the reduction in soil fertility (Fig. 9.8). The effectiveness of this strategy depends on the rates at which deep-rooting species colonise, the effectiveness with which these species tap nutrients at depth, and leaching rates. There is very little quantitative information on any of these important processes in practical restoration use.

Chemical amendment

Soil-nutrient availability changes in response to other factors such as pH and redox potential. As the leaching of many elements is increased in acidic conditions (Table 9.7), reducing pH artificially might enhance leaching losses. Acidic material (e.g. acidic peats, sulphur and acidic wastes containing pyritic material) could be incorporated into fertile soils to accelerate this process. Alternatively, rotational waterlogging/drainage may also have similar effects. However, the change in pH may cause its own difficulties.

Recent attempts to use chemical amendments have been promising (C.C. Stuckey, unpubl. data). Two chemicals (aluminium sulphate and ferric sulphate) have been added to soils, singly and in combination, with the aim of reducing the available phosphorus concentrations by chemical complexing. Initial results show that the available phosphorus concentrations have been halved within a month in the most effective treatment – ferric sulphate (Fig. 9.9). Care must be taken to ensure that the application rates are suitable, as too much can cause severe damage to the physicochemical structure of the soil (J. Tallowin, pers. comm.). More research in this area is needed.

Accumulation in organic matter

Organic matter is another pool that can be used to accumulate nutrients, which is known to build up in many boreal and temperate successions. This organic matter contains large amounts of nutrients,

Table 9.7. *Effects of irrigating a coniferous soil with acidic solutions (H_2SO_4) on base saturation*

				Quantity of base cations (keq ha^{-1})			
					Mineral soil		
Treatment	Acid added (keq H^+ ha^{-1})	pH of A_0	Base saturation of A_0 (%)	A_0	0–5 cm	5–10 cm	Total
Control	0	4.0	16.0	5.2	1.8	0.7	7.7
Acid 1	6.1	3.9	12.2	3.2	1.9	0.6	5.7
Acid 2	13.3	3.8	9.0	2.1	0.9	0.5	3.5

Source: Farrell *et al.* (1980).

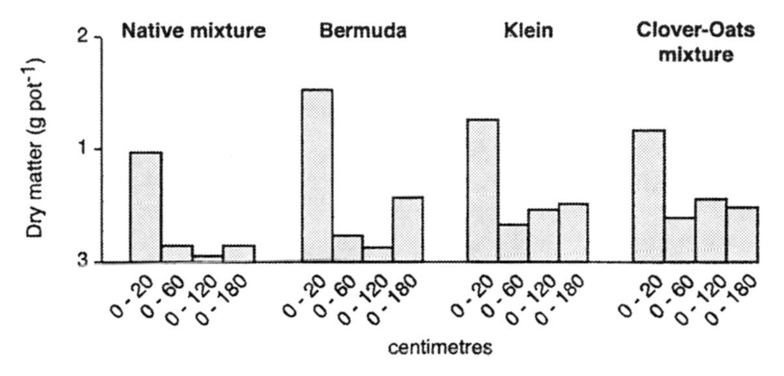

Fig. 9.8. The yield of four bioassay test plants on various mixtures of soil horizons. Soil profile mixes: 1 = 0–20 cm; 2 = 0–60 cm; 3 = 0–120 cm; 4 = 0–180 cm. Species: native mixture = *Bouteloua curtipendula* plus *Schizachyrium scoparium*; Bermuda = *Cynodon dactylon*; Klein = *Panicum coloratum*; clover–oats = *Trifolium vesiculosum* plus *Avena sativa*. Data from Chichester (1983).

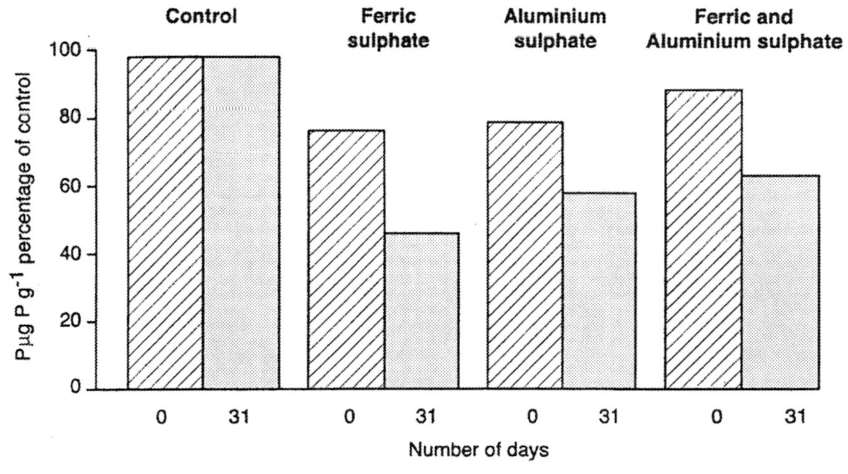

Fig. 9.9. Effects of chemical amendments on availability of soil phosphorus in a clay soil in eastern England: addition of ferric sulphate and aluminium sulphate, singly and in combination, markedly reduced the availability of soil phosphorus relative to untreated plots within one month. From C.C. Stuckey (unpubl. data).

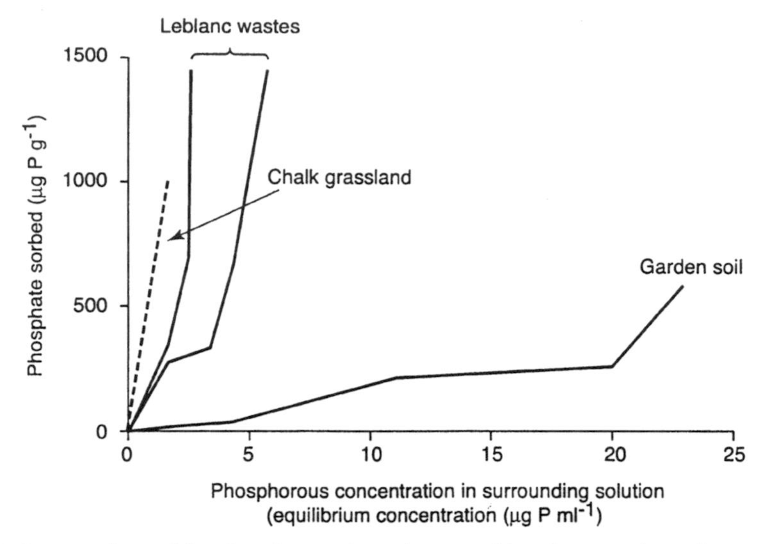

Fig. 9.10. A comparison of the phosphorus adsorption capacities of contrasting soil types; the calcareous and Leblanc soil are able to adsorb much more phosphorus than the normal garden soil and hence could be used as an amendment to reduce available soil phosphorus concentrations. From Marrs & Bradshaw (1993).

which are locked out of effective circulation at least in the short term. However, as we have already seen, nutrient mineralisation is an important process of soil nutrient cycling, and in many ecosystems the mineralisation supply will be in proportion to the capital: as the capital increases so too does the supply. For example, if nitrogen mineralises at about 2% yr^{-1} (Reuss & Innis, 1977), and if the capital increased from 1000 to 5000 kg N ha^{-1}, the available supply increases from 20 to 100 kg N ha^{-1} yr^{-1}. The turnover of all nutrients within soil organic matter will also be affected by resource quality, which in turn is affected by the quantity and quality of litter inputs, as well as the occurrence of soil flora and fauna (Swift *et al.*, 1979). Unfortunately, our knowledge of these processes is rudimentary.

Adsorption on soil minerals

It may also be possible to lock up nutrients, particularly phosphorus, through adsorption on the mineral matrix of the soil. Indeed phosphate adsorption curves are used to categorise soils for their ability to immobilise phosphorus added as fertiliser (Ozanne & Shaw, 1967). For instance, chalk grassland soil and Leblanc alkaline wastes have a much greater adsorption capacity than garden soil and hence require a much greater amount of phosphorus fertiliser to maintain the same phosphorus supply (Fig. 9.10). There is no reason why the phosphate adsorption capacity of a given soil cannot be manipulated to advantage in restoration schemes. Deep ploughing to incorporate fertile surface soils into the subsurface layers and addition of limestone or other substances, like the Leblanc wastes, are possibilities. Deep ploughing is particularly useful if subsurface layers have a high proportion of chalk, limestone or clay materials, all with a high adsorption capacity. Addition of limestone to surface layers might reduce the phosphorus supply, at least in the short term but, as calcium is easily leached, longer-term effects may be less predictable. There are, however, few published studies where the adsorption capacity of the soils has been deliberately manipulated to reduce the phosphorus availability for restoration purposes.

Acidification

A special case of the degrading approach has been highlighted in recent years, where the pH of the starting soil is too high for the desired ecosystem as a result of previous land use, and there is a need to acidify the soil (see Webb, volume 2). Usually this situation occurs on land that has been modified for agricultural purposes by liming or marling: both processes increase the pH of the soil but marling also affects the physical structure of the soil, by increasing the clay content and hence the cation exchange capacity. Under most temperate conditions the calcium is readily leached from the system by weakly acidic rainfall and hence agricultural treatments need to be repeated relatively frequently on most soils, the aim being to maintain soil pH between pH 6 and pH 7. In recent times, land that is marginal for agricultural use has being taken out of production and targeted for the re-creation of semi-natural ecosystems. The reasons for this are policy-driven: from an agricultural view – a need to reduce overall national production, and from a conservation view – a need to achieve Biodiversity Action Plan Targets developed as a response to the Rio Convention. Whilst we could wait for a very long time and allow natural processes to acidify the soil, restoration ecologists have attempted to develop more proactive methods, essentially by the addition of acidifying amendments.

There are well-known and predictable techniques for assessing the amount of lime needed to increase the pH of an acidic soil to that required for agricultural purposes (Jackson, 1958); there are no such equivalent approaches for acidifying soils. Thus, most attempts to acidify soils for restoration purposes have used an empirical approach to assess the amount of acidifying amendment needed for a given situation. Perhaps one of the most detailed that has been attempted was at the Minsmere RSPB reserve in Suffolk, UK (Owen *et al.*, 1999). Here, the aim was to develop techniques for the acidification on ex-arable soil in 11 fields in the centre of heathland fragments, so that heathland and acidic grassland communities could be established, with the ultimate aim of producing a single heathland block. The initial soil pH was *c.* pH 7, and the target pH was between pH 3.5 and pH 4, required to allow heathland and acid grassland species to grow, but at the same time restrict the growth of competitive ruderal species (Davy *et al.*, 1998; Owen & Marrs, 2000*a, b*). Experiments were set up in two fields and three potentially acidifying amendments were tested, at a range of application rates; the amendments were: (1) *Pinus sylvestris* chippings, (2) litter of bracken (*Pteridium aquilinum*), and (3) elemental sulphur. Elemental sulphur is a well-known soil acidifying agent for calcareous soils, it is relatively cheap and gives a long-lasting response because of its slow oxidation, which is enhanced by microbial activity (Nielsen *et al.*, 1993).

The results from this study showed that neither of the plant amendments reduced the pH to the target range (*Pinus* = pH 6–6.5; *Pteridium* = pH 4–4.5), but that elemental sulphur could achieve the target range if an appropriate application rate was applied. However, several important features emerged from this study. The soil pH changed through time depending on season and application rate, with most soils showing a complex relationship (Fig. 9.11) with a decline to a minimum followed by a slow recovery. Moreover, different results were found between the two fields sampled even although they were situated <1 km apart, suggesting that detailed empirical studies need to be done at a very local level. Owen *et al.* (1999) used the derived statistical relationships in their study to predict the pH after five and ten years and the theoretical equilibrium pH, as well as the amount of sulphur to be added to achieve a given pH. Thus, in the two different fields it was estimated that 2.1 and 3.4 t S ha^{-1} were needed to reduce soil pH to pH 4, and 4.6 and 5.3 t S ha^{-1} to reach pH 3.5. Owen & Marrs (2001) have suggested that there may be scope for mixing *Pteridium* litter with elemental S to reduce overall costs of acidification partly through a synergistic effect of the two treatments. The *Pteridium* litter produces a rapid acidification to a pH of *c.* pH 4 and the S, although slower acting, then reduces the pH below pH 4 and maintains this low level. Owen *et al.* (1999) also showed that there were few indirect effects of acidification on the chemical properties of the soil that might cause long-term problems for conservation, although considerable acidification was found at the lowest depth tested when high rates were applied. This acidification might cause problems if it were to leach into watercourses.

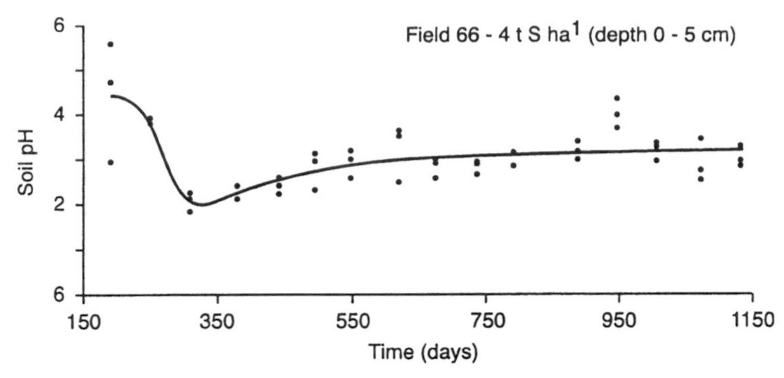

Fig. 9.11. An example of the temporal response of arable soils to addition of elemental sulphur (4 t S ha^{-1}) during an acidification scheme at Minsmere, Suffolk, UK; note the rapid decline in soil pH followed by a slow recovery to an asymptote. From Owen *et al.* (1999).

However, the amount of sulphur required in different situations still has to be calculated empirically, essentially because the results vary by order of magnitude in different situations. On the one hand, Williams *et al.* (1996) needed just 0.1 t S ha^{-1} to reduce soil pH by <1 pH unit to recreate heathland on a freely drained acid podzol, whilst Nielsen *et al.* (1993) required 4.5 t S ha^{-1} to effect a pH reduction of *c.* 1.5 pH units on a calcareous soil with a starting pH between 7.3 and 8.5. Chambers *et al.* (1996) in a heathland re-creation scheme in the Brecklands of Suffolk required similar amounts to those at Minsmere, but the amount required depended on the starting soil material. Thus, 9 and 18 t S ha^{-1} reduced soil pH from pH 8 to pH 3.4 and pH2.8 respectively at their fertile site, but 3 t S ha^{-1} reduced soil pH from pH 5.7 to pH 2.9 at their infertile site. Essentially, different amounts are needed depending on the starting pH, the desired end point and the soil chemistry.

The most commonly used amendment for acidification has been elemental sulphur, but other materials containing sulphur may also be effective. Dunsford *et al.* (1998) described an experiment in which a naturally occurring deposit of pyritic peat was used to acidify agricultural soils with the aim of establishing lowland heath. Large amounts of the peat were used (incorporation to 0.5 m at 50% and 75% by volume or a 10-cm blanket overlying a 75% mixture) and there was rapid initial pH reduction to between pH 2.5 and pH 3, followed by a rise over five years to stabilise between pH 3 and pH 4. Good *Calluna* establishment occurred especially when the pH was between 3 and 4.

Other sulphur-containing materials are now available for acidifying treatments, even though many of them have not been tested in restoration schemes. One relatively new innovation is the use of pelletised sulphur, which has been shown to be effective in initial trials (C.C. Stuckey, unpubl. data). Pelletised sulphur may be slower to take effect than the elemental form, as it may take more time to break down and oxidise. However, pellets are much easier to apply from a health and safety viewpoint.

CONCLUDING REMARKS

This chapter has highlighted the importance of the soil in restoration schemes, taking two separate approaches to achieve the target ecosystem desired. It is implicit in this approach that the starting position and the final desired target must be known. Thereafter, knowledge of ecosystem processes and a proper quantification of the impacts of applied treatments are essential if a scheme is to be successful. Unfortunately, we are a long way from being able to predict the outcome of many management treatments on soil processes and nutrient cycling. Thus treatments often have to be applied using best judgement rather than a certainty. However, if restoration ecologists consistently quantify the effects of treatment when soils are being manipulated, then a collective benefit will be derived as the knowledge base is increased.

However, it would be foolish to believe that soil component can be isolated from other parts of the ecosystem and the suite of management practices applied. This was first discussed by Rorison (1971), where he argued that grazing and soil toxicity, both applied as management treatments, could counteract the effects of the increased soil nutrient supply. Grazing, for example, is often implemented routinely as part of restoration management to reduce the aboveground biomass, to prevent the ingress of late-successional species and to create microhabitats for regeneration (see Duffey *et al.*, 1974; Wells, 1980; Bakker, 1989). Their impact on the soil, although important, is not the prime purpose of its use. Hence, anyone involved in ecosystem restoration must take a whole ecosystem view, and attempt to consider the impacts of all management in a systematic way (see Bradshaw, this volume; Cairns, this volume).

ACKNOWLEDGMENTS

I thank the very many colleagues who have contributed wittingly or unwittingly in the development of the ideas in this paper; these include Professors A.D. Bradshaw FRS and P.J. Grubb and Drs M.W. Gough, K.M. Owen, R.D. Roberts, R.A. Skeffington and C.S.R. Snow. Clare McLoughlin and Sandra Mather helped prepare the typescript and diagrams respectively.

REFERENCES

Allen, S.E. (1964). Chemical aspects of heather burning. *Journal of Applied Ecology*, **1**, 347–367.

Baker, A.J.M., McGrath, S.P., Reeves, R.D. & Smith, J.A.C. (2000). Metal hyperaccumulators plants: a review of the ecology and physiology of a biological resource for phytoremediation of metal-polluted soils. In *Phytoremediation of Contaminated Soil and Water*, eds. N. Terry & G.S. Bañuelos, pp. 85–107. Boca Raton, FL: CRC Press.

Bakker, J.P. (1989). *Nature Management by Grazing and Cutting*. Dordecht: Kluwer.

Bloomfield, H.E., Handley, J.F. & Bradshaw, A.D. (1981). Top soil quality. *Landscape Design*, **135**, 32–34.

Bobbink, R. & Willems, J.H. (1987). Increasing dominance of *Brachypodium pinnatum* (L.) Beauv. in chalk grasslands: a threat to a species-rich ecosystem. *Biological Conservation*, **40**, 301–314.

Boorman, L.A. & Fuller, R.M. (1982). Effects of added nutrients on dune swards grazed by rabbits. *Journal of Ecology*, **70**, 345–355.

Bradshaw, A.D. (1983). The reconstruction of ecosystems. *Journal of Applied Ecology*, **20**, 1–18.

Bradshaw, A.D. & Chadwick, M.J. (1980). *The Restoration of Land*. Oxford: Blackwell.

Bradshaw, A.D., Humphries, R.N., Johnson, M.S. & Roberts, R.D. (1978). The restoration of vegetation on derelict land produced by industrial activity. In *The Breakdown and Restoration of Ecosystems*, eds. M.W. Holdgate & M.J. Woodman, pp. 249–274. New York: Plenum Press.

Brasher, S. & Perkins, D.F. (1978). The grazing intensities and productivity of sheep in the grassland ecosystem. In *Production Ecology of British Moors and Montane Grassland*, eds. O.W. Heal & D.F. Perkins, pp. 354–374. Berlin: Springer-Verlag.

Brimblecombe, P. & Stedman, D.H. (1982). Historical evidence for a dramatic increase in the nitrate component of acid rain. *Nature*, **298**, 460–461.

Britton, A.J. (1998) Modelling invasions on heathlands. PhD Thesis, University of Liverpool, UK.

Bülow-Olsen, A. (1980). Nutrient cycling in a grassland dominated by *Deschampsia flexuosa* (L.) Trin. and grazed by nursing cows. *Agro-Ecosystems*, **6**, 209–220.

Chambers, B.J., Cross, R.B. & Pakeman, R.J. (1996). Recreating lowland heath on ex-arable land in the Breckland Environmentally Sensitive Area. *Aspects of Applied Biology*, **44**, 393–400.

Chapin, F.S. III (1980). The mineral nutrition of wild plants. *Annual Review of Ecology and Systematics*, **11**, 236–260.

Chapman, S.B. (1967). Nutrient budgets for a dry heath ecosystem in the south of England. *Journal of Ecology*, **55**, 677–689.

Chapman, S.B., Rose, R.J. & Clarke, R.T. (1989). A model of phosphorus dynamics of *Calluna* heathlands. *Journal of Ecology*, **77**, 35–48.

Chichester, F.W. (1983). Premining evaluation of forage grass growth on mine soil materials from an east-central Texas lignite site. 2: Soil profile horizons. *Soil Science*, **135**, 236–244.

Clements, F.E. (1916). *Plant Succession: An Analysis of the Development of Vegetation*. New York: Carnegie Institute.

Commission of the European Community (1986). Council directive of 12th June 1986 on the protection of the environment and in particular of the soil, when sewage sludge is used in agriculture. *Official Journal of the European Communities L181 (86/278/EEC)*, 6–12.

Connell, J.H. & Slatyer, R.D. (1977). Mechanisms of succession in natural communities and their role in community stability and organization. *American Naturalist*, **111**, 1119–1144.

Costigan, P.A., Bradshaw, A.D. & Gemmell, R.P. (1981). The reclamation of acidic colliery spoil. 1: Acid production potential. *Journal of Applied Ecology*, **18**, 865–878.

Crocker, R.L. & Major, J. (1955). Soil development in relation to surface age at Glacier Bay, Alaska. *Journal of Ecology*, **43**, 427–448.

Crompton, G. & Sheail, J. (1975). The historical ecology of Lakenheath Warren in Suffolk, England: a case study. *Biological Conservation*, **8**, 299–314.

Daft, M.J. & Hacskaylo, E. (1976). Arbuscular mycorrhizas in the anthracite and bituminous coal wastes of Pennsylvania. *Journal of Applied Ecology*, **13**, 523–531.

Dancer, W.S. (1975). Leaching losses of ammonium and nitrate in the reclamation of sand spoils in Cornwall. *Journal of Environmental Quality*, **4**, 499–504.

Davy, A.J., Dunsford, S.J. & Free, A.J. (1998). Acidifying peat as an aid to the reconstruction of lowland heath on arable soil: lysimeter experiments. *Journal of Applied Ecology*, **35**, 649–659.

Diemont, W.H. & Heil, G. (1984). Some long-term observations on cyclical and seral processes in Dutch heathlands. *Biological Conservation*, **30**, 283–290.

Digby, P.G.N. & Kempton, R.A. (1987). *Multivariate Analysis of Ecological Communities*. London: Chapman & Hall.

Duffey, E., Morris, M.G., Sheail, J., Ward, L.K., Wells, D.A. & Wells, T.C.E. (1974). *Grassland Ecology and Wildlife Management*. London: Chapman & Hall.

Dunsford, S.J., Free, A.J. & Davy, A.J. (1998). Acidifying peat as an aid to the reconstruction of lowland heath on arable soil: a field experiment. *Journal of Applied Ecology*, **35**, 660–672.

During, H.J. & Willems, J.H. (1986). The impoverishment of the bryophyte and lichen flora of the Dutch chalk heaths over the thirty years 1953–1983. *Biological Conservation*, **36**, 143–158.

Dutton, R.A. & Bradshaw, A.D. (1982). *Land Reclamation in Cities*. London: HMSO.

Dyke, G.V., George, B.J., Johnston, A.E., Poulton, P.R. & Todd, A.D. (1983). The Broadbalk Wheat Experiment 1968–1978: yields and plant nutrients in crops grown continuously and in rotation. *Report of Rothamsted Experimental Station 1982, Part 2*, 5–44.

Ebbs S.D., Lasat, M.M., Brady, D.J., Cornish, J., Gordon, R. & Kochian, L.V. (1997). Phytoextraction of cadmium and zinc from a contaminated soil. *Journal of Environmental Quality*, **26**, 1424–1430.

Egler, F.E. (1954). Vegetation science concepts. 1: Initial floristic composition, a factor in old-field vegetation development. *Vegetatio*, **14**, 412–417.

Elias, C.O., Morgan, A.L., Palmer, J.P. & Chadwick, M.J. (1982). *The Establishment, Maintenance and Management of Vegetation on Colliery Spoil Sites*. York, UK: Derelict Land Research Unit, University of York.

Epstein, E. (1972). *Mineral Nutrition of Plants: Principles and Perspectives*. New York: John Wiley.

Farrell, E.P., Nilsson, I., Tamm, C.O. & Wiklander, G. (1980). Effects of artificial acidification with sulfuric acid on soil chemistry in a Scots Pine forest. In *Ecological Impact Growth of Acid Precipitation*, eds. D. Drablos & A. Tollan, pp. 186–187. Oslo: SNSF.

Fowler, D. (1987). Rain cloud chemistry and acid deposition on mountains. In *Annual Report of the Institute of Terrestrial Ecology, 1986–7*, pp. 61–63. Swindon, UK: NERC.

Friedel, H. (1938*a*). Die Pflanzenbesiedlung im Vorfelde des Hintereisfereners. *Zeitschrift für Gletscherkunde*, **26**, 215–239.

Friedel, H. (1938*b*). Bodem und Vegetationsentwicklung im Vorfelde des Rhonegletschers. *Berichte des Geobotanischen Forschungs Institutes Rübel, Zurich 1937*, pp. 65–76.

Gimingham, C.H. & de Smidt, J.T. (1983). Heaths as natural and semi-natural vegetation. In *Man's Impact on Vegetation*, ed. M.J.A. Werger & I. Ikusima, pp. 185–199. The Hague: Dr W. Junk.

Gorham, E., Vitousek, P.M. & Reiners, W.A. (1979). The regulation of chemical budgets over the course of terrestrial ecosystem succession. *Annual Review of Ecology and Systematics*, **10**, 53–84.

Gough, M.W. & Marrs, R.H. (1990*a*). A comparison of soil fertility between seminatural and agricultural plant communities: implications for the creation of floristically rich grassland on abandoned agricultural land. *Biological Conservation*, **51**, 83–96.

Gough, M.W. & Marrs, R.H. (1990*b*). Trends in soil chemistry and floristics associated with the establishment of a low-input meadow system on an arable clay soil in Essex, England. *Biological Conservation*, **52**, 135–146.

Green, B.H. (1972). The relevance of seral eutrophication and plant competition to the management of successional communities. *Biological Conservation*, **4**, 378–384.

Green, B.H. (1980). Management of extensive amenity areas by mowing. In *Amenity Grasslands: An Ecological Perspective*, eds. I.H. Rorison & R. Hunt, pp. 151–161. Chichester, UK: John Wiley.

Grime, J.P. (1979). *Plant Strategies and Vegetation Processes*. Chichester, UK: John Wiley.

Grime, J.P. & Hunt, R. (1975). Relative growth rate: its range and adaptive significance in a local flora. *Journal of Ecology*, **63**, 393–422.

Grubb, P.J. (1986). The ecology of establishment. In *Ecology and Design in Landscape*, eds. A.D. Bradshaw, D.A. Goode & E. Thorp, pp. 83–98. Oxford: Blackwell.

Grubb, P.J. & Key, B.A. (1975). Clearance of scrub and re-establishment of chalk grassland on the Devil's Dyke. *Nature in Cambridgeshire*, **18**, 18–22.

Harper, J.L. (1971). Grazing, fertilizers and pesticides in the management of grasslands. In *The Scientific Management of Animal and Plant Communities for Conservation*, eds. E. Duffey & A.S. Watt, pp. 15–32. Oxford: Blackwell.

Heil, G. & Diemont, W.H. (1983). Raised nutrient levels change heathland into grassland. *Vegetatio*, **53**, 113–120.

Hirose, T. (1986). Nitrogen uptake and plant growth. 2: An empirical model of vegetation growth and partitioning. *Annals of Botany*, **58**, 487–496.

Hutson, B.R. (1980*a*). Colonization of industrial reclamation sites by Acari, Collembola and other invertebrates. *Journal of Applied Ecology*, **117**, 255–275.

Hutson, B.R. (1980*b*). The influence of soil development on the invertebrate fauna colonizing industrial reclamation sites. *Journal of Applied Ecology*, **117**, 277–286.

Hyder, M.B. & Barrett, G.W. (1986). Effects of nutrient enrichment on the producer trophic level of a six-year old-field. *Ohio Journal of Science*, **86**, 10–14.

Jackson, M.L. (1958). *Soil chemical analysis*. London: Prentice Hall.

Jefferies, R.A. (1981). Limestone amendments and the establishment of legumes on pyritic colliery spoil. *Environmental Pollution (A)*, **26**, 167–172.

Jefferies, R.A., Bradshaw, A.D. & Putwain, P.D. (1981*a*). Growth, nitrogen accumulation and nitrogen transfers by legume species established on mine spoils. *Journal of Applied Ecology*, **18**, 945–956.

Jefferies, R.A., Willson, K. & Bradshaw, A.D. (1981*b*). The potential of legumes as a nitrogen source for the reclamation of derelict land. *Plant and Soil*, **59**, 175–177.

Jenkinson, D.S. (1988). Soil organic matter and its dynamics. In *Russell's Soil Conditions and Plant Growth*, 11th edn, ed. A. Wild, pp. 564–607. Harlow, UK: Longman.

Jenkinson, D.S. & Johnston, A.E. (1977). Soil organic matter in the Hoosfield Continuous Barley experiment. *Report of Rothamsted Experimental Station 1976, Part 2*, 87–102.

Jenny, H. (1941). *Factors of Soil Formation*. New York: McGraw-Hill.

Jenny, H. (1980). *The Soil Resource*. Berlin: Springer-Verlag.

Jenny, H., Gessel, S.P. & Bingham, F.T. (1949). Comparative study of decomposition rates of organic matter in temperate and tropical ecosystems. *Soil Science*, **68**, 419–432.

Johnston, A.E. (1970). The value of residues. *Report of Rothamsted Experimental Station 1969, Part 2*, 5–90.

Johnston, A.E. & Poulton, P.R. (1977). Yields on the Exhaustion Land and changes in the NPK content of the soils due to cropping and manuring. *Report of Rothamsted Experimental Station 1976, Part 2*, 53–86.

Jones, I.L. (1967). Studies on hill land in Wales. *Technical Bulletin of the Welsh Plant Breeding Station*, **2**, 1–179.

Kenworthy, J.B. (1964). A study of the changes in plant and soil nutrients associated with moorburning and grazing. PhD thesis, University of St Andrews, UK.

Lawrence, D.B., Schoenike, R.E., Quispel, A. & Bond, G. (1967). The role of *Dryas drummondii* in vegetation development following ice recession at Glacier Bay, Alaska, with special reference to its nitrogen fixation by root nodules. *Journal of Ecology*, **55**, 793–813.

Lee, J.A., Press, M.C., Woodin, S. & Ferguson, P. (1987). Responses of acidic deposition in ombrotrophic mires in the UK. In *Effects of Acidic Pollution on Forests, Wetlands and Agricultural Ecosystems*, eds. T.C. Hutchinson & K.M. Meema, pp. 549–560. Heidelberg, Germany: Springer-Verlag.

Leisman, G.A. (1957). A vegetation and soil chronosequence on the Mesabi iron range spoil banks, Minnesota. *Ecological Monographs*, **27**, 221–245.

Maly, M.S. & Barrett, G.W. (1984). Effects of two types of nutrient enrichment on the structure and function of contrasting old-field communities. *American Midland Naturalist*, **111**, 342–357.

Marrs, R.H. (1985). Techniques for reducing soil fertility for nature conservation purposes: a review in relation to research at Roper's Heath, Suffolk, England. *Biological Conservation*, **34**, 307–332.

Marrs, R.H. (1993). Soil fertility and nature conservation in Europe: theoretical considerations and practical management solutions. *Advances in Ecological Research*, **24**, 241–300.

Marrs, R.H. & Bradshaw, A.D. (1980). Ecosystem development on reclaimed china clay wastes. 3: Leaching of nutrients. *Journal of Applied Ecology*, **117**, 727–736.

Marrs, R.H. & Bradshaw, A.D. (1993). Primary succession on man-made wastes: the importance of resource acquisition. In *Primary Successions*, eds. J. Miles & D.W.H. Walton, pp. 221–248. Oxford: Blackwell.

Marrs, R.H. & Bradshaw, A.D. (1998). The importance of nitrogen fixation in the reclamation of mine wastes: some general principles with applications in the monsoon tropics. In *Problems of Wasteland Development and Role of Microbes*, ed. A.K. Misra, pp. 47–64. Bhubaneswar, India: Amifen Publications.

Marrs, R.H. & Lowday, J.E. (1992). Control of bracken and the restoration of heathland. 2: Regeneration of the heathland communities. *Journal of Applied Ecology*, **29**, 204–211.

Marrs, R.H., Granlund, I.H. & Bradshaw, A.D. (1980). Ecosystem development on reclaimed china clay wastes. 4: Recycling of above-ground plant nutrients. *Journal of Applied Ecology*, **117**, 803–813.

Marrs, R.H., Roberts, R.D., Skeffington, R.A. & Bradshaw, A.D. (1983). Nitrogen and the development of ecosystems. In *Nitrogen as an Ecological Factor*, eds. J.A. Lee, S. McNeill & I.H. Rorison, pp. 113–136. Oxford: Blackwell.

Marrs, R.H., Gough, M.W. & Griffiths, M. (1991). Soil chemistry and leaching losses of nutrients from semi-natural grassland and arable soils on three contrasting parent materials. *Biological Conservation*, **57**, 257–271.

McGrath, S.P. (1995). The use of hyperaccumulator plants to clean up metal polluted soils: a four year feasibility study. *Current Topics in Plant Biochemistry, Physiology and Molecular Biology*, **14**, 126.

McGrath, S.P., Sidoli, C.M.D., Baker, A.J.M. & Reeves, R.D. (1993). The potential for the use of metal-accumulating plants for the *in situ* decontamination of metal-polluted soils. In *Integrated Soil and Sediment Research: A Basis for Proper Protection*, eds. H.J.P. Eijsackers & T. Hamers, pp. 673–676. Dordecht: Kluwer.

Miles, J. (1979). *Vegetation Dynamics*. London: Chapman & Hall.

Miles, J. (1981). *Effect of Birch on Moorland*. Cambridge: Institute of Terrestrial Ecology.

Miles, J. (1987). Succession: past and present perceptions. In *Colonization, Succession and Stability*, eds. A.J. Gray, M.J. Crawley & P.J. Edwards, pp. 1–30. Oxford: Blackwell.

Mountford, J.O., Lakhani, F.H. & Kirkham, F.W. (1993). Experimental assessment of the effects of nitrogen addition under hay-cutting and aftermath grazing on the vegetation on a Somerset peat moor. *Journal of Applied Ecology*, **30**, 321–332.

Nielsen, D., Hogue, E.J., Hoyt, P.B. & Drought, B.G. (1993). Oxidation of elemental sulfur and acidification of calcareous orchard soils in southern British Columbia. *Canadian Journal of Soil Science*, **73**, 103–114.

Odum, E.P. (1971). The strategy of ecosystem development. *Science*, **164**, 262–270.

Olsen, J.S. (1963). Energy storage and the balance of producers and decomposers in ecological systems. *Ecology*, **44**, 322–331.

Oomes, M.J.M. (1990). Changes in dry matter and nutrient yields during the restoration of species-rich grasslands. *Journal of Vegetation Science*, **1**, 333–338.

Owen, K.M. & Marrs, R.H. (2000*a*). Creation of heathland on former arable land at Minsmere, Suffolk: the effect of soil acidification on the establishment of *Calluna* and ruderal species. *Biological Conservation*, **93**, 9–18.

Owen, K.M. & Marrs, R.H. (2000*b*). Acidifying arable soils for the restoration of acid grasslands. *Applied Vegetation Science*, **3**, 105–116.

Owen, K.M. & Marrs, R.H. (2001). The use of mixtures of sulphur and bracken litter to reduce pH of arable soils and control arable weeds. *Restoration Ecology*, **9**, 397–409.

Owen, K.M., Marrs, R.H., Snow, C.S.R. & Evans, C. (1999). Soil acidification: the use of sulphur and acidic litters to acidify arable soils for the recreation of heathland and acidic grassland at Minsmere, UK. *Biological Conservation*, **87**, 105–122.

Ozanne, P.G. & Shaw, T.C. (1967). Phosphate sorption by soils as a measure of the phosphate requirement for pasture growth. *Australian Journal of Agricultural Research*, **18**, 601–612.

Petgel, D.M. (1987). Soil fertility and the composition of semi-natural grassland. In *Disturbance in Grasslands*, ed. J. van Andel, pp. 51–66. Dordrecht: Dr W. Junk.

Pitcairn, C.E.R., Fowler, D. & Grace, J. (1991). *Changes in Species Composition of Semi-Natural Vegetation associated with the Increase in Atmospheric Inputs of Nitrogen*. Huntingdon, UK: Institute of Terrestrial Ecology.

Putwain, P.D. & Gilham, D.A. (1990). The significance of the dormant viable seed bank in the restoration of heathland. *Biological Conservation*, **52**, 1–16.

Reuss, J.O. & Innis, G.S. (1977). A grassland nitrogen flow simulation model. *Ecology*, **58**, 379–388.

Richard, J.L. (1968). Les groupment vegetaux de la reserve d'Aletsch. *Beitraege Geobotanischen Landesaufnahme der Schweiz*, **51**, 305.

Roberts, R.D., Marrs, R.H. & Bradshaw, A.D. (1980). Ecosystem development on reclaimed china clay wastes. 2: Nutrient compartmentation and nitrogen mineralisation. *Journal of Applied Ecology*, **17**, 719–725.

Roberts, R.D., Marrs, R.H., Skeffington, R.A. & Bradshaw, A.D. (1981). Ecosystem development on naturally colonised china clay wastes. 1: Vegetation changes and overall accumulation of organic matter and nutrients. *Journal of Ecology*, **69**, 153–162.

Robertson, G.P. & Vitousek, P.M. (1981). Nitrification potentials in primary and secondary successions. *Ecology*, **62**, 376–386.

Robinson, B.H., Leblanc, M., Petit, D., Brooks, R.D., Kirkman, J.H. & Glegg, P.E.H. (1998). The potential of *Thlaspi caerulescens* for phytoremediation of contaminated soils. *Plant and Soil*, **203**, 47–57.

Rorison, I.H. (1971). The use of nutrients in the control of the floristic composition of grassland. In *The Scientific Management of Animal and Plant Communities for Conservation*, eds. E. Duffey & A.S. Watt, pp. 65–78. Oxford: Blackwell.

Singh, J.S., Raghubashani, A.S, Singh, R.S. & Srivastave, S.C. (1989). Microbial biomass acts as a source of nutrients in dry tropical savanna. *Nature*, **338**, 499–500.

Skeffington, R.A. & Bradshaw, A.D. (1980). Nitrogen fixation by plants growing on reclaimed china clay wastes. *Journal of Applied Ecology*, **17**, 469–477.

Smith, C.J., Elston, J. & Bunting, A.H. (1971). The effects of cutting and fertiliser treatments on the yield and botanical composition of chalk turf. *Journal of the British Grassland Society*, **26**, 213–219.

Smith, R.A.H. & Bradshaw, A.D. (1979). The use of metal-tolerant plant populations for the reclamation of metalliferous wastes. *Journal of Applied Ecology*, **16**, 595–612.

Smith, R.E.N., Webb, N.R. & Clarke, R.T. (1991). The establishment of heathland on old fields in Dorset, England. *Biological Conservation*, **57**, 221–234.

Smith, R.S. (1988). The effect of fertilisers on the conservation interest of traditionally managed upland vegetation. In *Agriculture and Conservation in the Hills and Uplands*, eds. M. Bell & R.G.H. Bunce, pp. 38–43. Huntingdon, UK: Institute of Terrestrial Ecology.

Stahl, P.D., Williams, S.E. & Christensen, M. (1988). Efficacy of native vesicular–arbuscular mycorrhizal fungi after severe soil disturbance. *New Phytologist*, **110**, 347–354.

Swift, M.J., Heal, O.W. & Anderson, J.M. (1979). *Decomposition in Terrestrial Ecosystems*. Oxford: Blackwell.

Tilman, D. (1988). *Plant Strategies and the Dynamics and Structure of Plant Communities*. Princeton, NJ: Princeton University Press.

Traynor, C.H. (1995). The management of heathland by turf cutting: historical perspective and application to conservation. PhD thesis, University of Liverpool, UK.

van der Woude, B.J., Pegtel, D.M. & Bakker, J.P. (1994). Nutrient limitation after long-term fertiliser application in cut grasslands. *Journal of Applied Ecology*, **31**, 405–412.

Viereck, L.A. (1966). Plant succession and soil development on gravel outwash of the Muldrow Glacier, Alaska. *Ecological Monographs*, **36**, 181–199.

Warren Springs Laboratory (1987). *Acid Deposition in the United Kingdom 1981–1985*. Stevenage, UK: Warren Springs Laboratory.

Wedin, D.A. & Tilman, D. (1990). Species effects on nitrogen cycling: a test with perennial grasses. *Oecologia*, **84**, 433–441.

Wells, T.C.E. (1980). Management options for lowland grassland. In *Amenity Grassland: An Ecological Perspective*, eds. I.H. Rorison & R. Hunt, pp. 175–196. Chichester, UK: John Wiley.

Whiting, S.N., Reeves, R.D. & Baker, A.J.M. (in press). Mining, metallophytes and land reclamation: conservation of biodiversity. *Mining Environmental Management*.

Williams, C.M., Ford, M.A. & Lawson, C.S. (1996). The transformation of surplus farmland into semi-natural habitat. 2: On the conversion of arable land to heathland. *Aspects of Applied Biology*, **44**, 185–192.

Williamson, N.A., Johnson, M.S. & Bradshaw, A.D. (1982). *Mine Waste Reclamation: The Establishment of Vegetation on Metal Mining Wastes*. London: Mining Journal Books.

Willis, A.J. (1963). Braunton Burrows: the effects on the vegetation of the addition of mineral nutrients to the dune soils. *Journal of Ecology*, **51**, 353–374.

10 • Chemical treatment of water and sediments with special reference to lakes

MARTIN SØNDERGAARD, KLAUS-DIETER WOLTER AND WILHELM RIPL

INTRODUCTION

The impact of human activities on the aquatic environment has increased during the past century. Chemical pollutants have increased in rivers, lakes and coastal areas due to rising population densities, farming and industrialisation. The effects have included acid rain and acidification of surface water over large areas where catchment soils as well as bedrock are poor in limestone, and increased deposition of heavy metals and other chemicals causing contamination and bioaccumulation of toxic products. A marked increase in the use of pesticides and the enhanced production of organic substances used in various industries have led to increased pollution by a wide variety of organic micropollutants (Kristensen & Hansen, 1994).

Measures to combat industrial sources of pollutants have been implemented at least in some parts of the world, although improvements in many areas are still needed, just as the environmental impact of many organic micropollutants remains to be elucidated (Kristensen & Hansen, 1994). However, the influence from nutrient-rich wastewater from cities or aquaculture and the use and leaching of fertilisers in agriculture still constitute significant problems that often overshadow other environmental problems. Apart from more local industrial influences, increased nutrient loading, resulting in eutrophication and a loss of the natural functionality of many ecosystems, is considered to be the most important and widespread environmental problem of lentic and coastal waters. In lakes, one of the most important factors is the increased availability of nutrients, especially phosphorus which – via its limiting effect on the growth of phytoplankton and thus indirectly on the community of higher organisms – has had a very important influence on lake metabolism and lake water quality (Ohle, 1953; Thomas, 1969; Jeppesen *et al.*, 1999). For decades, a reduction of external nutrient loading, especially phosphorus, has therefore been of paramount importance in lake management for counteracting undesired eutrophication effects and for improving water quality.

Multiple measures have been employed in the catchments to diminish nutrient loading. The first step has often been to establish sewage works in communities above a certain size to remove organic pollution and to avoid low oxygen concentrations in rivers. Biological sewage treatment has, however, only minor effects on nutrient loading and sewage works have often been expanded to incorporate varying degrees of phosphorus stripping and nitrogen removal. The effort to reduce phosphorus loading has often been supplemented with increased use of phosphate-free detergents. More recently, measures have also been taken to reduce nutrient loading from arable soils such as increased storage capacity of animal manure on farms, establishment of uncultivated buffer strips along streams and rivers, maintenance of green cover of fields in winter and retirement of agricultural land (Jeppesen *et al.*, 1999). Finally, improved nutrient removal and retention may also be achieved through new constructed wetlands, remeandering of channelised streams and biomanipulation of lakes.

Besides reducing the external nutrient and organic loading, restoration of inland freshwaters has mainly focused on lakes where chemical restoration techniques have been widely used. Rivers and streams are open systems and have a greater potential for natural recovery as they are flushed and noxious substances more easily diluted. In rivers, the general trend is thus simply to stop the input of materials and not to establish internal control by chemical means.

The reason why lakes have long been subjects of in-lake restoration measures is that they often respond slowly to a reduction in external phosphorus loading, and lake water phosphorus concentrations are not reduced to the same extent as external loading (Marsden, 1989; Sas 1989; Søndergaard *et al.*, 1999). Correspondingly, the desired improvement in water quality fails to occur even if phosphorus loading has been reduced to a level where improvements were expected. The reason for this resilience is internal loading of phosphorus from the sediment where phosphorus was accumulated when external loading was high. For a period following an external load reduction, part of this phosphorus pool is released concurrently with the establishment of a new dynamic equilibrium between the sediment and water phase. This internal phosphorus loading may be of great significance in both shallow, unstratified lakes where nutrients are added to the photic zone all summer (Jeppesen *et al.*, 1991; Phillips *et al.*, 1994; Søndergaard *et al.*, 1999), and in deep lakes where phosphorus is accumulated in the bottom layer of water during summer stratification. The duration and intensity of the internal phosphorus release after an external loading reduction depend on the degradability of sedimentary organic matter as an energetic basis for micro-organisms and on the size of the releasable sediment phosphorus pool. The release rates depend on the intensity of the microbial metabolism and transport mechanisms within the sediment. Internal phosphorus release is recorded especially in eutrophic lakes, but in summer even lakes with relatively low nutrient concentrations may experience a short-term net internal loading (Fig. 10.1). In shallow eutrophic lakes, summer phosphorus concentrations may rise to values more than twice as high as the concentrations derived from external loading (Jeppesen *et al.*, 1997; Søndergaard *et al.*, 1999). The release may be very persistent and endure for many years, with a conservative estimate of at least ten years after an external load reduction (Welch & Cooke, 1999). In some lakes, phosphorus retention has, in fact, remained negative for more than 15 years after the nutrient loading reduction (Søndergaard *et al.*, 1999).

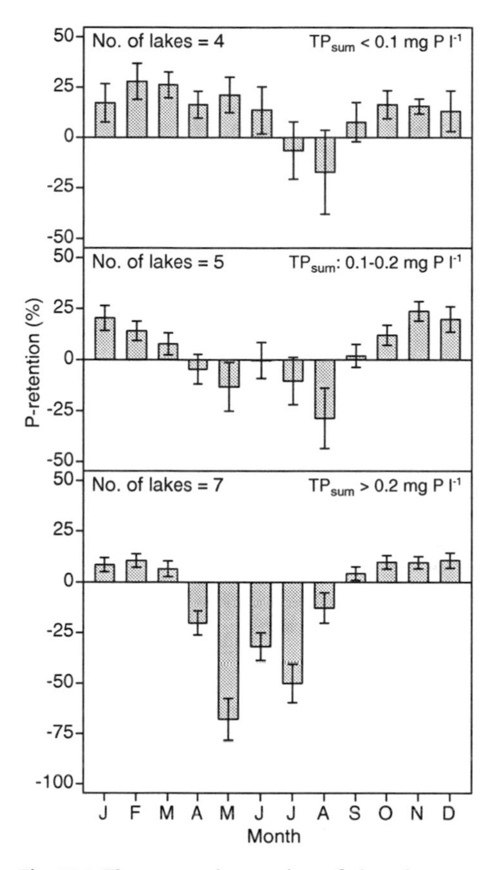

Fig. 10.1. The seasonal retention of phosphorus at different nutrient concentrations (mean summer concentration of total phosphorus, TP_{sum}) based on mass balance calculations for eight years (20 annual inlet–outlet samplings) in 16 Danish lakes. Phosphorus retention is given as percentage of external loading. From Søndergaard *et al.* (1999).

In an attempt to reduce internal phosphorus loading and accelerate lake recovery after a decrease in external loading, numerous experiments and lake restoration projects have been undertaken using various methods aimed at decreasing the sediment phosphorus release. Many methods have been chemical and have focused on reducing the effects of eutrophication by influencing phosphorus availability. The methods for chemical restoration of lakes have been applied to both stratified and shallow lakes with the objectives of influencing bioactivity and redox-dependent phosphorus fixation.

This chapter describes the background and the chemical and biological relations in lakes affecting the design and evaluation of the different chemical restoration tools. We focus mainly on the many chemical processes in which phosphorus may be part, and on the mechanisms controlling the exchange between the sediment and water phase. The last part of the chapter briefly describes the application of different types of chemical restoration tools and the underlying hydrological, chemical and biological principles and techniques, and possible problems that may arise.

INTERACTIONS BETWEEN SEDIMENT AND WATER IN LAKES AND THEIR IMPLICATIONS FOR LAKE RESTORATION

Stratification and oxygen supply

Due to the temperature-dependent water density (maximum at 4°C), most lakes deeper than 5–10 metres exhibit thermal stratification during part of the season. However, the depth required for stratification to occur varies considerably, depending on the surface area and surrounding topography. In temperate regions where most restoration projects have been implemented, dimictic lakes are the most common lake type. These lakes circulate freely twice a year: in spring when surface layer temperatures increase above 4°C, and in autumn when surface stratum temperatures decrease and approach 4°C. In summer, stratification divides dimictic lakes into three zones: a warm and less dense upper stratum (epilimnion), usually being more or less completely mixed, a cold and dense bottom stratum with more quiescent water (hypolimnion), and a transient zone with a steep temperature gradient (metalimnion), separating and minimising the exchange of nutrients and other substances between the epilimnion and hypolimnion.

Because of the summer temperature stratification in deep lakes, the chemical environment, including redox conditions, undergoes a significantly different development from that of shallow lakes. In shallow and completely mixed lakes, water tends to be saturated with oxygen except immediately above the sediment surface all summer. In calm periods, interim stratification may occur, this being, however, quickly eliminated as soon as the wind rises or thermal homogeneity is achieved at night.

In the deeper and more permanently summer-stratified lakes, oxygen concentrations in the hypolimnion decrease following the onset of thermal stratification in early summer, with stratification minimising the input of oxygenated water from the epilimnion. The rate of hypolimnetic oxygen depletion depends on the volume of hypolimnion, the water movement across the metalimnion, and on sediment oxygen consumption. In eutrophic lakes, where high primary production and sedimentation normally create an organically rich sediment with a proportionately high oxygen consumption, hypolimnion oxygen concentrations will be depleted sooner than in more nutrient-poor lakes with less production as well as sedimentation of organic matter. The thickness of the oxygen-depleted bottom layer increases concurrently with the consumption of oxygen in the hypolimnion in nutrient-rich lakes, which, as a consequence, become more and more undersaturated. When stratification sets in, oxygen is first depleted in the layers nearer the bottom sediments. Over the course of the summer the oxygen-depleted bottom layer includes more of the hypolimnion.

Following oxygen depletion, the hypolimnion concentrations of nitrate and sulphate typically decrease as they are alternative electron acceptors to oxygen. In contrast, the concentrations of phosphate, ammonium, iron, alkalinity and pH increase due to the metabolic processes in the sediment. In nutrient-poor lakes, sediment oxygen consumption may be so insignificant that hypolimnion oxygen concentrations may be more or less saturated all summer. In extremely nutrient-poor lakes, summer oxygen concentrations may even increase towards the bottom (orthograde oxygen profile) since oxygen is dissolved in higher concentrations at low temperatures during the spring overturn. Minimum oxygen concentrations are expected in the metalimnion in such lakes.

The different availability and consumption of oxygen implies that the concentration of various substances at the sediment/water interface differs widely between both shallow and deep as well as nutrient-rich and nutrient-poor lakes. In shallow, well-oxidized lakes, the gradients are established close to the sediment/water interface where a diffusive boundary layer of variable thickness, depending on mixing conditions, is established. In shallow nutrient-poor lakes, oxygen may penetrate the sediment by as much as a few centimetres, while penetration is limited to a few millimetres in nutrient-rich lakes, where the abundance of biodegradable organic matter is higher. Correspondingly, the penetration depth of other terminal electron acceptors into the sediment (nitrate and sulphate), depleted during the course of decomposition, is considerably less in nutrient-rich lakes. Different penetration depths, expressed by the classical redox stratification of various electron acceptors (Boström *et al.*, 1982), divide the sediment into an upper oxidised layer of variable depth (depending on eutrophication level) and a lower, reduced sediment layer. In eutrophic stratified lakes, steep concentration gradients of the numerous organic and inorganic compounds, involved in oxidation–reduction reactions, are found in the water phase established close to, or below, the thermocline where marked decreases in oxygen concentrations occur.

Many of the chemical and biochemical processes in water and sediment are redox processes, i.e. involve electron transfer. The extent of oxidative or reduced conditions in a solution can be described by the redox potential (Eh), the capability to oxidise or reduce the surrounding environment. The redox potential decreases 58 mV for each pH unit increase and strongly depends on oxygen concentrations. The theoretical redox potential in water saturated with oxygen (pH = 7 and T = 25 °C) is 800 mV, and well-oxidised water will normally have a potential ranging between 400 and 600 mV. Not until oxygen concentrations become very low (<0.1 mg l^{-1}) will the redox potential decrease to *c.* 200 mV. This is the level (200–300 mV) where iron is reduced from Fe^{3+} to Fe^{2+}, which is of crucial importance for the sorption of phosphorus. Oxidised iron oxides and hydroxides such as $Fe(OH)_3$ have a high affinity to adsorb phosphorus. In contrast, reduced iron is mostly incapable of adsorbing phosphorus. In the complete absence of oxygen, even lower redox potentials are established, and may reach −100 to −200 mV in the sediment. The presence of nitrate is, however, able to maintain the redox potential at a relatively high level, thus preventing the reduction of Fe^{3+}.

The sediment and phosphorus fixation

The sediment is to a large extent the terminal site for the particulate substances added to or produced in the water column and which are not mineralised during sedimentation before reaching the sediment surface. Therefore, sediments usually act as nutrient sinks for autochthonous and allochthonous particulate material. In nutrient-rich and productive lakes this net accumulation adds several millimetres or more to the sediment annually. Some lakes and reservoirs also receive significant input from river inlets rich in suspended solids. The phosphorus reaching the sediment is mostly in the particulate form inorganically bound to active surfaces of iron, aluminium or calcium, or as organic debris.

The fixation of phosphorus in the sediment varies depending on four processes: (1) transport of soluble phosphate between solid components; (2) adsorption–desorption mechanisms; (3) chemosorption; and (4) biological assimilation (Jacobsen, 1978). Chemosorption normally signifies chemical fixation of soluble compounds subsequently unaffected by changes in solute concentrations. Adsorption, on the other hand, is a physical fixation of soluble compounds on surfaces in constant equilibrium with solute concentrations. Both adsorption and chemosorption of phosphate by sediments involve numerous compounds, the most important being iron, calcium, aluminium, manganese, clay and organic matter. Apart from concentrations, adsorption and chemosorption processes are often dependent on both pH and redox potentials, which themselves are consequences of bacterial metabolism.

After a long period of high nutrient loading, a lake's processes of production and respiration will be out of equilibrium and the P/R quotient will stay well below 1, causing accumulation of nutrients and biodegradable organic matter in the sediment. Because of the high affinity of iron to bind phosphorus, total phosphorus concentrations in the surface sediment not only depend on the external phosphorus loading, but also on the concentrations of iron (Søndergaard *et al.*, 1996). With increased loading, the sediment undergoes significant concurrent changes in structure and function. With the onset of anoxic conditions and increased activity of anaerobic bacteria at the sediment surface, the sediments become anaerobic and sapropelic. Poisonous hydrogen sulphide may also be liberated with the consequent destruction of higher fauna.

To define sediment characteristics, sequential extractions with various chemical compounds have been developed for fractionation and description of the sediment phosphorus pool (Williams *et al.*, 1971; Hieltjes & Lijklema, 1980; Psenner *et al.*, 1988). The sediment phosphorus pool is often divided into inorganic and organic fractions, with the latter comprising a loosely sorbed and easily releasable form and a more tightly fixed, refractory form. The inorganic fraction is often subdivided into a loosely sorbed fraction and fractions bound to iron, aluminium and calcium.

From a management perspective, the sediment's pool of releasable phosphorus determines the magnitude and duration of internal phosphorus loading that continues following a reduction in nutrient loading. Often, both the loosely sorbed organic and inorganic fractions, as well as the iron-bound and redox sensitive sorption of phosphorus, are considered potentially mobile and releasable. However, as yet, it has not been possible to establish any simple and reliable relationships between the different sediment phosphorus fractions and the ultimate pool of releasable phosphorus. Although such knowledge may provide information on the overall and long-term conditions expected to prevail concerning the sorption of phosphorus in the sediments, such information on static phosphorus binding gives only limited insight into actual changes in phosphorus forms released under dynamic conditions. Another problem associated with the use of static parameters is determining the sediment depths from which phosphorus release may be expected. Traditionally, phosphorus in the upper 10 cm is considered potentially mobile. However, some studies indicate that phosphorus may be transported upwards from depths up to 20–25 cm (Søndergaard *et al.*, 1993, 1999).

Phosphorus release from sediments

Although the interstitial water normally contains <1% of the sediment's total phosphorus pool (Boström *et al.*, 1982), this pool, nevertheless, has a significant bearing on the phosphorus transport between sediment and water. This is because the interstitial water's phosphate content constitutes the direct link between the particulate phosphorus pool and the water phase above. The transport of phosphorus between the sediment and water phase results from a diffusion-mediated concentration gradient, normally appearing just below the sediment surface. Bioturbation from benthic invertebrates or through gas bubbles produced in deeper sediment layers during the microbial decomposition of organic matter may, however, significantly enhance the upward transport of phosphorus. Ohle (1958, 1978) reported that released methane gas is a significant transport process for further mixing of excessive phosphate concentrations from the interstitial into the overlying water.

Biodegradability of an organic substrate is necessary for bacterial degradation and for the release of phosphorus. At low sedimentation rates in oligotrophic lakes, organic matter is so resistant that further decomposition of the settled material does not occur. In contrast, organic matter settling at high rates in the upper sediment layers in eutrophic lakes is usually abundant and easily degraded. Bacterial metabolism in these lakes is therefore only limited by the delivery of the electron acceptors, oxygen, nitrate and sulphate, in order to oxidise organic matter. High rates of phosphorus redissolution are dominated by the process-conditioned modifications of the redox potential and pH, caused by this increased metabolism of the micro-organisms.

A number of factors influence the exchange of phosphorus between water and sediments, including redox conditions, pH, iron:phosphorus ratio and resuspension (Boström *et al.*, 1982; Søndergaard, 1988; Jensen *et al.*, 1992; Søndergaard *et al.*, 1992). The solid/liquid phase boundary between water and sediment, or between sediment particles and interstitial water, as well as the different possibilities of transport of matter across these boundary layers, are of crucial importance for the understanding of the dynamic release of phosphorus from sediments. On one hand, the actively metabolising bacteria in the interstitium cannot be supplied with the necessary electron acceptors (oxygen, nitrate and sulphate) by molecular diffusion from the supernatant water alone (Duursma, 1967). On the other hand, the inhibitory metabolic final products (e.g. hydrogen sulphide) cannot be removed by diffusion alone. Thus, individual velocity gradients between water and the solid phase are potential controls for processes and biota. Beyond the scale of diffusion at the boundary water/sediment layer, water flow, microturbulences, eddies, waves, a slowly circulating hypolimnion and seiches influence the processes.

Phosphorus retention and release depend on temperature. During the cold season, with low sedimentation rates and sufficient supply of oxygen or nitrate to the sediments, high redox potentials develop, maintaining the sedimentary iron in its oxidised form. The combined ferric oxides and hydroxides available in the sediment may effectively bind phosphate. Oxidised iron and/or molecular sulphur from H_2S oxidation usually colour the surface sediments (approx. 1–10 cm) light brown to orange-yellow (Gorham, 1958). Even the reduced iron sulphide can be oxidised again to Fe(III) with its high phosphate-binding capacity, via nitrate. Thus, even eutrophic lakes, suffering from a significant net annual internal phosphorus loading, are capable of retaining phosphorus during the winter season (Fig. 10.1).

At higher temperatures during spring, and high supply of biodegradable organic matter, oxygen and nitrate, which are transported into the sediment from the water above, are consumed in the highest millimetres to centimetres of the sediments as well as immediately above the sediment surface. Thus, at slow water flow, oxygen consumption and mineralisation of organic matter are the result of coupling between nitrification and denitrification (Ripl & Lindmark, 1978). If nitrate is consumed at sufficiently high sulphate concentrations and with a sufficient supply of biodegradable organic matter, sulphate reduction becomes the dominant sediment process. This phase can be determined from a definite decrease in sulphate concentrations in interstitial water and from steep sulphate gradients into the sediment. Hydrogen sulphide formed from sulphate reduction causes the reduction of Fe(III) and the formation of iron sulphide (FeS) by the following formula:

$$2\,FeO(OH) + 3\,H_2S \rightarrow 2\,FeS + S + 4\,H_2O$$

In iron-rich sediments, the colour changes from brown (oxidised status) to black (reduced status). In contrast to Fe(III), the reduced Fe(II) cannot efficiently fix phosphate. Therefore, the process leads to redissolution of phosphate into the interstitial water and, finally, into the water.

From the stoichiometry of the reactions it follows that sulphate reduction usually takes place in the sediments only up to a molar sulphur: iron ratio of about 1.5. Hydrogen sulphide, formed after reaching this ratio, can no longer be detoxified, implying that even reduced sulphur products are restricted by negative feedback due to H_2S as the final product. Thus, in the sediments of the Schlei estuary (Northern Germany), sulphate concentrations in the interstitial water increased after exhaustion of the dissolved Fe(II). This was a visible indication of inhibited sulfate reduction activity. Not until the exhaustion of the sediment iron buffer was almost complete, did a release of iron-bound phosphorus occur. This, in turn, led to an increase in the phosphorus concentration in the interstitial water (Ripl, 1986*a*, *b*).

In some shallow highly eutrophic lakes, the phosphorus release may be so high that the release over a few weeks amounts to the entire phosphorus content of the upper 1–2 mm of sediment. In hypertrophic Lake Søbygaard, several periods of low phytoplankton biomass and low net sedimentation rates resulted in a net sediment release of phosphorus as high as 100–200 mg P m^{-2} day^{-1} (Søndergaard *et al.*, 1990). In agreement with these processes, a negative correlation was found between sediment phosphorus (as % of the acid-soluble fraction) and the molar sulphur: iron ratio in the sediments of Lake Tegel, Berlin (W. Ripl *et al.*, unpubl. data) (Fig. 10.2). If the organic matter is still further degradable, methane fermentation may occur below the sulphate reduction zone.

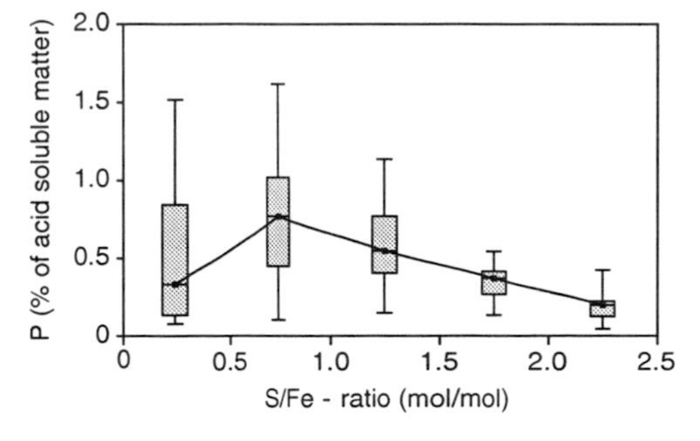

Fig. 10.2. Distribution of phosphorus as percentage of acid-soluble matter depending on the molar sulphur: iron ratio in sediments (0–25 cm) of Lake Tegel, Berlin, 1985–9. 5%, 25%, 50% (line with squares), 75% and 95% percentiles for each sulphur: iron ratio classes. Total number $n = 508$.

Consequently, the depletion of oxygen and the subsequent decrease in the redox potential may not be the principal reason (*sensu* Mortimer 1941, 1942) behind the phosphorus release from highly enriched sediments. Instead, the formation of hydrogen sulphide and its reaction with iron (Hasler & Einsele, 1948) after complete consumption of the nitrate appears to be a significant factor. The internal dynamic phosphorus release from lake sediments can, in most cases, be understood as a ligand exchange of phosphate versus sulphide with iron. Additionally, these processes are accelerated mainly by sedimentation of fresh organic matter after, or during, a planktonic algal bloom. In a H_2S-free anoxic environment, phosphate would be dissolved simultaneously with iron, but immediately reprecipitate as iron phosphate in an oxic environment.

CHEMICAL METHODS FOR LAKE RESTORATION

As pointed out earlier, a sufficient reduction of the external phosphorus loading is of paramount importance for achieving a high water quality in lakes, and every lake restoration project should start by examining and, if necessary, controlling the external loading of nutrients. In some instances, reduction of external loading may suffice to achieve a satisfactory water quality, while in other cases the internal loading is so high that in-lake measures are required. The possibilities of establishing permanent effect via in-lake restoration techniques are, however, poor if the external nutrient has not been brought to a level so low that equilibrium phosphorus concentrations can ensure the desired lake quality.

Either diversion, eliminating totally the anthropogenic source, or advanced wastewater treatment normally removing about 90% or more of the phosphorus content, have been most frequently employed to limit external nutrient loading. In principle, sewage plant removal of phosphorus is based on the same chemical principles as those used in in-lake restoration techniques (see below), i.e. phosphorus is precipitated from the wastewater solution by addition of metal salts, where the metal phosphate is insoluble and subsequently removable. Most often used are salts of alum, iron or calcium (Table 10.1). In some instances also treatment of river inflows with iron salts, e.g. $FeCl_3$, is used to precipitate phosphorus.

As for in-lake measures, there are two kinds of fundamental chemical restoration techniques to counteract eutrophication and these will be described below: (1) improvement of the phosphorus sorption of the substances already present in the lake, or (2) supply of new chemical sorption capacity to the lake (Table 10.1). Phosphorus control by improving existing sorption potentials is usually obtained using oxygen or, less often, nitrate to improve the redox-dependent phosphorus sorption. Phosphorus control by increasing the sorption capacity is usually obtained using alum or iron, or, more rarely, calcium. Although the different techniques are described in isolation below, the maximum effect may be achieved by a combination of techniques.

Besides counteracting eutrophication, chemical techniques may be used to restore soft water lakes from the effects of acid rain. In some countries acidification is the most serious environmental problem (Sandøy & Romundstad, 1995). In such cases, calcium is used to increase pH and restore a suitable environment for several organisms.

Hypolimnetic oxygenation

Aim and chemical background

Hypolimnetic oxygenation or aeration normally aims to increase the oxygen concentration and input of oxygen to the hypolimnion, in order to increase the sorption capacity of phosphorus through increased sorption to oxidised iron components in iron-rich lakes (McQueen *et al.*, 1986; Cooke *et al.*, 1993). Increased availability of oxygen may also decrease the gas formation in the sediment and dim- inish the resuspension of sediment particles and phosphorus release resulting from gas ebullition (Matinvesi, 1996). Long-term oxygenation may decrease the organic content, total nitrogen and the biological oxygen demand in the uppermost sediment (Matinvesi, 1996), which, provided that the external phosphorus loading has been sufficiently reduced, may produce a more permanent improvement of lake water quality.

Table 10.1. *Chemical measures used to restore lakes*

Chemical compound used	Aim	Techniques
Catchment measures		
Iron, alum, calcium	To precipitate and remove phosphorus with iron or alum hydroxides and calcium	Addition of iron ($FeCl_3$ and $FeSO_4$), alum ($Al_2(SO_4)_3$) or calcium ($Ca(OH)_2$) in wastewater treatment
Iron	To reduce loading by precipitation of phosphorus with iron hydroxides	Addition of iron ($FeCl_3$) in river inflows
In-lake measures		
Oxygen	To improve redox potential and sorption of phosphorus on iron; to enhance the distribution of fish and invertebrates	Injection of pure oxygen or atmospheric air into the hypolimnion or use of full-lift aerators
Nitrate	To oxidise organic matter and to improve redox potential and sorption of phosphorus on iron; to decrease hypolimnetic oxygen deficit	Injection of nitrate into the sediment or the hypolimnion
Iron	To increase phosphorus sorption capacity	Dosing of iron to water or sediment
Alum	To increase phosphorus sorption capacity in aluminium compounds	Dosing of alum to water
Calcium	Increased sorption of phosphorus in calcium–phosphorus compounds	Dosing of calcium to water

Oxygenation may also be used to improve and expand living conditions for cold-water fish (Prepas *et al.*, 1997) and other fauna in the hypolimnion (Dinsmore & Prepas, 1997*a*, *b*; Field & Prepas, 1997). For example, in the oxygen-treated basins of Amisk Lake, Alberta, cisco (*Coregonus artedii*) were able to feed throughout the water column, whereas in untreated basins, hypoxia restricted cisco to epilimnetic and metalimnetic waters (Aku & Tonn, 1997). Hypolimnetic aerators have also been installed in water supply reservoirs to improve raw water quality by lowering the concentrations of iron, manganese and sulphide (Cooke *et al.*, 1993; Burris & Little, 1998). Hypolimnetic aeration may also be combined with iron addition to facilitate phosphate precipitation and retention in the sediment (McQueen *et al.*, 1986; Jaeger, 1994).

Hypolimnetic oxygenation should be distinguished from complete mixing techniques that do not retain thermal stratification. Among the advantages of avoiding destratification are prevention of the transport of nutrient-rich bottom water to the epilimnion and maintenance of suitable habitats for cold-water fish and other species adapted to cold water. Artificial circulation or destratification (Cooke *et al.*, 1993; Simmons, 1998), hypolimnetic withdrawal (Nurnberg, 1987; Livingstone & Schanz, 1994) and other techniques manipulating the physical environment are not addressed in this chapter.

Techniques

Over the years, numerous different restoration methods and aerator designs have been implemented to increase hypolimnetic oxygen concentrations (Cooke *et al.*, 1993). Often either pure oxygen or atmospheric air is pumped into deep parts of the lake where it is dissolved into the hypolimnion via diffusers (Fig. 10.3). Oxygen can be supplied from a storage tank at the shore containing liquid oxygen (Prepas *et al.*, 1997; Gächter & Wehrli, 1998). Other

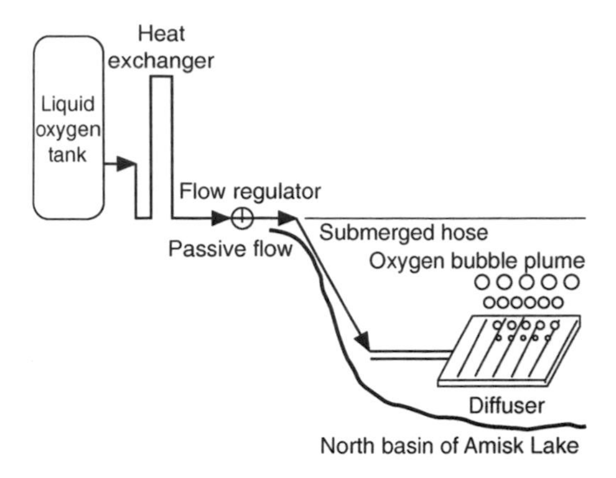

Fig. 10.3. Schematic illustration of the oxygenation system used in Amisk Lake, Alberta. From Prepas *et al.* (1997).

techniques include deep-water aeration where oxygen-depleted bottom water is brought to the surface via a full-lift aerator where it is aerated and returned to the bottom (Ashley *et al.*, 1987; Jaeger, 1994; Burris & Little, 1998).

Hypolimnetic oxygenation during summer stratification can be supplemented with winter aeration, thus ensuring complete mixing of the water column. Pressurised air is released in deep parts of the lake to enhance vertical mixing and dissolved oxygen concentrations and to prolong the oxic period following stratification (Gächter & Wehrli, 1998). However, disturbance of winter stratification during very cold periods may decrease deep-water temperatures, which may cause more extensive freezing of the lake.

Possible problems in connection with treatment and their effects

If the installation configuration is not designed properly, hypolimnectic aeration may lead to destriction of the thermocline and destratification of the lake, particularly in weakly stratified lakes (Lindenschmidt, 1999). Aeration also often leads to increased hypolimnetic temperatures even when destratification does not occur, but this is usually restricted to a few degrees (Prepas *et al.*, 1997).

Compared to untreated lakes, the hypolimnetic oxygen demand often increases during oxygenation. Several factors may be involved, including enhanced oxygen consumption due to induced circulation currents above the sediment, diminished thickness of the diffusive sublayer adjacent to the sediment and fewer transport-limited processes (Sweerts *et al.*, 1989; Moore *et al.*, 1996; Nakamura & Inoue, 1996). This implies that more oxygen may be needed than that previously calculated using the oxygen demand at stagnant conditions as a basis.

An anoxic hypolimnion and high phosphorus release rates may not be 'cause–effect' related, but two parallel symptoms of common cause are: (1) excessive organic matter sedimentation exhausting dissolved oxygen and (2) high sedimentation rates of phosphorus exceeding the phosphorus retention capacity of the anoxic sediment (Gächter & Wehrli, 1998). Even if oxygenation affects the transitory binding of phosphorus, it is questionable whether hypolimnetic oxygenation causes the permanent burial of phosphorus and, in turn, produces permanent effects on the trophic state of a lake. This may be important if external loading is not reduced before, or simultaneously with, the oxygenation (Gächter, 1987; Gächter & Wehrli, 1998).

Nitrate treatment

Aim and chemical background

Nitrate treatment of anaerobic sapropelic sediments aims to reduce the high potential reactivity of sediments by the oxidation of biodegradable organic matter. *In situ* oxidation of sedimentary biodegradable organic matter has its highest potential to control phosphorus in iron-rich systems, enabling a fixation of phosphorus with iron, and an increase of the iron buffer in the sediment (Ripl, 1976, 1978).

Following oxygen consumption, denitrification denotes the first step in the bacterially mediated oxidation of organic matter. Denitrification proceeds during consumption of nitrate, organic matter being oxidised with nitrate to carbon dioxide and water. Although dissimilatory reduction of nitrate to ammonia in a reducing environment has been suggested (Priscu & Downes, 1987; Søndergaard *et al.*, 2000), nitrogen is usually thought to be released as molecular, gaseous nitrogen:

$$5\,CH_2O + 4\,NO_3^- + 4\,H^+ \rightarrow 2\,N_2 + 5\,CO_2 + 7\,H_2O$$

For a non-equilibrium system, a definite redox potential cannot be specified, but the redox potential of denitrification is always in the positive range where iron is oxidised to the Fe^{3+}, which binds phosphate very effectively. Trivalent iron binds phosphorus, either directly as iron phosphate or via adsorption to ferric oxide hydroxides (Stumm & Morgan, 1981; Wetzel, 1983).

Nitrate treatment increases the activity of ubiquitous bacteria and their natural processes at the sediment surface. Through nitrate treatment, oxygen uptake of sediments is reduced, the anaerobic sapropelic sediments are stabilised and putrefaction, sulphate reduction and methane fermentation is diminished. Hydrogen sulphide formation is lowered and sulphides already present can be oxidised bacterially by nitrate to sulphate:

$$5\ S^{2-} + 8\ NO_3^- + 3\ H^+ + H_2O \rightarrow 5\ SO_4^{2-} + 4N_2 + 5\ OH^-$$

The detoxification of anaerobic sapropelic sediments, accompanying this process, allows renewed colonisation of the sediments with benthic fauna. Iron present as sulphide is also transformed into oxidised, trivalent iron by denitrification. As described earlier, the increased phosphorus binding capacity of sediments causes lower internal phosphorus release and lower nutrient supply to the pelagic zone in eutrophic lakes (Ripl, 1976, 1978).

In a carefully planned nitrate treatment, a single or double treatment is sufficient to achieve permanent improvement. However, the quantity of nitrate (and iron) should be adjusted to ensure that the phosphorus concentration in the water declines to such a low level that renewed algal blooms and sedimentation of biodegradable organic matter do not result in renewed feedback establishing phosphorus release and planktonic production.

Techniques

Since nitrate treatment significantly affects the metabolism of a lake, pre-treatment and accompanying investigations are necessary to plan and control for the extent of the intervention required. Sediment experiments are needed to calculate the preparatory quantities of nitrate to be used for oxidation of biodegradable organic matter and of iron available for phosphorus binding. If nitrate treatment is to be effective, the iron content of the sediment must be high enough, i.e. well above the lithospheric average of approximately 50 mg g^{-1} of mineral substance (Mackereth, 1966).

Nitrate treatment can be used in both deep and shallow lakes and is normally performed during the last phase of spring circulation and should be finished within a maximum of two months. If combined with iron, iron addition should precede that of nitrate. A solution of calcium nitrate is used, since calcium has a stabilising effect on sediments. This can be dosed into deep water under slow water circulation (Fig. 10.4).

Commercially produced nitrate as well as the outflow from sewage treatment plants may also be used. With the latter, nitrified and clarified water with low phosphorus should be added above anaerobic sapropelic sediments over a longer period (Ripl, 1985). Solid calcium nitrate has also been used (Søndergaard *et al.*, 2000), but this can be difficult to distribute evenly and may lead to negative effects. Since nitrate is generally used up within a few weeks, the water exchange should preferably last from weeks to months. If water residence time is short, the exchange time may be prolonged (e.g. by physical inclusion within a sheet pile wall or rubber apron) or nitrate may be injected directly into the sediment (Fig. 10.5).

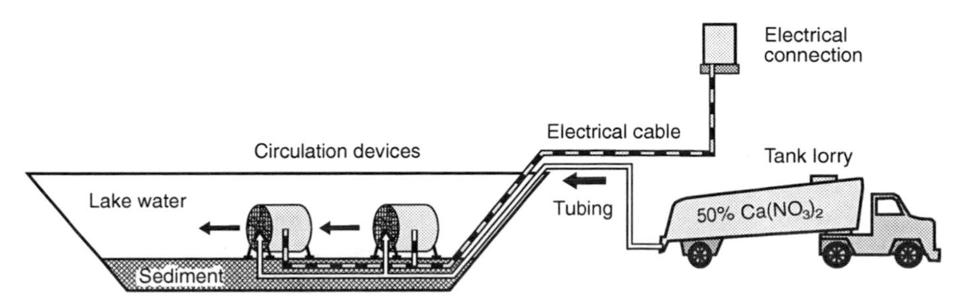

Fig. 10.4. An Example of treatment with calcium nitrate: distribution of nitrate by circulation devices.

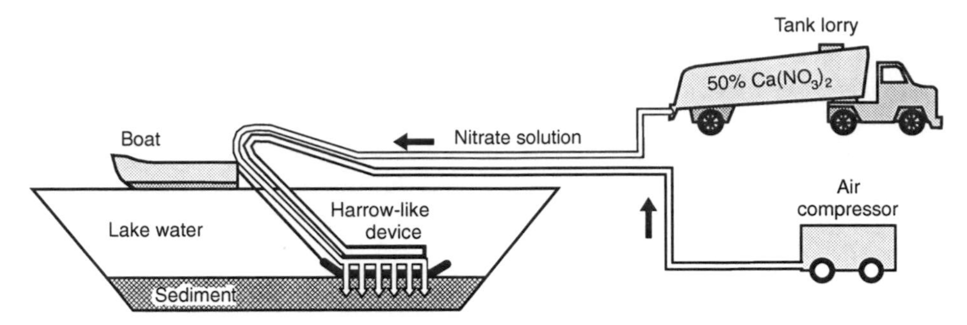

Fig. 10.5. An example of treatment with calcium nitrate: injection into the sediment.

Possible problems arising from nitrate treatment

The nitrate used might also act as a nutrient for phytoplankton and lead to increased growth if nitrogen were limiting. However, when phosphate is the target limiting factor for algal growth, nitrate should not increase phytoplankton biomass. A sufficient reduction of external phosphorus loading is crucial for successful nitrate treatment to avoid planktonic algal blooms. If sedimentation of fresh organic matter is not reduced, phosphorus release from the sediments via bacterial metabolism may recur causing feedback reinforcement of planktonic production. Furthermore, release of ammonium from nitrate ammonification should not occur, because ammonium usually originates from reduction of organic matter (Gottfreund & Schweisfurth, 1982). In fact, it has been reported that high ammonium concentrations in interstitial water decrease during and after nitrate treatment (Ripl, 1978, 1986*a*, *b*).

Where lake water is used for a potable supply, water-quality standards for nitrate may be exceeded by nitrate treatment. However, there is virtually no danger of nitrate export into groundwater, since hydraulic gradients usually enable only infiltration of groundwater into the lake and not vice versa, and efficient denitrification occurs in the lake sediment.

In some cases, there may be a short-term increase in nitrite and gaseous nitrogen oxide concentrations as intermediate products of the nitrification–denitrification process. However, gaseous nitrogen oxides are metabolised in the aqueous environment so quickly that major release is avoided. Consequently, damage to sensitive fish fauna has never been observed. Rapid metabolism of any trace metals released also occurs. For example, through the oxidation of sulphides, even sulphur-bound trace metals can be dissolved. Moreover, pH rises during the process of denitrification and since nitrate treatment is undertaken in iron-rich systems, binding of trace metals occurs in hydroxides or iron-trace-metal hydroxides.

Iron addition

Aim and chemical background

Iron addition is used to increase the iron buffer within the sediment and is often used in combination with nitrate (see above) (Ripl, 1976; Donabaum *et al.*, 1999; Dokulil *et al.*, 2000). In lakes with low iron content and low biodegradable organic matter in the upper sediments, iron addition may suffice without simultaneous nitrate treatment (see above). Such conditions are often found in very shallow lakes in which water movement and oxygen supply of the sediment surface are sufficient all year.

The objectives of iron treatment are: (1) precipitation of phosphorus from the water body; (2) increase of the sediment's phosphorus-binding capacity; and (3) decontamination or precipitation of surplus hydrogen sulphide. In many eutrophic lakes, sulphate reduction in the sediment plays a substantial role in the oxidation of organic substances. The H_2S generated is on the one hand toxic for benthic fauna but, on the other hand, performs a ligand exchange with phosphate when binding to iron (Hasler & Einsele, 1948; Brümmer, 1974). At increasing sediment iron concentrations, the buffer capacity for decontamination and H_2S binding increases, with negative consequences only becoming apparent.

In contrast to other adsorbents (e.g. aluminium), iron forms a dynamic phosphate- and sulphide-binding redox buffer at the sediment/water boundary layer. Dissolution of iron in deeper sediment layers occurs at small redox potentials and, in the absence of H_2S, increases its concentration in interstitial water. When iron is transported upwards along the concentration gradient, it accumulates in the boundary layer between the reductive and oxidative zone. This zone is also particularly active in the processes of phosphorus release and binding in which iron may constitute a dynamic sink for phosphorus.

When iron(III) chloride is introduced, hydrolysis occurs with the formation of gelatinous flocs of ferric hydroxide, which may go on to form a mixture of iron oxide and hydroxide with dewatering. Phosphorus binds to iron either by adsorption to these iron flocs or as iron phosphate (Stumm & Morgan, 1981):

$$\text{hydrolysis: } FeCl_3 + 3H_2O \rightarrow Fe(OH)_3 + 3H^+ + 3\,Cl^-$$

$$\text{dewatering: } Fe(OH)_3 \rightarrow FeO(OH) + H_2O$$

formation of iron phosphate:

$$FeO(OH) + H_3PO_4 \rightarrow FePO_4 + 2\,H_2O$$

adsorption to iron oxide–hydroxide:

$$FeO(OH) + PO_4^{3-} \rightarrow FeO(OH) \sim PO_4^{3-}(aq)$$

Orthophosphate is best precipitated by this reaction. In contrast to treatment by concentrated aluminium (Cooke *et al.*, 1986), co-precipitation of organic particles does not usually occur during iron treatment. Therefore, iron treatment during an expanded planktonic algal bloom may not successfully precipitate phosphorus.

Trivalent ferric oxide–hydroxide reacts with 1.5 mole hydrogen sulphide per mole iron. First, trivalent iron is reduced by H_2S to Fe(II) and 0.5 S^{2-}, which then precipitates with sulphide ions:

binding of hydrogen sulphide:

$$2\,FeO(OH) + 3\,H_2S \rightarrow 2\,FeS + S^0 + 4\,H_2O$$

A combined technique of phosphorus precipitation by hypolimnetic injection of a $FeCl_2$ solution in combination with transport of hypolimnetic water rich in free carbon dioxide into the upper layers to reduce *Microcystis* blooms has also been used (Deppe *et al.*, 1999).

Techniques

To understand iron treatment attempts, pre-treatment, accompanying and post-treatment monitoring has to be undertaken. This must include spatial and temporal distribution of phosphorus fractions (phosphate, particulate and organically bound phosphorus). The most suitable time for iron treatment can be determined from the annual circulation pattern and it should be effected when the phosphate fraction reaches its relative maximum (Cooke *et al.*, 1986), usually in late autumn to early spring. The dose should contain sufficient iron for phosphorus binding even if sulphate reduction should occur. In relation to phosphorus in water, iron should always be utilised over-stoichiometrically.

Most frequently, iron(III) chloride ($FeCl_3$) has been used as the iron salt, although iron sulphate has also been used (Daldorph, 1999). The latter may be less suitable because of its negative effect on sediment sulphate reduction. As a consequence of hydrolysis of acid iron bonds and the associated formation of protons, the buffer capacity (alkalinity) of the water is particularly important. With hydrolysis of iron(III) chloride, about 3 moles of protons per mole iron chloride are formed. At low alkalinity, this acid must be buffered or neutralised by the addition of fine particles of lime (calcium carbonate) with high reactivity:

$$\text{hydrolysis: } FeCl_3 + 2\,H_2O \rightarrow FeO(OH) + 3\,H^+ + 3\,Cl^-$$

$$\text{neutralisation: } 6\,H^+ + 3\,CaCO_3 \rightarrow 3\,CO_2 + 3\,H_2O + 3\,Ca^{2+}$$

Acid-forming iron bonds should be used with caution if dosing is to take place from a helicopter or aeroplane, since they can be transported as fine dust which may damage the surroundings.

Apart from adding iron as soluble iron salts, ferric oxide–hydroxides in a solid paste can be used. The latter may originate from the processing of iron-rich ground and mine waters for drinking supply, and may be dispersed by machine on winter ice cover, if this is thick enough. However, since the structure of the ice changes after the treatment, dispersal should take place rapidly and no further visits on to the ice should be undertaken. Ferric

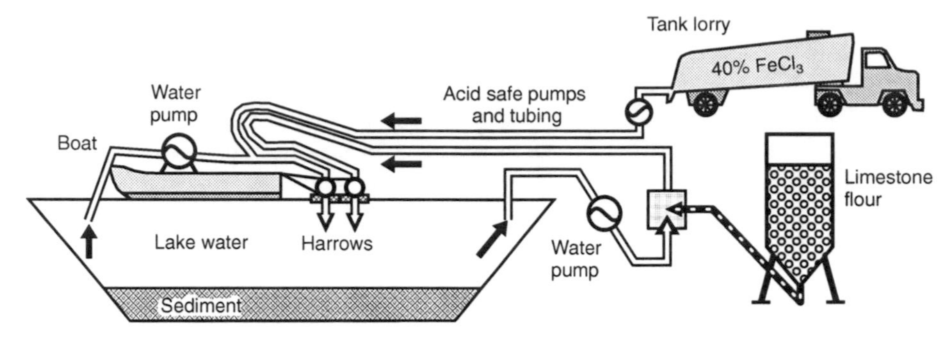

Fig. 10.6. Exemplary scheme of iron treatment with $FeCl_3$ and lime.

oxide–hydroxides should not contain strongly aged, drained material which transforms to oxides that cannot ensure phosphate and sulphide binding. In addition, undesirably high concentrations of phosphorus can be found in some preparations. Solid ferric oxide–hydroxide should therefore always be tested as to its suitability.

Usually iron is added in dissolved form. It can be brought to the lake as a concentrated solution, be mixed ashore and subsequently mixed with lake water on barges or in the boat, immediately prior to dosing. If added from land, the preparation can be pumped to a boat with, for example, acid-resistant equipment (pumps and flexible high-pressure polyethylene tubings). From the boat it is injected over several harrow-like arranged injection nozzles below the water surface (Fig. 10.6). The laborious direct injection of iron into the sediment, as was used in Lake Lillesjön, Sweden and Lake Schlei, Germany (Ripl, 1976, 1978, 1985) has since been proved unnecessary. The distribution of iron over the sediment should be as even as possible. If uneven, the method is less efficient. During precipitation of phosphorus, a clear reduction of total phosphorus to a level below 30-40 $\mu g\ l^{-1}$ should be the target. At higher levels the subsequent algal bloom could jeopardise the restoration attempt.

Possible problems in connection with iron treatment

As with the other restoration techniques, iron treatment should be preceded by a significant reduction of catchment nutrient loading to achieve effective and long-lasting results. Otherwise, the positive effects of the treatment may disappear in a few months as a result of continued external loading (Boers *et al.*, 1994).

The lake under treatment should be closely monitored to respond to any adverse effects of treatment. The most important of these is lowering of pH, which could change biological structure. However, toxic effects on fish and the benthic community are rarely observed if the pH does not drop below 6. Reduction of pH may also be prevented by the addition of lime (Ripl, 1976; Dokulil *et al.*, 2000) (see below). Other minor effects include the potential for brown staining of the water column during and, for a few days, after treatment, and a short-term increase in the occurrence of iron floc. Bathing should be directed to other areas for the sake of operational safety.

When adding iron chloride, the chloride concentration in the water column rises to 50 to several hundred milligrams per litre depending on hydraulic retention time and dose(s) used. This sort of increase is generally thought to be unimportant. For example, drinking-water limit values for chloride are *c.* 250 mg l^{-1}, and even an increase to 500 mg l^{-1} or more will probably not result in any biological damage (Schönborn, 1992).

Alum treatment

Aim and chemical background

Aluminium sulphate ($Al_2(SO_4)_3$) or alum has been used for decades to precipitate and increase the sorption capacity of phosphorus and to remove it from internal cycling (Dunst *et al.*, 1974; Cooke & Kennedy, 1978). Alum treatment may be used in both stratified and unstratified lakes. Also, but less frequently, combined iron–aluminium additions in the form of ferric aluminium sulphate have been used (Foy, 1986; Foy & Fitzsimons, 1987). Aluminium complexes and polymers have the advantage over iron of requiring a low redox potential for the reduction of insoluble Al^{3+} to soluble Al^{2+}, meaning that adsorbed phosphorus will not be released from the sediment during periods of anoxia (Foy, 1986; Welch *et al.*, 1988). Alum treatment aiming to reduce the amount of natural organic matter has also been investigated in reservoirs used for drinking supply (Chow *et al.*, 1999).

When added to water, alum forms an aluminium hydroxide complex ($Al(OH_3)$), which has a cotton-like appearance called 'floc' (Dunst *et al.*, 1974; Soltero & Nichols, 1981; Cooke *et al.*, 1993):

$$Al^{3+} + H_2O \rightarrow Al(OH)^{2+} + H^+ + 2\,H_2O \rightarrow Al(OH)_3 + 3H^+$$

Phosphorus adsorbs to the floc and sinks to the bottom where it can be permanently removed from the phosphorus cycle and fixed and buried in the sediment. If alum treatment is capable of transforming loosely sorbed and iron-bound phosphorus to aluminium-bound phosphorus, it may reduce the internal phosphorus loading caused by anoxia in the hypolimnion (Ryding & Welch, 1998). The floc also tends to physically entrap algae and other particulate matter (Soltero & Nichols, 1981; Connor & Martin, 1989).

Techniques

Alum is usually applied as concentrated liquid alum which is dispersed into the lake from a small boat or pontoon barges (Soltero & Nichols, 1981; Foy, 1986; Cooke *et al.*, 1993). Alum may be injected at prescribed depths and into different parts of the lake to facilitate complete coverage and obtain maximum effect. In most cases, aluminium is added in quantities ranging from 5 to 100 g Al m^{-2} or 5 to 25 g Al m^{-3} (Welch & Cooke, 1999; Ryding *et al.*, 2000). The amount added may be adjusted according to alkalinity in the lake and mobile sediment phosphorus concentrations.

Aluminium in the sediment of alum-treated lakes is usually indistinguishable and the alum floc is believed to settle gradually through the usually low-density sediments of most lakes and become buried by newly formed sediment (Welch & Cooke, 1999). In some cases, aluminium has been detected in the sediment at a depth corresponding to the time of treatment (Ryding *et al.*, 2000).

Possible problems in connection with treatment and their effects

Alum is normally added as a single treatment based on the current water and sediment phosphorus content, implying that the capacity to adsorb further phosphorus will eventually cease. Thus, a single alum treatment usually does not have a long-term effectiveness. If the external phosphorus loading remains high or is not reduced sufficiently, only a short-term effect on lake trophic state can be expected. In most cases the longevity of an effective treatment has been reported to last for about ten years, fluctuating between one and 20 years (Welch & Cooke, 1999). Treatments have had greater longevity and been more successful in stratified rather than unstratified lakes (Foy, 1986; Welch *et al.*, 1988; Welch & Cooke, 1999). However, treatments in shallow lakes are more certain to affect phosphorus availability in the photic zone than in stratified lakes where sediment-released phosphorus to the hypolimnion is unavailable.

Aluminium hydroxy complexed phosphorus is sensitive to pH, and phytoplankton or macrophyte-induced photosynthetically elevated pH has been blamed for the failure of one alum treatment in a shallow lake (Welch & Cooke, 1999). Dense macrophyte beds may also diminish the effectiveness of the treatment as they may cause uneven floc distribution or sediment phosphorus recycling from below the floc layer through plant senescence and decay (Welch & Cooke, 1999). Depending on the dosage, alum treatment may elevate sulphate levels and thereby lead to increased hydrogen sulphide production eventually reversing the lake back to eutrophy (Soltero & Nichols, 1981).

Acidification of alum-treated lakes to below pH 6 may result in increased aluminium concentrations and adverse toxic effects associated with enhanced metal solubility (Soltero & Nichols, 1981; Cooke *et al.*, 1993). At pH 4 to 6, various soluble intermediate forms occur, while at a pH below 4, soluble Al^{3+} dominates. This form is particularly toxic to biota (Cooke *et al.*, 1993). Because hydrogen ions are liberated when alum is added, pH decreases in the lake water at a rate depending on alkalinity. To avoid toxic effects, the maximum dosage of alum has been defined as the maximum amount of aluminium which, when added to lake water, would ensure a dissolved aluminium concentration below 50 $\mu g\ l^{-1}$ (Cooke & Kennedy, 1981; Kennedy & Cooke, 1982). Buffering agents, such as sodium aluminate and sodium bicarbonate, have been added in treatments of soft-water lakes to maintain pH above 6. Adverse effects in terms of reduced invertebrate populations have usually not been observed and no fish kills have been reported (Cooke *et al.*, 1993). Usually several days are required to treat a 300-ha lake so there is ample time for fish to avoid areas of water disturbance.

Lime treatment to reduce eutrophication

Aim and chemical background

Slaked lime (calcium hydroxide ($Ca(OH)_2$)) has been added to eutrophic lakes to diminish phosphorus availability by the formation of calcite (calcium carbonate, $CaCO_3$) and the precipitation of phosphate into insoluble Ca–PO_4 complexes (hydroxyapatite):

$$10\,CaCO_3 + 6\,HPO_4^{2-} + 2\,H_2O \rightarrow Ca_{10}\,(PO_4)_6(OH)_2 + 10\,HCO_3^-$$

In the short term (<15 days), calcium hydroxide treatment may also directly decrease phytoplankton biomass and chlorophyll *a* through the precipitation of phytoplankton cells or colonies (Zhang & Prepas, 1996).

Co-precipitation of inorganic phosphorus with calcite in hard-water lakes is a natural process usually triggered by an increase in pH caused by photosynthesis (Otsuki & Wetzel, 1972; Murphy *et al.*, 1983; Hartley *et al.*, 1997). It is believed that phosphate initially adsorbs to the surface of calcite crystals and later becomes incorporated into the crystal during crystal growth (Kleiner, 1988; House, 1990). Calcite sorbs phosphate especially when pH exceeds 9 and hydroxyapatite has its lowest solubility at high pH (>9.5).

Techniques

Fundamentally, lime application involves the same techniques as those developed for sewage treatment plants. *In situ* lake treatment, however, cannot be controlled similarly (i.e. by adjusting pH) and one of the problems is to find an application technique promoting carbonate precipitation at an acceptable pH remaining within its natural range (Murphy *et al.*, 1988).

Lime is usually added from a boat as a slurry of hydrated lime mixed with water, which is then sprayed over the lake surface or injected at a depth of a few metres. Alternatively, lime has been injected into the hypolimnion in combination with aeration, in order to shift the equilibrium of the calcite–carbonic acid systems towards calcite precipitation in the hypolimnion (Dittrich *et al.*, 2000). Repeated low-dose treatments or a single high-dose treatment have been used. Calcium hydroxide dosage normally ranges from 25 to 300 mg l^{-1}. When the calcium hydroxide slurry has been added, large particles will sink through the water column while small particles dissolve in the water and form calcite (Zhang & Prepas, 1996).

Less frequently, calcite or calcite-rich lake marl taken from natural deposits in the littoral zone or from the lake sediment and then spread over the lake surface has been used as an alternative to slaked lime in order to co-precipitate phosphorus (Stuben *et al.*, 1998; Hupfer *et al.*, 2000). Calcite is generally believed to have a lower capacity to sorb phosphorus than lime, where freshly nucleated calcite crystals are generated in the presence of phosphate.

Possible problems in connection with treatment and their effects

Turbidity increases after the lime treatment, but usually only for a few days at most. Lime treatment and the following pH shock may, however, have a negative impact on the macroinvertebrate community and other animals, and may last for a year or more after the treatment (Miskimmin *et al.*, 1995; Yee *et al.*, 2000). The extent of pH elevation after the addition depends on the buffering capacity of the lake and the dosage applied. In hard-water lakes, it is usually possible to keep pH below 10, while in soft-water lakes pH may increase to above 11 (Zhang & Prepas,

1996), this having severe implications for most organisms.

The affinity of calcium to sorb phosphorus in natural systems is relatively low compared with elements like iron. Several studies have thus shown that there is no relationship between the calcium carbonate content in the sediment of lakes and the amount of calcium carbonate-bound phosphorus (Søndergaard *et al.*, 1996; Rzepecki, 1997; Gonsiorczyk *et al.*, 1998).

Phosphorus precipitated with calcium carbonate may redissolve and thus prevent permanent effects of the treatment. Redissolved calcite may reprecipitate later as conditions change, establishing more long-term mechanisms (Murphy *et al.*, 1988). The solubility of precipitated phosphate increases in the hypolimnion and close to the sediment, where bacterial respiration causes lowered pH (Driscoll *et al.*, 1993).

Other chemical methods combating eutrophication

A number of other chemicals have been used to increase the binding capacity of phosphorus, many of them being relatively inexpensive industrial by-products.

Gypsum ($CaSO_4 * 2\,H_2O$) has been used in a few cases in a parallel manner to the use of calcium hydroxide to establish calcium–phosphorus compounds (hydroapatite) and reduce phosphorus release from the sediment (Wu & Boyd, 1990; Salonen & Varjo, 2000). However, the addition of sulphate may, in the longer term, lead to increased internal loading via the ligand exchange of sulphide and iron-bound phosphorus.

Slag, a by-product in the refining process of iron ore with caustic lime (Yamada *et al.*, 1986), has also been used. It contains large amounts of calcium and other elements like aluminium and iron that can be used to adsorb dissolved inorganic phosphate. Clay and fly ash have also been considered as sorption agents for phosphorus (Dunst *et al.*, 1974).

Liming to adjust pH

Aim and chemical background

Liming or base addition of acidified lakes is used to counteract and mitigate the decrease of pH in lakes where the acid deposition exceeds buffering capacity. Liming is thus used to enhance or prevent a decrease in species richness and species diversity, and to ensure that the natural fauna and flora can survive or recolonise (Stenson & Svensson, 1995; Nyberg, 1998). Normally, the aim is to raise pH to above 6 and the alkalinity to 0.1 meq l^{-1}, in order to establish an acceptable buffering capacity (Svenson *et al.*, 1995).

Most often, limestone powder or gravel (calcite, $CaCO_3$) and, less often, other buffering agents such as magnesium (dolomite, $CaMg(CO_3)_2$), are added to increase the cation pool. Wood ash from forest residues has been used alternatively in catchments, adding, besides calcium, also a number of other buffering elements (Bramryd & Fransman, 1995; Fransman & Nihlgård, 1995).

Apart from restoring faunal diversity, liming can also cause a net precipitation of phosphorus equivalent to the effects seen after the addition of slaked lime (Smayda, 1990). Liming also promotes the precipitation of aluminium, iron and manganese (Andersen & Pempkowiak, 1999), which may influence phosphorus availability. The reverse process can also be seen, as increased pH in the catchment may lead to increased phosphorus availability by stopping the acidification process which tends to increase the precipitation of phosphorus with aluminium in the soil matrix (Broberg & Persson, 1984). Increased pH may also increase the mineralisation rate in the sediment and the release of nutrients (Dickson *et al.*, 1995; Roelofs *et al.*, 1995).

Techniques

Liming using limestone powder or dolomite powder is mostly applied directly to the lake, but liming can also be conducted as a watershed treatment depending on hydrology (Svenson *et al.*, 1995). In-lake treatment is often conducted from a boat connected with a pipeline to a container on shore. Small lakes can be limed manually from a boat or on the ice during winter. Lakes situated in remote areas may be limed from a helicopter. Wetland liming can be used as a supplement to lake liming. Rivers can be limed by continuous automatic dosers (Sandøy & Romundstad, 1995).

Possible problems in connection with treatment and their effects

Liming is usually regarded as a temporary solution where no permanent effects are established. Contrarily, if the loading of acids to the lake continues, pH will eventually decrease again unless reliming is conducted.

Increased nutrient mobilisation effected by liming can cause internal eutrophication in shallow lakes followed by changes in the macrophyte and plankton community (Brandrud & Roelofs, 1995; Dickson *et al.*, 1995). Some plants (e.g. *Juncus bulbosus*) may benefit from higher nutrient concentrations in limed lakes when the carbon dioxide concentrations in the water are relatively high, as is the case after reacidification (Lucassen *et al.*, 1999). When liming in wetlands, undesirable effects, such as changes in mosses and lichens (Svenson *et al.*, 1995), may occur. Liming may lead to decreased transparency, but usually the lake water returns to the pre-treatment situation (Pulkkinen, 1995).

CONCLUDING REMARKS

Numerous chemical restoration measures have been developed to combat resilience in lake recovery during the past decades. In many lakes, internal loading of phosphorus from lake sediments prevents improvements in water quality despite a reduction of the external loading. In the case of acidification, liming has been used to counteract a pH decrease in lakes where the acid deposition exceeds the buffering capacity.

For both types of restoration measures, an important prerequisite for obtaining success and long-term effects is elimination of the underlying reasons for the undesirable water quality; i.e. a sufficient reduction of external phosphorus loading in the case of eutrophication, and a decline in the deposition of acids in the case of acidification.

Restoration measures to neutralise eutrophication effects focus on either increasing the phosphorus sorption capacity of compounds (especially of iron by improving redox conditions) already present in the sediment, or on increasing the sorption capacity by the addition of new sorption capacity (mainly alum, iron and calcium).

Five categories of chemical restoration measures can be summarized:

1. Hypolimnetic oxygenation, with pure oxygen or atmospheric air being injected into the hypolimnion with various types of equipment, to improve the redox sensitive sorption of phosphorus to iron and the living conditions of benthic animals.
2. Oxidation of the hypolimnion and the sediment using nitrate as an electron acceptor to oxidise organic matter in the sediment and improve the sorption of phosphorus to iron by preventing the formation of iron sulphide.
3. Addition of iron to increase the phosphorus sorption capacity of the sediment, this often being used as a supplement to oxidation with nitrate or oxygen.
4. Alum treatment to increase the phosphorus sorption capacity by increasing the non-redox sensitive binding of phosphorus to aluminium hydroxides.
5. Addition of slaked lime to increase the formation of calcite and the precipitation of phosphorus into hydroxyapatite.

ACKNOWLEDGMENTS

The technical staff of the National Environmental Research Institute are thanked for their assistance. Layout and manuscript assistance was provided by K. Møgelvang and A.M. Poulsen. We thank Eugene Welch and Martin Perrow for valuable comments on the manuscript.

REFERENCES

Aku, P.M.K. & Tonn, W.M. (1997). Changes in population structure, growth, and biomass of cisco (*Coregonus artedi*) during hypolimnetic oxygenation of a deep, eutrophic lake, Amisk Lake, Alberta. *Canadian Journal of Fisheries and Aquatic Sciences*, **54**, 2196–2206.

Andersen, D.O. & Pempkowiak, J. (1999). Sediment content of metals before and after lake water liming. *Science of the Total Environment*, **244**, 107–118.

Ashley, K.I., Hay, S. & Scholten, G.H. (1987). Hypolimnetic aeration: field test of the empirical sizing method. *Water Research*, **21**, 223–227.

Boers, P., van der Does, J., Quaak, M. & van der Vlught, J. (1994). Phosphorus fixation with iron(III)chloride: a new method to combat internal phosphorus loading in shallow lakes? *Archiv für Hydrobiologie*, **129**, 339–351.

Boström, B., Jansson, M. & Forsberg, C. (1982). Phosphorus release from lake sediments. *Archiv für hydrobiologische Ergebnisse der Limnologie*, **18**, 5–59.

Bramryd, T. & Fransman, B. (1995). Silvicultural use of wood ashes: effects on the nutrient and heavy metal balance in a pine (*Pinus sylvestris*, L.) forest soil. *Water, Air and Soil Pollution*, **85**, 1039–1044.

Brandrud, T.E. & Roelofs, J.G.M. (1995). Enhanced growth of the macrophyte *Juncus bulbosus* in South Norwegian limed lakes: a regional survey. *Water, Air and Soil Pollution*, **85**, 913–918.

Broberg, O. & Persson, G. (1984). External budgets for phosphorus, nitrogen and dissolved organic carbon for the acidified Lake Gårdsjön. *Archiv für Hydrobiologie*, **99**, 160–175.

Brümmer, G. (1974): Phosphatmobilisierung unter reduzierenden Bedingungen: Ein Beitrag zum Problem der Gewässereutrophierung. *Mitteilungen deutsche bodenkundliche Gesellschaft*, **18**, 175–177.

Burris, V.L. & Little, J.C. (1998). Bubble dynamics and oxygen transfer in a hypolimnetic aerator. *Water Science and Technology*, **37**, 293–300.

Chow, C.W.K., van Leeuwen, J.A., Drikas, M., Fabris, R., Spark, K.M. & Page, D.W. (1999). The impact of the character of natural organic matter in conventional treatment with alum. *Water Science and Technology*, **40**, 97–104.

Connor, J.N. & Martin, M.R. (1989). An assessment of sediment phosphorus inactivation, Kezar Lake, New Hampshire. *Water Research Bulletin*, **4**, 845–853.

Cooke, G.D. & Kennedy, R.H. (1978). Effects of a hypolimnetic application of aluminium sulfate to an eutrophic lake. *Verhandlungen Internationale Vereinigung für theoretische und angewandte Limnologie*, **20**, 28–39.

Cooke, G.D. & Kennedy, R.H. (1981). *Precipitation and Inactivation of Phosphorus as a Lake Restoration Technique*, EPA-600/3-81-012. Washington, DC: US Environmental Protection Agency.

Cooke, G.D., Welch, E.B., Peterson, S.A. & Newroth, P.R. (1986). *Lake and Reservoir Restoration*. Boston, MA: Butterworth.

Cooke, G.D., Welch, E.B., Peterson, S.A. & Newroth, P.R. (1993). *Restoration and Management of Lakes and Reservoirs*, 2nd edn. Boca Raton, FL: Lewis Publishers.

Daldorph, P.W.G. (1999). A reservoir in management-induced transition between ecological states. *Hydrobiologia*, **395/396**, 325–333.

Deppe, T., Ockenfeld, K., Meybohm, A., Opitz, M. & Benndorf, J. (1999). Reduction of microcystic blooms in a hypertrophic reservoir by a combined ecotechnological strategy. *Hydrobiologia*, **409**, 31–38.

Dickson, W., Borg, H. Ekström, C., Hörnström, E. & Grönlund, T. (1995). Reliming and reacidification effects on lakewater. *Water, Air and Soil Pollution*, **85**, 919–924.

Dinsmore, W.P. & Prepas, E.E. (1997*a*). Impact of hypolimnetic oxygenation on profundal macroinvertebrates in a eutrophic lake in central Alberta. 1: Changes in macroinvertebrate abundance and diversity. *Canadian Journal of Fisheries and Aquatic Sciences*, **54**, 2157–2169.

Dinsmore, W.P. & Prepas, E.E. (1997*b*). Impact of hypolimnetic oxygenation on profundal macroinvertebrates in a eutrophic lake in central Alberta. 2: Changes in *Chironomus* spp. abundance and biomass. *Canadian Journal of Fisheries and Aquatic Sciences*, **54**, 2170–2181.

Dittrich, M., Casper, P. & Koschel, R. (2000). Changes in the porewater chemistry of profundal sediments in response to artificial hypolimnetic calcite precipitation. *Archiv für hydrobiologische Ergebnisse der Limnologie*, **55**, 421–432.

Dokulil, M.T., Teubner, K. & Donabaum, K. (2000). Restoration of shallow, ground-water fed urban lake using a combination of internal management strategies: a case study. *Archiv für hydrobiologische Ergebnisse der Limnologie*, **55**, 271–282.

Donabaum, K., Schagerl, M. & Dokulil, M.T. (1999). Integrated management to restore macrophyte domination. *Hydrobiologia*, **395/396**, 87–97.

Driscoll, C.T., Effler, S.W., Auer, M.T., Doerr, S.M. & Penn, M.R. (1993). Supply of phosphorus to the water column of a productive hardwater lake: controlling mechanisms and management considerations. *Hydrobiologia*, **253**, 61–72.

Dunst, R., Born, S., Uttormark, P., Smith, S., Nichols, S., Peterson, J., Knauer, D., Serns, S., Winter, D. & Wirth, T. (1974). *Survey of Lake Rehabilitation Technique and Experiences*, Technical Bulletin no. 75. Madison, WI: Department of Natural Resources.

Duursma, E.K. (1967). The mobility of compounds in sediments in relation to exchange between bottom and supernatant water. In *Chemical Environment in the Aquatic Habitat*, eds. H. L. Goltermann & R.S. Clymo, pp. 288–296. Amsterdam: Noord-Hollandsche Uitgevers Maatsahappij.

Field, K.M. & Prepas, E.E. (1997). Increased abundance and depth distribution of pelagic crustacean zooplankton during hypolimnetic oxygenation in a deep, eutrophic Alberta lake. *Canadian Journal of Fisheries and Aquatic Sciences*, **54**, 2146–2156.

Foy, R.H. (1986). Suppression of phosphorus release from lake sediments by the addition of nitrate. *Water Research*, **11**, 1345–1351.

Foy, R.H. & Fitzsimons, A.G. (1987). Phosphorus inactivation in a eutrophic lake by the direct addition of ferric aluminium sulphate: changes in phytoplankton populations. *Freshwater Biology*, **17**, 1–13.

Fransman, B. & Nihlgård, B. (1995). Water chemistry in forested catchments after topsoil treatment with liming agents in south Sweden. *Water, Air and Soil Pollution*, **85**, 895–900.

Gächter, R. (1987). Lake restoration: why oxygenation and artificial mixing cannot substitute for a decrease in the external phosphorus loading. *Schweiziche Zeitschrift für Hydrologie*, **49**, 170–185.

Gächter, R. & Wehrli, B. (1998). Ten years of artificial mixing and oxygenation: no effect on the internal phosphorus loading of two eutrophic lakes. *Environmental Science and Technology*, **32**, 3659–3665.

Gonsiorczyk, T., Casper, P. & Koschel, R. (1998). Phosphorus-binding forms in the sediment of an oligotrophic and an eutrophic hardwater lake of the Baltic Lake District (Germany). *Water Science and Technology*, **37**, 51–58.

Gorham, E. (1958). Oberservations on the formation and breakdown of the oxidized microzone at the mud surface in lakes. *Limnology and Oceanography*, **3**, 291–298

Gottfreund, J. & Schweisfurth, R. (1982). Über die Herkunft von Ammonium in Wasser. *Vom Wasser*, **58**, 187–205.

Hartley, A.M., House, W.A., Callow, M.E. & Leadbeater, S.C. (1997). Coprecipitation of phosphate with calcite in the presence of photosynthesizing green algae. *Water Research*, **31**, 2261–2268.

Hasler, A.D. & Einsele, W. (1948). Fertilization for increasing productivity of natural inland waters. *Transactions of the North American Wildlife Conference*,**13**, 527–555.

Hieltjes, A.H.M. & Lijklema, L. (1980). Fractionation of inorganic phosphates in calcareous sediments. *Journal of Environmental Quality*, **9**, 405–407.

House, W.A. (1990). The prediction of phosphate coprecipitation with calcite in freshwaters. *Water Research*, **24**, 1017–1023.

Hupfer, M., Pöthig, R., Brüggemann, R. & Geller, W. (2000). Mechanical resuspension of autochthonous calcite (seekreide) failed to control internal phosphorus cycle in a eutrophic lake. *Water Research*, **34**, 859–867.

Jacobsen, O.S. (1978). Sorption, adsorption and chemosorption of phosphate by Danish lake sediments. *Vatten*, **4**, 230–243.

Jaeger, D. (1994). Effects of hypolimnetic water aeration and iron-phosphate precipitation on the trophic level of Lake Krupunder. *Hydrobiologia*, **275/276**, 433–444.

Jensen, H.S., Kristensen, P., Jeppesen, E. & Skytthe, A. (1992). Iron:phosphorus ratio in surface sediment as an indicator of phosphate release from aerobic sediments in shallow lakes. *Hydrobiologia*, **235/236**, 731–743.

Jeppesen, E., Kristensen, P., Jensen, J.P., Søndergaard, M., Mortensen, E. & Lauridsen, T. (1991). Recovery resilience following a reduction in external phosphorus loading of shallow, eutrophic Danish lakes: duration, regulating factors and methods for overcoming resilience. *Memorie dell'Istituto italiano di Idrobiologia*, **48**, 127–148.

Jeppesen, E., Jensen, J.P., Søndergaard, M. & Lauridsen, T. (1997). Top-down control in freshwater lakes: the role of nutrient state, submerged macrophytes and water depth. *Hydrobiologia*, **342/343**, 151–164.

Jeppesen, E., Søndergaard, M., Kronvang, B., Jensen, J.P., Svendsen, L.M. & Lauridsen, T. (1999). Lake and catchment management. In *Ecological Basis for Lake and Reservoir Management*, eds. D. Harper, A. Ferguson, B. Brierley & G. Phillips. *Hydrobiologia*, **408/409**, 419–432.

Kennedy, R.H. & Cooke, G.D. (1982). Control of lake phosphorus with aluminium sulfate: dose determination and application techniques. *Water Research Bulletin*, **18**, 389–395.

Kleiner, J. (1988). Coprecipitation of phosphate with calcite in lake water: a laboratory experiment modelling phosphorus removal with calcite in Lake Constance. *Water Research*, **22**, 1259–1265.

Kristensen, P. & Hansen, H.O. (eds.) (1994). *European Rivers and Lakes*. Copenhagen: European Environment Agency.

Lindenschmidt, K.E. (1999). Controlling the growth of *Microcystis* using surged artificial aeration. *Internationale Revue der gesamten Biologie*, **84**, 243–254.

Livingstone, D. & Schanz, F. (1994). The effects of deep-water siphoning on a small, shallow lake: a long-term case study. *Archiv für Hydrobiologie*, **132**, 15–44.

Lucassen, E., Bobbink, R. & Oonk, M.M.A. (1999). The effects of liming and reacidification on the growth of *Juncus bulbosus*: a mesocosm experiment. *Aquatic Botany*, **64**, 95–103.

Mackereth, F.J.H. (1966). Some chemical observations on post-glacial lake sediments. *Philosophical Transactions of the Royal Society London B*, **250** (**765**), 165–213.

Marsden, M.W. (1989). Lake restoration by reducing external phosphorus loading: the influence of sediment phosphorus release. *Freshwater Biology*, **21**, 139–162.

Matinvesi, J. (1996). The change of sediment composition during recovery of two Finnish lakes induced by waste water purification and lake oxygenation. *Hydrobiologia*, **335**, 193–202.

McQueen, D.J., Lean, D.R.S. & Charlton, M.N. (1986). The effects of hypolimnetic aeration on iron–phosphorus interactions. *Water Research*, **9**, 1129–1135.

Miskimmin, B.M., Donahue, W.F. & Watson, D. (1995). Invertebrate community response to experimental lime ($Ca(OH)_2$) treatment of an eutrophic pond. *Aquatic Sciences*, **57**, 20–30.

Moore, B.C., Chen, P.H., Funk, W.H. & Yonge, D. (1996). A model for predicting lake sediment oxygen demand following hypolimnetic aeration. *Water Research Bulletin*, **32**, 723–731.

Mortimer, C.H. (1941). The exchange of dissolved substances between mud and water in lakes. 1. *Journal of Ecology*, **29**, 280–329.

Mortimer, C.H. (1942). The exchange of dissolved substances between mud and water in lakes. 2. *Journal of Ecology*, **30**, 147–201.

Murphy, T.P., Hall, K.G. & Yesaki, I. (1983). Coprecipitation of phosphate with calcite in a naturally eutrophic lake. *Limnology and Oceanography*, **28**, 58–69.

Murphy, T.P., Hall, K.G. & Northcote, T.G. (1988). Lime treatment of a hardwater lake to reduce eutrophication. *Lake and Reservoir Management*, **4**, 51–62.

Nakamura, Y. & Inoue, T. (1996). A theoretical study on operation conditions of hypolimnetic aerators. *Water Science and Technology*, **34**, 211–218.

Nurnberg, G. (1987). Hypolimnetic withdrawal as lake restoration technique. *Journal of Environmental Engineering*, **113**, 1006–1016.

Nyberg, P. (1998). Biotic effects in planktonic crustacean communities in acidified Swedish forest lakes after liming. *Water, Air and Soil Pollution*, **101**, 257–288.

Ohle, W. (1953). Der Vorgang rasanter Seenalterung in Holstein. *Naturwissenschaften*, **40**, 153–162.

Ohle, W. (1958). Die Stoffwechseldynamik der Seen in Abhängigkeit von der Gasausscheidung ihres Schlammes. *Vom Wasser*, **25**, 127–149.

Ohle, W. (1978). Ebullition of gases from sediment, condition, and relationship to primary production of lakes. *Verhandlungen internationale Vereinigung der Limnologie*, **20**, 957–962.

Otsuki, A. & Wetzel, R.G. (1972). Coprecipitation of phosphate with carbonates in a marl lake. *Limnology and Oceanography*, **17**, 763–766.

Phillips, G., Jackson, R., Bennet, C. & Chilvers, A. (1994). The importance of sediment phosphorus release in the restoration of very shallow lakes (The Norfolk Broads, England) and implications for biomanipulation. *Hydrobiologia*, **275/276**, 445–456.

Prepas, E.E., Field, K.M., Murphy, T.P., Johnson, W.L., Burke, J. M. & Tonn, W. (1997). Introduction to the Amisk Lake Project: oxygenation of a deep, eutrophic lake. *Canadian Journal of Fisheries and Aquatic Sciences*, **54**, 2105–2110.

Priscu, J.C. & Downes, M.T. (1987). Microbial activity in the surficial sediments of an oligotrophic and eutrophic lake, with particular reference to dissimilatory nitrate reduction. *Archiv für Hydrobiologie*, **108**, 385–409.

Psenner, R., Boström, B., Dinka, M., Petterson, K., Pucsko, R. & Sager, M. (1988). Fractionation of phosphorus in suspended matter and sediment. *Archiv für hydrobiologische Ergebnisse der Limnologie*, **30**, 98–110.

Pulkkinen, K. (1995). Measuring movement and settling of limestone powder after liming using acoustics, beam attenuation and conductivity. *Water, Air and Soil Pollution*, **85**, 1021–1026.

Ripl, W. (1976). Biochemical oxidation of polluted lake sediment: a new lake restoration method. *Ambio*, **5**, 132–135.

Ripl, W. (1978). *Oxidation of Lake Sediments with Nitrate: A Restoration Method for Former Recipients*. Lund, Sweden: Institute of Limnology, University of Lund.

Ripl, W. (1985). Oxidation of sapropelic sediments by nitrified effluents from a treatment plant. In *Lake and Reservoir Management: Practical Applications*, NALMS Symposium EPA, pp. 153–156. Merrifield, VA: North American Lake Management Society.

Ripl, W. (1986*a*). Internal phosphorus recycling mechanisms in shallow lakes. In *Lake and Reservoir Management*, vol. 2,

Proceedings of the 5th Annual Conference and International Symposium on Applied Lake and Watershed Management, 13–16 November 1985, Lake Geneva, pp. 138–142. WI. Merrifield, VA: North American Lake Management Society.

Ripl, W. (1986*b*). Restaurierung der Schlei: Bericht über ein Forschungsvorhaben. In *Auftrag des Landesamtes für Wasserhaushalt und Küsten, Kiel*, Schriftenreihe des Landesamtes für Wasserhaushalt und Küsten D 5. Berlin: Technische Universität Berlin, Fachgebiet Limnologie.

Ripl, W. & Lindmark, G. (1978). Ecosystem control by nitrogen metabolism in sediment. *Vatten*, **34**, 135–144.

Roelofs, J.G.M., Smoldersm A.J.P., Brandrud, T.-E. & Bobbink, R. (1995). The effect of acidification, liming and reacidification on macrophyte development, water quality and sediment characteristics of soft-water lakes. *Water, Air and Soil Pollution*, **85**, 976–972.

Ryding, E. & Welch, E.B. (1998). Dosage of aluminium to absorb mobile phosphate in lake sediments. *Water Research*, **32**, 2969–2976.

Ryding, E., Huser, B. & Welch, E.B. (2000). Amount of phosphorus inactivated by alum treatments in Washington lakes. *Limnology and Oceanography*, **45**, 226–230.

Rzepecki, M. (1997). Bottom sediments in a humic lake with artificially increased calcium content: sink or source for phosphorus? *Water, Air and Soil Pollution*, **99**, 457–464.

Salonen, V.-P. & Varjo, E. (2000). Gypsum treatment as a restoration method for sediments of eutrophied lakes: experiments from southern Finland. *Environmental Geology*, **39**, 353–369.

Sandøy, S. & Romundstad, A.J. (1995). Liming of acidified lakes and rivers in Norway: an attempt to preserve and restore biological diversity in the acidified regions. *Water, Air and Soil Pollution*, **85**, 997–1002.

Sas, H. (co-ordinator) (1989). *Lake Restoration by Reduction of Nutrient Loading: Expectations, Experiences, Extrapolations*. St Augustin, Germany: Academia Verlag Richarz.

Schönborn, W. (1992): *Fließgewässerbiologie*. Jena, Germany: Gustav Fischer.

Simmons, J. (1998). Algal control and destratification at Hanningfield Reservoir. *Water Science and Technology*, **37**, 309–316.

Smayda, T. (1990). The influence of lime and biological activity on sediment, pH, redox and phosphorus dynamics. *Hydrobiologia*, **192**, 191–203.

Soltero, R.A. & Nichols, D.G. (1981). Lake restoration: Medical Lake, Washington. *Journal of Freshwater Ecology*, **2**, 155–165.

Søndergaard, M. (1988). Seasonal variations in the loosely sorbed phosphorus fraction of the sediment of a shallow and hypereutrophic lake. *Environmental Geology and Water Sciences*, **11**, 115–121.

Søndergaard, M., Jeppesen, E., Kristensen, P. & Sortkjær, O. (1990). Interactions between sediment and water in a shallow and hypertrophic lake: a study on phytoplankton collapses in Lake Søbygård, Denmark. *Hydrobiologia*, **191**, 139–148.

Søndergaard, M., Kristensen, P. & Jeppesen, E. (1992). Phosphorus release from resuspended sediment in the shallow and wind exposed Lake Arresø, Denmark. *Hydrobiologia*, **228**, 91–99.

Søndergaard, M., Kristensen, P. & Jeppesen, E. (1993). Eight years of internal phosphorus loading and changes in the sediment phosphorus profile of Lake Søbygaard, Denmark. *Hydrobiologia*, **253**, 345–356.

Søndergaard, M., Windolf, J. & Jeppesen, E. (1996). Phosphorus fractions in the sediment of shallow lakes as related to phosphorus load, sediment composition and lake chemistry. *Water Research*, **30**, 992–1002.

Søndergaard, M., Jensen, J.P. & Jeppesen, E. (1999). Internal phosphorus loading in shallow Danish lakes. *Hydrobiologia*, **408/409**, 145–152.

Søndergaard, M., Jeppesen, E. & Jensen, J.P. (2000). Hypolimnetic nitrate treatment to reduce internal phosphorus loading in a stratified lake. *Journal of Lake and Reservoir Management*, **16**, 195–204.

Stenson, J.A.E. & Svensson, J.-E. (1995). Changes of planktivore fauna and development of zooplankton after liming of the acidified Lake Gårdsjön. *Water, Air and Soil Pollution*, **85**, 979–984.

Stuben, D., Walpersdorf, E., Voss, K., Ronicke, H., Schimmele, M., Baborowski, M., Luther, G. & Elsner, E. (1998). Application of lake marl at Lake Arendsee, NE Germany: first results of a geochemical monitoring during the restoration experiment. *Science of the Total Environment*, **218**, 33–44.

Stumm, W. & Morgan, J.J. (1981). *Aquatic Chemistry: An Introduction Emphasising Chemical Equilibria in Natural Waters*. New York: John Wiley

Svenson, T., Dickson, W., Hellberg, J., Moberg, G. & Munthe, N. (1995). The Swedish liming programme. *Water, Air and Soil Pollution*, **85**, 1003–1008.

Sweerts, J.-P.R.A., St Louis, V. & Cappenberg, T.E. (1989). Oxygen concentration profiles and exchange in sediment cores with circulated overlying water. *Freshwater Biology*, **21**, 401–409.

Thomas, E.A. (1969). The process of eutrophication in Central European lakes. In *Eutrophication: Causes, Consequences, Correctives*, pp. 29–49. Washington, DC: National Academy of Science.

Welch, E.B. & Cooke, G.D. (1999). Effectiveness and longevity of phosphorus inactivation with alum. *Journal of Lake and Reservoir Management*, **15**, 5–27.

Welch, E.B., DeGasperi, L., Spyrikadis, D.E. & Belnick, T. (1988). Internal phosphorus loading and alum effectiveness in shallow lakes. *Journal of Lake and Reservoir Management*, **4**, 27–33.

Wetzel, R.G. (1983). *Limnology*. Philadelphia, PA: W.B. Saunders.

Williams, J.D.H., Syers, J.K., Harris, R.F. & Armstrong, D.E. (1971). Fractionation of inorganic phosphate in calcareous lake sediments. *Soil Science Society of America Proceedings*, **35**, 250–255.

Wu, R. & Boyd, C.E. (1990). Evaluation of calcium sulfate for use in aquaculture ponds. *Progressive Fish-Culturist*, **52**, 26–31.

Yamada, H., Kayama, M., Saito. K. & Hara, M. (1986). A fundamental research on phosphate removal by using slag. *Water Research*, **20**, 547–557.

Yee, K.A., Prepas, E.E., Chambers, P.A., Culp, J.M. & Scrimgeour, G. (2000). Impact of $Ca(OH)_2$ treatment on macroinvertebrate communities in eutrophic hardwater lakes in the Boreal Plain region of Alberta: *in situ* and laboratory experiments. *Canadian Journal of Fisheries and Aquatic Sciences*, **57**, 125–136.

Zhang, Y. & Prepas, E.E. (1996). Short-term effects of $Ca(OH)_2$ additions on phytoplankton biomass: a comparison of laboratory and *in situ* experiments. *Water Research*, **30**, 1285–1294.

11 • Atmospheric chemistry

PETER BRIMBLECOMBE

INTRODUCTION

The impact of the atmosphere on life has been recognised since the earliest times. We need to breathe and this has always been obvious. The relation of plants to the atmosphere is more subtle. Pliny the Elder's *Natural History* (XXXI.29) shows early interest in damage to vegetation by salty rain in coastal environments. Roman agriculturists thought rainfall carried nutrients for vegetation, notions which persisted to the Middle Ages and fuelled debates on the spontaneous generation of life within rainwater. Stephen Hales recognised that plant leaves absorb air and a portion of this is used in nutrition. By the nineteenth century vast tracts of land had been destroyed by hydrogen chloride vapours from the alkali industry and Liebig and the workers at the fledgling experimental station at Rothamsted argued over the amount of nitrogen delivered to plants in rainfall. Despite this long history of the relationship between the atmosphere and ecosystems, the acid rain of recent decades has show how much remains to be resolved.

Here we will look at the way in which atmospheric trace composition is influenced by a delicate balance between sources and sinks. The underlying oxidative chemistry of the atmosphere means that compounds are often removed as soluble acidic compounds within rainfall. Changes imposed by human activities have been both profound and difficult to ameliorate.

SOURCES OF TRACE MATERIALS IN THE ATMOSPHERE

The chemistry of the atmosphere tends to be more the product of its trace components than the bulk elemental nitrogen and oxygen that make up so much of its composition. On long time-scales this matrix has been the product of geochemical and biological activity and gases such as oxygen and nitrogen have long lifetimes. The oxygen and carbon dioxide in the air are intimately related to life through photosynthesis and respiration.

Trace gases in the atmosphere are typically regarded to be in steady state such that their sources by and large balance the sinks. The short-lived gases also tend to be more variable in concentration than their stable counterparts that are well mixed in the atmosphere.

Geochemical sources

There are a number of geochemical sources of materials for the atmosphere. These are regional and sporadic in nature. Volcanoes are the most dramatic and can profoundly affect the global budget in years of major eruptions (Fig. 11.1). The forces can be so great that materials are pushed into the stratosphere. In the past, massive eruptions have had great ecological impact, through the deposition of dust and ash and noxious materials such as sulphur oxides, hydrogen chloride and hydrogen fluoride that create an acidic environment. More continuous emissions from fumaroles also characterise volcanically active regions. These release reduced gases such as hydrogen sulphide.

Lightning is a source of nitrogen oxides that ultimately come down as nitrates in rain. The global magnitude of this source has been notoriously difficult to establish, but probably lies around 4–5 $\times$ 10^{12} g (N)yr^{-1} (Warneck, 1999).

Wind-blown dusts and sea-sprays make a further contribution. Dusts tend to resemble the material from which they are blown, but will age in the

Fig. 11.1. A volcanic smog (locally termed 'vog') lying against the slope of Mauna Loa in Hawaii. Such natural acidic pollution typifies the continuous release of volcanic gases.

atmosphere. Sea-spray is mostly sodium chloride, although it can become depleted as it ages and is replaced with nitrate or sulphate ions.

Biological sources

Although the release of carbon dioxide during respiration hardly needs mentioning, there are numerous complex transformations of carbon compounds. A surprising amount of carbon monoxide is produced by the oxidation of methane, isoprene and terpenes (Granier *et al.*, 2001) along with the degragadation of litter and organic material in the soil, providing 2–8% of the global budget (Schade & Crutzen, 1999). Trace gases also arise from biological sources, with microbiological organisms playing an especially important role. Methane is released from wet soils in processes that contribute about 100 $Tg\ yr^{-1}$, perhaps a quarter of global emissions. Large quantities of more complex hydrocarbons derive from vegetation, the best known of which are isoprenes (Geron *et al.*, 2001) from deciduous vegetation along with the terpenes. Monoterpenes such as pinene and limonene are particularly important from coniferous forests.

Soils are rich in nitrogen compounds. Ammonium salts are frequent products of the degradation of urea and or amino acids and ammonia is readily degassed under alkaline conditions. Volatile oxides or nitrogen, most typically N_2O, are produced in soils from the reduction of nitrate. Legrand *et al.* (1998) showed that at coastal Antarctic sites, bacterial decomposition of uric acid is a source of ammonium and aerosols which contain oxalate.

The oceans are not especially nitrogen rich, but have high concentrations of sulphate ions and are an important source of sulphur compounds. The activity of phytoplankton in the oceans provides a major source of dimethylsulphide (DMS) a volatile compound that is rapidly lost from seawater to the atmosphere. Along with it there are smaller emissions of methylmercaptan, dimethyl disulphide, carbonyl sulphide and carbon disulphide. Carbonyl sulphide is relatively stable in the troposphere, and thus accumulates to become the dominant reduced sulphur gas. There are also some terrestrial biological sources for the same sulphur compounds, often related to waterlogged sediments.

Halogen compounds are released into the atmosphere through biological activity, for example. The fungal degradation of wood is an important source of methyl chloride, the most abundant halocarbon in the atmosphere. Chloroform is produced through the activities of fungi such as *Caldariomyces fumago* at a rate of 0.15–20 $ng\ m^{-2}\ s^{-1}$ (Hoekstra *et al.*, 1998a) and soils may also yield brominated trihalomethanes (Hoekstra *et al.*, 1998b). Understandably there are also marine sources for a wide range of halocarbon compounds including methyl chloride, methyl bromide and carbon tetrachloride (Huntersmith *et al.*, 1983).

Agricultural sources

Agriculture can alter many of the biological processes outlined above. High concentrations of nitrogen in the soil can mean a more active denitrification process. Animal feedlots also yield increased ammonia emission from urine saturated soils.

Agricultural dressings, pesticides and herbicides now form an important part of the anthropogenic contribution to the atmosphere, particularly as organohalogen and organometallic compounds. The best known of these compounds was DDT which achieved a peak annual production of 8×10^{10}g in the 1960s. It was found at high concentrations in soils and lost with a mean lifetime of about five years, being gradually removed by evaporation, harvesting of the crops, in drainage water and through degradation. The atmospheric concentration is now much reduced. Recent measurements of the atmospheric concentrations of a variety of pesticides around Paris find lindane 0.3–6.3, atrazine < 0.03–2 and simazine < 0.03–3 ng m^{-3} and in Wisconsin, average concentrations of σ-chlordane 0.035, DDT 0.0087, DDE 0.015 and toxaphene 0.059 ng m^{-3} (Chevreuil *et al.*, 1996; McConnell *et al.*, 1998). Although pesticides tend to be at highest concentrations during application, less volatile ones such as lindane tend to be present throughout the year (van Dijk & Guicherit, 1999).

Fires are a large source of atmospheric pollution. This can come from wildfires or as the product of agricultural activities. They make a large contribution to the global particulate carbon (some teragrams per year). European law now limits the extent of stubble-burning which used to lead to high atmospheric particulate concentrations in the summer months. Particular concern has begun to focus on the forest fires of Southeast Asia and the tropics where air pollution can spread great distances altering particle load and regional chemistry.

Wood can release about 2 mg of methyl chloride for every gram burnt, along with significant quantities of hydrogen cyanide and hydrogen chloride. Some combustion is only partial and leads to residual or pyrolised hydrocarbons which include the derivatives of retene (1-methyl, 7-isopropyl phenanthrene) from the thermal transformation of resin-derived compounds such as abietic acid (Standley & Simoneit, 1994).

Perturbed ecosystems

Ecosystems that have been disturbed in the past may shift their atmospheric emissions. Loss of vegetation cover or a failure in care for soils can result in erosion and an enhanced release of dust. In arid regions this can also mean high concentrations of wind-blown salts. Some soils can be nitrogen rich, perhaps from intense occupation from livestock, others will have an altered pH balance. Fire damage can lead to large amounts of ash with readily leachable ions, particularly potassium and the alkaline earth elements.

Where the soils have low oxygen concentrations, typically when they are waterlogged, the environment becomes reducing. This is likely to enhance gaseous emissions, most typically methane if there are high organic loads, but also reduced sulphur gases: methylmercaptan, dimethyl disulphide, carbonyl sulphide and carbon disulphide. Biological processes are capable of producing more exotic compounds and as an example the atmospheres of riverine environments contain dimethylselenide ($(CH_3)_2Se$), produced in ways similar to DMS (Zhang & Frankenberger, 2000).

Methylation of metals is also a common process in sediments, with the production of methylmercury compounds. These have broad environmental impacts in many sites of mercury discharge around the world. Methylation is not restricted to mercury and other elements may be mobilised this way. The transformation of arsenic by the mould *Scopulariopsis brevicaulis*, which produces trimethyl arsine ($(CH_3)_3As$), has long been of interest.

Will-o'-the-wisps, the luminous phenomenon found over marshy ground, have sometimes been attributed to the ignition of methane by traces of phosphane (PH_3) or diphosphane (P_2H_4). Some bacteria are able to produce these phosphorus gases, along with methane, under the strongly reducing conditions (Gassmann & Glindemann, 1993).

Landfill sites are also a source of reduced gases, with methane the best known although hydrogen sulphide (Fairweather & Barlaz, 1998) and nitrous oxide (Bogner *et al.*, 1999) are also released in substantial amounts. The biochemistry of contaminated land can lead to the production of a range of exotic compounds. Organic transformations are also known within landfill sites and contaminated land. We hear much of leachates, but the odour alone reminds us that many volatile compounds (organosulphides) are also produced. There appears to be little modification in polychlorinated biophenyls (PCB) by plants although there can be vapour transport of them from contaminated land to other plants. In addition polycyclic aromatic hydrocarbons (PAH) and many other organic compounds appear to be bound into the organic matter in composts (Semple *et al.*, 2001).

Urban processes

Urban areas contain small ecological elements, but often have an importance for us that far outweighs their actual area or diversity. They are confronted by a wide array of anthropogenic pollutants. Most typically air pollution has resulted from fuel combustion. This is a primary source of major pollutants, such as carbon monoxide, sulphur dioxide, nitric oxide and hydrogen fluoride among others. Combustion in furnaces or engines can be represented very simply as:

$$\text{'CH'} + O_{2(g)} \rightarrow CO_{2(g)} + H_2O_{(g)}$$

Restricted amounts of oxygen in the combustion process means:

$$\text{'CH'} + O_{2(g)} \rightarrow CO_{(g)} + H_2O_{(g)}$$

$$\text{'CH'} + O_{2(g)} \rightarrow C_{(s)} + H_2O_{(g)}$$

The combustion, particularly where coal is burnt at moderate temperatures, is likely to give rise to PAH such as the carcinogen benzoalphapyrene. Other processes in combustion can tranform organic compounds into carcinogenic materials such as 1,3-butadiene.

The formation of nitric oxide in flames is simply written in terms of the Zeldovich cycle:

$$O + N_2 \rightarrow NO + N$$

$$N + O_2 \rightarrow NO + O$$

which leads to the net production of NO.

In some fuels, most notably coal or fuel oil, sulphur is present at high concentrations and can be oxidised on combustion:

$$S + O_{2(g)} \rightarrow SO_{2(g)}$$

The twentieth century saw a transition from coal to oil, which in effect meant solid fuels in stationary sources were replaced by liquid fuels burnt in mobile sources. This is much the result of the ever-increasing demands transport places on our energy needs. Many cities have experienced a decline in concentrations of classical pollutants such as sulphur dioxide and smoke, but increases in nitrogen oxide and photo-oxidants.

Volatile liquid fuels are responsible for increasing photochemical smog. This has important differences from classical primary pollution and is a product of reactions in the atmosphere. Such secondary air pollutants were first noticed in Los Angeles around the time of the Second World War and discussed under atmospheric chemistry. In addition to the production of smog, the organic compounds in fuel, most especially some of the aromatic compounds such as benzene, are of concern as carcinogens.

In the last decades of the twentieth century, particularly in Europe, diesel fuel has become increasingly important. Diesel-powered vehicles have introduced large amounts of fine soot and pyrolised materials into the air.

Chlorine compounds are common in the urban atmosphere. In the past the combustion of high chlorine coal has released substantial amounts of hydrochloric acid. Hydrogen chloride is also generated during the incineration of plastics such as polyvinylchloride, such that municipal incineration represents a potential source of hydrochloric acid. This demands careful control of stack gases. Chlorine is of further concern because of reactions that can lead to the production of carcinogenic chlorinated compounds such as the polychlorinated dibenzodioxins and dibenzofurans (PCDD and PCDF).

Metals are also found as impurities in fuels. The concentrations of metallic trace elements in fuels are highly variable, but some elements are notably enriched over their background levels. In oils, vanadium and nickel are found at enhanced concentrations because oil is able to extract these elements from rock very efficiently often as metalloporphyrin chelates. This means that oil fly ashes often contain vanadium and nickel oxides. Many elements are also found in coal ash. Some of these such as boron can be volatilised in combustion, but are ultimately washed out as boric acid, with notable increases in boron concentration in soils downwind of such sources.

These pictures of the enhancement of metals through vapour transport suggest increased concentrations on the surface of the particles, making the elements more readily available in ecosystems. Trace metals also arise from contamination during fuel manufacture or as additives. Tetraethyl lead was long used as an anti-knock agent in petrol and caused high lead concentrations at the roadside. In most countries automotive fuel is now unleaded, so emissions have declined, but a substantial lead burden near major highways remains.

Mining and extractive processes

Industrial activities, often placed in remote areas, can give rise to large quanitities of dust. Even the use of unsealed road surfaces is a potential source that can contaminate nearby soils and, even if inert, cover plants with thick coatings of dust, clogging stomata and decreasing photosynthetic efficiency.

Many mineralogical operations can give large amounts of air pollution, frequently as metalliferous dusts. For example sulphide ores have been roasted in the past with uncontrolled emissions:

$$Ni_2S_3 + 4O_2 \rightarrow 2NiO + 3SO_2$$

This sulphur dioxide often destroyed large tracts of vegetation downwind from smelters (Treshow & Anderson, 1989).

Aluminium smelting and brick-making are important sources of fluoride pollution. In aluminium smelting the fluoride comes from the cryolite ($AlF_3.3NaF$) used as a solvent for alumina. Brick production requires that clays are baked to drive off water. Fluorine can substitute for the OH group in clays, so hydrogen fluoride can be driven off at high temperature. This often leads to fluoride contamination downwind of brick-smelters, often recognised in terms of an increase in fluorosis among grazing animals.

REACTIONS IN THE ATMOSPHERE

The atmosphere is a highly oxidising medium, so the oxidation becomes the fate of many reactive compounds.

Hydrocarbons are frequently oxidised and broken down to smaller molecules with formaldehyde and then carbon monoxide as important and relatively stable intermediates, although carbon dioxide is the end product. Carbonyl compounds are very common in the atmosphere as oxidation products. Acetone, for example, although having direct biogenic emissions, is also produced from the oxidation of monoterpenes, isoprene and 3-methyl-butenol. Monoterpenes, such as pinene, oxidise to relatively stable derivatives of cyclobutane such as pinonaldehyde, pinonic and pinic acid.

Reduced sulphur compounds of biological origin such as DMS are readily oxidised in the atmosphere. Some can be converted to sulphur dioxide and then further oxidised to sulphuric acid, although there are stable organosulphur compounds such as dimethylsulphoxide and methanesulphonic acid, which are useful tracers of DMS emissions.

Pesticides will also undergo transformation in the atmosphere. To take malathion as an example, it is oxidised to malaoxon which seems to have a greater toxicity to mammals than the initial insecticide (Findlayson-Pitts & Pitts, 2000). Some, such as the soil fumigant 1,3-dichloropropene, are very volatile and readily react with OH radicals in the atmosphere to give formyl chloride and chloroacetaldehyde (FP).

The best-known oxidative process in the atmosphere is the formation of photochemical pollutants within smogs. It leads to the production of ozone. The problem of photochemical smog was studied in Los Angeles in the 1950s where it had become a severe problem. The oxidation of fuel hydrocarbons is initiated by the hydroxyl (OH) radical. The process can be simplified using the simplest hydrocarbon methane (CH_4) to represent the volatile emissions:

$$OH + CH_4 \rightarrow H_2O + CH_3$$
$$CH_3 + O_2 \rightarrow CH_3O_2$$
$$CH_3O_2 + NO \rightarrow CH_3O + NO_2$$
$$CH_3O + O_2 \rightarrow HCHO + HO_2$$
$$HO_2 + NO \rightarrow NO_2 + OH$$

These reactions can be summed, where $h\nu$ is a photon, such that:

$$CH_4 + 2O_2 + 2NO + h\nu \rightarrow H_2O + HCHO + 2NO_2$$

We see that the hydrocarbon is oxidised to an aldehyde, and nitric oxide to nitrogen dioxide. Methane is rather unreactive, so a poor example of the larger hydrocarbons typically involved, but the reaction sequences are rather similar, leading to aldehydes or ketones as products.

The nitrogen dioxide produced in such sequences absorbs light and dissociates:

$$NO_2 + h\nu \rightarrow O + NO$$

This reforms the nitric oxide, but also gives an oxygen atom that can react to form ozone:

$$O + O_2 + M \rightarrow O_3 + M$$

where M is a third body, such as a nitrogen molecule, to remove excess kinetic energy. Ozone and other oxidation products characterise photochemical smog. The intricacy of the chemistry often means that ozone and oxidant concentrations are higher away from cities than in their core. On continental scales there has been a significant increase in ozone concentrations over the last century.

Ozone is an obvious phytotoxin, but a range of other oxidation products are also toxic, perhaps most notably various peroxides produced in the atmosphere. Hydrogen peroxide is common, but the oxidation of alkenes in ozone-rich atmospheres can yield phytotoxic hydroxyhydroperoxides, e.g hydroxymethylhydroperoxide:

$$CH_2CR_2 + O_3 \rightarrow HCHOO + R_2CO$$

$$HCHOO + H_2O \rightarrow HOCH_2O_2H$$

Some of the oxidised organic material becomes associated with particles in the atmosphere. These will include long-chain (*c.* 10 carbons) aldehydes and some short-chain dicarboxylic acids (oxalic, malonic and succinic acid). Aromatic materials can end up as phenols, phthalic acid and after ring cleavage, as unsaturated aldehydes and glyoxal (CHO.CHO).

The nitrogen compounds are oxidised in smog, for example:

$$OH + NO_2 \rightarrow HNO_3$$

The most characteristic organo nitrogen compounds are the peroxyacyl and peroxyaryl nitrates. These products of the interaction between aldehydes and nitrogen dioxide are well recognised at the lachrymators (tear-producing agents) within smog.

Rainfall chemistry and acid rain

Many of the oxidation products of gases in the atmosphere are soluble and dissolve in rainwater, which ultimately removes them from the atmosphere. Some are acidic and others undergo further oxidation, leading to an increase in acidity of the droplets, which is popularly thought of as 'acid rain'.

Dissolution of gases

The dissolution gases (here H_2O_2) can be represented as the equilibrium

$$H_2O_{2(g)} = H_2O_{2(aq)} \qquad K_H = mH_2O_2/pH_2O_2,$$

where mH_2O_2 is the aqueous phase concentration and pH_2O_2 the pressure of the trace gas and K_H the Henry's law constant, typically given the units mol l^{-1} atm^{-1} (or mol $kg^{-1}atm^{-1}$). If the Henry's law constants are greater than about 4×10^4 mol $kg^{-1}atm^{-1}$, the gas partitions predominantly into the liquid phase in clouds. In the case of H_2O_2, which has a Henry's law constant of about 10^5 mol $kg^{-1}atm^{-1}$, it is found in the liquid phase and is active in solution chemistry in atmospheric water.

Many gases undergo subsequent reactions in solution. Take for example the dissolution of sulphur dioxide:

$$SO_{2(g)} + H_2O = H_2SO_3 \qquad K_H = mH_2SO_3/pSO_2$$

$$H_2SO_3 = H^+_{(aq)} + HSO^-_{3\,(aq)}$$

$$K' = mH^+mHSO_3^-/mH_2SO_3$$

Under typical conditions in a remote area where pSO_2 is 5×10^{-9} atm, we can expect the dissolving SO_2 to impose an equilibrium pH of 4.85. Even the ubiquitous carbon dioxide in the atmosphere is at sufficient concentrations to maintain acidic pH values of about 5.5.

Acidification

Sulphur dioxide is oxidised to sulphuric acid, but the aqueous oxidation is typically slow. In atmospheric water pH values lie in the range 2–6, so most of the SO_2 is present as bisulphite anions, so the oxidation would be written:

$$0.5O_2 + HSO_3^- \rightarrow H^+ + SO_4^{2-}$$

The sulphur dioxide that dissolves above can oxidise under prevailing situations and produce very acidic droplets.

Because this simple oxidation by molecular oxygen is very slow, it is thought to be preceded by other mechanisms. Most typically, these involve catalysis by iron and manganese. However, in unpolluted rainfall the metal concentrations are likely to be low and the reaction proceeds using other oxidants such as dissolved hydrogen peroxide and ozone. The hydrogen peroxide route is a particularly significant one as the reaction is faster in acid solution. This means that it would not slow as the system became more acidic with the production of sulphuric acid: this oxidation can be complex (Hoffmann, 1986). Hydrogen peroxide and organoperoxides will react:

$$ROOH + HSO_3^- = ROOSO_2^- + H_2O$$

$$ROOSO_2^- \rightarrow ROSO_3^-$$

$$ROSO_3^- + H_2O \rightarrow ROH + SO_4^{2-} + H^+$$

Note how the final reaction produces a hydrogen ion. Similarly in the ozone driven processes:

$$HSO_3^- + O_3 \rightarrow H^+ + SO_4^{2-} + O_2$$

The two other important inorganic acids in the atmosphere, nitric acid and hydrogen chloride, are produced in the gas phase or have primary sources. This means that they often have different relationships with precipitation concentration. Large amounts of atmospheric water can dilute these acids, whereas with sulphuric acid the liquid water can encourage the production process.

Halogen acids

Hydrochloric acid can be produced from volcanoes, for example, or degassed from sodium chloride particles:

$$NaCl + H_2SO_4 \rightarrow HCl + NaHSO_4$$

This process can deplete chloride ions in atmospheric aerosols.

In recent years the production of novel refrigerants (e.g. hydrochlorofluorocarbons or HCHFs) has led to the formation of highly chlorinated acyl halides in the atmosphere. These are highly reactive in water and produce halogenated acids, the best-known of which is trifluoroactic acid:

$$CF3CClO + H2O \rightarrow HCl + CF3COOH$$

This has few degradation pathways and there are fears that it will accumulate in ecosystems (Tromp *et al.*, 1995).

Radicals

In the last few decades the photochemistry and radical of atmospheric water droplets has been seen as an increasingly important process (Herrmann *et al.*, 1999), but we have far less understanding of the balance of reactions in the liquid phase compared with the gas phase. The production and loss of hydrogen peroxide is a key to this liquid-phase chemistry. Hydrogen peroxide partitions very effectively into the liquid phase and is an important oxidant.

In addition to transfer into solution from the gas phase, hydrogen peroxide can be produced in solution via iron-mediated photoproduction. This has been observed in simulated cloudwater experiments. Potential electron donors for these types of processes (e.g. oxalate, formate or acetate) are known to exist in atmospheric cloudwater (Arakaki & Faust, 1998).

Radical species found in the gas phase (OH, HO_2, NO_3 and CH_3O_2) can be absorbed into solution. This most important dissolution process may be that of the HO_2 radical. Dissolved hydroperoxide ion is a moderately strong acid (pKa 4.88) and gives O_2^-.

$$HO_2 = O_2^- + H^+$$

$$O_2^- + Fe^{2+} \rightarrow H_2O_2 + FeOH^{2+}$$

The hydroperoxide radical can also be produced in solution. This can be by reaction with OH:

$$H_2O_2 + OH \rightarrow HO_2 + H_2O$$

The O_2^- from the peroxide system can react rapidly with dissolved ozone to give O_3^-:

$$O_2^- + O_3 \rightarrow O_2 + O_3^-$$

$$O_3^- \rightarrow O_2 + O^-$$

$$O^- + H^+ \rightarrow OH$$

Methyleneglycol ($CH_2(OH)_2$, i.e. hydrated formaldehyde) is an important sink of OH and leads to the formation of oxidised products such as formic acid. This can react further with OH and yield carbon dioxide.

Stratospheric chemistry

Although the chemistry of the stratosphere may seem remote from ecological interactions it can be important because it controls ozone concentrations. Any thinning of the ozone layer leads to increased UV flux at ground level. The enhanced rate of ozone destruction comes as the result of chlorine-containing compounds, mostly refrigerants and spray propellants. This can be represented as:

$$CFCl_3 + h\nu \rightarrow CFCl_2 + Cl$$

where the process is initiated photochemically and ozone is destroyed through a radical chain reaction:

$$O_3 + Cl \rightarrow O_2 + ClO$$
$$O + ClO \rightarrow O_2 + Cl$$

The most substantial destruction of the ozone layer has occurred over Antarctica during spring, but it is widespread. Fortunately a number of international protocols have reduced the production of ozone-depleting substances.

Biological utilisation as a sink

Biological processes represent an important sink in the removal of compounds from the atmosphere. Obviously compounds such as ammonia and urea, or nutrient minerals deposited at the earth's surface, can be readily utilised. The removal of carbon monoxide by vegetation and soil microbes has long been known, and it is utilised in much the same way as carbon dioxide, although the efficiency varies widely. This is in marked contrast to mammals where carbon monoxide is toxic as it combines with haemoglobin.

Vegetation can also remove sulphur gases from the atmosphere and represents an important sink for tropospheric OCS. The fate of the oxidation products of DMS have not always been clear, but recently there have been some biochemical studies that suggest they may degrade biologically. Facultatively methylotrophic species of *Hyphomicrobium* and *Arthrobacter* can apparently produce the enzymes necessary for a reductive–oxidative pathway (Borodina *et al.*, 2000) for dimethylsulphoxide and dimethylsulphone (although the former appears to be more generally utilisable). Methanesulphonic acid is stable to photochemical decomposition in the atmosphere and its fate on land has been puzzling. A range of terrestrial methylotrophic bacteria that appear common in soils may mineralise methanesulphonic acid to carbon dioxide and sulphate (De Marco *et al.*, 2000).

Utilisation of the halogen compounds is obviously very important as they are so widely dispersed from industry. The soil rhizosphere provides a favourable environment for microbial degradation of organic compounds such as trichloroethylene (Anderson *et al.*, 1993) although the fate of the chlorine is not always clear. There is little clarity on the biological degradation of fluorinated compounds such as trifluoroacetic acid. As a general rule one might postulate that the dehalogenation of halocarbons would lead to mineralisation as halide salts in soils. By contrast the decarboxylation of compounds such as trichloroacetic acid would yield chloroform, but this decarboxylation is slow.

Relatively little is known about biochemically mediated reactions in atmospheric water. It is most likely to be important in dew, or rather guttation exuded from the stomata and enriched in plant-derived materials. This is often more alkaline than rainfall and has high carbon dioxide concentrations, so offers the potential for a rich and complex chemistry. This would not be without broader impacts as leaf wetness serves as an important sink for atmospheric gases (Brimblecombe, 1978).

IMPACTS OF AIR POLLUTANTS ON PLANTS

Vegetation is particularly sensitive to air pollutants. Lichens rapidly die out in polluted areas, often shown as a decline in diversity. The gradual reduction of sulphur dioxide concentrations in Europe has encouraged the reinvasion of lichens into urban areas. The air is often still acidic, so it is often acidophilic lichens that colonise first.

The effect of air pollutants on vascular plants has been much studied because of their commercial importance. Alfalfa is especially sensitive to sulphur dioxide and can be affected by just a few hours exposure at 300–500 p.p.b. Air pollutants change the water balance of plants which is expressed as structural changes. High sulphur dioxide or fluoride

Fig. 11.2. Conifers in southern Sweden, showing loss of needles especially from the tops of the trees.

concentrations, for instance, cause collapse and death of cells throughout the leaf. The areas of damage often start as a somewhat bruised or apparently water-soaked patches. These necrotic markings on the leaves arise from a rapid breakdown of chlorophyll and chlorosis is common.

Sulphur dioxide pollution remains in some industrialising nations and remote areas where industry or mining causes high deposition. The general relationship between forest diseases and sulphur depositon is complex. Leaf and stem rusts and mildew tend to decrease with air pollution, while frost damage increases (Worrell, 1994).

Nitrogen oxides can cause leaf damage, but only at very high concentrations close to sources. Concentrations of about 30 $\mu g\ m^{-3}$ NO_x affords some protection to vegetation. There has been considerable interest in the effects of enhanced nitrogen deposition, which can exceed that required for plant growth. In The Netherlands this has been related to increased stress. It is a likely explanation for the type of crown thinning and reduced growth observed on the northern German plain (Fig. 11.2.)

Ozone is the primary damaging agent within photochemical smog. It appears an important factor in the decline of lichen species, in the San Bernadino Mountains, not far from Los Angeles (Treshow & Anderson, 1989). In sensitive plants, such as tobacco, damage is apparent as white or tan flecks after some hours exposure at 200 p.p.b. The almost continental increases in ozone have the broadest of impacts on plant growth, reducing yields in crops and resistance to biotic and abiotic stress. Ozone enters through the stomata and produces oxygen species that are highly reactive, e.g. HO_2. These overload natural plant protection mechanisms and lead to necrotic markings on the leaf surface or shorter leaf lifetimes. Such damage is important throughout the growing season (PORG, 1997). There is a loss of yield at concentrations that are regularly found in Europe and North America. Corn exposed to 50–100 p.p.b. ozone, for six hours a day, through the growing season shows reduced yields.

Descriptions of plant exposure need to consider the concentration of ozone and the conductance of the leaf to the gas to define an effective dose. A simple way is to integrate exposure over time, i.e. daily mean concentrations over a seven-hour period (daylight hours typically 0900–1600 hrs). The idea of thresholds is embodied in the AOT40 standard. This is the accumulated exposure above the concentration of 40 p.p.b. over a three-month growing season. It is suggested that 3 and 6 p.p.m. hours give rise to a 5% and 10% yield reduction in crops. Trees probably require an AOT40 expressed over a 6-month period and here 10 p.p.m. hours has been suggested. Short-term critical exposure standards also need to be developed.

The difficulty with generalised standards is that they do not account for issues such as stomatal closure during drought which might reduce the effective dose of ozone. In Mediterranean areas some of the standards, which derive from northern European data, might not be appropriate. The co-occurrence of pollutants is also an important consideration. The best-known interaction is that between sulphur dioxide and ozone. There is evidence in tobacco that the synergism does not occur below the threshold for damage by ozone. The mechanisms for synergisms are uncertain, but it is possible that sulphur dioxide may decrease stomatal resistance (Treshow & Anderson, 1989).

Impact of acid rain

Precipitation in much of Europe and North America was close to pH 4.0 in the 1980s. This marked decrease over that of natural rainfall prompted wide public concern over acid rain. Since that time the emissions of sulphur dioxide have declined. In eastern Britain and some parts of Germany there has been a pronounced reduction in deposited sulphur. For the first time in decades, sulphur is being added to agricultural land as a nutrient. However, acidity may not have decreased as substantially, because the decrease in sulphur is easily offset by nitric acid from increasing nitrogen oxide emissions. Additionally, long-term records show evidence of a decline in the concentration of alkalis, most notably calcium, in rainfall (Hedin *et al.*, 1994). This may arise from a reduction in airborne alkaline dusts from unpaved roads or industrial emissions, thus lowering the capacity of the atmosphere to neutralise airborne acidity.

Although the effects of gaseous air pollutants have long been evident, it has proved much harder to show the harmful effects from acid precipitation on vegetation. Nevertheless, acid rain in the broadest sense of the word, has been seen as a key agent in forest decline. In some million hectares of forests in central Europe trees progressively died in the 1980s. The worst hit areas in Bavaria show a loss of about 30–40% of the firs on south-facing, semi-arid slopes.

There are many explanations for the mechanisms of forest decline, such as the effect of acids on the mobility of metals. Magnesium deficiency is thought to explain the needle yellowing in the Black Forest. Multiple stress is often suggested as a cause of increased sensitivity to drought, nutrient deficiency and pathogens. Increases in soil acidity have been related to enhanced aluminium mobilisation and explain the view that aluminium toxicity can cause the damage. This has been hard to establish, but it does seem that high aluminium concentrations and low ratios of calcium:aluminium and magnesium: aluminium lead to an enhanced sensitivity to acidic soils, but the picture is complex.

Impact of air pollutants on animals

Animals are generally less sensitive to pollutants than plants, although carbon monoxide and hydrogen sulphide are exceptions. There has been an obvious interest in the direct effect of pollutants on human health, but the epidemiological studies so often undertaken to understand the exposures of human populations to air pollutants are not readily adaptable to animals in their natural surroundings. Historically, there was evidence that cattle brought into London during severe winter fogs often died, seemingly at higher incidences than people, but it may simply have been that humans were protected by being indoors.

Acid mists, sprays and emissions from factories can be serious locally. The most notable problem may be with grazing animals that can accumulate large quantities of deposited pollutants from grass. Fluoride emitted from high fluoride coal, aluminium production or brick-making, but also important near volcanic emissions, is a special problem. Particulate materials also contain toxic metals and these may be found at high levels on the surfaces of plants near pollution sources. Concern again centres on fluorosis and similar diseases in grazing animals. There is also the potential for exposure to carcinogens, most notably PAH such as benzoalphapyrene in combustion-derived particles. There has been a change in the general spectrum of PAH compounds following the decrease in the domestic use of coal. However, diesel emissions have contributed new and sometimes nitrated PAH to atmospheric aerosols.

In both Europe and in North America the 1980s revealed widespread decrease in fish and amphibian populations of lakes. A pH range of 6–9 is desirable to support a good fish stock and below pH 5 lakes do not seem to support healthy fish, and this level of acidity also harms amphibian eggs. Salmonids are especially sensitive species of significant economic importance. The toxicity of acid waters may be related to aluminium as this reaches a maximum in brown trout at pH 5.0. Aluminium and hydrogen ions both affect the gill, by altering the active uptake of sodium ions. Aluminium precipitates at the gill filament and clogs it with mucus. Acid waters enhance this toxic effect of aluminium because the lower pH values increase aluminium solubility. The toxicity of aluminium is lessened if calcium is relatively abundant in the water.

Many aquatic organisms are affected by pulses of acid water. This may be in times of heavy storms or at snowmelt. As snowpacks melt acids tend to leave the pack first and rapid water flow means that there is little time for contact with soils which buffer against the acidity. These events have been responsible for many fish kills. Often fish can tolerate a reduction in pH if given time to acclimatise, but are killed by rapid changes.

Impact of air pollutants on ecosystems and evolution

The impact of environmental pollutants on ecosystems as a whole is clearly important although tended to be neglected in comparison with studies of single species, especially humans. The impacts of pollutants on ecosystems have usually been studied using biomarkers and looking at toxic effects on target species, but much of the interest tends to focus on soils and water pollution (Linhurst *et al.*, 1995).

Bioaccumulation of deposited substances, most particularly metals and chlorinated organic compounds, has been studied and arguments are often made on the basis of bioconcentration that can be related to the water to octanol partition of organic compounds. There has also been much debate about the subsequent concentration of materials up food chains, although Moriarty (1999) cautions against accepting such biomagnification without really firm evidence.

Increased nitrogen deposition (potentially both as nitrogen oxides or ammonia) can alter the nutrient status of an ecosystem. Forest ecosystems are also affected by ozone and the broader considerations of the 'acid rain' problem. Forest decline ultimately alters water balance and protection against soil erosion. Increasing levels of haze from forest fires in Southeast Asia have raised concerns about the growth and survival of understorey plants. Tang *et al.* (1996) showed a decrease of leaf carbon gain under haze conditions. There are similarly broad impacts of air pollution within acidified lakes. Damage to insect populations by acid waters can mean a decline in the birds that feed on them.

Exposure to air pollutants has continued over long periods of time in some places. Here some species that have evolved tolerances to pollution (Bradshaw & McNeilly, 1981). Rye grass populations in near-urban locations have developed a tolerance to sulphur dioxide (Bell & Mudd, 1976). The best-known effect on animals is the selection of different coloured types of the peppered moth *Biston betularia*. A dark form was relatively uncommon before the industrial revolution, but then became common in many parts of England and North America. Recent declines in air pollution, particularly sulphur dioxide and its concomitant smoke, have been correlated with declines in the frequency of melanic forms (Grant *et al.*, 1998).

Reduction of air pollutants

Air pollution is not easy to reduce at the receptor and is usually tackled by reduction of the sources. It generally proceeds by defining tolerable concentrations, doses or critical loads for individuals or ecosystems.

Emission reduction in industrialised countries has followed a path of first lowering the emissions of primary pollutants, such as sulphur dioxide and smoke. Smoke and particulate material is removed by grit arrestors, bag house and electrostatic precipitators. A reduction in sulphur emissions has been easiest to achieve by changing to fuels of low sulphur content, but also more recently through flue gas desulphurisation. Regenerative desulphurisation processes such as the Wellman–Lord procedure absorb sulphur dioxide into sodium sulphite solutions converting them to sodium bisuphite. The sulphur dioxide is later degassed and can be used as a feedstock for the chemical industry. Lime slurries are also used to absorb sulphur dioxide. However, the main product calcium sulphate, can become contaminated with trace metals. Furthermore, the amounts of lime required can be extremely large and it is often complained that these are mined from attractive sites of great ecological and recreational value.

The automobile is now such a major contributor to air pollution that catalytic converters have been widely added to exhaust systems. Three-way catalysts remove nitrogen oxide, carbon monoxide and hydrocarbons. The hydrocarbons and nitrogen oxide contribute to the broad photochemical production of ozone at a continental scale. However, the reduction in emission required to ensure low tropospheric ozone concentrations are quite substantial and it may be some time before they can be widely met. Hydrocarbons from industry, where they are used as solvents, have fallen out of favour and have been increasingly replaced by aqueous solvents. Chlorinated compounds have given very wide concern because of their potential toxicity and harm to the ozone layer. As a consequence, industrial processes have again begun to move away from these in an age of greener chemistry.

Once an ecosystem is exposed to air pollution it is difficult to cope with at the site. It is true in some extreme cases, more typically with artistic monuments, that protection against air pollution is possible by making a cover for the building. Some ecosystems have been covered also, but these are experimental rather than practical undertakings, where microcatchments have been isolated to establish the effects of acid rain.

One widely used treatment is the use of lime to counteract the effects of acid deposition. It has been widely used in Scandinavia and although it is not always desirable to treat at the receptor, it does have some merits. Not only does lime neutralise the acid, but it also raises calcium concentrations (see Søndergaard *et al.*, this volume). This can be important in systems where aluminium has enhanced mobility. Aluminium toxicity is decreased for many species (e.g. trees, fish) by increases in the calcium: aluminium ratio.

Trees may also reduce air pollution. Large forest expanses have a degree of self-protection against gases such as sulphur dioxide, which deposit out rapidly. Trees in urban areas can serve to remove both gaseous pollutants and particles, although such a remedial approach needs to be adopted with some care. This is necessary because trees planted in cities are also excellent sources of hydrocarbons that induce photochemical smog.

On the broader scale pollutants are reduced through a range of strategies. The oldest and most direct are emission reduction, but these are more difficult to apply when reactions take place in the atmosphere. Air-quality management strategies that account for the interactions between pollutants are an increasing part of most national plans to control air pollutants. Economic instruments, such as pollution taxes, emissions trading and cost–benefit analysis are also applied, although they often draw disapproval from those who do not believe the natural elements of the environment are properly valued.

In the twentieth century agreements had to be developed at an international level to combat the long-range transport of air pollutants and then pollution on a global scale that caused damage to the ozone layer and raised concentrations of the greenhouse gas, carbon dioxide. These international agreements while difficult to implement on a practical level were useful in establishing underlying principles, upon which more specific policies could be developed. Sustainability, ensuring what we do today does not limit our future, has become an important element of environmental planning. The precautionary principle requires us to pay particular attention to issues that provoke potential risks, even when we do not understand their magnitude.

CONCLUDING REMARKS

The relationship between ecosystems and air chemistry may be less tangible than interactions with climate, water or soils. Nevertheless, few would deny its importance in decades that have followed concern over acid rain. The subtlety and complexity of the interactions once limited our understanding of their importance. The atmosphere changed markedly in the twentieth century, but we have a greater level of understanding, so that the twenty-first century should see more carefully chosen policies that will ensure the restoration of an acceptable atmospheric environment.

REFERENCES

Anderson, T.A., Guthrie, E.A. & Walton, B.T. (1993). Bioremediation in the rhizosphere. *Environmental Science and Technology*, **27**, 2630–2636.

Arakaki, T. & Faust, B.C. (1998). Production of H_2O_2 sources, sinks, and mechanisms of hydroxyl radical ((OH)-O- photoproduction and consumption in authentic acidic continental cloud waters from Whiteface Mountain, New York: the role of the Fe(r) (r = II, III) photochemical cycle. *Journal of Geophysical Research – Atmospheres*, **103**, 3487–3504.

Bell, J.N.B. & Mudd, C.H. (1976). Sulphur dioxide resistance in plants: a study of *Lolium perenne*. In *Effects of Air Pollution on Plants*, ed. T.A. Mansfield, pp. 87–103. Cambridge: Cambridge University Press.

Bogner, J.E., Spokas, K.A. & Burton, E.A. (1999). Temporal variations in greenhouse gas emissions at a midlatitude landfill. *Journal of Environmental Quality*, **28**, 278–288.

Borodina, E., Kelly, D.P., Rainey, F.A., Ward-Rainey, N.L. & Wood, A. P. (2000). Dimethylsulfone as a growth substrate for novel methylotrophic species of *Hyphomicrobium* and *Arthrobacter. Archives of Microbiology*, **173**, 425–437.

Bradshaw, A.D. & McNeilly, T. (1981). *Evolution and Pollution.* London: Edward Arnold.

Brimblecombe, P. (1978). Dew as a sink for SO_2. *Tellus*, **30**, 151.

Chevreuil, M., Garmouma, M., Teil, M.J. & Chesterikoff, A. (1996). Occurrence of organochlorines (PCBs, pesticides) and herbicides (triazines, phenylureas) in the atmosphere and in the fallout from urban and rural stations of the Paris area. *Science of the Total Environment*, **182**, 25–37.

De Marco, P., Murrell, J.C., Bordalo, A.A. & Moradas-Ferreira, P. (2000). Isolation and characterization of two new methanesulfonic acid-degrading bacterial isolates from a Portuguese soil sample. *Archives of Microbiology*, **173**, 146–153.

Fairweather, R.J. & Barlaz, M.A. (1998). Hydrogen sulfide production during decomposition of landfill inputs. *Journal of Environmental Engineering-ASCE*, **124**, 353–361.

Findlayson-Pitts, B.J. & Pitts, J.N. (2000). *Chemistry of the Upper and Lower Atmosphere.* San Diego, CA: Academic Press.

Gassmann, G. & Glindemann, D. (1993). Phosphane (PH_3) in the biosphere. *Angewandte Chemie – International Edition*, **32**, 761–763.

Geron, C., Harley, P. & Guenther, A. (2001). Isoprene emission capacity for US tree species. *Atmospheric Environment*, **35**, 3341–3352.

Granier, C., Petron, G., Muller, J.-F. & Brasseur, G. (2001). The impact of natural and anthropogenic hydrocarbons on the tropospheric budget of carbon monoxide. *Atmospheric Environment*, **34**, 5255–5271.

Grant, B.S., Cook, A.D. Clarke, C.A. & Owen, D.F. (1998). Geographic and temporal variation in the incidence of melanism in peppered moth populations in America and Britain. *Journal of Heredity*, **89**, 465–471.

Hedin, L.O., Granat, L., Likens, G.E., Buishand, T.A., Galloway, J.N, Butler, T.J. & Rodhe, H. (1994). Steep declines in atmospheric base cations in regions of Europe and North America. *Nature*, **367**, 351–354.

Herrmann, H., Ervens, B., Nowacki, P., Wolke, R. & Zellner, R. (1999). A chemical aqueous phase radical mechanism for tropospheric chemistry. *Chemosphere*, **38**, 1223–1232.

Hoekstra, E.J., Verhagen, F.J.M., Field, J.A., de Leer, E.W.B. & Brinkman, U.A.T. (1998*a*). Natural production of chloroform by fungi. *Phytochemistry*, **49**, 91–97.

Hoekstra, E.J., de Leer, E.W.B. & Brinkman, U.A.T. (1998*b*). Natural formation of chloroform and brominated trihalomethanes in soil. *Environmental Science and Technology*, **32**, 3724–3792.

Hoffmann, M.R. (1986). On the kinetics and mechanisms of oxidation of aquated sulfur dioxide by ozone. *Atmospheric Environment*, **20**, 1145–1154.

Huntersmith, R.J., Balls, P.W. & Liss, P.S. (1983). Henry law constants and the air–sea exchange of various low-molecular-weight halocarbon gases. *Tellus, Series B – Chemical and Physical Meteorology*, **35**, 170–176.

Legrand, M., Ducroz, F., Wagenbach, D., Mulvaney, R. & Hall, J. (1998). Ammonium in coastal Antarctic aerosol and snow: role of polar ocean and penguin emissions. *Journal of Geophysical Research – Atmospheres*, **103**, 11043–11056.

Linhurst, R.A., Bourdeau, P. & Tardiff, R.G. (eds.) (1995). *Methods to Assess the Effects of Chemicals on Ecosystems.* Chichester, UK: John Wiley.

McConnell, L.L., Bidleman, T.F., Cotham, W.E. & Walla, M.D. (1998). Air concentrations of organochlorine insecticides and polychlorinated biphenyls over Green Bay, WI, and the four lower Great Lakes. *Environmental Pollution*, **101**, 391–399.

Moriarty, F. (1999). *Ecotoxicology.* London: Academic Press.

PORG (1997). *Ozone in the United Kingdom.* London: Department of Environment, Transport and Regions.

Schade, G.W. & Crutzen, P.J. (1999). CO emissions from degrading plant matter.2: Estimate of the global source strength. *Tellus*, **51B**, 909-918.

Semple, K.T., Reid, B.J. & Fermor, T.R. (2001). Impact of compositing strategies on the treatment of soils contaminated with organic pollutants: a review. *Environmental Pollution*, **112**, 269–283.

Standley, L.J. & Simoneit, B.R.T. (1994). Resin diterpenoids as tracers for biomass combustion aerosols. *Journal of Atmospheric Chemistry*, **18**, 1–15.

Tang, Y.H., Naoki, K., Akio, F. & Awang, M. (1996). Light reduction by regional haze and its effect on simulated leaf photosynthesis in a tropical forest of Malaysia. *Forest Ecology and Management*, **89**, 205–211.

Treshow, M. &. Anderson, F.K (1989). *Plant Stress from Air Pollution*. Chichester, UK: John Wiley.

Tromp, T.K., Ko, M.K.W., Rodriguez, J.M. & Sze, N.D. (1995). Potential accumulation of CFC replacement degradation product in seasonal wetlands. *Nature*, **376**, 327–330.

van Dijk, H.F.G. & Guicherit, R. (1999). Atmospheric dispersion of current-use pesticides: a review of the evidence from monitoring studies. *Water, Air and soil Pollution*, **115**, 21–70.

Warneck, P. (1999). *Chemistry of the Natural Atmosphere*. San Diego, CA: Academic Press. World Health Organization (1999). *Ecotoxic Effects: Ozone on Vegetation, Air Quality Guidelines for Europe*, 2nd edn. Copenhagen: World Health Organisation 1999.

Worrell, J.J. (1994). Relationships of acid deposition and sulfur dioxide with forest diseases. In *Effects of Acid Rain on Forest Processes*, eds. D.L. Godbold & A. Huttermann, pp. 163-182. New York: Wiley–Liss.

Zhang, Y.Q. & Frankenberger, W.T. (2000). Formation of dimethylselenonium compounds in soil. *Environmental Science and Technology*, **34**, 776–783.

Part 4 • Manipulation of the biota

12 • Establishment and manipulation of plant populations and communities in terrestrial systems

ANTHONY J. DAVY

INTRODUCTION

The reconstruction of an appropriate plant community is a *sine qua non* for the restoration of any degraded ecosystem. Clearly, the plant communities of any ecosystem have an element of intrinsic distinctiveness that represents the biodiversity of the system. Furthermore, attempts to restore most other aspects of ecosystem structure and function cannot succeed, partially or wholly, without the authentic primary producers. The physical structure and chemical composition of the stands of plants that are established, combined with the specificity of many trophic relationships, strongly influence the potential for restoration of animal and microbial communities.

Plant communities are essentially dynamic, being the product of a combination of historical and current successional processes that involve interactions both between species and between the vegetation and the abiotic environment (Connell & Slatyer, 1977; Davy, 2000). Disturbance or damage to an ecosystem is likely to affect all aspects of its successional status, including soil development, accumulated nutrient and biomass capital, and nutrient cycling. Practical approaches to the establishment and manipulation of plant communities ignore this at their peril. For the purposes of restoration, landform and the properties of the soil environment are determinants of plant communities in two senses. Primarily, they are integrated attributes of successional status that are part of the functional specification of any 'target' ecosystem and hence the vegetation it can support. These aspects are dealt with in detail elsewhere in this volume: landform and the physical properties of soil by Whisenant, the chemical environment of soil by Marrs and the role of soil micro-organisms by Allen *et al.* However, as fundamental determinants of plant survival and growth, these aspects of the environment also represent tools that may be used indirectly to predetermine, or manipulate, populations of particular species and whole plant communities. Such use is considered in this chapter.

The starting-point for the restoration of plant communities must therefore be the restoration of physical and soil environments appropriate to them, or their successional precursors. Even where highly satisfactory emulations of a desired physicochemical environment can be achieved, it will usually be necessary to introduce populations of desired plant species, to regulate their relative abundance and to remove or discourage unwanted, invasive species. Indeed, such manipulations may become the mainstay of restoration when the initial disturbance is primarily the result of the removal of crucial (or keystone) species or of invasion by alien species. Despite some superficial similarities with gardening, plant manipulations for restoration are likely to be most successful when based on the best possible knowledge of the population and community ecology of the species involved. This chapter will briefly examine the background to the synthesis of plant communities for restoration and then focus on the various means by which populations of particular species may be established, regulated and removed.

TARGET PLANT COMMUNITIES AND THEIR ASSEMBLY

The manipulation of plant populations for restoration purposes implies that a defined target or reference community is being aimed at. The objectives can range from re-establishment of lost species to complete reconstruction of complex communities from a more or less known pool of species. When it is known beforehand that damage is inevitable, the communities at a site can sometimes be quantitatively described before any degradation takes place. In other cases, examination of similar ('reference') sites nearby, or even fragmentary historical records, may have to suffice. The commonness of difficulties both in the initial identification of appropriate targets and with subsequent attempts to emulate them are symptomatic of our generally very incomplete understanding of how plant communities are structured.

An insight into the ways in which particular species interact with each other and their roles in communities may be derived from examination of the syndromes of traits, or strategies, that they have evolved. This led Grime (1977, 2001) to recognise three primary strategies of established plants based on tolerance of stress (defined as external constraints on the rate of dry matter production), tolerance of disturbance (defined in terms of the partial or complete destruction of biomass) and competitive ability (the tendency of neighbouring plants to utilise the same resources). The relevance of these issues to the manipulation of species and communities in restoration has been reviewed by Grime (1986). Another, highly significant, dimension of community structure derives from the regenerative behaviour (both from seed and vegetative spread) of individual species (Grubb, 1977, 1986; Grime, 2001); conditions that favour regeneration of any species may be entirely different from those that pertain most of the time or over most of the area occupied by a community and this can be crucial for species coexistence.

The idea of species coexistence is central to the mechanisms that maintain species diversity in natural communities is inevitably a key to their reconstruction or restoration. Current models include both stable and unstable coexistence (Chesson, 2000). Stable coexistence implies that populations of individual species can recover from episodes of low density, whereas unstable coexistence means that there is no tendency to recovery and thus species will be lost. The extent to which successful restoration depends critically on the initial absolute and relative abundances of the species established will reflect the strength of the stabilising mechanisms in a community.

In an initially uniform environment, competitive and synergistic interactions between species ought to result in a series of community assembly rules (Diamond, 1975). These represent generalised restrictions on species presence or abundance that depend on the presence or abundance of one or several other species, or types of species (Wilson & Whittaker, 1995). Evidence for the operation of assembly rules in plant communities is still surprisingly limited. In order to identify such rules, it is necessary to test distributional data against a null model that assumes a free association of species, but the statistical methods for seeking such non-randomness have been only developed relatively recently (e.g. Wilson & Roxburgh, 1994). However there is good evidence for non-random structure and species guilds, or functional types, in communities as different as saltmarshes (Wilson & Whittaker, 1995) and temperate rainforest (Wilson *et al.*, 1995).

Wilson (1992) proposed that an understanding of assembly rules would be an important starting-point for studies that would 'chart the future of ecosystems in the face of the human onslaught'. More immediately, knowledge of relevant assembly rules would provide a sounder theoretical basis for plant introductions than simply aiming at widespread random or uniform distributions of species from the pertinent species pools; such guidance in the construction of communities should also have practical advantages, particularly by allowing more efficient utilisation of propagules and a more rapid convergence on the target community structure. Assembly rules should also define the invasibility of a community exposed to immigration from a specific pool of alien species; hence, they could facilitate the prediction of destructive invasions and provide a focus for species-removal programmes. Investigation of assembly rules and their mechanisms in the target communities could be an invaluable precursor to almost any vegetation restoration programme. Such research may amply repay the cost involved, especially in large-scale projects.

ESTABLISHMENT OF NEW POPULATIONS

At a mechanistic level, the survival and abundance of the various species comprising a target community are functions of their respective intrinsic population dynamics. Success in the establishment of self-maintaining populations could be affected critically by many aspects of population biology, such as density-dependent fecundity or mortality, mode and distance of pollination and dispersal, numerous features of germination biology, capacity for and mode of clonal growth, and interactions with other species (Silvertown & Charlesworth, 2001). The main impetus for understanding how to design founding populations has, not surprisingly, come from the need to reintroduce endangered species to their former habitats (Falk *et al.*, 1996; Guerrant, 1996), although there is no reason why the same principles should not apply to the establishment of self-sustaining populations generally. Indeed, rare and endangered species are more likely to be both genetically depauperate and exacting in their requirements than more common ones (Fischer & Stöcklin, 1997; Fischer & Matthies, 1998; Kéry *et al.*, 2000). The demographic processes that determine the long-term viability of a population can be projected many years into the future using size/stage matrix models, provided that realistic transition probabilities for the species in question are available from empirical population studies in the same type of habitat (Guerrant, 1996). Such stochastic simulations allow estimation of population growth rates and the risk of extinction when founding populations are composed of plants of particular sizes or stages (e.g. dormant seed, fresh seed, plants of different sizes). The size of a founding population is of crucial importance for its ability to withstand chance variations in its environment. Pavlik (1996) discusses the dependence of minimum viable population (MVP) size on a variety of life-history characteristics: long-lived, woody, self-fertile species with high fecundity could to have an MVP in the range 50 to 250 individuals; for short-lived, herbaceous outcrossers, it might be in the range of 1500 to more than 2500 individuals (see Table 12.1). Sensibly, restoration involving non-endangered species should use founding populations of at least the

Table 12.1. *Selection of an objective for minimum viable population (MVP) size depends on the life-history characteristics of the target taxon*

Life-history characteristic	Minimum viable population[a] 50	2500
Longevity:	perennial ⟶	annual
Breeding system:	selfing ⟶	outcrossing
Growth form:	woody ⟶	herbaceous
Fecundity:	high ⟶	low
Ramet production:	common ⟶	rare or none
Survivorship:	high ⟶	low
Seed duration:	long ⟶	short
Environmental variation:	low ⟶	high
Successional status:	climax ⟶	seral or ruderal

[a]Long-lived, woody, self-fertile plants with high fecundity would have an MVP in the range of 50 to 250 individuals. MVP for short-lived, herbaceous outcrossers, however, would be in the range of 1500 to 2500+individuals.
Source: Pavlik (1996).

MVP size. Knowledge of the population biology of individual plant species can hardly be overvalued in designing restoration programmes to manipulate their abundances (Schemske *et al.*, 1994). The following practical approaches to establishing species stress the relevant aspects of population biology.

Establishment from natural seed banks

Types of soil seed bank

Most plant communities include populations of viable seed buried in the soil (Leck *et al.*, 1989; Thompson *et al.*, 1996; Thompson, 2000). A proportion of such seeds is usually capable of germinating as soon as they are exposed to suitable conditions. The remainder – often relatively recent recruits to the soil seed bank – may exhibit a variety of types of dormancy. Seed banks of different species differ in their persistence, depending on seed longevity in the soil and their propensity to germinate (Thompson & Grime, 1979). The persistence of seed banks has been

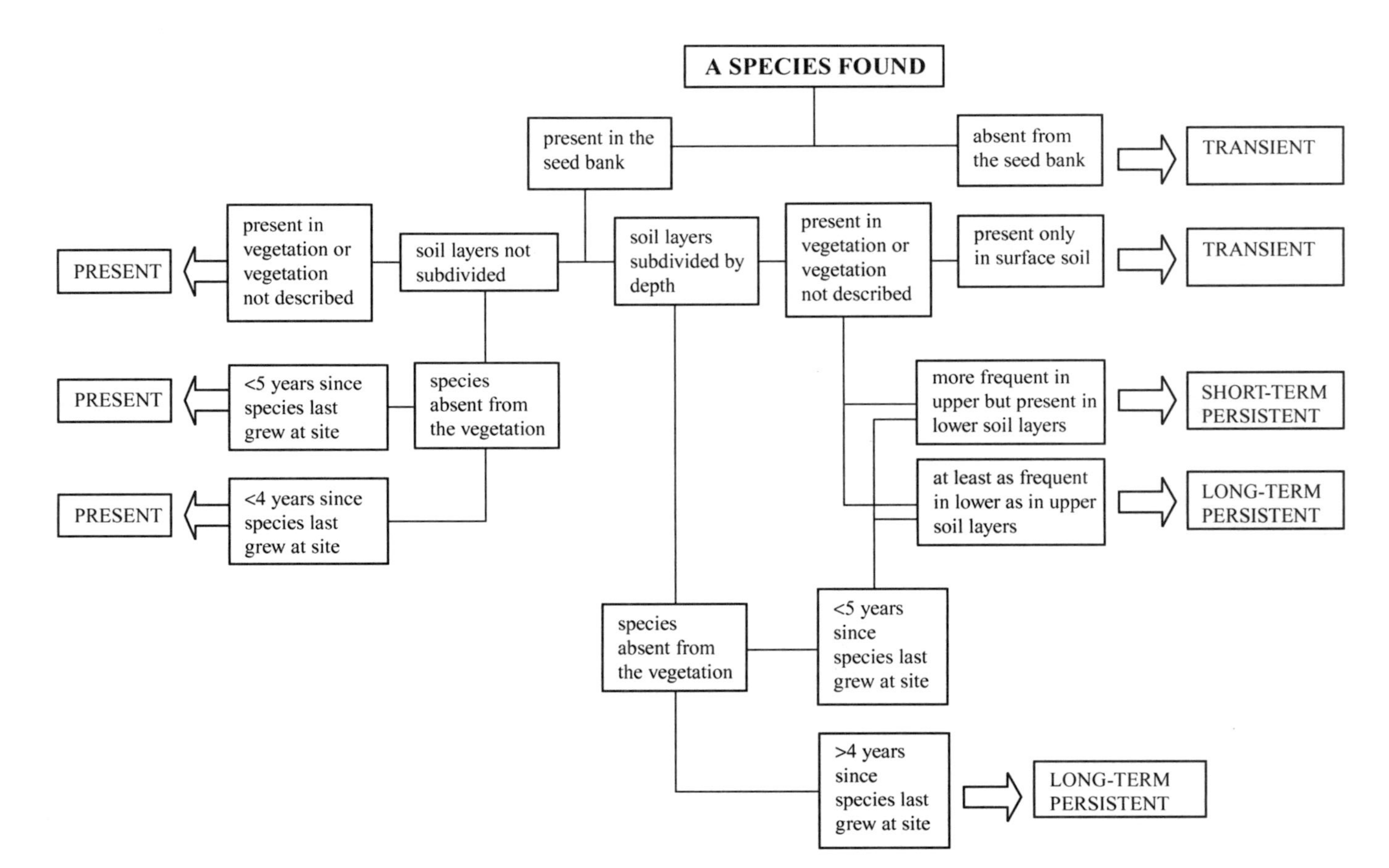

Fig. 12.1. Dichotomous key for assigning species to transient, short-term persistent and long-term persistent seed bank types used in a comprehensive database of the soil seed banks of northwest Europe. From Thompson *et al.* (1996).

classified in several ways but the classes used by Thompson *et al.* (1996) are easily related to their likely value in restoration. The transient type describes species with seeds that persist in the soil for less than a year (often much less); they have rather little practical value for restoration. The short-term persistent type includes species with seeds that persist for at least one, but less than five years. The long-term persistent type describes species of which the seeds persist in the soil for longer than five years (often much longer). Persistent seed banks may often be available for use in restoration and the more persistent they are the more likely they are to be usable. However, the determination of persistence of a species' seed bank can require considerable research effort and previous seed-banks studies have been exceptionally variable, both in approach and findings. The criteria of Thompson *et al.* (1996) for classifying species into seed bank types, using a variety of data, are described fully in Fig. 12.1.

The use of soil seed banks

In principle, the idea of exploiting the naturally occurring soil seed bank for restoration is attractive. Indigenous populations might be re-established from relict propagules in topsoil and litter retained from the same site, or by importing equivalent materials from nearby similar sites; furthermore, the genetic structure of such populations might be expected to closely resemble the pre-disturbance populations (see Gray, this volume). Imported soil could also provide the most appropriate rooting environment for the species concerned and, at the same time, 'seed' the site with suitable micro-organisms and invertebrates. Where the use of seed banks is feasible, it is likely to be a highly cost-effective method of re-establishing natural vegetation (van der Valk & Pederson, 1989). Even if authentic surface soil is not used deliberately as a source of desirable propagules, its restitution to a site might have considerable physical and chemical advantages for ecosystem restoration (see Whisenant, this volume; Marrs, this volume). In this case, the possible adverse consequences of a seed bank in allowing weedy species to become established also need to be considered.

The most fundamental questions raised by the seed-bank approach concern the extent to which the composition of the seed bank reflects that of the established vegetation prior to disturbance, or even that of an earlier successional stage that could be an appropriate target for restoration. Many studies have found large disparities between the composition of vegetation and the seed bank in the underlying soil, even in the absence of any disturbance. This is partly because seed banks may accumulate species that have been prominent over a series of successional stages at the site and partly because of differential survival of species in the seed bank. Smaller-seeded, shorter-lived and weedy species all tend to be more persistent in the soil. Susceptibility to predation is associated with large seeds but very small seeds, with limited resources, are less likely to become established successfully. Succession of lowland heath in Dorset, UK to vegetation of lower conservation value dominated by *Pinus sylvestris*, *Pteridium aquilinum*, *Betula* spp., *Rhododendron ponticum* or *Ulex europaeus* in each case gave rise to seed banks that were different from that under the original heath, with significantly lower numbers of heathland species represented; the soil beneath the latter three vegetation types also had significantly higher numbers of non-heathland species (Mitchell *et al.*, 1998).

Disturbance or damage significant enough to require restoration may further alter the character of a seed bank. The effects of different disturbances on natural seed banks are not well documented. It appears that conversion to forestry plantation, which may cause little relative disturbance to the soil, allows good persistence of native seeds (Hill & Stevens, 1981). Repeated cultivation of land leads to the destruction of the indigenous seed bank, either through deep burial, or by bringing seeds to the surface to germinate and meet their doom. At the same time, the seeds of weedy species tend to establish, allowing them to reproduce and augment their seed banks. Twenty years of arable cultivation in the Sandlings area of Suffolk, UK, was sufficient to remove all vestiges of the seedbank of ancient, lowland *Calluna* heath, even though the seeds of the dominant *Calluna vulgaris* can have considerable longevity in undisturbed heathland soil; at the time of restoration experiments, the seed bank comprised mainly arable weeds (Davy *et al.*, 1998; Dunsford *et al.*, 1998).

The storage of topsoil as a seed bank for subsequent restoration purposes also raises practical problems. Storage in piles exposes seeds to the same hazards as deep burial, particularly hypoxia and high concentrations of carbon dioxide, and accelerates loss of viability. On the other hand, storage in a relatively shallow layer would require a greater area than is likely to be available and would guarantee significant contamination with a weed flora, by dispersal of seeds from elsewhere. Emergence and establishment from the seed bank can be modified within limits by manipulation of the environment to promote desirable species or discourage weeds, for instance through effects on water availability (van der Valk & Pederson, 1989) or pH (Dunsford *et al.*, 1998; Marrs, this volume).

The uncertainties of using seed banks in the restoration of vegetation more than justify experimental assessment of the material for use in each case. Standard methods are available to determine the composition and viability of seed banks (Leck *et al.*, 1989; Thompson *et al.*, 1996). These should at least show whether there are sufficient propagules of the desired species and whether weed seeds are likely to represent a problem; they may also allow prediction of the changes in storage and of the probable composition of the resulting vegetation.

Establishment from seed sown *in situ*

In the absence of an adequate relict soil seed bank, seed that has been harvested and stored is a virtual necessity. Potentially the least labour-intensive way of using such seed is to sow directly into the restoration site. This, however, exposes the target population to the considerable hazards associated with seed germination and seedling establishment, particularly if conditions are suboptimal; it may not ultimately be the most economical or cost-effective use of seed. The prime requirements for successful establishment from seed are: a source of suitable seeds; methods for testing their quality; the means of storing them without unacceptable loss of viability; knowledge of the conditions required to break any dormancy and promote germination; and knowledge of the environmental conditions conducive to seedling establishment.

Sources of seed

Seed of many wild species is available commercially. Nevertheless it should be used with caution, because most plant species show considerable genetic variation and are found in a range of habitats, perhaps over a wide geographical range; seed may well have been collected from a different habitat, or a different part of the species' range from the site to be restored (see Gray, this volume). Hence it may be an ecotype poorly adapted for the restoration site, or at least genetically different from the indigenous population. The problem may have been exacerbated by cultivation for unknown generations, using semi-agricultural methods to increase production. Although this may protect wild populations and reduce the cost of seed purchased in bulk, during this process there will inevitably have been selection of genotypes favoured by the cultivation conditions. Ecologically, a better restoration might be expected with material collected locally, using methods aimed at sampling the range of genetic variation present (e.g. Walmsley & Davy, 1997*a*) but this is not necessarily so (see Gray, this volume). Seed should be collected at maturity, bearing in mind that this may leave a narrow window of opportunity before it is shed and dispersed. When local seed availability is limited by marginal climatic and soil conditions, as Greipsson & Davy (1997) found for the perennial, dune-building grass *Leymus arenarius* on the coast of Iceland, judicious application of nitrogen and, to a lesser extent, phosphorus and potassium fertiliser, can improve the yield of seed dramatically.

Seed quality

Commercially produced seed is normally subject to stringent, specified quality assurance. Seed collected *ad hoc* for restoration purposes may be highly variable, and so its quality should be examined under controlled conditions. The proportion of unfilled and insect-predated seeds can be assessed by dissection or by the use of X-ray techniques. Seeds placed on or between moist filter papers should be checked for gain in mass over 12 h, as an indication of imbibition; failure to imbibe water may indicate an impermeable seed or fruit coat that will require scarification before it will germinate (Baskin & Baskin, 1998).

Both viability and germination tests can be carried out. Viability can be tested by vital staining with tetrazolium. Imbibed seeds may be cut longitudinally, or their seed coats ruptured, to promote penetration of the stain. A variety of staining methods has been used (Moore, 1985), but typically seeds are incubated in a 1% solution of 2,3,5-triphenyl tetrazolium chloride for 24 h at 30 °C, in darkness. Red or pink staining indicates respiratory activity and hence viability; seeds are usually classified as: (1) sound; (2) viable but weak; (3) non-viable and weak but stained; or (4) dead and unstained.

Germination tests on moist filter paper in petri dishes, over a range of temperatures and under light and dark conditions, are normally used to test for germinability (proportion of seeds germinating and rate of germination). Baskin & Baskin (1998) discuss methodology and the many possible combinations of conditions in detail. In combination with tetrazolium testing, these germination tests are also likely to reveal whether there is significant seed dormancy.

The range of seed size found within a species is typically rather narrow (Westoby *et al.*, 1996). Nevertheless, seed size or mass may be a good predictor of quality for restoration purposes, particularly if the seed has been collected from *in situ* or wild populations, which are more likely to be subject to resource limitations than those produced under agricultural conditions. Greipsson & Davy (1995) examined the consequences of seed mass for germination in 34 natural populations of *Leymus arenarius* and found remarkable plasticity. A doubling of seed mass was associated with more than doubling of the final germination percentage, and a fivefold reduction in the median time to germination, both within and between populations. Establishment of the threshold seed size for good germination in a population and then screening the seeds before use will reduce the risk of failure in direct sowing programmes.

Seed storage

Storage usually becomes an issue if seed is collected more than a year before it is needed for use. It presents little problem for the great majority of species, although it can be costly if long-term storage is required. Modern gene-bank facilities can preserve viability for decades or centuries in a wide range of orthodox species. The Royal Botanic Gardens, Kew expects 80% of species stored there to survive for at least 200 years (Millennium Seed Bank Project, 2001). Long-term storage normally involves drying seeds gradually at 15 °C and 15% relative humidity, before drying them further at 18 °C and 10% relative humidity to equilibrium moisture content (typically 3–7% dry mass); then they are placed in hermetically sealed packages and stored at −18 to −20 °C. A minority of species with recalcitrant seeds cannot survive desiccation and, therefore, alternative means of storage or propagation must be sought. Such species are found in a wide range of plant families, suggesting that this trait must have evolved many times; wetland and aquatic plant species are particularly noted for the incidence of recalcitrant seeds but there are many examples of recalcitrance among trees, both temperate and tropical, and legumes (Baskin & Baskin, 1998; Farnsworth, 2000). In practice, limitations on the ability to store certain seeds can be a serious consideration for restoration work. Species of *Carex*, which are important as physiognomic dominants for the restoration of many types of wetlands, are often best stored under wet, cold conditions to maintain viability (Budelsky & Galatowitsch, 1999); van der Valk *et al.* (1999) advocate the use of freshly collected seed of *Carex* to maximise establishment.

Regular viability testing of stored seed is advisable to detect deterioration (Roberts & Ellis, 1983; Ellis *et al.*, 1985). In practice, Walmsley & Davy (1997*a*) found that storage over silica gel in sealed polypropylene boxes at 2 °C maintained good viability for seven years in most shingle-beach species investigated. Although there was some deterioration in the range of environmental conditions under which subsequent germination could occur, such conditions may be sufficient to store most species for several years.

Seed pre-treatment

Seed pre-treatments may improve establishment or its timing sufficiently to be economic. In dry climates, simple seed 'priming' with water can be hugely effective at virtually no cost. Priming involves soaking seed for several hours (typically

12–24) to allow imbibition but not germination immediately before sowing; the safe period of treatment needs to be determined experimentally for each species. Seeds can subsequently be allowed to surface-dry in order to facilitate sowing (Harris *et al.*, 1999).

Species from environments subject to unpredictable disturbances have often evolved long-term dormancy, providing a persistent seed bank as insurance against local extinction. Where a significant proportion of mature seeds is dormant, direct sowing in the field would result in sporadic, slow germination, as the natural loss of dormancy occurred. This may not be on an acceptable time-scale for restoration purposes and might not even achieve sufficient plant density for a self-sustaining population. Pre-treatments can relieve dormancy and give more or less synchronous, rapid germination. Dormancy resulting from a hard or impermeable seed coat can be removed by mechanical or chemical scarification (Baskin & Baskin, 1998). Only about 10% of mature seeds of *Lathyrus japonicus* freshly collected for the restoration of shingle-beach communities germinated under normally favourable conditions; this germinability did not change appreciably over seven years of dry storage at 2 °C. However, the same seed samples showed nearly 100% germination after a 45-minute treatment with concentrated sulphuric acid (96% v/v) to soften the testas and this germinability had only declined to about 90% after seven years of dry storage (Walmsley & Davy, 1997*a*). Similarly, only about 20% of seeds of *Crambe maritima* from the same site germinated from fresh fruits and none germinated from fruits stored dry for up to seven years; in contrast, seeds removed manually from the hard fruits had a germinability of 40–90% that was largely independent of storage time. Similar effects could be achieved in other species with mechanical abrasion. There are many kinds of physiological dormancy but a common mechanism of breaking them is stratification, a more or less prolonged exposure of imbibed seeds to low temperature. In many cases, treatment with gibberellic acid or other growth regulators can substitute, at least partially, for the requirement for cold stratification (Bewley & Black, 1994; Baskin & Baskin, 1998).

Coatings applied to individual seeds can be designed to improve water retention for inbibition but they can also incorporate locally targeted nutrients, growth regulating compounds and even the spores of beneficial arbuscular mycorrhizal fungi. Greipsson (1999), for instance, reported significantly improved establishment of *Poa pratensis* on an Icelandic barren when seeds had been coated with diatomaceous earth, and significant further improvement when cytokinin was incorporated into the coating. Coating technology is relatively expensive and might be economic only for large-scale restoration.

Direct sowing

The sowing of seeds raises many practical questions that do not lend themselves well to generalisation across ecosystems. Schiechtl & Stern (1996) have provided a range of practical seed-sowing protocols aimed primarily at soil stabilisation. In a seasonal climate, spring, autumn or before the beginning of a wet season are generally propitious for sowing, depending also on the phenology of the species. Soil physical preparation normally follows agricultural practice but a minimum of tillage consistent with good subsequent establishment is probably the best practice. Sowing density should be determined by experimentation, depending on the population biology of the species concerned, the likely competitive relationships with weed species and the distributions of mature plants in the target community.

Much ecological theory and experimentation can be applied to the determination of optimal depth of sowing. The essential compromise is simple: insufficient coverage with soil carries with it increased risks of desiccation and predation, but deeper burial risks exceeding the shoot's capacity to reach the surface and photosynthesise after germination. For any batch of seed, the probability of emergence declines with depth of burial to zero (Harris & Davy, 1986; Greipsson & Davy, 1996). Even deeper burial may produce conditions that inhibit germination, perhaps beyond the survival time of the seed. The capacity to emerge from burial is highly dependent on seed size, as larger seeds have more stored resources. This has been demonstrated repeatedly on coastal shores and dunes, where burial is an important phenomenon, because of potentially rapid rates of accretion (Davy & Figueroa, 1993; Davy *et al.*, 2001). Even within a single population the maximum length of shoot produced by etiolated seedlings (i.e. grown in continuous darkness) was directly related to their individual seed masses (Greipsson & Davy, 1996). Germination responses to light and temperature are also relevant to the way in which seeds should be handled. Certain small-seeded species, for which suitable (safe) microsites are readily available at the surface and which cannot emerge from significant burial, have evolved a light requirement for germination. Other species of open habitats appear to use such a light requirement as a 'gap-sensing' mechanism, to prevent ill-fated germination under the canopy of competitors (Pons, 2000). Likewise, the frequent germination requirement for diurnally alternating temperatures may have evolved as a gap- or burial-sensing mechanism to maintain dormancy in unfavourable conditions, because temperature variations are damped by the canopy and soil layers. Conversely, germination is inhibited by light in some larger-seeded species that have little chance of a safe microsite at the surface and appreciable ability to emerge from burial (Greipsson & Davy, 1994). A good knowledge of the germination responses to environmental factors is desirable for any species to be sown in a restoration programme.

The scale of a restoration project also constrains the methodology of sowing. Local restorations, employing spot or broadcast sowing, can seek to imitate the biotic heterogeneity of the target community (Walmsley & Davy, 1997*b*). Larger-scale projects may, perforce, use modified agricultural seed drills or, where there are steep slopes and rocky surfaces, pumped hydro-seeding (e.g. Cullen *et al.*, 1998) and must rely more on environmental heterogeneity and stochastic processes to help structure the developing community. Very large-scale projects, such as in Iceland, require specialised aerial seeding from low-flying aircraft (Greipsson & El-Mayas, 1999).

Establishment of vegetative plants

It is sometimes necessary or advantageous to bypass the hazards of germination in the field and use more mature plants in order to establish populations. Rapid physical stabilisation of the environment by plant roots may be a prerequisite for further restoration (Whisenant, this volume), and physically adverse environments may limit the success possible with directly sown seed. In either case, the restoration process may be greatly accelerated by the rapid establishment of important or keystone species.

Perennial species that readily produce adventitious roots are frequently propagated and planted directly as clonal offsets. The classic example of this is in the stabilisation and restoration of dunes with dune-building perennial grasses (Ranwell & Boar, 1986; Greipsson, volume 2). Likewise, cuttings of some trees and shrubs, such as *Salix* spp., can be planted and rooted *in situ* (Schiechtl & Stern, 1996). Large numbers of individuals can often be produced relatively rapidly. Clonal material, however, may lack genetic variation and this could present a number of potential dangers for long-term persistence. Where possible, the founder plants for the clones should be selected to represent the range of natural variation and the planting pattern should reflect the distribution of that variation. Unfortunately, the genetic variation present in either target or donor populations is rarely known. Perhaps the only mitigation is that naturally occurring clones can also be very extensive.

An alternative approach is to germinate seed and raise seedlings under laboratory or horticultural conditions. The resulting plants are transplanted to individual containers and allowed to grow under glasshouse or otherwise protected conditions until sufficiently large for transplanting into the field. Ideally, the rooting medium should mimic the texture of that at the eventual restoration site, although enhanced nutrient supply and water retention may be necessary for efficient production of high-quality plants. A hardening-off period in the open, with watering as necessary, often aids the transition to the field. By the time plants reach the field, they have substantial above- and belowground reserves of biomass and nutrients, and may even be approaching reproductive activity.

Plantings are advantageous in many environments. In physically adverse or unstable environments, such as on a shingle beach, container-grown plants could be planted with a high probability of survival (Walmsley & Davy, 1997*c*; Walmsley, volume 2). The only aftercare required was watering shortly after planting. Similarly, in the shade of deciduous woodland, where seedling growth can be slow, container-grown individuals of the distinctive perennial species were the only effective solution to restoration of the ground-flora (Packham *et al.*, 1995). Nursery-reared seedling transplants into newly burnt areas of its typical habitat were found to be more efficient than seed sowing in an endemic South African conifer (*Widdringtonia cedarbergensis*) since the former resulted in greater survival (Mustart *et al.*, 1995). Seedling survival in transplants was also improved by shading.

The proportions of species used and the pattern of planting should generally reflect their spatial distributions in the target communities, particularly with regard to any micro-environments or spatial heterogeneity present, although the density of planting may be necessarily lower than in the target community. There is a tendency for plantings of discrete individuals to take on a regular pattern, which gives an unnatural appearance, at least for the first few years. It is virtually impossible to achieve random arrangements in practice, and a small-scale, irregular pattern that is repeated across the site may be the best option for a more natural appearance in the initial stages (Walmsley & Davy, 1997*c*). Clearly, such methods are labour-intensive and potentially expensive; in consequence, they are especially suitable for long-lived perennial species with a distinctive role in their communities or ecosystems.

The translocation of mature individuals is even more labour-intensive. Most plants, even large trees, can be moved if measures are taken to minimise water stress. These include taking a root-ball sufficiently large to avoid damage to the root system, translocating when plants are dormant or during the cool/wet season and, subsequently, reducing transpiration; reducing the area of leaf canopy, tying it together or wrapping it, and the application of anti-transpirant chemicals are all options for reducing transpiration. Translocation allows limited numbers of keystone species, especially the long-lived, woody ones, to be established much more rapidly than waiting for successional development. The disadvantage is that a donor site is correspondingly damaged but where this is inevitable for other reasons, it is clearly valuable to salvage as much as possible.

Large-scale translocation

Salvaging plant communities from a donor site that is facing destruction raises the possibility of wholesale translocation. The use of suitable machinery can allow relatively undisturbed turves with intact vegetation and the rooting layer of soil to be cut and transported to a restoration site. This potentially circumvents many of the problems of establishing the right surface soil conditions, which is particularly useful when normal pedogenesis depends on hundreds or thousands of years of leaching. The most authentic restoration requires the turves to be abutted together in their original configuration, yielding approximately the same area of vegetaton. A much larger area can be reconstructed by spacing the turves and using them as nuclei for colonisation of the intervening parts of the restoration site. Not only should turves and the accompanying soil represent the composition of a functioning plant community accurately, including any natural seed banks, but they would include components that have rarely been the subject of separate restoration projects, such as lichens and bryophytes. Presumably, similar advantages accrue from integrated translocation of the microbial and invertebrate communities in the soil (Allen *et al.*, this volume).

In practice translocation to a different substrate may compromise the restoration achieved. Good *et al.* (1999) reported a pilot study at an opencast coal-mining site in Wales, in which herb-rich, mesic grassland turf was translocated. After three years, the cover and species composition were similar in plots that had received whole turves and in plots where turf had been spread over twice its area and rotovated in. More than half the original species were re-established but the plant communities were altered substantially by either technique of translocation, probably as a consequence of altered soil hydrology and nutritional status, and the substitution of a cutting treatment for grazing.

SPECIES REMOVAL

The problem of removing unwanted, weedy species can be more intractable than that of establishing the characteristic, desired ones. Invasion by weeds is an inherent danger associated with any disturbance or damage to ecosystems and with any attempt to manipulate local plant communities. Unfortunately the frequency of problems arising is becoming much greater. As barriers to dispersal have been progressively broken down, with deliberate and accidental introduction of propagules into new environments by human activity, the invasion of ecosystems by aggressive, alien plants has become an increasingly important cause of damage. Such invasion may be facilitated by many kinds of disturbance, both natural ones such as hurricanes, and those from exploitation by humans, such as the felling of forest trees or grazing. The resulting damage may involve no changes to the physical landform and little long-term adverse effect on soil conditions. Nevertheless, changes to the whole character of the plant communities, with the loss of keystone (perhaps endemic) species, can affect biodiversity and ecosystem function. The knock-on effects inevitably include habitat loss for many other groups of organisms. Consequently, the control of alien plant species is not infrequently the key to ecological restoration and, almost by definition, invasive species are the most difficult to control. Removal of such species must be highly selective. There are essentially three strategies by which it can be achieved: physical removal, the use of chemical herbicides and biological control.

Physical removal

Removal by uprooting or cutting down the offending individuals is highly focused and hence extremely successful. It is very labour-intensive but, in the case of some woody species at least, may not need to be repeated. Where there is continuing germination from the soil seed bank, a few successive removals may suffice to allow the target species to achieve dominance and the invasive species to be largely controlled or suppressed. Cutting down individual woody plants, whether it involves the removal of canopy trees, understorey species or scrub, allows the possibility of potentially rapid regrowth of sprouts. Even when tree stumps are killed by felling, or subsequent treatment of the cut face with herbicide, decomposition of the residue can cause local eutrophication or may encourage fungi harmful to desirable species. Where practicable, the slower and more expensive approach of uprooting is preferable, using mechanical devices appropriate to the scale of the problem. Removal by hand is one of the mainstays of work to restore limited areas of native ebony forests in the Black River Gorges National Park of Mauritius that have been degraded by invasions of rapidly growing species of guava (*Psidium cattleianum*) and privet (*Ligustrum walkeri*) (Dulloo et al., 1996). Selective methods require sufficient sophistication in the work-force to recognise the relevant species in potentially diverse communities. Even with selective removal, additional manipulations may be necessary to re-establish target species. The obvious requirement for the restoration of sand-dune pine plantations to natural sand dune communities is clear-felling of the pine trees. However, Sturgess & Atkinson (1993) found that removal of the deep layer of needle litter greatly accelerated the change to a near-natural dune vegetation.

Less selective methods are easier to mechanise. Tillage kills herbaceous weeds and repeated tillage depletes their soil seed banks; it can therefore be employed where none of the existing vegetation is to be retained, before sowing or planting the species to be restored. Cutting off the aerial parts of invasive, clonal plants that have underground organs rarely kills them. However, repeated cutting can have an important role in the long-term control of invasive densely clonal species, such as bracken (*Pteridium aquilinum*) (see section 'Manipulation of plant interactions', below).

Herbicides

There are numerous herbicides, from chemically diverse families, that differ in their formulation, mode of action, species specificity, toxicity to other organisms and persistence in the environment (Hance & Holly, 1990; Anderson, 1996). Chemical control is probably not the most attractive option for use by many restoration ecologists, because of uncertainties about side-effects, collateral damage to desirable species and persistence in environments other than those in which chemicals have been tested. However, certain non-selective and selective compounds with relatively low toxicities to other organisms and low persistence in the soil have found fairly general acceptance. Amongst these are various formulations of glyphosate, a non-selective post-emergence herbicide. This can be used to destroy annual and perennial weed populations without disturbance to the soil and hence also without the risk of stimulating further germination from its seed bank, prior to the introduction of desired plant species. It can also be painted on to the cut ends of woody plants to prevent regrowth. Asulam is a widely used, selective post-emergence herbicide that is particularly effective on bracken (and other ferns).

In the short term, selective herbicides can be highly effective in reducing the abundance of unwanted species. However, eradication of the unwanted species is rarely achieved and in the longer term, invasive, weedy species tend to recolonise and increase in abundance again unless herbicide is used as part of an integrated control programme with other control measures. A classic example is the not-infrequent necessity of controlling bracken for the restoration of lowland Atlantic heathland. Marrs *et al.* (1998*a*) used multiple treatments with asulam in combination with annual cutting and the introduction of seeds of *Calluna vulgaris*; after 18 years an acceptable form of breckland grass heath had been re-established but *Calluna* itself had not survived. No treatment had eradicated bracken after 18 years of continuous application (Marrs *et al.*, 1998*b*, *c*), and in plots where treatment had ceased after six years, there was appreciable recovery of bracken in the subsequent 12 years. Similarly, Hurst & John (1999) attempted to control the invasive dominant grass *Brachypodium pinnatum* with glyphosate, where it was forming monoculture patches in species-diverse chalk grassland. Again, initial results were promising but *Brachypodium* recolonised all treated stands within a few years and an integrated control programme would be required for the restoration of characteristic chalk-grassland species.

Herbicides are generally not labour-intensive to use, regardless of whether they are applied from a hand sprayer, a tractor-mounted sprayer or weed-wiper, or by aerial spraying.

Biological control

Biological control, using pathogenic micro-organisms and phytophagous insects, has had a chequered history, with relatively few unqualified successes worldwide (Simberloff & Stiling, 1996). In principle it is most appropriate where an alien species is invasive because it has been introduced into an environment where it lacks its normal complement of predators and pathogens. The dangers of introducing potential pathogens and predators are considerable, however, because their behaviour in new environments, faced with a new range of potential hosts or prey, cannot necessarily be predicted. Indigenous or desirable species may prove to be more susceptible or attractive than the intended target species. The direct effects on the populations of unintended victims are bad enough, but the knock-on effects arising from impacts on their trophic relationships may be more far-reaching. There are already too many examples of conservation disasters arising from misguided attempts at biological control: some of the major problems requiring ecological restoration have been created by species introductions themselves. Introductions of predators in particular should not be carried out without exhaustive experimental work to examine the possible trophic consequences. This should first be done in the laboratory; only if this raises no problems, should work be extended to heavily quarantined conditions in the field.

Biological control of plants is more likely to succeed in constrained, isolated areas such as on oceanic islands. *Cordia curassavica*, one of the worst introduced weeds, has been effectively controlled on Mauritius. The invasive impact of *Lantana camara* and *Opuntia vulgaris* has also been successfully limited there by biological control (C. Jones, pers. comm.). Introductions of specialist herbivorous insects have been proposed for the biological control of bracken in Britain, with possible candidate species from Papua-New Guinea (Kirk, 1982) and South Africa (Fowler, 1993).

The use of fungal pathogens may be less controversial and fraught with danger, perhaps because of their generally high host specificity. 'Mycoherbicides' are preparations containing the spores of particular pathogenic fungi that can be applied as required, in the same way as selective chemical herbicides. There are, as yet, few examples of their use in restoration ecology. A bracken pathogen (the imperfect fungus *Ascochyta pteridis*) has been suggested as a potential biological control agent (Webb & Lindow, 1981) and spores of *A. pteridis* have been successfully formulated as a mycoherbicide. However, the effects after spraying on bracken have so far been rather disappointing (Womack & Burge, 1993; Womack *et al.*, 1996).

MANIPULATION OF PLANT INTERACTIONS

Neither the establishment of indispensable plant species nor the removal of deleterious ones can immediately restore target plant communities. Whether the target is a successional climax or an artificially managed earlier stage, some degree of development is likely to be necessary. The reconstruction of ecosystems in physically degraded environments can be seen as the initiation and acceleration of successional processes. Where the damage is due to the effects of invading species, either as a result of management or species introductions, it is itself a successional process. Admittedly, the latter case may involve a novel succession that would have been impossible without human intervention. Such change is intimately bound up with the restoration of ecosystem function, including the reinstatement of appropriate nutrient pools in soil and vegetation and their fluxes, especially microbially mediated ones (Marrs, this volume). However, most models of succession (Connell & Slatyer, 1977) are driven by interactions between species, both positive (facilitation) and negative (competition and inhibition), in various combinations. Hence, anything that affects the balance of interactions between plant species can be used to manipulate the composition of communities and thus influence their development or regression.

Direct effects of manipulations are generally associated with disturbance and the removal of biomass, classically by grazing, cutting or burning. At its simplest, such selective removal of biomass reduces the competitive or inhibitory potential of species, thus releasing others from a constraint. Even if the removal is not in itself selective, species may differ in their tolerance to the challenge and therefore in their recovery from it. Disturbance may also create germination microsites for the regeneration and proliferation of target species that have already been established. The indirect consequences may include positive interactions, as the release of one species may facilitate others.

Facilitation and nurse species

The importance of positive effects in plant communities has only become fully apparent in recent work. The range of mechanisms includes various modifications of resource availability, modification of the substrate, protection from herbivores, promotion of pollination, dispersal or concentration of seeds, and subsidies of nutrients via root grafts or mycorrhiza (Callaway, 1995; Callaway & Walker, 1997). 'Nurse' species have been used for a considerable time, however, in silviculture and restoration to facilitate the establishment and growth of other species. They may be appropriately selected early-successional species that are sown or planted to ameliorate the physical environment of a degraded site, thus enabling the subsequent establishment of the target community. For example, Vieira *et al.* (1994) examined the role of the shrub *Cordia multispicata* in facilitating the establishment of woody species in abandoned Amazon pastures. The rain of bat- and bird-dispersed seeds of woody species and the density of woody seedlings were both much greater in the areas occupied by *C. multispicata*. In addition soil nutrient availability and light (photosynthetic photon flux density) at the soil surface were generally more favourable for the growth of rainforest tree seedlings in the *C. multispicata* patches, as compared to the grass-dominated zones. This shrub apparently acts both as a magnet for flying seed vectors and as a nurse plant.

Alternatively, nurse species may be entirely alien to the target community but selected for restricted longevity and lack of competitive ability in the longer term. Annual or short-lived perennial grasses, such as *Lolium multiflorum* and *Agrostis castellana*, can help to provide moist microsites for the germination of the tiny seeds of *Calluna vulgaris* and the protection of its young seedlings but do not persist once the *Calluna* is established (Environmental Advisory Unit, Liverpool, 1988).

Grazing by herbivores

Grazing by vertebrate herbivores and its exclusion are both long-established tools in the management of species composition for conservation. Herbivores differ in their selectivity and plants have evolved diverse defences against them, but the tissues of fast-growing, competitive species tend to be the most available and are often the most palatable. Crawley (1988) has pointed out that where facilitation is the dominant successional process, herbivore feeding would be expected to slow down or reverse succession, because it would reduce the rate of environmental amelioration by the dominant plants; conversely, where inhibition is more important, herbivores should accelerate succession by weakening the current dominants and allowing the invasion of later-successional species. Herbivores also tend to recycle nutrients locally and provide regeneration niches by physical disturbance to the vegetation and the soil surface.

Grazing has been employed much less for explicit ecological restoration than for general conservation management. One somewhat counter-intuitive use is in the restoration of forest in areas that have been cleared and converted to pasture. When such pastures are abandoned, the introduced grasses may impede the re-establishment of woody species and hence forest regeneration. Posada *et al.* (2000) described an interesting example, where they sought methods to re-establish tropical montane rainforest on Colombian upland pastures dominated by the African grasses *Pennisetum clandestinum* and *Melinis minutiflora*. A low stocking density with cattle facilitated the establishment of shrubs that created microhabitats more suitable for the subsequent establishment of montane forest trees. Similarly, simulated grazing treatments in abandoned Brazilian Amazonian pastures significantly improved germination and establishment of the 'successional facilitator' *Cordia multispicata* (Vieira *et al.*, 1994).

Grazing has a more obvious role in the restoration of grassland and prairies, where its effect in preventing the establishment of shrubs and trees arrests succession at the stage desired. However successional trajectories can be very sensitive to the initial conditions. The Rockefeller Experimental Tract is a 40-year old restoration experiment in the prairie–forest ecotone of Kansas with native tallgrass prairie and oak–hickory forest. Tracts of prairie were reseeded in 1957 and have been managed in different ways since (Kettle *et al.*, 2000). The effect of grazing in preventing afforestation depended crucially on biotic and edaphic conditions at the time of restoration; afforestation was 6% on land that had been in cultivation prior to seeding, 20% on former pastureland and 98% on land deforested before seeding. The presence of various prairie and forest species resulted from persistence rather than colonisation following restoration. The character and biodiversity of many grasslands are intrinsically linked with grazing (Hutchings & Stewart, volume 2) and it would be surprising if continuous or seasonal grazing treatments did not often play a major role in their restoration. Selective removal of biomass by herbivores, such as rabbits, may also have a role in protecting certain nutrient-poor, species-diverse communities from loss of biodiversity in response to eutrophication, and this may become more significant as the aerial deposition of nitrogen from atmospheric pollution increases.

Grazing, particularly overgrazing, by domestic animals is a major cause, world-wide, of ecosystem degradation. Huge damage has also been inflicted by the introduction of alien herbivores to previously isolated environments, especially on islands. Hence the abatement of grazing, by appropriate fenced exclosures or by removal of herbivore populations, is a potentially powerful tool for restoration.

Cutting

Cutting or mowing is generally used as a proxy for grazing. It can usually be mechanised by adapting agricultural equipment and can be attractive because it obviates the many complexities and costs associated with owning and managing herbivore populations (e.g. Bokdam & Devries, 1992). Cutting can have similar effects to grazing in reducing the dominance of the taller, faster growing species and, in this sense, it is species-selective. Nevertheless, it lacks the selectivity associated with palatability to particular herbivores. If the cuttings are removed, cutting also emulates grazing in depleting the nutrient capital of the system and in reducing the accumulation of litter. Mechanical cutting or mowing is likely to be much more damaging to invertebrate populations than grazing. On the other hand, it is particularly valuable in the management of generally unpalatable or poisonous species. Hence, cutting once or twice a year has been an integral part of programmes to restore grass heaths and grasslands that are challenged with invasive dominants such as bracken (Marrs *et al.*, 1998*a*, *b*) and *Brachypodium pinnatum* (Bobbink & Willems, 1993). In fact, mowing precluded forest establishment in the Rockefeller Experimental Tract prairie restoration more consistently than grazing (Kettle *et al.*, 2000).

Fire

Fire is another form of management that essentially reverses or arrests succession. It mobilises and reduces the nutrient capital of the system, some of which is lost in smoke and by subsequent leaching from the ash (Debano *et al.*, 1998). It is also potentially selective, because species differ in their sensitivity to fire and because the duration and temperature of fires can be controlled. Fire is an integral part of the ecology of certain ecosystems and plants in them have evolved considerable resistance to its damaging effects, including thick, insulating bark and serotinous fruits that only release their seeds after exposure to high temperatures (e.g. Enright *et al.*, 1998*a*, *b*). Burning in such communities can be used to the disadvantage of less-tolerant aliens and invaders. As with cutting, burning precluded forest establishment more consistently than grazing in the Rockefeller Experimental Tract prairie restoration (Kettle *et al.*, 2000). Oak savannas maintained by fire essentially disappeared from silt-loam soils in midwestern North America soon after European settlement, because of fire suppression and agriculture. Bowles & McBride (1998) showed that restoration of a 2-ha remnant of savanna in Illinois would require a return to high fire frequencies and reduction of the subcanopy to its former grub layer, which should increase abundance of light- and fire-adapted species and reduce that of alien species. There would be practical difficulties in doing this, because existing maturing oaks are too fire-resistant, there is a lack of grasses for fuel and the scale of the fire on this small site might be inadequate to open canopy gaps. Examination of vegetational change on a remnant barrens area in southern Illinois, over 25 years, showed that fire management temporarily reversed a trend to increasing dominance of woodland species and declining prairie species (Anderson *et al.*, 2000). Reintroduction of prescribed burning in 1990–3, after 16 years without burning, altered the vegetation trajectory but not back toward a species composition like that on the site before cessation of fire management. Fire management on the site apparently would not recover the barrens that occurred on the site previously. Nevertheless, consistent fire management would drive vegetation changes toward increasing abundance of prairie and open woodland species that would otherwise be lost without burning.

Controlled burning also has a role in preventing the accumulation of exceptionally large amounts of fuel, with its accompanying risk of promoting a catastrophic fire. A restoration programme to reinstate the declining, endemic conifer *Widdringtonia cedarbergensis* to the fynbos vegetation of the Cederberg Mountains, South Africa, includes burning (Mustart *et al.*, 1995); loss of the fire-sensitive adult trees has been reduced by practising frequent, low-intensity burning, which precludes the occurrence of intense wildfires.

CONCLUDING REMARKS

It is easy to assume that if a reasonable facsimile of the abiotic environment of a reference ecosystem can be constructed then the recolonisation of plant communities appropriate to that ecosystem is just a matter of time. While in certain cases this may be true, it is rarely so, at least on an acceptable timescale. Ecological restoration usually demands the deliberate re-establishment of plant species and subsequent management of them to direct or suppress successional processes. This is true whether the establishment is aimed immediately at creating the target communities or whether it is designed to accelerate succession towards them. The reference communities will vary in time and space and so may be regarded as a moving target. The achievement of spatial and temporal heterogeneity, both biotic and abiotic, on the same scales as reference sites would be an informative criterion for success. The ability to remove or control invasive, alien species is as important as the ability to establish desirable ones. The damage and destruction that necessitate restoration work, and the potential for world-wide dissemination of propagules, create many unbridled opportunities for catastrophic invasions. Although restored communities should have considerably greater resistance to invasion than degraded ones, because resources and niches will have been pre-empted by the established species, continued vigilance once an acceptable state of restoration has been achieved is likely to be important.

The principles underlying a wide range of approaches to establishing and removing plant species have been reviewed. Not all will be applicable to any restoration and sometimes the choice of approach will be influenced by economics, time-scale or other constraints. However, more than one is likely to be needed at a particular site. The establishment of a matrix of common, long-lived or keystone species might require a more rapid technique than the subsequent interpolation of rarer but possibly more distinctive components of a community. In practice, virtually any technique or ecological process whose consequences for plant abundance are sufficiently well understood may be appropriated as an artificial manipulation.

Community development and composition together represent only one facet of ecosystem structure and function. The restoration of plant community composition is inextricably bound up with the restoration of ecosystem function and needs to be considered in parallel with all of the processes that comprise ecosystem function. Similarly, we recognise that a plant community will interact with the communities of all other organisms present. The restoration is likely to be more successful and more rapid if the population dynamics of as many as possible of the species introduced, and their interactions with other species (negative and positive), are already understood in similar habitats. Populations within communities will only be self-sustaining if their regeneration niches are also catered for, explicitly or implicitly. Restoration ecology, although arguably in its infancy, has been exceedingly successful in highlighting how little such information exists in the majority of cases. For the time being, pragmatism and informed guesswork are in practice likely to remain important ingredients of many restoration projects.

REFERENCES

Anderson, R.C., Schwegman, J.E. & Anderson, M.R. (2000). Micro-scale restoration: a 25-year history of a southern Illinois barrens. *Restoration Ecology*, **8**, 296–306.

Anderson, W.P. (1996). *Weed Science: Principles and Applications*, 3rd edn. St Paul, MN: West Publishing Company.

Baskin, C. & Baskin, J. (1998). *Seeds: Ecology, Biogeography, and Evolution of Dormancy and Germination*. San Diego, CA: Academic Press.

Bewley, J.D. & Black, M. (1994). *Seeds: Physiology of Development and Germination*, 2nd edn. New York: Plenum Press.

Bobbink, R. & Willems, J.H. (1993). Restoration management of abandoned chalk grassland in The Netherlands. *Biodiversity and Conservation*, **2**, 616–626.

Bokdam, J. & Devries, M.F.W. (1992). Forage quality as a limiting factor for cattle grazing in isolated Dutch nature-reserves. *Conservation Biology*, **6**, 399–408.

Bowles, M.L. & McBride, J.L. (1998). Vegetation composition, structure, and chronological change in a decadent midwestern North American savanna remnant. *Natural Areas Journal*, **18**, 14–27.

Budelsky, R.A. & Galatowitsch, S. M. (1999). Effects of moisture, temperature, and time on seed germination of five wetland Carices: implications for restoration. *Restoration Ecology*, **7**, 86–97.

Callaway, R.M. (1995). Positive interactions among plants. *Botanical Review*, **61**, 306–349.

Callaway, R.M. & Walker, L.R. (1997). Competition and facilitation: a synthetic approach to interaction in plant communities. *Ecology*, **78**, 1958–1965.

Chesson, P. (2000). Mechanisms of maintenance of species diversity. *Annual Review of Ecology and Systematics*, **31**, 343–366.

Connell, J.H. & Slatyer, R.O. (1977). Mechanisms of succession in natural communities and their role in community stability and organization. *American Naturalist*, **111**, 1119–1144.

Crawley, M.J. (1988). Herbivores and plant population dynamics. In *Plant Population Ecology*, eds. A.J. Davy, M.J. Hutchings & A.R. Watkinson, pp. 367–392. Oxford: Blackwell.

Cullen, W.R., Wheater, C.P. & Dunleavy, P.J. (1998). Establishment of species-rich vegetation on reclaimed limestone quarry faces in Derbyshire, UK. *Biological Conservation*, **84**, 25–33.

Davy, A.J. (2000). Development and structure of salt marshes: community patterns in time and space. In *Concepts and Controversies in Tidal Marsh Ecology*, eds. M.P. Weinstein & D.A. Kreeger, pp. 137–156. Dordrecht, The Netherlands: Kluwer.

Davy, A.J. & Figueroa, M.E. (1993). The colonization of strandlines. In *Primary Succession on Land*, eds. J. Miles & D.W.H. Walton, pp. 113–131. Oxford: Blackwell.

Davy, A.J., Dunsford, S.J. & Free, A.J. (1998). Acidifying peat as an aid to the reconstruction of lowland heath on arable soil: lysimeter experiments. *Journal of Applied Ecology*, **35**, 649–659.

Davy, A.J., Willis, A.J. & Beerling, D.J. (2001). The plant environment: aspects of the ecophysiology of shingle species. In *Ecology and Geomorphology of Coastal Shingle*, eds. J.R. Packham, R.E. Randall, R.S.K. Barnes & A. Neal, pp. 191–201. Otley, UK: Westbury.

Debano, L.F., Neary, D.G. & Ffolliott, P.F. (1998). *Fire's Effects on Ecosystems*. New York: John Wiley.

Diamond, J.M. (1975). Assembly of species communities. In *Ecology and Evolution of Communities*, eds. M.L. Cody & J.M. Diamond, pp. 342–444. Cambridge, MA: Harvard University Press.

Dulloo, M.E., Jones, C., Strahm, W. & Mungroo, Y. (1996). Ecological restoration of native plant and animal communities in Mauritius, Indian Ocean. In *The Role of Restoration in Ecosystem Management*, eds. D.L. Pearson & C. V. Klimas, pp. 83–91. Madison, WI: Society for Ecological Restoration.

Dunsford, S.J., Free, A.J. & Davy, A.J. (1998). Acidifying peat as an aid to the reconstruction of lowland heath on arable soil: a field experiment. *Journal of Applied Ecology*, **35**, 660–672.

Ellis, R.H., Hong, T.D. & Roberts, E.H. (1985). *Handbook of Seed Technology for Genebanks*, vol. 2, *Compendium of Specific Germination Information and Test Recommendations*. Rome: International Board for Plant Genetic Resources.

Enright, N.J., Marsula, R., Lamont, B.B. & Wissel, C. (1998*a*). The ecological significance of canopy seed storage in fire-prone environments: a model for non-sprouting shrubs. *Journal of Ecology*, **86**, 946–959.

Enright, N.J., Marsula, R., Lamont, B.B. & Wissel, C. (1998*b*). The ecological significance of canopy seed storage in fire-prone environments: a model for resprouting shrubs. *Journal of Ecology*, **86**, 960–973.

Environmental Advisory Unit, Liverpool (1988). *Heathland Restoration: A Handbook of Techniques*. London: British Gas.

Falk, D.A., Millar, C.I. & Olwell, M. (eds.) (1996). *Restoring Diversity: Strategies for Reintroduction of Endangered Plants*. Washington, DC: Island Press.

Farnsworth, E. (2000). The ecology and physiology of viviparous and recalcitrant seeds. *Annual Review of Ecology and Systematics*, **31**, 107–138.

Fischer, M. & Matthies, D. (1998). Effects of population size on performance in the rare plant *Gentianella germanica*. *Journal of Ecology*, **86**, 195–204.

Fischer, M. & Stöcklin, J. (1997). Local extinctions of plants in remants of extensively used calcareous grasslands 1950–1985. *Conservation Biology*, **11**, 727–737.

Fowler, S.V. (1993). The potential for control of bracken in the UK using introduced herbivorous insects. *Pesticide Science*, **37**, 393–397.

Good, J.E.G., Wallace, H.L., Stevens, P.A. & Radford, G.L. (1999). Translocation of herb-rich grassland from a site in Wales prior to opencast coal extraction. *Restoration Ecology*, **7**, 336–347.

Greipsson, S. (1999). Seed coating improves establishment of surface seeded *Poa pratensis* used in revegetation. *Seed Science and Technology*, **27**, 1029–1032.

Greipsson, S. & Davy, A.J. (1994). Germination of *Leymus arenarius* and its significance for land reclamation in Iceland. *Annals of Botany*, **73**, 393–401.

Greipsson, S. & Davy, A.J. (1995). Seed mass and germination behaviour in populations of the dune-building grass *Leymus arenarius*. *Annals of Botany*, **76**, 493–501.

Greipsson, S. & Davy, A.J. (1996). Sand accretion and salinity as constraints on the establishment of *Leymus arenarius* for land reclamation. *Annals of Botany*, **78**, 611–618.

Greipsson, S. & Davy, A.J. (1997). Responses of *Leymus arenarius* to nutrients: improvement of seed production and seedling establishment for land reclamation. *Journal of Applied Ecology*, **34**, 1165–1176.

Greipsson, S. & El-Mayas, H. (1999). Large-scale reclamation of barren lands by aerial seeding in Iceland. *Land Degradation and Development*, **10**, 185–193.

Grime, J.P. (1977). Evidence for the existence of three primary strategies in plants and its relevance to ecological and evolutionary theory. *American Naturalist*, **111**, 1169–1194.

Grime, J.P. (1986). Manipulation of plant species and communities. In *Ecology and Design in Landscape*, eds. A.D. Bradshaw, D.A. Goode & E.H.P. Thorp, pp. 175–194. Oxford: Blackwell.

Grime, J.P. (2001). *Plant Strategies, Vegetation Processes, and Ecosystem Properties*, 2nd edn. Chichester, UK: John Wiley.

Grubb, P.J. (1977). The maintenance of species-richness in plant communities: the importance of the regeneration niche. *Biological Reviews*, **52**, 107–145.

Grubb, P.J. (1986). The ecology of establishment. In *Ecology and Design in Landscape*, eds. A.D. Bradshaw, D.A. Goode & E.H.P. Thorp, pp. 83–97. Oxford: Blackwell.

Guerrant, E.O. (1996). Designing populations: demographic, genetic, and horticultural dimensions. In *Restoring Diversity: Strategies for Reintroduction of Endangered Plants*, eds. D.A. Falk, C.I. Millar & M. Olwell, pp. 171–207. Washington, DC: Island Press.

Hance, R.J. & Holly, K. (eds.) (1990). *Weed Control Handbook: Principles*, 8th edn. Oxford, UK: Blackwell Scientific Publications.

Harris, D. & Davy, A.J. (1986). The regenerative potential of *Elymus farctus* from rhizome fragments and seed. *Journal of Ecology*, **74**, 1057–1067.

Harris, D., Joshi, A., Khan, P.A., Gothkar, P. & Sodhi, P.S. (1999). On-farm seed priming in semi-arid agriculture: development and evaluation in maize, rice and chickpea in India using participatory methods. *Experimental Agriculture*, **35**, 15–29.

Hill, M.O. & Stevens, P.A. (1981). The density of viable seeds in soils of forest plantations of upland Britain. *Journal of Ecology*, **69**, 693–709.

Hurst, A. & John, E. (1999). The effectiveness of glyphosate for controlling *Brachypodium pinnatum* in chalk grassland. *Biological Conservation*, **89**, 261–265.

Kéry, M. Matthies, D. & Spillmann, H.-H. (2000). Reduced fecundity and offspring performance in small populations of the declining grassland plants *Primula veris* and *Gentiana lutea*. *Journal of Ecology*, **88**, 17–30.

Kettle, W.D., Rich, P.M., Kindscher, K., Pittman, G.L. & Fu, P. (2000). Land-use history in ecosystem restoration: a 40-year study in the prairie–forest ecotone. *Restoration Ecology*, **8**, 307–317.

Kirk, A.A. (1982). Insects associated with bracken fern *Pteridiun aquilinum* (Polypodiaceae) in Papua-New Guinea and their possible use in biological control. *Acta Oecologia–Oecologia Applicata*, **3**, 343–359.

Leck, M.A, Parker, V.T. & Simpson, R.L. (1989). *Ecology of Soil Seed Banks*. San Diego, CA: Academic Press.

Marrs, R.H., Johnson, S.W. & Le Duc, M.G. (1998*a*). Control of bracken and restoration of heathland. 8: The regeneration of the heathland community after 18 years of continued bracken control or 6 years of control followed by recovery. *Journal of Applied Ecology*, **35**, 857–870.

Marrs, R.H., Johnson, S.W. & Le Duc, M.G. (1998*b*). Control of bracken and restoration of heathland. 6: The response of bracken fronds to 18 years of continued bracken control or 6 years of control followed by recovery. *Journal of Applied Ecology*, **35**, 479–490.

Marrs, R.H., Johnson, S.W. & Le Duc, M.G. (1998*c*). Control of bracken and restoration of heathland. 7: The response of bracken rhizomes to 18 years of continued bracken control or 6 years of control followed by recovery. *Journal of Applied Ecology*, **35**, 748–757.

Millennium Seed Bank Project (2001). http://www.rbgkew.org.uk/seedbank

Mitchell, R.J., Marrs, R.H. & Auld, M.H.D. (1998). A comparative study of the seed banks of heathland and successional habitats in Dorset, Southern England. *Journal of Ecology*, **86**, 588–596.

Moore, R.P. (1985). *Handbook on Tetrazolium Testing*. Zürich, Switzerland: International Seed Testing Association.

Mustart, P., Juritz, J., Makua, C., Vandermerwe, S.W. & Wessels, N. (1995). Restoration of the clanwilliam cedar *Widdringtonia cedarbergensis*: the importance of monitoring seedlings planted in the Cederberg, South Africa. *Biological Conservation*, **72**, 73–76.

Packham, J.R., Cohn, E.V.J., Millett, P. & Trueman, I.C. (1995). Introduction of plants and manipulation of field layer vegetation. In *The Ecology of Woodland Creation*, ed. R. Ferris-Kaan, pp. 129–148. Chichester, UK: John Wiley.

Pavlik, B.M. (1996). Defining and measuring success. In *Restoring Diversity: Strategies for Reintroduction of Endangered Plants*, eds. D.A. Falk, C.I. Millar & M. Olwell, pp. 127–155. Washington, DC: Island Press.

Pons, T.L. (2000). Seed responses to light. In *Seeds: The Ecology of Regeneration in Plant Communities* 2nd edn, ed. M. Fenner, pp. 237–260. Wallingford, UK: CAB International.

Posada, J.M., Aide, T.M. & Cavelier, J. (2000). Cattle and weedy shrubs as restoration tools of tropical montane rainforest. *Restoration Ecology*, **8**, 370–379.

Ranwell, D.S. & Boar, R. (1986). *Coast Dune Management Guide.* Abbots Ripton, UK: Institute for Terrestrial Ecology.

Roberts, E.H. & Ellis, R.H. (1983). The implications of the deterioration of orthodox seeds during storage for genetic resources conservation. In *Crop Genetic Resources: Conservation and Evaluation*, eds. J.H.W. Holden & J.T. Williams, pp. 18–37. London: George, Allen & Unwin.

Schemske, D.W., Husnabd, B.C., Ruckelshaus, M.H., Goodwillie, C., Parker, I.M. & Bishop, J.G. (1994). Evaluating approaches to the conservation of rare and endangered plants. *Ecology*, **75**, 584–606.

Schiechtl, H.M. & Stern, R. (1996). *Ground Bioengineering Techniques for Slope Protection and Erosion Control*, transl. L. Jaklitsch. Oxford: Blackwell.

Silvertown, J.W. & Charlesworth, D. (2001). *Introduction to Plant Population Biology*, 4th edn. Oxford: Blackwell Scientific Publications.

Simberloff, D. & Stiling, P. (1996). How risky is biological control? *Ecology*, **77**, 1965–1974.

Sturgess, P. & Atkinson, D. (1993). The clear-felling of sand-dune plantations: soil and vegetational processes in habitat restoration. *Biological Conservation*, **66**, 171–183.

Thompson, K. (2000). The functional ecology of soil seed banks. In *Seeds: The Ecology of Regeneration in Plant Communities*, 2nd edn, ed. M. Fenner, pp. 215–235. Wallingford, UK: CAB International.

Thompson, K. & Grime, J.P. (1979). Seasonal variation in the seed banks of herbaceous species in ten contrasting habitats. *Journal of Ecology*, **67**, 893–921.

Thompson, K, Bakker, J. & Bekker, R.M. (1996). *The Soil Seed Banks of North West Europe: Methodology, Density and Longevity.* Cambridge: Cambridge University Press.

van der Valk, A.G. & Pederson, R.L. (1989). Seed banks and the management and restoration of natural vegetation. In *Ecology of Soil Seed Banks*, eds. M.A. Leck, V.T. Parker & R.L. Simpson, pp. 329–346. San Diego, CA: Academic Press.

van der Valk, A.G., Brehmholm, T.L. & Gordon, E. (1999). The restoration of sedge meadows: seed viability, seed germination requirements, and seedling growth of *Carex* species. *Wetlands*, **19**, 756–764.

Vieira, I.C.G., Uhl, C. & Nepstad, D. (1994). The role of the shrub *Cordia multispicata* Cham. as a succession facilitator in an abandoned pasture, Paragominas, Amazonia. *Vegetatio*, **115**, 91–99.

Walmsley, C.A. & Davy, A.J. (1997*a*). Germination characteristics of shingle-beach species, effects of seed ageing and their implications for vegetation restoration. *Journal of Applied Ecology*, **34**, 131–142.

Walmsley, C.A. & Davy, A.J. (1997*b*). The restoration of coastal shingle vegetation: effects of substrate composition on the establishment of seedlings. *Journal of Applied Ecology*, **34**, 143–153.

Walmsley, C.A. & Davy, A.J. (1997*c*). The restoration of coastal shingle vegetation: effects of substrate composition on the establishment of container-grown plants. *Journal of Applied Ecology*, **34**, 154–165.

Webb, R. & Lindow, E. (1981). Evaluation of of *Ascochyta pteridium* as a potential biological control agent of bracken fern. *Phytopathology*, **71**, 911.

Westoby, M., Leishman, M. & Lord, J. (1996). Comparative ecology of seed size and dispersal. *Philosophical Transactions of the Royal Society of London B*, **351**, 1309–1318.

Wilson, E.O. (1992). *The Diversity of Life.* Cambridge, MA: Harvard University Press.

Wilson, J.B. & Roxburgh, S.H. (1994). A demonstration of guild-based assembly rules for a plant community, and determination if intrinsic guilds. *Oikos*, **69**, 267–276.

Wilson, J.B. & Whittaker, R.H. (1995). Assembly rules demonstrated in a saltmarsh community. *Journal of Ecology*, **83**, 801–807.

Wilson, J.B., Allen, R.B. & Lee, W.G. (1995). An assembly rule in the ground and herbaceous strata of a New Zealand rain-forest. *Functional Ecology*, **9**, 61–64.

Womack, J.G. & Burge, M.N. (1993). Mycoherbicide formulation and the potential for bracken control. *Pesticide Science*, **37**, 337–341.

Womack, J.G. Eccleston, G.M. & Burge, M.N. (1996). A vegetable oil-based invert emulsion for mycoherbicide delivery. *Biological Control*, **6**, 23–28.

13 • Ecology and management of plants in aquatic ecosystems

STEFAN E.B. WEISNER AND JOHN A. STRAND

INTRODUCTION

Recently, research on aquatic plants has received considerable attention, both from a basic research point of view – to understand aquatic ecosystem behaviour and plant population dynamics – but also because of the importance of aquatic plants in restoration of shallow lakes, wetland management and nutrient removal from aquatic systems. Furthermore, research on aquatic plants is quite often derived from problems arising when exotic aquatic species invade ecosystems (Madsen *et al.*, 1991), or when human impact makes native species into a nuisance. During the last decade there has been a shift in focus from factors affecting macrophytes, to how macrophytes affect the ecosystem (Carpenter & Lodge, 1986; Søndergaard & Moss, 1998).

Despite a considerable amount of research, we still lack knowledge on the relative importance of mechanisms regulating and structuring aquatic plant populations, and how vegetation affects the ecosystem. Basic knowledge regarding, for example, life-history characteristics, germination requirements or dispersal ecology of different species is also often lacking. It is also still surprisingly common for researchers to ignore or not fully appreciate the influence and importance of aquatic plants in theoretical as well as applied aspects of limnology (e.g. modelling and biomanipulation).

This chapter is an attempt to outline the basic forces structuring the distribution of aquatic vegetation, and to give examples on how knowledge about structuring forces can be used in management.

Definitions

The diverse and systematically heterogeneous group that comprise aquatic plants has challenged definition and classification. The term 'macrophytes' is typically used to exclude microscopic plankton and benthic algae. The distinction between aquatic and terrestrial plants is problematic as a number of terrestrial species can live in water for a considerable part of their life cycle while some aquatic species can withstand prolonged dry periods. A useful definition of macrophytes is 'plants that normally live in water and that must spend at least some part of their life cycle in water' (Sculthorpe, 1967).

This chapter will follow Hutchinson's (1975) classification of macrophytes, based on the position of the leaves in relation to the water surface: with a division into submerged, floating-leafed and emergent species. The submerged and floating-leafed plants can then be subdivided into rooted and free-floating species.

Role of plants in aquatic ecosystems

The role of macrophytes in aquatic ecosystems is extremely diverse and it is difficult to assess their most important contribution for the ecosystem as a whole or for a specific group of organisms, without taking different spatial and temporal scales into consideration. An important aspect of aquatic vegetation is the seasonal moderation of abiotic factors (Wetzel & Søndergaard, 1998). During the growth period, nutrients derived from both the water column and interstitial water in the sediments are incorporated in the tissue of the plants in competition with other primary producers (i.e. phytoplankton and periphyton). Aquatic vegetation is exceedingly important in the regulation of the productivity and metabolism of aquatic systems (Wetzel, 1979), and freshwater plant communities are among the most productive in the world. The littoral flora has been viewed as a metabolic sieve (Mickle & Wetzel, 1978), since water from the drainage basin, containing inorganic nutrients and organic compounds, passes through the vegetation which leads to marked alterations in the chemical composition of the influent water (Wetzel, 1979).

Macrophytes affect the chemical and physical composition of the surrounding water (see Carpenter & Lodge, 1986 for review). Water movements, temperature, light climate, oxygen and carbon dioxide concentration, pH, sedimentation rates, resuspension and turbidity are affected by aquatic vegetation (e.g. Barko & James, 1998). Sediment characteristics such as redox, oxygen concentration and organic content are also strongly influenced by the presence of macrophytes.

Aquatic vegetation is also of great importance for a wide range of other organisms as a food source (fish, birds, mammals, invertebrates), breeding sites (birds, fish, molluscs, invertebrates) or refuge against predation (fish, zooplankton).

The concept of cascading trophic interactions and alternative stable states, together with the practical implementation of that knowledge (i.e. biomanipulation: Jeppesen & Sammalkorpi, volume 2), has led to an appreciation of the importance of submerged aquatic macrophytes for other organisms and on an ecosystem level (Carpenter & Lodge, 1986; Søndergaard & Moss, 1998). Indeed, much current aquatic plant research is focused on how submerged macrophytes affect phytoplankton communities directly or indirectly via modulating the higher trophic levels. Also, the effects of plants on nutrient dynamics in littoral zones, wetlands and shallow eutrophic lakes are being studied (see Perrow *et al.*, this volume).

The importance of macrophytes for the structure and functioning of most aquatic systems makes them a relevant target for management. If we understand the mechanisms that structure aquatic vegetation we greatly increase our scope for managing aquatic systems effectively.

ESTABLISHMENT

Establishment ecology of most species and groups of aquatic plants is rather poorly investigated. The sequence: dispersal – seed banks – germination – seedling establishment is often broken by gaps in knowledge, even for common species. This is perhaps caused by the difficulty in adopting experimental approaches (e.g. for dispersal investigations) and the vast number of interacting mechanisms and factors affecting each step in the establishment sequence.

Dispersal

Aquatic macrophyte species often cover a larger geographic area than terrestrial species, implying effective dispersal (Cook, 1987). Both biotic and abiotic vectors for the dispersal of diaspores (seeds, fruits and vegetative parts) exist. Abiotic vectors are split between water (hydrochoric dispersal) and wind (anemochoric dispersal), whereas biotic vectors include dispersal on (epizoochoric) or inside (endozoochoric) animals, and dispersal by humans.

Wind dispersion is rare among aquatic plants. It is thought that dispersal by wind is disadvantageous because the diaspores are more likely to be blown to terrestrial than to aquatic sites (Cook, 1987). There are, however, some common and important emergent species that are dispersed as seeds by wind, e.g. *Phragmites* and *Typha*. No submerged or free-floating species have wind-dispersed seeds. Wind can however play an important secondary role in water-dispersed seeds and blow floating material over large distances.

Water is the most common vector for dispersal of aquatic plant diaspores, both within (Smits *et al.*, 1989) and between water bodies, the latter by means of rivers (Johansson *et al.*, 1996). Flooding of wetlands (Cellot *et al.*, 1998) and tidal currents (Bakker *et al.*, 1996) are also important for water-borne dispersal. Dispersal distance for seeds spread by water varies significantly among species (Johansson *et al.*, 1996). In some species (e.g. *Zannichelia*) the diaspores sink immediately upon release, thus leading to very short dispersal distances. In contrast, other species have diaspores that float for long periods of time (more than one year), thereby greatly increasing the potential dispersal distance (Cook, 1987).

Dispersal by animals can explain the occurrence of aquatic plants in isolated water bodies (Sculthorpe, 1967). Seeds and vegetative parts stick to animals (mainly waterfowl) and may be transported over long distances (Hutchinson, 1975). Disc-shaped, slightly hydrophobic seeds in particular can readily stick to feathers and fur (Cook, 1987).

Seeds may also be abundant in the guts of waterfowl. For example, Austin *et al.* (1990) showed that 50% of the individuals of the duck *Aytha valisneria* had seeds of *Potamogeton pectinatus* in their stomachs. Seeds of some species (e.g. *Potamogeton* species) can pass through the digestive system of birds and fish without damage (de Vlaming & Proctor, 1968; Agami & Waisel, 1986; Middleton *et al.*, 1991), while other seeds (e.g. *Nuphar* and *Nymphea*) are digested (Smits *et al.*, 1989). Moreover, germination of seeds of some species is enhanced after passage through the digestive system of animals (Agami & Waisel, 1988; Smits *et al.*, 1989).

Seed banks

Most seed-bank studies have been done in terrestrial environments, but emergent species have been reasonably well investigated in aquatic environments (e.g. van der Valk & Davis, 1976, 1978, 1979; Smith & Kadlec, 1983; Leck & Simpson, 1987; Ekstam, 1995). In contrast, only few investigations on seed banks of submerged macrophytes have been undertaken (e.g. Haag, 1983; Grillas *et al.*, 1992, 1993).

Correlations between abundance of submerged species in the seed bank and abundance in the present vegetation are sometimes strong (Grillas *et al.*, 1993) but often weak or non-existent (Haag, 1983; Grillas *et al.*, 1993). The correlation is stronger for annuals than for perennials (Westcott *et al.*, 1997). Annual angiosperms, such as *Z. palustris* and *Najas flexilis*, are often abundant in the seed bank, whereas seeds of perennial submerged macrophytes are rare (Westcott *et al.*, 1997). Annuals are dependent on successful regeneration from seeds for their survival (Grime, 1979). Most perennial submerged macrophytes rely on vegetative propagation for survival and dispersal (Sculthorpe, 1967), which may explain the low occurrence in the seed bank (Haag, 1983).

Most studies on aquatic macrophyte seed banks have only concerned the top 5 or 10 cm of the sediment (e.g. Haag, 1983; Grillas *et al.*, 1993). However, the proportion of seeds in different sediment layers can have implications for long-term development of vegetation, and the possibility of re-establishment of submerged macrophytes long after the disappearance of the vegetation (Bonis & Lepart, 1994). Stochastic disturbances (e.g. heavy storms), bioturbating animals (livestock, fish, birds and invertebrates) and invasive (drastic) management actions such as suction-dredging can affect vertical seed distribution in the sediment, and possibly bring deeply buried propagules to the surface or expose a long-buried seed bank.

Germination

Factors that control and initiate germination of aquatic plants include temperature, light, oxygen concentrations and desiccation (Baskin & Baskin, 1982; Pons & Schröder, 1986; Madsen & Adams, 1988; Ekstam, 1995). Generally, vegetative diaspores have higher and faster germination rates than seeds, and dormancy (if any) is more easily broken (Rogers & Breen, 1980).

It has been shown that temperature fluctuations initiate germination in some emergent species (Ekstam, 1995) and it has been suggested that a site, water depth and time sensor operates to indicate that the seeds are located at a favourable site in spring (i.e. moist mudflats without vegetation and standing water) (Pons & Schröder, 1986; Ekstam, 1995). For submerged plants, on the other hand, it appears that fluctuating temperatures may inhibit germination (Stross, 1989). This may be an adaptation to avoid germination in very shallow water where seedling survival would be uncertain.

Germination of seeds of some submerged macrophytes (e.g. *Potamogeton* species) is light induced (Spence *et al.*, 1971). Red light (725 nm) may initiate germination (Sokol & Stross, 1986; Coble & Vance, 1987). However, under some conditions, germination is independent of light conditions (Hartleb *et al.*, 1993).

Seedling establishment

The timing of establishment is of great importance for seedling survival both for submerged (Kimbel, 1982) and emergent macrophytes (Weisner & Ekstam, 1993). Physical disturbances are more likely to have impact on seedlings than on established plants. Early establishment in the growth season, giving the plants time to reach a certain size or store energy, may be vital for survival of emergent macrophytes if the water level increases towards the end of the growth season or the next year. In submerged macrophytes, survival and growth probably are higher if the seedlings are established early during the season due to higher temperatures during summer (Barko & Smart, 1981).

REGULATION OF ESTABLISHED VEGETATION

Vegetative propagation is the dominant form of reproduction in aquatic plants (Grace, 1993) which means that, to understand how plant distribution is regulated, we have to consider two very different life stages: the establishment stage and the stage of the established vegetation. The requirements of aquatic plants generally differ completely between these two stages, representing two very different sets of environmental conditions. For example, most emergent macrophytes (e.g. *Phragmites*, *Typha*, *Scirpus*) spread within water bodies by means of clonal expansion through rhizomatous growth. Much deeper water is needed to prevent clonal expansion than to prevent seedling establishment (Ekstam & Weisner, 1991).

Abiotic regulating factors

Water depth obviously affects vegetation in aquatic habitats. For submerged macrophytes, light availability is a product of water depth and water transparency (Chambers & Kalff, 1985). Water transparency can be affected by phytoplankton (see section 'Planktonic algae' below) but can also be strongly influenced by physical factors, e.g. suspended solids or humic substances. Thus, submerged vegetation is often lacking in wind-exposed as well as humic lakes. Further, in oligotrophic lakes with a high water transparency, submerged macrophytes may be able to colonise such water depths that pressure becomes the limiting factor (Hutchinson, 1975).

For emergent macrophytes, the influence of water depth is more direct, determining the distance between the rooting medium (the substrate or sediment) and the water surface (above which light and availabilty of carbon dioxide and oxygen drastically increase). Emergent macrophytes depend on an efficient transfer of oxygen from shoots to the roots, which are often located in an oxygen-consuming substrate. The efficiency of this transfer is negatively affected by deep water (Weisner, 1988; Vretare & Weisner, 2000).

Emergent vegetation is able to expand into deeper water when rooted in a firm compared to a soft substrate (Weisner, 1991). This is probably because the anchorage ability of emergent macrophytes decreases with water depth and with substrate softness (Weisner, 1991). Thus, the distribution of emergent vegetation within a wetland or a lake can, to some degree, be predicted from water depth and substrate characteristics. This means that emergent vegetation can spread to deeper water in wave-exposed minerogenic substrates more easily than at sheltered sites with soft sediments, although a strong wave exposure may cause damage and affect the depth penetration of the vegetation (Weisner, 1991). Moreover, sediments at sheltered sites are generally highly organic with a low redox potential. This increases the need for an efficient oxygen transport from shoots to roots and may thus also contribute to limiting the maximum depth of the emergent vegetation (Vretare & Weisner, 2000).

Submerged vegetation may also exhibit better growth at relatively wave-exposed sites than at sheltered ones (Weisner *et al.*, 1997). However, the underlying mechanisms differ from those determining the distribution of emergent vegetation. Observations in shallow eutrophic lakes suggest that the hampered growth of submerged macrophytes at sheltered sites is caused by intense grazing by waterfowl and competition from epiphytic algae (Weisner *et al.*, 1997). Further, comparisons between lakes with different sediment characteristics have suggested that wave action may be an important factor uprooting plants growing in soft sediments, acting differently on different species depending on shoot architecture (Schutten & Davy, 2000).

Competition

Planktonic algae

In deep eutrophic lakes, where nutrients are less likely to limit plant growth, light is considered to be the most important factor regulating the distribution of submerged macrophytes (Spence, 1982; Hough *et al.*, 1989). The depth distribution of submerged macrophytes is mainly regulated by light attenuation through the water column (Chambers & Kalff, 1985), which, in turn, is often determined by the concentration of phytoplankton.

However, in shallow lakes, submerged vegetation exhibits an 'escape effect' where the plants may reach to, or close to, the water surface, thus avoiding a poor light climate below water (e.g. Scheffer *et al.*, 1992). The plants can thus improve their access to light by displaying plasticity in their morphology (Strand & Weisner, 2001*a*). Therefore, in shallow eutrophic lakes, the light availability in the water column may be less important than other regulating factors (Weisner *et al.*, 1997).

Epiphyton

Phillips *et al.* (1978) suggested that light inhibition from periphyton attached to the submerged macrophytes ('epiphyton') is the mechanism initiating the decline of submerged macrophytes during eutrophication. Brönmark & Weisner (1992) argued that changes in the fish community resulting from eutrophication may have cascading effects through the benthic food chain, causing a decline of submerged macrophytes through light competition from epiphyton.

Epiphyton production was shown to be higher on sheltered sites compared to more wave-exposed sites within lakes (Strand & Weisner, 1996; Weisner *et al.*, 1997). The biomass of submerged vegetation was however lower on sheltered than wave-exposed sites. These patterns suggest that a low macrophyte biomass at sheltered sites might be caused by epiphyte shading. The influence of epiphytes on macrophytes may, besides shading, involve mineral nutrient or inorganic carbon competition, and effects of increased boundary layers (Sand-Jensen & Borum, 1991). Snail grazing may alleviate the impact of epiphyton on submerged macrophytes (Brönmark & Weisner, 1992).

A neglected mechanism is nutrient competition between periphyton and phytoplankton, which is likely to increase in the presence of submerged macrophytes since the available surface area for periphyton increases (Strand & Weisner, 2001*b*). Both nutrient uptake by periphytic algae and nitrogen transformations (nitrification/denitrification) by bacteria in the periphyton layers (Eriksson & Weisner, 1999) are important mechanisms for removing nutrients from the water column. The importance of these processes for the nutrient dynamics of shallow lakes and wetlands should not be underestimated.

Other plants

In eutrophic systems, aquatic macrophytes mainly compete for light (Spence, 1982; Chambers & Kalff, 1985; Hough *et al.*, 1989), and it is obvious that an emergent life-form is an advantage in this aspect.

Competition within species or between species of the same functional group of macrophytes is more complicated and it is important to take effects of competition at different life stages into consideration. Spatial heterogeneity on a small scale in submerged macrophyte communities is probably the result of competition (Titus & Stephens, 1983; Jefferies & Rudmik, 1991). However, the observed distribution pattern may not reflect current competitive ability but rather be the result of competitive interactions in the past.

Those submerged plants that can increase their light interception by concentrating the photosynthetic biomass near or at the water surface (Adams *et al.*, 1974; Strand & Weisner, 2001*a*) may outcompete plants without this capability (Madsen *et al.*, 1991). If nutrient availability is sufficient, it is likely that weakly rooted submerged plants (e.g. *Elodea* and *Ceratophyllum*) may have an advantage over those more strongly rooted in this aspect.

'Shade plants' have been used as a biological control of submerged aquatic weeds. Species such as *Salvinia* and *Eleocharis* can effectively outcompete invasive submerged weeds (Pieterse, 1990).

Grazing

Herbivory is important in structuring aquatic ecosystems (Lodge, 1991; Cyr & Pace, 1993; Lodge *et al.*, 1998). The impact of grazing on aquatic macrophytes is dependent on the life stage of the macrophyte when grazed, and it is likely that grazing affects seedlings and younger plants more than older, well-established plants (Crawley, 1983; Nyström & Strand,

1996). It also appears that effects of grazing on submerged macrophyte biomass early in the growth period can be compounded through the growth period (Mitchell & Perrow, 1998). In emergent macrophytes, grazing can have dramatic effects if stems are broken and the gas transport between shoots and roots is interrupted (Weisner & Granéli, 1989; Vretare & Weisner, 2000). Thus, in both submerged and emergent macrophytes, the effects on the vegetation of the grazing can be dramatic even if the amount of biomass consumed by the grazers is small.

Birds

Some studies on the effects of waterfowl grazing on submerged macrophytes indicate that the effects are insignificant because grazing occurs mainly during autumn when plant growth has decreased (Kiørboe, 1980; Mitchell, 1989, Perrow *et al.*, 1997*a*). Recent studies, however, suggest that waterfowl grazing can have pronounced direct effects on submerged macrophytes in shallow eutrophic temperate lakes (Lauridsen *et al.*, 1993; Lodge *et al.*, 1998), and probably also affects the long-term vegetation dynamics (Perrow *et al.*, 1997*a*; Mitchell & Perrow, 1998). More intense waterfowl grazing at sheltered rather than at wind-exposed sites may be the reason for the generally low abundance of macrophytes found at sheltered sites in shallow, eutrophic lakes (Scheffer *et al.*, 1992; Weisner *et al.*, 1997). The grazing pressure on submerged plants is also higher closer to marginal reedbelts, probably because waterfowl such as coot (*Fulica atra*) prefer to stay close to a refuge from predators, particularly in the breeding season (Weisner *et al.*, 1997).

Fish and crayfish

Fish can negatively affect aquatic macrophytes by either direct consumption (Prejs & Jackowska, 1978; Tanner *et al.*, 1990; Lodge *et al.*, 1998) or by uprooting and increased resuspension of sediments by bottom-feeding species (ten Winkel & Meulemans, 1984; Hansson *et al.*, 1987). Macrophyte-feeding fish are common in Europe, Asia, Africa and South America but absent from North America (Lodge *et al.*, 1998).

Adult crayfish feed mostly on vegetation, detritus and seeds (Hessen & Skurdal, 1986). Both field and laboratory studies have shown that crayfish can have a negative impact on macrophyte biomass (Feminella & Resh, 1989; Nyström & Strand, 1996) and species richness (Lodge & Lorman, 1987; Lodge *et al.*, 1994). Crayfish affect macrophytes directly through consumption, but also by cutting stems near the bottom when consuming the lowermost part of the stems (Lodge & Lorman, 1987). Crayfish grazing may structure vegetation dynamics by removing young, newly established plants of both emergent and floating-leafed species, as well as severely damaging mature emergent plants to some extent (Nyström & Strand, 1996).

Other grazers

Several studies have revealed the importance of insect grazers (Lodge, 1991; Newman, 1991; Jacobsen & Sand-Jensen, 1992). Many studies on herbivory on macrophytes have focused on insects, but grazing impact of insects seems to be relatively small (Lodge *et al.*, 1998).

Snails can graze directly on the macrophytes but snails have however mainly been shown to have a positive effect on submerged macrophytes by grazing on epiphyton, thus releasing the macrophytes from light competition (Brönmark & Weisner, 1992; Underwood *et al.*, 1992)

Introductions of muskrat (*Ondanata zibethicus*) to Europe (Hengeveld, 1989) and nutria (*Myocastor coypus*) into North America (Wilsey & Chabreck, 1991) are striking examples of the potential impact of mammals on macrophyte biomass. In exclosure experiments, grazing was shown to affect biomass of perennial emergent macrophytes (*Spartina patens* and *Scirpus olneyi*) negatively while annual species (*Cyperus*) were positively affected in a North American oligohaline marsh (Taylor *et al.*, 1994). These effects were attributed to nutria which was the dominant grazer in the marsh.

MANAGEMENT

Establishing aquatic vegetation

Establishment of aquatic vegetation can be promoted in shallow water bodies through several different methods. Controlled establishment of aquatic vegetation can be used to obtain a desired vegetation

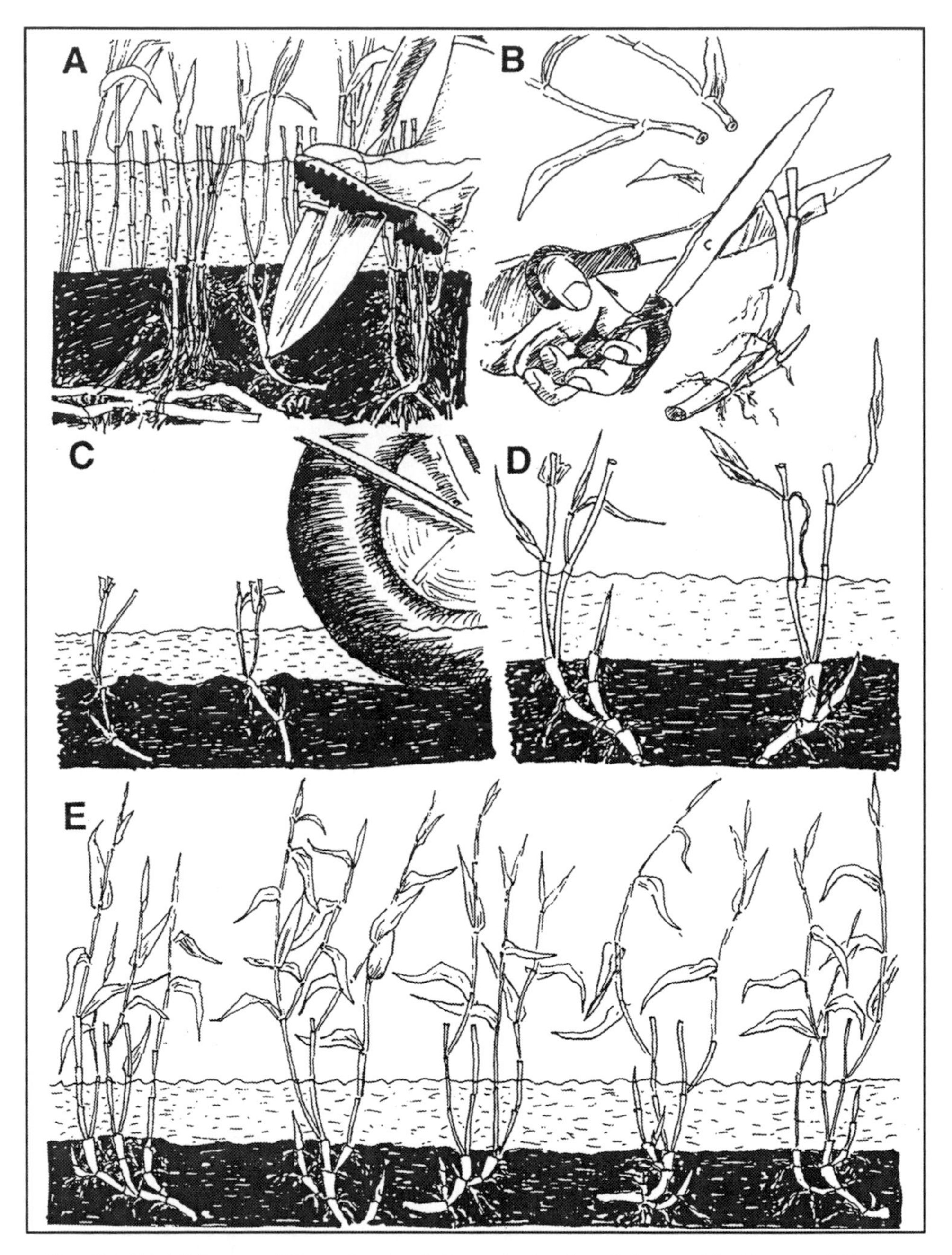

Fig. 13.1. Establishment of *Phragmites* stands through planting of shoots with rhizomes. (A) In an existing stand of *Phragmites*, young shoots with a piece of adherent rhizome are cut below the substrate surface in the beginning of the growth period. (B) Leaf area is reduced to decrease transpiration. (C) The shoots are immediately planted out at the new location. (D) New shoots will form at nodes. (E) At the end of the growth period, new shoot buds for the next growth period will form from rhizomes below the substrate surface. From Ekstam *et al.*, (1992).

structure. However, it must be remembered that the vegetation composition which gradually develops is a function of the environmental conditions. Thus, the desired vegetation composition initially obtained through successful controlled establishment may subsequently change drastically if the environmental conditions are more suitable for other species.

Emergent vegetation

The seed bank in the sediment, or plant propagules transported naturally from external sources or from internal plant populations, seem often to be sufficient for a rapid vegetation establishment if environmental conditions are right. The most common method for promoting establishment from existing sources is lowering of the water level (e.g. van der Valk & Davis, 1978). It will promote the growth of young emergent macrophytes since these are largely dependent on the physical distance from the sediment to the water surface (Weisner & Ekstam, 1993). Also, if the water level is lowered so much that the sediment surface is more or less exposed, this will induce temperature and redox changes in the sediment, promoting germination of many emergent species (Smith & Kadlec, 1983).

More controlled establishment methods include sowing and planting of precultivated plants or plant cuttings. These methods differ from the seed bank method in that we can control the plant material more closely and choose exactly what species or even genotype we want to establish. The exact designs of the methods also differ, depending on physiology and life-history characteristics of the species involved. Common reed (*Phragmites australis*) has been established in wetlands by several methods (Ekstam *et al.*, 1992). Sowing or simply spreading of fragmented panicles requires a close control of the water level since germination and successful seedling establishment depend on the water level being kept very close to the sediment surface. As the young reed plants grow the water level should be carefully increased to prevent establishment of terrestrial weeds. The use of precultivated plants speeds up the establishment. Planting of rhizomes, shoot cuttings and shoots with rhizomes has been performed with some success. Reeds have also been propagated by layering, i.e. placing shoots horizontally along the soil surface and thus promoting formation of side shoots. The multiplicity of methods used for establishing reed exemplifies how far we are from having elaborated standardised methods for the establishment of aquatic vegetation. Which method to choose is probably most often dependent on the available amount of funding, labour supply and how fast we want to achieve the desired vegetation composition (Ekstam *et al.*, 1992). However, planting of shoots with adherent rhizomes seems promising as a cost-effective method resulting in relatively fast establishment of *Phragmites* stands (Fig. 13.1).

Submerged vegetation

Decreased water depth often results in an increased light availability below water thus promoting growth of submerged macrophytes. An alternative method to increase light availability is to increase water transparency through biomanipulation (Jeppesen *et al.*, 1990; Hansson *et al.*, 1998). In theory, the removal of a large fraction of zooplanktivorous fish will lead to increased zooplankton biomass and a subsequent increase in grazing pressure on phytoplankton (Perrow *et al.*, this volume; Jeppesen & Sammalkorpi, volume 2). The ultimate goal of biomanipulation (clear water) is thus reached by altering the biomass of higher trophic levels. An underlying assumption is also that sufficient disturbance will cause the ecosystem to stabilise itself in the clearwater state through positive feedback mechanisms (Scheffer *et al.*, 1993).

Biomanipulation has been used frequently in recent decades, but a number of biomanipulation projects have not been successful or only partially successful (in effect or in time) (Jeppesen *et al.*, 1990; Perrow *et al.*, 1997*b*; Hansson *et al.*, 1998). Whether a lake will respond positively and permanently to biomanipulation seems to depend largely on the development of the submerged vegetation (Strand, 1999). In shallow, eutrophic lakes the clearwater state is largely stabilised by submerged macrophyte growth through a number of mechanisms (Scheffer, 1998). There is still some confusion about the relative importance of the various mechanisms by which submerged macrophytes stabilise the lake in the clearwater state, but the most often mentioned include: nutrient competition, periphyton–phytoplankton interactions, improved possibilities for zooplankton to avoid fish predation, decreased sediment resuspension and allelopathy (Jeppesen *et al.*, 1998).

Although the importance of re-establishment of submerged macrophytes has been emphasised for a some time, surprisingly few long-term studies have been undertaken with the aim of clarifying the structuring forces for colonisation of submerged vegetation after biomanipulation (but see Ozimek *et al.*, 1990; Lauridsen *et al.*, 1993), particularly in large lakes (Strand, 1999).

The results from two case studies in Sweden (Lake Finjasjön and Lake Ringsjön) show that submerged macrophytes may rapidly re-establish if conditions improve, even in large lakes such as Lake Finjasjön (11 km^2) (Strand & Weisner, 2001*b*), whereas no improvement at all in the submerged vegetation was observed in Lake Ringsjön (Strand, 1999).

It is important that the fish reduction is successful, so that Secchi depth increases substantially, thus allowing the submerged macrophytes to recolonise. Results from a number of biomanipulations suggest that at least 75% of the fish population should be removed if long-lasting effects on the ecosystem are to be obtained (Perrow *et al.*, 1997*b*). In Lake Finjasjön, 80% of the cyprinid fish was removed, whereas in Lake Ringsjön 49–60% was removed (Hansson *et al.*, 1998).

It has been suggested that if a lake is to be successfully restored by biomanipulation, it must be small and maximum water depth less than 4 m (Reynolds, 1994), or even 3 m (Phillips & Moss, 1994). The successful biomanipulation of Lake Finjasjön and elsewhere clearly illustrates that these restrictive values are irrelevant. Maximum and mean water depth is not as important as the percentage of the lake area that is shallow. This must also be related to a realistic Secchi depth. Thus, in Lake Finjasjön, the Secchi depth increased from 0.2–0.3 m prior to biomanipulation to more than 2 m afterwards, with a subsequent expansion of vegetation to 3 m water depth (Strand & Weisner, 2001*b*). This corresponds to roughly 30% of the lake area. In Lake Ringsjön, the increase in Secchi depth after biomanipulation (from 0.75 to 1.2 m: Hansson *et al.*, 1998) did not suffice for large areas of the lake to be covered by submerged vegetation (Strand, 1999). These areas must also be suitable for submerged macrophyte growth. Unsuitable areas include very stony substrates (Strand, 1999), very wave-exposed areas (Strand & Weisner, 1996; Chambers, 1987) as well as very sheltered areas (Strand & Weisner, 1996; Weisner *et al.*, 1997).

In cases where we cannot rely on an existing seed bank or other existing plant propagule sources (e.g. in newly created wetlands), a seed bank can be supplied using donor seed banks (van der Valk *et al.*, 1992). Vegetation has been successfully established in newly created water bodies by pumping up sediment from a nearby water body, transporting the sediment (including seeds and propagules) by truck to the receiving water body and spreading the sediment over the water surface by compressed air (Fig. 13.2). The timing of sediment/seed sampling and the subsequent spreading into the new habitat is probably important. If this is done too late into the growth season, the propagules may have entered secondary dormancy and will not germinate (at least not until next spring). Generally, vegetative propagules such as turions are more effective with a more rapid germination compared to seeds. Submerged macrophytes have also been successfully established in newly constructed wetlands simply by spreading of plant fragments over the water surface.

Control of aquatic vegetation

Aquatic weed problems have been reported from all continents of the world, and a number of different approaches have been used to reduce and control the undesired vegetation (Pieterse, 1990). Aquatic vegetation can be removed or temporarily controlled through the use of herbicides, by drawdowns killing the vegetative biomass of submerged vegetation (e.g. Poovey & Kay, 1998). Emergent vegetation can be controlled by cutting the shoots below water, thus inhibiting the transport of oxygen from shoots to the roots. In *Phragmites*, it seems that this treatment is most effective if performed in early summer and on reed growing in an organic substrate (Weisner & Granéli, 1989). Clipping of the emergent macrophytes *Ipomea aquatica* and *Paspalum distichum* below water, growing in a monsoonal wetland in India, was effective in killing

Fig. 13.2. Vegetation can be established in newly created water bodies by transferring sediment from a nearby water body. In experimental ponds in Sweden, this method resulted in rapid establishment of a number of species, including *Chara vulgaris, Potamogeton natans, P. pectinatus, P. filiformis* and *Elodea canadensis*. Data was kindly provided by Börje Ekstam and Karin Tonderski. Photo by Susanne Thuresson.

the plants (Middleton, 1990). Similar clipping was however less effective on a floating-leaved species, *Nymphoides christatum*.

However, to obtain a more long-term effect we must control environmental or biological factors that regulate the vegetation. If water depth can be increased sufficiently, this is generally a very effective measure to prevent expansion of emergent vegetation (Björk, 1988).

The biological control of aquatic weeds can be divided into two classes: (1) the use of competitive plant species or (2) the use of organisms that directly attack the weeds (Pieterse, 1990). Grazers are most common but fungal infections have also been used (Charudattan, 1990). Grazers can be selective, thus only attacking the weed species, or non-selective thus attacking all species present.

Selective species are, for example, the beetle *Agasicles hygrophila* (against alligator weed, *Alternanthera philoxeroides*) or the weevils *Neochetina eichhorniae* (against water hyacinth, *Eichhornia crassipes*) and *Cyrtobagus salviniae* (against *Salvinia molesta*) (Pieterse, 1990).

The most important non-selective organism for the control of aquatic weeds is grass carp (*Ctenopharyngon idella*), which has been used to eradicate all plants in aquatic systems, thus improving conditions for the indigenous species that can re-establish from the seed banks as opposed to invasive species that often only spread vegetatively, e.g. *Egeria densa* in New Zealand (Tanner *et al*., 1990). Grass carp has been used to control vegetation in water bodies in many countries around the world, e.g. in channel systems in Egypt, and it has also been tested for use in irrigation channels in Argentina (Dall Armellina *et al*., 1999).

However, the results have not always been satisfactory. One problem is that eradication of the macrophytes may result in algal blooms. Management of vegetation in wetlands traditionally often includes grazing by domestic animals, preventing dominance by a few competitive species (Rosenzweig, 1995). Overgrowth of shallow lakes with reed vegetation has caused a decline in the value of these lakes for nature conservation and biodiversity (Björk, 1988). These changes have often been attributed to eutrophication of the lakes. However, the historical development of the vegetation in shallow eutrophic lakes in southern Sweden illustrates that decreased grazing by domestic animals in combination with lowered water levels to obtain more arable land were the causes behind the degradation of these ecosystems. Reintroduction of grazing by domestic animals has restored biological values in lakes with excessive growth of reed vegetation.

CONCLUDING REMARKS

The central role of macrophytes for the functioning of aquatic systems means that the most effective way of managing these systems is often through vegetation management. This requires an understanding of the mechanisms regulating the distribution of vegetation. For submerged vegetation these are mainly water depth, water transparency and epiphytic growth. The distribution of emergent vegetation can largely be predicted from water depth and substrate characteristics. Moreover, the effects of grazing can be dramatic for both submerged and emergent macrophytes.

Management should aim at providing environmental conditions favouring the desired ecosystem state, rather than methods directly aimed at the vegetation. For example, the best method for promoting establishment of emergent vegetation is often lowering of the water level. To establish submerged vegetation, water transparency can be increased through biomanipulation (the removal of zooplanktivorous fish leading to increased zooplankton grazing pressure on phytoplankton). Conversely, in the case of nuisance vegetation (aquatic weeds) changes in water depth and introduction of grazers are often effective control measures.

ACKNOWLEDGMENTS

This study was done within the Swedish Water Management Research Program (VASTRA) which is financed by the Swedish Foundation for Strategic Environmental Research (MISTRA).

REFERENCES

Adams, M.S., Titus, J. & McCracken, M.D. (1974). Depth distribution of photosynthetic activity in a *Myriophyllum spicatum* community in Lake Wingra. *Limnology and Oceanography*, **19**, 377–389.

Agami, M. & Waisel, Y. (1986). The role of mallard ducks (*Anas platyrhynchos*) in distribution and germination of seeds of the submerged hydrophyte *Najas marina* L. *Oecologia*, **68**, 473–475.

Agami, M. & Waisel, Y. (1988). The role of fish in distribution and germination of seeds of the submerged macrophyte *Najas marina* L. and *Ruppia maritima* L. *Oecologia*, **76**, 83–88.

Austin, J.E., Serie, J.R & Noyes, J.H. (1990). Diet of canvasbacks during breeding. *Prairie Naturalist*, **22**, 171–176.

Bakker, J.P., Poschlod, P., Strykstra, R.J., Bekker, R.M. & Thompson, K. (1996). Seed banks and seed dispersal: important topics in restoration ecology. *Acta Botanika Neerlandica*, **45**, 461–490.

Barko, J.W. & James, W.F. (1998). Effects of submerged aquatic macrophytes on nutrient dynamics, sedimentation, and resuspension. In *The Structuring Role of Submerged Macrophytes in Lakes*, eds. E. Jeppesen, M. Søndergaard, M. Søndergaard, K. Christofferson, pp. 197–214. New York: Springer-Verlag.

Barko, J.W. & Smart, R.M. (1981). Comparative influences of light and temperature on the growth and metabolism of selected submersed macrophytes. *Ecological Monographs*, **5**, 219–235.

Baskin, J.M. & Baskin, C.C. (1982). Effects of wetting and drying cycles on the germination of seeds of *Cyperus inflexus*. *Ecology*, **63**, 248–252.

Björk, S. (1988). Redevelopment of lake ecosystems: a case-study approach. *Ambio*, **17**, 90–98.

Bonis, A. & Lepart, J. (1994). Vertical structure of seed banks and the impact of depth of burial on recruitment in two temporary marshes. *Vegetatio*, **112**, 127–139.

Brönmark, C. & Weisner, S.E.B. (1992). Indirect effects of fish community strucure on submerged vegetation in shallow eutrophic lakes: an alternative mechanism. *Hydrobiologia*, **243/244**, 293–301.

Carpenter, S.R. & Lodge, D.M. (1986). Effects of submersed macrophytes on ecosystem processes. *Aquatic Botany*, **26**, 341–370.

Chambers, P.A. (1987). Nearshore occurrence of submersed aquatic macrophytes in relation to wave action. *Canadian Journal of Fisheries and Aquatic Sciences*, **44**, 1666–1669.

Chambers, P.A. & Kalff, J. (1985). Depth distribution and biomass of submersed aquatic macrophyte communities in relation to secchi depth. *Canadian Journal of Fisheries and Aquatic Sciences*, **42**, 701–709.

Charudattan, R. (1990). Biological control of aquatic weeds by means of fungi. In *Aquatic Weeds: The Ecology and Management of Nuisance Aquatic Vegetation*, eds. A.H. Pieterse & K.J. Murphy, pp. 186–200. Oxford: Oxford Science Publications.

Cellot, B., Mouillot, F. & Henry, C.P. (1998). Flood drift and propagule bank of aquatic macrophytes in a riverine wetland. *Journal of Vegetation Science*, **9**, 631–640.

Coble, T.A. & Vance, B.D. (1987). Seed germination in *Myriophyllum spicatum* L. *Journal of Aquatic Plant Management*, **25**, 8–10.

Cook, C.D.K. (1987). Dispersion in aquatic and amphibious vascular plants. In *Plant Life in Aquatic and Amphibious Habitats*, ed. R.M.M. Crawford, pp. 179–190. Oxford: Blackwell.

Crawley, M.J. (1983). *Herbivory: The Dynamics of Animal–Plant Interactions*. Oxford: Blackwell.

Cyr, H. & Pace, M.L. (1993). Magnitude and pattern of herbivory in aquatic and terrestrial ecosystems. *Nature*, **361**, 148–150.

Dall Armellina, A.A., Bezic, C.R. & Gajardo, O.A. (1999). Submerged macrophyte control with herbivorous fish in irrigation channels of semiarid Argentina. *Hydrobiologia*, **415**, 265–269.

de Vlaming, V. & Proctor, V.W. (1968). Dispersal of aquatic organisms: viability of seeds recovered from the droppings of captive killdeer and mallard ducks. *American Journal of Botany*, **55**, 20–26.

Ekstam, B. (1995). Regeneration traits of emergent clonal plants in aquatic habitats. PhD thesis, Lund University, Sweden.

Ekstam, B. & Weisner, S.E.B. (1991). Dynamics of emergent vegetation in relation to open water of shallow lakes. In *Wetlands Management and Restoration*, Workshop eds. C.M. Finlayson & T. Larsson, pp. 56–64. Stockholm: Swedish Environmental Protection Agency.

Ekstam, B., Granéli, W. & Weisner, S. (1992). Establishment of reedbeds. In *Reedbeds for Wildlife*, ed. D. Ward, pp. 3–19. Bristol, UK: University of Bristol.

Eriksson, P.G. & Weisner, S.E.B. (1999). An experimental study on effects of submersed macrophytes on nitrification and denitrification in ammonium-rich aquatic systems. *Limnology and Oceanography*, **44**, 1993–1999.

Feminella, J.W. & Resh, V.R. (1989). Submersed macrophytes and grazing crayfish: an experimental study of herbivory in a Californian freshwater marsh. *Holarctic Ecology*, **12**, 1–8.

Grace, J.B. (1993). The adaptive significance of clonal reproduction in angiosperms: an aquatic perspective. *Aquatic Botany*, **44**, 159–180.

Grillas, P., van Wijk, C. & Boy, V. (1992). Transferring sediment containing intact seed banks: a method for studying plant community ecology. *Hydrobiologia*, **228**, 29–36.

Grillas, P., Garcia-Murillo, O., Geertz-Hansen, N., Montes, C., Duarte, C.M., Tan Ham, L. & Grossman, A. (1993). Submerged macrophyte seed bank in a Mediterranean temporary marsh: abundance and relationship with established vegetation. *Oecologia*, **94**, 1–6.

Grime, J.P. (1979). *Plant Strategies and Vegetation Processes*. New York: John Wiley.

Haag, R.W. (1983). Emergence of seedling of aquatic macrophytes from lake sediment. *Canadian Journal of Botany*, **61**, 148–156.

Hansson, L.-A., Johansson, L. & Persson, L. (1987). Effects of fish grazing on nutrient release and succession of primary producers. *Limnology and Oceanography*, **32**, 723–729.

Hansson, L.-A., Annadotter, H., Bergman, E., Hamrin, S.F., Jeppesen, E., Kairesalo, T., Luokkanen, E., Nilsson, P.-Å., Søndergaard, M. & Strand, J.A. (1998). Biomanipulation as an application of food-chain theory: constraints, synthesis, and recommendations for temperate lakes. *Ecosystems*, **1**, 558–574.

Hartleb, C.F., Madsen, J.D. & Boylen, C.W. (1993). Environmental factors affecting seed germination in *Myriophyllum spicatum* L. *Aquatic Botany*, **45**, 15–25.

Hengeveld, R. (1989). *Dynamics of Biological Invasion*. New York: Chapman & Hall.

Hessen, D.O. & Skurdal, J. (1986). Analysis of food utilized by the crayfish *Astacus astacus* in Lake Steinsfjorden. *Freshwater Crayfish*, **6**, 187–193.

Hough, R.A., Fornwall, M.D., Negele, B.J., Thompson, R.L. & Putt, D.A. (1989). Plant community dynamics in a chain of lakes: principal factors in the decline of rooted macrophytes with eutrophication. *Hydrobiologia*, **173**, 199–217.

Hutchinson, G.E. (1975). *A Treatise on Limnology*, vol. 3, *Limnological Botany*. New York: John Wiley.

Jacobsen, D. & Sand-Jensen, J.P. (1992). Herbivory of invertebrates on submerged macrophytes from Danish freshwaters. *Freshwater Biology*, **28**, 301–308.

Jefferies, R.L. & Rudmik, T. (1991). Growth, reproduction and resource allocation in halophytes. *Aquatic Botany*, **39**, 3–16.

Jeppesen, E., Jensen, J.P., Kristensen, P., Søndergaard, M., Sortkjœr, O. & Olrik, K. (1990). Fish manipulation as a lake restoration tool in shallow, eutrophic, temperate lakes. 2: Threshold levels, long-term stability and conclusions. *Hydrobiologia*, **200/201**, 219–227.

Jeppesen, E., Søndergaard, M., Søndergaard, M. & Christoffersen, K. (eds.) (1998). *The Structuring Role of Submerged Macrophytes in Lakes*. New York: Springer-Verlag.

Johansson, M.E., Nilsson, C. & Nilsson, E. (1996). Do rivers function as corridors for plant dispersal? *Journal of Vegetational Science*, **7**, 593–598.

Kimbel, J.C., (1982). Factors influencing potential intralake colonisation of *Myriophyllum spicatum* L. *Aquatic Botany*, **14**, 295–307.

Kiørboe, T. (1980). Distribution and production of submerged macrophytes in Tipper Grund (Ringkøbing Fjord, Denmark), and the impact of waterfowl grazing. *Journal of Applied Ecology*, **17**, 675–688.

Lauridsen, T.L., Jeppesen, E. & Østergard Andersen, F. (1993). Colonisation of submerged macrophytes in shallow fish manipulated Lake Væng: impact of sediment composition and waterfowl grazing. *Aquatic Botany*, **46**, 1–15.

Leck, M.A. & Simpson, R.L. (1987). Seed bank of a freshwater tidal wetland: turnover and relationship to vegetation change. *American Journal of Botany*, **74**, 360–370.

Lodge, D.M. (1991). Herbivory on freshwater macrophytes. *Aquatic Botany*, **41**, 195–224.

Lodge, D.M. & Lorman, J.G. (1987). Reductions in submerged macrophyte biomass and species richness by the crayfish *Orconectes rusticus*. *Canadian Journal of Fisheries and Aquatic Sciences*, **44**, 591–597.

Lodge, D.M., Kershner, M.W., Aloi, J.E. & Covich, A.P. (1994). Effects of an omnivorous crayfish (*Orconectes rusticus*) on a freshwater littoral food web. *Ecology*, **75**, 1265–1281.

Lodge, D.M., Cronin, G., van Donk, E. & Froelich, A.J. (1998). Impact of herbivory on plant standing crop: comparisons among biomes, between vascular and nonvascular plants, and among freshwater herbivore taxa. In *The Structuring Role of Submerged Macrophytes in Lakes*, eds. E. Jeppesen, M. Søndergaard, M. Søndergaard & K. Christofferson, pp. 149–174. New York: Springer-Verlag.

Madsen, J.D. & Adams, M.S. (1988). The germination of *Potamogeton pectinatus* tubers: environmental control by temperature and light. *Canadian Journal of Botany*, **66**, 2523–2526.

Madsen, J.D., Sutherland, J.W., Bloomfield, J.A., Eichler, L.W. & Boylen, C.W. (1991). The decline of native vegetation under dense Eurasian watermilfoil canopies. *Journal of Aquatic Plant Management*, **29**, 94–99.

Mickle, A.M. & Wetzel, R.G. (1978). Effectiveness of submersed angiosperm–epiphyte complexes on exchange of nutrients and organic carbon in littoral systems. 1: Inorganic nutrients. *Aquatic Botany*, **4**, 303–316.

Middleton, B.A. (1990). Effect of water depth and clipping frequency on the growth and survival of four wetland plant species. *Aquatic Botany*, **37**, 189–196.

Middleton, B.A., van der Valk, A.G., Mason, D.H., Williams, R.L. & Davis, C.B. (1991). Vegetation dynamics and seed bank of a monsoonal wetland overgrown with *Paspalum distichum* L. in northern India. *Aquatic Botany*, **40**, 239–259.

Mitchell, S.F. (1989). Primary production in a shallow eutrophic lake dominated alternatively by phytoplankton and by submerged macrophytes. *Aquatic Botany*, **33**, 101–110.

Mitchell, S.F. & Perrow, M.R. (1998). Interactions between grazing birds and macrophytes. In *The Structuring Role of Submerged Macrophytes in Lakes*, eds. E. Jeppesen, M. Søndergaard, M. Søndergaard & K. Christofferson, pp. 175–196. New York: Springer-Verlag.

Newman, R.M. (1991). Herbivory and detrivory of freshwater macrophytes by invertebrates: a review. *Journal of the North American Benthological Society*, **10**, 89–114.

Nyström, P. & Strand, J.A. (1996). Crayfish grazing on aquatic macrophytes, and the importance of macrophyte life stage, temperature and crayfish species. *Freshwater Biology*, **36**, 673–682.

Ozimek, T., Gulati, R.D. & van Donk, E. (1990). Can macrophytes be useful in biomanipulation of lakes? The Lake Zwemlust example. *Hydrobiologia*, **200/201**, 399–407.

Perrow, M.R., Schutten, J.H., Howes, J.R., Holzer, T., Madwick, F.J. & Jowitt, A.J.D. (1997*a*). Interactions between coot (*Fulica atra*) and submerged macrophytes: the role of birds in the restoration process. *Hydrobiologia*, **342/343**, 241–255.

Perrow, M.R., Meijer, M.-L., Dawidowicz, P. & Coops, H. (1997*b*). Biomanipulation in shallow lakes: state of the art. *Hydrobiologia*, **342/343**, 355–365.

Phillips, G.L. & Moss, B. (1994). *Is Biomanipulation a Useful Technique in Lake Management? A Literature Review.* Bristol, UK: National Rivers Authority.

Phillips, G.L., Eminson, D. & Moss, B. (1978). A mechanism to account for macrophyte decline in progressively eutrophicated freshwaters. *Aquatic Botany*, **4**, 103–126.

Pieterse, A.H. (1990). Biological control of aquatic weeds. In: *Aquatic Weeds: The Ecology and Management of Nuisance Aquatic Vegetation*, eds. A.H. Pieterse & K.J. Murphy, pp. 174–177. Oxford: Oxford Science Publications.

Pons, T.L. & Schröder, H.F.J.M. (1986). Significance of temperature fluctuations and oxygen concentration for germination of the rice field weeds *Fimbristylis littoralis* and *Scirpus juncoides*. *Oecologia*, **68**, 315–319.

Poovey, A.G. & Kay, S.H. (1998). The potential of a summer drawdown to manage monoecious *Hydrilla*. *Journal of Aquatic Plant Management*, **36**, 127–130.

Prejs, A. & Jackowska, H. (1978). Lake macrophytes as the food of roach (*Rutilus rutilus* L.) and rudd (*Scardinus erythrophthalmus* L.) 1: Species composition and dominance relations in the lake and food. *Ekologia Polska*, **26**, 429–438.

Reynolds, C.S. (1994). The ecological basis for the successful biomanipulation of aquatic communities. *Archiv für Hydrobiologie*, **130**, 1–33.

Rogers, K.H. & Breen, C.M. (1980). Growth and reproduction of *Potamogeton crispus* in a South African lake. *Journal of Ecology*, **68**, 561–571.

Rosenzweig, M.L. (1995). *Species Diversity in Space and Time.* Cambridge: Cambridge University Press.

Sand-Jensen, K. & Borum, J. (1991). Interactions among phytoplankton, periphyton and macrophytes in temperate freshwaters and estuaries. *Aquatic Botany*, **41**, 137–175.

Scheffer, M. (1998). *Ecology of Shallow Lakes*. London: Chapman & Hall.

Scheffer, M., de Redelijkheid, M.R. & Noppert, F. (1992). Distribution and dynamics of submerged vegetation in a chain of shallow eutrophic lakes. *Aquatic Botany*, **42**, 199–216.

Scheffer, M., Hosper, S.H., Meijer, M.-L., Moss, B. & Jeppesen, E. (1993). Alternative equilibria in shallow lakes. *Trends in Ecology and Evolution*, **8**, 275–279.

Schutten, J. & Davy, A.J. (2000). Predicting the hydraulic forces on submerged macrophytes from current velocity, biomass and morphology. *Oecologia*, **123**, 445–452.

Sculthorpe, C.D. (1967). *The Biology of Aquatic Vascular Plants*. London: Edward Arnold.

Smith, L.M. & Kadlec, J.A. (1983). Seed banks and their role during drawdown of a North American marsh. *Journal of Applied Ecology*, **20**, 673–684.

Smits, A.J.M., van Ruremonde, R. & van der Velde, G. (1989). Seed dispersal of three nymphaeid macrophytes. *Aquatic Botany*, **35**, 167–180.

Sokol, R.C. & Stross, R.G. (1986). Annual germination window in oospores of *Nitella furcata* (Charophyceae). *Journal of Phycology*, **22**, 403–406.

Spence, D.H.N. (1982). The zonation of plants in freshwater lakes. *Advances in Ecological Research*, **12**, 37–125.

Spence, D.H.N., Milburn, T.R., Ndawula-Senuimba, M. & Roberts, E. (1971). Fruit biology and germination of two tropical *Potamogeton* species. *New Phytologist*, **70**, 197–212.

Strand, J.A. (1999). Development of submerged macrophytes in Lake Ringsjön after biomanipulation. *Hydrobiologia*, **404**, 113–121.

Strand, J.A. & Weisner, S.E.B. (1996). Wave exposure related growth of epiphyton: implications for the distribution of submerged macrophytes in eutrophic lakes. *Hydrobiologia*, **325**, 113–119.

Strand, J.A. & Weisner, S.E.B. (2001*a*). Morphological plastic responses to water depth and wave exposure in an aquatic plant (*Myriophyllum spicatum*). *Journal of Ecology*, **89**, 166–175.

Strand, J.A. & Weisner, S.E.B. (2001*b*). Dynamics of submerged macrophyte populations in response to biomanipulation. *Freshwater Biology*, **46**, 1397–1408.

Stross, R.G. (1989). The temporal window of germination in oospores of *Chara* (Charophyceae) following primary dormancy in the laboratory. *New Phytologist*, **113**, 491–495.

Søndergaard, M. & Moss, B. (1998). Impact of submerged macrophytes on phytoplankton in freshwater lakes. In *The Structuring Role of Submerged Macrophytes in Lakes*, eds. E. Jeppesen, M. Søndergaard, M. Søndergaard & K. Christofferson, pp. 115–133. New York: Springer-Verlag.

Tanner, C.C., Wells, R.D.S. & Mitchel, C.P. (1990). Re-establishment of native macrophytes in Lake Parkinson following weed control by grass carp. *New Zealand Journal of Marine and Freshwater Research*, **24**, 181–186.

Taylor, J.N., Grace, J.B., Guntenspergen, G.R. & Foote, E.L. (1994). The interactive effects of herbivory and fire on an oligohaline marsh, Little Lake, Louisiana, USA. *Wetlands*, **14**, 82–87.

ten Winkel, E.H. & Meulemans, J.T. (1984). Effects of fish upon submerged vegetation. *Hydrobiological Bulletin*, **18**, 157–158.

Titus, J.E. & Stephens, M.D. (1983). Neighbor influences and seasonal growth patterns for *Vallisneria americana* in a mesotrophic lake. *Oecologia*, **56**, 23–29.

Underwood, G.J.C., Thomas, J.D. & Baker, J.H. (1992). An experimental investigation of interactions in snail–macrophyte–epiphyte systems. *Oecologia*, **91**, 587–595.

van der Valk, A.G. & Davis, C.B. (1976). The seed bank of prairie glacial marshes. *Canadian Journal of Botany*, **54**, 1832–1838.

van der Valk, A.G. & Davis, C.B. (1978). The role of seed banks in the vegetation dynamics of prairie glacial marshes. *Ecology*, **59**, 322–335.

van der Valk, A.G. & Davis, C.B. (1979). A reconstruction of the recent vegetational history of a praire marsh, Eagle Lake, Iowa, from its seed bank. *Aquatic Botany*, **6**, 29–51.

van der Valk, A.G., Pederson, R.L. & Davis, C.B. (1992). Restoration and creation of freshwater wetlands using seed banks. *Wetlands Ecology and Management*, **1**, 191–197.

Vretare, V. & Weisner, S.E.B. (2000). Influence of pressurised ventilation on performance of an emergent macrophyte (*Phragmites australis*). *Journal of Ecology*, **88**, 978–987.

Weisner, S.E.B. (1988). Factors affecting the internal oxygen supply of *Phragmites australis* (Cav.) Trin. ex Steudel in situ. *Aquatic Botany*, **31**, 329–335.

Weisner, S.E.B. (1991). Within-lake patterns in depth penetration of emergent vegetation. *Freshwater Biology*, **26**, 133–142.

Weisner, S.E.B. & Ekstam, B. (1993). Influence of germination time on juvenile performance of *Phragmites australis* on temporarily exposed bottoms: implications for the colonization of lake beds. *Aquatic Botany*, **45**, 107–118.

Weisner, S.E.B. & Granéli, W. (1989). Influence of substrate conditions on the growth of *Phragmites australis* after a reduction in oxygen transport to below-ground parts. *Aquatic Botany*, **35**, 71–80.

Weisner, S.E.B., Strand, J.A. & Sandsten, H. (1997). Mechanisms regulating abundance of submerged vegetation in shallow eutrophic lakes. *Oecologia*, **109**, 592–599.

Westcott, K., Whiles, T.H. & Fox, M.G. (1997). Viability and abundance of seeds of submerged macrophytes in the sediment of disturbed and reference shoreline marshes in Lake Ontario. *Canadian Journal of Botany*, **75**, 451–456.

Wetzel, R.G. (1979). The role of the littoral zone and detritus in lake metabolism. *Archive für Hydrobiologie*, **13**, 145–161.

Wetzel, R.G. & Søndergaard, M. (1998). Role of submerged macrophytes for the microbial community and dynamics of dissolved organic carbon in aquatic ecosystems. In *The Structuring Role of Submerged Macrophytes in Lakes*, eds. E. Jeppesen, M. Søndergaard, M. Søndergaard & K. Christofferson, pp. 133–148. New York: Springer-Verlag.

Wilsey, B.J. & Chabreck, R.H. (1991). Nutritional quality of nutria diets in three Louisiana wetland habitats. *Northeast Gulf Science*, **12**, 67–72.

14 • Micro-organisms

MICHAEL F. ALLEN, DAVID A. JASPER AND JOHN C. ZAK

INTRODUCTION

Most restoration ecologists have concerned themselves with the re-establishment of desirable plants and animals but generally overlook the resource base upon which the animal and plant communities are maintained, the soils. Reconstructing the soils is difficult because a soil comprises both organic and inorganic components that have weathered *in situ* for a very long time. As such, despite the best efforts, soils that have been reconstructed always contain early-successional characteristics (see Marrs, this volume). In most situations, the plants that colonise early-successional sites and are capable of surviving on these soils are not those that are considered desirable for restoration 'success'. Restoration managers are primarily responsible for placing late seral vegetation into early seral soils.

Existing at the interface of plant and soil are soil organisms: the most diverse group of organisms at any site. Current estimates using molecular diversity measurements suggest that there are from 4 000 to 40 000 distinct genetic populations of soil organisms per gram of soil, of which only less than 5% potentially may be cultured (e.g. Collins *et al.*, 1995; Allen *et al.*, in press)! No accurate estimates of soil organism richness across a landscape exist but it certainly is in the thousands to millions. These organisms change with successional time (many form highly specific associations with a limited number of plant species) in addition to their activities being altered due to discontinuous resource inputs (in the form of organic matter) and soil structural and compositional changes. Clearly, one cannot 'restore' in the theoretical sense the microbial communities that existed prior to disruption.

Despite the difficulty of reconstructing the composition of soil organisms, restoring the functions catalysed by soil organisms is essential to restoration success. Virtually all species comprising mid and late seral vegetation are facultatively or obligately mycorrhizal. Early or mid seral plants are often nitrogen-fixing and thus inject a large fraction of the nitrogen that is continually recycled in late seral ecosystems. Establishing an ecosystem that cycles nutrients intrabiotically and turns over organic matter appropriately requires a complete soil trophic structure with both immobilising microflora and grazing soil fauna. Indeed, in most 'stable' ecosystems, the vast majority of energy flows not through the grazing vertebrates or even insects, but through belowground food webs that recycle the majority of nutrients required for new plant production (Allen *et al.*, 1999).

Although there is likely a high degree of redundancy amongst soil micro-organisms, much of that 'redundancy' is due to our inability to measure accurately the rates of the critical ecosystem processes (e.g. Collins *et al.*, 1995). Just as most plants photosynthesise (a single important function), so do most mycorrhizal fungi enhance phosphorus uptake. However, as differing plants photosynthesise at differing rates and under differing conditions, so do differing mycorrhizal fungi have different abilities to enhance phosphorus uptake, as well as perform a large diversity of other physiological processes important to the growth and survival of a plant (Allen, 2000). Nitrogen-fixing prokaryotes represent a highly diverse group of organisms ranging from symbiotic bacteria (such as *Rhizobium*) to actinomycetes (*Frankia*) and cyanobacteria (*Azolla*). All of the differences in rates and conditions are probably the result of the characteristics of the organisms separate from their nitrogen-fixing capacity. Therefore, it is critical to recognise that there is not just a black box of 'soil organisms' but that this group comprises an extremely highly diverse collection of taxa each requiring different resources and processing available resources differently. In short, each species of these organisms has a separate niche.

What are the organisms comprising this component? The soil community consists of organisms from virtually all phyla. On a simple count basis, the prokaryotes, primarily eubacteria, are predominant. One can expect to find 10^6 to 10^9 culturable colony-forming units of these organisms and a diversity of several thousand taxa per gram of dry soil. Given certain conditions such as extreme salinity, pH, temperatures, archaebacteria may also be present. Actinomycetes and cyanobacteria are also common in soils. Various protists can also be present including slime moulds and algae. Probably the largest biomass group in most ecosystems (excluding the plants themselves) are the fungi. In highly organic soils, there can be up to 500 m or approximately 25 mg of fungal hyphae per gram of dry soil. Christensen (1981) noted that a species increment curve of culturable fungi for a single site continued to increase well after 1000 isolates. Finally, there is a wide diversity of soil animals that graze on the plants directly or on the micro-organisms living in the soil. These vary highly among sites and biomes but primarily consist of nematodes, mites, collembolans and other arthropods (see Majer *et al.*, this volume).

The soil organisms can also be divided into free-living organisms and those organisms living symbiotically with plants. We use the original definition for symbiotic, that is an organism living in intimate association with a host (deBary, 1887). Each of the different groups utilises differing resources provided by the dominant organismal biomass group, the primary producers. The free-living organisms are primarily responsible for decomposition, some nitrogen fixation, and most nitrogen transformations. They may also be responsible for the mineralisation of many essential elements including phosphorus, calcium and iron. Symbiotic organisms include both mutualists and parasites. Important mutualistic groups include nitrogen-fixing bacteria, mycorrhizal fungi and lichens. Parasites include fungal and bacterial parasites which can reduce production and plant survival. Plant parasitic nematodes are common and can consume as much as 25% of the net primary production (Stanton *et al.*, 1981).

In summary, soil micro-organisms are the most diverse group of organisms encountered when attempting to restore a site. The vast majority cannot be identified, cultured or replaced. One must set up the appropriate conditions for restoration and allow colonisation to occur naturally.

Despite our inability to quantify soil organisms and directly restore them, we can add many critical groups artificially and can manipulate the conditions important for the establishment of others. The remainder of this chapter will be to assess the nature of disturbance and its effects on soil organisms, to describe the importance of different groups of organisms to the restoration process, and finally to describe remedies that may be undertaken to restore soil organisms and the processes that they catalyse. It is important that the reader remember that each site has its own distinctive grouping of these organisms; thus we can only provide general recipes that include groups of organisms, rarely species. It is also important to restore a diversity of species that perform similar tasks, as the conditions and rates of different species for a given function can vary dramatically.

FUNCTIONAL GROUPS OF SOIL MICRO-ORGANISMS

With the other chapters of this volume, little effort must be made to describe the organisms that are of concern. However, few ecologists and land managers recognise the incredible diversity of organisms that are present in soils. In order to acquaint the reader with the diversity of processes of concern, we describe briefly some of the organisms that catalyse the important processes in soils. The reader should understand that this is merely a brief overview and we recommend basic texts in soil microbiology (e.g. Paul & Clark, 1996) and other volumes pertaining to the particular groups of interest.

Primary producers

Although rarely considered, many soil micro-organisms are important contributors to the primary production of a site. Cyanobacteria and green algae as well as mosses and other cryptogams can be found in many soil surfaces (Eldridge & Rosentreter, 1999). To our knowledge, no estimates have been made of their contribution to the overall carbon balance of a restored site but they can fix large amounts of carbon (e.g. Lange *et al.*, 1998). Following precipitation events, almost any soil can be found to turn a slight to dark green colour, indicative of cyanobacteria or green algae. In arid regions, these organisms may be the predominant form of carbon input important for both carbon and nitrogen dynamics and for control of soil erosion (Belnap & Gillette, 1998; Evans & Belnap, 1999).

Lichens, important for weathering rocks, are a symbiosis between fungi and either cyanobacteria or green algae. Little is known of their specific roles in restoring any site but they may be extremely important in many areas. In the Arctic, reindeer lichens can be the dominant primary producers and are critical for caribou (*Rangifer tarandus*) and other animals (Nash, 1996). In the tropics, lichens can form dense canopy epiphyte cover providing up to 100 kg of carbon to the ecosystem (Forman, 1975). Again, to our knowledge, they have never been assessed for their importance in restoration dynamics.

Nitrogen-cycling microbes

The importance of re-establishing a conservative nitrogen cycle in any restoration effort cannot be underestimated. The importance of nitrogen-fixing organisms in early-successional ecosystems is well known (e.g. Sprent, 1987). Nitrogen-fixing micro-organisms can be divided into symbiotic and asymbiotic fixers. Symbiotic nitrogen-fixing organisms tend to be of major importance to ecosystem functioning, as these associations are the most effective nitrogen-fixing systems. In order for nitrogen-fixation to occur, nitrate concentrations must be low and the localised conditions must be anaerobic. Plants, such as legumes, form specific structures in response to the presence of the bacteria that scavenge oxygen, using an enclosed unit (the nodule) with a circulating compound, leghaemoglobin, which binds and removes oxygen from within the nodules. The plant is also required to provide energy directly, an important function, as nitrogen-fixation is extremely energy-expensive.

Of the symbiotic fixers, there are two primary types. The best known are the legume–rhizobial associations. These associations include members of the legume family and several genera of bacteria including *Rhizobium*, *Bradyrhizobium* and *Azorhizobium*. Importantly, the taxonomy of the Rhizobiaceae has traditionally been based on the host range, e.g. *Rhizobium meliloti* (associating with *Melilotus*, and *Medicago*). It is now known that the gene that regulates the host range, the *nod* gene, is carried on a plasmid that is transferable amongst the differing strains (Atlas & Bartha, 1993). The importance of this fact is that the traditional taxonomy is not viable and we must learn the new basis of taxonomy for these critical groups. These bacteria form symbiotic associations with legumes that nodulate. The second group are the actinorhizal associations. These are symbioses between actinomycetes in the genus *Frankia* and a wide array of host plants, primarily shrubs or small trees in several families. Common plants include those in the families Casuarinaceae, Myricaceae, Betulaceae, Eleagnaceae, Rhamnaceae and Rosaceae (e.g. Sprent, 1987).

There is a large variety of asymbiotic prokaryotes capable of nitrogen-fixation, given appropriate conditions. These include organisms such as *Azospirillum* and *Azotobacter*, which commonly live in a loose association with grass roots, sulphur bacteria such as *Chromatium* that live in severe conditions, and free-living cyanobacteria such as *Nostoc* that form crusts and eubacteria such as *Beijerinckia*. Some of these organisms are known to enhance the nitrogen nutrition of plants in diverse habitats.

Cyanobacteria can often be found colonising early-successional sites as well as open areas in later successional sites. The importance of this is that open patches are required for establishment. Shading, such as caused by weed cover, inhibits establishment of cyanobacteria. In wet areas, cyanobacteria can often be found colonising soil surfaces and fixing nitrogen. Little is known about the importance of cyanobacteria in restoration. In undisturbed arid and semi-arid ecosystems, they are common and extremely important in many ecosystems, forming 'desert crusts' (West & Skujins, 1978). These crusts may provide considerable nitrogen to these ecosystems (Evans & Belnap, 1999). However, they are extremely sensitive to disturbance and only recolonise slowly or not at all. In the many reclamation efforts that have been studied in the Great Basin, to our knowledge, none has yet been found to have re-formed cyanobacterial crusts (Belnap, 1993). Nonetheless, there is a high diversity of these cyanobacteria in most sites and they are important to nitrogen enrichment in ecosystems ranging from wetlands to deserts. Lichens formed with cyanobacteria as the producer component can also be found in many ecosystems including deserts, grasslands and tropical forests. These organisms have been poorly studied but they can fix up to 8 kg N ha^{-1} yr^{-1} in the tropics (Forman, 1975) and are found in every terrestrial environment (Nash, 1996).

In addition to nitrogen inputs to sites regulated by soil microbes, these organisms are responsible for much of the nitrogen transformation within a soil. One of the difficulties in understanding and manipulating the nitrogen balance of a site is the range of forms that nitrogen will take and the vast array of means whereby microbes transform that nitrogen. Greater details of the nitrogen cycle have been published elsewhere (e.g. Paul & Clark, 1996). It is important to review the critical processes to understand what may be altered in the disturbance and restoration processes.

Soil micro-organisms catalyse virtually every step in the nitrogen cycle. Microbes are responsible for converting detrital organic nitrogen to ammonia (NH_3) which is rapidly transformed into ammonium ions (NH_4^+) in the presence of an excess of H^+ ions in soils. NH_4^+ can be fixed or taken up by plants, a common pathway in ecosystems dominated by forested and shrubland plant communities. Importantly, the mycorrhizal associations play an important part in the uptake of NH_4^+−N (e.g. Smith & Read, 1997). In all moderate (pH 4–6, organic matter content of 0.5% to 3%) soils, but especially those highly organic soils with neutral to high pH, the NH_4^+ is transformed into nitrite ions (NO_2^-) and then to nitrate ions (NO_3^-) via nitrification. This is a two-phase oxidation process that co-occurs chemoautotrophically with the prokaryotes using carbon dioxide or carbonates as a carbon source (Paul & Clark 1996). Importantly, as NO_3^-, the nitrogen is subject to leaching, uptake by plants or mycorrhizal fungi, immobilisation by decomposer micro-organisms, or denitrification. Uptake and immobilisation are desirable as the nitrogen is retained in the system and recycled. Immobilisation by soil microbes can initially reduce plant nitrogen uptake but this changes with time. Leaching and denitrification are not desirable processes in a restoration site as the nitrogen is lost from the system. Leaching is purely a physical process but one aggravated by irrigation, a common practice in many reclamation efforts. Denitrification is a set of microbial-reducing processes whereby the NO_3^- is reduced to nitrogen dioxide (NO_2), nitrous oxide (N_2O) and then N_2. Both N_2O and N_2 are lost to the atmosphere as gases. In denitrification, a group of bacteria utilise the oxides of nitrogen compounds as terminal electron acceptors to complete their own energy conversion. As such, this process is affected by soil NO_3^- content, carbon availability, and soil-water content (the process is inhibited linearly by oxygen and denitrification generally does not occur at water potentials more negative than −0.01 MPa [Paul & Clark, 1996]). Soil pH and temperature also affect the rates of denitrification by regulating the growth of the responsible micro-organisms. Many of the micro-organisms deemed desirable for plant protection or nitrogen-fixation (e.g. *Pseudomonas*, *Azospirillum* and *Rhizobium*) also have the capacity to denitrify NO_3^-.

Nutrient uptake associations

Probably the group best known to alter nutrient uptake patterns of plants is the mycorrhizal fungi. These fungi invade plant roots and transport soil resources to the host plants in exchange for plant carbon. This association is formed between most plants and fungi in all of the true fungal groups. There is an extensive literature on the importance of mycorrhizae in ecosystems and land restoration specifically (e.g. Allen 1991; Allen *et al.*, in press).

There are several types of mycorrhizae. The most common is called an arbuscular mycorrhiza (AM).

Most plants form this association with Zygomycetes in the Glomales. None of these fungi can yet be cultured and mass-produced for large inoculation efforts although inoculated individuals for outplanting are available. This symbiosis is characterised by arbuscules, structures of the fungi that penetrate individual cortical plant cells. The second most common mycorrhizal type is the ectomycorrhiza (EM) formed between a fungus and many forest and chaparral trees and shrubs. The fungi forming these associations include Zygomycetes (in the Endogonales), as well as Ascomycetes and Basidiomycetes (Smith & Read, 1997). In this association, the fungi do not penetrate the plant cells but surround them, forming a Hartig net, and also surround individual roots with a mantle. These associations include fungi such as *Pisolithus tinctorius* used in reclamation efforts world-wide (e.g. Marx *et al.*, 1984) and truffles, important for human as well as plant restoration. Other mycorrhizal types are important in the growth of particular plant species, which need to be incorporated into restoration planning for specific species. For example, orchids form orchidaceous mycorrhizae with *Rhizoctonia* spp. and ericaceous plants form ericoid or arbutoid mycorrhizae depending on the soil conditions (Smith & Read, 1997).

Decomposers

Decomposition is a complex activity consisting of both immobilisation and mineralisation. In immobilisation, plant litter is decomposed and nutrients are absorbed by the decomposers. These organisms represent the groups with the highest diversity in any ecosystem. Moreover, they are always found, whether a site is disturbed or undisturbed, restored or not restored. It is obtaining the complex of groups best able to catalyse carbon release at a desirable rate that is an important goal of restoration. Mineralisation occurs as the immobilised nutrients are released into soil and made available for plant uptake. This can occur via cell death and protoplasm release or, more commonly, grazing of the immobilising fungi or bacteria. Thus, the other groups of organisms essential to the decomposition process are the soil invertebrates, primarily amoebas, nematodes, mites and collembolans. More details on the basic processes of decomposition and its linkages to nutrient cycling can be found in Curry & Good (1992) and Whitford (1988).

Bacterial and fungal activities are essential to the functioning of all terrestrial ecosystems through their roles in organic matter decomposition and nutrient cycling. Soil generally contains a vast number of microbial species with a wide diversity of metabolic versatility. Through their acquisition of carbon and nutrients for growth, fungal and bacterial biomass comprises an important source and sink of essential plant nutrients (Paul & Clark, 1996). The root region in particular has been shown to be an intense centre of microbial activity that is crucial for plant growth. The immediate area around the root, termed the rhizosphere, represents a major interface between the root and the soil environment. It is within this region that root exudates, containing carbohydrates and amino acids, interact with the soil microflora and stimulate the activity of bacteria and fungi. Increased rhizosphere activity will in turn affect nutrient availability and nutrient uptake. By forming a network of hyphae that can reach spatially heterogeneous nutrient pools, root-region fungi can assist plants in overcoming the restrictions to plant growth imposed by the spatial and temporal heterogeneity in nutrient availability.

Most efforts at describing soil microbe distribution concentrate on the rhizosphere (soil surrounding the root) and rhizoplane (the root–soil interface) region. However, it is important to remember that the mycorrhizal fungi dramatically affect the exudate rate and quality, provide a source of carbon and nutrients, and interact with other mycorrhizal fungi in a myriad of ways (Azcon-Aguilar & Barea, 1992). Bethlenfalvay (1993) describes the importance of the 'mycorrhizosphere' which contains vastly different conditions than does the bulk soil. Together, the rhizosphere and mycorrhizosphere constitute a relatively high fraction of the soil volume in an undisturbed habitat. Indeed, as root density commonly exceeds 2–4 cm cm^{-3} of soil and there is up to 100 cm of mycorrhizal hyphae cm^{-3} soil (Friese & Allen, 1991), it is arguable whether there is any bulk soil in an undisturbed ecosystem. The true 'bulk soil' may exist only in agricultural systems and in early-successional sites, prior to restoration. One goal in restoration may be to eliminate the bulk soil!

Fungi constitute the majority of the microbial biomass associated with decomposing litter (Allen *et al.*, 1999). Fungal diversity is also very high (Zak & Rabatin, 1997). In a shrub steppe ecosystem, isolate diversity curves indicated that as many as 1500 isolates would not enumerate all of the fungal taxa present (Allen *et al.*, in press) and Allen & MacMahon (1985) noted that the isolate-diversity curve began to level off at over 300 isolates. Importantly, all of these techniques fail to isolate a vast array of soil fungi, especially the basidiomycetes that may constitute the majority of the microbial mass.

Prokaryotic decomposers represent the highest diversity of organisms at most sites and, in many systems, are responsible for mineralising the majority of litter carbon. In fact, the diversity of these organisms in soils is so high that it is an almost impossible task to estimate richness with any technique. However, numbers of 4 000 to 40 000 have been discussed (e.g. Collins *et al.*, 1995; Allen *et al.*, in press). The bacteria tend to have a smaller standing crop than do the fungi but have a more rapid turnover. When conditions are appropriate, the rapid growth allows them to exploit organic carbon rapidly.

The soil fauna is the third major group involved in decomposition. The importance and dynamics of the soil fauna have been almost completely overlooked in restoration processes. Organisms ranging in size from protozoa to collembola are responsible for a large fraction of the mineralisation of nutrients by grazing on the bacteria and fungi that mineralise the litter carbon. In addition, these organisms are crucial to decomposition by fractionating the litter, dispersing decomposer organisms and changing the mycofloral composition (e.g. Curry & Good, 1992).

There is a relatively high diversity in the types of soil fauna in soils but most can be divided into functional groups based on their mouthparts. These structures dictate the types of organisms on which they feed, ranging from bacteria to other soil fauna. The wide array of these organisms is important from their size structure as well as for their feeding preferences. For example, in fine-textured soils or in sites with patches of small-pored soil, nematodes, mites and collembolans may be too large to penetrate those pores providing an escape mechanism for bacteria. In these cases, the protozoa, because of their small size, are important to the mineralisation of the bacterial immobilised nutrients (Curry & Good, 1992).

Food web structure

Soil organisms, just as the larger easily visualised organisms, aggregate into food webs that contain important structural elements. The manner in which they are organised is of critical importance because these webs regulate nutrient availability and carbon accumulation, and the rates and amounts determine the soil organic matter characteristics. Laboratory studies have shown many possible pathways and regulatory points but unfortunately the range of scenarios that could be important to restoring disturbed habitats is not well known. However, we will describe some of the salient features of soil food webs and their importance in nutrient and carbon cycling.

In general, bacteria and fungi utilise the carbon provided by plants and soil animals graze on the soil microflora. The presence of the differing groups of invertebrates allows for the grazing of the various groups of soil microflora thereby mineralising the nutrients bound in the microbial biomass. Generally, this grazing is presumed to reduce the microfloral mass (see Collins *et al.*, 1995).

By grazing on the soil microbial mass, these organisms release carbon dioxide back into the atmosphere, which in turn reduces the carbon-to-nutrient ratio. They then excrete the nutrients directly or by their own deaths thereby releasing the now available nutrients. Thus, these organisms are obviously crucial to restoring an active nutrient cycle.

Parasites and pathogens

Organisms parasitic on plants have generally been ignored in the restoration/reclamation literature. In general, there have been few descriptions of diseases that affected the restoration of a site. However, when the conditions are appropriate, diseases can dramatically or subtly alter the establishment rate of vegetation. Kowalski (1982) reported that the antagonistic interactions between *Cylindrocarpon*, a parasitic fungus, and some ectomycorrhizal fungi, mediated by the soil saprophytic community, determined the reforestation ability of some Polish forests. M.F. Allen *et al.* (1987) found that snow mould (consisting of a number of species) reduced the survival and growth of *Artemisia tridentata* transplants and that high nematode and mite activity dramatically reduced the mycorrhizal fungi which, in turn, lowered mycorrhizal benefits. Stanton & Krementz (1982) also reported high densities of plant parasitic nematodes in an early-successional shrub steppe, which could result in the consumption of a high amount of the carbon that is needed for root growth and soil stabilisation. E.B. Allen & M.F. Allen (unpubl. data) recently observed that a species of *Fusarium* was killing one of the planted trees in a reconstructed tropical seasonal forest.

BIOMES AND DOMINANT MICRO-ORGANISMS

Despite the work of Waksman (1927) and subsequent microbiologists, many ecologists and land managers still are under the impression that soil microbes are just soil microbes. However, differing organisms occupy different habitats, and even functional groups change with latitude, altitude and precipitation.

Read and colleagues probably developed the most comprehensive scheme relating the types of mycorrhizae to different habitats (e.g. Read, 1983). He suggested that as one moves up in altitude or latitude, there is a shift from mineral soils to highly organic soils and a concomitant shift in mycorrhizal types from dominance by AM to EM to ericoid mycorrhizae. This scheme is relatively accurate and useful for restoring functional groups of mycorrhizae, although it should be cautioned that these groupings are not exclusive (Allen, 1991). For example, AM were common in plants used for reclamation of mines in the Beartooth Plateau in Montana (elevation 4000 m), which also is a relatively arid site (E.B. Allen *et al.*, 1987).

When assessing mycorrhizal activity, the patterns of diversity become important. At the upper latitudes and altitudes where conifers predominate, the diversity of EM fungal symbionts becomes very high. Trappe (in Allen *et al.*, 1995) estimates that there can be more than 2000 species of ectomycorrhizal fungi associated with Douglas-fir alone. Alternatively, in the seasonal tropical forest near Chamela on the west coast of Mexico, to date we have found approximately 25 species of AM fungi in a forest with over 900 plant species catalogued in a 2000-ha reserve. This corresponds with a richness of 15 species of AM fungi found on a single patch of ground dominated by sagebrush and only eight other plant species. These data and many others in the literature support the notion that there is an inverse general relationship between the low diversity of AM fungi and high plant diversity at low latitudes shifting to a high fungal diversity and low plant diversity at the higher latitudes (Allen *et al.*, 1995).

Dynamics of nitrogen-fixing symbionts appear also to display a latitudinal gradient. At lower latitudes, legumes and their associated rhizobial bacteria comprise a high proportion of the nitrogen input. At Chamela, there are over 140 species of legume trees and, in natural disturbances, legumes such as *Phaseolus* often predominate, associating with vesicular arbuscular (VA) mycorrhizal fungi and rhizobia. *Phaseolus* is found in forest gaps whereas many of the legume trees, such as *Caesalpinia eriostachys*, are common in the surrounding undisturbed forests. At high-latitude coastal sites, such as in mined lands in Alaska, plants such as *Alnus* and *Dryas* form actinorhizal associations with *Frankia* and various ectomycorrhizal associations with several fungi including species of *Inocybe* and *Leccinum*. In inland areas of Alaska, legumes such as *Hedysarum alpinum* form EM and rhizobial nodules. These tripartite symbioses predominate in natural disturbance zones such as stream banks and glacial outwashes. Such associations have been found to enhance plant establishment in restored mined lands at these latitudes (Helm & Carling, 1993). At intermediate sites, a mix of types exists. On Mount St Helens (Washington, USA), the legumes *Lupinus latifolius* and *L. lepidus* invaded the ash within two years of the volcanic eruption. Both had active nodules when first found and became AM following the invasion of animals carrying inoculum

Box 14.1 Succession in the restoration of an open-pit coal mine in western Wyoming

HYPOTHESES AND EXPERIMENTAL DESIGN

A large interdisciplinary project was initiated in the autumn of 1981 to study the applicability of succession theory to restoration of an open-pit coal mine in western Wyoming, near the town of Kemmerer, Wyoming, USA. A major focus was the interactions of microbes, plants, animals and soils in creating a self-perpetuating ecosystem. The site is a cold-desert sagebrush steppe at 2200 m in elevation located at 42° N 100° W. Further details can be found in Allen & MacMahon (1985) and Allen (1993). Several groups of organisms were targeted for assessment in this project. Methods were appropriate to each process and can be found in the relevant citations.

Our general hypothesis was that the invasion and establishment of soil micro-organisms is a limiting factor in the recovery of soil processes, which, in turn, can limit the establishment of acceptable plant diversity and productivity. In this context, we assessed microbial mass and immobilisation, mycorrhizal fungal activity and diversity, nitrogen-fixing prokaryotes and plant pathogen activity.

INITIAL CONDITIONS

The soils disturbed were high in phosphorus and low in nitrogen. Bicarbonate-extractable phosphorus ranged from 20–30 mg kg^{-1} and the total phosphorus was 800 mg kg^{-1}. The average soil nitrogen was 10–20 mg kg^{-1} with the available soil nitrogen of 10 mg kg^{-1} of both NO_3^- and NH_4^+. However, an important feature was the dispersion pattern. In the undisturbed areas, nutrients were concentrated around the base and canopy edge of the individual shrubs, with low concentrations of available nutrients in the intervening spaces (interspaces). In the site to be restored, nutrients were relatively evenly dispersed across the site. Both variance:mean ratios and semi-variograms showed no co-correlation in any parameter assessed. Nor were any patterns observed whether the scale of analysis was at the cm level or across the entire site.

The microbial composition was dramatically altered by the mining and topsoil storage. Bioassays showed no symbiotic nitrogen-fixing prokaryotes. Spores of AM fungi survived the storage but only marginal infection was reinitiated upon planting. AM fungal hyphal length studies showed that little of the mycelium survived. Microbial mass declined significantly comparing the adjacent undisturbed soils and the newly respread topsoils. Microbial mass was also evenly distributed across the site, and highly organised around the shrubs in the undisturbed site. The fungal composition was highly altered with disturbance, with a Sorensen's overlap index of only 0.09. In the undisturbed area, the predominant fungi were characteristic of tundra–taiga regions but were desert–grassland forms in the disturbed region (Allen & MacMahon, 1985; Allen, 1993).

ESTABLISHMENT OF MICROBES

While many microbes invaded the site following the respreading of the topsoil, the different groups invaded using differing vectors and at different rates. There was no apparent establishment of *Rhizobium*. In planted *Hedysarum boreale*, nodules that increased plant performance were present only on artificially inoculated plants (Carpenter & Allen, 1988). Both fungal and bacterial mass increased with time, especially under shrubs (Allen, 1993). Plant pathogens also were present and dramatically reduced plant growth in the high precipitation years that would have been expected to enhance plant establishment and production (Allen *et al.*, 1987).

The recovery of mycorrhizal fungal populations was associated with the patterns of shrub replanting. We found that there were more than two orders of magnitude more propagules of AM fungi immigrating via wind than by animals (Allen, 1988*a*). Both animal activity and wind concentrated inoculum with shrubs. The majority of those animals that served as vectors utilised planted shrubs for cover or food. For example, chipmunks (*Eutamius amoenus*) were found to deposit spores (Allen, 1988*a*). Wind not only moved spores but preferentially deposited those spores on the lee side of shrubs (Allen, 1988*b*). With multiple shrubs, if they were planted within 10 × their height, many more propagules would be deposited between them than would be deposited around either independently (Allen *et al.*, 1997).

Surface-applied inoculum was found to be ineffective in forming mycorrhizae (Friese & Allen, 1993). However, inoculum that was buried artificially or was placed in the soil by ants was found to be an effective inoculum source (Friese & Allen, 1993). After six years, mycorrhizal infection was found to be both of a higher density and in a more effective spatial pattern when shrubs were planted in patches than either with no planting or with the grass plantings that were the standard reclamation practice (Allen & Friese, 1992). With no topsoil preservation, there was little re-establishment of mycorrhizae. Even after 30 years, with no replaced topsoil, the mycorrhizal infection and spore counts were lower than those in the topsoiled sites (between one and six years old) and neither was up to the level of activity of the undisturbed areas (Waaland & Allen, 1987; Allen & Friese, 1992). Thus, it appears that careful planting in a patchy pattern, coupled with an adjacent source area, wind, and animals all interacted to facilitate the re-establishment of mycorrhizae.

The planting pattern was also important to the saprophytic activity. The soil microbial mass increased between 1981 and 1987. Interestingly, the microbial phosphorus increased proportionately more than did the microbial mass; we do not know what the cause may be. Microbial mass increased more under the shrubs than in the interspaces, probably associated with the increasing organic matter from decomposing roots, as the leaf litter did not begin to accumulate until 1987 (Allen, 1993).

The implications from these experiments are clear. The microbial component of a restored site is complex and dependent on the treatment practices that disturb the ecosystem, the care with which the topsoil is handled, and the location and retention of a diverse source area.

(Allen, 1987; Allen & MacMahon, 1988). Alders subsequently invaded and were initially AM and later EM (Allen *et al.*, 1992). At the Kemmerer restoration site, a mined arid shrub steppe in Wyoming (see Box 14.1), early-successional sites were often colonised by *Hedysarum boreale*, a legume, whose growth was enhanced by forming a dual symbiosis with AM fungi and *Rhizobium* (Carpenter & Allen, 1988). Undisturbed surrounding vegetation consisted of a high proportion of *Purshia tridentata*, an actinorhizal shrub forming AM. In chaparral vegetation, AM–rhizobial symbiotic *Lotus scoparius* predominates in early-successional sites such as road sides (Stylinski & Allen, 1999) and *Ceanothus*, forming actinorhizae and AM, is found in late successional stands. In many areas of the world, *Acacia* spp. that form associations with AM fungi and rhizobia predominate.

The diversity of saprophytic fungi is exceedingly high at all sites (Zak & Rabatin, 1997). Moreover, different species certainly have very distinct ranges. In fact, this large resilience in the decomposer composition and its implications should be a major concern but has not been addressed. Allen & MacMahon (1985) noted that there was a shift in the soil saprophytic fungal community from a predominance of tundra–taiga species to desert grassland forms following strip-mining and topsoil replacement in a shrub steppe. The tundra–taiga forms presumably remained since the Pleistocene but, upon severe disturbance and direct exposure of the soil to solar radiation, were not able to survive or tolerate competition from invading desert forms.

SHIFTS IN POPULATIONS OF ORGANISMS WITH DISTURBANCE

Symbiotic organisms

Shifts in populations of symbiotic organisms as a result of disturbance may be measured in terms of changes either in total levels of activity of the organism or as shifts in the relative proportions of species forming the symbiosis. If either change occurs the net benefit to the host may be reduced, either because of the lower incidence of formation or association with less effective species.

Changes in populations will occur as a result of three possible factors. First, direct physical damage may occur to the micro-organisms; this is most applicable for organisms such as mycorrhizal fungi whose soil hyphal network may extend relatively long distances in soil and whose infectivity may depend on an intact hyphal network. Second, the soil micro-organism population may be exposed to unfavourable chemical or physical factors in the soil as a result of disturbance. The third important result of ecosystem disturbance is removal of the host.

Each of these factors may be important for the survival of mycorrhizal fungi, for example, in soil disturbance during mining. The connecting network of hyphae in the soil is directly disrupted, the host plant is removed and the soil is likely to reach extremes of temperature and dryness. If soil is also stockpiled then further damage, from waterlogging or exposure to heightened microbial activity, is likely to decrease the survival of the mycobiont (Jasper, 1994).

Disturbance of an ecosystem is likely to lead to death of a proportion of the population of symbiotic soil micro-organisms thereby indirectly affecting plants, even though the microbes themselves may remain. For example, individual propagules of AM fungi may survive soil disturbance but destruction of the network of soil hyphae dependent on an intact plant can reduce the nutrient uptake capacity (Jasper *et al.*, 1989).

Mycorrhizal fungi

Soil disturbance during mining is the most important impact in terms of its scale and severity, and has also been the subject of most research. During mining, topsoil layers are usually removed separately and subsequently respread. As propagules of AM fungi occur mainly in this surface soil (Allen, 1991), the numbers and infectivity of propagules of AM fungi in topsoils are decreased (Allen & Friese, 1992).

Disturbance of forests by logging is likely to have the biggest impact on mycorrhizal fungi through mechanical destruction of the host plants or by burning. No loss of infectivity of mycorrhizal fungi in the soils has been found as a result of clear-cutting the mixed conifer forests of the Pacific Northwest of the United States, where invading plants are typically AM, in which the fungal symbionts are typically found deeper in the mineral soil (Parke *et al.*, 1983*a*; Pilz & Perry, 1984). However, the loss of the litter layer could have major effects on EM inoculum (Parke *et al.*, 1983*b*; Gardner & Malajczuk, 1988).

Burning can potentially affect the infectivity of AM fungi through direct effects of high temperatures on the propagules in the soil or indirectly through reduced numbers of mycorrhizal host plants (Klopatek *et al.*, 1988; Wicklow-Howard, 1989).

Any mechanical disruption of soils appears to reduce the infectivity of mycorrhizal inoculum in soils (e.g. Jasper *et al.*, 1987, 1989). Recovery of infectivity following mechanical disturbance, such as strip-mining, may take several years if it occurs at all (Allen & Allen, 1980; Waaland & Allen, 1987). For example, the formation of AM by *Agropyron smithii*, four months after discing of soil, was reduced from 95% root length colonised in the undisturbed soil to 62% in the disced soil (Allen & Allen, 1980). However, after 15 months, there was no difference in colonisation between the two soils.

The effect of disturbance must be considered in relation to the nature of the propagule. Spores are relatively robust and unlikely to be greatly affected by disturbance. Mycorrhizal root fragments are likely to be more vulnerable to disturbance than spores because the root cortex containing the fungal tissue is likely to be easily displaced or damaged. Disturbance of soil will detach the network of hyphae from host roots and will break the network into smaller components. The hyphae of AM fungi can remain infective even in dry soil if remaining attached to old roots. However, when separated from the host plant this infectivity is lost with soil disturbance (Jasper *et al.*, 1989). The mycelial network in soil is an important propagule in undisturbed soils (Smith & Read, 1997) and losses in infectivity after soil disturbance may be at least partly due to damage to this network, rather than damage to the relatively robust structures of spores and mycorrhizal roots.

A spatial dilution in inoculum could have the same effect on the effective inoculum as an absolute loss. Allen & MacMahon (1985) noted that these fungi were highly concentrated around individual shrubs in a cold desert, with the spores concentrated at the shrub bases, such that new infections developed with spring root regrowth. This pattern was lost as the spores and hyphae were randomly dispersed throughout the soil following topsoil respreading. In a grassland site, however, the inoculum was evenly distributed across the stand. When the topsoil was stripped, respread and uniformly seeded with grasses, it remained diluted but rapidly recovered uniformly across the site (Allen & Allen, 1980). In this case, the structure rapidly recovered. In most instances there is a reduction in inoculum due to the vertical mixing of the deeper soils with the surface soils that contain the majority of inoculum (e.g. Schwab & Reeves, 1981).

In summary, a loss of mycorrhizal inoculum will occur with almost any disruption of an existing soil. That loss can be dramatic when soil disturbance is followed by long-term storage as the propagules die without new plants; alternatively the loss can be minimised by careful replacement of surface soil and litter layers, and rapid replanting of species compatible with the existing inoculum.

Nitrogen-fixing organisms

Survival of the symbiotic nitrogen-fixing organisms *Rhizobium* (including *Bradyrhizobium*) and *Frankia* during and after ecosystem disturbance has received little attention. Therefore, this section will be limited to research that has examined, in various contexts, the effect on these organisms of individual factors associated with ecosystem disturbance.

Unlike the hyphal network of mycorrhizal fungi, the discrete cells of these organisms are unlikely to be directly affected by soil disturbance. The most important constraints on the survival of infective organisms are likely to be the absence of host plants and changes to soil condition. It appears that many symbionts are able to survive for long periods in the absence of host plants. *Frankia* was shown to be well distributed in soils from a range of sites in which actinorhizal plants had been absent for periods ranging from 20 to 80 years (Smolander & Sundman, 1987). Following logging and fire, infective *Frankia* were found in soils in which *Ceanothus velutinus* had not been growing for approximately 300 years (Wollum *et al.*, 1968). For rhizobia, Bergerson (1970) reported observations of good survival of *Rhizobium meliloti* for 11 years in completely bare soil with surface temperatures in excess of 65 °C. Other major factors that are likely to limit the survival of rhizobia or *Frankia* in disturbed soils are high temperatures and possible desiccation. Removal of the vegetation can expose the soil surface to extremely high temperatures, which in some bare soils may exceed 60 °C. The resulting below-surface temperatures in dry soil are not necessarily lethal but are likely to cause a decrease in the numbers of rhizobia (Chatel & Parker, 1973). The effect of high temperatures may be more severe in moist soil, with severe reductions being recorded at 40 °C, although *Frankia* has been shown *in vitro* to maintain some growth at temperatures as high as 40 °C (Burggraaf & Shipton, 1982). A further consequence of removal of vegetative cover and of disturbance of the soil at some sites may be that the soil becomes extremely dry. In general it appears that, while there will be some losses of these organisms in dry soil, the impact of this on the rate of formation of the symbiosis in restored vegetation is likely to be small (e.g. Shipton & Burggraaf, 1982).

At the opposite extreme, anaerobiosis due to prolonged waterlogging, as in flooded or in some stockpiled soils, is likely to decrease the numbers of *Rhizobium* and *Frankia* in a similar manner to that observed for other aerobic soil micro-organisms (Atlas & Bartha, 1993). It appears likely that in most situations sufficient rhizobia and *Frankia* would survive the extreme conditions to which disturbed soil may be exposed. However an understanding of the severity and causes of the losses would assist in optimising strategies for soil handling to ensure maximum survival. Inoculation may be needed if the rate of symbiont colonisation is slow. However, at Mount St Helens, lupine seedlings were already heavily nodulated in their first year even though the nearest inoculum source was over 5 km distant.

Free-living organisms

Unfortunately, there are few data describing the changes in soil autotrophs following perturbation. In semi-arid sites, fire, grazing, mining, off-road vehicles and other factors that disrupt the soil surface reduce or eliminate the algal and cyanobacterial components (Belnap & Gillette, 1998). Disturbance can disrupt or eliminate the lichen communities that predominate in northern latitudes. In Alaska and the Yukon, in areas disrupted by the Klondike gold rush of the 1890s, the lack of lichen cover is still evident.

In the case of decomposers, the compositional change is one of individual taxa, not broad functional groups (Allen *et al.*, 1999). In fact, no system on earth exists without decomposers. However, the composition and spatial arrangement of those decomposers can have dramatic effects on nutrient cycles and subsequently, on site stability.

Microbial biomass

Microbial mass tends to decline with immediate perturbation, then gradually increase following the input of organic matter in the newly dead plant tissue, decrease upon loss of that organic matter, followed by a gradual increase, as phytomass standing crop increases through succession. The time-frame of the drop is very rapid. At the Kemmerer mine restoration project (Box 14.1), the decomposer fungal mass immediately following topsoil replacement was less than one-third that of the adjacent native area. However, the time-scale of recovery is very slow largely because the recovery of phytomass and soil organic matter is slow. Over a seven-year period from topsoil replacement, the microbial mass never went beyond 50% of that found in the undisturbed site, reaching a maximum of 600 mg kg^{-1} microbial-mass carbon (Allen, 1993). Insam & Domsch (1988) noted that in a mesic system, microbial mass initially declined from a late-successional value of 800 mg kg^{-1} to a value of 200 mg kg^{-1} in a newly reconstructed reclamation site. The microbial mass then increased at a rate greater than the organic matter accumulation, reaching an equilibrium value of approximately 600–700 mg kg^{-1}. They estimated that equilibrium in microbial carbon/organic-matter carbon would take longer than 50 years to decline from a high of 0.15 (*c.* 15 years after revegetation) to a stable value of approximately 0.10. In an arid site, the pattern appeared to be very different in that there is little evidence that any stable value is reached.

In another interesting study, Klein *et al.* (1994) followed the fungal and bacterial mass following soil disturbance in a grassland. They noted that, although the total fungal mass increased with time, the increase consisted of dead hyphae. Total bacterial counts also did not change with time but the active mass declined. In this grassland site, dominated by bacterial mass in the undisturbed sites, the bacterial: fungal mass increased with time.

These three patterns indicate how little we know about the soil and microbial dynamics during restoration. To our knowledge, they represent the only three comparable data sets. One was a study of the recovery to an agricultural system (Insam & Domsch, 1988) whereas the others were undertaken in shrublands. Moreover, the Insam & Domsch (1988) presented data from a chronosequence but the Allen (1993) and Klein *et al.* (1994) were following the replanting of a single site.

Spatial array

Possibly of equal importance to the total mass is the change in the spatial distribution of the microbes. Allen & MacMahon (1985) noted that in a semi-arid shrub steppe, nutrients and microbial activity were tightly coupled to shrub distribution regulating the patterns of these 'islands of fertility'. Shrub roots concentrate nutrients by extending roots into the surrounding soil and concentrating those nutrients and organic matter directly under the canopy. The aboveground architecture also 'harvests' wind-blown material, like a snow fence, again concentrating soil organic matter and finer soil. This pattern was lost with severe disturbance (Allen, 1988*b*). Alternatively, in a desert grassland, Schlesinger *et al.* (1990) suggested that there was an increase in the patchiness of soil resources associated with invading shrubs resulting in desertification of the site. This would suggest that the spatial patterns shift from evenly distributed to patchily distributed between grass-dominated systems to shrub-dominated systems. In the Great Basin of the USA increasing grasses represent increasing desertification of the site, whereas in the Chihuahuan grasslands of New Mexico increased shrublands indicate increasing desertification.

Species composition

Assessing species shifts is difficult at best. The limited data that exist suggest that the diversity in micro-organisms is reduced with disturbance. Fresquez *et al.* (1986) noted that the richness in the fungi declined with disturbance. Allen & MacMahon (1985) reported a slight decline but they noted that 275 isolates were inadequate to assess the richness changes in this shrub–steppe system. They also reported no difference in the Shannon index (*H'*) because of the high richness of low-density species. Miller (1984) also noted that species richness did not decline in a disturbed cold desert site. Fresquez *et al.* (1986) also reported a decline in fungal diversity. Allen *et al.* (in press) noted that the species increment curves for fungi in restored areas level off faster than those in native shrublands.

Of equal or greater importance were the shifts in the species composition. All studies to date have shown a significant impact of disturbance on the composition of soil fungi. Miller (1984) reported that xerophyllic *Penicillium* spp. predominated in the disturbed soils. Allen & MacMahon (1985) also noted that there was a shift toward more xeric fungi following the disturbance. Fresquez & Dennis (1990) reported that the Sørensen's index of similarity, comparing an untreated grassland with the addition of sewage sludge, ranged from 0.11 to 0.38. Allen & MacMahon (1985) found that the similarity between a shrub steppe and a respread topsoil site was only 0.21. None of these would be considered as coming from a similar community.

Food webs

Several studies have demonstrated that the higher trophic levels of soil animals are affected by disturbance. Both numbers and species compositions of soil arthropods are known to be affected by subtle changes in soils, and especially by disturbance (e.g. Moldenke & Lattin, 1990). Stanton & Krementz (1982) found that densities of all nematode groups (microbivores, predators and plant parasites) were lower in reclaimed than undisturbed shrub–steppe systems. Elkins *et al.* (1984) described the activity in litter bags placed in mined versus unmined soils in northwestern New Mexico. Bacteria and fungal mass was not different in topsoiled versus unmined soils. No differences in the numbers of protozoans were observed. However, there was a decline initially in the numbers of many of the Acari. Whitford (1988) also noted that the numbers of arthropods were initially high following soil reconstruction but then declined in a stand of *Agropyron desertorum*. These studies consistently indicate that there was an initial burst in fungi following the respreading of topsoil or replanting (organic matter input) but that the fungi rapidly disappeared. It may take at least decades for the fungi and their associated trophic pathway to form. This is probably tightly coupled to changes in organic matter inputs (Insam & Domsch, 1988; Allen, 1993).

RESTORATION PRACTICES

Managing for survival

Managing disturbed soils for maximum survival of symbiotic micro-organisms can be considered from two perspectives: (1) ameliorating the disturbance effect to ensure the survival of as many micro-organisms as possible, and (2) managing the restoration of the soil to enhance their recovery.

Timing of disturbance

Amelioration strategies are obviously only possible if the disturbance event can be anticipated, such as in the case of mining or logging. In this context, there has been very little research investigating optimum timing of disturbance events.

The capacity of mycorrhizal fungi to survive disturbance events is likely to be a function of the number and type of propagules present (Jasper *et al.*, 1991). If initial numbers are high, then despite severe reductions, enough infective propagules may remain to initiate adequate colonisation in restored vegetation. While the number of propagules in an ecosystem is difficult to manipulate, disturbance events may be able to be co-ordinated with the period when most disturbance-tolerant propagules are present.

In seasonal climates, it is during the dry season that mycorrhizal fungi are likely to be in a resting or dormant phase. During the early growing season, propagules will germinate and new hyphae will be produced, colonising adjacent plants. At this stage the population of mycorrhizal fungi is likely to be most vulnerable to soil disturbance. As the growing season progresses, colonising mycorrhizal fungi will begin to form fruiting bodies or spores and thus should become more tolerant of disturbance, by virtue of the increasing number of these robust propagules. Therefore an important principle should be to co-ordinate disturbance events with periods when the soil is dry or the fungi and plants are dormant.

Managing soil to enhance recovery

The symbiotic soil micro-organisms considered in this chapter have, at best, limited saprophytic capability. After a disturbance event, the crucial factor affecting their survival is likely to be the length of time until plants are reintroduced to that soil. Therefore, another important principle in management of disturbed soils is to minimise the time that the soils spend without a plant cover.

During mining, soil may be stored in stockpiles for several years. Even when the soil is respread immediately after stripping, it may be several months before new plant growth occurs. Establishing a dense cover-crop of plants on soil stockpiles is likely to enhance the survival of all, especially symbiotic, micro-organisms. Similarly, benefits are likely from careful maintenance of existing plant cover during logging.

Actively growing roots not only enhance the survival of micro-organisms in soil through exudations and establishing symbiotic associations, but they also assist by using water in the soil. Excessive soil moisture can be an important factor limiting the survival of propagules in soil stored in stockpiles. Deep-rooted perennial species should be the most effective in reducing the water content of stockpiled soils. The proportion of soil in the stockpile that benefits from the presence of a dense plant cover will be greatest in shallow stockpiles. In large stockpiles, the active surface 'skin' of the stockpile forms a lower proportion of the total. Therefore, stockpiles should always be made as shallow as possible.

It is important to recall the immense diversity of saprophytic soil organisms when discussing recovery techniques. Every manipulation probably causes an irreversible change in the composition of the saprophytic community. Hence, it is unlikely that any manipulation will be effective in 'restoring' the saprophytic community. A more successful approach is probably to manipulate the diversity of processes or to enhance the diversity of organisms within a functional group.

Inoculation using natives or exotics

Even in the best of circumstances, the reinvasion of soil microbes can be limiting to the restoration of a given site. Where there is an inadequate source area or where conditions have changed so dramatically that the source-area microbes are not adapted to the altered conditions, the use of artificial propagation techniques for soil microbes may be critical. We are aware of no studies that have shown an enhancement of saprophytic activity using artificial inoculations. Microbial growth enhancements should be utilised with caution or, preferably, not at all. However, the use of exotic symbionts may be important to site recovery. Strain selection efforts using rhizobia for reclaiming mined lands and pastures has been ongoing for decades. Although commercial strains exist, they are often adapted for conditions favourable to agricultural crops and not native conditions.

The use of inoculation of ectomycorrhizal fungi has dramatically enhanced the forestation of differing habitats. When attempting to restore native species, however, the situation on the use of exotics becomes less clear. Mikola (1980) strongly advises that the local inoculum is more adapted to the local plant and climatic conditions. However, Marx *et al.* (1984) have advocated the use of exotic inoculations of ectomycorrhizal fungi when the conditions of the disturbance dramatically altered the soils to be restored. Another procedure is to utilise native fungal species as the inoculum source. Stewart & Pfleger (1985) used 'ecologically adapted' mycorrhizal fungi to inoculate red pine and jack pine in iron-mining sites and found these same fungi survived and spread from the inoculated to the uninoculated individuals over a period of several years.

The inoculation of a site or of plants by arbuscular mycorrhizal fungi is less well documented. Despite the wide array of pot-culture studies on mycorrhizae, there are few field studies that have studied the role of inoculation of arbuscular mycorrhizae in the field. Aldon (1975) found that inoculated seedlings had improved survival after outplanting when artificially inoculated with exotic mycorrhizal fungi. Davies & Call (1990) reported enhanced growth of several plant species following outplanting after inoculation. Alternatively, Call & McKell (1982) reported enhanced establishment and growth of *Atriplex canescens* when inoculated with local native mycorrhizal fungi and Allen & Allen (1986) found enhanced drought tolerance in mycorrhizal versus non-mycorrhizal plants using an inoculum from the neighbouring source area. Arbuscular mycorrhizae are almost ubiquitous. Because of this, establishment is crucial for initiating successful plant communities. Sand dunes have been an especially interesting test case since the work of Nicolson in the 1950s (see a detailed description in Greipsson, volume 2). Recent work has shown the importance of inoculation in these habitats and application to other habitats would be valuable.

However, the composition of that inoculum is also important. Weinbaum *et al.* (1996) also noted that local and introduced mycorrhizal fungi had different survival rates when inoculated in the field. Survival of the introduced inoculum in a region that is being restored to its original conditions is of concern. The long-term survival of introduced inoculum has not been well documented. *Pisolithus tinctorius* reportedly declines rapidly upon the regrowth of a forest following inoculation (Marx *et al.*, 1984). Friese & Allen (1991) found that *Gigaspora margarita* survived for two growing seasons when seedlings were inoculated and transplanted or when the spores were inoculated at 10 cm depth, but failed to establish when placed on the surface of the soil. Weinbaum *et al.* (1996) found that both *Scutellospora calospora* and *Acaulospora elegans* survived for three years following transplanting into a new habitat but failed to sporulate.

Managing for natural reinvasion

Soil organisms of all kinds will reinvade any disturbed site given adequate time. Saprophytes and nitrogen-fixers have been found in numbers approaching those of an undisturbed site in only months (Fresquez *et al.*, 1986). Free-living organisms such as soil mites, nematodes, fungi and bacteria all reinvade rapidly (Stanton & Krementz, 1982; Fresquez *et al.*, 1986). Even symbionts such as mycorrhizal fungi can invade in a few years given appropriate hosts and nearby sources (e.g. Allen *et al.*, in press). However, that reinvasion can be erratic in both space and time. In some sites where no topsoil had been added, mycorrhizal activity was low or nonexistent decades following abandonment (Allen & Allen, 1980; Waaland & Allen, 1987). Moreover, every measurement of microbial diversity indicates a reduction in taxon richness following disturbance (Allen *et al.*, in press).

Few soil micro-organisms needed for restoration can be cultured and used to replace the thousands that are lost with the disturbance. Of those most easily cultured, the saprophytes, there is little evidence that they will survive when replaced into the altered soils. These organisms also comprise most of the soil diversity and probably a majority of these have never been characterised. Of the symbionts, probably a majority cannot yet be grown in culture. Therefore, managing the site to enhance reinvasion of soil organisms is critical to the restoration process.

Despite this, there has been minimal research studying the natural dispersal mechanisms of micro-organisms. One general assumption is that, because microbes are extremely small and contamination of pure culture isolation is high, all micro-organisms must readily disperse. The other predominant approach to models of microbial migration is based on a log-distance relationship, where the organisms are deposited in exponentially decreasing numbers as the distance from the source areas increases. Other studies have assessed microbial invasions by plant pathogens as a function of disease symptoms, again using a log function (see discussion in Allen *et al.*, 1993).

Assessment of the dispersal of soil micro-organisms shows a much more complex situation but one that can be manipulated to enhance the recovery of soil organisms. Just as with many organisms, soil micro-organisms disperse using a variety of vectors that depend on the local conditions and the constraints imposed by the developmental biology of the organism. Soil micro-organisms can disperse by wind but do not always do so. This is because soil microbial structures come in a variety of sizes, some of which increase wind dispersal but others require very high winds to be entrained into the general wind flow for dispersal. Animals are effective vectors, either by accidental transport on the external body parts or via ingestion as food items. In some habitats, winds are adequate for entraining soils and the organisms inhabiting those soils. In others, the winds are inadequate or the soils are too moist and therefore not conducive to wind erosion.

Wind is probably the most common vector for the dispersal of soil micro-organisms. Mycorrhizal fungi and soil animals are frequently wind-dispersed. However, there are size and environmental factors that restrict the dispersal by wind (Allen *et al.*, 1993). The size limits are the first restriction on wind dispersal. As propagules increase in size, their terminal velocity increases. The terminal velocity is the speed at which particles fall in the atmosphere. The terminal velocity can be readily calculated based on propagule size. It takes a stronger wind to entrain the particle, as its terminal velocity increases. The other important factors are the air density and height above the ground at which the propagule resides. It takes a relatively high elevation to make enough of a change in air density to affect the terminal velocity so that for most instances, it can be ignored. In order for propagules to be wind-dispersed, they must be entrained into the turbulent upper airflows. A synthetic number (the Reynolds number) provides an indication of whether flow is laminar (straight) or turbulent. The two major factors critical to the Reynolds number are the height above the soil surface and the velocity of air movement. Generally, at Reynolds numbers above 2000, airflow is turbulent. This generally requires a relatively high wind at the soil surface to entrain particles. For example, soil particles containing spores in a desert surface must be 3–30 cm high for the Reynolds number to exceed 2000. In southwestern Wyoming where the winds commonly exceeded 10 m s^{-1}, only on the ridge tops was the wind found to move large numbers of fungal spores from the soil surface (Allen *et al.*, 1989) and the spores entrained from the valley bottom probably came from the surfaces of badger (*Taxidea taxus*) mounds that were raised 20–50 cm above the soil surface (Allen *et al.*, 1997). In habitats where the vegetation is dense, the wind is unable to penetrate the canopy and dispersal is negligible. Even in coniferous forest canopies where the understorey is minimal, there is rarely enough wind to disperse spores of mushrooms fruiting above ground (Allen, 1987; Allen *et al.*, 1993). In summary, wind dispersal of soil micro-organisms should not be assumed.

Animals are also important vectors for the dispersal of soil biota. In some cases, this dispersal may be the result of direct feeding of animals on soil organisms. The preference of many mammals for truffles is well known. The dispersal of mycorrhizal truffles is critical to the reforestation of many habitats (Allen *et al.*, 1997). In other cases, the dispersal may be indirect, caused by the movement of soil by the animal or by the transfer of rooting material with the microbial propagules. *Thomomys talpoides* has been shown to transport mycorrhizal fungi both horizontally and vertically and thereby initiate the re-establishment of mycorrhizal fungi (e.g. Allen & MacMahon, 1988). Harvester ants (*Pogonomyrmex* spp.) have been shown to clip and transport root segments with mycorrhizal propagules (Friese & Allen, 1993). Deer (*Cervus elaphus*) removed roots with propagules of mycorrhizal fungi while feeding on the plant shoots and transported these several kilometres into disturbed areas (Allen, 1987).

Animals also move soil organisms to locations from which they may disperse more readily. Murie (1962) observed that squirrels (*Sciurus* spp.) dried mushrooms in the crotches of trees and Allen (1991) observed a squirrel feeding on a mushroom several metres into a tree canopy. This results in the release of spores well into the turbulent air stream for longrange dispersal. Habitat characteristics thus appear to be critical to the dispersal vectors of soil micro-organisms. In mesic to wet habitats with wet, sticky soils and a high canopy cover, there is little evidence for wind dispersal. Most dispersal appears to be via animals. Alternatively, in arid habitats where wind erosion moves soils, there appears to be a relatively higher amount of wind dispersal (Allen, 1988*a*).

In a restoration project, one can either allow reinvasion to occur willy-nilly or manage the habitat to enhance the reinvasion of native microbes. Two facets should be evaluated: the locations (including topography) of source areas in relation to the site to be restored and the predominant vector(s) of the immigrating microbes. Manipulating these two factors can substantially enhance the rate of successful microbial community recovery. If the site is relatively arid with high natural wind erosion, spores often will be predominantly deposited on the lee slope of a ridge and the microbe will most probably establish there. Waaland & Allen (1987) found increasing mycorrhizal fungal activity on the lee slopes of abandoned western strip-mined lands. On the windward side, barriers such as snow fences can be planted to enhance the trapping of wind-blown propagules. As described shortly, Allen *et al.* (1997) enhanced the collection of fungal-sized propagules via manipulating the distances between plants and thereby creating patches whereby large groups of plants could act as islands of inoculum.

In more mesic sites, animal activity can also be manipulated to enhance microbial deposition. On Mount St Helens, mycorrhizal fungal spores were moved by mammals that keyed in on plant patches, predominantly lupine patches. They ignored individual plants (Allen *et al.*, 1992). Thus, the larger plants became the attractants for animal activity resulting in the establishment of mycorrhizae. The consequent growth enhancement of the plants in those patches subsequently increased the reproductive success of those plants and further increased the size and diversity of the patches. Thus, the manipulation of the vegetation can alter the activity of the animals resulting in greater microbial deposition, which in turn increases plant establishment.

Clearly then, the manipulation of the patch structure not only affects the spatial pattern of the plants (cf. Davy, this volume), but also results in the improved ecosystem dynamics that increase site stability. This approach uses no artificial inoculation process but can result in the rapid recovery of the soil organisms as well as of the plants and animals that most environmentalists are concerned with. Pattern manipulation is the basis of the large restoration project from Wyoming described in detail in Box 14.1.

CRITERIA FOR RECOVERY

Evaluation of recovery is difficult. This is because the same microbial processes that enhance restoration can reduce restoration potential. For example, it is critical for nitrification to proceed at a rate that resupplies nitrogen for plant growth. However, if nitrification is too fast, the result can be leaching or enhanced weed competition. If nitrification is too slow, the nitrogen becomes immobilised in dead litter or microbial mass and nitrogen deficiencies in plant growth result. Clearly, the end result of a stable plant community and appropriate animal activity is the goal of restoration. However, a diverse, stable and functioning microbial community is essential for the plants and animals to survive on a site.

Several studies have assessed particular groups or processes as important parameters of successful (or unsuccessful) restoration. These include soil enzymes, particular groups of organisms such as mycorrhizal and saprophytic fungi, microbial mass, and spatial distributions. Soil enzymes are used as important indicators of microbial functioning. Some investigators have suggested that soil enzymes can be used as a biological index to soil fertility and microbial activity. This approach has not been particularly successful. However, these enzymes do catalyze specific transformations such as phosphorus release, nitrogen fixation and microbial respiration (Klein *et al.*, 1998). This means that the enzymatic activity does represent specific microbially mediated processes and, as such, must occur at adequate rates to process soil nutrients and organic matter.

Looking at specific groups of organisms and their community characteristics has been suggested as an important indicator of microbial functioning. One indicator has been the community structure of the soil saprophytic fungi. Parameters such as the Shannon–Wiener index of diversity and dominance–diversity curves may be associated with the reclamation status of some mine-land soils (see Zak, 1992). Importantly, an adequate number of isolations needs to be made to generate a species increment curve (Allen *et al.*, in press).

Another component is the microbial mass or microbial mass:organic matter ratio. Microbial mass has generally not been a good indicator of the restoration status. Insam & Domsch (1988) found that microbial mass started low, increased rapidly and then levelled off with successional age in restored, strip-mined land. Alternatively, the soil organic matter continued to increase at a steady rate through the entire 50-year chronosequence. They suggested that 'recovery' occurred after the microbial carbon: organic-matter carbon was lowered to the levels in the surrounding forests and stabilised, at a rather low value. This means that the soil organic carbon (stable organic matter) is high relative to the microbial mass (the carbon turning over). That stabilisation occurred slowly in agricultural chronosequences but recovered relatively rapidly in forested sites. On the other hand, in a five-year study of a single site, Allen (1993) reported that the organic matter did not change significantly but the microbial mass and microbial phosphorus content increased. In this arid system, the limit may be the microbial activity, at least initially, and not the organic matter content.

Allen & MacMahon (1985) suggested that not only the mass of microbes but also the spatial dispersion is important to the recovery of a site. They found that the total mass and other parameters such as mycorrhizal fungal spore counts, hyphal lengths and fungal colony counts were not reduced by a large amount. However, in the respread topsoil, these components were evenly dispersed across the study area but, in the native areas, they were very patchy and distributed in association with shrubs. Allen & Friese (1992) reported that this pattern continued for at least five years following restoration; if the site was composed of evenly dispersed grasses or weedy vegetation, the total counts of mycorrhizal spores on a surface area basis were lower and evenly dispersed across a site. When patches of shrubs were planted, these fungi clumped around those shrubs, where they could be the most effective.

CONCLUDING REMARKS

Microbes are not the same everywhere. They are easily the most diverse elements of any ecosystem and do not always recover from a disturbance no matter what the time-frame. Just as with all other components of an ecosystem, the micro-organisms cannot be assumed to recover automatically from disturbance. Care in all phases of the management from the disturbance to the placement of plants to the retention of source areas is crucial to restoring a functioning ecosystem.

Help from inoculation efforts may help in some aspects of restoration. Using native nitrogen-fixers or mycorrhizal fungi may aid in the initial recovery of individual plants. However, just as plant production requires a diversity of plants to maintain sustainability, so do the functions that are catalysed by the soil micro-organisms. As these functions are numerous, recognition of the importance and care to manage them is a crucial step in a real restoration effort. These include several specific steps that should always be undertaken:

1. Maintenance of a source area. The vast majority of taxa of soil organisms cannot be replaced artificially. They must have a source area to colonise from. In addition, animals that immigrate are important both in acting as vectors of soil organisms and in serving to move inoculum locally or create conditions locally that enhance microbial reestablishment.
2. Careful management of the topsoil. The topsoil can serve as an important reservoir of soil organisms and also provides the appropriate habitat for invading microbes.
3. Careful planting of appropriate plant species and in appropriate spatial arrays. Understanding the ecosystem that is disrupted is crucial to restoring soil microbial functioning. For example, in a grassland, a uniform grass cover may be important to recovering the ecosystem. In the Great Basin, reestablishing patches of shrubs may serve as the catalyst for microbial recovery.
4. Recognition that all groups of micro-organisms contribute to the ecosystem functioning and community sustainability is crucial to restoration. Practices that disrupt any one group, e.g. parasitic nematodes and fungi, also cascade into disrupting groups such as bacterial-feeding nematodes and mycorrhizal fungi that are critical to the success of any restoration effort.

5. Finally, it should be understood that at this time we cannot measure microbial diversity but that retaining both functional and taxon diversity is crucial for restoring a sustainable ecosystem. Continued efforts to broaden our understanding of both microbial diversity and functioning is a high priority for future restoration research.

ACKNOWLEDGMENTS

We thank Christina Doljanin and Tracy Tennant for help in putting this manuscript together. This review was prepared with support from the National Science Foundation, Conservation and Restoration Biology and Biocomplexity Programs.

REFERENCES

Aldon, E.F. (1975). Endomycorrhizae enhance survival and growth of fourwing saltbush on coalmine spoils. *USDA Forest Service Research Note R.M.*, **294**, 1–5.

Allen, E.B. & Allen, M.F. (1980). Natural re-establishment of vesicular-arbuscular mycorrhizae following stripmine reclamation in Wyoming. *Journal of Applied Ecology*, **17**, 139–147.

Allen, E.B. & Allen, M.F. (1986). Water relations of xeric grasses in the field: interactions of mycorrhizae and competition. *New Phytologist*, **104**, 559–571.

Allen, E.B., Chambers, J.C., Conner, K.F., Allen, M.F. & Brown, R.W. (1987). Natural reestablishment of mycorrhizae in disturbed alpine ecosystems. *Arctic and Alpine Research*, **19**, 11–20.

Allen, E.B., Allen, M.F., Helm, D.J., Trappe, J.M., Molina, R. & Rincon, E. (1995). Patterns and regulation of mycorrhizal plant and fungal diversity. *Plant and Soil*, **170**, 47–62.

Allen, E.B., Brown, J.S., & Allen, M.F. (in press). Restoration and Biodiversity. In *Encyclopedia of Biodiversity*, ed. S. Levin. San Diego, CA: Academic Press.

Allen, M.F. (1987). Re-establishment of mycorrhizas on Mount St. Helens: migration vectors. *Transactions of the British Mycological Society*, **88**, 413–417.

Allen, M.F. (1988*a*). Re-establishment of VA mycorrhizae following severe disturbance: comparative patch dynamics of a shrub desert and a subalpine volcano. *Proceedings of the Royal Society of Edinburgh*, **94B**, 63–71.

Allen, M.F. (1988*b*). Belowground spatial patterning: influence of root architecture, microorganisms and nutrients on plant survival in arid lands. In *The Reconstruction of Disturbed Arid Lands: An Ecological Approach*, ed. E.B. Allen, pp. 113–135. Boulder, CO: Westview Press.

Allen, M.F. (1991). *The Ecology of Mycorrhizae*. Cambridge: Cambridge University Press.

Allen, M.F. (1993). Microbial and phosphate dynamics in a restored shrub steppe in southwestern Wyoming. *Restoration Ecology*, **1**, 196–205.

Allen, M.F. (2000). Mycorrhizae. In *Encyclopedia of Microbiology*, Vol. 3, ed. M. Alexander, pp. 328–336. San Diego, CA: Academic Press.

Allen, M.F. & Friese, C.F. (1992). Mycorrhizae and reclamation success: importance and measurement. In *Evaluating Reclamation Success: The Ecological Considerations*, General Technical Report no, NE-164, eds. J.C. Chambers & G.L. Wade, pp. 17–25. Radnor, PA: US Department of Agriculture Forest Service.

Allen, M.F. & MacMahon, J.A. (1985). Impact of disturbance on cold desert fungi: comparative microscale dispersion patterns. *Pedobiologia*, **28**, 215–224.

Allen, M.F. & MacMahon, J.A. (1988). Direct VA mycorrhizal inoculation of colonizing plants by pocket gophers (*Thomomys talpoides*) on Mount St Helens. *Mycologia*, **80**, 754–756.

Allen, M.F., Allen, E.B. & West, N.E. (1987). Influence of parasitic and mutualistic fungi on *Artemisia tridentata* during high precipitation years. *Bulletin of the Torrey Botanical Club*, **114**, 272–279.

Allen, M.F., Hipps, L.E. & Wooldridge, G.L. (1989). Wind dispersal and subsequent establishment of VA mycorrhizal fungi across a successional arid landscape. *Landscape Ecology*, **2**, 165–172.

Allen, M.F., Crisafulli, C., Friese, C.F. & Jeakins, S.L. (1992). Re-formation of mycorrhizal symbioses on Mount St Helens, 1980–1990: interactions of rodents and mycorrhizal fungi. *Mycological Research*, **96**, 447–453.

Allen, M.F., Allen, E.B., Dahm, C.N. & Edwards, F.S. (1993). Preservation of biological diversity in mycorrhizal fungi: importance and human impacts. In *Human Impact on Self-Recruiting Populations*, ed. G. Sundnes, pp. 81–105. Trondheim, Norway: Tapir Publishers.

Allen, M.F., Klironomos, J. & Harney, S.K. (1997). The epidemiology of mycorrhizal fungi during succession. In *The Mycota*, Vol. VB, eds. Cr. Carroll & P. Tudzynski, pp. 169–183. Berlin: Springer-Verlag.

Allen, M.F., Allen, E.B., Zink, T.A., Harney, S., Yoshida, L.C., Siguenza, C., Edwards, F., Hinkson, C., Rillig, M.,

Bainbridge, D., Doljanin, C., & MacAller, R. (1999). Soil microorganisms. In *Ecosystems of the World*, Vol. 16, *Ecosystems of Disturbed Ground*, ed. L. Walker, pp. 521–544. New York: Elsevier.

Atlas, R.M. & Bartha, R. (1993). *Microbial Ecology.* Redwood City, CA: Benjamin & Cummings.

Azcon-Aguilar, C. & Barea, J.M. (1992). Interactions between mycorrhizal fungi and other rhizosphere microorganisms. In *Mycorrhizal Functioning*, ed. M.F. Allen, pp. 163–198. London: Chapman & Hall.

Belnap, J. (1993). Recovery rates of cryptobiotic crusts: inoculant use assessment methods. *Great Basin Naturalist*, **53**, 89–95.

Belnap, J. & Gillette, D.A. (1998). Vulnerability of desert biological soil crusts to wind erosion: the influences of crust development, soil texture, and disturbance. *Journal of Arid Environments*, **39**, 133–142.

Bergerson, F.S. (1970). Some Australian studies relating to the long-term effects of the inoculation of legume seeds. *Plant and Soil*, **32**, 727–736.

Bethlenfalvay, G.J. (1993). Mycorrhizae in sustainable agriculture plant–soil system. *Symbiosis*, **14**, 413–425.

Burggraaf, A.J.P. & Shipton, W.A. (1982). Estimates of *Frankia* growth under various pH and temperature regimes. *Plant and Soil*, **69**, 135–147.

Call, C.A. & McKell, C.M. (1982). Vesicular-arbuscular mycorrhizae: a natural revegetation strategy for disposed processed oil shale. *Reclamation and Revegetation Research*, **1**, 337–347.

Carpenter, A.T. & Allen, M.F. (1988). Responses of *Hedysarum boreale* to mycorrhizas and *Rhizobium* plant and soil nutrient changes. *New Phytologist*, **109**, 125–132.

Chatel, D.L. & Parker, C.A. (1973). Survival of field-grown rhizobia over the dry summer period in Western Australia. *Soil Biology and Biochemistry*, **5**, 415–423.

Christensen, M. (1981). Species diversity and dominance in fungal communities. In *The Fungal Community: Its Organization and Role in the Ecosystems*, eds. D.T. Wicklow & G.C. Carroll, pp. 201–231. New York: Marcel Dekker.

Collins, H.P., Robertson, G.P. & Klug, M.S. (1995). *The Significance and Regulation of Soil Biodiversity.* Dordrecht, The Netherlands: Kluwer.

Curry, J.P. & Good, J.A. (1992). Soil faunal degradation and restoration. In *Soil Restoration, Advances in Soil Science*, eds. R. Lal & B.S. Stewart, pp. 171–216. New York: Springer-Verlag.

Davies, F.T., Jr & Call, C.A. (1990). Mycorrhizae, survival and growth of selected woody plant species in lignite overburden in Texas [USA]. *Agriculture Ecosystems and Environment*, **31**, 243–252.

deBary, A. (1887). *Comparative Morphology and Biology of the Fungi, Mycettozoa and Bacteria.* Oxford: Clarendon Press.

Eldridge, D.J. & Rosentreter, R. (1999). Morphological groups: a framework of monitoring microphytic crusts in arid landscapes. *Journal of Arid Environments*, **41**, 11–25.

Elkins, N.Z., Parker, L.W., Aldon, E. & Whitford, W.G. (1984). Responses of soil biota to organic amendments in stripmine spoils in Northwestern New Mexico. *Journal of Environmental Quality*, **13**, 215–219.

Evans, R.D., & Belnap, J. (1999). Long-term consequences of disturbance on nitrogen dynamics in an arid ecosystem. *Ecology*, **80**, 150–160.

Forman, R.T.T. (1975). Canopy lichens with blue-green algae: a nitrogen source in a Colombian rain forest. *Ecology*, **56**, 1176–1184.

Fresquez, P.R. & Dennis, G.L. (1990). Composition of fungal groups associated with sewage sludge amended grassland soils. *Arid Soil Research and Rehabilitation*, **4**, 19–32.

Fresquez, P.R., Aldon, E.F. & Lindemann, W.C. (1986). Microbial re-establishment and the diversity of fungal genera in reclaimed coal mine spoils and soils. *Reclamation and Revegetation Research*, **4**, 359–367.

Friese, C.F. & Allen, M.F. (1991). The spread of VA mycorrhizal fungal hyphae in the soil: inoculum types and external hyphal architecture. *Mycologia*, **83**, 409–418.

Friese, C.F. & Allen, M.F. (1993). The interaction of harvester ants and VA mycorrhizal fungi in a patchy semi-arid environment: the effects of mound structure on fungal dispersion and establishment. *Functional Ecology*, **7**, 13–20.

Gardner, J.H. & Malajczuk, N. (1988). Recolonization of rehabilitated Bauxite mine sites in Western Australia by mycorrhizal fungi. *Forest Ecology and Management*, **24**, 27–42.

Helm, D.J. & Carling, D.E. (1993). Use of soil transfer for reforestation on abandoned mined lands in Alaska. 2: Effects of soil transfers from different successional stages on growth and mycorrhizal formation by *Populus balsamifera* and *Alnus crispa*. *Mycorrhiza*, **3**, 107–114.

Insam, H. & Domsch, K.H. (1988). Relationship between soil organic carbon and microbial biomass on chronosequences of reclamation sites. *Microbial Ecology*, **15**, 177–188.

Jasper, D.A. (1994). Management of mycorrhizas in revegetation. In *Management of Mycorrhizas in Agriculture*,

Horticulture and Forestry, eds. A.D. Robson, L.K. Abbott & N. Malajczuk, pp. 211-219. Dordrecht, The Netherlands: Kluwer.

Jasper, D.A., Robson, A.D. & Abbott, L.K. (1987). The effect of surface mining on the infectivity of vesicular arbuscular mycorrhizal fungi. *Australian Journal of Botany*, **35**, 641–652.

Jasper, D.A., Abbott, L.K. & Robson, A.D. (1989). Hyphae of a vesicular-arbuscular mycorrhizal fungus maintain infectivity in dry soil, except when the soil is disturbed. *New Phytologist*, **112**, 101–107.

Jasper, D.A., Abbott, L.K. & Robson, A.D. (1991). The effect of soil disturbance on vesicular-arbuscular mycorrhizal fungi in soils from different vegetation types. *New Phytologist*, **118**, 471–476.

Klein, D.A., McLendon, T., Paschke, M.W. & Redente, E.F. (1994). Saprophytic fungal–bacterial biomass variations in successional communities of a semi-arid steppe ecosystem. *Biology and Fertility of Soils*, **19**, 253–256.

Klein, D.A., Paschke, M.W. & Redente, E.F. (1998). Assessment of fungal–bacterial development in a successional shortgrass steppe by direct integration of chloroform-fumigation extraction (FE) and microscopically derived data. *Soil Biology and Biochemistry*, **30**, 573–581.

Klopatek, C.C., DeBano, L.F. & Klopatek, J.M. (1988). Effects of simulated fire on vesicular-arbuscular mycorrhizae in pinyon-juniper woodland soil. *Plant and Soil*, **109**, 245–249.

Kowalski, S. (1982). Role of mycorrhiza and soil fungi in natural regeneration of fir (*Abies alba* Mill) in Polish Carpathians and Sudetes. *European Journal of Forest Pathology*, **12**, 107–112.

Lange, O.L., Belnap, J., & Reichenberger, H. (1998). Photosynthesis of the cyanobacterial soil-crust lichen *Collema tenax* from arid lands in southern Utah, USA: role of water content on light and temperature responses of CO_2 exchange. *Functional Ecology*, **12**, 195–202.

Marx, D.H., Cordell, C.E., Kenney, D.S., Mexal, J.G., Artman, J.D., Riffle, J.W. & Molina, R.J. (1984). Commercial vegetative inoculum of *Pisolithus tinctorius* and inoculation techniques for development of ectomycorrhizae on bare-root tree seedlings. *Forest Science Monograph*, **25**, 1–101.

Mikola, P. (1980). *Tropical Mycorrhiza Research*. Oxford: Clarendon Press.

Miller, R.M. (1984). Microbial ecology and nutrient cycling in disturbed arid ecosystems. In *Ecological Studies of Disturbed Landscapes: A Compendium of the Results of Five Years of Research Aimed at the Restoration of Disturbed Ecosystems*, DOE/NBM-5009372 (DE85009372), technical ed. A.J. Dvorak, pp. 3-1–3-29. Office of Scientific and Technical Information, US Department of Energy.

Moldenke, A.R. & Lattin, J.D. (1990). Density and diversity of soil arthropods as biological probes of complex soil phenomena. *Northwest Environmental Journal*, **6**, 409–410.

Murie, A. (1962). *Mammals of Denali*. Denali, AK: Alaska Natural History Association.

Nash, T.H. III (1996). *Lichen Biology*. Cambridge: Cambridge University Press.

Parke, J.L., Linderman, R.G. & Trappe, J.M. (1983*a*). Effect of root zone temperature on ectomycorrhiza and vesicular-arbuscular mycorrhiza formation in disturbed and undisturbed soils of southwest Oregon. *Canadian Journal of Forest Research*, **13**, 657–665.

Parke, J.L., Linderman, R.G. & Trappe, J.M. (1983*b*). Effects of forest litter on mycorrhiza development and growth of Douglas-fir and western red cedar seedlings. *Canadian Journal of Forest Research*, **13**, 666–671.

Paul, E.A. & Clark, F.E. (1996). *Soil Microbiology and Biochemistry*. San Diego, CA: Academic Press.

Pilz, D.P. & Perry, D.A. (1984). Impact of clearcutting and slash burning on ectomycorrhizal associations of Douglas-fir seedlings. *Canadian Journal of Forest Research*, **14**, 94–100.

Read, D.J. (1983). The biology of mycorrhiza in the Ericales. *Canadian Journal of Botany*, **61**, 985–1004.

Schlesinger, W.H., Reynolds, J.F., Cunningham, G.L., Huenneke, L.F., Jarrell, W.M., Virginia, R.A. & Whitford, W.G. (1990). Biological feedbacks in global desertification. *Science*, **247**, 1043–1048.

Schwab, S. & Reeves, F.B. (1981). The role of endomycorrhizae in revegetation practices in the semiarid west. 3: Vertical distribution of vesicular-arbuscular mycorrhiza inoculum potential. *American Journal of Botany*, **68**, 1293–1297.

Shipton, W.A. & Burggraaf, A.J.P. (1982). *Frankia* growth and activity as inluenced by water potential. *Plant and Soil*, **69**, 293–297.

Smith, S.E. & Read, D.J. (1997). *Mycorrhizal Symbiosis*, 2nd edn. San Diego, CA: Academic Press.

Smolander, A. & Sundman, V. (1987). *Frankia* in acid soils of forests devoid of actinorhizal plants. *Physiologia Plantarum*, **70**, 297–303.

Sprent, J.I. (1987). *The Ecology of the Nitrogen Cycle*. Cambridge: Cambridge University Press.

Stanton, N.L. & Krementz, D. (1982). Nematode densities

on reclaimed sites on a cold desert shrub-steppe. *Reclamation and Revegetation Research*, **1**, 233–241.

Stanton, N.L., Allen, M.F. & Campion, M. (1981). The effect of the pesticide carbofuran on soil organisms and root and shoot production in shortgrass prairie. *Journal of Applied Ecology*, **18**, 417–431.

Stewart, E.L. & Pfleger, F.L. (1985). *Selection and Utilization of Mycorrhizal Fungi in Revegetation of Iron Mining Wastes*, A Mining Research Contract Report, Contract No. J0225008, project officer K. Bickel. St Paul, MN: Bureau of Mines, US Department of the Interior.

Stylinski, C. & Allen, E.B. (1999). Lack of native species recovery following severe exotic disturbance in southern California shrublands. *Journal of Applied Ecology*, **36**, 544–554.

Waaland, M.E. & Allen, E.B. (1987). Relationships between VA mycorrhizal fungi and plant cover following surface mining in Wyoming. *Journal of Range Management*, **40**, 271–276.

Waksman, S.A. (1927). Microbiological analysis of soil as an aid to soil characterization and classification. *Journal of the American Society of Agronomy*, **19**, 297–311.

Weinbaum, B.S., Allen, M.F. & Allen, E.B. (1996). Survival of arbuscular mycorrhizal fungi following reciprocal transplanting across the Great Basin, USA. *Ecological Applications*, **6**, 1365–1372.

West, N.E. & Skujins, J. (1978). Summary, conclusions, and suggestions for further research. In *Nitrogen in Desert Ecosystems*, eds. N.E. West & J. Skujins, pp. 244–253. Stroudsburg, PA: Dowden, Hutchinson & Ross, Inc.

Whitford, W.G. (1988). Decomposition and nutrient cycling in disturbed arid ecosystems. In *The Reconstruction of Disturbed Arid Ecosystems*, ed. E.B. Allen, pp. 136–161. Boulder, CO: Westview Press.

Wicklow-Howard, M. (1989). The occurrence of vesicular-arbuscular mycorrhizae in burned areas of the Snake River Birds of Prey area, Idaho [USA]. *Mycotaxon*, **34**, 253–258.

Wollum, A.G., Youngberg, C.T. & Chichester, F.W. (1968). Relation of previous timber stand age to nodulation of *Ceanothus velutinus*. *Forest Science*, **14**, 114–118.

Zak, J.C. (1992). Responses of soil fungal communities to disturbance. *Mycological Research*, **9**, 403–425.

Zak, J.C. & Rabatin, S.C. (1997). Organization and description of fungal Communities. In *The Mycota*, vol. 4, eds. K. Esser & P.A. Lemke, pp. 33–46. Berlin: Springer-Verlag.

15 • Terrestrial invertebrates

JONATHAN D. MAJER, KARL E. C. BRENNAN AND LUBOMIR BISEVAC

INTRODUCTION

In this chapter we outline some of the invertebrate-related issues that need to be considered during the restoration of ecosystems. Since a major factor in relation to this involves knowing what is present in the area, we also discuss the sampling protocols that need to be followed when surveying the invertebrate fauna. Readers are referred to the book *Animals in Primary Succession: The Role of Fauna in Reclaimed Lands* (Majer, 1989) for a comprehensive account of the importance of invertebrates in restored areas.

Horwitz *et al.* (1999) point out that restoration of a system requires more than just recreating the vegetation assemblage. Restoration may be deemed to have failed unless the services of nutrient retention and cycling, purification of air and water, detoxification and decomposition of wastes, pollination, dispersal of seeds, and other ecosystem services are recovered. Invertebrates dominate the functions and processes of most ecosystems. The goods and services provided by these systems to humanity rely on their invertebrate communities and the ecological processes that are driven by these animals. Anyone who is concerned with restoration of ecosystems must therefore consider this important component of the biota, both in terms of its role and its diversity. Although not necessarily linearly related (Schwartz *et al.*, 2000), the links between biodiversity and ecosystem functioning are demonstrable, and reflect the increased functional roles that are possible in ecosystems that contain more species (see Tilman, 1997). Diversity may increase overall ecosystem productivity, although diversity may be greatest at intermediate levels of productivity (Rosenzweig, 1971; Tilman, 1982). Increased diversity stabilises the functioning of the total ecosystem by increasing the resistance of ecosystems to perturbation, thus providing an 'insurance value' (Tilman, 1997) to minimise costs associated with unpredictable events.

Figure 15.1 illustrates the various ecosystem functions and processes that need to be re-established in the restored area and indicates some of typical organisms that are involved. Although micro-organisms, plants and vertebrates are of unquestionable importance and often interact with invertebrates, invertebrates feature prominently as drivers of these functions and processes. In the following section, we outline the importance of invertebrates in some of these ecosystem functions and processes – issues related to some of the other functions and processes are raised elsewhere in the chapter. Where appropriate, we also include anecdotal information that should be taken into account when considering restoration of ecosystems.

ROLES OF INVERTEBRATES IN ECOSYSTEM FUNCTIONING

Soil structure

Abbott (1989) has reviewed the various ways in which invertebrates contribute to the maintenance of soil structure in restored lands. Under the general umbrella of the term 'soil structure', there are various soil attributes which are important to the wellbeing of the ecosystem. Included here are the distribution of pores in the soil – important for soil aeration, drainage and root penetration; soil friability – important for root penetration and burrowing by animals; and soil stratification – important for ensuring that nutrients are available at a depth where they are available to plants. Animals

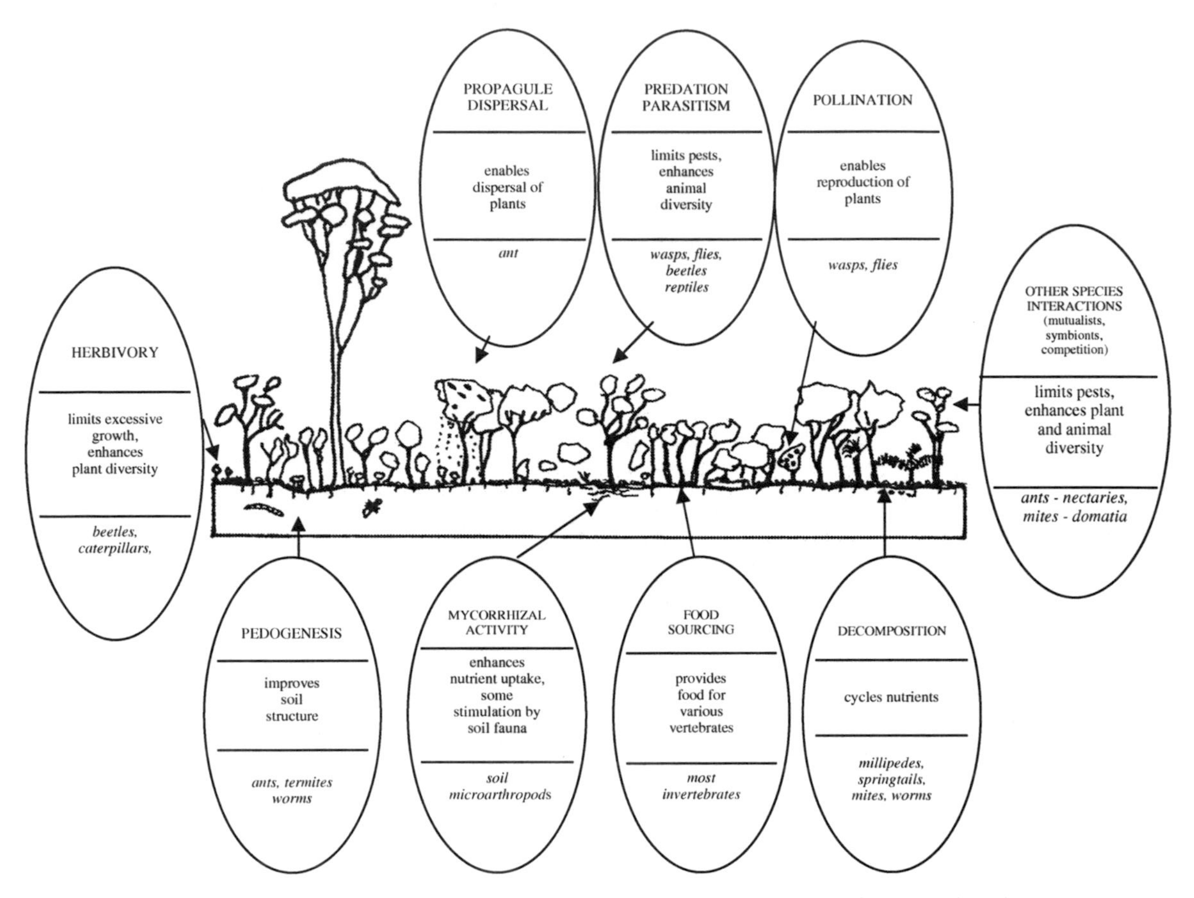

Fig. 15.1. The ecosystem functions and processes that need to be established in restored land.

have an important influence on all three structural attributes. Earthworms, termites, ants and beetles all create cavities in the soil by their burrowing activities and also increase the friability of the soil. Earthworms, termites, ants and certain vertebrates also bring soil from lower strata to the surface and hence increase nutrient availability to certain plants.

The wheat belt of Western Australia has been largely cleared for agriculture, resulting in a large shift in soil conditions. In comparison with virgin soils, cultivated soils are less tractable, they often have a compacted zone beneath the surface and are more prone to surface runoff after heavy rains (Abbott *et al.*, 1979). The structure of these soils must be improved if sustainable agriculture is to continue (Parker, 1989). Abbott *et al.* (1979) describe a 'passive' restoration attempt at Kodj-Kodjin, Western Australia in which ploughing and stock were excluded from fields for seven years. They compared this 'formerly cultivated' soil with cultivated soil and soil under virgin woodland. Physical and chemical variables were measured and the density of large soil animals estimated by taking soil cores. Although cultivated soil was 'inferior' to virgin soil in terms of pH, organic matter content, size of water-stable aggregates, compactability and ease of water infiltration, the 'formerly cultivated' soil had substantially recovered towards virgin soil levels. This was associated with a recovery of the large soil invertebrates, such as ants, termites and beetles, and the creation of pores and holes within the soil. Continuing this work at nearby Kellerberrin, Lobry de Bruyn & Conacher (1994 *a*,*b*, 1995) quantified the influence of soil invertebrates on soil structure. They specifically examined the role of ants and termites in soil turnover (bioturbation) and the influence of ant biopores on water infiltration. Ant biopores were found to conduct water four to six times faster than the control soil. The density of ant biopores in farmland soils was comparable to the naturally vegetated habitat, and varied according to soil type, suggesting that they are potentially capable of maintaining soil structure, even in disturbed environments such as a wheat field (Lobry de Bruyn, 1993).

These studies highlight how soil invertebrates rejuvenate degraded agricultural soils and there is no reason not to believe that they would not play an equally important role in soils of restored areas, such as mine sites.

Nutrient turnover

Hutson (1989) has reviewed the role of fauna in decomposition and nutrient turnover in restored lands, and has examined literature on reclaimed pit heaps in England (e.g. Hutson & Luff, 1978), open-cast coal mines in the United States (e.g. Elkins *et al.*, 1984), peat extractions in Ireland (Curry *et al.*, 1985) and polders in Holland (van Rhee, 1963). In addition, Dunger (1989) reviewed the extensive work on the soil and litter fauna of coal-mined areas in the former German Democratic Republic. Micro-organisms such as fungi and bacteria play an enormous part in this process. Nevertheless, animals such as earthworms, millipedes, termites and springtails play an overriding regulatory role. They break up the litter so that micro-organisms can act on it, redistribute the material to strata where nutrient release can take place more rapidly, or directly stimulate the activity of micro-organisms. It follows that the presence of these animals is necessary if nutrient cycles are to be established in restored areas.

A study on restored bauxite mines was conducted in the forested south-west part of Western Australia. It examined the development of the biota on a series of 30 bauxite mines, representing a wide range of restoration prescriptions and ages, and three jarrah (*Eucalyptus marginata*) forest controls. Ward *et al.* (1991) quantified the rate of decomposition of eucalypt litter using litter bags. Decomposition was assessed by weight loss of leaves and also by carbon dioxide production from leaves. Greenslade & Majer (1993) also sampled the Collembola fauna by pitfall trapping and by heat extraction of soil and litter samples. A range of environmental variables was measured in each plot (see Majer *et al.*, 1984) and their relationships with decomposition and collembolan variables were examined using multivariate statistics. The collembolan data included total species and numbers of individuals and were further divided into species associated with litter, shrubs and grasses, as well as introduced species. Many of the collembolan variables were significantly associated with particular environmental parameters, with a species-rich fauna building up in areas with a rich flora and tree cover. Decomposers built up in areas with a dense litter layer and also high shrub and tree cover. Grass-associated species were most numerous in areas of good shrub cover, while introduced species declined as the restoration matured with age. Decomposition was often higher in moist areas with higher litter and shrub cover (Ward *et al.*, 1991). These data suggest that collembolan populations are mediating decomposition. The need to conclusively differentiate between correlation and causation still exists so further studies, using microcosms, are required to confirm the association between collembolan build-up and decomposition activity in these restored areas.

A study at the Ranger uranium mine in northern Australia has found that recolonisation by ant species is correlated with other insect groups. Ant community parameters were also correlated with soil microbial biomass, a measure of nutrient-cycling potential (Andersen & Sparling, 1997). Thus, there is the potential to use surveys of certain invertebrates as surrogates for this important ecosystem function.

Mycorrhizal associations

The nutrition and vigour of many plants is enhanced by the presence of mycorrhizal fungal associations with their roots. This is because the fungus contributes to nutrient uptake by the plant. Prior to restoration, the soils of a degraded area may have little or no microbial activity. There is therefore a need to build up the microbial status so that these important plant–mycorrhizal associations may be established.

There are many records of vertebrate animals consuming the nutritious fruiting bodies of higher fungi. The spores are often present in their faeces, so it is likely that these mycophagous animals may act as their agents of dispersal and inoculation. One example of this relationship, however, involves earthworms, which can act as vectors of mycorrhizal fungi (Reddell & Spain, 1991). It therefore follows that if mycorrhizae are to become established, it is desirable that animal vectors first colonise or move through the restored area.

The case of earthworms

This chapter would not be complete without making special reference to the particularly important role of earthworms in ecosystem restoration. Indeed, their role in soils was pointed out by Charles Darwin in a book published over 120 years ago (Darwin, 1881). In mesic environments they can attain densities of 1 million per hectare, so it is not surprising that their importance in soil can outweigh that of many other invertebrate groups. Bradshaw (1983) cites a striking example in which brick rubble from an urban wasteland had become covered by several centimetres of soil as a result of earthworm activity. Although earthworms have generally been attributed the major role in soil structuring, an Australian study has shown that ants can bring 841 $g\,m^{-2}\,yr^{-1}$ of soil to the surface while earthworms only turn up 133 $g\,m^{-2}\,yr^{-1}$ (Humphreys, 1981). Readers should therefore not regard earthworms as a universal panacea for soil improvement; they should first obtain a basic understanding of the relative composition of soil faunas in the environments in which they are working.

Herbivory

It is well known that selective grazing by mammals can increase or decrease the species diversity of vegetation. More recently, it has been suggested that differential grazing by insects on particular *Eucalyptus* species may actually increase the diversity of vegetation (Morrow, 1977). The reason is that an otherwise dominant *Eucalyptus* species can be suppressed by herbivore activity, and this in turn allows other plant species to thrive. The implications of this are particularly important to land restoration since the early stages may be characterised by the abundance of a few species of plants. Herbivore activity may play an important role in reducing the dominance of these plants and thus provide opportunities for plants that are normally less able to compete with them.

Casotti & Bradley (1991) have investigated tree herbivory rates in restored Western Australian bauxite mines. They found that between 0.9 and 1.9% of leaf area could be lost to herbivores per month and that the level of herbivory was positively correlated with leaf nitrogen of the current or previous month. The rates of leaf loss were significantly different between the three eucalypt species which they investigated, lending support to the possibility that herbivores could differentially affect the vigour of certain plant species and, in the longer term, plant diversity.

Pollination

Pollination is an important ecosystem function as it provides a crucial step in the perpetuation of plant species through sexual reproduction. It thus seems desirable to re-establish pollination relationships in restored communities to produce self-sustaining plant populations. Tepedino (1979) has estimated that 67% of flowering plants rely upon insects for their pollination. The role of insects in this function is therefore critical to consider in restoration, but the effect of pollinator limitation at a community scale has rarely been studied and conclusions remain largely inferential (Whelan, 1989). However, recent reviews by Rathcke & Jules (1993) and Kearns & Inouye (1997) document an accumulation of evidence that demonstrates the relationship between insects and the persistence and resilience of a plant community through the process of pollination.

Pollination interactions are vulnerable to many disturbances related to ecosystem fragmentation and alteration (Kearns & Inouye, 1997), and may therefore provide good models for the role of invertebrates in ecosystem functions. There are very few studies of pollination dynamics in restored ecosystems. However, several trends apparent from studies in fragmented and isolated habitats are relevant, since restored communities often share biogeographic characteristics with fragments.

Insect pollinator diversity and abundance have been found to decline with decreasing habitat area and increasing isolation (Jennersten, 1988; Aizen & Feinsinger, 1994*a*). These, and other, studies (e.g. Aizen & Feinsinger, 1994*b*; Buchmann & Nabhan, 1996), have found a corresponding decline in seed set with increasing fragmentation of habitat. Plants relying on one, or a few types of pollinators, suffer lower seed set in small isolated populations if the pollinators disappear and fail to return (Jennersten, 1988; Houston *et al*., 1993; Pavlik *et al*., 1993).

Generalisation of pollination systems may buffer both plants and pollinators (Bond, 1994; Waser *et al*., 1996), but are open to exploitation by super-generalists such as the honey bee. In some circumstances, competition with this bee is detrimental to native bee abundance and diversity, and may lead to lowered seed set of some plants (Aizen & Feinsinger, 1994*a*; Paton, 1996). Generalised pollination systems can also lead to the exclusion of locally rare plant species from the community through competition for shared pollinators (Waser, 1978).

Cascade effects, in which the loss of one partner in a mutualistic relationship indirectly affects other species in the community, may follow pollination disruptions, particularly if the partner lost is a keystone species (e.g. Lambeck, 1992). Janzen (1974) has recorded such a situation in central American tropical forests. Male euglossine bees in these forests are highly specific to orchid species, but the females rely on a variety of scattered woody plants for nectar. Habitat clearing and development has led to a decline in bee habitat and numbers, fewer native woody plants, the use of weedy species in clearings by the female bees, and reduced visits to the more widely scattered native woody plants. The fate of the bees, orchids and native woody plants are thus all intimately connected. Cascade effects obviously have great potential to affect restoration efforts which may suffer pressures relevant to pollinators such as isolation, colonisation by exotic plant and animal species, altered community structure and loss of suitable nesting sites. The fact that more cases like this have not been documented probably reflects our poor understanding of pollinator dynamics at different spatial and temporal scales, and time-lags between the disappearance of pollinators and its effects on the community (e.g. Bond, 1994).

As a result of this evidence from fragmented ecosystems, it is apparent that abundance and diversity of insect pollinators do have the capacity to significantly influence the process of pollination and seed set, both at the individual plant and community level. If the dynamics of the original plant community are to be restored, the process of pollination, and the insects responsible for it, must be restored as well. Since the processes of pollination and seed set are related to a wide variety of biological and physical factors within an ecosystem, they are recommended as direct, easily measurable indicators of community health (Aizen & Feinsinger, 1994*b*).

Propagule disperal

The majority of mining companies in Australia use direct seeding to revegetate mine sites. However, the dispersal and survival of many Australian plant seeds is intimately linked with the ant fauna of the region. The removal of seeds or diaspores by ants conforms to three categories: nest decoration – removal and incorporation of the diaspore into the nest structure; granivory – removal of the diaspore and subsequent consumption of the embryo; and myrmecochory – dispersal of the diaspore while leaving the embryo intact. Nest decoration is not a widespread phenomenon, granivory is detrimental to the plant, but myrmecochory is particularly prevalent in arid countries such as Australia. In this last category, seeds possess an elaiosome, often an oil- or fat-bearing body, which is attractive to ants. After the diaspore is transported, the ants eat the elaiosome and discard the seed. This may benefit the seed by: (1) dispersing it; (2) isolating it from seed predators; (3) isolating it from wildfires while exposing it to germination-inducing temperatures; (4) enhancing longevity; (5) placing it in a nutrient enriched environment; or (6) a combination of these (Andersen, 1990; Majer, 1990).

The beneficial effects of myrmecochory to seed survival are probably not all relevant to the initially-seeded mine site. Seeds are sown in a dispersed pattern, into fertilised soils and they have often been pre-treated by heat or smoke to accelerate germination. However, as the restoration matures, it is desirable that myrmecochorous relationships are restored for improved seed survival. This is important because restored areas are often subjected to accidentally or deliberately lit fires – if the seeds are not buried they may be incinerated. Seed removal rates in 3-year-old restored areas in Western Australian bauxite mines (Majer, 1980) and 2.5-year-old mineral sand mines in Queensland (Majer, 1985) were similar to those in the original vegetation, suggesting that relationships have been restored. The relative rates of seed removal by granivores and elaiosome collectors were similar in the maturing regrowth and in the forest, suggesting that the relationship was recovering. However, in Queensland, although seed was taken in approximately equal proportions by granivores and myrmecochores in the forest, only the latter were involved in the regrowth. Andersen & Morrison (1998) provide further evidence that full restoration of this relationship may be delayed in restored areas. Working at a uranium mine in the Northern Territory, they found that seed dispersal distances on restored waste rock sites were less than 0.5 m, compared with 2.2 m in undisturbed sites. This has important implications for the survival of seedlings produced by mature plants within the restoration.

Food resources for vertebrates

One trend that frequently shows through from studies in native ecosystems is the link between the biodiversity of arthropod communities of vegetation and site fertility (e.g. Recher *et al.*, 1996*a*). On the basis of the data from eucalypt forests, the most abundant and rich arthropod communities have been found to occur in habitats with high foliage nutrient levels. There is also evidence that foliar nutrients are related to site fertility (Recher *et al.*, 1996*a*), although this relationship does not always hold. Embellished over these trends, is the tendency for particular tree species to have higher nutrient levels than co-occurring species. This relationship can have a follow-on effect to the insectivorous avifauna, with bird usage of trees reflecting the invertebrate loads which those tree species support. A similar response has been suggested in relation to some arboreal marsupials occurring in southern Australia (Braithwaite *et al.*, 1983).

This relationship has important implications for restoration. Seeding or planting trees and shrubs that have low-nutrient foliage may be an ineffective way of encouraging invertebrates and the associated insectivorous vertebrates. Thus, if restoration is to provide for a rich fauna, it should ideally include some plant species which support high invertebrate densities.

The need for diversity in restored areas

By now it should be evident that it is desirable to encourage the return of a rich invertebrate fauna, containing species that are associated with the full range of ecosystem functions and processes. But invertebrates are incredibly diverse! The question therefore arises: is it necessary for all of the original invertebrate species to recolonise the restored area, or are some of the species redundant in the ecological sense? This is a highly contentious and much-discussed issue (Walker, 1992). Ehrlich & Ehrlich (1981) have likened the situation to rivets on an aeroplane – maybe there is no problem if one or even a few are lost, but there comes a point when so many are missing that the plane crashes. The resilience of an ecosystem, such as an area of restoration, may similarly be affected by an absence of species, but at what point does this reach problematic levels? Walker's (1992) paper on 'Biodiversity and ecological redundancy' was understood by some to suggest that in those functional groups where there was some redundancy, we could afford to lose some of its members. Walker (1995) subsequently clarified this misunderstanding by pointing out that where one of the member species declines or disappears due to species-specific effects, ecological equivalence allows functional compensation by the other member species that are not so affected. In cases where species in a functional group are virtually equivalent, they probably differ in their environmental adaptations and are thus each able to compensate for species loss under different sets of environmental conditions. We therefore conclude that, wherever possible, restorers of disturbed land should strive to maximise the return of the full range of species.

RESTORATION OF INVERTEBRATES: PRACTICAL ISSUES

Facilitation of invertebrate recolonisation

Although the majority of invertebrates have excellent powers of dispersal, either by flight (most insects), ballooning (e.g. some juvenile spiders) or by travelling on the bodies of other more mobile animals (e.g. phoretic mites), others are more limited in their capacity to colonise newly restored areas. This has frequently led managers to consider the possibility of reintroducing certain invertebrates in to the area.

Earthworms have been the focus of considerable interest in this regard. Cocoons can be dispersed on the feet and feathers of birds, as evidenced by two species of earthworms being found close to a colony of gulls on an isolated area of reclaimed sea (a polder) in the Netherlands (Meijer, 1989). Both Meijer (1989) and Ma & Eijsackers (1989) canvass the possibility of translocating earthworms into reclaimed polders, but express reservations. Meijer (1989) points out that this may destroy the possibility of monitoring the natural development of the fauna and may infringe the regulations of the relevant nature conservation agency. Ma & Eijsackers (1989) cite several examples where translocations have been successful (e.g. Vimmerstedt, 1983), but point out that the process may fail if the soil is too saline or has an unacceptable level of toxicity.

Another instance where invertebrates have inadvertently been introduced is where revegetation has involved the introduction of turves of vegetation (see Davy, this volume). Although not directly stating that they came in with the plant, Leong & Bailey (2000) found numbers of the thrips *Frankliniella minuta* on translocated patches of the vernal pool plant *Blennosperma bakeri*. Although this thrips is pestiferous, it does flag the potential for translocated turves to bring in beneficial soil and litter invertebrates.

Although there is the potential to speed up the colonisation of invertebrates, it should also be remembered that many species have specific habitat and food requirements. For instance, a particular millipede may require litter of sufficient depth, while another ant species may require an appropriate mix of shaded and unshaded ground. Given the high richness of invertebrates, it is highly likely that many of them have highly specific ecological niches and that a translocation will fail if the appropriate conditions are not available. It is our opinion that encouragement of invertebrate diversity may be better served by creating the appropriate habitat conditions, a process that we discuss below.

Relationships with restoration procedures

A number of studies have shown that the type of restoration can have a profound effect on the rate of colonisation by invertebrates and also the type of animals that occupy the area. Majer (1981) compared the rate of ant and epigaeic invertebrate recolonisation in bauxite mines that had either been unvegetated, planted with a single species of *Eucalyptus* or seeded with mixed native species. Within the first three years, the ant fauna in the seeded area had become extremely diverse and had started to resemble that of the forest control. By contrast, the ant and ground invertebrate fauna of the planted plot had not developed much more that it had in the unvegetated plot. A subsequent study (Majer *et al.*, 1984) measured the ant fauna and a full set of environmental parameters in 30 restored bauxite mines and three forest controls. Analysis of the factors that contributed to the return of a rich and diverse ant fauna indicated that plant richness and diversity, plant stratification and the presence of logs all contributed to the development of a forest-like ant community. Subsequently, restoration procedures were modified to incorporate these factors into the prescriptions.

This is not the only study that has demonstrated the importance of restoration procedures. Fox & Fox (1982) have reached similar conclusions for ants in eastern Australian sand mines, as have Hawkins & Cross (1982) for insects in Alabama surface mines, Johnson *et al.* (1983) for ground and foliage arthropods on Arizona transmission line corridors and Cullen & Wheater (1997) for ground invertebrates in British limestone quarries. So great is the potential for restoration option to influence the colonisation of invertebrates, that we believe that achieving the most appropriate restoration prescription is the most economical way of encouraging the return of invertebrates.

Species-centred restoration

In some instances, the issue is not the invertebrate community in general, but restoration of the area for the benefit of a rare or threatened species. Invertebrates have not featured widely in such endeavours, partly because of the 'taxonomic impediment' (New, 1984) – we don't have described names for the majority of terrestrial invertebrates, let alone know whether they are rare or threatened. Nevetheless, the situation is rapidly changing, as evidenced by the escalation in publication of *Red Data Books* concerning various invertebrate groups (see Samways, 1994, chapter 7, for references). In our own state of Western Australia, few invertebrates were gazetted as rare or likely to become extinct until the last ten years. Now, they number 38 (plus five 'protected fauna' and 39 listed as priority fauna in need of monitoring) (Mawson & Majer, 1999), compared with a total count of 2288 plant and 208 vertebrate species.

In many countries, a 'recovery' or 'action' plan is required for species that are categorised as having the highest degree of threat. This may well lead to steps being taken to restore the animals' habitat. Because of the superior knowledge about their degree of threat, and also because of their charismatic lure, butterflies lead the list of terrestrial invertebrates for which recovery plans have been drawn up.

Pullin (1996) has reviewed the literature on restoration of butterfly populations in Britain, while New (1987) has provided equivalent information for Australia. Many of the earlier attempts involved restoration or conservation of the remaining habitat of the species, followed by breeding up and release of the butterfly, often from stock collected outside of the area or even the country! More of these releases have failed than have succeeded, often because managers have failed to take into account information on the autecology and habitat requirements of the species. In his review of British case studies, Pullin (1996) has drawn up a series of recommendations for enhancing the success of attempts to restore butterfly populations. These include the need to thoroughly study the specific habitat requirements and life cycles of the species before the restoration attempt. Then a series of hypotheses about how the species can survive should be drawn up and experimentally tested. The outcomes of these experiments may then lead to acceptance of the plans or to their rejection or modification. Although time-consuming, this should be more economical in the long term than funding restoration projects that are destined to fail. Additionally, release of animals is not enough; management of the habitat may also be necessary, particularly in habitats that have been substantially reduced in extent or been modified by human activity. Examples of such management include: regular coppicing of woodland to provide sunny clearings for the heath fritillary (*Mellicta athalia*); the clearing of gorse scrub and careful management of grazing on what was formerly calcareous grassland to provide a mosaic of short, grazed turf and patches of thyme foodplant (*Thymus praecox*) for the large blue (*Maculinea arion*); and the manipulation of grazing to provide areas of bare ground and sheep's fescue food plant (*Festuca ovina*) of the height that is preferred by the silver spotted skipper (*Hesperia comma*) (Pullin, 1996).

A potential flow-on effect from the protection of a designated species is that this may help with the conservation of other rare or threatened species: indeed the conservation effort may lead to an invertebrate fauna that is characteristic of the previously undisturbed ecosystem. This is certainly the case in some instances; the site of the endangered Piceatus jewel butterfly (*Hypochrysops piceatus*) in Queensland, Australia is now a designated nature reserve and is known to contain other, less charismatic arthropod species (Kitching, 1999). We do not advocate the restoration of habitat for individual species as a panacea for the conservation of biodiversity in general, indeed there could be instances where conservation of a species does not favour biodiversity in general. However, species-centred restoration is probably a useful adjunct to efforts to conserve and restore biodiversity in general.

Pest problems

At least during the early stages, the plant community of revegetated areas can differ from native ecosystems in a number of important ways. It can consist of fewer species – some of which may dominate the regrowth – it can contain species of plants that are not native to the area, and the vegetation may be stressed due to it growing in open conditions or on hostile substrates. Each feature can render the revegetation vulnerable to pest attack.

Revegetation that is dominated by one or a few species of plant tends towards a monoculture, much like that of an agro-ecosystem. Root (1973) has suggested that such communities may be more prone to heavy predation and insect outbreaks because of the concentration of a single resource (plant species) in the area. By contrast, species-rich communities should be less vulnerable to damage because the greater environmental complexity makes it more difficult for specialist herbivores to find their food resource (Samways, 1994). While we are not aware of reports of this contributing to pest outbreaks in revegetation, one of us (JDM) has observed grasshopper outbreaks in revegetated Wyoming coal-mines, where a simplified prairie has been recreated.

The introduction of non-native plants into an area can result in them being encountered by invertebrate herbivores with which they have no co-evolved defence mechanism. A classic example of this is where Australian *Eucalyptus* species have been used for revegetating areas in Brazil, where large tracts of these trees have been defoliated by native Lepidoptera (Pedrosa-Macedo, 1993; Zanúncio, 1993; Fagundes *et al.*, 1996), Coleoptera (Pedrosa-Macedo, 1993; Zanúncio *et al.*, 1993) and also leaf-cutting ants (*Atta* and *Acromyrmex* spp.) (Vilela, 1986; Anjos *et al.*, 1993), some of which have become significant pests.

Often when a plant is grown under environmental stress, there follows a complex change in the quantity, distribution and composition of its nitrogen content as it responds to the conditions that it is facing. Such changes can increase the amount of nitrogenous food that is available to sucking and chewing herbivores, such as psyllids and beetles respectively. Nitrogen induced by stress has been proposed as a major reason for psyllid outbreaks (White, 1969) and may be a contributing reason for pest outbreaks in restored areas (Louda, 1988). Another factor that can contribute to the elevation of plant nitrogen is the application of fertilisers, a common practice during the initial stages of revegetation. This is also known to result in pest problems in such areas (Boyer & Zedler, 1996).

INVERTEBRATES AS INDICATORS OF RESTORATION

Potential for using invertebrates

By now, it should be evident that it is desirable for managers of restored land to have an understanding of the invertebrates that are present in the area. This, however, is not generally the case. A frequently cited reason for not considering invertebrates is that they are simply too diverse to study; there are so many species that a proper consideration of them would be overwhelming. It is true that on a global scale they are diverse – estimates of terrestrial arthropod diversity range around 12.5 million (Hammond, 1992) to 30 million-plus (Erwin, 1982), although Stork (1999) believes that 10–15 million appears to be the more probable total. This richness can also be reflected at the individual site level. For instance, studies in the southwest of Australia revealed 607 arthropod species in the canopies of two eucalypt species at a single locality (Majer *et al.*, 2000), 1100 arthropod species on the bark of four eucalypt species situated along a transect through the entire forest (J.D. Majer & B.E. Heterick, unpubl. data), and 290 species of soil and litter arthropods under a small area of the same forest (Postle *et al.*, 1991). These figures are not necessarily at the highest end of the scale; tropical sites can yield more spectacular figures (e.g. Stork, 1993) and even temperate areas, commonly regarded as considerably less rich in species, can support high arthropod diversities (Platnick, 1991).

McGeoch (1998) has reviewed the issues concerning the use of terrestrial invertebrates as bio-indicators. There is obviously a need to rationalise the approach to sampling and identifying invertebrates in order to keep investigations within sensible financial and logistic limits. One approach is to consider invertebrates at 'ordinal' levels, such as slaters (Crustacea, Isopoda), springtails (Collembola) and beetles (Coleoptera). This is indeed the strategy that has been adopted by many people who have assessed invertebrate recolonisation of restored land. The disadvantage, however, is that it often fails to separate species that possess totally different ecological requirements or roles in the ecosystem. It may be possible to use springtail abundance as an indicator of the decomposer activity, but the lumping of all beetles together can result in mixing of data on herbivores, predators and decomposers. Even the lumping of springtails together is dubious – some species are cosmopolitan while others have very narrow habitat requirements, some are litter-dwellers, while others are associated with the soil. So, the presence of large numbers of springtails does not necessarily indicate that restorers of land have been successful in, say, re-establishing the original litter fauna and an adequate level of nutrient cycling.

Another approach is to sort a range of organisms into distinguishable morphospecies, often referred to as recognisable taxonomic units (RTUs), as described by Beattie (1993) and Oliver & Beattie (1993). The advantage of this approach is that the sorting may be carried out by parataxonomists, also referred to as biodiversity technicians, who have received a relatively minimal amount of training to carry out the task. This circumvents the bottleneck of access to specialised taxonomists. Although this procedure is cheaper because it uses less specialised personnel, it can lead to underestimates, or even overestimates, of the true variety of organisms within a particular taxonomic group (Cranston & Hillman, 1992). Scientists are of differing opinion as to whether this margin of error is an impediment to the comparison of invertebrates in restored areas with those in undisturbed benchmark areas.

There is also considerable interest in using one or more taxa as surrogates for a range of other groups. The reasoning is that an area that is found to support a diverse grasshopper fauna, for instance, should also support a high diversity of other invertebrate groups. Various workers have examined this idea in relatively pristine environments (e.g. Yen *et al.*, 1989; Trueman & Cranston, 1997) or a range of mildly disturbed sites (e.g. grazing: Hadden & Westbrooke, 1999; Landsberg *et al.*, 1999). They have sometimes concluded that variations in richness of certain invertebrate groups across sites are not necessarily correlated, and may not be correlated with floristic diversity. The most common reason advanced for this discrepancy is that certain taxa may be responding to different components of the environment.

Trends across sites tend to be more harmonious when highly disturbed areas, such as restored minesites, are compared with each other and with undisturbed areas (Majer, 1983), suggesting that surrogate taxa have more applicability to this type of situation. Probably the most satisfactory option is to survey a range, or 'shopping basket' (Hammond, 1994) of taxonomic groups, representing organisms that are associated with a complementary range of ecological processes. Potential groups include termites – soil structuring; springtails – nutrient cycling; hemipterans (sucking bugs) – herbivory; or ants as indicators of several processes.

Sampling protocols

In addition to knowing what invertebrates are present, managers may also use these animals to monitor the progress of restoration of disturbed ecosystems. We cannot overestimate the importance of following a sampling protocol for monitoring fauna in restored areas. By doing this, the researcher is saved the task of designing the programme, and the data that are generated should be comparable with those from other areas where the protocol has been followed. This concept has been widely adopted for the monitoring of river quality (e.g. RIVPACS: see Richardson & Jackson, this volume), and an analogous scheme, known as SOILPACS (Soil Invertebrate Prediction and Classification Scheme) has been proposed for monitoring of contaminated land (Spurgeon *et al.*, 1996). A sampling protocol that has been widely used by the authors for evaluating the success of mine-site restoration in various parts of the world is shown in Box 15.1. An important feature of this protocol is that it allows the integration of

Box 15.1. Example of a sampling protocol

This protocol was originally described by Allen (1989) and provides an appreciation of the level of baseline fauna surveys which operators of major land disturbance projects can practically attain. The approach can also be considered as a possible template for other organisations seeking to standardise their survey programmes.

It was designed for the comprehensive baseline survey commissioned by a bauxite and gold mining company (Worsley Alumina Pty Ltd, 1985, 1999). The broad objectives of the company for restoration were to regenerate a self-sustaining forest ecosystem, planned to maintain recreation, conservation and other nominated forest values. The overriding aim of the original surveys was to ensure that data were collected on a systematic and equal-effort basis, to enable legitimate comparisons between faunal richness and abundance in the restoration and the undisturbed vegetation communities.

At each intensive study location, an invertebrate sampling transect, a bird observation transect, together with a vertebrate fauna trapping grid, and a vegetation plot were established (see Fig. B15.1).

The invertebrate sampling transect was located along one of the bird observation transects at each intensive study location, and formed the focal point for the pitfall trapping, vegetation sweeping and tree beating sampling techniques employed for invertebrates. Along each transect, 10 or 20 pitfall-traps were located at 10 m intervals. At each trap location a PVC pitfall trap holding cylinder was sunk into the ground. At the commencement of each sampling period, pitfall traps comprising 18 mm internal

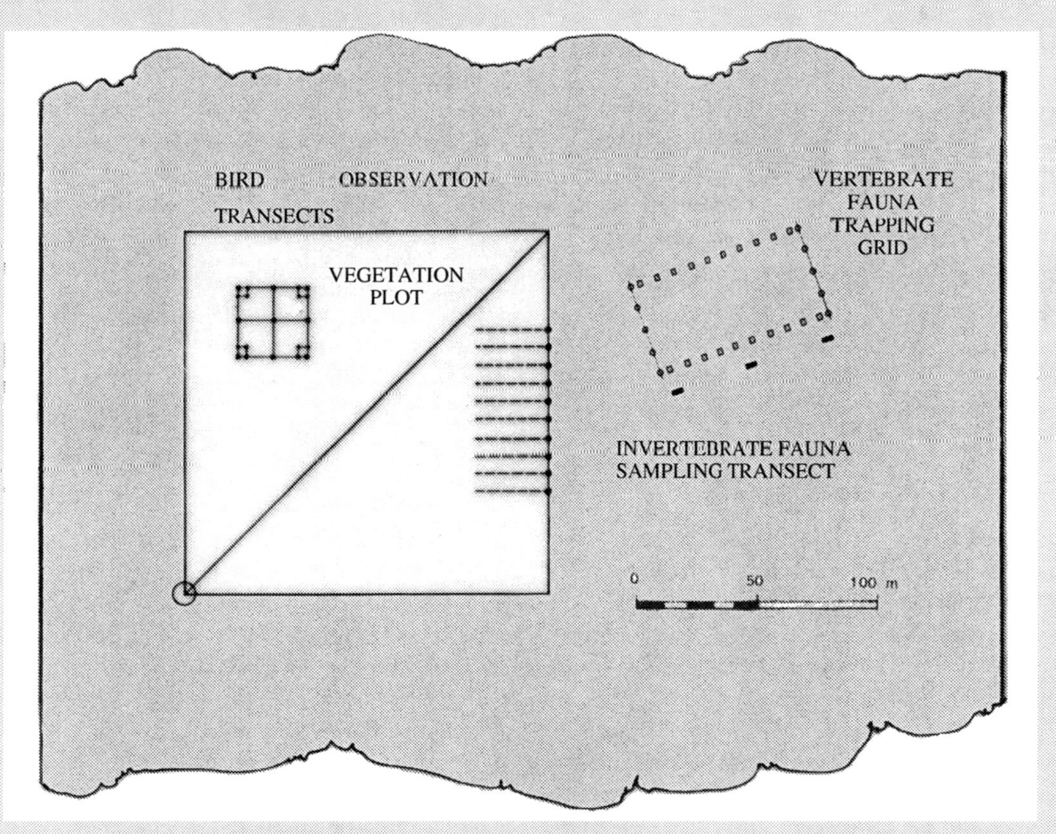

Fig. B15.1. Schematic diagram of the intensive study locations used for systematic flora, vertebrate and invertebrate fauna surveys in native vegetation and restored mine sites. Adapted from Allen (1989).

diameter (43 mm now the preferred size) Pyrex® test tubes half full of 70/30 v/v alcohol/glycerol preservative (the glycerol reducing the rate of evaporation of the alcohol) were placed in each holding cylinder. They were then left open for five consecutive days and nights. Each pitfall trap site also formed the starting-point for each vegetation sweep sample, which ran at right angles to the transect and into the core area of habitat being sampled. Herb and shrub sweeps were performed using a quadrilateral-shaped sweep-net. Each sweep consisted of a 40 m walk away from the pitfall trap site and a 40 m return walk, the latter aligned 2 m to the side of the original traverse. Eighty sweeps were performed over the entire 2 × 40 m distance and the contents of the net were placed in containers of 70% alcohol for later sorting. A simple and inexpensive suction trap that is adapted from a petrol-driven garden leaf bagger (Smith, 1999) may be used in place of a sweep-net. Up to 20 small trees or large shrubs were flagged for beat sampling in each study location. Generally, the tree nearest to each pitfall trap was selected, although in all cases a representative mix of tree species was sampled for each location. Beating was performed by jarring each tree with a stout stick, the dislodged animals being collected on a 1 m^2 calico beating tray, which was moved four times around the tree canopy during the exercise in order to ensure adequate sampling. Animals were hand-picked from the tray and placed in containers of 70% alcohol for sorting.

The three sampling methods provided relative estimates of the abundance and species richness of invertebrates in the ground (pitfall traps), herb and shrub (sweep samples) and tree (beat samples) strata. If resources permit, the leaf litter fauna may be sampled using either Tullgren funnels or the relatively inexpensive and portable Winkler sacks (Besuchet *et al.*, 1987). The advantage of the latter is that they do not require electricity and may be packed and taken on field trips. A tutorial for using these devices may be found at the web address of the American Museum of Natural History (2002).

The strength of this protocol is that it allows the integration of vegetation data with information on vertebrates and invertebrates. By following this protocol, investigators are provided with the opportunity to compare data from other areas within the country, or even between countries, just as Majer has been able to do with components of the invertebrate fauna in restored areas in Australia (e.g. Bisevac & Majer, 1999*a*), South Africa (e.g. Majer & de Kock, 1992) and Brazil (e.g. Majer, 1995).

data on invertebrates, vertebrates and plants – a feature that is missing from many restoration studies.

Experiment and monitoring designs

Of fundamental importance when applying a sampling protocol is determining an appropriate monitoring or experimental design that can accurately evaluate restoration success. Considerable literature exists on designing monitoring programmes to assess the impact and recovery of ecological communities following disturbances. Brevity does not permit us to review this literature here and, for detailed information, the reader is referred to the following publications (Green, 1979; Hurlbert, 1984; Underwood, 1991, 1992, 1997; Wiens & Parker, 1995; Michener, 1997). Instead, two commonly utilised approaches to document restoration success using terrestrial invertebrates are described: long-term regular sampling and chronosequence approaches.

Long-term regular sampling

This involves repeated sampling of impacted and control sites over a long time period (usually years). After which, at some time in the future, an assessment is made as to how well the passage of time has allowed the invertebrate community in the impacted sites to resemble communities in control sites. For example Majer & Nichols (1998) considered mine restoration success and tracked successional changes in ant communities over 14 years in three mine pits by repeatedly sampling the same mine pits and a forest control in 1977, 1978, 1979, 1980, 1981, 1984 and 1989. Using this approach, they were able to conclude that mine-pit ant community composition converged towards the forest site as time since restoration increased. However, distinct differences between the mined and unmined site still persisted.

The major drawback with long-term regular sampling is that it takes a very long time (usually many years, sometimes decades) before enough data points have been gathered for trends to become apparent. Hence, there has to be a considerable long-term commitment by either an individual or a company to safeguard the data, preserve a reference collection of species collected, protect the study sites (from additional disturbances such as wildfires), maintain consistency in sampling methods between time periods and set aside adequate funding in future years for sampling to be conducted (Strayer *et al.*, 1986; Pickett, 1991; Michener, 1997). Moreover, as restoration technology has advanced (with changes in seed mixes, topsoil return procedures, etc.), the particular sites studied may eventually end up representing outdated restoration technology. Consequently, feedback from the sites studied may be of an outdated restoration technique.

Chronosequence (space-for-time substitution)

This is a more rapid approach whereby a temporal trend is inferred by using sites of different ages (Pickett, 1989, 1991). Instead of repeatedly sampling the same sites, control sites and a series of different sites representing a chronological sequence of increasing age since disturbance are sampled. This approach is therefore only valid when the following premises are met:

1. All sites were the same prior to the disturbance.
2. The same magnitude of disturbance (*sensu* Sousa, 1984) was applied to the disturbed sites.
3. All treatments applied to sites after the disturbance were the same (e.g. topsoil return procedures, seed mixes).
4. All disturbed sites follow the same pattern or trajectory of recovery (i.e. spatial and temporal variation is equivalent) (Majer & Nichols, 1998; Pickett, 1989).

An example of a study utilising a chronosequence approach is that of Jackson & Fox (1996), who sampled ants in control sites and mine pits with age classes that represented a chronosequence of ages since restoration: 1, 5, 11, and 17 years. By adopting this approach, they were able to determine that the trajectory of recovery of ant communities following mining was moving towards (but did not match) a species composition similar to unmined forest.

Although both the study designs detailed above are widely used, none is sufficiently rigorous that differences between control and disturbed sites or their developmental trajectories are wholly attributable to the disturbance they measured rather than to other unmeasured factors. In the studies described above, no data were collected at impact sites prior to the disturbance. Differences between control and disturbed sites, therefore, may have always been present. For this reason BACI and Beyond BACI study designs have been developed (see Underwood, 1991, 1992). These are the only study designs that can reliably detect impact and recovery from disturbance against the background variation experienced in nature.

Currently, we are not aware of many terrestrial invertebrate studies that have adopted BACI or Beyond BACI designs to monitor the restoration of degraded ecosystems. However, with increasing monitoring standards, government regulation and the need to have unambiguous interpretations of data, it is foreseeable that more rigorous monitoring designs, which adopt Beyond BACI principles, will be required. Given the high abundances of terrestrial invertebrates that can be obtained during a sampling programme, their sensitivity to disturbance, their functional importance in ecosystems and their lower costs compared to sampling vertebrates, we advocate strongly their use where these advanced and extremely powerful monitoring designs are required.

Spatial and temporal heterogeneity of faunas

Considerable temporal variation is a characteristic of invertebrate populations in many ecosystems. Both species richness and community composition (the abundance of individual species within invertebrate communities) change from season to season and from year to year. Variation between seasons is driven primarily by changing climatic conditions and becomes more pronounced as temperature and rainfall differences between season increase. As such, temporal variation increases with latitude

Box 15.2. Use of invertebrates in completion criteria

Bisevac & Majer (1999*b*) evaluated the possibility of using selected invertebrate taxa as part of Completion Criteria schedules, using a Western Australian mineral sand mine as a case study. A chronosequence of ten restored areas and four heath controls was selected and sampled by a version of the plant, invertebrate and vertebrate sampling protocol described in Box 15.1. Invertebrate samples were sorted in the laboratory to broad taxonomic (ordinal) levels and, for a selection of groups representing different trophic levels, to species. Plants and vertebrates were surveyed in the same plots by independent consultants. The entire procedure, including sampling, was timed, and the period of time allocated to sampling, ordinal level sorting, sorting of each individual taxonomic group and data processing was recorded.

The mean time required to sample the entire invertebrate material was 2.3 h per plot, comprising: 1.1 h for pitfall traps, 1 h for suction samples and 0.2 h for litter samples. The time needed to identify material to ordinal level was in total 3.5 h per plot, comprising: 2.5 h for pitfall traps, 0.4 h for suction samples and 0.6 h for litter samples. The ratio of time spent in collecting material to time spent in the laboratory identifying and tabulating material to ordinal level was, on average, 1/1.5. The highest ratio was for litter samples (1/3), since much of the material was microscopic. The ratio for pitfall traps was 1/2.3 but, because relatively few animals were obtained in suction samples, the time spent in the laboratory processing this material was less than the time in the field (1/0.4).

Table B15.1 shows the number of species of plants, selected invertebrates and vertebrates obtained. It also shows the mean field time taken to sample and sort/identify these taxa from a plot. Although the most diverse group was the plants, the beetles, spiders and ants were in the same league. Birds were reasonably diverse, but reptiles, amphibians and mammals were represented by few species. These trends were also represented in the number of species obtained per hour of effort.

The results of this survey indicate that collection of invertebrate material can be undertaken as rapidly as that for plant and much more rapidly than for vertebrate material. The time to process invertebrate material is in the same order as for plants, although it is longer than the time taken to process vertebrate data. The reason for this is probably that once vertebrates are observed and recorded in the field, there is little to do. Nevertheless, the cumulative time taken to obtain and process invertebrate material is generally less than for vertebrates. The information yield for invertebrates, in terms of number of species or 'orders' per plot, is almost as high as for plants and considerably higher than for amphibians, reptiles and mammals. Taking ants as an example for the basis of comparison, only birds come anywhere near in terms of species per unit of effort. If additional groups of invertebrates had been sorted to species level, the information content of the invertebrate samples would be even higher.

This high invertebrate diversity has implications for the types of statistical analyses that may be performed on the data. Collections that contain high numbers of species lend themselves to robust data analyses by such techniques as classification, ordination and other multivariate analyses. Trends in diversity indices also tend to be more meaningful in cases where high numbers of species are involved; variations between sites in low-diversity taxa can yield serendipitous results. Thus the invertebrate data prove themselves to be cost-effective to gather and potentially high in information content. Being the most diverse members of the animal kingdom, their inclusion in surveys can contribute to the data on physical factors and plant and vertebrate communities in restored areas. As well as strengthening the conclusions reached from a study of these aspects alone, invertebrate data can provide indication of the degree of re-establishment of ecosystem functioning. When the processes of invertebrate recolonisation are sufficiently well understood, they will provide a valuable addition to the more widely accepted physical and biological measures which are currently being considered for use as Completion Criteria (Tacey & Treloar, 1994).

Table B15.1. *Numbers of species or orders sampled during five seasons in ten restored plots and four heathland control plots at the Iluka mineral sand mine, near Eneabba, Western Australia, with mean time to sort/identify each group to morphospecies or species level*

Taxon	Total morphospecies or orders	Time to sample one plot (h)[a]	Time to sort one plot (h)[b]	Total time (h)	Species or orders (h^{-1})
Plants	194	3.0	2.4	5.4	35.9
Arthropoda – Orders	27	2.3	3.5	5.8	4.7
Crustacea – Isopoda	3	2.3	0.2	2.5	1.2
Chilopoda	3	2.3	0.2	2.5	1.2
Collembola	22	2.3	11.1	13.4	1.6
Chelicerata – Araneae	96	2.3	4.3	6.6	14.6
Insecta – Coleoptera	172	2.3	6.3	8.6	20.0
Insecta – Formicidae	86	2.3	7.1	9.4	9.1
Vertebrata – Aves	47	4.0	0.0	4.0	11.8
Vertebrata – Amphibia	9	7.0	1.0	8.0	1.1
Vertebrata – Reptilia	15	7.0	2.5	9.5	1.6
Vertebrata – Mammalia	4	7.0	1.0	8.0	0.5

[a]Within the invertebrate groups and vertebrate groups, the field times collectively cover all groups, rather than being cumulative times.
[b]Time to sort invertebrate groups assumes that they have already been sorted to order.

and, in arid ecosystems, greatest variation may occur not between seasons or years but between periods of rain and no rain.

Not all groups of invertebrates respond to seasonal changes in the same manner. Recher *et al.* (1996*b*) followed changes in tree canopy arthropods sampled at three-monthly intervals for two years in Western Australia. They found that members of different functional groups peaked in abundance during different seasons. Herbivores (beetle larvae, bugs, butterflies and grasshoppers) peaked in abundance during spring, predators and/or parasites (ants, spiders and wasps) peaked in autumn, whilst decomposers (springtails) and fungivores (psocids and thrips) peaked during winter (Recher *et al.*, 1996*b*). They explained these differences as herbivores responding to periods of high leaf production, decomposers and fungivores responding to periods of high moisture availability and predators/parasites responding to prey abundance. Additionally, they documented that differences in arthropod numbers between years can be as great as between seasons.

Ideally, where seasonal variation is likely to be present, we recommend sampling once during the middle of each season. However, where financial resources are limited we suggest sampling only twice per year. This should be done when herbivores attain maximum abundance and secondly when decomposers are most active. In what seasons this occurs will differ between climatic regions. In Mediterranean climates, for instance, herbivores generally peak in abundance during spring to early summer, whilst decomposers generally peak during late winter to early spring. Seasonal trends in other climatic zones can generally be elucidated by consulting the local literature. Finally, where financial resources are extremely limited we suggest sampling only once per year during the season in which the invertebrate group/s being targeted are considered to attain highest abundance and species richness. This final suggestion is based on the premise that collecting more animals and species is likely to provide a data set in which the elucidation of similarities and differences between sites/treatments is facilitated.

Invertebrates as completion criteria indicators

The mining industry is increasingly using Completion Criteria, for assessing the end product of restoration attempts (Mills *et al.*, 1992). Such criteria allow government and other agencies, as well as mining companies, to evaluate the quality of restoration techniques employed. They also permit the success of the restoration in reaching a self-sustaining ecosystem that is suitable for the agreed final land use to be evaluated. The criteria that do exist tend to be site-specific, and include physical and biological factors, as well as water quality and safety topics. Box 15.2 gives an example of an evaluation of the use of invertebrates in schedules of Completion Criteria, and concludes that they compare favourably with plants and vertebrates in terms of cost and information yield.

CONCLUDING REMARKS

This chapter has attempted to raise some of the issues concerning invertebrates in restored lands. Some of the comments, such as those concerning sampling protocols and sampling design, apply to vertebrates as well. Others are more specific. The inclusion of invertebrates in pre-development environmental surveys and in post-development monitoring is no longer in its infancy (see Rosenberg *et al.*, 1986). It is encouraging to see that the importance of this significant component of the biota is now being recognised. Hopefully, the issues raised in this chapter will enable readers to apply our recommendations to their own projects and to gain worthwhile data that will assist with the entire restoration program.

REFERENCES

Abbott, I. (1989). The influence of fauna on soil structure. In *Animals in Primary Succession: The Role of Fauna in Reclaimed Lands*, ed. J.D. Majer, pp. 39–50. Cambridge: Cambridge University Press.

Abbott, I., Parker, C.A. & Sills, I.D. (1979). Changes in the abundance of large soil animals and physical properties of soils following cultivation. *Australian Journal of Soil Research*, **17**, 343–353.

Aizen, M.A. & Feinsinger, P. (1994*a*). Habitat fragmentation, native insect pollinators, and feral honey bees in Argentine Chaco Serrano. *Ecological Applications*, **4**, 378–392.

Aizen, M.A. & Feinsinger, P. (1994*b*). Forest fragmentation, pollination, and plant reproduction in a chaco dry forest, Argentina. *Ecology*, **75**, 330–351.

Allen, N.T. (1989). A methodology for collecting standardised biological data for planning and monitoring reclamation and rehabilitation programmes. In *Animals in Primary Succession: The Role of Fauna in Reclaimed Lands*, ed. J.D. Majer, pp. 179–205. Cambridge: Cambridge University Press.

American Museum of Natural History (2002). http://research.amnh.org/entomology/social_insects/-winkler_demo.html.

Andersen, A.N. (1990). Seed harvesting ant pests in Australia. In *Applied Myrmecology: A World Perspective*, eds. R.K. Vander Meer, K. Jaffe & A. Cedeno, pp. 34–39. Boulder, CO: Westview Press.

Andersen, A.N. & Morrison, S.C. (1998). Myrmecochory in Australia's seasonal tropics: effects of disturbance on dispersal distance. *Australian Journal of Ecology*, **23**, 483–491.

Andersen, A.N. & Sparling, G.P. (1997). Ants as indicators of restoration success: relationship with soil microbial biomass in the Australian seasonal tropics. *Restoration Ecology*, **2**, 109–114.

Anjos, N., Moreira, D.D.O. & Della Lucia, T.M.C. (1993). Manejo integrado de formigas cortadeiras em reflorestamentos. In *As Formigas Cortadeiras*, ed. T.M.C. Della Lucia, pp. 212–241. Brazil: Folha de Viçosa.

Beattie, A.J. (ed.) (1993). *Rapid Biodiversity Assessment. Proceedings of the Biodiversity Assessment Workshop.* Macquarie, NSW: Macquarie University.

Besuchet, C., Burckhardt, D.H. & Loble, I. (1987). The 'Winkler/Moczarski' eclector as an efficient extractor for fungus and litter Coleoptera. *Coleopterists' Bulletin*, **41**, 392–394.

Bisevac, L. & Majer, J.D. (1999*a*). Comparative study of ant communities of rehabilitated mineral sand mines and surrounding heathland, Western Australia. *Restoration Ecology*, **7**, 117–126.

Bisevac, L. & Majer, J.D. (1999*b*). An evaluation of invertebrates for use as success indicators for minesite rehabilitation. In *The Other 99%: The Conservation and Biodiversity of Invertebrates*, eds. W. Ponder & D. Lunney, pp. 46–49. Mosman, NSW: Royal Zoological Society of New South Wales.

Bond, W.J. (1994). Do mutualisms matter? Assessing the impact of pollinator and disperser disruption on plant extinction. *Philosophical Transactions of the Royal Society of London B*, **344**, 83–90.

Boyer, K.E. & Zedler, J.B. (1996). Damage to cordgrass by scale insects in a constructed marsh: effects of nitrogen additions. *Estuaries*, **19**, 1–12.

Bradshaw, A.D. (1983). The reconstruction of ecosystems. *Journal of Applied Ecology*, **20**, 1–17.

Braithwaite, L.W., Dudzinski, M.L. & Turner, J. (1983). Studies on the arboreal marsupial fauna of eucalypt forests being harvested for woodpulp at Eden, New South Wales. 2: Relationship between fauna densities, richness and diversity, and measured variables of habitat. *Australian Wildlife Research*, **10**, 231–247.

Buchmann, S.L. & Nabhan, G.P. (1996). *The Forgotten Pollinators*. Washington, DC: Island Press.

Casotti, G. & Bradley, J.S. (1991). Leaf nitrogen and its effects on the rate of herbivory on selected eucalypts in the jarrah forest. *Forest Ecology and Management*, **41**, 167–177.

Cranston, P.S. & Hillman, T. (1992). Rapid assessment of biodiversity using 'biological diversity technicians'. *Australian Biologist*, **5**, 144–154.

Cullen, P. & Wheater, R. (1997). The flora and invertebrate fauna of abandoned limestone quarries in Derbyshire, United Kingdom. *Restoration Ecology*, **5**, 77–84.

Curry, J.P., Kelly, M. & Bolger, T. (1985). Role of invertebrates in the decomposition of *Salix* litter in reclaimed cutover peat. In *Ecological Interactions in Soil: Plants, Microbes and Animals*, eds. A.H. Fitter, D. Atkinson, D.J. Read & M.B. Usher, pp. 355–365. Oxford: Blackwell.

Darwin, C. (1881). *The Formation of Vegetable Mould through the Action of Worms, with Observations on their Habits.* London: Faber & Faber.

Dunger, W. (1989). The return of soil fauna to coal mined areas in the German Democratic Republic. In *Animals in Primary Succession: The Role of Fauna in Reclaimed Lands*, ed. J.D. Majer, pp. 307–337. Cambridge: Cambridge University Press.

Ehrlich, P.R. & Ehrlich, A.H. (1981). *Extinction: The Causes and Consequences of the Disappearance of Species.* New York: Random House.

Elkins, N.Z., Parker, L.W., Aldon, E. & Whitford, W.G. (1984). Responses of soil biota to the organic amendments in strip-mine spoils in northwest New Mexico. *Journal of Environmental Quality*, **13**, 215–219.

Erwin, T.L. (1982). Tropical forests: their richness in Coleoptera and other species. *Coleopterists' Bulletin*, **36**, 74–75.

Fagundes, M., Zanúncio, J.C., Lopes, F.S. & Marco P. Jr (1996). Comunidades de Lepidópteros noturnos desfolhadores de eucalípto em três regiões do cerrado de Minas Gerais. *Revista Brasileira de Zoologia*, **13**, 763–771.

Fox, M.D. & Fox, B.J. (1982). Evidence for interspecific competition influencing ant species diversity in a regenerating heathland. In *Ant–Plant Interactions in Australia*, ed. R.C. Buckley, pp. 99–110. The Hague: Dr W. Junk.

Green, R.H. (1979). *Sampling Design and Statistical Methods for Environmental Biologists.* New York: John Wiley.

Greenslade, P. & Majer, J.D. (1993). Recolonization by Collembola of rehabilitated bauxite mines in Western Australia. *Australian Journal of Ecology*, **18**, 385–394.

Hadden, S.A. & Westbrooke, M.E. (1999). A comparison of the Coleoptera, Araneae and Formicidae fauna in a grazed native grassland remnant of Victoria. In *The Other 99%: The Conservation and Biodiversity of Invertebrates*, eds. W. Ponder & D. Lunney, pp. 101–106. Mosman, NSW: Royal Zoological Society of New South Wales.

Hammond, P.M. (1992). Species inventory. In *Global Biodiversity: Status of the Earth's Living Resources*, ed. B. Groombridge, pp. 17–39. London: Chapman & Hall.

Hammond, P.M. (1994). Practical approaches to the estimation of the extent of biodiversity in speciose groups. *Philosophical Transactions of the Royal Society of London* B, **345**, 119–136.

Hawkins, B.A. & Cross, E.A. (1982). Patterns of refaunation of reclaimed strip-mine spoils by nonterricolous arthropods. *Environmental Entomology*, **11**, 762–775.

Horwitz, P., Recher, H.F. & Majer, J.D. (1999). Putting invertebrates on the agenda: political and bureaucratic challenges. In *The Other 99%: The Conservation and Biodiversity of Invertebrates*, eds. W. Ponder & D. Lunney, pp. 398–406. Mosman, NSW: Royal Zoological Society of New South Wales.

Houston, T.F., Lamont, B.B., Radford, S. & Errington, S.G. (1993). Apparent mutualism between *Verticordia nitens* and *V. aurea* (Myrtaceae) and their oil-ingesting bee pollinators (Hymenoptera: Colletidae). *Australian Journal of Botany*, **42**, 369–380.

Humphreys, G.S. (1981). The rate of ant mounding and earthworm casting near Sydney, New South Wales. *Search*, **12**, 129–131.

Hurlbert, S.H. (1984). Pseudoreplication and the design of ecological experiments. *Ecological Monographs*, **54**, 187–211.

Hutson, B.R. (1989). The role of fauna in nutrient turnover. In *Animals in Primary Succession: The Role of Fauna in Reclaimed Lands*, ed. J.D. Majer, pp. 51–70. Cambridge: Cambridge University Press.

Hutson, B.R. & Luff, M.L. (1978). Invertebrate colonization and succession on industrial reclamation sites. *Scientific Proceedings, Royal Dublin Society, Series A*, **6**, 165–174.

Jackson, G.P. & Fox, B.J. (1996). Comparison of regeneration following burning, clearing or mineral sand mining at Tomago, NSW. 2: Succession of ant assemblages in a coastal forest. *Australian Journal of Ecology*, **21**, 200–216.

Janzen, D.H. (1974). The deflowering of Central America. *Natural History*, **83**, 49–53.

Jennersten, O. (1988). Pollination in *Dianthus deltoides* (Caryophyllaceae): effects of habitat fragmentation on visitation and seed set. *Conservation Biology*, **2**, 359–366.

Johnson, C.D., Beley, J.R., Ditsworth, T.M. & Butt, S.M. (1983). Secondary succession of arthropods and plants in the Arizona Sonoran Desert in response to transmission line construction. *Journal of Environmental Management*, **16**, 125–137.

Kearns, C.A. & Inouye, D.W. (1997). Pollinators, flowering plants, and conservation biology. *BioScience*, **47**, 297–307.

Kitching, R.L. (1999). Adapting conservation legislation to the idiosyncrasies of the arthropods. In *The Other 99%: The Conservation and Biodiversity of Invertebrates*, eds. W. Ponder & D. Lunney, pp. 274–282. Mosman, NSW: Royal Zoological Society of New South Wales.

Lambeck, R.J. (1992). The role of faunal diversity in ecosystem function. In *Biodiversity of Mediterranean Ecosystems in Australia*, ed. R.J. Hobbs, pp. 129–148. Chipping Norton, NSW: Surrey Beatty.

Landsberg, J., Morton, S. & James, C. (1999). A comparison of the diversity and indicator potential of arthropods, vertebrates and plants in arid rangelands across Australia. In *The Other 99%: The Conservation and Biodiversity of Invertebrates*, eds. W. Ponder & D. Lunney, pp. 111–120. Mosman, NSW: Royal Zoological Society of New South Wales.

Leong, J.M. & Bailey, E.L. (2000). The incidence of a generalist thrips herbivore among natural and translocated patches of an endangered vernal pool plant, *Blennosperma bakeri*. *Restoration Ecology*, **8**, 127–134.

Lobry de Bruyn, L.A. (1993). Ant composition and activity in naturally vegetated and farmland environments on contrasting soils at Kellerberrin, Western Australia. *Soil Biology and Biochemistry*, **25**, 1043–1056.

Lobry de Bruyn, L.A. & Conacher, A.J. (1994*a*). The effect of ant biopores on water infiltration in soils in undisturbed bushland and in farmland in a semi-arid environment. *Pedobiologia*, **38**, 193–207.

Lobry de Bruyn, L.A. & Conacher, A.J. (1994*b*). The bioturbation activity of ants in agricultural and naturally vegetated habitats in semi-arid environments. *Australian Journal of Soil Research*, **32**, 555–570.

Lobry de Bruyn, L.A. & Conacher, A.J. (1995). Soil modification by mound-building termites in the central wheatbelt of Western Australia. *Australian Journal of Soil Research*, **33**, 179–193.

Louda, S.M. (1988). Insect pests and plant stress as considerations for revegetation of disturbed ecosystems. In *Rehabilitating Damaged Ecosystems*, Vol. 2, ed. J. Cairns, pp. 51–67. Boca Raton, FL: CRC Press.

Ma, W.-C. & Eijsackers, H. (1989). The influence of substrate toxicity on soil macrofauna return in reclaimed land. In *Animals in Primary Succession: The Role of Fauna in Reclaimed Lands*, ed. J.D. Majer, pp. 223–244. Cambridge: Cambridge University Press.

Majer, J.D. (1980). The influence of ants on broadcast and naturally spread seeds in rehabilitated bauxite mines. *Reclamation Review*, **3**, 3–9.

Majer, J.D. (1981). *The role of invertebrates in bauxite mine rehabilitation*, Forests Department of Western Australia Bulletin no. 93. Perth, WA: Forests Department.

Majer, J.D. (1983). Ants: bioindicators of minesite rehabilitation, land use and land conservation. *Environmental Management*, **7**, 375–383.

Majer, J.D. (1985). Recolonization by ants of rehabilitated mineral sand mines on North Stradbroke Island, Queensland, with particular reference to seed removal. *Australian Journal of Ecology*, **10**, 31–48.

Majer, J.D. (ed.) (1989). *Animals in Primary Succession: The Role of Fauna in Reclaimed Lands.* Cambridge: Cambridge University Press.

Majer, J.D. (1990). The role of ants in Australian land reclamation seeding operations. In *Applied Myrmecology: A World Perspective*, eds. R.K. Vander Meer, K. Jaffe & A. Cedeno, pp. 544–554. Boulder, CO: Westview Press.

Majer, J.D. (1995). Ant recolonisation of rehabilitated bauxite mines at Trombetas, Pará, Brazil. *Journal of Tropical Ecology*, **11**, 1–17.

Majer, J.D. & de Kock, A.E. (1992). Ant recolonisation of sand mines near Richards Bay, South Africa: an evaluation of progress with rehabilitation. *South African Journal of Science*, **88**, 31–36.

Majer, J.D. & Nichols, O.G. (1998). Long-term recolonisation patterns of ants in rehabilitated bauxite mines, Western Australia. *Journal of Applied Ecology*, **35**, 161–182.

Majer, J.D., Day, J.E., Kabay, E.D. & Perriman, W.S. (1984). Recolonization by ants in bauxite mines rehabilitated by a number of different methods. *Journal of Applied Ecology*, **21**, 355–375.

Majer, J.D., Recher, H.F. & Ganesh, S. (2000). Diversity patterns of eucalypt canopy arthropods in eastern and western Australia. *Ecological Entomology*, **25**, 295–306.

Mawson, P.R. & Majer, J.D. (1999). The Western Australian Threatened Species Committee: lessons from invertebrates. In *The Other 99%: The Conservation and Biodiversity of Invertebrates*, eds. W. Ponder & D. Lunney, pp. 369–373. Mosman, NSW: Royal Zoological Society of New South Wales.

McGeoch, M.A. (1998). The selection, testing and application of terrestrial insects as bioindicators. *Biological Review*, **73**, 181–201.

Meijer, J. (1989). Sixteen years of fauna invasion and succession in the Lauwerszeepolder. In *Animals in Primary Succession: The Role of Fauna in Reclaimed Lands*, ed. J.D. Majer, pp. 339–369. Cambridge: Cambridge University Press.

Michener, W.K. (1997). Quantitatively evaluating restoration experiments: research design, statistical analysis, and data management considerations. *Restoration Ecology*, **5**, 324–337.

Mills, C., Chandler, R. & Caporn, N. (1992). *Completion Criteria*, Proceedings of Conference on Management and Rehabilitation of Mined Lands. Perth, WA: Curtin University of Technology.

Morrow, P.A. (1977). Host specificity in a community of three co-dominant *Eucalyptus* species. *Australian Journal of Ecology*, **2**, 89–106.

New, T.R. (1984). *Insect Conservation: An Australian Perspective.* Dordrecht: Dr W. Junk.

New, T.R. (1987). *Butterfly Conservation.* Melbourne, Victoria: Entomological Society of Victoria.

Oliver, I. & Beattie, A.J. (1993). A possible method for the rapid assessment of biodiversity. *Conservation Biology*, **7**, 562–568.

Parker, C.A. (1989). Soil biota and plants in degraded agricultural soils. In *Animals in Primary Succession: The Role of Fauna in Reclaimed Lands*, ed. J.D. Majer, pp. 423–438. Cambridge: Cambridge University Press.

Paton, D. (1996). *Overview of the Impacts of Feral and Managed Honeybees in Australia.* Canberra, ACT: Australian Nature Conservation Council.

Pavlik, B.M., Ferguson, N. & Nelson, M. (1993). Assessing limitations on the growth of endangered plant populations. 2: Seed production and seed bank dynamics of *Erysimum capitatum* ssp. *angustatum* and *Oenothera deltoides* ssp. *howellii*. *Biological Conservation*, **65**, 267–278.

Pedrosa-Macedo, J.H. (1993). *Manual de pragas em florestas*, Vol. 2, *Pragas florestais do sul do Brasil.* Viçosa, Brazil: Instituto de Pesquisas e Estudios Florestais/Sociedade de Investigações Florestais.

Pickett, S.T.A. (1989). Space-for-time substitution as an alternative to long-term studies. In *Long-Term Studies in Ecology: Approaches and Alternatives*, ed. G.E. Likens, pp. 110–135. New York: Springer-Verlag.

Pickett, S.T.A. (1991). Long-term studies: past experience and recommendations for the future. In *Long-term Ecological Research*, ed. P.G. Risser, pp. 71–88. Chichester, UK: John Wiley.

Platnick, N.I. (1991). Patterns of biodiversity: tropical vs. temperate. *Journal of Natural History*, **2**, 1083–1088.

Postle, A.C., Majer, J.D. & Bell, D.T. (1991). A survey of selected soil and litter invertebrate species from the northern jarrah (*Eucalyptus marginata*) forest of Western Australia, with particular reference to soil-type, stratum, seasonality and the conservation of forest fauna. In *Conservation of Australia's Forest Fauna*, ed. D. Lunney, pp. 193–203. Mosman, NSW: Royal Zoological Society of New South Wales.

Pullin, A.S. (1996). Restoration of butterfly populations in Britain. *Restoration Ecology*, **4**, 71–80.

Rathcke, B.J. & Jules, E.S. (1993). Habitat fragmentation and plant–pollinator interactions. *Current Science*, **65**, 273–277.

Recher, H.F., Majer, J.D. & Ganesh, S. (1996*a*). Eucalypts, arthropods and birds: on the relation between foliar nutrients and species richness. *Forest Ecology and Management*, **85**, 177–195.

Recher, H.F., Majer, J.D. & Ganesh, S. (1996*b*). Seasonality of canopy invertebrate communities in eucalypt forests of eastern and western Australia. *Australian Journal of Ecology*, **21**, 64–80.

Reddell, T. & Spain, A.V. (1991). Earthworms as vectors of viable propagules of mycorrhizal fungi. *Soil Biology and Biochemistry*, **23**, 767–774.

Root, R.B. (1973). Organization of a plant–arthropod association in simple and diverse habitats: the fauna of collards (*Brassica oeracea*). *Ecological Monographs*, **43**, 95–124.

Rosenberg, D.M., Danks, H.V. & Lehmkuhl, D.M. (1986). Importance of insects in environmental impact assessment. *Environmental Management*, **10**, 773–783.

Rosenzweig, M.L. (1971). Paradox of enrichment: destabilization of exploitation ecosystems in ecological time. *Science*, **171**, 385–387.

Samways, M.J. (1994). *Insect Conservation Biology.* London: Chapman & Hall.

Schwartz, M.W., Brigham, C.A., Hoeksema, J.D., Lyons, K.G., Mills, M.H. & van Mantgem, P.J. (2000). Linking biodiversity to ecosystem function: implications for conservation ecology. *Oecologia*, **122**, 297–305.

Smith, J.W.C. (1999). A new method for handling invertebrates collected using standard vacuum-sampling apparatuses. *Australian Journal of Entomology*, **38**, 227–228.

Sousa, W.P. (1984). The role of disturbance in natural communities. *Annual Review of Ecology and Systematics*, **15**, 353–391.

Spurgeon, D.J., Sandifer, R.D. & Hopkin, S.P. (1996). The use of macro-invertebrates for population and community monitoring of metal contamination: indicator taxa, effect parameters and the need for a soil invertebrate prediction and classification scheme (SOILPACS). In *Bioindicator Systems for Soil Pollution*, eds. N.M. Straalen & D.A. Krivolutsky, pp. 95–110. Dordrecht: Kluwer.

Stork, N.E. (1993). How many species are there? *Biodiversity and Conservation*, **2**, 215–232.

Stork, N.E. (1999). Estimating the number of species on Earth. In *The Other 99%: The Conservation and Biodiversity of Invertebrates*, eds. W. Ponder & D. Lunney, pp. 1–7. Mosman, NSW: Royal Zoological Society of New South Wales.

Strayer, D., Glitzenstein, J.S., Jones, C.G., Kolasa, J., Likens, G.E., McDonnell, M.J., Parker, G.G. & Pickett, S.T.A. (1986). *Long-Term Ecological Studies: An Illustrated Account of their Design, Operation, and Importance to Ecology.* New York: M.F. Cary Arboretum, New York Botanic Gardens.

Tacey, W. & Treloar, J. (1994). What do we want Completion Criteria to achieve? In *Proceedings of the 19th Australian Mining Industry Council Environmental Workshop*, pp. 246–256. Canberra, ACT: Australian Mining Industry Council.

Tepedino, V.J. (1979). The importance of bees and other insect pollinators in maintaining floral species composition. In *Great Basin Naturalist Memoirs*, No. 3, *The Endangered Species: A Symposium*, pp. 39–150. Provo, UT: Brigham Young University.

Tilman, D. (1982). *Resource Competition and Community Structure.* Princeton, NJ: Princeton University Press.

Tilman, D. (1997). Biodiversity and ecosystem functioning. In *Nature's Services: Societal Dependence on Natural Ecosystems*, ed. G.C. Daily, pp. 93–112. Washington, DC: Island Press.

Trueman, J.W.H. & Cranston, P.S. (1997). Prospects for the rapid assessment of terrestrial invertebrate biodiversity. *Memoirs of the Museum of Victoria*, **56**, 349–354.

Underwood, A.J. (1991). Beyond BACI: experimental designs for detecting human environmental impacts on temporal variations in natural populations. *Australian Journal of Marine and Freshwater Research*, **42**, 569–587.

Underwood, A.J. (1992). Beyond BACI: the detection of environmental impact on populations in the real, but variable, world. *Journal of Experimental Marine Biology and Ecology*, **161**, 145–178.

Underwood, A.J. (1997). *Experiments in Ecology: Their Logical Design and Interpretation Using Analysis of Variance.* Cambridge: Cambridge University Press.

van Rhee, J.A. (1963). Earthworm activities and the breakdown of organic matter in agricultural soils. In *Soil Organisms*, eds. J. Doeksen & J. van der Drift, pp. 54–59. Amsterdam: New Holland.

Vilela, E.F. (1986). Status of leaf-cutting ant control in forest plantations in Brazil. In *Fire Ants and Leaf-cutting Ants: Biology and Management*, eds. C.S. Lofgren & R.K. Vander Meer, pp. 399–408. Boulder, CO: Westview Press.

Vimmerstedt, J.P. (1983). Earthworm ecology in reclaimed opencast coal mining sites in Ohio. In *Earthworm Ecology*, ed. J.E. Satchell, pp. 229–240. London: Chapman & Hall.

Walker, B.H. (1992). Biological diversity and ecological redundancy. *Conservation Biology*, **6**, 18–23.

Walker, B.H. (1995). Conserving biological diversity through ecosystem resilience. *Conservation Biology*, **9**, 747–752.

Ward, S.C., Majer, J.D. & O'Connell, A.M. (1991). Decomposition of eucalypt litter on rehabilitated bauxite mines. *Australian Journal of Ecology*, **6**, 251–257.

Waser, N.M. (1978). Interspecific pollen transfer and competition between co-occuring plant species. *Oecologia*, **36**, 223–236.

Waser, N.M., Chittka, L., Price, M.V., Williams, N.M. & Ollerton, J. (1996). Generalization in pollination systems, and why it matters. *Ecology*, **77**, 1043–1060.

Whelan, R.J. (1989). The influence of fauna on plant species composition. In *Animals in Primary Succession: The Role of Fauna in Reclaimed Lands*, ed. J.D. Majer, pp. 107–142. Cambridge: Cambridge University Press.

White, T.C.R. (1969). An index to measure weather-induced stress on trees associated with outbreaks of psyllids in Australia. *Ecology*, **50**, 905–909.

Wiens, J.A. & Parker, K.R. (1995). Analysing the effects of accidental environmental impacts: approaches and assumptions. *Ecological Applications*, **4**, 1069–1083.

Worsley Alumina Pty Ltd (1985). *Worsley Alumina Project: Flora and Fauna Studies, Phase Two.* Perth, WA: Worsley Alumina Pty Ltd.

Worsley Alumina Pty Ltd (1999). *Boddington Gold Project: Flora and Fauna Studies.* Perth, WA: Worsley Alumina Pty Ltd.

Yen, A.L., Robertson, P. & Bennett, A.F. (1989). A preliminary analysis of patterns of distribution of plant and animal communities in the Victorian mallee. In *The Mallee Lands: A Conservation Perspective*, eds. J.C. Noble, P.J. Joss & G.K. Jones, pp. 54–60. Melbourne, Victoria: CSIRO.

Zanúncio, J.C. (1993). *Manual de Pragas em florestas*, Vol. 1, *Lepidoptera desfolhadores de eucalípto: Biologia, ecologia e controle.* Viçosa, Brazil: Instituto de Pesquisas e Estudos Florestais/Sociedade de Investigações Florestais.

Zanúncio, J.C., Bragança, M.A.L., Laranjeiro, A.J. & Fagundes, M. (1993). Coleópteros associados a eucaliptocultura em regiões de São Mateus e Aracruz, Espírito Santo. *Revista Ceres*, **40**, 583–589.

16 • Aquatic invertebrates

JOHN S. RICHARDSON AND MICHAEL J. JACKSON

INTRODUCTION

Aquatic invertebrates can be found in just about every imaginable freshwater habitat. They bridge the divide from glacial outwash to estuaries, torrential waterfalls to stagnant ponds and from water trapped in the tiny bracts of plants to the largest of our lakes and rivers. Across all of these habitats, invertebrates constitute the bulk of species diversity and account for most of the secondary productivity, as well as performing a multitude of ecological roles. They also create the vital link from primary producers and detrital materials to higher trophic levels. Not surprisingly then, the contribution of restoration activities to the structure and function of aquatic systems will be mediated largely through responses of invertebrates.

Aquatic invertebrates are represented by at least nine phyla in freshwater ecosystems, including the arthropods, molluscs, worms, rotifers and nematodes and many of the family groups are key players in food web structure and function. Various invertebrates may graze on algae, vascular plants and detritus or they may prey upon bacteria, protozoa, other invertebrates, and sometimes even fish or larval amphibians. Others also are parasitic.

Invertebrates influence nutrient cycling and rates of primary production in most aquatic environments. They play an important role in decomposition processes and in altering the transportation rates of organic and inorganic particles in streams (e.g. Wallace & Webster, 1996). Other invertebrates can have serious impacts on resource levels in freshwaters, for instance, the crustacean 'water fleas' *Daphnia* spp. that feed on planktonic algae are capable of visibly depleting the algal biomass of an entire lake.

There are many perturbations inflicted upon freshwater habitats that require active intervention in order to restore the lost structure and function. For example, water quality may be degraded by pollutants, and changes to hydrology (e.g. flow control, impounding, water extraction) can produce marked impacts on physical habitat, temperature regimes and other properties critical to particular species. Channelling, dredging and dyking have led to profound changes in many rivers around the world and the removal of vegetation and wood from lake margins or stream channels has often resulted in much diminished habitat complexity and stability. Finally, there are the problems caused by introductions of invertebrates, or the effects of introductions on native invertebrates. All of these insults have resulted in habitat alterations and impairment of freshwater ecosystems.

Rather than costly restoration, conservation of freshwater ecosystems by protective measures, for instance riparian buffers and watershed management, may be the best course of action (Downs *et al.*, volume 2). In some circumstances, riparian buffers have been shown to maintain water quality and habitat architecture for invertebrates (e.g. Newbold *et al.*, 1980; Castelle *et al.*, 1994) and in many parts of the world this is the primary line of defence for protection of aquatic habitats. Despite such efforts, restoration measures may still be required.

Among the many ecological objectives behind the restoration of aquatic environments is the enhanced capacity of a system to express its natural dynamics, i.e. through restoring functional attributes (Ebersole *et al.*, 1997). This will necessarily involve invertebrates, as most freshwater systems are dominated in terms of abundance, secondary production and numbers of species by invertebrates. Invertebrates are therefore frequently the key to maintaining diverse and productive ecosystems. Invertebrates also occupy every conceivable microhabitat, burrowing within plants and woody debris, clinging onto or wedging between mineral substrates, burying themselves in sediment and organic debris, migrating in and out of weedbeds, and even attaching themselves to other organisms. Maintenance of habitat complexity is often critical to the provision of these essential microhabitats, especially when animals need to seek refuge from predators or disturbance events (Gore *et al.*, 1998; Lancaster, 1999; Lake, 2000).

Restoration works are judged more often on the basis of physical structure or water quality, than on the basis of biological measures (e.g. Frissell & Nawa, 1992). Furthermore, the motivation for restoration activities often focuses on vertebrates, although invertebrates are increasingly being considered, at least in terms of their responses, and exploited for their role in restoring ecosystem function or as indicators of ecosystem condition. As more invertebrate species are listed as threatened or endangered, we should anticipate more projects directed specifically towards the recovery of invertebrate populations.

In this chapter we limit our discussion to freshwater habitats, but we have omitted treatment of wetlands and estuaries, since there are many books on those subjects already (e.g. Means & Hinchee, 2000; Zedler, 2001). We have chosen therefore to focus our attention on lakes, ponds and running waters. We concentrate here on the ecological interactions of invertebrates and the multiple roles that they may perform in the restoration of lakes and rivers together with their usefulness in monitoring restoration attempts.

ECOLOGICAL INTERACTIONS OF INVERTEBRATES

Invertebrates are critical links in all aquatic food webs. In most aquatic ecosystems the vertebrates are predaceous, and hence the invertebrates take up an intermediate position in the food web, processing algae or detritus into packages of food suitable for fish and other vertebrates. Fish often have specialist invertebrate diets and there are many examples of fish growth rates being limited by their supply of invertebrate prey (see review in Richardson, 1993). Some birds (e.g. dippers [*Cinclus mexicanus*], blue ducks [*Hymenolaimus malacorhynchus*], harlequin ducks [*Histrionicus histrionicus*]) and mammals (e.g. water shrews [*Neomys fodiens*, *Sorex bendirii*, *Sorex palustris*]) are also active, specialist foragers on aquatic invertebrates as are many juvenile and adult amphibians. In addition, the winged adult stages of aquatic insects often contribute to terrestrial food webs through predation by a host of animals including birds, bats, dragonflies and spiders.

Cummins (1973) described a functional feeding approach to the organisation of invertebrates in aquatic food webs (Fig. 16.1). Animals are classed on the basis of their food and means of acquisition. Those animals consuming detritus (dead organic matter, such as leaves) are divided into 'shredders' feeding on larger particles of detritus (>1 mm), and 'collectors' consuming smaller particles (<1 mm). Collectors are further classified by their mechanism of food capture as 'filterers' for all those taxa that use filters of any sort, and 'gatherers' that sequester particles using mouthpart brushes, etc. Some examples of filterers include blackfly larvae (Simuliidae), net-spinning caddisflies (e.g. Hydropsychidae and Philopotamidae), Cladocera, copepods, and mosquito larvae (Culicidae). 'Scrapers' or 'grazers' consume microscopic layers of periphyton (algae and bacteria) on rock and other surfaces, and a few invertebrates even pierce algal cells, e.g. larvae of the caddisfly family Hydroptilidae. Some invertebrates are parasites, for instance mermithid nematodes, horsehair worms (Nematomorpha), larval mites (Acari) and some Hymenoptera. Finally, there are many predaceous invertebrates, including all the dragonflies and damselflies (Odonata), dytiscid beetles, leeches (Hirundinaea) and representatives of many other groups.

Inputs of organic matter (detritus) fuel a considerable proportion of the energetics of aquatic food webs (e.g. Richardson, 1991; Wallace *et al.*, 1999). Detritivorous invertebrates can have a major influence on decomposition of detrital materials. In an experimental reduction of stream insects using insecticide, Wallace *et al.* (1991) were able to reduce detrital consumption by 56–59%. One estimate of consumption rates of leaf litter detritus by invertebrates was 25% of the input rate (Cuffney *et al.*, 1990).

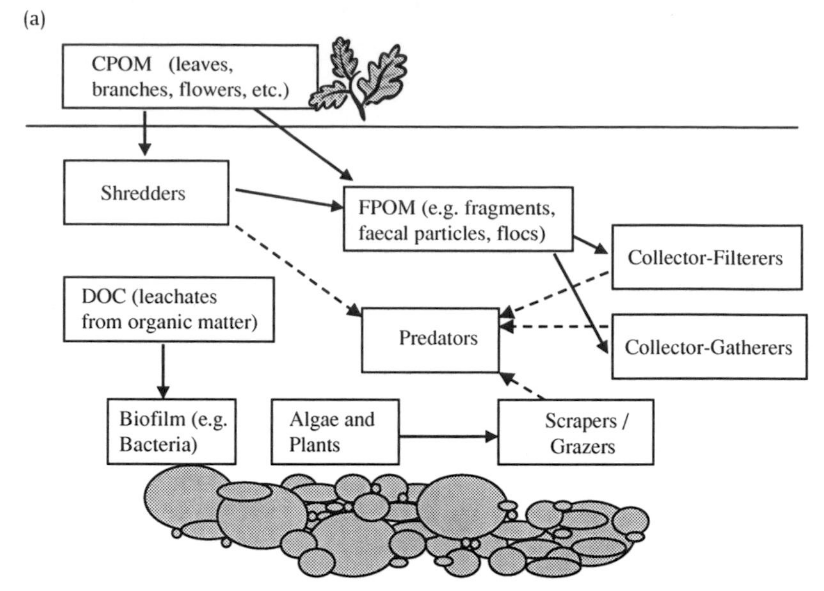

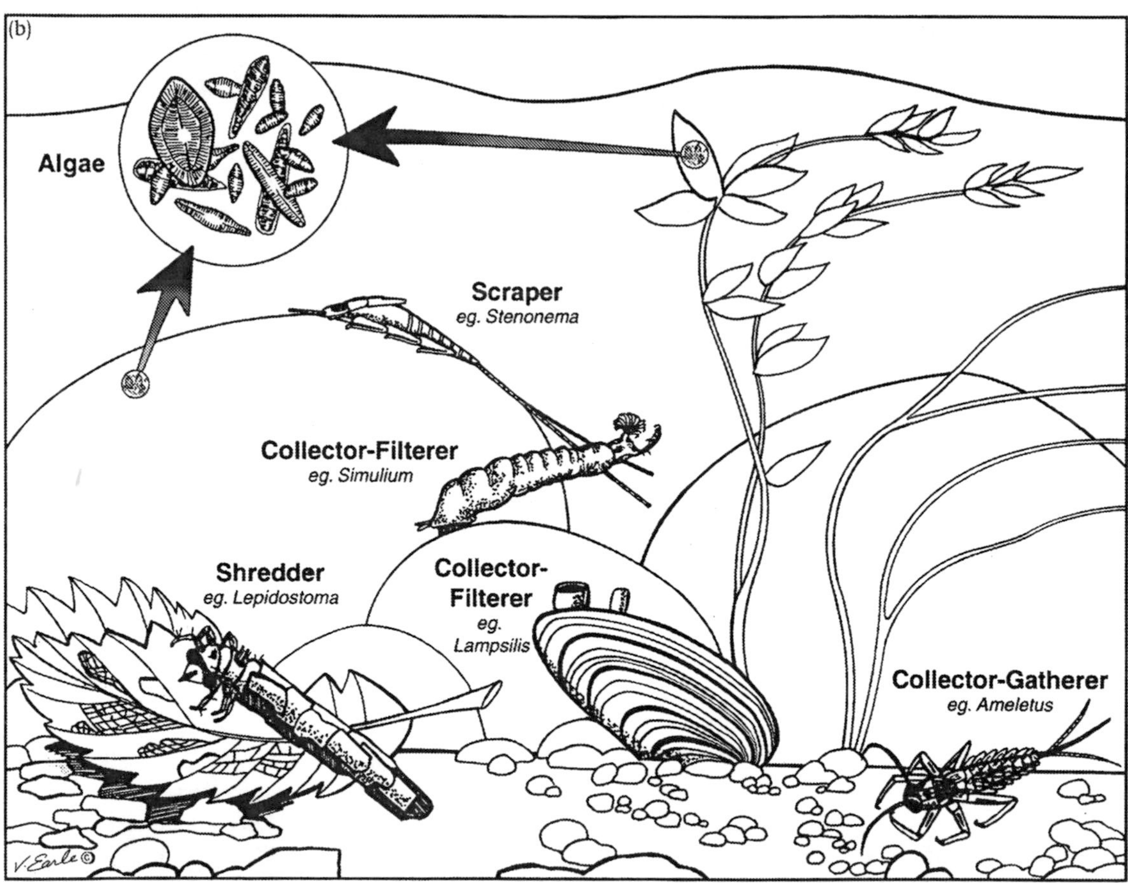

Fig 16.1. (a) Simplified food web of a stream based on major energetic pathways. CPOM enters the stream channel from the surrounding riparian areas contributing to the energy base of the stream food web. Organisms are grouped by their feeding mode or guild; see text for examples of species in each guild. Abbreviations: CPOM, coarse particulate organic matter (>1 mm); FPOM, fine particulate organic matter (0.63 μm–1 mm); DOC, dissolved organic matter (<0.63 μm). (b) Part of a stream bottom showing the major compartments into which lotic feeding guilds, or 'functional feeding groups' are classified, along with an example of each.

Aquatic food webs have a large number of omnivorous invertebrates. In particular many species feed on periphyton, which comprises not only algae, but also a large component of bacteria, protozoa and other microscopic animals. Periphyton can be largely composed of biofilms, bacteria and fungi feeding on dissolved organic carbon as part of the 'microbial loop' (Hairston & Hairston, 1993). The high degree of detritivory and omnivory, and the 'microbial loop' constitute major distinctions from terrestrial systems (Morin, 2000).

Running waters

Two of the key features of streams and rivers that set them apart from other aquatic environments are the disturbance caused by large variation in flow (floods and droughts), and the fact that moving water brings with it both food and assistance with motility. Flow variation can lead to profound changes in stream food webs and disturbance is widely acknowledged to be a key ecosystem process.

Refuge from flows and floods

High variation in discharge can result in floods capable of reshaping stream beds (see Newson *et al.*, this volume). Invertebrates respond to floods by moving to portions of the stream that are less hydraulically stressful, or they are killed. Evidence shows that some invertebrates take refuge from floods amongst the sheltered habitats ('refugia') within complex stream channels, such as behind and under large woody debris or large rocks (e.g. Lancaster, 1999), or deep into the substrate. Invertebrates also have been shown to move to shallow, newly inundated floodplains during high flow (e.g. Rempel *et al.*, 2000). Differential susceptibility to disturbance may reshape stream communities in the short term. For instance, the cased caddisfly *Dicosmoecus gilvipes* is fairly resistant to predators, but susceptible to mortality during floods (Wootton *et al.*, 1996). Following floods, losses of these large-bodied insects result in reduced competition for benthic algae and in large increases of small, faster-reproducing invertebrates that are more vulnerable to predation by fish. Stream complexity, either natural or resulting from restoration, can have a large influence on the provision of refuge from floods and the impact of disturbances on the stream community. Even under non-flood conditions invertebrates in streams have many means of holding on to avoid being swept away by the current, including hooks, claws, suckers and even silk threads.

One of the common means of restoration in running waters is the placement of woody materials to stabilise stream substrates and create heterogeneity of flow conditions and geomorphology. Stream habitats are classified in various ways based on flow velocity, depth, the bed materials, and the mechanism by which flow is locally affected (e.g. different kinds of pools), all of which contribute to the diversity of habitats available, and the organisms living therein. Wood is also an important habitat element for invertebrates in both streams and lakes. In low-gradient rivers where substrate is fine-grained, woody material forms an important substrate, and may support by far the largest portion of a river's production (e.g. Benke, 1998). Wood is also a substrate for the growth of biofilm, which is consumed directly by a number of invertebrates (Anderson *et al.*, 1978). There are even invertebrates that burrow into wood and complete their larval development there (e.g. *Lara* spp. and *Lipsothrix* spp.). Wood in streams may reduce the rate of export of detrital materials or nutrients contributing to productivity.

Invertebrates may also have an influence on the geomorphology of stream channels and influence sediment transport processes by their actions, sometimes called 'ecosystem engineering'. Some larger species of invertebrates have been shown to increase the movements of fine sediments stirred up by their foraging behaviour and general movements. For instance, crayfish moving around in artificial streams increased sand and gravel transport rates by more than four fold (Statzner *et al.*, 2000) and predaceous stoneflies increased fine sediment movements by about 25% over controls (Zanetell & Peckarsky, 1996). In an experimental insecticide treatment of a small stream the export rate of fine inorganic particles was decreased by 75% relative to a control stream, presumably due to the reduction in insect numbers (Wallace *et al.*, 1993). Some invertebrates may also enhance the stability of bottom substrate, for instance, net-spinning caddisfly larvae bind gravel and cobbles together with their silk nets thereby doubling the critical shear stress for movement (Statzner *et al.*, 1999).

Food delivery and movements in flowing water

Flowing water delivers food to a diverse group of filterer species in streams, which either construct nets to capture food using silk or mucus, or have some morphological structure to trap particles. These animals collect micro-organisms (e.g. bacteria), algae, small invertebrates, or particles of detritus from the passing water and they include a large variety of caddisfly species, black fly larvae, some midges (e.g. *Rheotanytarsus* spp.), mussels (Unionidae, Margaritiferidae and Sphaeriidae), and even a few mayflies (e.g. *Anepeorus* sp. and *Dolania* sp.). There is a great deal of variation between species or age classes within a species in the size of the filtering apparatus and the size of particles they collect. This food delivery mechanism can support enormous densities of stream invertebrates. The density of filterers reaches its highest level at lake outlets, where they can feed on the highly nutritious food particles exported from the lakes (Richardson & Mackay, 1991). Wotton (1987) documented over 1.2 million black fly larvae per square metre at an English lake outlet. Filterers may have a profound impact on water clarity and are able to consume from 5% to 100% of the suspended particles per kilometre of stream length (Heard & Richardson, 1995; Strayer *et al.*, 1999).

Animals also take advantage of flowing water as a means to move around streams as 'drift'. There are many non-exclusive explanations for this phenomenon. Invertebrates can use drift to escape from predators (often other species of invertebrates) in the benthos (e.g. Peckarsky, 1996; Huhta *et al.*, 1999). Drift may be used to move amongst food patches (Kohler, 1992) and as the flow increases or decreases it allows invertebrates to move to preferred microhabitats (Minshall & Winger, 1968). Drift is especially important for the numerous stream fishes, notably salmonids, which obtain their invertebrate prey directly from the water column.

Food web interactions in streams

In forested landscapes, the energetics of small streams is very largely based on detrital inputs from the forest canopy. Intermediate-sized streams have less canopy shading and are fuelled more by primary production, mostly algae. Large rivers are often deep and turbid, reducing the potential for primary production, and they rely more on the import of organic matter from upstream (Vannote *et al.*, 1980). Streams in non-forested settings (e.g., prairies, subalpine areas, tundra) are driven largely by primary production or dissolved carbon from groundwater. Some evidence of the importance of detritus to small, forested streams comes from energy budgets and experimental manipulations of detrital abundance (Richardson, 1991; Wallace *et al.*, 1999). Wallace *et al.* (1999) used an experimental exclusion of leaf litter inputs to a small stream and reduced production to 22% of pre-treatment levels after four years, including populations of detritivores as well as their invertebrate predators. The productivity of aquatic food webs is dependent upon conversion of algae or detritus to larger packages, i.e. invertebrate bodies, which only then become available to larger predators, such as fish or birds.

Invertebrates play a crucial role in food webs and trophic cascades in streams. Selective predation on particular species of invertebrates often results in considerable changes. Food web effects involve invertebrates 'transmitting' signals, generated by fish populations, through to the algal community (Power, 1992). Some invertebrates may also act as 'ecosystem engineers', for example, by clearing substrate of fine particles. In a tropical stream, algal biomass more than doubled in the presence of freshwater shrimps (Atyidae) that reduced silt relative to exclusion samples which, in turn, enhanced the survival of other invertebrates that consumed algae (Pringle *et al.*, 1993). Other species, such as crayfish, may reduce cover of filamentous algae, allowing for greater productivity of diatoms, and promoting invertebrates incapable of grazing filamentous types (e.g. Creed, 1994).

Stream invertebrates also contribute to a number of clear or putative facilitation interactions. Shredders break up larger particles of detritus that subsequently become available to collectors, although the evidence for the importance of this process in nature is limited (Heard & Richardson, 1995). Other species may repackage particles into larger bundles in their faeces making them available as food particles to other species (Wotton & Malmqvist, 2001). For instance, dense aggregations of blackfly larvae produced over 4.8 kg per day of larger particles in a 500 m reach of stream (Wotton *et al.*, 1998). The larvae of Hine's emerald dragonfly (*Somatochlora hineana*), an endangered species in the United States, appear to be frequent co-inhabitants of crayfish burrows, another facilitation interaction (D.A. Soluk, pers. comm.). In other cases invertebrates live on or near each other, for instance finding shelter among the shells of dead and living mussels in river beds (e.g. Strayer *et al.*, 1999). Invertebrates may also be facilitators for vertebrates. For instance, predaceous benthic invertebrates may predispose their fleeing prey to additional vulnerability from fish predation, resulting in faster fish growth (Soluk & Richardson, 1997).

Lakes and ponds

Aquatic invertebrates of standing waters range in size from microscopic zooplankton drifting more-or-less passively in the water column to sessile filter-feeding bivalves which may reach up to 20 cm or more in length and weigh almost 1 kg. In between these extremes there are all manner of worms, snails, mites and shrimps and thousands of species of insects, many of which have terrestrial as well as aquatic stages to their life cycles. Benthic invertebrates can reach very high densities; for example, chironomid larvae in lake sediments may reach upwards of 130 000 individuals m^{-2} (Mason, 1977) and *Daphnia hyalina* in weedbeds up to 1500 individuals l^{-1} (Lauridsen & Buenk, 1996).

Aquatic invertebrate communities vary across latitude, altitude, nutrient status, and lake size and depth. In each case very different communities occupy different zones of the lake (see Fig. 16.2), but in every case the invertebrate activity provides a vital nutrient-cycling mechanism and energy link within and between the zones, in all of which the invertebrates perform numerous functional roles.

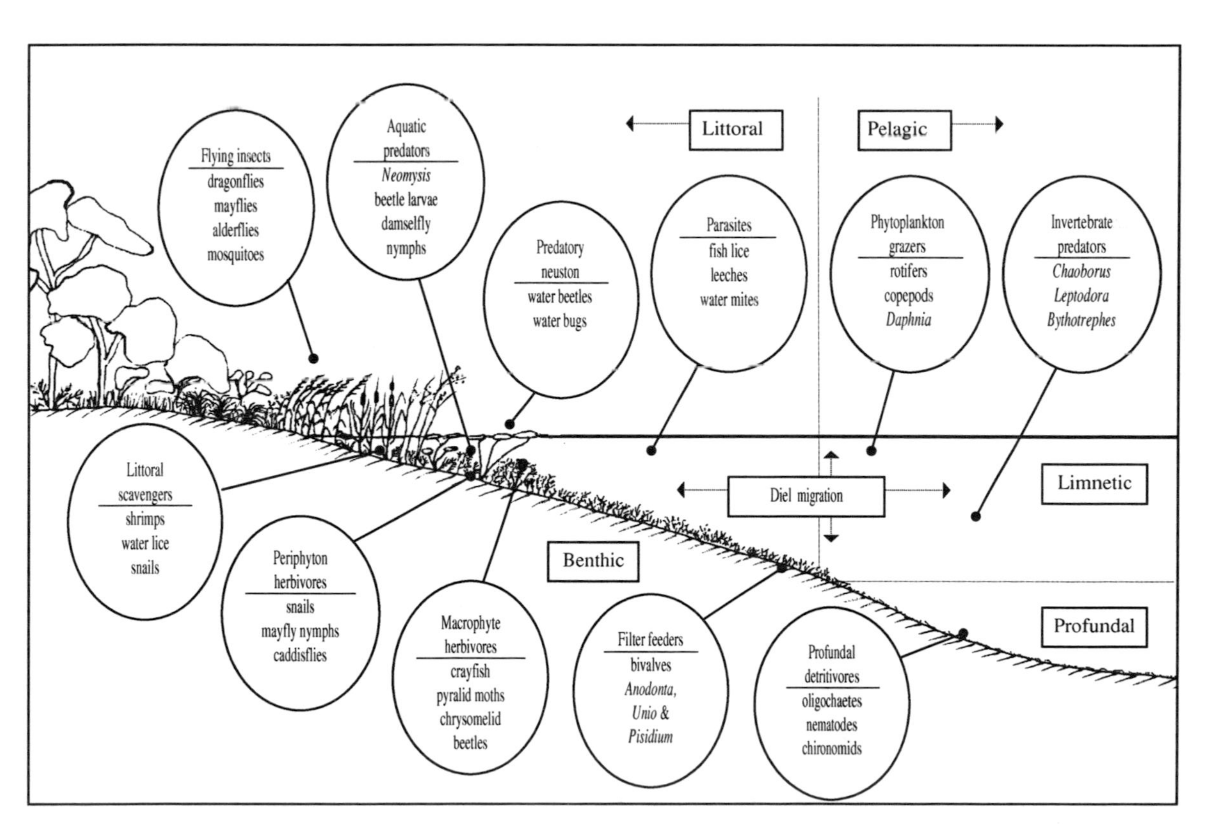

Fig 16.2. Zonation of standing waters showing major invertebrate groups associated with each zone.

The limnetic zone

In the limnetic zone, the main source of primary productivity comes from planktonic algae nurtured by nutrient inputs, derived from both external catchment processes, such as agricultural runoff and input of human and animal effluents, and internal cycling. The phytoplankton can quickly exploit these nutrients, often forming dense blooms, until an essential element is exhausted thereby limiting further algal growth. Zooplankton populations graze on the algae, resulting in peaks and troughs in water clarity over the growing season, as nutrients cycle and different species of algae and zooplankton come and go. The most distinctive of these peaks occurs during May or early June and is known as the 'spring clear-water phase' and often stimulates the spring growth of submerged macrophytes (Scheffer, 1998).

As well as temporal changes, there are spatial changes in distribution and abundance, which provide links with other zones. Movements of zooplankton vary between species, but many migrate daily either vertically or horizontally, vertically from the sediment surface or horizontally from macrophyte beds (e.g. Stansfield *et al.*, 1997). These movements are generally attributed to resource availability and predator-avoidance (Lampert, 1993; Lauridsen *et al.*, 1997) and can result in elaborate temporal patterns of 'hide-and-seek' that balance resource intake against predation risk (Diehl & Kornijów, 1998). Many aquatic invertebrates are partially nocturnal and use chemical stimuli to detect predators. Twilight is always a busy time for animals commuting in and out of refuges and can be particularly hazardous for invertebrates on bright moonlit nights: hence the evolution of lunar migration cycles (Helfman, 1993). Competition and habitat complexity play major roles in these behavioural patterns and create both seasonal and long-term cycles, sometimes involving morphological adaptations.

In the limnetic zone, the size and shape of prey relative to gape size and handling capabilities of predators can be critical. Invertebrates may make use of size refuges to avoid predation; for example *Daphnia* spp. may produce elongated helmets as an extension of their carapace under risk of predation from planktivores (O'Brien *et al.*, 1979). Such tactics may lead to sudden switching in predator–prey relationships as critical size thresholds are reached (Brooks & Dodson, 1965; Wooton, 1998). Nonetheless, predatory macroinvertebrates, such as *Chaoborus*, *Leptodora* and *Neomysis*, as well as planktivorous fish, may substantially reduce zooplankton populations.

The benthic zone

The benthic zone demarcates the lowermost profile of the waterbody, including animals living in and around the lake sediments – the zoobenthos. It stretches from the fringing (or littoral) communities to the deepest parts of the lake, beyond the limits of primary productivity, commonly known as the profundal. Because the sediments of the profundal lie below the thermocline, where physical and chemical conditions tend to be uniform, the littoral zone is usually far more diverse than the profundal (Palmer *et al.*, 1997). In very shallow lakes, the limnetic and profundal zones cease to exist, leaving only the littoral communities.

The profundal zone

The surface sediments of the profundal may be home to a number of different groups of benthic detritivores including oligochaetes, nematodes, isopods and chironomids. Many of these animals are capable of surviving low oxygen or anoxic conditions and they are known to have profound effects on the nature of the bottom sediments. For example, tubificids were found to alter the redox potential, pH and stratigraphy of lake sediments (Davies, 1974*a*, *b*,) and *Chironomus plumosus* influenced exchange of dissolved substances, in particular dissolved oxygen and nutrients, such as nitrogen and phosphorus, between water and sediments (Andersson *et al.*, 1978; Granéli, 1979*a*, *b*) mediated through complex interactions with the surface microbial community (Johnson *et al.*, 1989). Tube-building in worms and chironomids may also contribute greatly to the structural integrity of the sediments (Heines *et al.*, 1994). Algal production may be far in excess of zooplankton filtering capacities and much of the phytoplankton probably outlives predation. Whilst estimates vary, between 50% and 90% may remain ungrazed (Pomeroy, 1980). Dead and dying algal cells are partly decomposed *in situ* by heterotrophic bacterioplankton and the resulting detritus forms a constant 'rain' of material that fuels the benthic detritivore community (Barnes, 1980). Faecal material containing partially digested algae and gut symbiont bacteria and fungi may be crucial in preparing food for assimilation into the detritivore food chain (Pomeroy, 1980; Wotton & Malmqvist, 2001). Migrating *Daphnia* reach depths of 50 m or more and may graze these surface sediments (Parsons, 1980; Stich & Lampert, 1981). Filter-feeding prosobranch gastropods, such as *Valvata* spp., are also not uncommon at these extreme depths (Lodge & Kelly, 1985).

The littoral zone

The littoral zone marks the depth at which enough light reaches the bottom sediments to allow vascular plant growth (Scheffer, 1998). In shallow lakes (mean depth <3 m), most of the waterbody falls within these limits and the lake ought to be dominated throughout by a mix of macro-algae, submersed macrophytes and floating-leaved and emergent plants. However, the limit for submerged plants will depend also on water clarity for photosynthesis to be successful. In clear lakes, light intensity diminishes with depth on a roughly exponential basis but suspended sediments and phytoplankton blooms can lead to much higher light attenuation rate that can limit the growth of littoral vegetation (Scheffer, 1998). The extent of the littoral zone may also be determined by factors such as temperature, wind disturbance, herbivory, and water and sediment chemistry. In general terms, shallow lakes tend to be either completely devoid or brimfull of submerged plants as alternative stable states (Jeppesen & Sammalkorpi, volume 2), whilst deep lakes have only a narrow vegetated fringe.

In ponds and shallow lakes the structural complexity of the submerged vegetation plays a crucial role in supporting the benthic community (Gong *et al.*, 2000). The plants provide shelter from wave action, refuge from predation and a ready food source. Macroinvertebrate communities are not only more diverse in weedbeds than elsewhere, but some species show specificity towards particular macrophytes (Dvorák & Best, 1982). Adaptations for living in amongst plants are commonplace; for example, *Sida crystallina* possesses a gland at the back of its head for adhesion to macrophyte surfaces. Grazer life cycles often closely mirror the seasonal development of macrophyte beds; for example, gastropod distribution shows a strong correlation with plant cover (van den Berg *et al.*, 1997; Jackson, 1999). The zoobenthos benefits greatly from the littoral zone community above it and here too the plant structures hamper predation, particularly by benthivorous fish, such as carp and tench (Gilinsky, 1984; Brönmark & Vermaat, 1998). The littoral sediments may also be home to a variety of filter-feeding bivalves, ranging from the tiny pea mussels (*Pisidium* spp.) to the large duck and swan mussels (*Unio* and *Anodonta* spp.), which can occur in densities high enough to filter the entire water volume of a lake within a matter of days (Ogilvie & Mitchell, 1995).

In temperate regions, macrophytes appear to be fairly resistant to direct invertebrate attack, but groups such as the pyralid moths and chrysomelid beetles, that mine into the plant tissue, and large herbivores, such as crayfish, often cause extensive damage (Jacobsen & Sand-Jensen, 1992; Nyström & Strand, 1996). Many insects also oviposit into and pupate in the plant tissue. However, in tropical regions the impact of herbivorous arthropods can be remarkable and has been used as a valuable tool in the control of nuisance weeds. For example, a weevil, *Cyrtobagous salviniae* (Coleoptera: Curculionidae) was introduced into Lake Moondarra, Australia in 1980 to control the spread of water fern *Salvinia molesta*. By August 1981, the weevil had destroyed an estimated 18 000 tonnes of the weed. It has since been successfully used in the control of *Salvinia* on three continents. Similar invertebrate introductions have led to the successful control of alligator weed (*Alternanthera philoxeroides*), water hyacinth (*Eichhornia crassipes*) and water lettuce (*Pistia stratiotes*) (Harley & Forno, 1993).

Nonetheless, the periphyton attached to and surrounding the plants is believed to be the primary food source for grazing invertebrates. Periphyton consists of a mass of diatoms, filamentous algae, protozoa, detritus and bacteria. Periphytic algae is more productive even than the macrophytes it surrounds, and gastropods, mayfly larvae, chironomids and many other invertebrates capitalise on this rich food source (Kornijow *et al.*, 1995; Crowder *et al.*, 1997). The presence of grazing snails has been shown to enhance the growth of the macrophytes. For example, hornwort (*Ceratophyllum demersum*) grown in the presence of *Planorbis planorbis* produced significantly longer leaves and had more leaf nodes and growing tips than plants grown in the absence of the snails (Underwood, 1991).

THE ROLE OF INVERTEBRATE FUNCTIONS IN HABITAT RESTORATION

In general, there have been few restoration projects in freshwaters specifically designed for invertebrates. Invertebrates are used more often as a measure of the performance of restoration activities, especially those aimed at improving conditions for fish. Restoration may be divided broadly into various categories, such as physical rebuilding or modification of habitats, chemical remediation, and controlling watershed influences (e.g. Starnes, 1985). In lakes, invertebrate populations may be manipulated to attain a particular community structure. For example, biomanipulation may utilise the known filtering capacity of the generalist feeder *Daphnia* spp. In running waters, habitat alterations may be directed at producing more 'fish food', but are not typically based on manipulations of specific populations of invertebrates.

Lakes and ponds

Lake ecosystems have suffered greatly from human impacts and most lake managers are struggling to overcome the long-term effects of pollution or accidental introductions. Restoration responses vary from lake to lake, necessitating a range of techniques, but scale is paramount and large deep lakes respond very differently from small shallow ones, primarily because of differences in the zonation ratios. It is not possible to consider all the options here, so we shall concentrate on the restoration of shallow hypereutrophic lakes.

Early attempts to combat eutrophication in shallow lakes focused on internal and external nutrient control – the so-called 'bottom–up' approach (Moss, 1983). When this failed, biotic approaches using 'top–down' manipulation of the food web were employed (e.g. Shapiro *et al.*, 1975; Meijer *et al.*, 1994*a;* Carpenter *et al.*, 1995), although nutrient levels remain critical for success (Jeppesen & Sammalkorpi, volume 2). This 'ecosystem engineering' exploits the particular functions of key aquatic invertebrates as 'webmasters' (Coleman & Hendrix, 2000). Early efforts centred on zooplankton communities, but recently, particularly in view of the increasing success of biological control of weeds, attention has turned to the manipulation of macroinvertebrates. Much of the latter work is still at the theoretical stage, but initial management trials show promising signs of success.

Alternative stable states and stepwise restoration

In shallow lakes, a 'vegetation-dominated clear state' and a 'non-vegetated turbid state' are thought to be alternative equilibria over a range of nutrient levels (Scheffer *et al.*, 1993; Scheffer & Jeppesen, 1998). A number of switching mechanisms are thought to be responsible for the transition from one state to the other – forward switches include excessive nutrient enrichment, the mechanical loss of plants or the inappropriate use of biocides (Blindow *et al.*, 1998). Utilising invertebrates to revert from the turbid state is a stepwise process. The initial step must be to achieve clear water, first by reducing nutrient inputs and toxic contaminants and then by promoting zooplankton grazing of phytoplankton and filtering of suspended sediment. Even given sufficient light the sediments need to provide the correct chemical and physical conditions to promote seed germination. Finally, germinating seedlings need to be free of choking periphytic algae. Once a diverse assemblage of macrophytes has been established, a fully functional littoral macroinvertebrate community should follow and lasting stability ought to return to the system.

Step 1: Achieving clear water

Biomanipulation of turbid lakes seeks to alter community structure in order to stimulate a switch in the zooplankton community, usually by the removal of planktivorous fish (Shapiro *et al.*, 1975; Perrow *et al.*, this volume). Biomanipulation theory stems from the work of Hrbácek *et al.* (1961), which showed that once cyprinid fish populations were reduced, large *Daphnia* became abundant, algal blooms decreased and water clarity improved. Conversely, with increasing numbers of fish, smaller zooplankton prospered and algal blooms became commonplace. Brooks & Dodson (1965) postulated that at high fish densities the smaller, less efficient, zooplankters predominated because large zooplankton were more vulnerable to predation from planktivorous fish. Algal biomass decreased in the absence of fish because the larger cladocerans were superior competitors, fed on a wider range of food sizes and had a lower metabolic demand per unit mass. In practice, there were found to be other processes at work. Different sized zooplankton did select for particular planktonic algae, but some algae were more resistant to grazing than others (Leibold, 1989). Cyanobacteria, in particular, can proliferate in eutrophic lakes and can be highly toxic and aggregate into large colonies that are inedible to most herbivorous zooplankton. Further, research at Lake Ringstön in southern Sweden by Hansson *et al.* (1998) found a negative correlation between the ratio of the dominant rotifer, *Keratella cochlearis*, and macrozooplankton versus the ratio of 'small' (<30 μm) and 'large' (>30 μm) phytoplankton, but revealed that cyclopoid copepods could rapidly reduce rotifer abundance, suggesting that larger zooplankton affected algal size distributions both directly and indirectly. Provision of temporary artificial structures may be able to provide refuge for zooplankton against fish predation (Timms & Moss, 1984; Shapiro, 1990). However, reducing the zooplanktivorous fish population does not necessarily guarantee a predation-free environment for large zooplankton, because densities of predatory invertebrates, such as *Chaoborus*, *Leptodora* and *Neomysis*, can explode once planktivorous fish are removed, counteracting any advantage gained from the removal of the fish (see Box 16.1).

Manipulating the invertebrate community simultaneously with fish shows promise. For example, experiments with the introduction of zebra mussels (*Dreissena polymorpha*) to a hypereutrophic pond in The Netherlands reduced algal blooms by up to 50% (Reeders & Bij de Vaate, 1990). In the future, the manipulation of native bivalves could prove to be an equally powerful and more acceptable restoration tool than fish removal.

Step 2: Establishing macrophyte communities

Lakes undergoing restoration have often remained barren of submerged macrophytes for many years (George, 1992) and this may have a considerable bearing on the success of biomanipulation. The loss of plant structure increases the effects of wave action and destabilises the sediments. Detritivores become more vulnerable to predation from benthivorous fish (Gilinsky, 1984) and are starved of resources from periphytic grazers. Apart from cutting off carbon and nutrient recycling, the loss of bacteria, fungi and invertebrates may affect binding processes that normally help to structure the sediment, leaving only a fluid substrate that is too unstable for macrophyte colonisation (Pomeroy, 1980; Schutten, 2000). This problem might be solved by re-creating suitable conditions for the detritivore 'engineers' (Levinton, 1995). First, benthivorous fish, such as tench, carp and bream, would have to be removed (Fig. 16.3). Submerged artificial structures mimicking the plants could then be installed (e.g. plastic plants), which reduce fish predation on zooplankton and are rapidly colonised by periphyton and macroinvertebrates (Shapiro, 1990). Given a sufficient surface area of artificial substrate, the detritivore food chain should reactivate and begin to prepare the sediment conditions for macrophyte growth. In natural macrophyte beds the majority of the grazers die off with the plants (e.g. Lodge & Kelly, 1985). In a recent experiment M.J. Jackson & M.R. Perrow (unpubl. data) installed a submerged artificial refuge (approx. 30 m^2) made up of some 7500 plastic 'cobweb-brushes', with a total surface area of some 2900 square metres, in Alderfen Broad, Norfolk, UK to test the effectiveness of refuge provision in promoting macroinvertebrate communities. Preliminary results show that following the initial dominance of Diptera and Ephemeroptera larvae, there was a gradual increase in other groups leading to a diverse assemblage of Odonata, Trichoptera, Asellidae, Megaloptera and Neuroptera after five months (Fig. 16.4). Artificial refuges allow the

Box 16.1. Problems of introduced invertebrates

Two examples of introduced invertebrates and the troubles they can cause are the zebra mussel (*Dreissena polymorpha*) and opossum shrimps (*Mysis relicta* and *Neomysis integer*). The zebra mussel was first noted in eastern North America in the mid-1980s and has since spread over a considerable range. The damage caused by this highly invasive mussel to water supply facilities and native mussel populations (Strayer *et al.*, 1999) spurred an enormous effort to find methods to control its numbers, including chemical, physical and biological means of control (e.g. Nalepa & Schloesser, 1993; Claudi & Mackie, 1994) and removal by hand (Hallac & Marsden, 2001). All these attempts at control have been unsuccessful at stopping the spread of this species.

Mysis relicta was introduced to many lakes in North America as a mid-sized prey item for promoting fisheries production, especially where cladoceran or copepod zooplankters were too small for large 'forage' fish (Ashley *et al.*, 1997). In some systems the introduction was a success, but in others it was responsible for the apparent loss of some productive sports fisheries. Unfortunately, *Mysis* also competed effectively with fish for their cladoceran prey, reducing the food available for juvenile fish. *Mysis* migrates vertically, coming to the surface only at night when it can avoid predation from the very fish it was intended to feed (Johannsson *et al.*, 1994; Ashley *et al.*, 1997). In Kootenay Lake in British Columbia, Canada, the large reductions in numbers of kokanee (*Oncorhynchus nerka*) were attributed, in part, to introduction of *M. relicta*, and efforts to mitigate this impact are under way. The impact of *Mysis* is of sufficient concern that a government agency in Canada is experimenting with commercial-scale harvesting of *Mysis* in Okanagan Lake to reduce its impact on fish populations – a very costly restoration activity.

Another opposum shrimp, *Neomysis integer*, is commonly found in the brackish waters of Western Europe where it is known to tolerate salinities down to 150 mg Cl l^{-1} (0.8% seawater). *Neomysis* is an important predator on the zooplankton community, preying particularly on *Daphnia* spp. and the grazing copepod *Eurytemora affinis*, and experimental evidence has shown that, at high densities, it can significantly enhance the effects of eutrophication in nutrient-rich brackish waters, particularly in the absence of fish. *Neomysis* is an omnivore and also feeds on periphyton and detritus (Irvine *et al.*, 1993; Aaser *et al.*, 1995), so that even in the absence of animal prey it is able to maintain the potential of high predation on zooplankton. Waste products produced by *Neomysis* may even directly increase phytoplankton biomass (Aaser *et al.*, 1995). *Neomysis* is just one example of how large invertebrate omnivores may confound lake restoration attempts. Restoration efforts in Lake Wolderwijd in The Netherlands involving reductions of 75% of the fish population resulted in major increases in the density of *Neomysis integer* (Meijer *et al.*, 1994*b*). Similarly in the biomanipulated Cockshoot Broad, UK a combination of drought and an unusually high flood tide in 1992 increased the salinity of the lake and allowed *Neomysis* to invade (Moss *et al.*, 1996). In Cockshoot Broad the effect was somewhat short-lived as the brackish conditions soon returned to fresh water. Numbers of these animals were grossly underestimated until it was discovered that they were mainly active at night (Aaser *et al.*, 1995). During the day many macroinvertebrates take refuge from predation in amongst the littoral vegetation and in the bottom sediments and some may be having far greater effects on the system than 'daytime' researchers might otherwise imagine.

grazers to survive and continue the induction processes over the winter period. Some benthic macroinvertebrates, such as snails, can be slow to colonise and may have to be introduced onto the refuges to further facilitate the process, particularly if the lake has been dredged and colonisation sources are remote. Terrestrial areas may also have to be managed to create suitable habitat for adult stages of aquatic larvae.

Once plants have germinated there is always a danger that they will be choked by periphytic algae and detritus. In mature systems, periphyton is

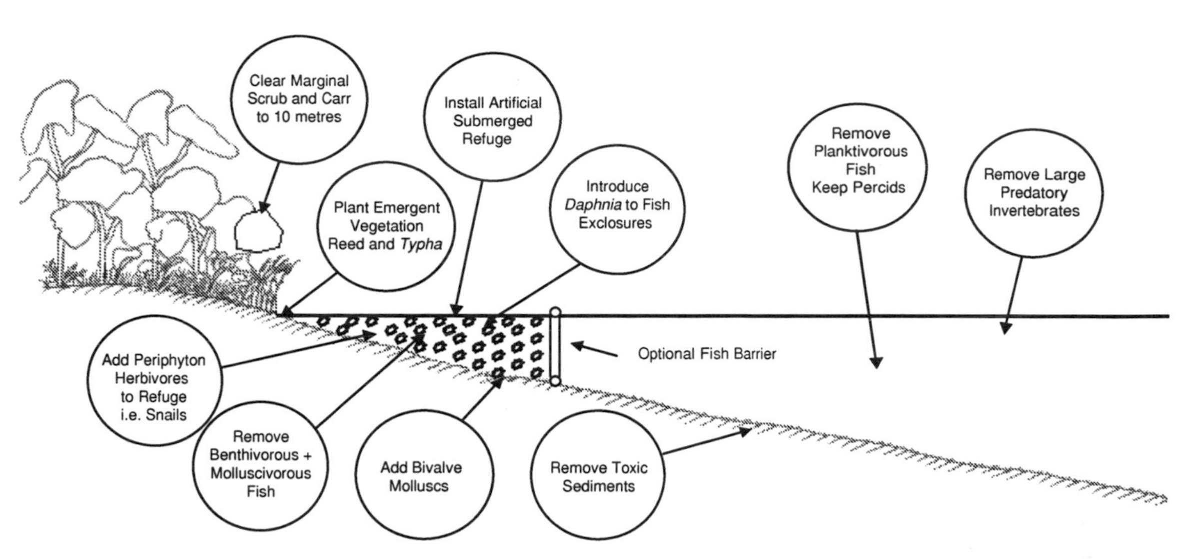

Fig 16.3. Possible management options in shallow lake restoration using invertebrates to improve water clarity, enhance growth of submerged plants and to encourage stability of the system. Note that artificial refuges and fish barriers can be removed and fish stocks readjusted once macrophytes recover.

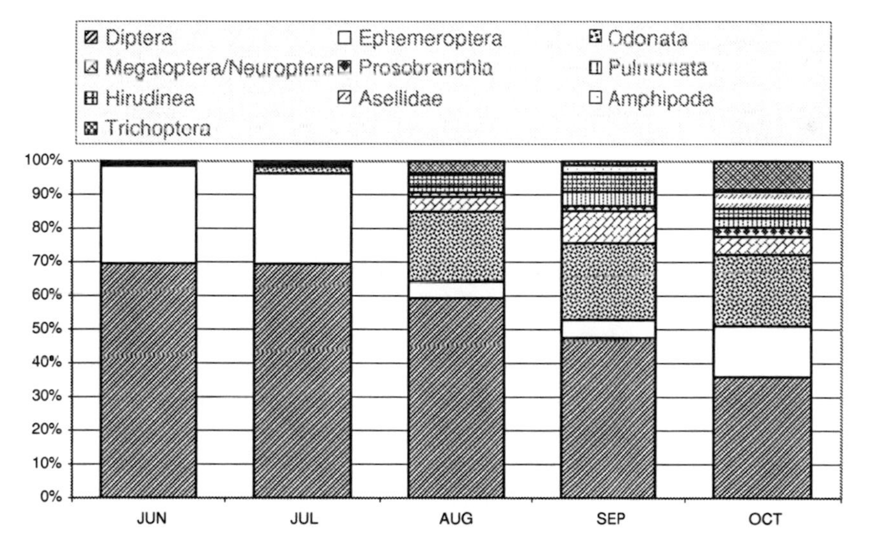

Fig 16.4. Preliminary results showing macroinvertebrate colonisation of 30 m^2 of plastic brush refuge at Alderfen Broad, UK, since installation in May 2000. Proportions of different groups of taxa taken from 60 brushes on each sampling occasion. M.J. Jackson & M.R. Perrow, unpublished data.

usually grazed off by gastropods and insect larvae (Brönmark & Vermaat, 1998) promoting photosynthetic activity in the macrophytes (Jones *et al.*, 2000). In the absence of grazers the shading effect of periphyton reduces the growth rates of the plants (Brönmark, 1990; Underwood *et al.*, 1992). Large invertebrates are generally superior periphyton grazers to smaller ones, but if large grazers are removed the smaller species tend to compensate by increasing in number (Jones *et al.*, in press). Thus, large snails may prevent smaller chironomids, in particular, from tube-building on the macrophyte stems and leaves, which can otherwise be a major cause of shading (G.L. Phillips, pers. comm.). There are, however, considerable differences between the competitiveness of different periphyton grazers and their effectiveness in reducing shading will depend to a large degree on the presence of fish (Martin *et al.*, 1992; Jones *et al.*, in press). Macrophytes and invertebrates are thus entwined in a close commensal relationship that is central to achieving restoration objectives, but there are still many unknowns, for example, how might different plant architectures and stem densities benefit different macroinvertebrate species.

Step 3: Stabilising the system

It is often assumed that the more species there are in a food web, the more stable the system: but this may not necessarily be the case (May, 1972; Lawler, 1993) and diversity alone may not be the best gauge of success. In any developing system, successive species may come and go many times before the *status quo* is reached (Morin, 2000), implying that numerous attempts at biomanipulation may be required before the desired result is accomplished. Invertebrate responses are liable to have inherent time lags rooted in seasonal and long-term population cycles.

Maintaining the stability of lentic systems may be dependent particularly on the presence of generalised omnivores (Diehl, 1995; Nyström *et al.*, 1996). Omnivores form the basis of reticulated food webs and understanding the processes by which these species are able to switch from one food source and/or habitat to another may hold the key to restoration success (Diehl & Eklöv, 1995). Recent developments in metapopulation theory imply that patchily distributed habitats and food resources offer a greater chance of survival to both predator and prey (Holyoak & Lawler, 1996). There are numerous omnivorous macroinvertebrates in lentic systems, and although little is known about their food preferences, age and size structure is likely to be critical (see Box 16.1).

Streams and rivers

Considerable time and energy is being devoted to restoring streams and rivers around the world. Many projects are aimed at enhancing fish habitat, including for example, woody debris additions, boulder placements, riffle reconstruction, riparian planting, changes in flow regulation, and removal of concrete and training devices. The majority of these activities are, however, not focused on invertebrates, nor evaluated for invertebrate responses, but rather for creation of fish or wildlife habitat. In the cases where invertebrates are specifically considered it is usually in their role as fish food or as a component of biodiversity. As an exception, the restoration of the Kissimmee River in Florida made very explicit predictions about the roles and responses of invertebrates in the recovery process (see Box 16.2). The functions that invertebrates serve have been little utilised in stream and river restoration in contrast to the situation in lakes.

Artificial reconstruction of streambeds by adding materials to create riffles has become common in streams where the channels have been geomorphologically simplified. There is a strong positive relation between the habitat heterogeneity in aquatic environments and the diversity of aquatic communities (Power, 1992; Lancaster & Hildrew, 1993; Wallace & Webster, 1996; Rempel *et al.*, 2000). Most of that heterogeneity is derived from diversity in flow conditions and substrate types. Most evidence points to riffle habitats as the most productive channel unit in streams, and there has been a large effort at reconstruction of riffles for the purpose of increasing productivity of invertebrates. Riffle habitats may already exist but be left dysfunctional by flow regulation or channelisation, making a stream broad and shallow. Addition of materials according to simple rules about the relative spacing of natural riffles has been successful in recreating habitat for rheophilic invertebrates (e.g. Cobb *et al.*, 1992; Gore *et al.*, 1998, Harper *et al.*, 1998).

Efforts to adjust flow patterns to increase the heterogeneity of flow conditions and stream substrates have been applied in many situations (e.g. Gore, 1985). Placements of boulders, large woody debris or snags can result in flow variation across a channel

Box 16.2. The Kissimmee River Ecosystem restoration project

One of the most ambitious restoration projects in a freshwater system to date is the Kissimmee River Ecosystem restoration project in south Florida (Toth & Anderson, 1998). In that project, approximately 35 out of 161 km of channelised river will be restored to its former configuration as a floodplain-river system (Merritt *et al.*, 1996). The previous channelisation resulted in alienation of the floodplain and loss of habitat heterogeneity within the river, especially the absence of fast flow areas suitable for rheophilic invertebrates (Harris *et al.*, 1995). It was also estimated to have resulted in the loss of around 6 billion freshwater shrimp (*Palaemonetes paludosus*), an important prey species to many predatory birds and fish (Toth, 1993).

The importance of invertebrates as key linkages in aquatic ecosystems was clearly recognised by the project team (Harris *et al.*, 1995). The project leaders made predictions about ecosystem responses, and in particular, for riverine and floodplain invertebrates, although, notably, these were not quantitative (Harris *et al.*, 1995). Predictions about invertebrate responses were based on knowledge of the associations between particular invertebrates and the physical environments they require, especially substrate and flow conditions (Dahm *et al.*, 1995; Harris *et al.*, 1995). Additionally, expected changes in the trophic basis of the river food web allowed for other predictions for the invertebrates based on their known food requirements. The integrity of an ecosystem is often based on comparison to some reference condition (Rosenberg & Resh, 1993); however in the Kissimmee there were no remnant, unmodified reaches for comparison. The primary reason for making specific predictions of how the system would change was to judge the success and effectiveness of the restoration activities (Anderson & Dugger, 1998).

A demonstration project was launched in the Kissimmee system in 1987 to determine if it was truly feasible to restore the river. The trial restoration of a nearly 20-km-long section of the river proceeded by allowing the flow to follow the original channel and the provision of flow variation (Toth, 1993). The inundation of the river quickly resulted in expansion of the floodplain wetlands and colonisation of crayfish (*Procambarus fallax*) and shrimp (*Palaemonetes paludosus*) (Toth, 1993). The reintroduction of flow to the remnant, old channels resulted in rapid increases in invertebrate numbers, particularly the rheophilic groups, such as the mayfly (*Stenacron* spp.), the filter-feeding caddisfly (*Cheumatopsyche* spp.) and the filtering midge (*Rheotanytarsus* spp.) (Toth, 1993). As a consequence of the demonstration project, the larger restoration programme proceeded.

and enhance the proportion of streambed in riffles and inundated at most flow rates (Gore, 1985). Agencies responsible for flow regulation have begun to consider how their flows affect macroinvertebrates in streams and adjust them to provide suitable habitats (Gore *et al.*, 1998).

Fluvial systems depend on natural dynamics to express heterogeneity at many scales (e.g. Ebersole *et al.*, 1997; Swanson *et al.*, 1998; Newson *et al.*, this volume). The re-engineering of structure for fish habitat in some streams is not based on how it affects invertebrate microhabitats or basal resources, and often attempts to reduce natural dynamics (Frissell & Nawa, 1992), in contrast to the goals of restoration. Despite the enormous effort at stream restoration, there is still a lack of clear predictions for the responses of invertebrate community structure or biological processes. As a result of their diverse roles invertebrates may be one of the best measures of success (see below).

The placement of large woody debris into streams to re-create structural complexity is a common technique in habitat enhancement for fish. There are few studies of the effect of large woody debris placements on invertebrates or ecosystem processes such as organic matter retention and nutrient dynamics. Further, two such studies of the impacts of large wood additions on invertebrates and ecosystem processes, both in similarly-sized streams in similar terrain in the eastern USA, found somewhat different results. In one study, additions of logs across the channel resulted in greater than 17-fold increases in the abundance of particulate organic matter, with an eightfold increase in invertebrate densities and a 51% increase in secondary production relative to the reference reaches (Wallace *et al.*, 1995). The production of collectors increased by 680% over the control (but not for all species), the production of scrapers decreased, and surprisingly shredder production decreased by 33% relative to the reference reaches (Wallace *et al.*, 1995). In the other study, by Lemly & Hilderbrand (2000), additions of logs resulted in increased amount of pool area, thus enhancing the habitat heterogeneity. However, there were no significant increases in the amounts of organic matter, nor changes in their invertebrate communities at the stream reach scale. These two contrasting results suggest that there are still many unknowns about the mechanisms by which structural features may affect stream ecosystems and the scales at which such effects should be evaluated. This uncertainty suggests the need for a conceptual framework for restoration of ecosystems.

In many parts of the world increased nutrient levels are problematic. However, in some places fertilisation of streams and lakes with inorganic nutrients has been employed to enhance productivity of algae, and hence invertebrates, usually with the aim of increasing fish food production. One government agency used nutrient additions at low levels in an oligotrophic stream to increase algal production by five to ten times and invertebrate numbers about twofold (Johnston *et al.*, 1990). The invertebrates showing the biggest increases were midge larvae (Chironomidae) and hydropsychid caddisflies.

Pollution from metals, acids and organic compounds is widespread and may even be delivered by long-distance transport mechanisms in the atmosphere. One treatment used in stream restoration is the addition of lime to counter problems of acidification, and, to a limited extent, metal contamination. Experimental studies of the addition of lime have shown that many streams recover their diversity of invertebrate assemblages as a result (Ormerod *et al.*, 1990; Lingdell & Engblom, 1995). These results indicate the potential for successful chemical means of restoration from acidification. However, addition of lime to streams that had soft water prior to acidification resulted in loss of characteristic species (Ormerod *et al.*, 1990), showing that such restoration techniques cannot be applied without a broader consideration of the nature of the ecosystem.

In some cases, restoration of habitat for fish may indirectly contribute to the recovery of invertebrate populations. The larval stage (glochidia) of many mussels (Unionidae) is dependent upon a limited subset of fish species on which they are parasitic (Ricciardi *et al.*, 1998). Reintroduction of particular fish species or restoration of the habitat for those fish may consequently result in the recovery of some species of mussels (e.g. Smith, 1985).

SPECIES-CENTRED RESTORATION

There are many species of invertebrates that have become endangered through loss of habitat. Freshwater mussels are probably the most imperilled group of invertebrate organisms in fresh waters (see Box 16.3). Despite the large, and increasing, number of invertebrates that are listed as endangered, or are candidates for listing, there appear to be no particular restoration activities currently in place for them. There are conservation plans for many of these listed species, but none that currently involve habitat manipulations. Whilst there are habitat restoration plans for species such as the pearl mussel (*Margaritifera margaritifera*), there are few projects, if any, currently under way (e.g. Cosgrove & Hastie, 2001). This is in stark contrast to restoration for other aquatic organisms such as fish (Perrow *et al.*, this volume).

Restoration activities may also call for removal or control of introduced invertebrates that have wreaked havoc on some species or whole ecosystems. Invertebrates have been introduced wilfully or accidentally into aquatic environments around the world, sometimes with no measurable impact and sometimes with detrimental changes. There are many examples of invertebrate introductions (see Box 16.1), including zebra mussel and Asiatic clam (*Corbicula fluminea*). The opposum shrimps (*Mysis* spp. and *Neomysis* spp.) and other invertebrates have been intentionally introduced to lakes to provide an

Box 16.3. Endangerment of native unionid mussels

Recovering endangered species through restoration is a challenging exercise partly because of the extent of habitat alterations, but also because many invertebrates have complex life cycles and we lack specific information on their biology. Disruption of the habitat for any single stage may be sufficient to limit or eliminate the species, regardless of the condition of the habitat for other stages. Here we use one group, the freshwater mussels (family Unionidae), to illustrate some of the difficulties one would have to overcome to develop restoration options.

Freshwater mussels are one invertebrate group that is vulnerable to large-scale extirpations and even extinctions. In the United States, about 192 species or subspecies (about 60% of described taxa) are considered threatened or endangered (Ricciardi *et al.*, 1998; Vaughn & Taylor, 1999). Nineteen species or subspecies are recorded as extinct (Ricciardi *et al.*, 1998). These species are facing many threats, including loss of fish hosts for their parasitic larval stages, water control through impoundments, increased fine sediments, overharvesting and competitive exclusion by introduced species (Strayer *et al.*, 1999; Vaughn & Taylor, 1999). Moreover, an individual mussel may live many decades, but may not begin reproducing until it is more than ten years old, and thus populations are predisposed to respond slowly to any restoration efforts.

Native mussels have often been displaced by competition from introduced species, such as the zebra mussel and the Asiatic clam (Ricciardi *et al.*, 1998), either through direct fouling of shells or through their highly efficient reduction of suspended food particles leading to the starvation of the native species (Strayer *et al.*, 1999). Efforts to remove the zebra mussels from endangered unionids by hand have met with only limited success (Hallac & Marsden, 2001).

Mussels were formerly very common in larger rivers and lakes, and probably had a tremendous influence on ecosystem structure and function. In some cases, introduced zebra mussels have resulted in even higher clearing rates of suspended particles, resulting in increased light penetration, increased macrophyte growth, and increased shallow-water benthos population densities (Strayer *et al.*, 1999). In systems where native mussels have been lost by flow regulation and sedimentation, there may be a positive feedback since decreased mussel numbers could also result in increased densities of suspended particles, thus exacerbating problems associated with turbidity. Restoration towards natural flow conditions for the sake of mussels may have many other positive benefits for water quality.

Native mussels are harvested as raw material for producing buttons, or for creating inocula for other mussels to produce pearls. The European freshwater pearl mussel (*Margaritifera margaritifera*) was heavily harvested in the past, which accounts for much of the 90% population decline (Strayer *et al.*, 1999). Overharvesting has reduced the numbers and range of many other mussels (e.g. Anthony & Downing, 2001). Cessation of harvesting alone will do little to allow populations to recover if habitat is not restored. Protection of existing populations seems the most prudent course of action, but there appears to have been no efforts to reverse the trends for this group of molluscs using habitat restoration. Given the complexity of mussel life histories, it poses a complex challenge to use habitat restoration to bring freshwater mussels back from the brink.

additional prey source for fishes, typically sports fishes. The planktonic crustacean *Bythotrephes longimanus* was first observed in the Laurentian Great Lakes in 1982 and has since expanded its range and been implicated in changes to the relative species composition and production in some lake food webs (Dumitru *et al.*, 2001). There are many other examples of exotic invertebrates that have been introduced to fresh waters outside their former range. In general these introduced invertebrates have high reproductive rates, are relatively tolerant, and are aided in their dispersal by human activities, all features that make them difficult to eradicate locally or regionally. Unfortunately, one of the conclusions that can be drawn is that it is doubtful that these ecosystems can be restored to their former composition simply by the removal or control of the introduced species.

MONITORING THE RESPONSES OF INVERTEBRATES TO RESTORATION

Invertebrate assemblages have become widely used as a measure of the condition of aquatic systems, commonly referred to as biomonitoring. Restoration managers may use biomonitoring to judge their success (see below) and many agencies from around the world have developed tools to use invertebrate community composition as a way to compare restored sites to the communities in reference sites (Rosenberg & Resh, 1993). Invertebrate communities have many advantages as biological indicators, including their ubiquity, high density and species diversity, and they demonstrate a broad range of tolerances to perturbations (e.g. chemical pollution, suspended sediments, temperature changes, species introductions). Invertebrates also respond on a convenient time-scale for monitoring, unlike fish, which may take more than a year, or algal composition, which can change in a matter of days. Invertebrates are generally easy to sample and identify and provide an integrative measure of ecosystem condition that could be missed by periodic, direct sampling for water quality or physical variables.

The success of restoration might be judged on the basis of how similar a restored site is to a reference site that has not been seriously perturbed by human activities. The use of particular species or invertebrate community structure as indicators of the condition of aquatic systems has exploded in the last decade (e.g. Rosenberg & Resh, 1993). There are a variety of indices that have been developed for particular regions, and a more general approach using a reference condition method. The latter uses minimally disturbed sites as predictors of the expected invertebrate communities given a particular set of physical and chemical conditions. Deviations from that reference condition can then be detected at the sample site. There are now many such models from around the world for rivers, lakes and ponds (e.g. Wright, 1995; Reynoldson *et al*., 1997). Thus, there is a great potential for using invertebrates as indicators to gauge the progress of restoration.

There have been several other methods developed for monitoring of the state of aquatic systems, including a variety of indices, one of the most common being variations on the Benthic Index of Biotic Integrity (e.g. Karr & Chu, 1999). These indices rely on detailed knowledge of the sensitivity of particular species to a specific perturbation. The potential of these indices, or metrics as they are sometimes called, is in providing a simple numerical representation of the condition of the system, with obvious advantages to regulatory personnel (Biggs *et al*., 2000). An extension of this approach has resulted in the Rapid Bioassessment Protocols based on sampling for benthic macroinvertebrates used by government and non-government organisations to screen for water quality problems (Barbour *et al*., 1999).

The use of macroinvertebrate community composition has been used to detect impacts from pollution, but similarly can be used to determine when an ecosystem has been successfully restored. In the Laurentian Great Lakes, a multivariate model was developed to identify deviations in the lake benthos community from the reference state, and successfully discriminated sites that had been contaminated (Reynoldson *et al*., 1995).

CONCLUDING REMARKS

In many systems we know very little about the role of particular invertebrates and especially how restoration activities may affect their populations. In lakes and ponds, an understanding of the roles of certain invertebrates, such as *Daphnia* spp., has been exploited in biomanipulation, and in some cases used to augment prey availability to fish through species introductions or nutrient additions. In running waters there has been no clear move towards using invertebrate responses directly as a 'tool' to promote restoration processes. The apparent lack of utilisation of invertebrates as a restoration tool in streams may reflect the predominance of the physical habitat determining species occurrences and abundance. This leaves much scope for making advances in more focused use of invertebrate responses to aid in restoration.

What is the capacity for invertebrates to respond to restoration? And do we need to think about the responses of invertebrates differently from those of vertebrates? Invertebrates are often considered to be relatively resilient to perturbation, but some invertebrates are poor colonisers. For instance, many of the large mussels that are endangered are dependent on particular fish to disperse their larvae, fish that may also be in peril. The aquatic insects vary in their capacity to disperse, from strong flyers such as dragonflies, to weak flyers such as some of the stoneflies, which will affect colonisation by oviposition. These differences between species mean that not all species will respond on the same temporal scale, and recolonisation rates will depend upon the proximity of 'seed' populations (Gore, 1985). Dispersal in streams may come from upstream areas (tributaries or the main stream) via drifting, or may occur through animals crawling along the bottom. Propagules arriving in lakes and ponds may have dispersed passively or on animal vectors or they may have been washed in through flooding or from lentic habitats upstream. Invertebrates in general have comparatively high reproductive rates and some species have the capacity for an extremely rapid numerical response to changes in habitat. However, it is rare that specific predictions are made for how stream invertebrates will respond to restoration, and if predictions are made at all, they are usually community-level measures, not at the species level.

We still face many methodological problems with considering invertebrates, not the least of which is the state of our taxonomic and ecological knowledge of many species. Many species are difficult to identify beyond genus in their larval stages; however, there are sometimes large differences in the tolerances or functions of species in the same genus. There are sampling questions that result in some uncertainty about the responses of invertebrates in aquatic ecosystems (Jackson, 1997). These methodological problems confound our ability to interpret the functions of particular species and how they contribute to, or respond to, restoration.

There are many challenges to the restoration of aquatic systems, especially given how little we know about many of the invertebrates that are so critical to ecosystem function. Aquatic systems are dynamic and have a high degree of resilience, provided that organisms can indeed recolonise. We need to acknowledge the uncertainty inherent in manipulating any system and the vagaries of the outcomes of restoration given the patchwork of sites from which invertebrate animals may propagate. Moreover, the roles of most invertebrates in the processes of recovery are largely unknown. The actual practice of restoration has yet to embrace a conceptual framework derived from principles of conservation biology or community ecology. An ecosystems approach, which takes into account all the processes and elements, including invertebrates, may be a more effective way to structure plans for restoration activities.

The biggest challenge may simply be: how do we know when we have succeeded at restoring an ecosystem? Most aquatic ecosystems have been severely impacted by human activity. It may be that we are unable to find 'pristine' systems to act as our reference points for judging the success of restoration activities. The goal of restoring a self-maintaining system may still be compromised if some of the original pieces are missing, i.e. extirpation of some of the invertebrates (or any species) and the processes they drive. Theory predicts that a diverse ecosystem is less sensitive to perturbation by virtue of having some alternative pathways or species that can compensate for each other (e.g. Naeem, 1998), thus maintaining ecosystem functions. These theories are not sufficiently tested to determine how far we have to recover the original structure of an ecosystem for it to operate as expected. The issue of choosing the correct target model for our restoration objectives and actions, especially for the invertebrate components, will not be easily resolved.

REFERENCES

Aaser, F.H., Jeppesen, E. & Søndergaard, M. (1995). Seasonal dynamics of the mysid *Neomysis integer* and its predation on the copepod *Eurytemora affinis* in a shallow hypereutrophic brackish lake. *Marine Ecology Progress Series* **127**, 47–56.

Anderson, D.H. & Dugger, B.D. (1998). A conceptual basis for evaluating restoration success. *Transactions of the North American Wildlife and Natural Resources Conference*, **63**, 111–121.

Anderson, N.H., Sedell, J.R. Roberts, L.M. & Triska, F.J. (1978). The role of aquatic invertebrates in processing of wood debris in coniferous forest streams. *American Midland Naturalist*, **100**, 64–82.

Andersson, A., Berggren, H. & Cronberg, G. (1978). Effects of planktivorous and benthivorous fish on organisms and water chemistry in eutrophic lakes. *Hydrobiologia*, **59**, 9–15.

Anthony, J.L. & Downing, J.A. (2001). Exploitation trajectory of a declining fauna: a century of freshwater mussel fisheries in North America. *Canadian Journal of Fisheries and Aquatic Sciences*, **58**, 2071–2090.

Ashley, K., Thompson, L.C., Lasenby, D.C., McEachern, L. Smokorowski, K.E. & Sebastian, D. (1997). Restoration of an interior lake ecosystem: the Kootenay Lake fertilization experiment. *Water Quality Research Journal of Canada*, **32**, 295–323.

Barbour, M.T., Gerritsen, J., Snyder, B.D. & Stribling, J.B. (1999). *Rapid Bioassessment Protocols for Use in Streams and Wadeable Rivers: Periphyton, benthic macroinvertebrates and fish*, 2nd edn, EPA 841-B-99-002. US Washington, DC: Environmental Protection Agency, Office of Water.

Barnes, R.S.K. (1980). The unity and diversity of aquatic systems. In *Fundamentals of Aquatic Ecosystems*, eds. R.S.K. Barnes & K.H. Mann, pp. 5–23. Oxford: Blackwell.

Benke, A.C. (1998). Production dynamics of riverine chironomids: extremely high biomass turnover rates of primary consumers. *Ecology*, **79**, 899–910.

Biggs, J., Williams, P., Whitfield, M., Fox, G. & Nicolet, P. (2000). *Biological Techniques of Still Water Quality Assessment, Phase 3, Method Development*, R&D Technical Report no. E110. Bristol, UK: The Environment Agency.

Blindow, I., Hargeby, A. & Andersson, G. (1998). Alternative stable states in shallow lakes: what causes a shift? In *The Structuring Role of Submerged Macrophytes in Lakes*, eds. E. Jeppesen, Ma. Søndergaard, Mo. Søndergaard & K. Christoffersen, pp. 353–360. New York: Springer-Verlag.

Brönmark, C. (1990). How do herbivorous freshwater snails affect macrophytes? a comment. *Ecology*, **71**, 1212–1215.

Brönmark, C. & Vermaat, J.E. (1998). Complex fish–snail–epiphyton interactions and their effects on submerged freshwater macrophytes. In *The Structuring Role of Submerged Macrophytes in Lakes*, eds. E. Jeppesen, Ma. Søndergaard, Mo. Søndergaard & K. Christoffersen, pp. 47–68. New York: Springer-Verlag.

Brooks, J.L. & Dodson, S.I. (1965). Predation, body size and composition of plankton. *Science*, **150**, 28–35.

Carpenter, S.R., Christensen, D.L., Cole, J.C., Cottingham, K.L., He, X., Knight, S.E., Pace, M.L., Post, D.M., Schindler, D.E. & Voichick, N. (1995). Biological control of eutrophication in lakes. *Environmental Science and Technology*, **29**, 784–786.

Castelle, A.J., Johnson, A.W. & Conolly, C. (1994). Wetland and stream buffer size requirements: a review. *Journal of Environmental Quality*, **23**, 878–882.

Claudi, R. & Mackie, G.L. (1994). *Practical Manual for Zebra Mussel Monitoring and Control*. Boca Raton, FL: CRC Press.

Cobb, D.G., Galloway, T.D. & Flannagan, J.F. (1992). Effects of discharge and substrate stability on density and species composition of stream insects. *Canadian Journal of Fisheries and Aquatic Sciences*, **49**, 1788–1795.

Coleman, D.C. & Hendrix, P.F. (2000). *Invertebrates as Webmasters in Ecosystems*. Wallingford, UK: CAB International.

Cosgrove, P.J. & Hastie, L.C. (2001). Conservation of threatened freshwater pearl mussel populations: river management, mussel translocation and conflict resolution. *Biological Conservation*, **99**, 183–190.

Creed, R.P. (1994). Direct and indirect effects of crayfish grazing in a stream community. *Ecology*, **75**, 2091–2103.

Crowder, L.B., McCollum, E.W. & Martin, T.H. (1997). Changing perspectives on food web interactions in lake littoral zones. In *The Structuring Role of Submerged Macrophytes in Lakes*, eds. E. Jeppesen, Ma. Søndergaard, Mo. Søndergaard & K. Christoffersen, pp. 240–49. New York: Springer-Verlag.

Cuffney, T.F., Wallace, J.B. & Lugthart, G.J. (1990). Experimental evidence quantifying the role of benthic invertebrates in organic matter dynamics of headwater streams. *Freshwater Biology*, **23**, 281–299.

Cummins, K.W. (1973). Trophic relations of aquatic insects. *Annual Review of Entomology*, **18**, 183–206.

Dahm, C.N., Cummins, K.W., Valett, H.M. & Coleman, R.L. (1995). An ecosystem view of the restoration of the Kissimmee River. *Restoration Ecology*, **3**, 225–238.

Davies, R.B. (1974*a*). Tubificids alter profiles of redox potential and pH in profundal lake sediment. *Limnology and Oceanography*, **19**, 342–346.

Davies, R.B. (1974*b*). Stratigraphic effects of tubificids in profundal lake sediments. *Limnology and Oceanography*, **19**, 466–488.

Diehl, S. (1995). Direct and indirect effects of omnivory in a littoral lake community. *Ecology*, **76**, 1727–1740.

Diehl, S. & Eklöv, P. (1995). Effects of piscivore-mediated

habitat use on resources, diet and growth of perch. *Ecology*, **76**, 1712–1726.

Diehl, S. & Kornijów, R. (1998). Influence of submerged macrophytes on trophic interactions among fish and invertebrates. In *The Structuring Role of Submerged Macrophytes in Lakes*, eds. E. Jeppesen, Ma. Søndergaard, Mo. Søndergaard & K. Christoffersen, pp. 24–46. New York: Springer-Verlag.

Dumitru, C., Sprules, W.G. & Yan, N.D. (2001). Impact of *Bythotrephes longimanus* on zooplankton assemblages of Harp Lake, Canada: an assessment based on predator consumption and prey production. *Freshwater Biology*, **46**, 241–251.

Dvorák, J. & Best, P.H. (1982). Macroinvertebrate communities associated with the macrophytes of Lake Vechten: structural and functional relationships. *Hydrobiologia*, **95**, 115–126.

Ebersole, J.L., Liss, W.J. & Frissell, C.A. (1997). Restoration of stream habitats in the western United States: restoration as re-expression of habitat capacity. *Environmental Management*, **21**, 1–14.

Frissell, C.A. & Nawa, R.K. (1992). Incidence and causes of physical failure of artificial habitat structures in streams of western Oregon and Washington. *North American Journal of Fisheries Management*, **12**, 182–197.

George, M. (1992). *The Land Use, Ecology and Conservation of Broadland*. Chichester, UK: Packard Publishing.

Gilinsky, E. (1984). The role of fish predation and spatial heterogeneity in determining benthic community structure. *Ecology*, **65**, 455–468.

Gong Z., Xie, P. & Wang, S. (2000). Macrozoobenthos in two shallow, mesotrophic Chinese lakes with contrasting sources of primary production. *Journal of the North American Benthological Society*, **19**, 709–724.

Gore, J.A. (1985). Mechanisms of colonization and habitat enhancement for benthic macroinvertebrates in restored river channels. In *The Restoration of Rivers and Streams: Theories and Experience*, ed. J.A. Gore, pp. 81–101. Toronto, Canada: Butterworth.

Gore, J.A., Crawford, D.J. & Addison, D.S. (1998). An analysis of artificial riffles and enhancement of benthic community diversity by physical habitat simulation (PHABSIM) and direct observation. *Regulated Rivers: Research and Management*, **14**, 69–77.

Granéli, W, (1979*a*). The influence of *Chironomus plumosus* larvae on the exchange of dissolved substances between sediments and water. *Hydrobiologia*, **66**, 149–159.

Granéli, W. (1979*b*). The influence of *Chironomus plumosus* larvae on the oxygen uptake of sediment. *Archiv für Hydrobiologie*, **87**, 385–403.

Hairston, N.G. Jr & Hairston, N.G. Sr (1993). Cause–effect relationships in energy flow, trophic structure, and interspecific interactions. *American Naturalist*, **142**, 379–411.

Hallac, D.E. & Marsden, J.E. (2001). Comparisons of conservation strategies for unionids threatened by zebra mussels (*Dreissena polymorpha*): periodic cleaning vs. quarantine and translocation. *Journal of the North American Benthological Society*, **20**, 200–210.

Hansson, L.-A., Bergman, E. & Cronberg, G. (1998). Size structure and succession in phytoplankton communities: the impact of interactions between herbivory and predation. *Oikos*, **81**, 337–345.

Harley, K.L.S. & Forno, I.W. (1993). Biological control of aquatic weeds by means of arthropods. In *Aquatic Weeds: The Ecology and Management of Nuisance Aquatic Vegetation*, eds. A.H. Pieterse & K.J. Murphy, pp. 177–186. Oxford: Oxford University Press.

Harper, D., Ebrahimnezhad, M. & Cot, F.C.I. (1998). Artificial riffles in river rehabilitation: setting the goals and measuring the successes. *Aquatic Conservation: Marine and Freshwater Ecosystems*, **8**, 5–16.

Harris, S.C., Martin, T.H. & Cummins, K.W. (1995). A model for aquatic invertebrate response to Kissimmee River restoration. *Restoration Ecology*, **3**, 181–194.

Heard, S.B. & Richardson, J.S. (1995). Shredder-collector facilitation in stream detrital food webs: is there enough evidence? *Oikos*, **72**, 359–366.

Heines, F., Sweerts, P.J. & Loopik, E. (1994). The micro-environment of chironomid larvae in the littoral and profundal zone of Lake Maarsseveen I, The Netherlands. *Archiv für Hydrobiologie*, **130**, 53–67.

Helfman, G.S. (1993). Fish behaviour by day, night and twilight. In *Behaviour in Teleost Fishes*, 2nd edn, ed. T. J. Pitcher, pp. 479–512. London: Chapman & Hall.

Holyoak, M. & Lawler, S.P. (1996). Persistence of an extinction-prone predator–prey interaction through metapopulation dynamics. *Ecology*, **77**, 1867–1879.

Hrbácek, J., Dvorakova, M., Korinek, V. & Procházková, L. (1961). Demonstration of the effect of the fish stock on the species composition of zooplankton and the intensity of metabolism of the whole plankton association. *Verhandlungen der Internationalen Vereinigung für Limnologie*, **14**, 192–195.

Huhta, A., Muotka, T., Juntunen, A. & Yrjönen, M. (1999). Behavioural interactions in stream food webs: the case of drift-feeding fish, predatory invertebrates and grazing mayflies. *Journal of Animal Ecology*, **68**, 917–927.

Irvine, K., Bales, M.T., Moss, B. & Snook, D. (1993). Trophic relationships in the ecosystem of a shallow, brackish lake - Hickling Broad, Norfolk, with special reference to the role of *Neomysis integer* Leach. *Freshwater Biology*, **29**, 119–139.

Jackson, M.J. (1997). *Sampling Methods for Studying Macroinvertebrates in the Littoral Vegetation of Shallow Lakes*, Broads Authority Research Series no. 17. Norwich, UK: Broads Authority.

Jackson, M.J. (1999). The aquatic macroinvertebrate fauna of the littoral zone of the Norfolk Broads 1977–1995. *Transactions of the Norfolk and Norwich Naturalists Society*, **32**, 27–56.

Jacobsen, D. & Sand-Jensen, K. (1992). Herbivory of invertebrates on submerged macrophytes from Danish freshwaters. *Freshwater Biology*, **28**, 301–308.

Johannsson, O.E., Rudstam, L.G. & Lasenby, D.C. (1994). *Mysis relicta*: assessment of metalimnetic feeding and implications for competition with fish in Lakes Ontario and Michigan. *Canadian Journal of Fisheries and Aquatic Sciences*, **51**, 2591–2602.

Johnson, R.K., Boström, B. & van de Bund, W. (1989). Interactions between *Chironomus plumosus* (L.) and the microbial community in surficial sediments of a shallow, eutrophic lake. *Limnology and Oceanography*, **34**, 993–1003.

Johnston, N.T., Perrin, C.J., Slaney, P.A. & Ward, B.R. (1990). Increased juvenile salmonid growth by whole-river fertilization. *Canadian Journal of Fisheries and Aquatic Sciences*, **47**, 862–872.

Jones, J.I., Moss, B., Eaton, J.W. & Johnstone, O.Y. (2000). Do submerged aquatic plants influence periphyton community composition for the benefit of invertebrate mutualists? *Freshwater Biology*, **43**, 591–604.

Jones, J.I., Johnstone O.Y., Eaton J.W. & Moss, B. (in press). The influence of nutrient loading, dissolved inorganic carbon and higher trophic levels on the interaction between submerged plants and periphyton. *Journal of Ecology*.

Karr, J.R. & Chu, E.W. (1999). *Restoring Life in Running Waters: Better Biological Monitoring*. Washington, DC: Island Press.

Kohler, S.L. (1992). Competition and the structure of a benthic stream community. *Ecological Monographs*, **62**, 165–188.

Kornijow, R., Gulati, R.D. & Ozimek, T, (1995). Food preferences of freshwater invertebrates: comparing fresh and decomposed angiosperm and a filamentous alga. *Freshwater Biology*, **33**, 205–212.

Lake, P.S. (2000). Disturbance, patchiness, and diversity in streams. *Journal of the North American Benthological Society*, **19**, 573–592.

Lampert, W.H. (1993). Ultimate causes of diel vertical migration of zooplankton: new evidence for the predator-avoidance hypothesis. *Archiv für Hydrobiologie*, **39**, 79–88.

Lancaster, J. (1999). Small-scale movements of lotic macroinvertebrates with variations in flow. *Freshwater Biology*, **41**, 605–619.

Lancaster, J. & Hildrew, A.G. (1993). Flow refugia and the microdistribution of lotic macroinvertebrates. *Journal of the North American Benthological Society*, **12**, 385–393.

Lauridsen, T.L. & Buenk, I. (1996). Diel changes in the horizontal distribution of zooplankton in the littoral zone of two shallow lowland lakes. *Archiv für Hydrobiologie*, **137**, 161–176.

Lauridsen, T.L., Pedersen, L.J., Jeppesen, E. & Søndergaard, M. (1997). The importance of macrophyte bed size for composition and horizontal migration of cladocerans in a shallow lake. *Journal of Plankton Research*, **18**, 2283–2294.

Lawler, S.P. (1993). Species richness, species composition and population dynamics of protists in experimental microcosms. *Journal of Animal Ecology*, **62**, 711–719.

Leibold, M.A. (1989). Resource edibility and the effects of predators and productivity on the outcome of trophic interactions. *American Naturalist*, **134**, 922–949.

Lemly, A.D. & Hilderbrand, R.H. (2000). Influence of large woody debris on stream insect communities and benthic detritus. *Hydrobiologia*, **421**, 179–185.

Levinton, J.S. (1995). Bioturbators as ecosystem engineers: control of the sediment fabric, inter-individual interactions, and material fluxes. In *Linking Species and Ecosystems*, eds. C.G. Jones & J.H., Lawton, London: pp. 29–36. Chapman & Hall.

Lingdell, P.-E. & Engblom, E. (1995). Liming restores the benthic invertebrate community to 'pristine' state. *Water, Air and Soil Pollution*, **85**, 955–960.

Lodge, D.M. & Kelly, P. (1985). Habitat disturbance and the stability of freshwater gastropod populations. *Oecologia*, **68**, 111–117.

Martin, T.H., Crowder, L.B., Dumas, C.F. & Burkholder, J.M. (1992). Indirect effects of fish on macrophytes in Bays

Mountain Lake: evidence for a littoral trophic cascade. *Oecologia*, **89**, 476–481.

Mason, C.F. (1977). Populations and production of benthic animals in two contrasting shallow lakes in Norfolk. *Journal of Animal Ecology*, **46**, 147–172.

May, R.M. (1972). Will large complex systems be stable? *Nature*, **238**, 413–414.

Means, J.L. & Hinchee, R.E. (eds.) (2000). *Wetlands and Remediation: An International Conference, Salt Lake City, Utah.* Columbus, OH: Batelle Press.

Meijer, M.-L., Jeppensen, E., van Donk, E., Moss, B., Scheffer, M., Lammense, E., van Nes, E., van Berkum, J.A., de Jong, G.J., Faafeng, B.A. & Jensen, J.P. (1994*a*). Long-term responses to fish-stock reduction in small shallow lakes: interpretation of five-year results of four biomanipulation cases in The Netherlands and Denmark. *Hydrobiologia*, **275/276**, 457–466.

Meijer, M.-L., van Nes, E.H., Lammens, E.H.R.R. & Gulati, R.D. (1994*b*). The consequences of a drastic fish-stock reduction in the large and shallow Lake Wolderwijd, The Netherlands: can we understand what happened? *Hydrobiologia*, **276**, 31–42.

Merritt, R.W., Wallace, J.R., Higgins, M.J., Alexander, M.K., Berg, M.B., Morgan, W.T., Cummins, K.W. & Vandeneeden, B. (1996). Procedures for the functional analysis of invertebrate communities of the Kissimmee River-floodplain ecosystem. *Florida Scientist*, **59**, 216–274.

Minshall, G.W. & Winger, P.V. (1968). The effect of reduction in stream flow on invertebrate drift. *Ecology*, **49**, 580–582.

Morin, P.J. (2000). *Community Ecology.* Oxford: Blackwell.

Moss, B. (1983). The Norfolk Broadland: experiments in the restoration of a complex wetland. *Biological Review*, **58**, 521–561.

Moss, B., Stansfield, J.H., Irvine, K., Perrow, M.R. & Phillips, G.L. (1996). Progressive restoration of a shallow lake: a 12-year experiment in isolation, sediment removal and biomanipulation. *Journal of Applied Ecology*, **33**, 71–86.

Naeem, S. (1998). Species redundancy and ecosystem reliability. *Conservation Biology*, **12**, 39–45.

Nalepa, T.F. & Schloesser, D.W. (eds.) (1993). *Zebra Mussels: Biology, Impacts, and Control.* Boca Raton, FL: CRC Press.

Newbold, J.D., Erman, D.C. & Roby, K.B. (1980). Effects of logging on macroinvertebrates in streams with and without buffer strips. *Canadian Journal of Fisheries and Aquatic Sciences*, **37**, 1076–1085.

Nyström, P. & Strand, J.A. (1996) Grazing by a native and an exotic crayfish on aquatic macrophytes. *Freshwater Biology*, **36**, 673–682.

Nyström, P., Brönmark, C. & Granéli, W. (1996). Patterns in benthic food webs: a role for omnivorous crayfish? *Freshwater Biology*, **36**, 631–646.

O'Brien, W.J., Kettle, D. & Riessen, H. (1979). Helmets and invisible armour: structures reducing predation from tactile and visual planktivores. *Ecology*, **60**, 287–294.

Ogilvie, S.C. & Mitchell, S.F. (1995). A model of mussel filtration in a shallow New Zealand lake with reference to eutrophication control. *Archiv für Hydrobiologie*, **133**, 471–482.

Ormerod, S.J., Weatherley, N.S., Merrett, W.J., Gee, A.S. & Whitehead, P.G. (1990). Restoring acidified streams in upland Wales: a modelling comparison of the chemical and biological effects of liming and reduced sulphate deposition. *Environmental Pollution*, **64**, 67–86.

Palmer, M.A., Covich, A.P., Finlay, B.J., Gibert, J., Hyde, K.D., Johnson, R.K., Kairesalo, T., Lake, S., Lovell, C.R., Naiman, R.J., Ricci, C., Sabater, F. & Strayer, D. (1997). Biodiversity and ecosystem processes in freshwater sediments. *Ambio*, **26**, 571–577.

Parsons, T.R. (1980). Zooplankton production. In *Fundamentals of Aquatic Ecosystems*, eds. R.S.K. Barnes & K.H. Mann, pp. 46–66. Oxford: Blackwell.

Peckarsky, B.L. (1996). Alternative predator avoidance syndromes of stream-dwelling mayfly larvae. *Ecology*, **77**, 1888–1905.

Pomeroy, L.R. (1980). Detritus and its role as a food source. In *Fundamentals of Aquatic Ecosystems*, eds. R.S.K. Barnes & K.H. Mann, pp. 84–102. Oxford: Blackwell.

Power, M.E. (1992). Habitat heterogeneity and the functional significance of fish in river food webs. *Ecology*, **73**, 1675–1688.

Pringle, C.M., Blake, G.A., Covich, A.P., Buzby, K.M. & Finley, A. (1993). Effects of omnivorous shrimp in a montane tropical stream: sediment removal, disturbance of sessile invertebrates and enhancement of understory algal biomass. *Oecologia*, **93**, 1–11.

Reeders, H.H. & Bij de Vaate, A. (1990). Zebra mussels (*Dreissena polymorpha*) a new perspective for water quality management. *Hydrobiologia*, **200/201**, 437–450.

Rempel, L.L., Richardson, J.S. & Healey, M.C. (2000). Macroinvertebrate community structure along gradients of hydraulic and sedimentary conditions in a large gravel-bed river. *Freshwater Biology*, **45**, 57–73.

Reynoldson, T.B., Bailey, R.C., Day, K.E. & Norris, R.H. (1995). Biological guidelines for freshwater sediment based on Benthic Assessment of Sediment (the BEAST) using a multivariate approach for predicting biological state. *Australian Journal of Ecology*, **20**, 198–219.

Reynoldson, T.B., Norris, R.H., Resh, V.H., Day, K.E. & Rosenberg, D.M. (1997). The reference-condition: a comparison of multimetric and multivariate approaches to assess water-quality impairment using benthic macroinvertebrates. *Journal of the North American Benthological Society*, **16**, 833–852.

Ricciardi, A., Neves, R.J. & Rasmussen, J.B. (1998). Impending extinctions of North American freshwater mussels (Unionoida) following the zebra mussel (*Dreissena polymorpha*) invasion. *Journal of Animal Ecology*, **67**, 613–619.

Richardson, J.S. (1991). Seasonal food limitation of detritivores in a montane stream: an experimental test. *Ecology*, **72**, 873–887.

Richardson, J.S. (1993). Limits to productivity in streams: evidence from studies of macroinvertebrates. *Canadian Special Publication of Fisheries and Aquatic Sciences*, **118**, 9–15.

Richardson, J.S. & Mackay, R.J. (1991). Lake outlets and the distribution of filter feeders: an assessment of hypotheses. *Oikos*, **62**, 370–380.

Rosenberg, D.M. & Resh, V.H. (eds.) (1993). *Freshwater Biomonitoring and Benthic Macroinvertebrates*. New York: Chapman & Hall.

Scheffer, M. (1998). *Ecology of Shallow Lakes*. London: Chapman & Hall.

Scheffer, M. & Jeppesen, E. (1998). Alternative stable states. In *The Structuring Role of Submerged Macrophytes in Lakes*, eds. E. Jeppesen, Ma. Søndergaard, Mo. Søndergaard & K. Christoffersen, pp. 397–406. New York: Springer-Verlag.

Scheffer, M., Hosper, S.H., Meijer, M.-L., Moss, B. & Jeppesen, E. (1993). Alternative equilibria in shallow lakes. *Trends in Ecology and Evolution*, **8**, 275–279.

Schutten, J. (2000). Predicting the hydraulic forces on submerged macrophytes from current velocity, biomass and morphology. *Oecologia*, **123**, 445–452.

Shapiro, J. (1990). Biomanipulation: the next phase – making it stable. *Hydrobiologia*, **200/201**, 13–27.

Shapiro, J., Lamarra, V. & Lynch, M. (1975). Biomanipulation: an ecosystem approach to lake restoration. In *Proceedings of the Symposium on Water Quality Management through Biological Control*, eds. P.L. Brezonik & J.L. Fox, pp. 85–89. Gainsville, FL: University of Florida Press.

Smith, D.G. (1985). Recent range expansion of the freshwater mussel, *Anodonta implicata*, and its relationship to clupeid fish restoration in the Connecticut River system. *Freshwater Invertebrate Biology*, **4**, 105–108.

Soluk, D.A. & Richardson, J.S. (1997). The role of stoneflies in enhancing growth of trout: a test of the importance of predator–predator facilitation within a stream community. *Oikos*, **80**, 214–219.

Stansfield, J.H., Perrow, M.R., Tench, L.D., Jowitt, A.J.D. & Taylor, A.A.L. (1997). Submerged macrophytes as refuges for grazing cladocera against fish predation: observations on seasonal changes in relation to macrophyte cover and predation pressure. *Hydrobiologia*, **342/343**, 229–240.

Starnes, L.B. (1985). Aquatic community response to techniques utilized to reclaim eastern US coal surface mine-impacted streams. In *The Restoration of Rivers and Streams: Theories and Experience*, ed. J.A. Gore, pp. 193–222. Toronto, Canada: Butterworth.

Statzner, B., Arens, M.-L., Champagne, J.-Y., Morel, R. & Herouin, E. (1999). Silk-producing stream insects and gravel erosion: significant biological effects on critical shear stress. *Water Resources Research*, **35**, 3495–3506.

Statzner, B., Fiévet, E., Champagne, J.-Y., Morel, R., & Herouin, E. (2000). Crayfish as geomorphic agents and ecosystem engineers: biological behavior affects sand and gravel erosion in experimental streams. *Limnology and Oceanography*, **45**, 1030–1040.

Stich, H.B. & Lampert, W. (1981). Predator evasion as an explanation of diurnal vertical migration of zooplankton. *Nature*, **293**, 396–398.

Strayer, D.L., Caraco, N.F., Cole, J.J., Findlay, S. & Pace, M.L. (1999). Transformation of freshwater ecosystems by bivalves. *BioScience*, **49**, 19–27.

Swanson, F.J., Johnson, S.L., Gregory, S.V. & Acker, S.A. (1998). Flood disturbance in a forested mountain landscape. *BioScience*, **48**, 681–689.

Timms, R.M. & Moss, B. (1984). Prevention of growth of potentially dense phytoplankton populations by zooplankton grazing, in the presence of zooplanktivorous fish, in a shallow wetland system. *Limnology and Oceanography*, **29**, 472–486.

Toth, L.A. (1993). The ecological basis of the Kissimmee River restoration plan. *Florida Scientist*, **56**, 25–51.

Toth, L.A. & Anderson, D.H. (1998). Developing expectations for ecosystem restoration. *Transactions of the North American Wildlife and Natural Resources Conference*, **63**, 122–134.

Underwood, G.J.C. (1991). Growth enhancement of the macrophyte *Ceratophyllum demersum* in the presence of the snail *Planorbis planorbis*: the effect of grazing and chemical conditioning. *Freshwater Biology*, **26**, 325–334.

Underwood, G.J.C., Thomas, J.D. & Baker, J.H. (1992). An experimental investigation of interactions in snail–macrophyte–epiphyte systems. *Oecologia*, **91**, 587–595.

van den Berg, M.S., Coops, H., Noordhuis, R., van Schie, J. & Simons, J. (1997). Macroinvertebrate communities in relation to submerged vegetation in two *Chara*-dominated lakes. *Hydrobiologia*, **342/343**, 143–150.

Vannote, R.L., Minshall, G.W., Cummins, K.W., Sedell, J.R. & Cushing, C.E. (1980). The river continuum concept. *Canadian Journal of Fisheries and Aquatic Sciences*, **37**, 130–137.

Vaughn, C.C. & Taylor, C.M. (1999). Impoundments and the decline of freshwater mussels: a case study of an extinction gradient. *Conservation Biology*, **13**, 912–920.

Wallace, J.B. & Webster, J.R. (1996). The role of macroinvertebrates in stream ecosystem function. *Annual Review of Entomology*, **41**, 115–139.

Wallace, J.B., Cuffney, T.F., Webster, J.R., Lugthart, J.G., Chung, K., & Goldowitz, G.S. (1991). Export of fine particles from headwater streams: effects of season, extreme discharges, and invertebrate manipulation. *Limnology and Oceanography*, **36**, 670–682.

Wallace, J.B., Whiles, M.R., Webster, J.R., Cuffney, T.F., Lugthart, G.J. & Chung, K. (1993). Dynamics of inorganic particles in headwater streams: linkages with invertebrates. *Journal of the North American Benthological Society*, **12**, 112–125.

Wallace, J.B., Webster, J.R. & Meyer, J.L. (1995). Influence of log additions on physical and biotic characteristics of a mountain stream. *Canadian Journal of Fisheries and Aquatic Sciences*, **52**, 2120–2137.

Wallace, J.B., Eggert, S.L., Meyer, J.L. & Webster, J.R. (1999). Effects of resource limitation on a detrital-based ecosystem. *Ecological Monographs*, **69**, 409–442.

Wootton, J.T., Parker, M.S. & Power, M.E. (1996). Effects of disturbance on river food webs. *Science*, **273**, 1558–1561.

Wootton, R.J. (1998). *Ecology of Teleost Fishes*. London: Chapman & Hall.

Wotton, R.S. (1987). Lake outlet blackflies: the dynamics of filter feeders at very high population densities. *Holarctic Ecology*, **10**, 65–72.

Wotton, R.S. & Malmqvist, B. (2001). Feces in aquatic ecosystems. *BioScience*, **51**, 537–544.

Wotton, R.S., Malmqvist, B., Muotka, T. & Larsson, K. (1998). Fecal pellets from a dense aggregation of suspension-feeders in a stream: an example of ecosystem engineering. *Limnology and Oceanography*, **43**, 719–725.

Wright, J.F. (1995). Development and use of a system for predicting the macroinvertebrate fauna in flowing waters. *Australian Journal of Ecology*, **20**, 181–197.

Zedler, J.B. (ed.) (2001). *Handbook for Restoring Tidal Wetlands*. Boca Raton, FL: CRC Press.

17 • Fish

MARTIN R. PERROW, MARK L. TOMLINSON AND LUIS ZAMBRANO

INTRODUCTION

Fish exhibit far greater diversity than any other vertebrate group with 20 000 living species (Nelson, 1984), more than twice as many as birds or reptiles and amphibians and four times as many as mammals. Fish have colonised nearly all waters on the planet from the deepest oceans to temporary pools, with a diversity of form, ecology and biology to match. Around 1% of fish have even mastered the physiological stresses of migrating from salt to fresh waters and vice versa (Nelson, 1984). The fact that fish are supported in water has provided terrific scope for body size range. In the sea, only the great whales attain greater size than the 12.5 m long, 12.5 tonne whale shark (*Rhincodon typus*), making fish some of the largest vertebrates alive. At the other extreme the pygmy goby (*Pandaka pygmaea*) is 250 million times smaller, reaching 7.5–9.9 mm at adult size.

Fish may be carnivores, herbivores, detritivores and both external and internal parasites, specialising on all conceivable prey items from whole fish to parts of fish (scales, fins and blood), other vertebrates (including mammals), invertebrates, algae, fruit and plankton of various sorts. The radiation of fish into virtually every available aquatic niche has, in many cases, left them vulnerable to extinction as aquatic systems have been extensively modified and exploited of their resources. Several major river basins such as the Amazon are also likely to retain a relatively high proportion of unknown species. With continued habitat destruction and modification it is likely that many will disappear before even being described.

Lakes, rivers and seas have become deposits for human wastes and pollutants derived from industry and agriculture, with significant impact upon fish. The human need for water has resulted in regulation of many rivers, including damming (see Downs *et al.*, volume 2) and abstraction of lake waters. As some habitats have reduced in extent, others, such as water supply reservoirs, have increased and had fish introduced to them. Inevitably, some species have considerably expanded their range and abundance at the expense of others.

Moreover, fish are an important human food supply and over-exploited stocks of once-abundant species such as Peruvian anchoveta (*Engraulis ringens*) and Atlantic cod (*Gadus morhua*) have plummeted. Sharks are now a globally threatened group and a staggering, nearly 740 000 tonnes were slaughtered in 1994, primarily for the 'shark fin soup' trade (TRAFFIC, 1996). The enormous aquarium trade in coral reef fish is also a serious threat and a "responsible aquarists guide" has been produced by the Marine Conservation Society (2000).

The socio-economic importance of the exploitation of fish as food, for sport and as aquarium subjects has led to considerable investment in many aspects of fish biology, culture and ecology. The relative ease and extensive knowledge of fish culture, with enormous potential for production of individuals for reintroduction is a major asset in species-centred restoration efforts.

Moreover, as the habitat preferences of a few species, particularly salmonids, are well understood, fish habitat may be rather easily created. The desire for good fish stocks, particularly for sporting purposes, has provided the impetus for much habitat restoration, particularly in rivers. Whilst potentially beneficial to a range of species, the seeming preoccupation with fish from the perspective of ecologists interested in other groups, hydrologists and geomorphologists has been criticised (see Downs *et al.*, volume 2).

A number of factors aid fish-related control of available resources. With potential resource limitation, there is huge scope for intraspecific and interspecific competition, both within fishes and between fish and other groups (e.g. waterfowl: Giles, 1992). Predation and predation risk, both between fishes and particularly upon invertebrates, also have far-reaching structural and functional consequences. The introduction of alien fish species has thus often had catastrophic consequences. This has led to some attempts at control, a different form of species-centred restoration.

In this chapter we further consider the above issues in two main sections: (1) species-centred restoration, and (2) exploitation of the functionality of fish in habitat restoration, and offer our thoughts on the likely direction of future efforts. Whilst we concentrate on fresh waters, as this is the location of much restoration activity, the implications for marine systems are noted wherever possible.

SPECIES-CENTRED RESTORATION

Restoration of endangered species

The conservation and restoration of commercial species gained momentum towards the end of the last century. In the United States the concern over declining stocks of salmomids led to many projects, particularly by the US Fish and Wildlife service, but also by local citizens (Box 17.1). There are also examples of

Box 17.1. The restoration of the Mattole watershed and its salmonids

The Mattole River runs north for 65 miles along the Californian coast through the Humboldt and northern Mendocino counties before draining into the Pacific Ocean (Mattole Restoration Council, 2000). With 74 tributary streams the Mattole watershed is enormous at approximately 304 square miles. Ancestrally, the watershed contained dense old-growth forests of Douglas-fir (*Pseudotsuga menziesii*), redwood (*Sequoia sempervirens*) and other native hardwoods. After the Second World War, the development of steel-tracked bulldozers allowed timber companies to access the remote and steep forests. The logging boom from 1950 to 1970 left less than 9% of the old-growth forest intact (Mattole Salmon Group, 1999). The resulting erosion from logging, the myriad of 'skid trails' and road construction led to huge sediment loads entering the river, destroying the habitat utilised by the once abundant chinook (king) (*Oncorhynchus tshawytscha*) and coho (silver) salmon (*O. kisutch*) and steelhead trout (*O. mykiss*). The use of huge gillnets by the commercial fishery in the Pacific and the Mattole estuary provided additional pressure on declining stocks (Bernard & Young, 1997).

Concern grew among the citizens of the Mattole region, as the once-fabled salmon runs continuously reduced. In 1980, they formed the Mattole Watershed Salmon Support Group (now called the Mattole Salmon Group) and pledged to restore the run. The Mattole Salmon Group implemented a hatch box and rearing programme. During salmon runs, fish were captured using traps and eggs were stripped from the females, which were then fertilised by milt from males. The fertilised eggs were then placed in clean gravel inside wooden hatch boxes, fed by filtered water directly from the river. Filtered water and cleaned gravels from the Mattole were used to imprint the fry (Bernard & Young, 1997). It was estimated that the hatch boxes increased hatch success rates from 15% in the natural river to 80% (Mattole Restoration Council, 2000). Fry are then raised in creek holding areas. When smolts were released in May for their downstream migration, mortality was high. Now, smolts are held longer for release in autumn during higher river flows (Berger, 1992). Approximately 400 000 juvenile salmon have been released in this way over the years (Mattole Salmon Group, 1999). Initially, the hand-reared fish were not marked, preventing separation of hatched and 'natural' returning fish in visual surveys and a full evaluation of the success of the scheme. However, reared fish are now fin-clipped.

It became clear the hatch box and rearing programme alone was failing to increase the number of salmon returning to spawn and that the whole watershed required attention. As a result, the Mattole Restoration Council was formed in 1983, to protect and restore natural systems, including forests, fisheries, soils and other native plant and animal communities,

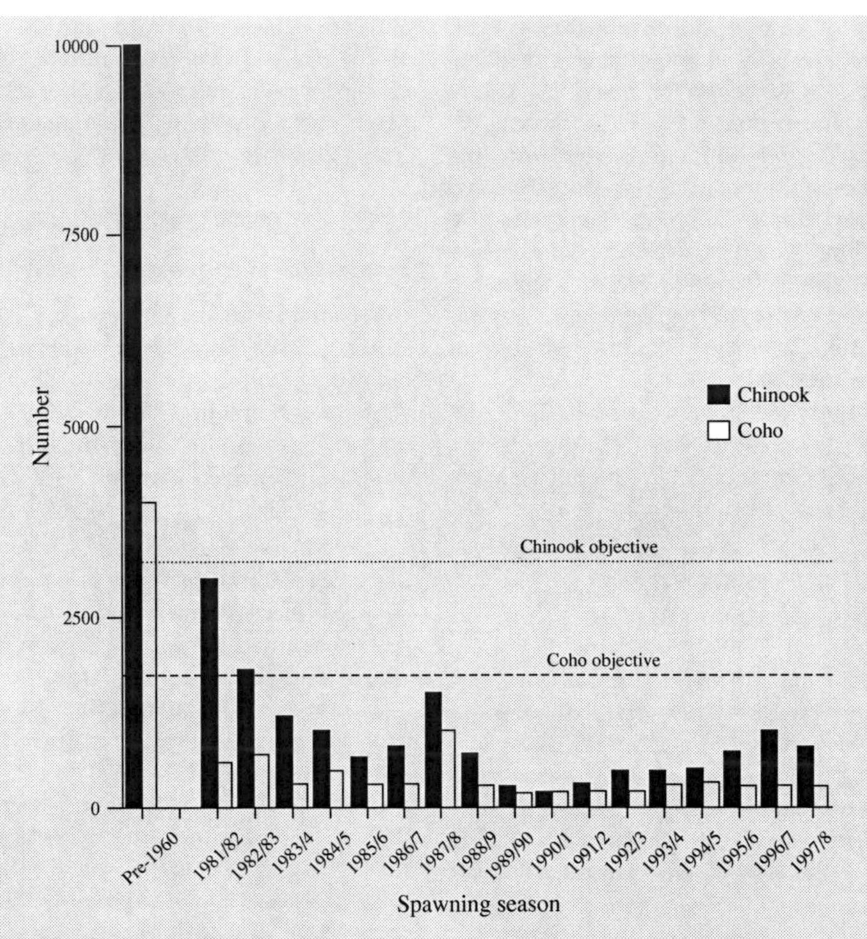

Fig. B17.1. Estimated number of adult chinook and coho salmon returning to spawn in the Mattole River watershed. Redrawn from Mattole Salmon Group (1999).

in the watershed (Mattole Restoration Council, 2000). Together with the Mattole Salmon Group, the Mattole Restoration Council identifies sections of the Mattole and its tributaries in need of enhancement. Works such as excavations to deepen the channel, input of woody debris and the planting of riparian trees to re-establish the bankside vegetation and improve habitat for smolt have continued. Roads and trails throughout the watershed were improved as part of the restoration after surveys and mapping revealed they were responsible for 76% of all the erosion in the area (Mattole Restoration Council, 2000). The ongoing work is funded by monies raised by the Mattole Restoration Council, by grants from government agencies and donations in kind by local landowners.

In 1990, fewer than 200 salmon spawned in the Mattole. In 2000/01 a total of 366 fish (80 chinook, 14 coho, 43 unidentified salmon, 178 steelhead and 51 unidentified fish) were seen in 72 miles of the main river and its tributaries. Just 158 redds were counted (Lingel & Fish, 2001). Moreover, the fact that a third of the fish seen were known to be hand-reared illustrates the continued importance of the hatchery. The goal of the Mattole Salmon Group to restore the salmon runs to their former glory (Fig. B17.1), making the hatching programme redundant, is thus far from realisation. However, it is widely accepted that without the citizen-driven restoration work, salmon would now be extinct in the Mattole watershed.

special cases from regions with particular problems, such as the Great Lakes (Selgeby *et al.*, 1995).

In contrast, the conservation of endangered species *per se* had received relatively little attention until very recently (Maitland, 1995). For example, in Britain, a country with a long history of conservation, especially of birds, one formerly common species, the burbot (*Lota lota*) became extinct only 30 or so years ago and several *Coregonus* species teeter on the edge of extinction (Box 17.2). The inclusion of eight species on the species Annex II of the European Community's Habitats Directive has brought fish conservation on to the agenda for governmental conservation bodies and triggered steps towards species centred-restoration projects. Special Areas of Conservation (SACs) are now required for a range of poorly studied species such as lampreys (river [*Lampetra fluviatilis*], brook [*L. planeri*] and sea [*Petromyzon marinus*]) and spined loach (*Cobitis taenia*).

The latter is particularly intriguing as it is known to exist in a variety of forms in a species complex and may also display polyploidy. It is currently

Box 17.2. Towards the restoration of common whitefish (*Coregonus lavaretus*) in Haweswater, a reservoir in the Lake District, UK

Four whitefish species are recorded from the UK: common whitefish (*Coregonus lavaretus*), vendace (*C. albula*), Arctic cisco (*C. autumnalis*) and houting (*C. oxyrinchus*). In a European context only the latter two are considered endangered/vulnerable, although within Britain they all are and are protected under Schedule 5 of the Wildlife and Countryside Act 1981. In fact, houting is probably extinct, Arctic cisco is restricted to Lough Neagh in Ireland where it is vulnerable to increasing eutrophication and exploitation, vendace only occurs in two lakes in Cumbria, and common whitefish is known from a small number of widely separated lakes in England, Scotland and Wales where it is known by different names. These include 'schelly' in its four English populations in the Lake District (Ullswater, Haweswater, Red Tarn and Brotherswater); 'powan' in Lochs Lomond and Eck in Scotland and 'gwyniad' in Lake Bala in Wales (Giles, 1994; Beaumont *et al.*, 1995).

In the case of common whitefish, studies by Winfield *et al.* (1996) revealed the population at Haweswater – a locally important potable water supply managed by the private company North West Water – to be in serious decline, as a result of inconsistent and poor recruitment. Whitefish spawn in mid winter (October to January) over stone and gravel beds, when the water temperature is approximately 6 °C (Giles, 1994). The females usually lay between 2000 and 11 000 eggs, which incubate for up to 100 days amongst the stones and gravels (Maitland & Campbell, 1992). Winfield *et al.* (1998) suggested that during the incubation period from January to March, fluctuating water levels at Haweswater could have left eggs exposed, possibly resulting in the elimination of an entire year class. In addition, a breeding colony of cormorants (*Phalacrocrax carbo*) established at Haweswater in 1992, with potential impact on the adult population.

In order to combat the decline, a series of conservation/restoration actions were suggested: (1) improvement of spawning grounds; (2) reduction in water-level fluctuation; (3) management of the cormorant population; and (4) the establishment of 'refuge' populations at two further suitable water bodies within the Haweswater catchment.

The introduction of artificial substrata had been used to improve fish stocks, predominantly in streams (e.g. Gray & Cameron, 1987), and as a biomanipulation tool in lakes (Tomlinson *et al.*, 2002), although the approach was novel for *Coregonus* spp. In January 1997, five 1.8 × 4.3 m mats of artificial grass (Model RA1: Nordon Enterprises Ltd, Altham, UK) (Fig. B17.2), were introduced on the only known spawning ground. The main advantage of the mats over natural gravel substrata was that they could be drawn further in to the lake during periods of low water level thus ensuring the eggs would remain submerged during the incubation period.

An underwater video camera revealed eggs were successfully laid on the substrata although, ironically, high rainfall during the study period caused the reservoir to overflow, thereby preventing a full

Fig. B17.2. The introduction of artificial spawning substrate at Haweswater.

evaluation of the success of the mats. Even so, during a period of low water levels in September 1997, a visual survey was conducted to identify potential whitefish spawning grounds and suitable areas where larger scale artificial substrata could be deployed by North West Water, if required. A total of 21 locations were identified.

North West Water also agreed to operate the reservoir, when it was practical to due so, to maintain water levels during the incubation period. Schemes aimed at reducing the fluctuations in water levels have included work to reduce pipe leakage and the promotion of efficient water use by customers. However, sympathetic management is still heavily reliant on rainfall.

With respect to the cormorant population, measures were taken by North West Water to prevent the birds nesting on the reservoirs' island in 1999. Various methods to attempt to scare the birds away were used, but the most effective was frequent disturbance by humans. With the successful prevention of nesting in 1999 the same measures were repeated in 2000. The foot-and-mouth disease outbreak in the UK prevented access and any further action in 2001.

On the issue of establishing refuge populations, after exploratory surveys and discussions with English Nature, Blea Water and Small Water were selected. In February 1997, *c.* 48 000 eggs and milt were obtained from five female whitefish, gillnetted from the known spawning ground within Haweswater. The eggs from each female were kept separate until fertilisation by milt from three and five males. The fertilised eggs were stored in equal volume in two containers lined with the aforementiomed grass mats, until transportation, in stainless steel vacuum flasks, to the recipient lakes. The eggs were then introduced into six containers lined with grass mats at each lake. The fertilised eggs were monitored by underwater video camera and periodic physical removal of samples of the mats. This revealed egg survival rates were very high (Winfield *et al.*, 2002) at both lakes. Visual searches at both lakes for young whitefish during the summer of 1997 were then undertaken, but were unsuccessful. However, eight young whitefish were captured in gillnets (10 mm mesh size) in Small Water in September 1997, although none were captured at Blea Water. In 2000, echo-sounding revealed 'appropriate' signals in Small Water, although no whitefish were captured in fyke nets. The same methods utilised in Blea Water found no evidence of whitefish presence.

In conclusion, more sensitive management appears to have the potential of boosting recruitment of whitefish within Haweswater. Further, action can be implemented to sustain recruitment whenever environmental conditions are not suitable. The potential also exists for the establishment of further populations in the catchment as a 'safety net', although further work will be required to determine the best strategy for establishing and maintaining populations. With continued support of the water company, the longer-term future of whitefish at Haweswater seems bright.

unknown whether the fish in the five catchments in which it is found in the UK are genetically distinct from those in Continental Europe or even each other (Perrow & Jowitt, 2000). A precautionary principle of protecting populations in each catchment has thus been suggested.

Whilst maintenance of genetic diversity is a key issue in species-centred restoration (see Gray, this volume), it is difficult to justify restoration action such as reintroduction of particular populations on the grounds of genetic diversity, where the species remains common in its wider range (see Macdonald *et al.*, Chapter 4, this volume).

Perhaps the single most important issue in species-centered restoration is to tackle the causes of the initial decline (Brown & Evans, 1986), which may be manifold (see above). Water quality and habitat restoration and management are often key issues (Maitland, 1995). Once corrected, natural recovery of populations within a reasonable time-scale may then prove possible, although promising signs such as the colonisation of a few individuals may be far from a self-sustaining population. If the latter is unlikely, reintroduction, perhaps using the powerful tool of aquaculture technology (Box 17.1), can then be justified.

This logical stepwise process has rarely been undertaken in a fully co-ordinated manner and reintroductions and habitat restoration have run concurrently or, at worse, reintroductions have been undertaken with little apparent regard for whether the environment can support the introduced stock and will be sustainable over the longer term (Jokikokko, 1990).

Habitat restoration

It is probably true to say that a desire to produce more fish, particularly 'game' species (especially salmons and trouts), for sport and the table propagated much of the early interest in stream and river restoration (better termed enhancement and rehabilitation in this case). The territoriality and distinct habitat preferences of salmonids provides a straightforward link between an increased abundance of particular habitat and more fish. As defined in the US National Research Council (1992) quoting Raleigh & Duff (1980), essential habitat features include 'clear cold water, a rocky substrate, an approximate pool to riffle ratio of 1 to 1 with areas of slow deep water, a relatively stable flow regime, well vegetated stream banks and abundant in-stream cover'.

A style of promoting or simply installing particular habitat elements, using both natural and artificial materials was typically adopted (for review see National Research Council, 1992). A habitat element approach is still used in large-scale restoration schemes (Box 17.1) and the means of habitat enhancement for fish are also documented for estaurine and marine habitats (e.g. Adams & Whyte, 1990; Clark, volume 2; Hawkins *et al.*, this volume).

The number of schemes illustrating beneficial impacts of habitat enhancements over the longer term and beyond the generation time is increasing (e.g. Ptolemy, 1997; van Zyll de Jong *et al.*, 1997), as are attempts to provide habitat for poorly studied endangered species (Fuselier & Edds, 1995; Winemiller & Anderson, 1997).

For the purposes of this chapter, we adopt a life-history stage approach to illustrate the range of species-centred restoration efforts.

Spawning passage

Anadromous and catadromous species provide the most dramatic examples of spawning migrations.

However, many other species undergo migrations to spawning grounds. Obstructions to free passage of fish in the form of dams and weirs is a serious issue in many river catchments, accounting for the decline in many species (see Adams, volume 2 for wider-scale impacts of dams).

A number of types of passage structures including Denil fish passes, pool fish passes, fish elevators and natural bypass channels are in use particularly in the developed world. As noted by Jungwirth *et al.* (1998) in an international conference on the topic, '[fish pass] research reached perhaps its zenith in cost, scale and complexity with the passage facilities on a series of dams on the Columbia River in the north-western United States. Here, one can boast of passing adult salmon through a staircase of concrete and metal baffles, orifices and louvres with an elevation gain of 35 m across 1300 m of length in under 4 hours.'

In the absence of passes, migrating fishes may even be captured and physically moved upstream (Hendricks, 1995). Moreover, whilst many pass structures may be suitable for large, fast-swimming salmonid species, adapted to leap natural obstructions such as falls and debris dams, their use by smaller and/or less vigorous species remains questionable. Even what may be deemed to be minor obstructions may be difficult to navigate even for large species (Lucas & Frear, 1997). For small species such as the <15 cm bullhead *Cottus gobio*, even a barrier of 10 cm in height may be impassable, leading to population fragmentation (Utzinger *et al.*, 1998).

The removal of dams has considerable consequences for flow and sediment dynamics with associated flooding risks for human interests. Even so, dam removal has been undertaken in a number of countries, such as Denmark (e.g. Rasmussen, 1999). Especially where undertaken with other habitat works, the effect on fish stocks may be dramatic (Scruton *et al.*, 1998).

Spawning substrate

Fish utilise a range of spawning substrates from bed material to overhanging and emergent vegetation. Many species are lithophilous (gravel-spawning) in rivers, exerting considerable choice in the size of the substrate in which to cut redds (salmonids) or broadcast their eggs. Deposition of fine sediments in spawning gravels as a result of modifications in land practice in the upper catchment is a major problem for many species. Clogging of interstitial spaces may reduce egg habitat or oxygen concentrations thereby reducing egg and larval survival.

Cleaning of gravels may be undertaken through natural, sustainable means by restoring flushing flows. This has been undertaken on the Colorado river for the Colorado pikeminnow (*Ptychocheilus lucius*) amongst other endangered fish species (see Newson *et al.*, this volume). Where the ultimate cause of sediment deposition cannot be tackled, manual disturbance of spawning gravels can be undertaken with regular ploughing, raking or jetting (Mih & Bailey, 1981). Suitable spawning substrate may also be introduced, although a thorough understanding of flow and sediment dynamics will be required for this to be sustainable. A particularly satisfactory outcome was reported in East Fork Lobster Creek in Oregon, following introduction of 15 rock-filled gabion baskets filled with gravel after winter freshets in successive winters. The quality of gravels impounded by gabions equalled or exceeded the quality of gravels in unmodified areas and over 50% of coho salmon (*Oncorhynchus kisutch*) and steelhead (*O. mykiss*) spawning over an 11-year period used the gabions (House, 1996).

Water-level fluctuation is a frequent limiting factor for littoral spawning species in lakes, with vegetation becoming unavailable immediately prior to spawning, or eggs that have already been laid being left high and dry (Box 17.2). Water-level management may thus be a critical aspect in fish restoration programmes. Otherwise, management of the littoral margin may be readily undertaken to establish the right sort of vegetation for spawning fishes. Where suitable substrate is still unavailable, artifical substrates may be introduced (Bry, 1996) (Box 17.2).

Larval and fry habitat

Habitats within the riparian zone/floodplain operate as essential larvae and fry nurseries for many species (Copp & Peňáz, 1988; Schiemer & Waidbacher, 1992; Kurmayer *et al.*, 1996). All too frequently, even if the habitats themselves, such as backwaters, meander loops, side channels and even artificially dug drainage channels, are still present, connectivity with the main channel may have been much reduced or eliminated altogether (see Downs *et al.*, volume 2). Isolation of lakes from their catchments is also commonplace.

Much of the need for floodplain habitats stems from the limited swimming ability of young fish and their consequent need for flow refugia and a dependence on zooplankton, which may proliferate in calm, warm waters.

In lakes, Skov & Berg (1999) successfully introduced spruce trees as a habitat for larval pike (*Esox lucius*), a potentially important piscivore stabilising lake food webs (see below).

Reintroduction

Introduction of hatchery-bred fish to restore declining stocks is not new. For example, the initial attempt to restore Atlantic salmon (*Salmo salar*) to the Connecticut River occurred over 100 years prior to the beginning of the current programme in 1967 (Jones, 1978).

The relative ease of fish aquaculture has meant that eggs and young fish may be produced on a massive scale and stocked at whatever stage or size is suitable. However, success of production is not guaranteed as disease may be prevalent (Harrell *et al.*, 1987). Moreover, all too frequently, insufficient attention has been given to the optimal size of stocked fish and the timing and means of introduction (e.g. spot or scatter-stocking). In the case of the latter, spot stocking may result in density-dependent mortality and scatter-stocking may be the better option (Berg & Jørgensen, 1994).

The genetic composition of stocked individuals is also a major issue and genetic pollution should be avoided. In an ironic twist, since the 1960s the native Atlantic salmon was thought to be extinct from Danish waters and stocks were maintained with Scottish-born fish. However, recent DNA analysis of old scale material revealed that small populations of indigenous fish still existed in a number of rivers (Nielsen *et al.*, 1997, 2001). Parent fish taken from the river are now subject to genetic analysis and assigned to the most likely population of origin before being stripped of eggs and milt for supportive breeding of juvenile fish for introduction. Any alien fish are destroyed. Thus, there is hope the alien stock will be diluted over time.

Post-project appraisal of stocking is essential (Hickley, 1994). Where eggs were introduced and some native stock was present it may be impossible to determine the success of the reintroduction strategy, particularly where several other means of improving stocks are undertaken (Box 17.1). For young fish however, the introduced individuals can be separated from natives by means of tags, fin clips, dyes or other markers (for example, the antibiotic tetracycline produces a mark in the otolith).

Where properly undertaken, stocking of hatchery fish may have a great impact. Hendricks (1995) showed that the hatchery contribution (of 5–20 million per annum: Saint-Pierre, 1996) to the juvenile population of endangered American shad (*Alosa sapidissima*) population in the Susquehanna River ranged from 79% to 99% from 1985 to 1991. Coupled with transplant of thousands of returning adults above barriers and the use of innovative measures to direct outmigrating juveniles away from turbine intakes at dams, the population increased from a few hundred to over 60 000 over an 11-year period (Saint-Pierre, 1996). The success of the scheme was a major factor in the decision of several hydroelectric companies to construct fish passage facilities at a series of large dams, opening up over 700 km of historic spawning habitat to American shad and other anadromous fishes such as river herring (*A. aestivalis* and *A. pseudoharengus*), for the first time since 1910.

Control of aliens

Exotic fish have been introduced to lakes and rivers throughout the world as food, for sporting purpose, as bait, biological control or even by accident. In rural communities of developing countries, aquaculture has been widely promoted to provide a cheap source of animal protein to combat nutritional problems. This activity is mostly based on species with a standard technology for production at reduced costs. Consequently, a few species of proven productivity, such as common carp and *Tilapia*, have been widely dispersed (Box 17.3).

The 'tens' rule considers the probability of an exotic species becoming a pest by the probability of

Box 17.3. The ecological and sociological impact of common carp *Cyprinus carpio* aquaculture in Mexico

Following international trends around 30 years ago, the Mexican government started a programme promoting culture of common carp (*Cyprinus carpio*) in rural regions (Contreras-Balderas, 2000). Carp production was encouraged in any type of freshwater system, which has resulted in the presence of this fish in 95% of Mexican lakes (Mujica, 1987). Carp grows well in Mexican shallow ponds as a result of its ability to survive and grow in environments with low food supply and high turbidity (Maitland & Campbell, 1992). The well-known negative effects of carp on fresh waters (Cahn, 1929; Crowder & Painter, 1991; Bales, 1992; Carvalho & Moss, 1995; Roberts *et al.*, 1995) is of real conservation concern as these could lead to the disappearance of a large proportion of the endemic Mexican fish diversity. To date, approximately 30% of the Mexican freshwater fish species that are classified as 'in danger', 'rare' or 'extinct' is related to the introduction of an exotic fish in their native systems (Contreras-Balderas, 2000).

Perhaps inevitably, there is a potential trade-off between conservation and social concerns, and, at face value, the social benefits that carp production (30 000 tonnes per year on average) may provide poor people in rural zones, may be argued to override biodiversity conservation. However, this is highly simplistic and there was a need to understand the actual benefits of carp culture to the local people. Studies at Acambay within the Lerma–Santiago Basin in Central Mexico proved useful in analysing the real balance of the trade-off between stocking carp and preserving native fishes and other species.

The Lerma–Santiago Basin contains the largest and richest lakes in the country, hosting endemic freshwater organisms that are economically and culturally important, such as the white fish (pescado blanco) (*Chirostoma estor*), the crayfish (acocil) (*Cambarellus montezumae*) and axolotl (*Ambystoma* spp.).

Moroever, there are a great number of small (<10 ha), shallow (<3 m deep) artificial irrigation ponds in the region, which, despite being highly influenced by the needs of the farmers, have a similar water regime to the natural lakes and rivers of the region. Namely, they are filled during the rainy season that starts in June and ends in November. Close to Easter, the ponds partially or totally dry out as a result of irrigation of the surrounding agricultural land in the middle of the dry season. Probably as a result of the hydrological similarity, ponds contain a similar suite of native organisms to the natural systems.

Carp culture starts each year at the beginning of the rainy season with the introduction of fingerlings provided by the government. They grow for around five months until Easter, when they are easily harvested. Benthivory in that time may deplete benthic organisms, whilst promoting an increase in water turbidity through resuspension of solids (Zambrano *et al.*, 1999) (Fig. B17.3), thereby depleting water quality. Carp can thus initiate the switch from a clear to a turbid water system (Fig. B17.3). An increase in turbidity further modifies many relationships within the trophic web (Scheffer *et al.*, 1993), including a detrimental impact upon the abundance of submerged macrophytes

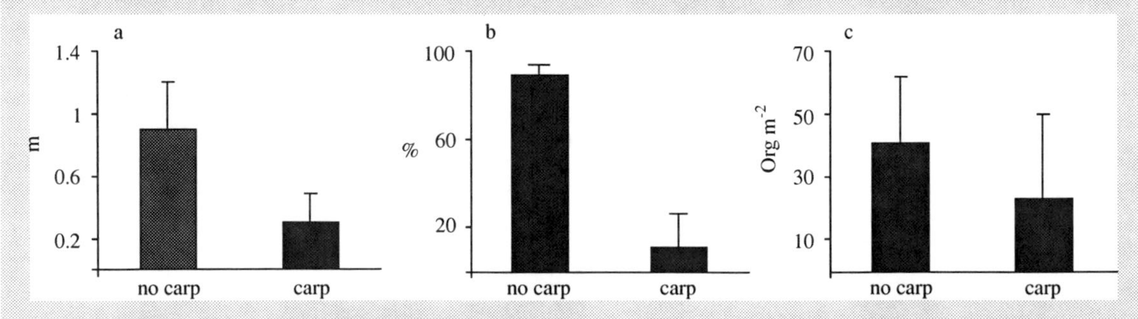

Fig. B17.3. Variables in different ponds with and without carp. (a) Secchi depth; (b) percentage of macrophyte cover; and (c) abundance of benthic organisms. Bars are means with standard error.

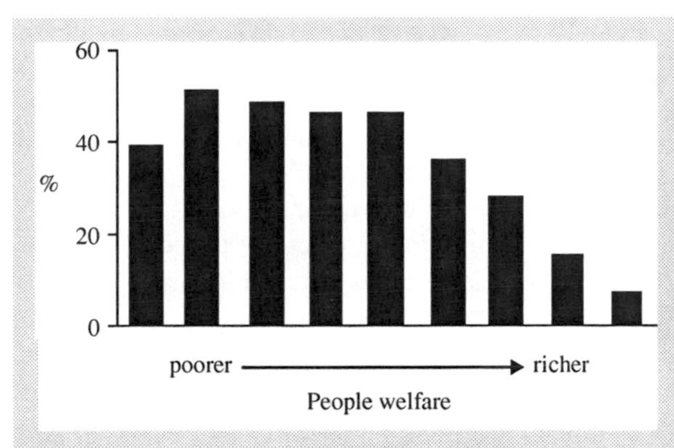

Fig. B17.4. Percentage of the consumption of native species of fish in Mexico, considering the region welfare index.

(Zambrano & Hinojosa, 1999), that provide spatial and food resources for most of the fauna in shallow lakes (Scheffer, 1998).

A notable indirect effect of carp is a decrease in fish abundance and diversity. Crayfish are also particularly affected (Hinojosa unpubl. data). Native fishes and crayfish are exploited by native people, and to compensate their loss, there should be clear nutritional benefits of introducing carp. However, this does not appear to be the case.

The main beneficiaries of the carp introduction policy are those who own a pond; in other words, those from wealthy families with enough money to build one in the first place. Without carp in their ponds, owners allow people to fish freely on their property, since they do not hold native species in high esteem and a source of further wealth. As a result of the open fishing rights, poorer members of society have better access and control over native species, which is reflected in the higher proportion of poor people that eat native species (Fig. B17.4). The introduction of carp provides a saleable commodity to the owners and the poor people lose their right to fish for native species, which are diminished by the presence of carp. Carp stocking does not, therefore, directly benefit the poorest members of society as it was supposed to do, and it also generates a considerable threat to native endemic species.

Natural resource use in rural societies is frequently pushed by ideas promoting the development of an industrial society in one way and by the desire for conservation in another. This case study illustrates that the stocking of exploitable fish must be carefully researched beforehand and appropriate aquaculture policies developed to balance both ways. The balance must consider the potential close links between ecological values of the system and cultural factors from the local society, in order to achieve the successful management of exploited fresh waters.

one-tenth in each of three steps: (1) a 10% probability that an introduced species will appear in the wild; (2) 10% of these will become established in the new environment; and (3) 10% of these will become a pest (Williamson & Fitter, 1996). Intentional introduction for aquaculture increases these proportions to a 'twos' or 'threes' rule (Minns & Cooley, 2000), as producers generate conditions for survival of the exotic fish (Zambrano & Macías-García, 2000).

Negative impacts of aliens are manifold, depending on the nature of the fish involved. The primary impacts upon native species are typically competition and predation (see below). The introduction of the voracious predator Nile perch (*Lates niloticus*) and five competing tilapiine species (*Oreochromis niloticus*, *O. leucostictus*, *O. rendalli*, *O. melanopleura* and *Tilapia zillii*) to Lake Victoria in the late 1950s led to massive decline in many of the 350 endemic haplochromine cichlids and the native *O. esculentus* (Okemwa & Ogari, 1994). From a fishery dominated by 250 haplochromine species, mormyrids, catfishes, cyprinids and *O. esculentus* and *O. variabilis*, only two exotics (*L. niloticus* and *O. niloticus*) and the native *Rastrineobola argentea* were important by the late 1980s (Ochumba *et al.*, 1994).

Further, as fish often function as keystone species in aquatic food webs (see below) the entire system may change dramatically for the worse (Ochumba *et al.*, 1994) (Box 17.3). In the case of Lake Victoria this extended to deforestation of the lake catchment, as firewood was required to fry the oily-fleshed Nile perch in its own fat to preserve it, unlike the native cichlids, which could be dried (Okemwa & Ogari, 1994).

Control measures

Rinne (1992) suggests: (1) research into the mechanisms of interactions between native and introduced fishes; (2) conservation and restoration of habitats for native species in an ecosystem or river basin concept; (3) incorporation of a value system for native fishes; and (4) stringent regulations for importation of non-native fishes, as measures towards control of alien species. However, where aliens are already established with obvious effect, a more direct form of action is required. Complete eradication of virtually any pest species is a daunting task and effective control is perhaps a more realistic aim. The current trend towards biomanipulation of even large lakes (e.g. the 2700-ha Lake Wolderwijd in the Netherlands: Backx & Grimm, 1994) and the reduction of many species in the oceans illustrate that anything is possible with sufficient ambition and resources. Cost–benefit ratios are critical, particularly where commercial interests are maintained and these need to be thoroughly evaluated. Even then, a desirable outcome may prove possible with little impact upon commercial interests. For example, simply changing mesh size in the Lake Victoria gillnet fishery to a 5-inch (12.5-cm) minimum mesh size was predicted to reduce predation by Nile perch on other fishes including haplochromine cichlids, and *Rastrineobola argentea* by as much as 44%, with little decrease (10%) in Nile perch yield (Schindler *et al.*, 1998).

In general, the means of controlling any species will be highly specific and multiple techniques targeting vulnerable life stages may be required. One of the few examples of successful control is that upon parasitic sea lamprey (ironically protected in Europe) in the Great Lakes. After a number of unsuccessful control attempts, a programme of treating nursery grounds with the chemical lampricide TFM (3-trifluoromethyl-4-nitrophenol) reduced lamprey abundance to 10% of its initial level (Christie, 1997), with obvious impact within three years at some sites (Johnson, 1988).

EXPLOITING THE FUNCTIONAL ROLES OF FISH IN RESTORATION

The functional role(s) of fish has probably been more widely exploited as a means of habitat restoration than in any other group, save perhaps invertebrates, through which the influence of fish is often expressed (Richardson & Jackson, this volume). It is unclear why this should be the case as it is difficult to argue that fish are better understood than, say, mammals or birds. Limnologists may have invested more effort in quantifying interactions in food webs than their terrestrial counterparts, which has led to limnology becoming a leader in combining ecological thinking with management practice (Harris, 1994). Nowhere is this more advanced than in lake restoration, with 'biomanipulation' of fish – the removal of zooplanktivorous/benthivorous species and/or the enhancement of piscivorous species – a key component of many lake restoration programmes (Jeppesen & Sammalkorpi, volume 2).

Here, we briefly consider some distinctive biological and ecological features of fish that provide the potential for a pervasive influence and then go on to illustrate the impacts fish have, particularly in freshwater lakes.

Key parameters

Diet and ontogenetic switching

A review of the diets of 600 species by Love (1980) revealed that the majority are carnivorous (85%), feeding on invertebrates of various types, other fishes and even upon mammals and birds in the case of some larger, obligate predators. However, most species are really rather generalist.

A high degree of carnivory is linked to the use of protein for energy production. Fish have been able to elevate their minimum protein requirement to >40% as a result of: (1) the relatively lower cost of maintaining body temperature and position in water; (2) ease of excretion of potentially toxic ammonia and nitrogenous waste through the gills; and (3) the capacity to digest protein irrespective of feeding habit (Pandian & Vivekanandan, 1985).

Diet may thus readily change according to life stage, i.e. ontogenetically. Eurasian perch (*Perca fluviatilis*), for example, is generally thought to switch from zooplankton to macroinvertebrates at the end of its first year of life at a length of 7–10 cm and then to fish at a length of 15–20 cm. In practice, the diet often changes according to the availability of different prey items, which is often seasonal and potentially influenced by competition with other species (Persson, 1983*a*, 1986*a*, 1987). Indeed, relationships may become complicated as a result (Box 17.4).

Fish may also exploit poor quality foodstuffs in the absence of preferred prey. In a southern Swedish lake blue-green algae and detritus constituted 75% of the diet of roach (*Rutilus rutilus*) and was sufficient for maintenance and growth (Persson, 1983*b*).

Box 17.4. Interspecific interactions between young-of-the-year roach and perch in a shallow eutrophic lake:competitor becomes prey?

The unrelated roach (*Rutilus rutilus*, Cyprinidae) and perch (*Perca fluviatilis*, Percidae) occur sympatrically in a number of habitats over much of Europe and northern Asia (Persson, 1986*b*). At least in the first few years of life there is scope for considerable resource overlap. Young roach enjoy a competitive advantage over young perch in shallow eutrophic waters, being better able to exploit prey in warmer (Persson, 1986*b*), structureless (Winfield, 1986) more turbid conditions. In the absence of submerged plants in shallow lakes, perch may be forced to abandon the more profitable open water zone and become restricted to the littoral margin (Persson, 1987), with implications for individual growth and survival. The intensity of the juvenile bottleneck may thus prevent the development of an age-structured perch population. The lack of older individuals, which may be functionally important piscivores, has further implications for trophic interactions and structure.

A two-year cycle in the recruitment of roach in Alderfen Broad (Townsend & Perrow, 1989; Perrow *et al.*, 1990; Townsend *et al.*, 1990), a small (5.1 ha), shallow (mean depth around 1m) lake in eastern England, provided an opportunity to explore further the nature of the interaction between the juveniles of both species. In good recruitment years the number of young-of-the-year roach was 19-fold higher on average than in poor recruitment years. In the former, roach severely depleted the shared zooplankton resource even leading to the elimination of larger cladoceran zooplankton by late summer (Fig. 17.1). This reduced growth and fecundity of older individuals and with high inter-annual mortality, the spawning population was always dominated by young fish. Competition with young-of-the-year in a good recruitment year thereby ultimately reduced the fecundity of older individuals coming to spawn the following year. The resultant poor year class released any older individuals from competition and good growth and fecundity resulted in a strong year class in the year after (and so on).

Several impacts of strong roach recruitment on the smaller perch population were expected:

- Perch would be forced to undergo their ontogenetic dietary switch to macroinvertebrates earlier than desired.
- Growth, and perhaps survival, of juvenile perch would be negatively affected.

As previously shown by Persson (1986*a*), perch did indeed switch to macroinvertebrates by the end of their first growing season in the presence of abundant roach (1985) compared to the following year when roach recruitment was limited (1986) (Fig. B17.5). Using data from the six years perch had reappeared in Alderfen (following a reduction in nutrient levels and an increase in submerged vegetation: Perrow *et al.*, 1994), no impact

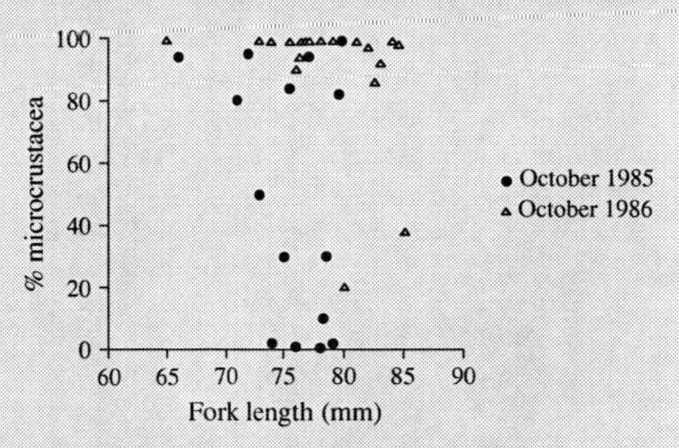

Fig. B17.5. Relationship between the percentage of microcrustaceans by volume in the stomachs of underyearling perch and their size (mm fork length) After Perrow (1989).

of roach recruitment on perch abundance was noted. However, there was a significant positive effect on the length attained by perch by the end of the growing season. In contrast, there was no obvious intraspecific effect (Fig. B17.6).

This unexpected result suggested individual perch benefited by switching early. But if this was true, this meant perch foraged suboptimally in the absence of roach, which seems unlikely given the general ability of fish to forage in an optimal manner (Townsend & Winfield, 1985). Moreover, even though there was no obvious variation in individual diet in relation to body size (Fig. B17.5), it is plausible that the trend was an artifact of the survival of a few larger perch while all smaller individuals perished. Larger perch with larger mouths have greater potential to take macroinvertebrates and other large prey items such as fish.

A number of authors have now illustrated that perch may become piscivorous at an early age and small size: 17–28 mm in the case of Spanovskaya & Grygorash (1977) and 10.5 mm in the case of Brabrand (1995). In the latter study, perch were cannibalistic on conspecifics in the same cohort. Predation on young-of-the-year cyprinids by young-of-the-year perch is always possible as perch are typically born a month or so earlier and thus enjoy a size advantage. Borcherding *et al.* (2000) showed young-of-the-year perch of 30 mm were effective predators of 0+ bream of 14 mm and 19 mm.

The fact that fish were not represented in the diet of perch at Alderfen is not unusual given the small number of fish examined and the likely relative infrequency of fish in the diet, belying their potential importance. Thus, in conclusion, perhaps the best explanation of the observed positive relationship between the size achieved by perch and the abundance of roach is that rather than suffer from the effects of competition at least some perch were able to benefit by consuming the smaller juvenile roach.

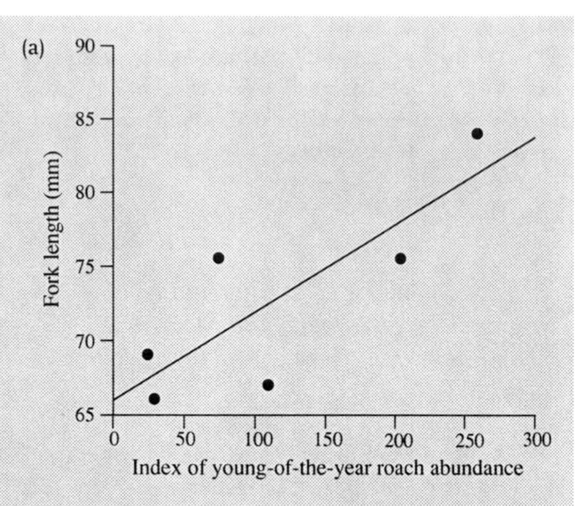

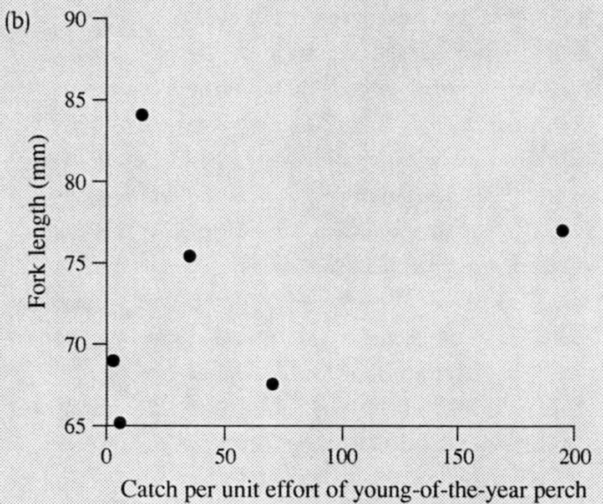

Fig. B17.6. Relationships between size attained by underyearling perch (mm fork length in October) and abundances of (a) underyearling roach and (b) underyearling perch from 1979 to 1986. After Perrow (1989).

Patterns of foraging

Studies illustrate that fish generally forage optimally, in close agreement with predictive models of maximal intake (Townsend & Winfield, 1985). However, a common pattern is that fish continue to sample other habitats or prey types, presumably to continually assess whether to persist with a strategy or change. Fish may readily switch between different prey items, utilising widely different foraging strategies in the process (Lammens & Hoogenboezem, 1991). Fish are able to achieve such results by behavioural adaptations, whilst retaining a relatively

Box 17.5. Explaining the benthic foraging efficency and diet of three sympatric European cyprinids: the effect of body size and mouth protrusion

The cyprinids roach (*Rutilus rutilus*), common bream (*Abramis brama*) and silver bream (*Abramis bjoerkna*) occur sympatrically in lakes and slow-flowing rivers over much of Europe. All three species take a range of prey, including planktonic crustacea (such as *Daphnia* spp.) and benthic items such as chironomid larvae. The protrusibility of the mouth, the aperture size of the gill rakers and relative strength of pharyngeal teeth are key factors influencing the efficiency of foraging on particular groups of invertebrates (Lammens & Hoogenboezem, 1991).

Roach, with the strongest teeth, use their forceps-like relatively non-protrusible mouths to feed preferentially on molluscs, whilst the protrusible mouth of bream allows them to forage efficiently on chironomid prey in the sediments. Prey items are winnowed from the sediments by the use of the moving gill raker basket, with actual aperture size modified in the process. The mechanism of prey selection appears to be olfactory, and local bulging of the palatal organ surface traps food against the palate before being transferred to the pharyngeal teeth for crushing, whilst non-food particles are ejected through the operculum (Sibbing, 1991).

The greater mouth protrusibility of bream over silver bream and roach allows more suction pressure and more effective penetration into bottom sediments (Lammens *et al.*, 1987). Overall body size influences the relationship, with larger fish being able to push deeper into the sediments with their larger mouths, which may handle prey more efficiently, particularly in relation to increasing sediment particle size (Fig. B17.7). Consequently, the intake rate of chironomids was higher for large bream (25–35 cm) compared to smaller ones (10–20 cm) (Fig. B17.7). This is thought to have been partly responsible for larger bream taking a larger proportion of chironomids in their diet compared to smaller conspecifics and also roach and silver bream (Fig. B17.7).

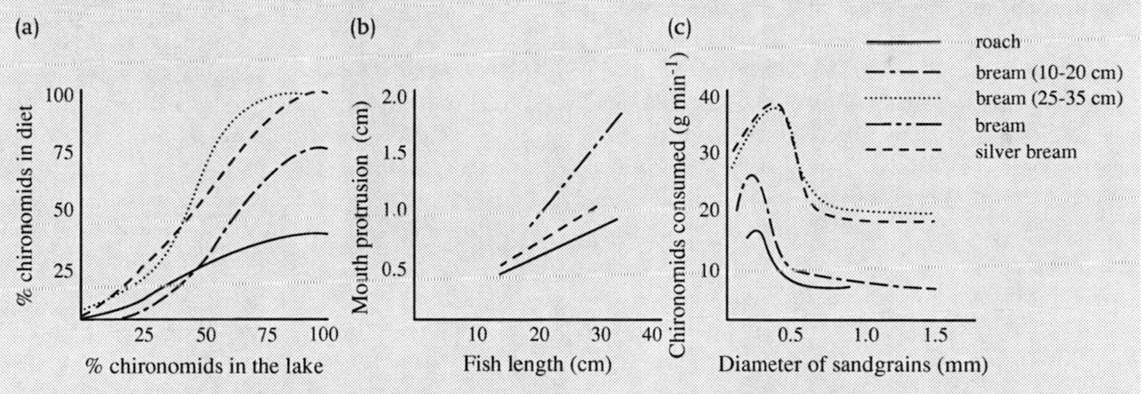

Fig. B17.7. Relationships between: (a) the proportion of chironomids in the diet versus the proportion in the benthos of the lake; (b) mouth protrusion and fish length; and (c) rate of chironomid consumption and the diameter of sandgrains in the substrate; for roach, two size classes of bream and silver bream, three sympatrically occurring cyprinids. Redrawn from Lammens *et al.* (1987).

simple body plan. Thus, many species closely resemble each other and even very subtle changes to body form can result in a real advantage for one species over another when foraging on a particular prey item (Box 17.5).

Body size and related parameters

An increase in body size leads to an increase in gape size and swimming ability (particularly speed), both of which are key determinants of foraging success and ontogenetic diet shifts (see above). Even small differences may provide a distinct advantage. Individuals reaching key body size (gape size) thresholds may go on to enjoy increased growth rates, and ultimately survival and fitness (also see below). This is perhaps at its keenest in flowing environments where small size differences may allow fish to forage over a greater area of channel (Mills, 1991) whilst exploiting a greater range of prey.

However, the advantages do not always lie with larger individuals. One of the fundamentals of body size relationships is that whilst smaller individuals require relatively more food to support each unit of body mass, being smaller, they need absolutely less. With limited resources small individuals may be at an advantage and exhibit better growth (Hamrin & Persson, 1986).

Reaching a particular size can also expose the individuals concerned to a 'bottleneck', where one limitation or another restricts growth and survival resulting in an asymmetric age/size structure in the population (Box 17.4).

Fecundity and age class strength

The number of eggs potentially laid by a female is typically a function of body size and body condition. Where competition for resources in any one growing season is intense, the number of eggs laid by a female may be orders of magnitude lower than where resources were abundant. Herein lies the potential for feedback, as where growth was good and fecundity high, a high number of young fish may be recruited into the population, assuming other controls on growth and survival were relaxed. A strong year class may then dominate the population and even the system for many years (e.g. Hodgson *et al.*, 1993).

With suppression of other year classes and the preponderance of egg potential within individuals of the dominant year class, strong year classes may then beget other strong year classes. This is one mechanism for population cycles (Townsend & Perrow, 1989).

Growth and longevity

Unlike higher terrestrial invertebrates and in common with plants, fish display essentially indeterminate growth (Weatherley & Gill, 1987). Moreover, individuals trapped for some time at a particular size may grow rapidly should conditions change for the better.

The essence of growing until death, coupled with the long life spans of many species, leads to the potential for storage of a considerable amount of the available nutrient within fish bodies, particularly those at the top of the food chain such as largemouth bass (*Micropterus salmoides*) (Kitchell & Carpenter, 1993).

Competition

Notwithstanding the influence of temperature upon the poikilothermic body of fishes, the intense competition for resources is typically a major determinant of growth of individuals within a population. Many studies have illustrated density-dependent growth rates as a result of intraspecific (e.g. Perrow & Irvine, 1992) and interspecific competition (see Box 17.4), which may have far-reaching consequences (Hamrin & Persson, 1986; Perrow *et al.*, 1994).

Impacts

Zooplanktivory

The mechanics

Zooplankton is probably the most important food source for young fish, including larvae and fry. Even the fry of facultative piscivores such as northern pike (*Esox lucius*) typically consume zooplankton in their early life stages (Bry, 1996). As fish mature, they typically switch to larger, more profitable items (see above). However, this really depends on the size of the zooplankton involved and whether the fish concerned are obligate zooplanktivores or not. With relatively large and abundant zooplankton such as krill (*Euphausia superba*) and the young life stages of fish, crabs, corals and other invertebrates in the seas, even the largest fish such as whale and basking sharks (*Cetorhinus maximus*) can be supported. Large species are invariably filter-feeders in order to meet the required ratio of intake rate to metabolic demand. Specialised and elaborate filtering structures to consume zooplankton are relatively rare. It is much more common that fish use rather generalist mouthparts. For example, amongst the Cyprinidae, one of the largest families of vertebrates in the world, inhabiting every conceivable freshwater and brackish habitat, there are few specialist plankton feeders with associated mouthparts, the Siberian genus *Hypophthalmicthys* – containing the phytoplankton specialist silver carp (*H. molitrix*) and the zooplankton specialist bighead carp (*H. nobilis*) – being the best known.

The means of consuming zooplankton by cyprinids lies in the adoption of multiple foraging strategies. Lammens & Hoogenboezem (1991) recognise several modes:

1. Particulate feeding, whereby the fish detects an individual prey (the reactive distance), pursues and attacks it with fast, directed suction.
2. Pump filter-feeding (gulping), where a slow-swimming or stationary fish takes a long series of suctions/snaps especially directed at local areas of higher plankton density.
3. Tow-net filter-feeding, which involves a fish swimming quickly with the mouth agape and opercula abducted, thereby engulfing a cylinder of water containing food particles.

Fish may switch between different modes according to the prevailing conditions including the density of food particles, light level and the escape efficiency of the prey concerned. Moreover, the size of the fish and thus the intake required may limit the modes that can be adopted. For example, roach typically adopt a true particulate strategy when young (Winfield *et al.*, 1983), achieving a high intake rate (*c.* 0.7 individuals s^{-1}) upon optimally sized *Daphnia* spp. (Perrow & Irvine, 1992). Fish above 135 mm were only able to approach this intake rate by adopting a gulping strategy (Perrow, 1989). Large bream also use pump filter-feeding, especially in turbid lakes and at night, relying on retention of small prey – smaller than the diameter between rakers – in the moving gill raker basket (Lammens & Hoogenboezem, 1991). With any sort of filter-type feeding, foraging efficiency is primarily determined by the retention abilities of the filtering system.

The same cannot be said for true particulate feeding, whether it is metabolically viable being determined by search time (encounter rate), prey visibility and handling time (Werner & Hall, 1974; Zaret & Kerfoot, 1975) and the escape ability of the prey (see below). Fish typically exhibit size-selective predation, taking the larger, more profitable items. However, this may be constrained by an increased handling time of the largest prey for the smallest fish. Selectivity is also influenced by a number of factors such as prey density and water turbidity (Gardner, 1981).

Prey responses to predation risk

Different zooplankton groups differ widely in their abilities to avoid predation. Copepods make fast evasive jumps when attacked, which makes them significantly less vulnerable to some fish predators than cladoceran zooplankton (Winfield *et al.*, 1983). However, this advantage may virtually disappear against fish with protrusible mouths such as bream (Winfield *et al.*, 1983). Most planktonic cladocerans, such as *Daphnia*, *Ceriodaphnia* and *Bosmina* spp. on the other hand, are straightforward to catch.

Cladocerans may attempt to reduce predation by the production of different morphological structures such as spines and helmets (see Richardson & Jackson, this volume) as well as through behavioural means. The most obvious of these is the use of refuges. This leads to the use of deep, potentially anoxic waters in deep lakes by day with the rapid ascent into the surface waters at night to feed: diel vertical migration (DVM) (Lampert, 1993). In shallow lakes, although *Daphnia* may sit very close to the bottom sediments to avoid predation in the water column by day, they are more likely to exploit vegetation at the littoral margins of the lake and undergo diel horizontal migration (DHM) (see Jeppesen *et al.*, 1998*a* for review). The use of submerged macrophytes as refuges has received much attention in recent years with the truly pelagic *Daphnia* spp. trading off reduced foraging efficiency amongst macrophytes for their relative safety from fish predation.

The use of any refuge is determined by the potential risks, which *Daphnia* appear to assess by detecting kairomones. For example, Lauridsen & Lodge (1996) instigated the use of macrophytes by *Daphnia* in laboratory experiments, through the introduction of water that had contained fish. However, in a lake situation where the scent of fish is likely to be all-pervading and zooplankton and fish may simultaneously use the same habitat (Venugopal & Winfield, 1993) as both attempt to avoid their predators (larger fish and birds in the case of small fish), it is suggested the most relevant cue for *Daphnia* to utilise macrophytes is that of crushed *Daphnia* subject to actual predation.

The impacts

The seminal work of Hrbacek *et al.* (1961) illustrated that fish were the major structuring force configuring zooplankton communities. Brooks & Dodson (1965) then showed that size-selective predation of the larger-bodied zooplankton was the key mechanism. Under predation pressure, the typical response of zooplankton is to reproduce at a smaller size and then produce a sexual generation leading to ephippia (the resting stage) production. Cryer *et al.* (1986) were among the first to suggest that young-of-the-year fish could influence the dynamics of favoured large cladocerans and lead to their extinction in the water column by late summer (Fig. 17.1).

More recent data suggests that whilst submerged macrophyte refuges may delay the decline of large cladoceran zooplankton, this is still inevitable if a critical fish density is exceeded (Stansfield *et al.*, 1997; Perrow *et al.*, 1999*a*). In the absence of macrophytes, even a density of 0.2 m^{-2} of efficient zooplanktivores (roach in the case of Perrow *et al.*, 1999*a*) could cause the collapse of cladoceran populations. This threshold increased to a density of nearer 1 m^{-2} where refuges were present. Also paramount is the quality of the refuges themselves. Whereas Schriver *et al.* (1995) suggested a macrophyte PVI (per cent volume of the water column infested by macrophytes) of as little as 15–20% could provide a refuge, Stansfield *et al.* (1997) suggested 75% was preferred.

The reason behind the inevitable decline of *Daphnia* was argued by Perrow *et al.* (1999*a*) to be that *Daphnia* were obliged to expose themselves to predators by migrating out of refuges to feed. Even under the cover of darkness, predation by young cyprinids adapted to forage in very low light conditions (Townsend & Risebrow, 1982) could still occur. The potential for competition between *Daphnia* and species better able to exploit the resources within macrophyte beds such as *Ceriodaphnia* spp. and *Simocephalus* spp. may also be a factor.

For pelagic species subject to intense predation pressure the only true refuge may be one of size, with smaller individuals and species becoming unprofitable even to young-of-the-year fish. An abundance of small zooplankters is generally insufficient to control algal populations however, as filtering capacity is directly linked to body size (Mourelatos & Lacroix, 1990).

Predation of macroinvertebrates

Fish predation as a structuring force in macroinvertebrate communities

Macroinvertebrates are a particularly important prey item for fish in rivers (Richardson & Jackson, this volume). Selective predation by fish on invertebrate predators such as damselflies and caddis can have ramifications for the whole invertebrate community (Hildrew *et al.*, 1984; Power, 1992). In the case of the study by Power (1992) this also had an ecosystem effect, with increased populations of algivores and ultimately a lower biomass of periphytic algae.

In lakes, fish have a pronounced impact upon community structure, abundance and biomass of benthic macroinvertebrates in relation to their own abundance and biomass (Tatrai *et al.*, 1994) (Box 17.3). The ecosystem implications of such impacts are growing (Jones *et al.*, 1998; Richardson & Jackson, this volume) particularly in relation to effects through grazing. The most widely accepted of these is the fish–mollusc–macrophyte interaction.

The fish–mollusc–epiphyte–macrophyte interaction and implications for restoration

Several unrelated fish families may include a high proportion of molluscs in their diet. In North America this includes the centrarchid sunfishes, notably pumpkinseed (*Lepomis gibbosus*) and redear sunfish (*L. microlophus*), whereas in Europe many cyprinids of a range of sizes including roach and tench (*Tinca tinca*) may do so with great impact. For example, in the survey of Brönmark & Weisner (1996) ponds without fish had a high density and biomass of snails and a low biomass of periphyton, whereas the others had the opposite.

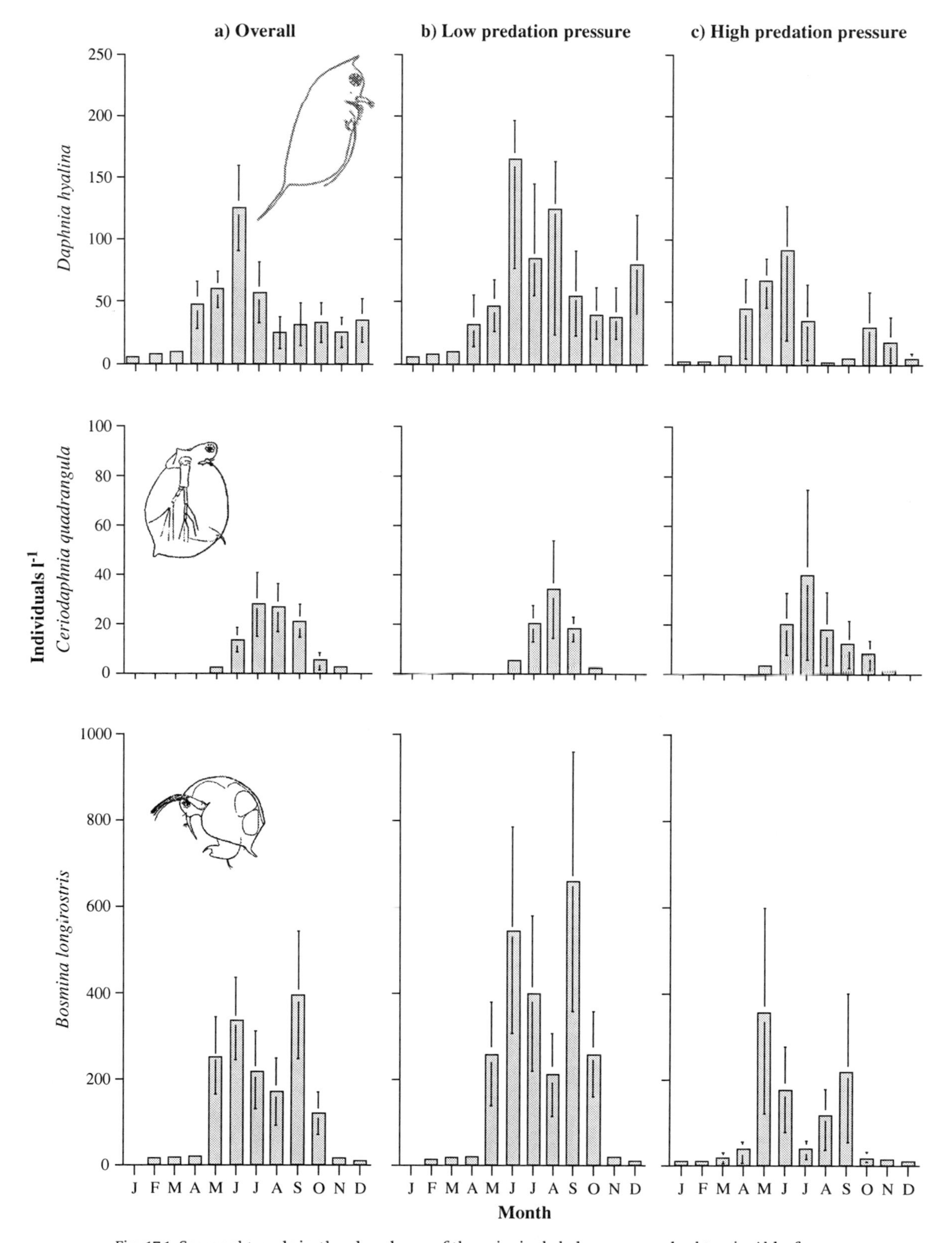

Fig. 17.1. Seasonal trends in the abundance of the principal cladoceran zooplankton in Alderfen Broad: *Daphnia hylina*, *Ceriodaphnia quadrangula* and *Bosmina longirostris* comparing the (a) overall pattern with year of (b) low predation pressure and (c) high predation pressure. Monthly means and standard errors are shown. After Perrow (1989).

Several experimental studies have shown that snail grazing may reduce the abundance of periphytic algae (e.g. Underwood *et al.*, 1992; see Brönmark & Vermaat, 1998 for review) and epiphytes may have an adverse effect on the growth and perhaps fitness of their macrophyte host (see Brönmark & Vermaat, 1998 for review), possibly ultimately leading to macrophyte decline with eutrophication (Phillips *et al.*, 1978). Recent experimental studies using both centrarchids and tench have shown that the impact of predation of fish may cascade to the growth performance of the macrophytes concerned (Brönmark *et al.*, 1992; Martin *et al.*, 1992; Brönmark, 1994).

Whilst the strength of the fish–mollusc–epiphyte–macrophyte interaction will clearly vary according to nutrient concentrations and a suite of factors including the density and biomass and type of the fish predators, snail grazers, periphyton community and the type of macrophyte, there is clear scope for its use in the restoration of aquatic plant communities. At its simplest, this may involve controlling molluscivorous fish at least in the early stages of plant development.

Benthivory

The mechanics

Foraging amongst bottom sediments for the plants and especially animals living there is a common pattern amongst fishes. Insect larvae, especially larval chironomids, are important prey and these may be buried from a few millimetres to several centimetres or more deep.

With anti-predator behaviours and crypsis of potential prey and poor water clarity, fish often rely on senses other than vision to detect prey. These include proprioceptors in the lateral line, electrolocation (e.g. in the knifefishes of Africa [*Gymnotus* spp.]) or most commonly olfaction.

Where prey are small in relation to the predator, making individual detection inefficient, direct assessment of the location of high prey densities may be adopted through sampling the sediment and sieving it (Boxes 17.4 and 17.5). Fish then concentrate on food-rich patches (e.g. Lammens *et al.*, 1987). Different species differ in their abilities to penetrate into the sediment in the pursuit of prey, which in part depends on body bulk and musculature as well as mouth structure. Large carp (which may reach 1.5 m in length) can dig to >10 cm. In cyprinids using a gill-raker sorting mechanism, the size of the apertures in relation to the prey and the size of the prey in relation to the substratum are important determinants of feeding efficiency (Box 17.4).

The impacts

As with zooplanktivory, the ecosystem impact of benthic foraging and how it may relate to food-web interactions and thus be exploited in restoration activity, are best understood in lakes. As outlined above, there is currently little evidence for a cascading impact of direct predation upon benthic invertebrates and it is the indirect impacts of benthivory that have received more attention. These include: (1) resuspension of bottom sediments; (2) changes in sediment chemistry, especially the release of phosphorus; and (3) uprooting of submerged and emergent macrophytes.

In pond experiments, Breukelaar *et al.* (1994) showed a biomass of bream (*Abramis brama*) greater than 100 kg ha^{-1} had a noticeable effect, with every further 100 kg ha^{-1} causing an increase of 46 g sediment m^{-2} day^{-1} and a corresponding 38 cm decrease in Secchi disc depth, over and above that expected from the increases in chlorophyll (9 $\mu g\ l^{-1}$), total phosphorus (30 $\mu g\ l^{-1}$) and inorganic nitrogen (0.48 mg l^{-1}). The same biomass of carp had about half the impact on resuspension (but see Box 17.3).

The removal of the disturbance of benthivorous fish also appeared to be the key to the colonisation of submerged vegetation macrophytes in the biomanipulated Ormesby Broad, UK (Tomlinson *et al.*, 2002) and the large Lakes Wolderwijd and Veluwe in The Netherlands (Meijer, 2000).

Whilst direct disturbance is relatively easily quantified, the issue of nutrient release is more complex as increased nutrient levels may be linked directly to excretion by the fish themselves (Tatrai & Istvanovics, 1986; Breukelaar *et al.*, 1994). Cline *et al.* (1994) suggested that trophic guild was a clue to the mechanism of increased nutrient levels, with omnivores operating through physiological processes and benthivorous fish operating through sediment resuspension.

Herbivory

There are relatively few truly herbivorous fish, although many include some plant material in their diets. One of the more easily utilisable forms of plant matter is that of periphytic algae, coating the substratum of lakes and streams. Here, specialists such as the stone roller (*Campostoma anomalum*) of North America and locariid catfish of central America may limit algal standing crop (Power, 1992).

Fewer species specialise on submerged macrophytes, with the notable exception of grass carp (*Ctenopharyngodon idella*). This large (to 120 cm) cyprinid, originally from China, has been introduced all over the world to control nuisance growths of aquatic plants (Weisner & Strand, this volume). In some circumstances, grass carp may perform the job for which they were intended with beneficial consequences for other species (Petridis, 1990), but when stocked at too high a density, they may eliminate submerged macrophytes (Petridis, 1990) conceivably with ecosystem consequences.

Selectivity is always likely to be an issue, with potential avoidance of nuisance macrophyte species and selective grazing of other, less problematic ones. Van Donk & Otte (1996) suggested that selective grazing by rudd (*Scardinius erythrophthalmus*) caused a shift from the softer, more palatable *Elodea nuttallii* spp. to the calcareous *Ceratophyllum demersum*. Rudd and roach, which may also take a reasonable proportion of macrophytes in its diet (Prejs, 1984), are common European species and may, in fact, structure macrophyte community composition in many lakes and slow-flowing rivers.

Piscivory

The mechanics

Piscivorous fish range considerably in size and structure. Two common body forms are an elongate, fusiform shape associated with extreme acceleration (e.g. pike, gars [*Lepisosteus* spp.], and marlin [*Tetrapturus* spp.]) and a more stocky shape with considerable muscle bulk often associated with more sustained swimming (e.g., tuna [*Thunnus* spp.]), but also capable of fast burst speed (e.g. largemouth bass). Still others rely on stealth or crypsis and even lures (e.g. anglerfish [*Lophiiformes*]) and electricity (e.g. electric eel [*Electrophotus electricus*]).

In general, ambush predators tend to be solitary with open-water species foraging within shoals, although this pattern may change with the age and size of the predator (Eklöv, 1992). Ambush predators, including pike and wels catfsh (*Silurus glanis*), tend to take single, relatively large prey (to 50%) in relation to their body size (van Densen, 1994). Sequential digestion may then move the prey, the end of which may still be protruding from the mouth, down the digestive tract.

Obligate piscivores often have sizeable teeth to injure, capture and manoeuvre prey in their jaws. More facultative piscivores may have no obvious adaptations and include smaller fishes in their diet as they become available. Virtually all predatory and omnivorous species are likely to take fry of their own species as well as others.

Selectivity of particular prey is a common theme, but is obviously partly determined by the availability of prey in the water body concerned. For example, although escosids such as pike may prefer cyprinids over spiny-finned percids, the latter may comprise virtually the entire diet when other prey are not available (van Densen, 1994).

Many piscivores rely on vision to orientate towards their prey, although even largely visual predators may forage effectively in turbid conditions or become active at low light levels (i.e. at dawn and dusk), when their prey are more susceptible to attack (e.g. Dobler, 1977). Fish operate in a sensory world that is difficult to perceive and it is only recently that we have begun to unravel what may be intricate predator–prey interactions as each seeks an advantage. For example, pike may defaecate away from their ambush sites to reduce the chances of detection by their potential prey (Brown *et al.*, 1996).

Gape size and size refuges

Gape size is a key factor determining the efficiency of any piscivore (van Densen, 1994). Prey may thus avoid predation by growing beyond exploitable size. In the study of Brönmark & Weisner (1996), large molluscivorous tench reached equivalent or greater size than their potential predators with consequences for molluscan grazers. Indeed, many cyprinids in Eurasia may achieve a size refuge from all but the largest piscivores, which are relatively rare and patchily distributed. Tonn *et al.* (1991) suggested this was an important difference between fish communities in North America and Europe accounting for the greater incidence of piscivore control in the former, where the dominant cyprinids (e.g. *Notemigonus*, *Phoxinus* and *Pimephales* spp.) only reach a few centimetres at adult size and are always within the scope of predators such as largemouth bass.

Such is the sensitivity of the gape-limitation factor that individuals within crucian carp populations may adapt to the presence of predators by accumulating tissue on the dorsal surface to widen their bodies and make them less vulnerable to attack (Poleo *et al.*, 1995). Thus, fish may demonstrate a similar ability to invertebrates to respond in physiological terms to the threat of predation.

Impact of direct predation

Despite the long history of management of piscivores to promote populations of commercially valuable or desirable species (e.g. Stroud & Clepper, 1979), there are relatively few convincing examples of direct predation being of functional consequence in food webs, although the number is growing with more larger-scale, replicated experiments supported by detailed empirical and theoretical studies.

Using a reciprocal transfer design in two experimental lakes, and supported by detailed dietary work, the studies of Carpenter & Kitchell (1993) clearly demonstrated that largemouth bass controlled populations of forage fish which cascaded to lower trophic levels. This confirmed the suggestions of Spencer & King (1984), amongst others, of the potential impact of largemouth bass.

In northwestern Europe, through survey of a large number of lakes, Brönmark *et al.* (1995) showed that in the presence of piscivorous perch, populations of crucian carp and tench were dominated by a low density of large individuals. Without piscivores, a high density of small individuals was typical.

Although Brönmark & Weisner (1996) have suggested that the loss of piscivores may have trophic consequences resulting in switches between macrophyte and algal dominance in shallow lakes, there is little unequivocal evidence to support this argument to date. The addition of piscivores both for sporting purposes and as a restoration tool provides opportunity to test the theory. Recent ongoing studies (e.g. Skov *et al.*, in press) show promise.

Predation risk

The habitat use and behaviour of different species within a community makes them more vulnerable to attack by one type of predator over another. This is manifested as predator-specific escape responses (Christensen & Persson, 1993). In a similar fashion to zooplankton vulnerable fish may respond dramatically to the risk of predation, changing behaviour and even morphology (see above). In pond experiments, Jacobsen & Perrow (1998) showed young roach abandoned their diel movement between macrophytes and open water when pike were present, preferring to stay in open water during the day, as pike lay hidden amongst the vegetation. Such diel patterns by small zooplanktivorous fish have been frequently observed in lakes and often attributed to predation risk from both piscivorous fish and birds (Gliwicz & Jachner, 1992). Here, littoral emergent and submerged vegetation may offer an effective refuge against some sorts of predators, although the density of vegetation may be a critical factor in its efficacy as a refuge (Bean & Winfield, 1995).

Such is the magnitude of the response to predation risk, by small zooplanktivorous fish in particular (which are more vulnerable to predation than the larger benthivorous species), there is great potential for cascading effects to other trophic levels. The space vacated by zooplanktivores or occupied by piscivores, or both, may be available to zooplankton (Perrow *et al.*, 1999*a*) or zooplankton populations may become more abundant and contain larger individuals, where their predators are concentrated on avoiding their own (Turner & Mittelbach, 1990). Jacobsen *et al.* (1997) illustrated a cascading risk interaction in enclosure, but not pond, experiments, illustrating the likely vagaries of the effect, with the type and relative density of predators, prey and zooplankton all potentially determining the strength of the interaction.

Nutrient cycling

Fish are integrally involved in nutrient dynamics, via a number of direct and indirect effects, which may be difficult to separate in practice, demanding rigorous logical design (Threlkeld, 1994). When this is undertaken, the results may be surprising. In an elegant experiment, Threlkeld (1987) showed dead fish induced a cascading effect down to phytoplankton dynamics, an effect generally thought to be mediated through reduced zooplankton grazing when live fish were present.

The release of nutrients from fish bodies and its impact upon the ecosystem is well known from studies on rivers subject to migrations of salmonids. Where the stock and the magnitude of the run have declined historically, the general productivity of the system may have been seriously affected (Larkin & Slaney, 1997). To mimic the input of nutrients from fish mortality following spawning, nutrient additions have been undertaken with beneficial impact on primary productivity, invertebrates and fish growth (Deegan & Peterson, 1992; Harvey *et al.*, 1998). In a similar fashion, nutrients may have to be introduced into lakes whose nutrient supply has been cut off (e.g. by regulation of inflows for hydroelectric schemes: Milbrink, 1988) in order to ultimately support fish.

In lakes, the deliberate removal of fish to induce a shift to clear-water conditions through zooplankton grazing or a reduction in sediment resuspension has resulted in dramatic changes in nutrient dynamics. In several Danish lakes, total nitrogen concentrations decreased and the percentage of nitrogen retained in the sediment or lost by denitrification increased substantially (Jeppesen *et al.*, 1998*b*). Conversely, in Ormesby Broad in the UK, phosphorus concentrations increased substantially following fish removal (Tomlinson *et al.*, 2002), perhaps linked to an increase in benthic chironomids, which may pump phosphorus within interstitial spaces in the sediment into the water column (Phillips *et al.*, 1994).

Implications for lake restoration

The development of biomanipulation

Shapiro *et al.* (1975) coined the term 'biomanipulation' for a series of management measures aimed at reducing algal biomass and promoting clear water in the battle against eutrophication.

By the late 1980s, biomanipulation had become synonymous with the 'top–down' removal of zooplanktivorous and/or benthivorous fish. The development of theoretical models clearly illustrate the potential for an impact of fish and how biomanipulation may be used (e.g. Scheffer *et al.*, 1993; Scheffer, 1998; Zambrano *et al.*, 2000) (see below).

Biomanipulation should be seen as a tool for lake restoration, typically applied with a suite of other measures primarily directed at nutrient control (see Jeppesen & Sammalkorpi, volume 2). After 15 or so years of experimentation and management, Perrow *et al.* (1997) concluded that there was little doubt that biomanipulation worked, it was more a case of whether it was appropriate or not. In other words, whether fish were responsible for turbid water in the first place and whether fish were effectively controlled (see Jeppesen & Sammalkorpi, volume 2). The mechanics of biomanipulation are clearly outlined by Jeppesen & Sammalkorpi (volume 2).

Furthermore, it is now suggested that the cascading effect of fish through zooplankton to lower trophic levels exploited by biomanipulation is more likely to be stronger in eutrophic and hypertrophic lakes (Jeppesen *et al.*, 1997) than in oligotrophic ones, as initially argued by McQueen *et al.* (1986). The increased scope for the production of young-of-the-year in eutrophic lakes and the general dominance of young fish (as there are few resources for older ones: Townsend & Perrow, 1989) means that if recruitment of fry occurs to its full potential the predation pressure upon zooplankton is intense. Should this be released through manipulation of the fish stock, the corresponding scope for the development of zooplankton populations and the potential impact upon phytoplankton is also great. The strength of the young-of-the-year-zooplankton interaction is now exploited by targeting fry, although this may require removal of large adults or interference with spawning (Perrow *et al.*, 1997; Jeppesen & Sammalkorpi, volume 2).

The potential impact of benthivorous fish is also now more widely recognised (Meijer, 2000; Tomlinson *et al.*, 2002), particularly in shallow lakes where the per capita impact of zooplanktivorous/benthivorous species is argued to be more intense than in deeper ones (Jeppesen *et al.*, 1998*b*).

In recent times, there have also been numerous attempts to exploit the potential cascading impact from piscivores throughout the food web, by enhancing stocks of piscivores, particularly through stocking but also habitat manipulation (Skov & Berg, 1999). In Europe, stocking young-of-the-year pike at densities between 1000 and 4000 ha^{-1} (Prejs *et al.*, 1994; Berg *et al.*, 1997) showed considerable promise. As a result, the practice became a management technique before the establishment of a solid research basis for its use (Skov *et al.*, in press). Although further research is now under way, there are considerable doubts as to the efficacy of young-of-the-year pike (Skov *et al.*, in press) and pike in general (Perrow *et al.*, 1999*b*). Cannibalism of younger individuals by larger ones may be rife, limiting population size, which also seems to be limited by particular habitat requirements, especially when fish are young. Establishing a suite of piscivores, adapted to tackling a range of prey species and sizes seems to be the key and the importance of perch, an open-water shoaling piscivore, particularly in relation to predation of young-of-the-year planktivores, should not be underestimated (Perrow *et al.*, 1999*c*).

The development of fish communities after biomanipulation

The establishment of an appropriate fish community after biomanipulation is so important that it has been dubbed the 'cornerstone of restoration' (Meijer *et al.*, 1994). However, this has proved elusive, partly as the definition of what is appropriate has remained somewhat vague, despite the growing number of empirical studies illustrating what is expected at different nutrient levels and descriptions of community structure under different conditions (Perrow *et al.*, 1999*c*; Jeppesen & Sammalkorpi, volume 2). In general, the attainment and maintenance of a high proportion of piscivores is seen as essential, although this has proved difficult in practice (Meijer *et al.*, 1994, 1995; Walker, 1994; Perrow *et al.*, 1999*b*). Perrow *et al.* (1999*b*) argue that control by piscivores is more likely at lower nutrient concentrations, partly as the potential for production of enormous numbers of small zooplanktivores is reduced and piscivorous species typically comprise a higher proportion of the total fish biomass (see Jeppesen & Sammalkorpi, volume 2, Fig. 13.1).

Management of undesirable and desirable species, typically through selective removal and stocking respectively, is likely to prove to be an essential element of biomanipulation, ensuring the development of the fish community is in step with the development of the clear water state.

CONCLUDING REMARKS

Quite simply, the functional importance of fish in many aquatic food webs cannot be underestimated. This stems from a massive potential for reproduction, indeterminate and plastic growth and ability to exploit a variety of foodstuffs with relatively standard body form and gut design. Fish may exert an influence as zooplanktivores, benthivores, herbivores, piscivores and agents of nutrient release and regulation. The manipulation of fish communities is thus an integral part of lake restoration and 'top–down' rather than nutrient-limited 'bottom–up' thinking is in the ascendancy (Jeppesen & Sammalkorpi, volume 2). As our understanding grows, we may also learn to use the subtleties of predator–prey interactions through the use of refuges which may be exploited or created to induce particular beneficial effects. Ways of using fish to maintain lake food webs in the desired configuration are also required, although this can only follow greater understanding of the interrelation between fish and lake dynamics over the longer term.

In time and with further study, we may even come to see that the ecology of even large rivers and even the seas, which we perceive are under the control of physical, chemical and even climatic factors, may be heavily influenced by fish. At present, we are only just beginning to grasp the use of time and space by predatory fish in the oceans and their abilities to control their prey. It was only a short time ago our predecessors perceived the resources of the seas to be boundless and beyond over-exploitation. How wrong they were.

This introduces the notion that control of introduced alien species, whilst daunting, is not impossible, although it may be beyond the resources of developing nations where feeding people has priority over biodiversity conservation. Even in developed nations, fish conservation/restoration has focused on commercially important stocks of salmonids in particular. In contrast, the conservation and restoration of non-commercial endangered fish has lagged behind that of other vertebrate groups, despite its priority (Maitland, 1995).

Aquaculture technology is an enormously powerful tool for the production of individuals for reintroduction, where this is required. However, this seems not to have been used to its full potential as yet. Few, if any, projects exhibit the stringency normally associated with captive breeding and release schemes (see Macdonald *et al.*, Chapter 20, this volume). A high degree of wastage seems inherent, as does a focus on the technical rather than the ecological, both of which must change.

ACKNOWLEDGMENTS

We thank Christian Skov of the Danish Institute for Fisheries Reseach for a valuable literature search in the early phase of this project and for valuable comments on an earlier draft of this manuscript; and Ian Winfield for providing the information and figure for Box 17.2.

REFERENCES

Adams, M.A. & Whyte, I.W. (1990). *Fish Habitat Enhancement: A Manual for Freshwater, Estuarine and Marine Habitats.* Vancouver, Canada: Department of Fisheries and Oceans, Program Planning and Economics Branch.

Backx, J.J.G.M. & Grimm, M.P. (1994). Mass removal of fish from Lake Wolderwijd, the Netherlands. 2: Implementation phase. In *Rehabilitation of Freshwater Fisheries*, ed. I.G. Cowx, pp. 401–414. Oxford: Fishing News Books.

Bales, M. (1992). Carp and river environment deterioration: villain or innocent by-stander? *Newsletter of the Australian Society of Fisheries Biologists*, **22**, 26–27.

Bean, C.W. & Winfield, I. J. (1995). Habitat use and activity patterns of roach (*Rutilus rutilus* (L.)), rudd (*Scardinius erythrophthalmus* (L.)), perch (*Perca fluviatilis* (L.)) and pike (*Esox lucius* L.) in the laboratory: the role of predation threat and structural complexity. *Ecology of Freshwater Fish*, **4**, 37–46.

Beaumont, A.R., Bray, J., Murphy, J.M. & Winfield, I.J. (1995). Genetics of whitefish and vendace in England and Wales. *Journal of Fish Biology*, **46**, 880–890.

Berg, S. & Jørgensen, J. (1994). Stocking experiments with 0 + eels (*Anguilla anguilla* L.) in Danish streams: post-stocking movements, densities and mortality. In *Rehabilitation of Freshwater Fisheries*, ed. I.G. Cowx, pp. 314–325. Oxford: Fishing News Books.

Berg, S., Jeppesen, E. & Søndergaard, M. (1997). Pike (*Esox lucius* L.) stocking as a biomanipulation tool. 1. Effects on the fish population in Lake Lyng, Denmark. *Hydrobiologia*, **342/343**, 311–318.

Berger, J.J. (1992). Citizen restoration efforts in the Mattole River Watershed. In *Restoration of Aquatic Ecosystems: Science, Technology and Public Policy*, ed. National Research Council (US), pp. 457–463. Washington, DC: National Academy Press.

Bernard, T. & Young, J. (1997). *The Ecology of Hope: Communities Collaborate for Sustainability.* Gabriola Island, Canada: New Society Publishers.

Borcherding, J., Maw, S.K. & Tauber, S. (2000). Growth of 0+ perch (*Perca fluviatilis*) predating on 0+ bream (*Abramis brama*). *Ecology of Freshwater Fish*, **9**, 236–241.

Brabrand, Å. (1995). Intra-cohort cannibalism among larval stages of perch (*Perca fluviatilis*). *Ecology of Freshwater Fish*, **4**, 70–76.

Breukelaar, A.W., Lammens, E.H.R.R., Klein Breteler, J.G.P. & Tatrai, I. (1994). Effects of benthivorous bream (*Abramis brama* L.) and carp (*Cyprinus carpio* L.) on sediment resuspension and concentrations of nutrients and chlorophyll *a*. *Freshwater Biology*, **32**, 113–121.

Brönmark, C. (1994). Effects of tench and perch on interactions in a freshwater, benthic food chain. *Ecology*, **75**, 1818–1824.

Brönmark, C. & Vermaat, J.E. (1998). Complex fish–snail–epiphyton interactions and their effects on submerged freshwater macrophytes. In *The Structuring Role of Submerged Macrophytes in Lakes*, eds. E. Jeppesen, Ma. Søndergaard, Mo. Søndergaard & K. Christoffersen, pp. 47–68. New York: Springer-Verlag.

Brönmark, C. & Weisner, S.E.B. (1996). Decoupling of cascading trophic interactions in a freshwater, benthic food chain. *Oecologia*, **108**, 534–541.

Brönmark, C., Kloswiewski, S. P. & Stein, R. A. (1992). Indirect effects of predation in a freshwater, benthic food chain. *Ecology*, **73**, 1662–1674.

Brönmark, C., Paszkowski, C.A., Tonn, W.M. & Hargeby, A. (1995). Predation as a determinant of size structure in populations of crucian carp (*Carassius carassius*) and tench (*Tinca tinca*). *Ecology of Freshwater Fish*, **4**, 85–92.

Brooks, J.L. & Dodson, S.I. (1965). Predation, body size, and composition of plankton. *Science*, **150**, 28–35.

Brown, B.J. & Evans, D.R. (1986). Restoration of anadromous fish: not as simple as it sounds. In *Wild Trout, Steelhead and Salmon in the 21st Century*, ed. D. Guthrie, pp. 61–74. Portland, OR: Oregon State University Sea Grant Publication.

Brown, G.E., Chivers, D.P. & Smith, R.J.F. (1996). Effects of diet on localized defecation by Northern pike *Esox lucius*. *Journal of Chemical Ecology*, **22**, 467–475.

Bry, C. (1996). Role of vegetation in the life-cycle of pike. In *Pike: Biology and Exploitation*, ed. J.F. Craig, pp. 45–67. London: Chapman & Hall.

Cahn, A.R. (1929). The effect of carp on small lake, the carp as a dominant. *Ecology*, **10**, 271–274.

Carpenter, S.R. & Kitchell, J.F. (eds.) (1993). *The Trophic Cascade in Lakes.* Cambridge: Cambridge University Press. 385 pp.

Carvalho, L. & Moss, B. (1995). The current status of a sample of English sites of special scientific interest subject to eutrophication. *Aquatic Conservation: Marine and Freshwater Ecosystems*, **5**, 191–204.

Christensen, B. & Persson, L. (1993). Species-specific antipredator behaviours: effects on prey choice in different habitats. *Behavioural Ecology and Sociobiology*, **32**, 1–9.

Christie, G.C. (1997). Integrated management of sea lamprey: lessons for potential control of ruffe. In *International Symposium on Biology and Management of Ruffe*, Ann Arbor, Michigan, 21–23 March 1997, Symposium Abstracts, ed. D.A. Jensen, p.23. Ann Arbor, MI: Great Lakes Fisheries Commission.

Cline, J.M., East, T.L. & Threlkeld, S.T. (1994). Fish interactions with the sediment–water interface. *Hydrobiologia*, **275/276**, 301–311.

Contreras-Balderas, S. (2000). Annotated checklist of introduced invasive fishes in México, with examples of some recent introductions. In *Non-Indigenous Freshwater Organisms*, eds. R. Claudi & J.H.Leach, pp. 33–56. Boca Raton, FL: Lewis Publishers.

Copp, G.H. & Peňáz, M. (1988). Ecology of fish spawning and nursery zones in the flood plain, using a new sampling approach. *Hydrobiologia*, **169**, 209–224.

Crowder, A. & Painter, D.S. (1991). Submerged macrophytes in Lake Ontario: current knowledge, importance, threats to stability and needed studies. *Canadian Journal of Fisheries and Aquatic Sciences*, **48**, 1539–1545.

Cryer, M., Peirson, G. & Townsend, C.R. (1986). Reciprocal interactions between roach, *Rutilus rutilus*, and zooplankton in a small lake: Prey dynamics and fish growth and recruitment. *Limnology and Oceanography*, **31**, 1022–1038.

Deegan, L.A. & Peterson, B.J. (1992). Whole river fertilization stimulates fish production in an Arctic tundra river. *Canadian Journal of Fisheries and Aquatic Sciences*, **49**, 1890–1901.

Dobler, E. (1977). Correlation between the feeding time of the pike (*Esox lucius*) and the dispersal of a school of *Leucaspius delineatus*. *Oecologia*, **27**, 93–96.

Eklöv, P. (1992). Group foraging versus solitary foraging efficiency in piscivorous predators: the perch *Perca fluviatilis*, and pike *Esox lucius*, patterns. *Animal Behaviour*, **44**, 313–326.

Fuselier, L. & Edds, D. (1995). An artificial riffle as restored habitat for the threatened Neosho madtom. *North American Journal of Fisheries Management*, **15**, 499–503.

Gardner, M.B. (1981). Mechanisms of size selectivity by planktivorous fish: a test of hypotheses. *Ecology*, **62**, 571–578.

Giles, N. (ed.) (1992). *Wildlife after Gravel: Twenty Years of Practical Research by The Game Conservancy and ARC.* Fordingbridge, UK: Game Conservancy.

Giles, N. (1994). *Freshwater Fish of the British Isles: A Guide for Anglers and Naturalists.* Shrewsbury, UK: Swan Hill Press.

Gliwicz, Z.M. & Jachner, A. (1992). Diel migrations of juvenile fish: a ghost of predation past or present? *Archiv für Hydrobiologie*, **124**, 385–410.

Gray, R.W. & Cameron, J.D. (1987). A deep-substrate streamside incubation box for Atlantic salmon eggs. *Progressive Fish Culturist*, **49**, 124–129.

Hamrin, S.F. & Persson, L. (1986). Asymmetrical competition between age classes as a factor causing population oscillations in an obligate planktivorous fish species. *Oikos*, **47**, 223–232.

Harrell, L.W., Flagg, T.A. & Waknitz, F.W. (1987). *Snake River Fall Chinook Salmon Brood-Stock Program (1981–1986)*, Final Report of Contract A179-83BP39642. US Department of Energy.

Harris, G.P. (1994). Pattern, process and prediction in aquatic ecology: a limnological view of some general ecological problems. *Freshwater Biology*, **32**, 143–160.

Harvey, C.J., Peterson, B.J., Bowden, W.B., Hershey, A.E., Miller, M.C., Deegan, L.A. & Finlay, J.C. (1998). Biological responses to fertilization of Oksrukuyik Creek, a tundra stream. *Journal of the North American Benthological Society*, **17**, 190–209.

Hendricks, M.L. (1995). The contribution of hatchery fish to the restoration of American shad in the Susquehanna River. In *Uses and Effects of Cultured Fishes in Aquatic Ecosystems*, eds. H.L. Schramm Jr & R.G.Piper, pp. 329–336. Bethesda, MD: American Fisheries Society.

Hickley, P. (1994). Stocking and introduction of fish: a synthesis. In *Rehabilitation of Freshwater Fisheries*, ed. I.G. Cowx, pp. 247–254. Oxford: Fishing News Books.

Hildrew, A.G., Townsend, C.R. & Francis, J. (1984). Community structure in some southern English streams: the influence of species interactions. *Freshwater Biology*, **14**, 297–310.

Hodgson, J.R., He, X. & Kitchell, J.F. (1993). The fish populations. In *The Trophic Cascade in Lakes*, eds. S.R. Carpenter & J.F. Kitchell, pp. 43–68. Cambridge: Cambridge University Press.

House, R. (1996). An evaluation of stream restoration structures in a coastal Oregon stream, 1981–1993. *North American Journal of Fisheries Management*, **16**, 272–281.

Hrbacek, J., Dvorakova, M., Korinek, M. & Prochazkova, L. (1961). Demonstration of the effect of the fish stock on the species composition of zooplankton and the intensity of metabolism of the whole plankton association. *Verhandlungen itnernationale Vereinigung fur theoretische und angewandte Limnologie*, **14**, 192–195.

Jacobsen, L. & Perrow, M.R. (1998). Predation risk from piscivorous fish influencing the diel use of macrophytes by planktivorous fish in experimental ponds. *Ecology of Freshwater Fish*, **7**, 78–86.

Jacobsen, L., Perrow, M.R., Landkildehus, F., Hjørne, M., Lauridsen, T.L. & Berg, S. (1997). Interactions between piscivores, zooplanktivores and zooplankton in submerged macrophytes: preliminary observations from enclosure and pond experiments. *Hydrobiologia*, **342/343**, 197–205.

Jeppesen, E., Jensen, J.P., Søndergaard, W., Lauridsen, T.L., Pedersen, L.J. & Jensen, L. (1997). Top–down control in freshwater lakes: the role of nutrient state, submerged macrophytes and water depth. *Hydrobiologia*, **342/343**, 151–164.

Jeppesen, E., Lauridsen, T.L., Kairesalo, T. & Perrow, M.R. (1998*a*). Impact of submerged macrophytes on fish–zooplankton interactions in lakes. In *The Structuring Role of Submerged Macrophytes in Lakes*, eds. E. Jeppesen, Ma. Søndergaard, Mo. Søndergaard & K. Christoffersen, pp. 91–114. New York: Springer-Verlag.

Jeppesen, E., Jensen, J.P., Søndergaard, M. Lauridsen, T.L., Møller, P.H. & Sandby, K. (1998*b*). Changes in nutrient retention in shallow eutrophic lakes following a decline in density of cyprinids. *Archiv für Hydrobiologie*, **142**, 129–151.

Johnson, B.G.H. (1988). A comparison of the effectiveness of sea lamprey control in Georgian Bay and the North Channel of Lake Huron. *Hydrobiologia*, **163**, 215–222.

Jokikokko, E. (1990). Brown trout stockings in the restored rapids of the tributaries of the river Oulujoki system in eastern Finland. *Polskie Archiwum Hydrobiologii/Polish Archives of Hydrobiology*, **38**, 79.

Jones, J.I., Moss, B. & Young, J.O. (1998). Interactions between periphyton, nonmolluscan invertebrates, and fish in standing waters. In *The Structuring Role of Submerged Macrophytes in Lakes*, eds. E. Jeppesen, Ma. Søndergaard, Mo. Søndergaard & K. Christoffersen, pp. 69–90. New York: Springer–Verlag.

Jones, R.A. (1978). A history of trying success with salmon: breeding new stocks for the Connecticut rivers. *Oceans*, **11**, 59–60.

Jungwirth, M., Schmutz, S. & Weiss, S. (eds.) (1998). *Fish Migration and Fish Bypasses.* Oxford: Fishing News Books.

Kitchell, J.F. & Carpenter, S.R. (1993). Synthesis and new directions. In *The Trophic Cascade in Lakes*, eds. S.R. Carpenter & J.F. Kitchell, pp. 332–350. Cambridge: Cambridge University Press.

Kurmayer, R., Keckeis, H., Schrutka, S. & Zweimueller, I. (1996). Macro- and microhabitats used by 0+ fish in a side-arm of the River Danube. *Archiv für Hydrobiologie* (suppl. Large Rivers), **113**, 425–432.

Lammens, E.H.R.R. & Hoogenboezem, W. (1991). Diets and feeding behaviour. In *Cyprinid fishes: Systematics, Biology and Exploitation*, eds. I.J. Winfield & J.S. Nelson, pp. 353–376. London: Chapman & Hall.

Lammens, E.H.R.R., Geursen, J. & MacGillavry, P.J. (1987). Diet shifts, feeding efficiency and coexistence of bream *Abramis brama*, roach *Rutilus rutilus* and white bream *Blicca bjoerkna* in hypertrophic lakes. *In Proceedings of the 5th Congress of European Ichthyologists*, pp. 153–162. Stockholm: European Ichthyological Union.

Lampert, W. (1993). Ultimate causes of diel vertical migration of zooplankton: new evidence for the predator avoidance hypothesis. *Archiv für Hydrobiologie und Ergebnisse Limnologie*, **39**, 79–88.

Larkin, G.A. & Slaney, P.A. (1997). Implications of trends in marine-derived nutrient influx to south coastal British Columbia salmonid production. *Fisheries*, **22**, 16–24.

Lauridsen, T.L. & Lodge, D. (1996). Avoidance by *Daphnia magna* of fish and macrophytes: chemical cues and predator-mediated use of macrophyte habitat. *Limnology and Oceanography*, **41**, 794–798.

Lingel, R. & Fish, G. (2001). Mattole Salmon Group Update. *Mattole Restoration Newsletter*, Spring/Summer 2001, Issue 16.

Love, R.M. (1980). *The Chemical Biology of Fishes*. London: Academic Press.

Lucas, M.C. & Frear, P.A. (1997). Effects of a flow-gauging weir on the migratory behaviour of adult barbel, a riverine cyprinid. *Journal of Fish Biology*, **50**, 382–396.

Maitland, P.S. (1995). The conservation of freshwater fish: past and present experience. *Biological Conservation*, **72**, 259–270.

Maitland, P.S. & Campbell, R.N. (1992). *Freshwater Fishes*. London: HarperCollins.

Marine Conservation Society (2000) http://www.mtsuk.org/

Martin, T.H., Crowder, L.B. Dumas, C.F. & Burkholder, J.M. (1992). Indirect effects of fish on macrophytes in Bays Mountain Lake: evidence for a littoral trophic cascade. *Oecologia*, **89**, 476–481.

Mattole Restoration Council (2000). http://www.mattole.org/

Mattole Salmon Group (1999). http://www.humboldt.net/~salmon/

McQueen, D.J., Post, J.R. & Mills, E.L. (1986). Trophic relationships in freshwater pelagic ecosystems. *Canadian Journal of Fisheries and Aquatic Sciences*, **43**, 1571–1581.

Meijer, M.-L. (2000). *Biomanipulation in The Netherlands: 15 Years of Experience*. Lelystad, The Netherlands: Ministry of Transport, Public Works and Water Management, Institute for Inland Water Management and Waste Water Treatment.

Meijer, M.-L., Jeppesen, E., van Donk, E., Moss, B., Scheffer, M., Lammens, E.H.R.R., van Nes, E., van Berkum, J.A., de Jong, G.L., Faafeng, B.A. & Jensen, J.P. (1994). Long term responses to fish stock reduction in small shallow lakes: interpretation of five-year results of four biomanipulation cases in The Netherlands and Denmark. *Hydrobiologia*, **275/276**, 457–466.

Meijer, M.-L., Lammens, E.H.R.R., Raat, A.J.P., Klein Breteler, J.G.P. & Grimm, M.P. (1995). Development of fish communities in lakes after biomanipulation. *Netherlands Journal of Aquatic Ecology*, **29**, 91–101.

Mih, W.C. & Bailey, G.C. (1981). The development of a machine for the restoration of stream gravel for spawning and rearing of salmon. *Fisheries*, **6**, 16–20.

Milbrink, G. (1988). Restoration of fish population in impounded lakes in Scandinavia by locally applied fertilization. In *Eutrophication and Lake Restoration: Water Quality and Biological Impacts*, French–Swedish Limnological Symposium, Thonon-les-Bains, 10 June 1987, ed. G. Balvay, pp. 193–203. Thonon-les-Bains, France: Institut Nationale de la Recherche Agronomique, Institut de Limnologie.

Mills, C.A. (1991). Reproduction and early life history. In *Cyprinid Fishes: Systematics, Biology and Exploitation*, eds. I.J. Winfield & J.S. Nelson, pp. 483–508. London: Chapman & Hall.

Minns, K.C. & Cooley, J.M. (2000). Intentional introductions: are the incalculable risks worth it? In *Non-Indigenous Freshwater Organisms*, ed. R. Claudi & J.H. Leach, pp. 57–60. Boca Raton, FL: Lewis Publishers.

Mourelatos, S. & Lacroix, G. (1990). *In situ* filtering rates of Cladocera: effect of body length, temperature, and food concentration. *Limnology and Oceanography*, **35**, 1101–1111.

Mujica, C.E. (1987). Los cuerpos de agua continentales adecuados por el cultivo de carpa. *Revista Mexicana de Acuacultura*, **9**, 7–10.

National Research Council (1992). *Restoration of Aquatic Ecosystems: Science, Technology and Public Policy*. Washington, DC: National Academy Press.

Nelson, J.S. (1984). *Fishes of the World*, 2nd edn. New York: John Wiley.

Nielsen, E.E., Hansen, M.M. & Loeschcke, V. (1997). Analysis of microsatellite DNA from old scale samples of Atlantic salmon: a comparison of genetic composition over sixty years. *Molecular Ecology*, **6**, 487–492.

Nielsen, E.E., Hansen, M.M. & Bach, L. (2001). Looking for a needle in a haystack: discovery of indigenous salmon in heavily stocked populations. *Conservation Genetics*, **2**, 219–232.

Ochumba, P.B.O, Gophen, M. & Pollingher, U. (1994). Ecological changes in the Lake Victoria after the introduction of Nile perch (*Lates niloticus*): the catchment, water quality and fisheries management. In *Rehabilitation of Freshwater Fisheries*, ed. I.G. Cowx, pp. 338–347. Oxford: Fishing News Books.

Okemwa, E. & Ogari, J. (1994). Introductions and extinction of fish in Lake Victoria. In *Rehabilitation of Freshwater Fisheries*, ed. I.G. Cowx, pp. 326–337. Oxford: Fishing News Books.

Pandian, T.J. & Vivekanandan, E. (1985). Energetics of feeding and digestion. In *Fish Energetics: New Perspectives*, eds. P. Tytler & P. Calow, pp. 99–124. London: Croom Helm.

Perrow, M.R. (1989). Causes and consequences of a two-year cycle in recruitment of roach (*Rutilus rutilus* (L.)) in Alderfen Broad. PhD thesis, University of East Anglia, Norwich, UK.

Perrow, M.R. & Irvine, K. (1992). The relationship between cladoceran body size and the growth of underyearling roach (*Rutilus rutilus* (L.)) in two shallow lowland lakes: a mechanism for density-dependent reductions in growth. *Hydrobiologia*, **241**, 155–161.

Perrow, M.R. & Jowitt, A.J.D. (2000). On the trail of spined loach: developing a conservation plan for a poorly known species. *British Wildlife*, **11**, 390–397.

Perrow, M.R., Peirson, G. & Townsend, C.R. (1990). The dynamics of a population of roach (*Rutilus rutilus* (L.)) in a shallow lake: is there a two-year cycle in recruitment? *Hydrobiologia*, **191**, 67–73.

Perrow, M.R., Moss, B. & Stansfield, J. (1994). Trophic interactions in a shallow lake following a reduction in nutrient loading: a long-term study. *Hydrobiologia*, **275/276**, **94**, 43–52.

Perrow, M.R., Meijer, M.-L., Dawidowicz, P. & Coops, H. (1997). Biomanipulation in shallow lakes: state of the art. *Hydrobiologia*, **342/343**, 355–365.

Perrow, M.R., Jowitt A.J.D., Stansfield, J.H. & Phillips, G.L. (1999*a*). The importance of the interactions between fish, zooplankton and macrophytes in the restoration of shallow lakes. *Hydrobiologia*, **395/396**, 199–210.

Perrow, M.R., Jowitt, A.J.D., Leigh, S.A.C., Hindes, A.M. & Rhodes, J.D. (1999*b*). The stability of fish communities in shallow lakes undergoing restoration: expectations and experiences from Norfolk Broads (UK). *Hydrobiologia*, **408/409**, 85-100.

Perrow, M.R., Hindes, A.M., Leigh, S. & Winfield, I.J. (1999*c*). *Stability of Fish Populations after Biomanipulation*, Environment Agency R&D Technical Report no. W199. Swindon, UK: Water Research Council.

Persson, L. (1983*a*). Effects of intra- and interspecific competition on dynamics and size structure of a perch (*Perca fluviatilis*) and a roach (*Rutilus rutilus*) population. *Oikos*, **41**, 126–132.

Persson, L. (1983*b*). Food consumption and the significance of detritus and algae to intraspecific competition in roach *Rutilus rutilus* in a shallow eutrophic lake. *Oikos*, **41**, 118–125.

Persson, L. (1986*a*). Effects of reduced interspecific competition on resource utilisation in perch (*Perca fluviatilis*). *Ecology*, **67**, 355–364.

Persson, L. (1986*b*). Temperature-induced shift in foraging ability in two fish species, roach (*Rutilus rutilus*) and perch (*Perca fluviatilis*): implications for coexistence between poikilotherms. *Journal of Animal Ecology*, **55**, 829–839.

Persson, L. (1987). Effects on habitat and season on competitive interactions between roach (*Rutilus rutilus*) and perch (*Perca fluviatilis*). *Oecologia*, **73**, 170–177.

Petridis, D. (1990). The influence of grass carp on habitat structure and its subsequent effect on the diet of tench. *Journal of Fish Biology*, **36**, 533–544.

Phillips, G.L., Eminson, D.F. & Moss, B. (1978). A mechanism to account for macrophyte decline in progressively eutrophicated waters. *Aquatic Botany*, **4**, 103–125.

Phillips, G., Jackson, R., Bennet, C. & Chilvers, A. (1994). The importance of sediment phosphorus release in the restoration of very shallow lakes (the Norfolk Broads, England) and implications for biomanipulation. *Hydrobiologia*, **275/276**, 445–456.

Poleo, A.B.S., Oxnevad, S.A., Ostbye, K., Heibo, E., Andersen, R.A. & Vollestad, L.A. (1995). Body morphology of crucian carp *Carassius carassius* in lakes with or without piscivorous fish. *Ecography*, **18**, 225–229.

Power, M.E. (1992). Habitat heterogeneity and the functional significance of fish in river food webs. *Ecology*, **73**, 1675–1688.

Prejs, A. (1984). Herbivory by temperate freshwater fishes and its consequences. *Environmental Biology of Fishes*, **10**, 281–296.

Prejs, A., Martyniak, A., Borón, S., Hliwa, P. & Koperski, P. (1994). Food web manipulation in a small eutrophic Lake Wirbel, Poland: effect of stocking with juvenile pike on planktivorous fish. *Hydrobiologia*, **275/276**, 65–70.

Ptolemy, R.A. (1997). A retrospective review of fish habitat improvement projects in British Columbia: do we know enough to do the right thing? In *Symposium on Sea-Run Cutthroat Trout: Biology, Management and Future Conservation*, eds. J.D. Hall, P.A. Bisson & R.E. Gresswell, pp. 145–147. Corvallis, OR: American Fisheries Society.

Raleigh, R.F. & Duff, D.A. (1980). Trout stream habitat improvement: ecology and hydrology. In *Proceedings of Wild Trout II*, 24–25 September 1979, pp. 67–77. Vienna, VA: Trout Unlimited.

Rasmussen, J.B. (1999). *The Skjern River Restoration Project.* Copenhagen: Danish Ministry of the Environment and Energy, National Forest and Nature Agency.

Rinne, J.N. (1992). The effects of introduced fishes on native fishes in Arizona, southwestern United States. In *Protection of Aquatic Biodiversity, Proceedings of the World Fisheries Congress, Theme 3*, eds. D.P. Philipp, J.M. Epifanio, J.E. Marsden & J.E. Claussen, pp. 149–159. Enfield, NH: Science Publishers, Inc.

Roberts, J., Chick, A., Oswald, L. & Thompson, P. (1995). Effect of carp (*Cyprinus carpio* L.), an exotic benthivorous fish, on aquatic plants and water quality in experimental ponds. *Marine and Freshwater Research*, **46**, 1171–1180.

Saint-Pierre, R.A. (1996). *A Recipe for Success: Anadromous Fish Restoration in the Susquehanna River, USA*, Council Meeting Papers of the International Council for the Exploration of the Sea (ICES), Reykjavik, 27 September–4 October 1996. Copenhagen: ICES.

Scheffer, M. (1998). *Ecology of Shallow Lakes.* London: Chapman & Hall.

Scheffer, M., Hosper, S.H., Meijer, M.-L., Moss, B. & Jeppesen, E. (1993). Alternative equilibria in shallow lakes. *TREE*, **8**, 275–279.

Schiemer, F. & Waidbacher, H. (1992). Strategies for conservation of a Danubian fish fauna. In *River Conservation and Management*, eds. P.J. Boon, P. Calow & G.E. Petts, pp. 363–382. Chichester, UK: John Wiley.

Schindler, D.E., Kitchell, J.F. & Ogutu-Ohwayo, R. (1998). Ecological consequences of alternative gill net fisheries for Nile perch in Lake Victoria. *Conservation Biology*, **12**, 56–64.

Schriver, P., Bøgestrand, J., Jeppesen, E. & Søndergaard, M. (1995). Impact of submerged macrophytes on fish–zooplankton–phytoplankton interactions: large-scale enclosure experiments in a shallow lake. *Freshwater Biology*, **33**, 255–270.

Scruton, D.A., Anderson, T.C. & King, L.W. (1998). Pamehac Brook: a case study of the restoration of a river impacted by flow diversion for pulpwood transportation *Aquatic Conservation: Marine and Freshwater Ecosystems*, **8**, 145–157.

Selgeby, J.H., Eshenroder, R.L., Krueger, C.C., Marsden, J.E. & Pycha, R.L. (eds.) (1995). *Proceedings of the International Conference on Restoration of Lake Trout in the Laurentian Great Lakes*, Ann Arbor, Michigan, 10–14 January 1994. *Journal of Great Lakes Research*, **21**.

Shapiro, J., Lamacea, V. & Lynch, M. (1975). Biomanipulation: an ecosystem approach to lake restoration. In *Proceedings of Symposium on Water Quality Management through Biological Control*, eds. P.L. Brezonik & J.L. Fox, pp. 85–96. Gainesville, FL: University of Florida.

Sibbing, F.A. (1991). Food capture and oral processing. In *Cyprinid Fishes: Systematics, Biology and Exploitation*, eds. I.J. Winfield & J.S. Nelson, pp. 377–412. London: Chapman & Hall.

Skov, C. & Berg, S. (1999). Utilization of natural and artificial habitats by YOY pike in a biomanipulated lake. *Hydrobiologia*, **408/409**, 115–122.

Skov, C., Perrow, M.R., Berg, S. & Skovgaard, H. (in press). *Freshwater Biology.*

Spanovskaya, V.D. & Grygorash, V.A. (1977). Development and food of age-0 eurasian perch (*Perca fluviatilis*) in reservoirs near Moscow, USSR. *Journal of Fisheries Research Board of Canada*, **34**, 1551–1558.

Spencer, C.N. & King, D.L. (1984). Role of fish in regulation of plant and animal communities in eutrophic ponds. *Canadian Journal of Fisheries and Aquatic Sciences*, **41**, 1851–1855.

Stansfield, J.H., Perrow, M.R., Tench, L.D., Jowitt A.J.D. & Taylor, A.A.L. (1997). Submerged macrophytes as refuges for grazing Cladocera against fish predation: observations on seasonal changes in relation to macrophyte cover and predation pressure. *Hydrobiologia*, **342/343**, 229–240.

Stroud, R.H. & Clepper, H. (eds.) (1979). *Predator–Prey Systems in Fisheries Management, Proceedings of an International Symposium on Predator–Prey Systems in Fish Communities and their Role in Fisheries Management*, Atlanta, Georgia, 24–27 July 1978. Washington, DC: Sport Fishing Institute.

Tatrai, I. & Istvanovics, V. (1986). The role of fish in the regulation of nutrient cycling in Lake Balaton, Hungary. *Freshwater Biology*, **16**, 417–424.

Tatrai, I., Lammens, E.H.R.R., Breukelaar, A.W. & Klein Breteler, J.G.P. (1994). The impact of mature cyprinid fish on the composition and biomass of benthic macroinvertebrates. *Archive für Hydrobiologie*, **131**, 309–320.

Threlkeld, S.T. (1987). Experimental evaluation of trophic-cascade and nutrient mediated effects of planktivorous fish on plankton community structure. In *Predation: Direct and Indirect Impacts on Aquatic Communities*, eds. W.C. Kerfoot & A. Sih, pp. 161–173. Hanover, NH: University of New England Press.

Threlkeld, S.T. (1994). Benthic–pelagic interactions in shallow water columns: an experimentalist's perspective. *Hydrobiologia*, **275/276**, 293–300.

Tomlinson, M.L., Perrow, M.R., Hoare, D., Pitt J.-A., Johnson, S., Wilson, C. & Alborough, D. (2002). Restoration of Ormesby Broad through biomanipulation: ecological, technical and sociological issues. In *Management and Ecology of Lake and Reservoir Fisheries*, ed. I.G. Cowx, pp. 184–202. Oxford: Fishing News Books.

Tonn, W.M., Magnuson, J.J., Rask, M. & Toivonen, J. (1991). Intercontinental comparison of small-lake fish assemblages: the balance between local and regional processes. *American Naturalist*, **136**, 345–375.

Townsend, C.R. & Perrow, M.R. (1989). Eutrophication may produce population cycles in roach, *Rutilus rutilus* (L.), by two contrasting mechanisms. *Journal of Fish Biology*, **34**, 161–164.

Townsend, C.R. & Risebrow, A.J. (1982). The influence of light level on the functional response of a zooplanktivorous fish. *Oecologia*, **53**, 293–295.

Townsend, C.R. & Winfield, I.J. (1985). The application of optimal foraging theory to feeding behaviour in fish. In *Fish Energetics: New Perspectives*, ed. P. Tytler & P. Calow, pp. 67–98. London: Croom Helm.

Townsend, C.R., Sutherland, W.J. & Perrow, M.R. (1990). A modelling investigation of population cycles in the fish *Rutilus rutilus* (L.). *Journal of Animal Ecology*, **59**, 469–485.

TRAFFIC (1996). http://www.traffic.org/publications/pubs_sharks.html/

Turner, A.M. & Mittelbach, G.G. (1990). Predator avoidance and community structure: interactions among piscivores, planktivores, and plankton. *Ecology*, **71**, 2241–2254.

Underwood, G.J.C., Thomas, J.D. & Baker, J.H. (1992). An experimental investigation of interactions in snail–macrophyte–epiphyte systems. *Oecologia*, **91**, 587–595.

Utzinger, J., Roth, C. & Peter, A. (1998). Effects of environmental parameters on the distribution of bullhead *Cottus gobio* with particular consideration of the effects of obstructions. *Journal of Applied Ecology*, **35**, 882–892.

van Densen, W.L.T. (1994). Predator enhancement in freshwater fish communities. In *Rehabilitation of Freshwater Fisheries*, ed. I.G. Cowx, pp. 102–119. Oxford: Fishing News Books.

van Donk, E. & Otte, A. (1996). Effects of grazing by fish and waterfowl on the biomass and species composition of submerged macrophytes. *Hydrobiologia*, **340**, 285–290.

van Zyll de Jong, M.C., Cowx, I.G. & Scruton, D.A. (1997). An evaluation of instream habitat restoration techniques on salmonid populations in a Newfoundland stream. *Regulated Rivers: Research and Management*, **13**, 603–614.

Venugopal, M.N. & Winfield, I.J. (1993). The distribution of juvenile fishes in a hypereutrophic pond: can macrophytes potentially offer a refuge for zooplankton? *Journal of Freshwater Ecology*, **8**, 389–396.

Walker, P.A. (1994). Development of pike and perch populations after biomanipulation of fish stocks. In *Rehabilitation of Freshwater Fisheries*, ed. I.G. Cowx, pp. 376–89. Oxford: Fishing News Books.

Weatherley, A.H. & Gill, H.S. (1987). *The Biology of Fish Growth*. London: Academic Press.

Werner, E.E. & Hall, D.J. (1974). Optimal foraging and size selection of prey by the bluegill sunfish (*Lepomis macrochirus*). *Ecology*, **55**, 1042–1052.

Williamson, M. & Fitter. A. (1996). The varying success of invaders, *Ecology*, **77**, 1661–1666.

Winemiller, K.O. & Anderson, A.A. (1997). Response of endangered desert fish populations to a constructed refuge. *Restoration Ecology*, **5**, 204–213.

Winfield, I.J. (1986). The influence of simulated aquatic macrophytes on the zooplankton consumption rate of juvenile roach, *Rutilus rutilus*, rudd, *Scardinius erythrophthalmus*, and perch, *Perca fluviatilis*. *Journal of Fish Biology*, **29**, 37–48.

Winfield, I.J., Peirson, G., Cryer, M. & Townsend, C.R. (1983). The behavioural basis of prey selection by underyearling bream (*Abramis brama* (L.)) and roach (*Rutilus rutilus* (L.)). *Freshwater Biology*, **13**, 139–149.

Winfield, I.J., Cragg-Hine, D., Fletcher, J.M. & Cubby, P.R. (1996). The conservation ecology of *Coregonus albula* and *C. lavaretus* in England and Wales, UK. In *Conservation of Endangered Freshwater Fish in Europe*, eds. A. Kirchhofer & D. Hefti, pp. 213–223. Basel, Switzerland: Birkhauser.

Winfield, I.J., Fletcher, J.M. & Cubby, P.R. (1998). The impact on the whitefish (*Coregonus lavaretus* (L.)) of reservoir operations at Haweswater, UK. *Archiv für Hydrobiologie*, **50**, 185–195.

Winfield, I.J., Fletcher, J.M. & Winfield, D.K. (2002). Conservation of the endangered whitefish (*Coregonus lavaretus*) population of Haweswater, UK. In *Management and Ecology of Lake and Reservoir Fisheries*, ed. I.G. Cowx, pp. 232–241. Oxford: Fishing News Books.

Zambrano, L. & Hinojosa, D. (1999). Direct and indirect effects of carp (*Cyprinus carpio* L.) on macrophyte and benthic communities in experimental shallow ponds in central Mexico. *Hydrobiologia*, **408/409**, 131–138.

Zambrano, L. & Macías-García, C. (2000). Impact of introduced fish for aqua-culture in Mexican freshwater systems. In *Non-Indigenous Freshwater Organisms*, eds. R. Claudi & J.H. Leach, pp. 113–126. Boca Raton, FL: Lewis Publishers.

Zambrano L., Perrow, M., Aguirre-Hidalgo, V. & Macías-García, C. (1999). The impact of introduced carp (*Cyprinus carpio*) in subtropical shallow ponds. *Journal of Aquatic Stress Ecosystems and Recovery*, **6**, 281–288.

Zambrano, L., Scheffer, M. & Martinez-Ramos, M. (2001) Catastrophic response of lakes to benthivorous fish introduction. *Oikos*, **94**, 344–350.

Zaret, T.M., & Kerfoot, W.C. (1975). Fish predation on *Bosmina longirostris*: body-size selection versus visibility selection. *Ecology*, **56**, 232–237.

18 • Reptiles and amphibians

CARL G. JONES

INTRODUCTION

Amphibians and reptiles are important components of many ecosystems and reach high levels of biodiversity and biomass in many tropical areas, where most species are to be found. They occur at several trophic levels and have important interactions with other species. Besides the more obvious ones of being predators and prey, some lizards are important plant pollinators and tortoises may disperse seeds and maintain grazing climax communities.

World-wide there are about 12 000 species of reptiles and amphibians, about a quarter of known vertebrate species. Currently only 5.5% of amphibian and reptilian species are considered threatened or endangered, but this is probably an underestimate. Many populations have been disrupted by habitat change, destruction and fragmentation and by the impacts of exotic species and disease. Efforts to conserve them have mostly been passive by passing legislation and habitat protection, but there are increasing efforts to restore populations through habitat restoration, and by the intensive management of the threatened species. In many cases there have been efforts to replace extirpated populations with animals translocated from healthy wild populations or captivity.

The restoration of depleted or extirpated reptile and amphibian populations is a realistic goal even though there are, considering the number of species, relatively few successful cases. Nevertheless, even though our efforts may seem crude we have to pursue the goal of restoring populations and communities for the techniques being developed have wide currency for the conservation of reptiles and amphibians globally.

UNDERLYING PRINCIPLES

Understanding population restoration

The restoration of a population of an endangered species needs to start with a step-by-step evaluation of the problems it faces and to correct those factors that most compromise the population's health. It is helpful to go to basic principles and ask the question: Why is the population so rare? In any rare or declining population the species has to be declining due to: (1) poor productivity: few animals may be breeding, hatchability may be poor and/or juvenile survival may be depressed; (2) poor adult survival. It is likely that in many threatened reptile and amphibian populations the correction of one or a few limiting factors may result in a rapid increase in the population.

A successful restoration programme may follow five broad overlapping stages or steps that provide a conceptual framework in which to develop the restoration work (see Box 18.1). The components of a species restoration outlined below may seem self-evident but a surprisingly large number of restoration projects proceed without a clear coherent knowledge of the problems and a plan to address them.

Know your species

This is self-evident, but for most of the world's reptiles and amphibians we still know only cursory details of their life history and little about their population biology. The failure of several restoration projects for endangered reptiles and amphibians has occurred largely because not enough was known about their ecology and population biology to address the problems in an effective manner (Dodd & Seigal, 1991). The first step to conserving any species or population is to know the life history and its basic ecology together with details of its distribution and population size. Captive studies have often been used to compliment field studies on the life history of endangered species and for many, captive studies are the most effective way of collecting good quality data.

Understand limiting factors

Most vertebrate populations are limited by a relatively few variables. Provided the habitat still exists, some of the most important factors controlling population densities in reptiles and amphibians are: (1) food supply, (2) refugia and hibernacula, (3) breeding sites, (4) microhabitat variables (basking sites, water quality, etc.), (5) predators and competitors, and (6) disease.

The importance of these variables can, in some cases, be tested on a small scale. Sometimes a closely related species may be used as a model on which to test techniques, before any intensive attempt at the species' restoration, by the provisioning of food, providing suitable refugia, controlling predators, monitoring disease and trial translocations or reintroductions.

Intensive management

When every individual animal counts and the aim is maximising productivity and survival, intensive management is justified, but it is only applicable to critically endangered species and populations. It is usually supplemental to an extensive population management approach, correcting most, if not all, of the limiting factors identified in stage two. Intensive management may involve captive breeding and reintroductions, translocations, introduction onto predator-free islands, and the harvesting of eggs for head-starting (see below). Intensive management has been used for the restoration of some chelonian populations such as western swamp terrapin (Box 18.1) and Galapagos tortoises (Box 18.2), but, to date, has been rarely applied to other reptiles and amphibians since many techniques are still being developed.

Population management

This aims at increasing a population's growth by addressing the most important factors controlling the population and may require detailed research to identify these factor. This is not feasible with critically endangered species whose numbers are critically low.

Monitoring

It is important to carefully monitor species of conservation concern both during and after the restoration of the species or population. Consistent long-term population monitoring is most important: recording rates of productivity and survival that can be compared with other populations of the same species (or a closely related species).

Microhabitat considerations

Habitat destruction and degradation are the main causes of reptile and amphibian decline and extirpation. However large areas of apparently suitable habitat may have few reptiles and amphibians due to microhabitat deficiencies. Most species respond to microhabitat features and these are often species-specific and poorly understood. They include suitable refugia, basking sites, nesting sites, nursery areas and hibernacula. For aquatic species, water depth, flow, pH, nutrient levels etc may all be important. How microhabitat variables affect reptiles and amphibians is described below.

Basking and nesting sites

Since reptiles and amphibians are ectothermic they are largely dependent upon environmental temperatures and in temperate climates are only active in the warmer months. British reptiles are only able to achieve their selected body temperatures for 31–39% of all sunlight periods between April and September. Hence, basking sites are very important for many temperate species (Beebee & Griffiths, 2000). Evidence from artificially introduced exotic reptiles in the northern hemisphere suggests that the northerly range of oviparous reptiles is limited by the availability of suitable conditions for egg incubation as many can live normal life spans but are unable to breed successfully (Frazer, 1964).

Most temperate basking reptiles tend to become concentrated on south-facing sunny open areas. In suitable habitat, sand lizards (*Lacerta agilis*) are concentrated on south-facing slopes with bare sandy

Box 18.1. Restoration of the western swamp terrapin

Broad-based intensive management is restoring the population of this critically endangered western swamp terrapin (*Pseudemydura umbrina*) from Western Australia. The species initial decline was due to habitat loss. During the 1960s to 1980s most of the animals were in two blocks of habitat; one of 155 ha and the other of 65 ha. These were protected in 1962, but despite this the population declined from over 200 in the late 1960s to about 30 wild animals in the late 1980s (Kuchling & DeJose, 1989). The population declined due to reduced recruitment into the adult population due to increased juvenile mortality. The main disturbances and threats that impacted upon the population included road traffic, polluted water from intensive livestock farming, bush fires, lowering of the water table and predation by exotic mammals. Many of these problems have been addressed by management such as fencing along roads to reduce road kills and to exclude predators, diversion of drains with polluted water, fire management and control of water levels.

A captive colony of 25 western swamp terrapins was established in 1959 although the first captive breeding was not recorded until 1966. By 1978, 26 hatchlings had been produced. In 1979, a further two females and a male were taken into captivity from the wild. In 1980, eight additional hatchlings were produced, but these all subsequently died. Breeding in the captive population ceased and by 1987 the situation was critical with the wild population numbering 30 in one population. Only 17 animals remained in captivity, three of which were adult females. In 1988, the standards of captive management were improved, based on a greater understanding of the species reproductive biology. The captive animals resumed breeding and in the ten breeding seasons up to 1996/7 280 eggs had been laid and 13 eggs had been brought in from wild nests. From these, 228 hatched and 79 were released, 59 after head-starting until they weighed at least 95 g (Kuchling *et al.*, 1992; Kuchling, 1999).

By 1997, the world population had increased to over 250 individuals – including about 150 in captivity – although the number of breeding adults was only about 40. In 1991, the last wild population was fenced to exclude exotic red foxes (*Vulpes vulpes*). The second reserve that had held terrapins until the early 1990s was fenced in 1994 and captive-bred animals were introduced.

This case study shows the development of a project through the initial stages of understanding the biology of the animal and the factors that impacted upon the population. When these were understood an intensive conservation management programme began to address the main factors reducing the population and to boost productivity through captive breeding and reintroductions.

In summary, the main actions taken to restore the western swamp turtle population were (from Kuchling, 1999):

- Habitat management, including predator exclusion and control, fire management, changes of drainage patterns to reconstitute former conditions, and water supplementation during dry years.
- Population monitoring and reintroduction of captive-bred juveniles into managed habitat.
- Identification, protection and rehabilitation of former habitat to increase the carrying capacity of existing reserves and to establish new reserves and populations.
- Maintenance of a captive colony to produce juveniles for reintroduction and as a safeguard against problems in the wild.
- Education and awareness-raising in the Australian public about the problems of the species.

patches where they are able to bask and lay their eggs. Terrapins require logs and stones to bask on and snakes and lizards may need substrates such as logs and stones exposed to the sun. Towns (1996) suggested that when increasing in numbers shore skinks rapidly colonise areas with large rocks, which provide favourable basking sites. Some amphibians also bask, but because of excessive water loss by evaporation across permeable skin, can only be exposed to the sun for short periods and some species of frogs usually regulate temperature by changing microhabitats (Beebee & Griffiths, 2000).

Refugia

Densities of many species are related to the number of suitable refugia. The numbers of Mauritian forest day geckos (*Phelsuma guimbeaui* and *P. rosagularis*) are related to the number of cavities and cracks of suitable size that exist in mature trees for the geckos to shelter and lay eggs in (pers. obs.). The ability of many species to coexist with predators is dependent on the occurrence of secure refugia into which the smaller reptiles can escape. The New Zealand shore skink (*Oligosoma smithi*) persists on islands with the introduced Pacific rat (*Rattus exulans*), provided there are boulder fields with rocks of appropriate size that allow the skinks to penetrate the interstices between rocks but exclude rats (Towns, 1996). Similarly, in the Seychelles the bronze geckos (*Ailuronyx* spp.) survive on the main islands alongside introduced black rat (*R. rattus*) and *Pandanus* spp. in palms, where they can presumably escape into the acute interstices at the base of leaf axils (pers. obs.).

Substrate type

Substrate type influences the occurrence of many species that dig their own refugia. Natterjack toads (*Bufo calamita*) are found in light soil where they can burrow to reach moist soil to avoid desiccation and in which they can hibernate in winter, several toads often using the same hole (Beebee, 1983). Similarly many reptiles require loose well-draining soil in which to burrow. These include the gopher tortoise (*Gopherus polyphemus*), whose burrows protect it from temperature extremes, desiccation and predators as well as providing refugia for other species (Diemer, 1986).

Hibernacula

In temperate regions, reptiles hibernate beneath the ground to avoid killing frosts. Hibernacula may show species-specific characteristics as a result of some species hibernating alone whilst others choosing to hibernate in large groups perhaps with other species. The absence of suitable hibernation sites will preclude reptiles from establishing themselves in otherwise suitable habitat. Zappalorti & Reinert (1994) pioneered the construction of artificial hibernacula for use by free-ranging snakes in natural habitats. In an effort to improve snake habitat they constructed 25 subterranean refugia along a 5-km section of a sewer and electric line right-of-way in southern New Jersey, USA. At the end of three years, snakes had been observed at 17 of these sites. Nine different species had been recorded although most observations were of corn (*Elaphe guttata*) and pine snakes (*Pituophis melanoleucus*). These hibernacula, in addition to being used for winter hibernation, were also used for shelter, courting, nesting, shedding and basking.

Taxonomic and genetic considerations

Taxonomic and genetic considerations that are likely to affect restoration ecologists fall into two groups: (1) understanding the nature of natural taxonomic units such as the limits of racial and species variation, and (2) the genetic make-up of individual animals within a population.

There are many real problems in defining taxonomic units. What represents a distinct taxon is actively debated and developments in DNA techniques have highlighted some of these. For example, many subspecies from continental populations have been based on relatively minor morphological differences which are clinal in nature and would not stand rigorous reappraisal using modern techniques, since there may be gene flow across the population. However, allopatric and island populations that have long histories of isolation often show considerable genetic distance from their nearest populations, even though these differences may not have any obvious morphological or ecological expression. The significance of genetic difference may not always be clear since microsatellites used to determine genetic distance between populations utilise non-coding DNA. In other cases, variation may code for real, but undetected, differences which enhance the fitness of the respective populations.

Chromosomal studies of Hochstetter's frog (*Leiopelma hochstetteri*) have shown that the species is unique amongst vertebrates as a result of the large number of supernumerary chromosomes it may have, the number varying with location. Studies of this variation indicate that each regional frog population is distinctive. The significance of this chromosomal variation is not clear but the mixing of the different populations would compromise any evolutionary processes. The mixing of long-isolated populations of any species should be avoided.

Captive studies

These have proven to be of great use in a number of reptile and amphibian conservation programmes (Wiese & Hutchins, 1994). Captive studies have been useful in elucidating the life-history parameters of species that are also being studied in the field. Once the management and breeding techniques have been worked out captive breeding offers great opportunities for producing animals for reintroduction. Murphy *et al.* (1994) provide a useful review on the role of captive breeding in the conservation of reptiles and amphibians. Captive breeding and reintroduction programmes for amphibians and reptiles have certain advantages relative to those for mammals and birds. Many reptiles and amphibians have a relatively high reproductive potential allowing them to recover quickly from population bottlenecks and captive animals may have a much higher survival than wild individuals. The behaviour of reptiles and amphibians is relatively hard-wired (Chiszar *et al.*, 1994). Thus it is usually not necessary to teach captive-bred/reared animals to recognise food or habitats or how to interact appropriately with conspecifics and predators.

Captive reptiles and amphibians can often be maintained in a relatively small area and the captive requirements for many are relatively modest. Amphibians can often be more readily studied under captive conditions and most of the life-history data on which the conservation programmes for the Houston toad (*Bufo houstonensis*), Mallorcan midwife toad (*Alytes muletensis*) and the Puerto Rican crested toad (*Peltophryne Lemus*) are based have been collected from zoo-based studies (Tonge & Bloxam, 1989, 1991; Johnson, 1994).

Similarly, captive studies on the endangered endemic reptiles of Mauritius have clearly benefited work on the conservation of wild animals. Captive work on the Telfair's skink (*Leiolopisma telfairi*), Guenther's gecko (*Phelsuma guentheri*), Round Island boa (*Casarea dussumieri*) and night gecko (*Nactus* spp.) has provided life-history data. This has helped with the interpretation and focus of the fieldwork and has raised the public profile of these reptiles that has driven efforts to conserve these species in the wild.

For much species-centred restoration the main source of animals for reintroduction (see below) has been from captive breeding programmes. The most important examples are western swamp terrapin (Box 18.1), Galapagos giant tortoise (Box 18.2), Mallorcan midwife toad (Box 18.3), sand lizard (Box 18.5) and Puerto Rican crested toad (Johnson, 1994).

TECHNIQUES

Enhancing nesting success and juvenile survival

The protection of eggs and young to maximise hatchability and neonate survival has been a feature of several reptile conservation programmes especially those concerning chelonians and crocodilians. The physical protection of nest sites is a major feature of the Galapagos tortoise conservation programme (Cayot *et al.*, 1994) (Box 18.2) and has been widely applied to several species of sea turtles (Stancyk, 1982). In most cases the protection is usually the construction of physical barriers that protect the nest site from predation by exotic mammals.

Supplemental feeding

Supplemental feeding has rarely been used as a management tool with reptiles and amphibians and in most cases would be unnecessary, since food may be abundant. Moreover, for many species it is difficult to see how this technique could be applied because of the feeding ecology of the species concerned, but for large carnivorous and scavaging reptiles there are possibilities. Komodo dragons (*Varanus komodoensis*) were fed at a viewing site for these lizards so visitors could view them easily. Daily, 15–20 different dragons would turn up to feed. Supplemental feeding was continued for four years after which it was stopped because it may have

Box 18.2. Conservation of Galapagos giant tortoises

There are 14 named taxa of Galapagos tortoise (*Geochelone elephantopus*), three of which are extinct and, since the 1970s, the taxon *abingdoni* from Pinta island is only known from one individual. Of the remaining taxa, eight are threatened. The tortoise had declined primarily due to human exploitation with subsequent impaction by exotic mammals through predation upon eggs and young (by pigs, rats and dogs) and competition for food by herbivores (goats, burros and cattle) that also destroy tortoise habitat. As all the exotic mammals are not present on each island their impact varies from island to island depending on which species are present.

To save the endangered taxa, a captive-breeding programme for the Galapagos giant tortoises was established in 1965 at a dedicated facility on the Galapagos Islands,. This programme focused on the Espanola tortoises G. (*n.*) *hoodensis*. This was critically endangered and all tortoises found on the island were removed to the breeding centre. Between 1963 and 1974, 12 females and two males were found and there was no evidence of any recent reproduction in the population. An additional adult male was added to the population in 1977 from the San Diego Zoo. The first breeding was in 1971 when 20 young hatched and the first reintroduction was in 1975 when the 17 surviving young from the1971 production were released on Espanola. Incubation and rearing techniques were developed and improved between 1980–1987 (Cayot & Morillo, 1997). Incubation at 28 °C produces primarily females and 29.5 °C produces mainly males, which allows the luxury of determining the sex of the hatchlings. The eggs are usually hatched at 28.5–29 °C which gives the highest hatching success and a male-to-female ratio of 1:2. With better incubation management, hatchability of fertile eggs has increased from 20–30% to 60–70%. Annual survival of hatchlings has also improved dramatically from about 86% to more than 98%.

A total of 664 juvenile tortoises of between 1.5–4.5 years of age had been returned to their native island between 1975 and the end of 1994. Survival has been good and 55% of released tortoises have been located on the island during monitoring trips. The first nesting attempts and two dead hatchlings were noted in 1990, with live hatchlings discovered the following year. In addition to reintroducing captive-bred animals, feral goats (the only exotic mammal on Espanola) were completely eradicated from the island in 1978. Substantial regeneration of the vegetation followed.

For the other threatened taxa of Galapagos tortoise, captive breeding was not considered a priority. These taxa have been managed by the protection of nest sites, head-starting and the control or eradication of exotic mammals on the smaller islands and in critical tortoise areas. In two populations of Galapagos tortoise on Santa Cruz and Santiago, the main cause of nest failure was predation on the eggs, and to a lesser extent the hatchlings, by pigs. On these islands the nests have been protected with crude rock walls or the nest sites covered with chainlink fencing, preventing pigs digging up the nest and destroying the nest in one encounter. On San Cristobal Island nests were fenced against dogs until their eradication in the mid-1970s. On several islands exotic mammals are controlled around important tortoise areas (Cayot *et al*., 1994).

In some wild populations eggs and hatchlings were harvested towards the end of the incubation period and hatched and reared at the Tortoise Breeding Center. Between 1970 and 1992, 268 young head-started tortoises had been reintroduced to Pinzon island, where it has been estimated that survivorship is about 77%. Until the head-starting programme there had been no recruitment into the population because of hatchling predation by black rats. Although rats still exist on Pinzon, the presence of young introduced animals means that the critical status of this taxon has been reversed, giving more time for the eradication of the rats (Cayot *et al*., 1994).

Box 18.3. The importance of captive breeding in the restoration of Mallorcan midwife toad

Mallorcan midwife toad (*Alytes muletensis*) was only described in 1977 from subfossil remains and living animals were not discovered until 1980 when they were found in remote mountain gorges in the north of Mallorca, Spain. It then declined due to predation from the introduced snake *Natrix maura* and competition with the green frog (*Rana perezi*) and green toad (*Bufo viridis*). A survey revealed 13 natural breeding populations with only 1000–3000 animals with the largest subpopulations in seven separate gorges.

A captive-breeding programme was set up in 1985 at the Durrell Wildlife Conservation Trust which imported four toads and 16 tadpoles. The first captive breeding followed in 1988 and was subsequently highly successful. In the first three seasons (1988–1990) 1227 eggs were laid, of which 1156 were fertile and 926 hatched (Tonge & Bloxam, 1991). Toads were distributed to other institutions for research into the biology and behaviour of the species and for captive breeding, and animals have been returned to Mallorca for release. The first reintroduction from the Durrell Wildlife Conservation Trust was in 1989, when 76 tadpoles were returned to the island and released at two sites. Since then releases of both young toads and tadpoles have occurred on an annual basis up to 1997, using animals from several institutions. The toads from the different subpopulations have been kept separate since they may have been isolated for up to 2000 years, and genetically distinct.

Before reintroduction, faecal samples from the toads were screened for parasites and bacteria and more recently these results have been compared with the results from wild animals. Preliminary findings showed that the wild and captive animals had similar parasite faunas (Buley & Gonzalez-Villavicencio, 2000). The releases were all to sites that did not contain toads but were within the known historical range. Releases have been into natural sites and into old, man-made water holes. A number of criteria had to be met before a release site was chosen, including (summarised from Buley & Garcia, 1997):

- Permanent presence of water, as tadpoles will overwinter.
- Absence of *Natrix maura* and *Rana perezi*.
- Isolation from natural toad populations, to avoid mixing genetically distinct populations and to minimise the risk of disease transmission.
- Adequate refugia for adults (i.e. humid with rocky crevices).
- Correct amount of sunlight, for appropriate amounts of plant/algal growth.
- Accessibility to biologists for monitoring.
- Isolated from disturbance by the public and domestic livestock.

There is annual *in situ* monitoring of all natural and reintroduction sites, to establish the relative success of reintroductions and to determine the size and breeding status of the different subpopulations. There are ten release sites ready for reintroductions pending the availability of suitable captive-bred tadpoles and toads for release (Buley *et al.*, 1999).

Reintroduction has been highly successful and as a result 25% of the wild population originates from captive-bred toads. The distribution of the toads has increased from 100 km^2 in the early 1980s to about 200 km^2 in 2000, around 20% of what is believed to be the original range of the species. The number of sites that now have toads is 25, and 12 of these have been established with captive-bred animals. All reintroductions have apparently resulted in established populations (Buley & Garcia, 1997; Buley *et al.*, 1999; Buley & Gonzalez-Villavicencio, 2000).

The reintroduction of Mallorcan midwife toad is one of the most successful amphibian reintroduction projects. Captive breeding played a large part in obtaining life-history data, biological and behavioural data and in providing animals for release. The releases were carefully planned and executed and based on good habitat requirements derived from wild populations. Careful thought was given to maintaining the different subpopulations as genetically distinct entities and prior to reintroduction the animals were screened for pathogenic parasites and bacteria. The wild and reintroduced populations were carefully monitored to evaluate the causes of success or failure.

resulted in unmeasured ecological, behavioural and physiological changes, including habituation and health problems, although Walpole (2001), who conducted this study, does not suggest what these may have been, other than high density at the feeding site. Following the cessation of feeding numbers dropped down to pre-feeding levels and there were no detectable long-term effects upon the population. It is a pity that this interesting study was halted since it might have resulted in higher productivity and survival.

On Round Island, a satellite island off the north coast of Mauritius in the Mascarene group of islands in the Indian Ocean, the density of the large endemic Telfair's skink has increased up to tenfold around the camping area where there is an abundance of discarded food (pers. obs.). Similarly, the densities of skinks is very high around the main habitations and rubbish tips in on some of the offshore islands in the Seychelles, where the skinks *Mabuia wrightii* and *M. seychellensis* are taking discarded food (pers. obs.).

Eradication of predators/competitors: conservation of reptiles on small islands

Much of the conservation work on reptiles has been on oceanic islands, driven by high levels of endemicity and high extinction rates. While habitat destruction is an obvious cause of species loss, the gradual degradation of ecosystems by the invasion of exotic animals and plants is more insidious and is often difficult to evaluate.

The restoration of representative areas on large islands is possible but the long-term management of these areas is very labour-intensive due to the constant maintenance required to control the invasion of exotic organisms. A more achievable goal is the restoration of small islands where many of the invasive species can be completely eradicated and reinvasion rates can be kept very low by good quarantine protocols. In addition to this, many satellite islands have remained free of many of the most harmful aliens and are refuges for species that have long since become extirpated elsewhere. Consequently, uninhabited satellite islands are usually less ecologically damaged than larger inhabited islands and often have relatively intact reptile communities.

Rats (*Rattus* spp.) are among the most harmful exotic mammals on islands and have received more research and management attention than any other exotic mammal. Rats have been accidentally introduced to islands almost world-wide since the fifteenth century. The three most widespread species are the black rat, brown rat (*R. norvegicus*) and Pacific rat.

In New Zealand it was noted that the number of species and density of lizards was higher and tuatara (*Sphenodon* spp.) were generally commoner and larger, on rat-free New Zealand islands (Whitaker, 1978). Brown rats reached the 2-ha Whenuakura Island in about 1982 and the entire population of more than 130 tuatara disappeared (Newman, 1988). On Mauritius in the Indian Ocean it is believed rats were responsible for wiping out the mainland populations of the keel-scale boa (*Casarea dussimeri*), the large Guenther's gecko, the large skinks *Leiolopisma mauritiana* and *L. telfairi* (Telfair's skink) and the night gecko (*Nactus serpensinsula*). There is supporting evidence from satellite islands around Mauritius where some reptile populations persisted. For example, Telfair's skink persisted on two small islands until the middle of the nineteenth century, their disappearance coinciding with the introduction of rats. Telfair's skink, Guenther's gecko and keel-scale boa now only survive on rat-free Round Island. Some of the smaller species of Mauritian skinks and geckos coexist with exotic rats presumably because they are inaccessible to rats when in refugia (pers. obs.). Rats may also cause the extinction of amphibians. On New Zealand rats are believed to have been involved in the extinction of three species of *Leiopelma* frogs (Bell, 1994).

New Zealand biologists have pioneered work on the restoration of degraded island ecosystems for several decades (Box 18.4). This has been achieved by the control and eradication of exotic animals, the restoration of the vegetation communities and the introduction of missing vertebrates including 12 species of lizards (Towns *et al.*, in press). Over the last two decades the techniques for the eradication of exotic mammals have become increasingly sophisticated and the size of island that can be cleared of problem mammals has increased steadily.

Box 18.4. Conservation management of lizards on satellite islands off New Zealand

The native terrestrial reptile fauna of New Zealand is comprised of two species of tuatara and more than 60 species of lizards in two endemic genera of geckos (*Hoplodactylus* and *Naultinus*) and two endemic genera of skinks (*Cyclodina* and *Oligosoma*). The tuataras, 24% of the geckos and half of the skinks are in need of urgent conservation action. Tuatara and 37% of the New Zealand lizard species are now totally or mainly confined to islands (Daugherty *et al.*, 1994). Many of these have been extirpated from much of their range by exotic mammals and now have widely disjunctive distributions where they once had large continuous ones.

There are possibilities of reintroducing lizards to islands where they once occurred following the successful eradication of exotic predatory mammals, especially rats. Twelve species of lizards and both tuatara have been reintroduced to islands within their former range. However, this conservation management could not start until details on the distribution and taxonomy of the species had been clarified in a series of studies starting in the 1970s. The distributional and taxonomic work, including genetic studies, enabled the resolution of cryptic species complexes including dividing the tuatara into three taxa, of which there are two different species, *Sphenodon punctatus* and *S. guntheri*.

Tuatara were the flagship species and extensive laboratory studies on their reproductive biology were undertaken. These, and work on physiological effects of the nest environment in the field, enabled the development of artificial egg incubation methods that set the scene for conservation management. Egg-laying can be induced in gravid *S. punctatus* by injecting the hormone oxytocin, and the eggs can be hatched in incubators with a hatching success of 86–92%. Tuatara are also being bred in captivity in large naturalistic enclosures, although this work is still in its early stages and has yet to produce significant numbers of animals for release (Cree *et al.*, 1994; Towns *et al.*, in press). Since tuatara show temperature-dependent sex determination, the ratio of males to females can be carefully determined by hatching the eggs under the appropriate incubation regime. Young from several of the different island populations have been head-started and released on islands from which rats have been removed.

The removal of rats has also benefited some lizard species that were able to survive with rats, albeit at low densities. On one island the density of shore skinks (*Oligosoma smithi*) increased 3.6-fold over nine years (Towns *et al.*, 1990, in press). Twelve species of lizards have been re-established on islands as part of island restoration exercises. The best documented are for four species of skinks (followed by number reintroduced): marbled skink (*Cyclodina oliveri*), $n = 25$; Whitaker's skink (*Cyclodina whitakeri*), $n = 28$; robust skink (*C. alani*), $n = 14$; and the egg-laying skink (*Oligosoma suteri*), $n = 30$. These were released on Korapuki Island between 1988 and 1992, after the eradication of rats. The populations have been monitored for at least seven years and all have shown positive population growth despite the small number of animals released, although Whitaker's skink has shown a population growth of only 7% per annum. The main reason for this slow rate of increase was the low annual productivity of only one per female, but this was offset by the longevity of the lizards, some of which were at least 16 years old. Two of the other three species of skinks were monitored and *C. alani* showed an annual rate of increase of 28% and *O. suteri* showed an increase of 16%. Adult survival was excellent and varied between species from about 76% to 90% per annum (Towns & Ferreira, in press).

In conclusion, these re-establishments were successful, but the rate of recovery was limited by the life-history characteristics of the species with their relatively low breeding rates. With slow breeding species, during the initial reintroduction phase the populations are inevitably very vulnerable due to the small number of animals involved.

Translocation and reintroduction

The translocation of wild animals into an area where it is hoped to establish or restore a population and the reintroduction of captive-bred/reared reptiles and amphibians are increasingly important management techniques. The quality of many programmes that have used translocated and reintroduced animals varies considerably but the most carefully planned and professionally organised have generally been the most successful.

The translocation and reintroduction process falls into three parts: (1) pre-release training and/or conditioning with (2) the release or translocation process (hard vs. soft), and (3) post-release monitoring and management.

Pre-release training and conditioning

This is usually considered to be of most importance for captive-bred higher vertebrates that have to learn various survival skills before being released. It is usually assumed that the amount of learning necessary for a successful reintroduction of reptiles and amphibians is less than for birds and mammals. For animals that are being translocated it is usually not an important consideration since the animals are being moved from one area of habitat to another and will already possess the necessary survival skills. It is even less of a consideration for amphibians being moved as spawn since they will develop their various imprinting and survival skills *in situ* while they develop.

Early learning is likely to be important for larger long-lived species although little work has been done on this. The studies of Grassman *et al.* (1984) suggest that nestling Kemp's ridley turtles (*Lepidochelys kempi*) imprint upon the beach where they hatch (see Box 18.6). The apparent poor success of head-starting (see below) in sea turtles may be due to inappropriate early learning experiences during the head-starting process.

Release and subsequent monitoring and management

Higher vertebrates are released by hard or soft release techniques. Hard release is where the animals are just liberated without any habituation to the release area. Soft releases are typically carried out over an extended period where the animals are introduced gradually from a holding pen and may be given post-release support in the form of supplemental food, predator control etc., while the animals adjust to life at liberty. As we have learned more and more about reintroduction it has become obvious that soft-release techniques are essential for most captive-bred/reared higher vertebrates. The high levels of mortality that are often a feature of hard-release techniques have become unacceptable. In reptiles and amphibians hard release techniques are the norm, but we need to be aware of subtle learning and physiological adaptations that reptiles and amphibians may have to make when adapting to life at liberty. Some species may need post-release support and many species may need the opportunity to rehydrate after being released and for some of the larger, slower breeding species it may be appropriate to provide post-release support. Whitaker's skinks (*Cyclodina whitakeri*) introduced to Korapuki Island off New Zealand (Box 18.4) were provided with artificial home sites of stacked plywood sheets and artificial burrows made of plastic drainage pipe into which the released animals were individually placed (Towns, 1994). The reintroductions of natterjack toads in Britain were followed by support in the form of habitat management (Box 18.5).

Disease considerations

Quarantine has to be considered when moving animals from one area to another and are particularly important for source animals raised in captivity. Here they may have been exposed to exotic pathogens from different continents, which may not be detectable during routine screening.

The desert tortoise (*Gopherus agassizii*) was declining in the wild in California due to habitat destruction and predation, so a campaign of reintroductions from captive-breeding programmes has been undertaken by several United States state agencies. Unfortunately these were conducted without quarantine provision and have resulted in the introduction of an upper respiratory disease hitherto unknown in wild populations. This disease has reduced numbers of tortoises in some of the largest populations by up to 60%. The release of captive animals

Box 18.5. The restoration of the herpetofauna of British dunes and heathland

Two previously declining British species, the natterjack toad (*Bufo calamita*) and sand lizard (*Lacerta agilis*) have been restored by the management of their heathland and sand dune habitats. Where populations had become extirpated animals were reintroduced or translocated (Beebee & Denton, 1996; Moulton & Corbett, 1999). Sand lizards were released as juveniles or as adults, while natterjack toads have been moved as spawn or tadpoles taken directly from wild sites. A third rare British species, the smooth snake (*Coronella austriaca*) has also benefited by the restoration of heathland and some have also been translocated to restored areas.

NATTERJACK TOAD

Natterjack toad suffered a major decline during the first half of the twentieth century and disappeared from more than 75% of its historic sites, due to habitat destruction and successional changes. After broad-based ecological studies the problems faced by the species were assessed and a stepwise plan to correct them was suggested (from Denton *et al.*, 1997):

- The restoration of early successional habitats on heaths and dunes.
- The maintenance of these habitat stages by sustainable economic methods.
- Countering anthropogenic acidification of breeding ponds.
- Enhancement of natterjack populations, accounting for minimum viable population size and metapopulation theory.
- Reintroductions of natterjacks to restore the historically documented range.

The natterjack is adapted to early-successional stages, sand dunes, upper saltmarsh, heathlands and floodplains. The management of its heath and dune habitats has focused on the restoration and maintenance of early stages of seral succession, through the physical clearance of invasive scrub and woodland vegetation. Sheep and cattle were used to establish grazing regimes to maintain the successional stage. Population size in natterjacks is related to the availability of breeding habitat – shallow ephemeral ponds that dry out in midsummer (Beebee *et al.*, 1996) – and in colonies of less than 100 adult toads, additional ponds were built. Natterjacks need breeding ponds that are neutral or near neutral in pH and in some unsuitable acidic ponds the water was neutralised by adding powdered calcium hydroxide or removing the accumulated sulphate load when the pond dried out in midsummer. In some areas the common toad (*Bufo bufo*) has been a major competitor and the natterjack tadpoles do poorly in ponds where there are high densities of common toad tadpoles. Natterjacks have thus benefited from the removal of adult common toads and spawn (which were translocated to suitable ponds elsewhere with no toads) (Simpson, 2002). The captive rearing of natterjack spawn has been a useful way to boost productivity when population levels are low. The survival from egg to toadlet is 90% or more under good captive conditions compared with less than 5% in the wild (Beebee & Denton, 1996). An excellent account of the management and recovery of a population of natterjack toads is given by Simpson (2002).

Twenty reintroductions of natterjack into parts of its historic range where the species had become extirpated are documented by Denton *et al.* (1997). At least six of these resulted in new expanding populations and another eight have showed initial signs of success. The six failures were mainly early during the conservation programme when the needs of the species were poorly known and can be attributed to poor water quality and inadequate terrestrial microhabitats for the toads. At the 15 reintroduction sites since 1980, ancillary management has been implemented to improve habitats, including scrub clearance at eight sites, pond creation at all sites and grazing regimes at three sites (Denton *et al.*, 1997).

SAND LIZARD

Concomitant with the work being carried out on the natterjack there has been a programme for the restoration of sand lizard populations. In Britain the sand lizard is at the northwestern edge of its biogeographical range. About 90% of British colonies were lost in the last century due to habitat destruction and fragmentation. To conserve the species Moulton & Corbett (1999) suggested the following:

- Site protection, with the most important sites being made into nature reserves.
- Management of sites to maintain and, where necessary, restore suitable conditions for sand lizards.
- The reintroduction of captive-bred or the translocation of wild-caught sand lizards to suitable sites within their known or presumed historic range. Whilst the emphasis has been on restoring dune populations, heathland translocations are also being undertaken.

Sand lizards favour local variation in topography, banks, ridges, steep slopes and gullies that provide sunny aspects and microclimates. Of particular importance are bare sandy areas on south-facing slopes for basking and egg-laying and where concentrations of young sand lizards gather. The creation of basking and egg-laying sites for this species is one of the main conservation strategies in Britain since the northern range for oviparous reptiles is limited by the availability of suitable conditions for egg incubation. Studies on wild colonies showed that patches of bare sand occupying about 5% of the area are necessary for healthy populations. In managed areas vegetation was cleared to expose the bare sand up to an area of 2 × 1 m and these patches covered between 2% and 20% of the area. The managed areas where sand lizards are found vary in size between 1 and 50 ha and most are less than 5 ha.

Habitat restoration work for natterjacks (see above) also benefited sand lizards in those sites where both species occurred. Habitat management and restoration were carried out during the winter months when the reptiles and amphibians were hibernating, to minimise disturbance. Much of the scrub was removed to provide open sunny areas for the lizards and encourage mature heather (*Calluna* sp.) or marram grass (*Ammophila arenaria*).

As the result of management lizards were recorded breeding at all 25 managed sites. Relatively poor-quality unmanaged sites ($n = 10$) showed declining populations, good-quality unmanaged sites ($n = 17$) showed stable populations, while managed sites showed a annual doubling in sighting frequencies. Managed sites had on average nearly five times the density and productivity of lizards compared to unmanaged areas (Corbett & Tamarind, 1979).

Animals were reintroduced or translocated to restored areas within the historic range of the species to restore its former range both on sites where it occurred previously and to other areas in appropriate habitats, to compensate for irreversible habitat loss elsewhere (Moulton & Corbett, 1999). The majority of sand lizards used in reintroductions are captive-bred in outdoor vivaria that conform closely to their natural habitat. The eggs are artificially incubated and the young head-started for a few weeks before being released to hibernate in the wild. To establish a viable population with a good age structure, three annual releases of around 50 animals have been recommended (Moulton & Corbett, 1999). Wild-caught animals are sometimes translocated, but these are usually from sites due to be lost to development, although occasionally, a few animals have been available from large established populations. Where adults are released it has been recommended that 20 or more pairs are usually needed to produce a viable population (Moulton & Corbett, 1999). Lizards are released on warm days between mid-April and early September. Sand lizards have been re-established at sites in at least five English counties. To date, most reintroductions have been successful in establishing viable populations.

OTHER SPECIES

The restoration and management of dry heathland has also benefited the rare smooth snake; trends in sighting rates have followed those of sand lizard, doubling in managed areas. The grass snake (*Natrix natrix*), a species not normally associated with dry heaths, has also increased and in management areas encounter rates have also doubled (Corbett & Tamarind, 1979). The numbers of adders (*Vipera berus*), common lizards (*Lacerta vivipara*), slow worms (*Anguis fragilis*) and common toads have also increased on restored heathland.

may have done more damage to the population than the previous causes of decline.

If captive-bred animals from zoo or other mixed species collections are to be used for reintroduction they should be reared under carefully quarantined conditions and screened prior to release. Ideally, captive bred animals should come from dedicated facilities such as the one for Galapagos tortoises (Box 18.2). Some of the Mallorcan midwife toads that have been reintroduced were derived from a zoo population (Box 18.3). The latter were bred in dedicated facilities within the zoo and were screened for parasites and pathogenic bacteria before introduction, the results being compared with results obtained from wild animals (Buley & Gonzalez-Villavicencio, 2000). One of the aims of the Mallorcan Midwife Toad project is the 'comparative analysis of parasitological and bacteriological status of captive and wild populations' (Buley *et al.*, 1999).

In general, translocated animals from healthy wild populations should be preferred when stocking new populations.

Head-starting

Animals for reintroduction are typically captive-bred or are captive-reared from wild-harvested eggs or hatchlings. Head-starting is one of the most frequently used techniques to boost productivity and to obtain young for releases. Usually it involves the harvesting of wild-laid eggs and rearing these in captivity. Head-starting boosts the number of young that survive the first year of life by avoiding most, or all, of the high first-year mortality in the wild, which affects most reptiles and amphibians. The length of time a species is head-started for varies from species to species but there are usually minimum sizes that have to be reached to reduce the incidence of predation. Head-starting has the additional advantage that in those reptiles that show temperature-dependent sex determination it is possible to modify the proportions of male to females by careful control of the incubation temperature.

Marine turtles

Head-starting of marine turtles has been applied to several species in the USA, it was tried for green (*Chelonia mydas*) and loggerhead (*Caretta caretta*) turtles from 1959 to 1989, but the impact of this project has not apparently been evaluated. Other head-starting projects have been attempted for other marine turtles, but most of the work has been done on the endangered Kemp's ridley (Box 18.6). The rationale is that growing young turtles in captivity will protect them from high rates of natural predation that would otherwise have occurred. Critics have argued that it is not a proven technique and suggest that captive turtles suffer nutritional deficiencies and behavioural modification including inappropriate imprinting as well as disease and injury exacerbated by the crowded conditions of captivity.

Freshwater terrapins

Head-starting has been tried on several species of freshwater turtles in many parts of the world and is clearly a viable technique in boosting productivity. In Malaysia it has been used for *Batagur baska* and *Callagur borneoensis* (Moll, 1995) and in Brazil for the giant Amazonian turtle (*Podocnemis expansa*) (Cantarelli, 1997).

In Massachusetts, USA, studies on the redbelly terrapin (*Pseudemys rubriventis*) demonstrated high levels of mortality in the first year post-release of nine-month-old turtles. However, mortality was size-related and survival improved as the turtles became older and larger. Despite mortality, survival was sufficient to result in an increase in three out of four of the populations into which head-started turtles were released (Haskell *et al.*, 1996).

A particularly innovative approach to minimise the impact of road-kills on the northern diamond-back terrapin (*Malaclemys terrapin*) is recorded by Wood & Herlands (1997) who removed eggs from dead terrapins of the killed on roads in Cape May, New Jersey, USA. Between 1989 and 1995, 3690 eggs were salvaged from 4020 road-killed terrapins. A total of 1175 (32%) of these hatched and 782 (81%) young survived the ten-month head-starting period to be introduced into the saltmarsh habitat of their parents.

In parts of India, human bodies are cremated and then disposed of in holy rivers. The Ganga River, in Uttar Pradesh, suffers from the accumulation of partly cremated human bodies. To combat the problem, the Government of India launched a clean-up project which included the incubation,

Box 18.6. Head-starting Kemp's ridley turtle

The Kemp's ridley turtle (*Lepidochelys kempi*) is the most endangered of the marine turtles and virtually the entire species nested on a 15-km section of beach near Rancho Nuevo, Mexico. In 1947 it was estimated the nesting population was about 40 000 females. Egg exploitation followed and by the 1970s, the population had declined to about 200–500 nesting females.

In 1978 an experimental project was initiated with the aim of establishing a secondary nesting colony at Padre Island, Texas. Between 1978 and 1992 some 22 507 eggs and 9675 hatchlings were collected from Rancho Nuevo and transferred to Padre Island for incubation and/or rearing (Caillouet, 1995). Attempts were made to imprint the hatchlings to Padre Island so they would return there as adults.

Eggs were harvested at Rancho Nuevo as they were laid and placed in containers of Padre Island sand, and transported to Padre Island where they hatched. As it is believed that the hatchlings 'imprint' on the olfactory nature of the nesting beach and/or the adjacent waters as they leave the nest and travel out to sea, the young turtles are released on the beach and allowed to enter the surf and swim for a few minutes before being caught up for captive rearing. The turtles are then believed to store this information and use it to facilitate navigation back to the nesting beach (Klima & McVey, 1982). There is some evidence supporting the imprinting hypothesis since four-month-old head-started turtles spent more time in a solution made of Padre Island sand and sea-water than in control solutions, in a multiple-choice test in the laboratory (Grassman *et al.*, 1984).

The young turtles were then reared in captivity for (usually) seven to 15 months before release at sea. Although a large number of eggs were harvested they are not thought to have affected the productivity of the turtles at Rancho Nuevo since they are believed to represent just 2.8% of cumulative productivity (Caillouet, 1995).

Hatching and rearing rates in captivity were very high with over 85% of eggs hatching and a rearing rate of over 95%. Between 1978 and 1992, 23 102 head-started turtles were reared and most were released. Turtles were tagged to allow future identification. In 1996, two head-started turtles nested on Padre Island and in 1998, there were four confirmed nests in South Texas made by three Kemp's ridleys from the project. This project has thus demonstrated that experimentally imprinted and head-started turtles are able to join the wild population, find their way to nesting beaches, copulate successfully and produce viable offspring. The clutches were laid at or close to the beach where the turtles were imprinted as hatchlings and were comparable to those at Rancho Nuevo.

The Kemp's ridley head-starting programme has perfected many of the techniques necessary for maximising the productivity of sea-turtles. The techniques for transporting and hatching eggs, the captive rearing of young, diets and disease control have all been greatly improved. However, areas of uncertainty, such as the importance of early learning and imprinting and how these may be modified by captive rearing, remain.

Although small numbers of head-started turtles have successfully been recruited into the breeding population the monitoring techniques are not sensitive enough to compare their survival with that of wild turtles. This question will be answered when the percentage of head-started turtles returning to breed on Padre Island is known, but since they do not reach maturity until about ten years of age, only time will tell. However, the data do suggest that it is possible to use head-started turtles to establish new populations and as such, it is a potent technique.

head-starting and release of Indian soft-shell terrapin (*Aspideretes gangeticus*) which are efficient scavengers of the decomposing bodies. About 85 000 eggs have been harvested from nest sites along the Ganga, and with a hatch rate of 44%, about 25000 young have been released into a turtle sanctuary along the Ganga river (Whitaker & Andrews, 1997).

Tortoises

Head-starting of tortoises has been attempted on several species but the most extensive and perhaps the most successful work has been with the Galapagos tortoises (Box 18.2). Morafka *et al.* (1997) undertook some interesting trial studies on the Bolson tortoise (*Gopherus flavomarginatus*) and the desert tortoise

G. agassizii. They ran a three-year trial using Bolson tortoises as a model to develop the techniques of obtaining eggs, incubation and rearing. The presence of eggs inside gravid females was determined by inguinal palpation. Thereafter, whereas some females laid without any stimulation, others were induced to lay by an intramuscular injection of oxytocin at a dosage of 1.0 USP unit per kg of body weight. Eggs were incubated under three different regimes and hatchability (excluding infertile clutches) was 65–77% per year. Survivorship of the 107 hatchlings over their first year was 76%.

The work on the desert tortoise was more passive as wild-caught individuals were simply placed in a large (60 m × 60 m) natural, predator-proof pen. Here, they laid their eggs and were subsequently released. The eggs were left *in situ* and an estimated 90–94% hatched. Annual survivorship of all age classes (first to third years) was 86%. Trial releases were attempted with ten hatchling Bolson tortoises and nine desert tortoises (three and four years of age) but annual survival rates were not available.

Crocodilians

Head-starting has been used extensively for several species of crocodilians both as a means for ranching crocodiles for skins and meat, but also as a very powerful conservation tool. This technique has been used on several species in India: gharial (*Gavialis gangeticus*), estuarine crocodile (*Crocodylus porosus*) and mugger (*C. palustris*) (de Vos, 1984). Release occurs at two to four years old and most individuals have survived. Indeed the recovery of the gharial is attributed to the breeding of head-started animals (Rao, 1990).

Moreover, some of the most extensive and detailed work has been on the American alligator (*Alligator mississippiensis*), although this programme did not start releasing substantial numbers of animals until 1991 (Elsey *et al.*, 2000*a*). Minimum survival to four years post-release of 42 319 alligators was an impressive 85.3% (Elsey *et al.*, 1998). Comparisons in survival between released head-started and wild alligators revealed no difference (Elsey *et al.*, 1998). Released animals also grow as well as wild animals (Elsey *et al.*, 1992*a*), can forage for food efficiently (Elsey *et al.*, 1992*b*) and some produce eggs (Elsey *et al.*, 1998). Similarly, released head-started Nile crocodile (*C. niloticus*), Australian freshwater crocodile (*C. johnstoni*), Orinoco crocodile (*C. acutus*) and the Cuban crocodile (*C. rhombifer*) have all showed good survival rates (reviewed in Elsey *et al.*, 2000*b*).

Increasing genetic diversity

Small isolated populations may lose genetic variability as a consequence of inbreeding. The effects of inbreeding depression are well documented – decreased fertility and increase in malformed and stillborn young and depressed juvenile survival – although there are few studies that show evidence for this in wild reptile and amphibian populations. An important recent study on an isolated and inbred population of adders (*Vipera berus*) in Sweden (Madsen *et al.*, 1999) showed that they had low genetic variability, were producing many deformed and stillborn young and showed poor recruitment into the adult population. Adult males were counted annually (since many females were cryptic and difficult to find) and the population declined from 25 males in 1983 to just four adult males in 1995. In 1992, twenty marked males from a genetically variable but nearby population were introduced and left for four mating seasons, after which the eight surviving males were removed. The introduced snakes mated with the local females and these produced young. Adders reach maturity at about four years and from 1996 the number of adults increased dramatically so that by 1999 there were 32 adult males in the population. The genetic variability in the population also increased dramatically over the same time period. This study clearly shows the advantage of adding new animals to a failing population as it increased the genetic diversity and helped a small, inbred population recover.

Many continental reptiles and amphibians share a similar problem faced by the Swedish adders, as their populations become increasing fragmented and isolated. Small urban populations of common toads (*Bufo bufo*) and common frogs (*Rana temporaria*) breeding in garden ponds in Britain have become increasingly isolated and there is already evidence of inbreeding depression in toads (Hitchings & Beebee, 1998). The outbreaks of frog disease and an increase in colour varieties in common frogs have been attributed to the random fixing of deleterious genes due to inbreeding (Hitchings & Beebee, 1997). Beebee & Griffiths (2000) recommend mixing spawn between garden ponds to mitigate inbreeding effects.

The translocation of animals from nearby healthy populations is thus the strategy of choice when trying to restore extirpated or failing small populations of reptiles and amphibians. This approach makes sense since nearby populations are in most cases likely to be the closest genetically. In Britain, similar approaches have been widely practiced where they have complemented habitat improvement and have been applied to sand lizard (Box 18.5), slow worm (*Anguis fragilis*), smooth snake (*Coronella austriaca*) and natterjack toad (Box 18.5).

Use of analogue species

The introduction of species outside of their known range may be an appropriate conservation strategy for critically endangered species, especially when suitable habitat within its known range is not available or no longer suitable. Most often this type of introduction, usually called a 'conservation introduction' has been to an island where the population can be closely monitored. In some cases it may be worth considering the introduction of closely related or ecologically similar species to act as analogues or ecological replacements for species that have become extinct. The three main criteria that need to be considered while choosing an analogue species are: (1) taxonomic closeness to the extinct taxon; (2) ecologically equivalence to the extinct taxon; and (3) conservation importance.

If the analogue is taxonomically close to the extict form (i.e. a sister species, subspecies or in the same superspecies) and is also ecologically similar, then there is every possibility that the introduction will be successful. Taxonomic closeness alone, however, may not be enough in the choice of an analogue since in many radiations a closely related species may have a very different ecology.

Ecological closeness may override the need for taxonomic closeness since the purpose of using an analogue is to restore an ecological function that has been lost by an extinction (Box 18.7). Increasingly, it may be possible to find analogue species that are also rare and in need of conservation action. The case of

Box 18.7. The introduction of tortoises to Indian Ocean islands to replace the extinct tortoises of Mauritius, Rodrigues and the Seychelles

Giant tortoises were once widespread on islands in the western Indian Ocean with endemic taxa on islands in the Mascarenes and the Seychelles. These wild populations are now extinct except for the giant tortoises of Aldabra Island (a distant atoll in the Seychelles group). The giant tortoise populations are likely to have had a profound impact upon the vegetational structure and composition on the different islands. Early reports from the Mascarenes suggest that the tortoises maintained grazed areas – tortoise lawns – as they do on Aldabra. Some of the plants that would have formed these tortoise lawns may be adapted to high levels of grazing pressure. The Mascarene endemic tussock-forming grass *Vetevaria arguna* maintained its best population on Round Island off Mauritius, where it was the commonest grass, until exotic rabbits, which grazed the island heavily, were eradicated in 1986. Subsequently *V. arguna* declined and was replaced by other grasses. It is hypothesised that *Veteveria* survives in heavily grazed areas by having coarse unpalatable leaves which are avoided by herbivores that selectively feed on other grasses. Similarly, on Mauritius, captive Aldabra giant tortoises feed preferentially on softer exotic grasses ignoring the native grass *Zosyia*. Another species that may have been a plant of tortoise turf is *Aerva congesta*, which is now only found on open barren areas on Round Island. *Aerva* has a small prostrate form protecting it from grazing.

There are other possible examples of co-evolutionary plant–tortoise interactions in the Mascarenes. Many of the endemic plants show leaves with marked heterophyllly with characteristic lanceolate shape with marked red venation. The latter form is typically displayed by young leaves growing up to about 1.2 m above the ground. Feeding experiments offering heterophyllous leaves to Aldabra tortoises demonstrated that they would not eat these, but would readily browse adults leaves of the same species. This suggests that heterophylly may be an adaptation to deter browsing by giant tortoises. On both Rodrigues and Mauritius, which both had an endemic browsing saddle-backed tortoise, heterophylly is common, and recently constructed life-size models of the tortoises suggest that the largest animals would have been able to browse leaves up to about 1.2 m above the ground.

In the absence of giant tortoises it is only possible to speculate about the interactions that the extinct endemic tortoises would have had with the endemic flora. It may, however, be possible to reactivate some of the interactions by introducing an analogue species. It has been suggested that the Aldabra tortoise would be an appropriate analogue to introduce to Round Island, which had tortoises until 1842. It has been proposed to introduce six subadult animals to the island that would then be carefully monitored to check their health and condition, to note their feeding preferences and to evaluate the possible impact upon the vegetation if large numbers were introduced. As a trial, four Aldabra tortoises were released into a 1-ha pen on Ile aux Aigrettes, where their behaviour and impact upon the native plant community is being studied. It is planned to introduce more tortoises to Ile aux Aigrettes and to conduct replicated grazing studies that will be run in parallel with the release study on Round Island.

This initiative would not be the first use of analogue tortoises on Indian Ocean islands, since Aldabra tortoises have been introduced to two islands in the main Seychelles group, Fregate Island and Curieuse Island. The suitability of the Aldabra tortoise as an analogue to the extinct Seychelles tortoise is not entirely clear since the taxonomic and ecological affinities between the Aldabra and Seychelles species have yet to be clarified, although some feel that they were very close.

The translocation to Curieuse Island (280 ha) has been the most completely documented. Two hundred and fifty were released between 1978 and 1982, to provide an accessible population for ecotourism and scientific research (Stoddart *et al.*, 1982). These animals initially did very well and put on considerable weight, with some even doubling in weight in less than two years. The first hatchling was found in early 1980 and several more the following year. However the population subsequently declined due to human poaching of adults for food and hatchlings for pets, and extirpation was predicted by 1992 (Samour *et al.*, 1987). In 1990, only 117 animals could be accounted for, of which only six were juveniles that had hatched on the island (Hambler, 1994). From productivity data collected in 1990 it was estimated that 2100–3900 young had hatched on the island since 1980, and the very high rates of loss were attributed to predation by rats and feral cats. Surprisingly, the data also showed that reproduction and growth rates were depressed and suggested that the island was close to carrying capacity even though the tortoises had only a slight impact upon the native vegetation. Some animals had damaged dorsal scutes, although the cause of this was not known. It was not clear what was limiting carrying capacity and this is clearly a population that could benefit from careful population management, e.g. prevention of poaching, cat and rat eradication and health monitoring.

The translocated population on Fregate Island, (introduced around the late 1960s and early 1970s) has fared much better, although it has not been as well studied. There were about 160 adults in 2001 and many of the animals were very large and all size classes were represented, suggesting good breeding and recruitment. Many of the hatchlings and small animals are harvested and head-started for several years to avoid poaching by humans taking them for pets. Rats and cats have been eradicated and there are no major predators other than humans. On Fregate, the endemic magpie robin (*Copsychus sechellarum*) follows the adult tortoises and forages on invertebrates they disturb, which increases the foraging success of this endangered bird. The success of the translocation of tortoises to Fregate clearly shows the potential of analogues.

However, the use of analogues needs to proceed with great care to ensure there are no negative impacts upon native biota. The choice of tortoises as analogues allows the luxury of being able to closely monitor the released animals and, if necessary, to be able to remove them with relative ease. It is suggested that tortoises are used on Mauritian islands to maintain grazed areas, and to provide some browsing pressure, a role which in many restoration projects is fulfilled by domestic livestock. Tortoises will also allow us to evaluate more accurately the ecological role endemic tortoises once had on the pristine Mauritius.

Box 18.8. The introduction of Guenther's gecko and ornate day gecko to Rodrigues to replace extinct *Phelsuma* species

The introduction of Guenther's gecko (*Phelsuma guentheri*) and the ornate day gecko (*P. ornata*) to Rodrigues was suggested by Jones (1993). Rodrigues has lost all of its endemic reptiles and only the native, pantropical gecko *Lepidodactylus lugubris* survives. The introduction of the large nocturnal Guenther's gecko to one or more of the rat-free islets around Rodrigues would help to establish additional populations of this highly endangered gecko where it is highly unlikely to do any ecological damage. Rodrigues once had its own large noctural gecko (*Phelsuma gigas*) which was closely related to *P. guentheri* and judging from its morphology, may have been ecologically similar. Hence *P. guentheri* is likely to be a close ecological analogue for the species once present.

Similarly, Rodrigues once had a brightly coloured frugivorous day gecko, *P. edwardnewtoni*, which became extinct about 1914. Ecologically and taxonomically the ornate day gecko, a common species from Mauritius, may prove to be an appropriate analogue. Although the introduction of these two geckos to Rodrigues would be outside their natural range, this would be an area where close relatives were once present, for all Mascarene day geckos belong to a unique clade within the genus *Phelsuma*. With introduction, it may prove possible to resurrect species interactions that have been lost. *Phelsuma* geckos are known to be important pollinators of some endemic Mascarene plants and the introduction of analogue species from Mauritius to Rodrigues could reactivate lost plant–gecko interactions.

the suggested introduction of Guenther's gecko to Rodrigues in the Mascarenes to replace the extinct *Phelsuma gigas* meets all three criteria (Box 18.8).

CONCLUDING REMARKS

The restoration of reptiles and amphibians is an emerging discipline. Great strides have been made and there are notable successes. There are however the problems of 'halfway technologies', a phrase used by Frazier (1997), who wished to underline the fact that boosting the productivity of turtles by hatcheries, head-starting and captive breeding is only going to work if the original causes of the species decline have been addressed. As the use of half-way technologies emphasises the fact that boosting productivity may not be the whole answer, perhaps it would be more appropriate to view the problem as 'half-developed technologies'. Many of the techniques that have been used with reptiles and amphibians show great promise. Some which have not worked as well as expected, such as head-starting in marine turtles, should not be discarded but rethought and improved. It is more likely the technique and not the concept that is wrong, since head-starting works well with tortoises, crocodiles, tuatara, lizards and amphibians where it is effective in increasing productivity and increasing adult recruitment into the breeding population. However, in most studies the survival rate of the released young is not given, although where this has been measured it is usually close to or comparable to that of wild animals. What is poorly known is the rate of recruitment of head-started animals into the breeding population. Deficiencies need to be addressed and efforts made to develop this powerful technique further, since it has considerable potential for the conservation of many species of reptiles and amphibians.

Whilst Frazier (1997) draws reference to the techniques that boost productivity, management can also address problems that impact upon both juvenile and adult survival. The provision of refugia, hibernacula and other microhabitat manipulations may all enhance survival as will the control of disease and the control and eradication of competitors and predators.

In evaluating the relative success of restoration projects critics often point to the large number of failed projects. However, considerable caution needs to be exercised in comparing success among different studies because of the great difference in quality. Furthermore, different assumptions and perceptions of population characteristics often underlie the determination of success and failure. The successes to date are impressive but most have been the restoration of single species in relatively simple systems. There are even more fascinating challenges ahead with species restoration within complex habitats, restoration of reptile communities with multi-species introductions and restoring lost ecological interactions. Reptiles and amphibians will be important models in our efforts to rebuild vertebrate communities and provide suitable research subjects of competition and predation as successive species are introduced. When restoring communities it is desirable to start with the introduction of species at lower trophic levels, reinstating predators only after populations of the other species have been established.

It is often claimed that species recovery efforts alone are a waste of time since they do not address the real issues of habitat loss and degradation. However, there is increasing evidence to demonstrate that species recovery efforts drive habitat management. The examples from Britain of the recovery of the sand lizard and natterjack toad demonstrate that the efforts to restore these two species have resulted in the restoration of their dune and heathland habitats. In Mauritius, the efforts to restore the native habitat on Round Island have been driven by the conservation work on the threatened reptiles. Field studies and a captive-breeding project for the most endangered reptiles prompted a programme to restore the island vegetation, by eradicating goats and rabbits, some weed control and replanting native plants. Other islands around Mauritius are also being restored primarily to provide alternative sites for endangered Round Island reptiles.

ACKNOWLEDGMENTS

I am most grateful to Nick Arnold, Richard Gibson, Andrew Greenwood, Tom Fritts, Gerald Kuchling, Gordon Rodda and Dave Simpson who all provided information and helped shape my views on amphibian and reptile conservation.

REFERENCES

Beebee, T.J.C. (1983). *The Natterjack Toad.* Oxford: Oxford University Press.

Beebee, T.J.C. & Denton, J. (1996). *The Nattjack Toad Conservation Handbook.* Peterborough, UK: English Nature.

Beebee, T.J.C. & Griffiths, R. (2000). *Amphibians and Reptiles.* London: HarperCollins.

Beebee, T.J.C., Denton, J.S. & Buckley, J. (1996). Factors affecting population densities of adult natterjack toads *Bufo calamita* in Britain. *Journal of Applied Ecology* **33**, 263–268.

Bell, B.D. (1994). A review of the status of New Zealand *Leiopelma* species (Anura: Leiopelmatidae), including a summary of demographic studies in Coromandel and on Maud Island. *New Zealand Journal of Zoology*, **21**, 341–349.

Buley, K. & Garcia, G. (1997). The recovery programme for the Mallorcan midwife toad *Alytes muletensis*: an update. *Dodo, Journal of the Wildlife Preservation trust*, **33**, 86–90.

Buley, K.R. & Gonzalez-Villavicencio, C. (2000). The Durrell Wildlife Conservation Trust and the Mallorcan midwife toad, *Alytes muletensis*: into the 21st century. *Herpetological Bulletin*, **72**, 17–20.

Buley, K.R., Gibson, R.C. & Bloxam, Q.M.C. (1999). The recovery programme for the Mallorcan midwife toad (*Alytes muletensis*) In *Linking Zoo and Field Research to Advance Conservation*, eds. T.L. Roth, W.F. Swanson & L.K. Blattman, p. 249. Cincinnati, OH: Cincinnati Zoo.

Caillouet, C.W. (2000). Sea turtle culture: Kemp's ridley and loggerhead turtles. In *Encyclopedia of Aquaculture*, ed. R.R. Stickney, pp. 786–798. New York: John Wiley.

Cantarelli, V.H. (1997). The Amazon turtles: conservation and management in Brazil. In *Proceedings: Conservation, Management and Restoration of Tortoises and Turtles*, ed. J. Van Abbema, pp. 407–410. New York: New York Turtle and Tortoise Society.

Cayot, L.J. & Morillo, G.M. (1997). Rearing and repatriation of Galapagos tortoises: *Geochelone nigra hoodensis*, a case study. In *Proceedings: Conservation, Management and Restoration of Tortoises and Turtles*, ed. J. Van Abbema, pp. 178–183. New York: New York Turtle and Tortoise Society.

Cayot, L.J., Snell, H.L., Llerena, W. & Snell, H.M. (1994). Conservation biology of Galapagos Reptiles: twenty-five years of successful research and management. In *Captive Management and Conservation of Amphibians and Reptiles*, eds. J.B. Murphy, K. Adler & J.T. Collins, pp. 297–305. London: Society for the Study of Amphibians and Reptiles.

Chiszar, D., Smith, H.M. & Carpenter, C.C. (1994). An ethological approach to reproductive success in reptiles. In *Captive Management and Conservation of Amphibians and Reptiles*, eds. J.B. Murphy, K. Adler & J.T. Collins, pp. 147–173. London: Society for the Study of Amphibians and Reptiles.

Corbett, K.F. & Tamarind, D.L. (1979). Conservation of the sand lizard, *Lacerta agilis*, by habitat management. *British Journal of Herpetology*, **5**, 799–823.

Cree, A., Daugherty, C.H., Towns, D.R. & Blanchard, B. (1994). The contribution of captive management to the conservation of tuatara (*Sphenodon*) in New Zealand. In *Captive Management and Conservation of Amphibians and Reptiles*, eds. J.B. Murphy, K. Adler & J.T. Collins, pp. 377–385. London: Society for the Study of Amphibians and Reptiles.

Daugherty, C.H., Patterson, G.B. & Hitchmough, R.A. (1994). Taxonomic and conservation review of the New Zealand herpetofauna. *New Zealand Journal of Zoology*, **21**, 317–323.

Denton, J.S., Hitchings, S.P., Beebee, T.J.C. & Gent, A. (1997). A recovery program for the natterjack toad (*Bufo calamita*) in Britain. *Conservation Biology*, **11**, 1329–1338.

de Vos, A. (1984). Crocodile conservation in India. *Biological Conservation*, **29**, 183–189.

Diemer, J.E. (1986). The ecology and management of the gopher tortoise in the southeastern United States. *Herpetologica*, **42**, 125–133.

Dodd, C.K. & Seigal, R.A. (1991). Relocation repatriation and translocation of amphibians and reptiles: are they conservation strategies that work? *Herpetologica*, **47**, 336–350.

Elsey, R.M., Joanen, T., McNease, L. & Kinler, N. (1992*a*). Growth rates and body condition factors of *Alligator mississippiensis* in coastal Louisiana wetlands: a comparison of wild and farm-released juveniles. *Comparative Biochemistry and Physiology*, **103A**, 667–672.

Elsey, R. M., McNease, L., Joanen, T. & Kinler, N. (1992*b*). Food habits of native wild and farm-released juvenile alligators. *Proceedings of the Annual Conference of Southeastern Association of Fish and Wildlife Agencies*, **46**, 57–66.

Elsey, R.M., Moser, E.B., McNease, L. & Rebecca, G.F. (1998). Preliminary analysis of survival of farm-released alligators in southwest Louisiana. *Proceedings of the Annual Conference of Southeastern Association of Fish and Wildlife Agencies*, **52**, 249–259.

Elsey, R.M., Lance, V.A. & McNease, L. (2000*a*). Evidence of accelerated sexual maturity and nesting in farm-released alligators in Louisiana. In Grigg *et al*., pp. 244–255.

Elsey, R.M., McNease, L., & Joanen, T. (2000*b*). Louisiana's alligator ranching programme: a review and analysis of releases of captive-raised juveniles. In *Crocodilian Biology and Evolution*, eds. G.C. Grigg, F. Seebacher & C.E. Franklin, pp. 426–441. Chipping Norton, NSW: Survey Beatty.

Frazier, J.F.D. (1964). Introduced species of reptiles and amphibians in mainland Britain. *British Journal of Herpetology*, **3**, 145–150.

Frazier, N. (1997). Turtle conservation and halfway technology: what is the problem? In *Proceedings: Conservation, Management and Restoration of Tortoises and Turtles*, ed. J. Van Abbema, pp. 422–425. New York: New York Turtle and Tortoise Society.

Grassman, M.A, Owens, D.W., McVey, J.P. & Marquez, M. (1984). Olfactory based orientation in artificially imprinted sea turtles. *Science*, **224**, 83–84.

Hambler, C. (1994). Giant tortoise *Geochelone gigantea* translocation to Curieuse Island (Seychelles): success or failure? *Biological Conservation*, **69**, 293–299.

Haskell, A., Graham, T.E., Griffin, C.R. & Hestbeck, J.B. (1996). Size related survival of headstarted redbelly turtles (*Pseudemys rubriventris*) in Massachusetts. *Journal of Herpetology*, **30**, 524–527.

Hitchings, S.P. & Beebee, T.J.C. (1997). Genetic substructuring as a result of barriers to gene flow in urban common frogs *Rana temporaria*. *Heredity*, **79**, 117–127.

Johnson, R.R. (1994). Model programs for reproduction and management: *ex situ* and *in situ* conservation of toads of the family Bufonidae. In *Captive Management and Conservation of Amphibians and Reptiles*, eds. J.B. Murphy, K. Adler & J.T. Collins, pp. 243–254, London: Society for the Study of Amphibians and Reptiles.

Jones, C.G. (1993). The ecology and conservation of Mauritian skinks. *Proceedings of the Royal Society of Arts and Sciences, Mauritius*, **5**, 71–95.

Klima, E.F. & McVey, J.P. (1982). Headstarting the Kemp's ridley turtle *Lepidochelys kempi*. In *Biology and Conservation of Sea Turtles*, ed. K.A. Bjorndal, pp. 481–487. Washington, DC: Smithsonian Institution Press.

Kuchling, G. (1999). *The Reproductive Biology of the Chelonia*. Berlin: Springer-Verlag.

Kuchling, G. & DeJose, J.P. (1989). A captive breeding operation to rescue the critically endangered western swamp turtle *Pseudemydura umbrina* (Testudines: Chelidae) in the wild and in captivity. *International Zoo Yearbook*, **28**, 103–109.

Kuchling, G., DeJose, J.P., Burbidge, A.A. & Bradshaw, S.D. (1992). Beyond captive breeding: the western swamp tortoise *Pseudemydura umbrina* from extinction. *International Zoo Yearbook*, **31**, 37–41.

Madsen, T., Shine, R., Olsson, M. & Wittzell, H. (1999). Restoration of an inbred adder population. *Nature*, **402**, 34–35.

Moll, D. (1995). Conservation and management of river turtles: a review of methodology and techniques. In *Proceedings of the International Congress of Chelonian Conservation*, 6–10 July 1995, pp. 290–294. Gonfaron: Editions Soptom.

Morafka, D.J., Berry, K.H. & Spangenberg, E.K. (1997). Predator-proof field enclosures for enhancing hatching success and survivorship of juvenile tortoises: a critical evaluation. In *Proceedings: Conservation, Management and Restoration of Tortoises and Turtles*, ed. J. Van Abbema, pp. 147–165. New York: New York Turtle and Tortoise Society.

Moulton, N. & Corbett, K. (1999). *The Sand Lizard Conservation Handbook*. Peterborough, UK: English Nature.

Murphy, J.B., Adler, K. & Collins, J.T. (eds.) (1994). *Captive Management and Conservation of Amphibians and Reptiles*. London: Society for the Study of Amphibians and Reptiles.

Newman, D.G. (1988). Evidence of predation on a young tuatara, *Sphenodon punctatus*, by kiore, *Rattus exulans*, on Lady Alice Island. *New Zealand Journal of Zoology*, **15**, 443–446.

Rao, R.J. (1990). Recovered gharial population in the National Chambal Sanctuary. In *Crocodiles: Proceedings of the 9th Working Meeting of the Crocodile Specialist Group, IUCN* Gland, Switzerland: World Conservation Union.

Samour, H.J., Spratt, D.M.J., Hart, M.G., Savage, B. & Hawkey, C.M. (1987). A survey of the Aldabran giant tortoise population introduced on Curieuse Island, Seychelles. *Biological Conservation*, **41**, 147–158.

Simpson, D. (2002). The fall and rise of the Ainsdale natterjacks: the history of natterjack toad conservation at Ainsdale Sand Dunes National Nature Reserve. *British Wildlife*, **13**, 161–170.

Stancyk, S.E. (1982). Non-human predators of sea turtles and their control. In *Biology and Conservation of Sea Turtles*, ed. K.A. Bjorndal, pp. 139–152. Washington, DC: Smithsonian Institution Press.

Stoddart, D.R., Cowx, D., Peet, C. & Wilson, J.R. (1982). Tortoises and tourists in the Western Indian ocean: the Curieuse experiment. *Biological Conservation*, **24**, 67–80.

Tongue, S. & Bloxam, Q. (1989). Breeding the Mallorcan midwife toad *Alytes muletensis* in captivity. *International Zoo Yearbook*, **28**, 45–53.

Tonge, S. & Bloxam, Q. (1991). The breeding programme for the Mallorcan Midwife Toad *Alytes muletensis* at the Jersey Wildlife Preservation Trust. *Dodo, Journal of the Jersey Wildlife Preservation Trust*, **27**, 146–156.

Towns, D.R. (1994). The role of ecological restoration in conservation of Whitaker's skink (*Cyclodina whitakeri*), a rare New Zealand lizard (Lacertilia: Scincidae). *New Zealand Journal of Zoology*, **21**, 457–471.

Towns, D.R. (1996). Changes in habitat use by lizards on a New Zealand island following removal of the introduced Pacific rat *Rattus exulans*. *Pacific Conservation Biology*, **2**, 286–292.

Towns, D.R. & Ferreira, S.M. (in press). Conservation of New Zealand lizards (Lacertila: Scincidae) by translocation of small populations. *Biological Conservation*.

Towns, D.R., Atkinson, I.A.E. & Daughery, C.H. (1990). The potential for ecological restoration in the Mercury Islands. In *Ecological restoration of New Zealand Islands*, Conservation Sciences Publication no. 2. eds. D.R. Towns, C.H. Daugherty & I.A.E. Atkinson, pp. 91–108. Wellington: New Zealand Department of Conservation.

Towns, D.R., Daugherty, C.H. & Cree, A. (in press). Raising the prospects for a forgotton fauna: a review of ten years of conservation effort for New Zealand reptiles. *Biological Conservation*.

Walpole, M.J. (2001). Feeding dragons in Komodo National Park: a tourism tool with conservation complications. *Animal Conservation*, **4**, 67–73.

Whitaker, R. & Andrews, H.V. (1997). Captive breeding of Indian turtles and tortoises at the centre for herpetology/ Madras crocodile bank. In *Proceedings: Conservation, Management and Restoration of Tortoises and Turtles*, ed. J. Van Abbema, pp. 166–173. New York: New York Turtle and Tortoise Society.

Wiese, R.J. & Hutchins, M. (1994). The role of zoos and aquariums in amphibian and reptilian conservation. In *Captive Management and Conservation of Amphibians and Reptiles*, eds. J.B. Murphy, K. Adler & J.T. Collins, pp. 37–45. London: Society for the Study of Amphibians and Reptiles.

Wood, R.C. & R. Herlands (1997). Turtles and tires: the impact of roadkills on northern diamondback terrapin, *Malaclemys terrapin terrapin*, populations on the Cape May peninsula, southern New Jersey, USA. In *Proceedings: Conservation, Management and Restoration of Tortoises and Turtles*, ed. J. Van Abbema, pp. 46–53. New York: New York Turtle and Tortoise Society.

Zappalorti, R.T. & Reinert, H.K. (1994). Artificial refugia as a habitat-improvement strategy for snake conservation. In *Captive Management and Conservation of Amphibians and Reptiles*, eds. J.B. Murphy, K. Adler & J.T. Collins, pp. 369–375. London: Society for the Study of Amphibians and Reptiles.

19 • Birds

JOSÉ MARIA CARDOSO DA SILVA AND PETER D. VICKERY

INTRODUCTION

Birds form one of the most diversified groups of terrestrial vertebrates. With approximately 10 000 species and 300 billion individuals, birds are distributed world-wide, and occur on every continent (Gill, 1994). Because of their extraordinary array of adaptations in form and function – a product of at least 150 million years of evolution – birds inhabit an extraordinary array of different habitats, ranging from grasslands to tropical forests, from small ponds to the largest oceans, from searing deserts to the frigid climates found in the Arctic and Antarctic (Gill, 1994).

Because many species are sensitive to human-induced changes to the land, birds can be used as a valuable metric to determine the degree of habitat degradation, or conversely, the level of habitat restoration. For example, if one wants to assess the biodiversity status of a region, the number of threatened birds is generally regarded as a suitable indicator. Also, if one wants to follow the pace of vegetation recovery following any kind of man-made disturbance, bird diversity and composition are key variables to be monitored (Furness & Greenwood, 1993).

This chapter has three main parts. The first focuses on the restoration of bird populations, with special emphasis on endangered species. We discuss the methods that have been made to increase the severely depleted populations of particular bird species, to a condition similar to those prior to human interference, or at least to a population level that provides a high probability of long-term viability. The second part is concerned with the impacts of birds on the restoration of different ecosystems world-wide. By assuming that ecosystem restoration seeks to control in some way the natural successional sequence (MacMahon, 1997), then we ask how birds can accelerate or retard natural succession. Finally, in the third part of this chapter we demonstrate that birds are one of the best indicator animal groups to evaluate the success of large-scale, ecosystems restoration.

RESTORING BIRD POPULATIONS

Bird conservation in a changing world

A total of 1186 bird species is thought to be globally threatened. Three extant species have become extinct in the wild, 182 species have critically low populations, 321 species are endangered and another 680 species are vulnerable (BirdLife International, 2000). Threatened species are usually habitat specialists. They often inhabit tropical and subtropical forests, do not tolerate well habitat degradation, and have very restricted ranges (BirdLife International, 2000). To avoid further extinctions, conservation programmes should be created for each of these species, but resources are severely limited, especially in the tropical countries that harbour most endangered species. Although international collaboration can play a role in establishing well-designed conservation programmes in tropical countries, the best long-term strategy is to educate people within the country concerned to develop and implement these programmes locally.

Major causes of bird population declines

One of the major challenges in developing a conservation programme for a threatened species is the inadequate knowledge of the underlying reasons for a species' decline. Population decline is a consequence of both intrinsic (proximate) and extrinsic (ultimate) causes. Intrinsic causes are related to a species' demography, whereas extrinsic causes are environmental factors that lead to rapid changes in demographic parameters.

One strategy to identify intrinsic causes is to evaluate the net recruitment of a population. This method measures the difference between the growth of a population through birth or immigration, or the decrease of a population as a result of death and emigration. Decreasing bird populations occur when there are reductions in: (1) the proportion of the individuals in the population that reproduce; (2) the fecundity (clutch or brood size) of the individuals that reproduce; (3) the success rate of individuals that attempt to breed; or (4) the natural flux of immigrants from other populations (Temple, 1986; Brown, 1995).

Extrinsic causes that currently lead to bird populations declines may be traced directly or indirectly to human activities. The major extrinsic causes threatening birds world-wide are habitat loss/degradation, direct exploitation of species and introduction of invasive species (BirdLife International, 2000). Other factors such as natural disasters and pollution play a minor role. The relative importance of each of these factors in causing population declines has changed considerably. King (1981) calculated that 32% of the avian extinctions from 1600 to 1980 involved habitat loss/degradation, 91% involved the impacts of invasive species, and 25% involved excessive direct exploitation. As these estimates suggest, some species become extinct by a combination of two or more causes. Temple (1986) showed a different pattern by evaluating the present-day causes of bird endangerment: habitat loss/degradation threatens 82% of the world's endangered birds, excessive direct exploitation threatens 44%, invasive species threaten 35% and other less pervasive factors, such as toxic chemicals and natural events, threaten only 12%. The specific type of threats varies for different kinds of birds. Habitat loss and degradation primarily threatens birds that are habitat specialists and have small body size whereas exploitation and introduction of invasive species mostly threatens birds with large body size and long generation time (Owens & Bennett, 2000).

Conservation programmes for birds should include strategies that incorporate both intrinsic and extrinsic causes for population decline. Because intrinsic causes are generally more easily identified than extrinsic ones, there is a general trend to improve short-term success in the net recruitment of a species at a local scale, rather than identify the more complicated, difficult extrinsic causes. However, unless extrinsic causes are simultaneously treated, sole attention to intrinsic causes is unlikely to guarantee a secure future for any species (Temple, 1986).

Restoration as a strategy for bird conservation

Bird conservation usually requires the adoption of multiple strategies. These strategies are likely to vary among species, as well as among populations of a particular species. Because each species has a unique set of habitat requirements, its conservation status will largely reflect the extent to which the human-modified environment continues to meet these requirements (Brown, 1995). Variation among populations within a species is likely to be a consequence of the uneven geographic distribution of the threat factors. For example, within the range of a particular species, protection of extant populations in parks and reserves may be the primary conservation action, whereas in other regions, restoration will also play an important role. Thus, although conservation planning should always consider the entire species' range, long-term and field-oriented conservation or restoration action will probably be developed at a smaller spatial scale.

Protection of large natural areas continues to be the most cost-effective way to conserve birds and all biodiversity. However, at present, protected areas are insufficient to guarantee the conservation of a large part of the global biodiversity, and new areas must be created, primarily in tropical and subtropical regions (Myers *et al.*, 2000). Protected areas play a pivotal role in bird conservation world-wide and cannot be replaced by any other strategy. Although numerous large sites that have been officially declared as parks have not received adequate management, they are very important in halting deforestation and biodiversity loss (Bruner *et al.*, 2001). Guidelines for selecting natural reserves and establishing efficient systems of reserve networks are presented in Soulé & Terborgh (1999) and Margules & Pressey (2000).

Restoration should be viewed as a strategy to complement protection within parks. Restoration is generally adopted when populations for a particular species are so critically low that few viable and self-sustainable populations remain. Thus, restoration methods can be used to: (1) recover depleted extant populations; (2) reintroduce populations into a part of the species' native range from which it has disappeared or become extirpated; and (3) introduce populations outside the species' original range.

Tools for bird restoration

Legal measures

Legal measures provide important tools for conservation of endangered species because they help create public awareness and require social responsibility for protecting rare species and their habitats. Legal measures can be very effective in controlling hunting and trade. The list of threatened bird species produced by BirdLife International and IUCN (Birdlife International, 2000) has been adopted globally as a standard tool for guiding international conservation efforts. This list is also an important source for countries to establish their own endangered species lists and to develop species-oriented conservation actions, including recovery plans (policies in different regions are outlined in part 6, volume 2).

Habitat management

Because populations of most species have declined as a result of habitat loss or degradation, habitat management is usually a critical step in any endangered species conservation plan. One of the major challenges for bird conservation is the need to assess the availability of suitable habitat and the area requirements for endangered species.

Habitat preferences can be assessed from field-oriented studies that associate species' abundance and nesting success with measurements of habitat structure and floristic composition. As a result of these data, models of habitat structure can be derived and can provide a basis on which habitat restoration can be evaluated. This method has been successfully used by Kus (1998) to evaluate restored nesting habitat of the endangered least Bell's vireo (*Vireo bellii pusillus*).

A species' area requirements can be assessed by estimating density using a variety of methods, including spot-mapping (Robbins *et al.*, 1989; Vickery *et al.*, 1994). However, for birds with large area requirements, such as hawks, eagles, macaws, parrots, toucans and even bellbirds, reliable estimates will require other techniques, including radio telemetry. By combining habitat models with area requirement estimates, spatial models can be developed to guide habitat restoration efforts.

Although rehabilitation of natural habitats to an original condition may be essential to restore many endangered species, in some cases, even critically endangered species such as the Mauritius kestrel (*Falco punctatus*) can survive in suboptimal habitats 'enhanced' with exotic plants (Safford & Jones, 1998). Such cases afford an opportunity to develop alternative conservation strategies.

Manipulation of eggs, hatchlings and nests

One method used to increase fecundity of individuals within a depleted population is to manipulate eggs, hatchlings and nests (Barclay, 1987). Historically, birds have been classified as determinate layers (lay a fixed number of eggs), or indeterminate layers (lay extra eggs if some are removed). Although this classification should be evaluated with more experiments (Winkler & Walters, 1983), it has been useful as a guideline for restoration programmes. If the threatened species is an indeterminate layer, such as some birds of prey (Welty & Baptista, 1988), one can greatly increase the fecundity of the individual birds by removing clutches that are then incubated by foster parents (Drewien & Bizeau, 1977; Fyfe, 1977; Cade & Hardaswick, 1985) or in artificial incubators (Burnham, 1983). By using these techniques, managers have effectively enhanced fecundity of manipulated breeders (Barclay, 1987).

Another strategy related to egg manipulation is that some bird species lay two eggs (e.g. cranes), but usually only one of the nestlings survives to fledging

(Kuyt, 1981). Managers can increase populations by taking one of the eggs or nestlings for rearing by wild foster parents or hand-rearing. This has resulted in substantial increases in net recruitment of several species (Drewien & Bizeau, 1977).

Egg manipulation is also a good strategy to increase breeding success of certain raptors that have eggs thinned by pesticides. These fragile eggs can be removed from the nest and replaced with artificial eggs. The thin-shelled eggs can be successfully hatched in artificial incubators and then returned to the parents that have been incubating artificial eggs (Burnham, 1983).

Mortality rates are usually high among young birds less than one year old (Ricklefs, 1983). Most of this mortality is caused by factors that are not related to the intrinsic causes of the decline. Thus, increasing juvenile survivorship by protecting young birds from predation, competition, accidents or environmental calamities can increase bird populations (Barclay, 1987). This strategy seems to be very successful for secondary cavity nesters that can be induced to adopt safe artificial nest sites (Snyder, 1978).

Supplemental feeding

The availability of food supply is one of the most important natural factors controlling bird populations. Lack of high-quality food limits population growth and usually has a negative effect on population size (Lack, 1966; Newton, 1980; Grant, 1986). Food availability is often considerably reduced as a result of habitat modification (Temple, 1986). Food limitation increases mortality and therefore reduces net recruitment within a population. If natural food shortage is the only cause of population decline, then a suitable management strategy is to adopt supplemental feeding. This strategy is also recommended for those species whose food supply is contaminated with toxic chemicals.

Control of antagonistic organisms

The control of antagonistic organisms, usually predators, competitors or brood parasites, is one of the standard management strategies used to reverse population declines of endangered species (Temple, 1986). Control of predators, for example, is usually a critical step to the recovery of species that exist in very small numbers. For example, the greatest threat for the endangered tokoeka (*Apteryx australis*) is the impact of introduced predators such as brush-tailed opossum (*Trichosurus vulpecula*), stoat (*Mustela erminea*) and cats and dogs (McLennan *et al.*, 1996). Lack of predator control may have lead to the recent extinction of the last wild individual of the Spix's macaw (*Cyanopsitta spixii*).

Whilst it maybe critical to control nest predators, this can be a difficult task. Nest predation was an important factor in limiting net recruitment of several insular species in Mauritius (Temple, 1986), and special efforts to monitor and protect these nests were required before populations increased (Jones & Swinnerton, 1997).

Control of potential competitors has also been used as a strategy for enhancing nesting success rates of endangered species. To restore populations of red-cockaded woodpeckers (*Picoides borealis*), a cavity nester, it was necessary to remove southern flying squirrels (*Glaucomys volans*) from the restoration areas, because they competed for cavities (Franzreb, 1997). Elimination of colonies of aggressive large gulls (*Larus* spp.) by poisoning, shooting, egg and chick destruction, or human disturbance, has been used to re-establish breeding colonies of some species of terns (*Sterna* spp.) in North America and Europe (Kress, 1983; Anderson & Devlin, 1999).

On islands, exotic species can have especially serious negative impacts on birds. For example, the endangered Gould's petrel (*Pterodroma leucoptera leucoptera*) has been adversely affected by the introduction of the European rabbit (*Oryctolagus cuniculus*) on to Cabbage Tree Island, New South Wales, Australia, because rabbit foraging changed the structure of the rainforest where the petrels nest. To restore bird populations and native vegetation, rabbits were eradicated by sequential epizootics of myxomatosis and rabbit haemorrhagic disease, and a single application of cereal-based bait containing the anticoagulant brodifacoum (Priddel *et al.*, 2000).

Parasites and diseases can also cause serious declines in bird populations (Warner, 1968). Therefore, a carefully designed health monitoring system should be a part of any recovery programme, particularly if it involves translocations of either wild or captive-bred individuals (Cooper, 1977; Cunningham, 1996).

Translocation

Translocation is the movement of living organisms from one area to be released at another site. According to IUCN (1995), there are three main classes of translocation: (1) conservation introductions – an attempt to establish a species, for the purpose of conservation, outside its recorded distribution but within an appropriate habitat and ecogeographical area; (2) reintroduction – the intentional movement of an organism into a part of its native range from which it has disappeared or become extirpated in historical times as a result of human activities or natural catastrophe; and (3) supplementation – the addition of individuals to an existing population of conspecifics (see MacDonald *et al.*, chapter 4, this volume).

Bird translocations have been used to three main purposes: (1) to solve human–animal conflicts; (2) to supplement game populations; and (3) conservation (Fischer & Lindenmayer, 2000). Although translocations seemed to be a commonly employed management tool to solve human–animal conflicts, they were generally unsuccessful (Fischer & Lindenmayer, 2000). Translocation, seeking to supplement game bird populations, seems to be a popular tool among wildlife managers; between 1979 and 1998, 40.5% of the bird supplementation efforts were for game species (Fischer & Lindenmayer, 2000). Bird supplementation is based mostly on captive rather than wild populations. However, several efforts to supplement game bird populations have failed to build self-sustainable populations. For example, Heyl *et al.* (1988) supplemented populations of the Cape francolin (*Francolinus capensis*) with captive-reared birds, but there was no evidence of reproduction and within six months they had all died. The authors concluded that supplementing wild populations with captive-reared individuals could only have short-term benefits such as temporally increased hunting opportunities.

Translocations for conservation purposes have been used world-wide as a strategy to recover populations of endangered bird species (Cade, 1986; Temple, 1986). All three translocation classes have been used in bird conservation projects (see reviews in Temple, 1978).

Introduction of endangered species to predator-free islands has been effective in restoring populations. For example, several species of birds, including little spotted kiwi (*Apteryx owenii*), brown kiwi (*mantelli*), and takahe (*Porphyrio mantelli*), persist in the wild because populations of these three species were introduced to small offshore islands in New Zealand (BirdLife International, 2000).

Supplementing wild populations with captive-bred individuals was used to increase populations of the Mauritius kestrel in Mauritius (Cade & Jones, 1993). Seixas & Mourão (2000) supplemented populations of blue-fronted Amazon (*Amazona aestiva*) in Pantanal, Brazil, by releasing nestlings that were seized by police from illegal traders. Around 60% of the parrots survived at least 13 months after being released.

Reintroduction, the most important and frequently used method of translocation, has been used to restore many bird populations (Fischer & Lindenmayer, 2000). For example, reintroductions have restored populations of peregrine falcon (*Falco peregrinus anatum*) in North America (Cade, 1986); whooping crane (*Grus americana*) in the United States (Meine & Archibald, 1996); yellow-shouldered Amazon (*Amazona barbadensis*) on Margarita Island, Venezuela (Sanz & Grajal, 1998); and guans (*Penelope obscura* and *P. superciliaris*) in reforested areas in southeastern Brazil (Pereira & Wajntal, 1999). In addition, reintroduction of captive-bred individuals is now the only feasible strategy to restore wild populations of California condor (*Gymnogyps californianus*) in the western United States (Snyder & Snyder, 1989) and the Alagoas curassow (*Mitu mitu*) in northeastern Brazil (Collar *et al.*, 1992).

Although costs of translocation efforts cannot be easily assessed (Fischer & Lindenmayer, 2000), it is clear that translocation is the most expensive way to restore bird populations. Despite these costs, a high percentage of translocations have failed (Griffith *et al.*, 1989; Fischer & Lindenmayer, 2000; but see Reading *et al.*, 1997). Given the frequency of these failures, it is important to determine whether this approach is appropriate. Thus, a feasibility study should be conducted before initiating any translocation programme. A feasibility study should address not only ecological aspects of the translocation, but also biopolitical issues. Kleiman *et al.* (1994) proposed the following criteria to evaluate the feasibility of a translocation:

1. Augmentation of the wild population is necessary.
2. Stock is available.
3. The wild population is not jeopardised.
4. Causes of decline are removed.
5. The habitat is ecologically suitable.
6. The habitat is not saturated with members of the same species or other similar species.
7. The local human population is not negatively impacted.
8. Governmental and non-governmental organisations are supportive.
9. The programme complies with all laws and regulations.
10. Relocation technology is known or in development.
11. Sufficient information exists about the species' biology.
12. Sufficient resources exist for the programme.

Engelhardt *et al.* (2000) used these criteria to evaluate the feasibility and success of a translocation programme of trumpeter swans (*Cygnus buccinator*) in the United States.

Although there is no general agreement on what constitutes a successful translocation (Seddon, 1999), there have been several studies that have tried to identify the factors that predict the success of a translocation (Griffith *et al.*, 1989). In these studies, a translocation programme was considered as successful only if it resulted in a self-sustaining population. Generally, the success of translocations has been estimated by comparing individual species, without considering phylogenetic information. In a first effort to include phylogenetic information to analyse the success of translocations, Wolf *et al.* (1998) reported three factors that were good predictors of translocation success for both mammals and birds: (1) habitat quality of the release site; (2) release within the core of the species' historical range; and (3) the number of animals released. These factors may be used as guidelines to build a translocation proposal. Initially, a large-scale study of the species' original range should be conducted to identify the best sites (those with large suitable habitats) for translocation. Ideally, such a study should combine both satellite imagery and fieldwork. If several suitable sites are found, preference should probably be directed to sites within the core of the species' original range because restoration is more likely to succeed here than in sites located at the periphery of the species' range (Lawton, 1995; but see Macdonald *et al.*, chapter 4, this volume). After site selection, initial modelling incorporating both population and landscape data (e.g. Akçakaya & Atwood, 1997; but see recommendations by Beissinger & Westphal, 1998) should be used to estimate the minimum population size to be attained in the restoration.

THE ROLE OF BIRDS IN ECOSYSTEM RESTORATION

In general, the principles for restoration of ecosystems or landscapes are based on natural successional processes (Luken, 1990) and many case studies support this perspective (e.g. Cairns, 1995; MacMahon, 1997). Most management strategies used in ecosystem or landscape restoration seek to accelerate or inhibit natural succession. These strategies can be designed to speed up succession, by shortening the successional sequence, or to hold a community in a particular stage of succession that is considered to have some desirable attributes (MacMahon, 1997). Birds can be important agents controlling the pace of succession in an abandoned land, either by inhibiting or accelerating the rate of succession.

The impact of birds on the restoration process

Birds can play an important role in the restoration process of a given ecosystem or landscape. The role will be positive if birds accelerate natural vegetation succession. Conversely, if birds act in a way which delays natural vegetation succession, this role could be viewed as negative.

Positive impacts

Birds may have a positive impact on vegetation succession in three ways: (1) by controlling populations of herbivorous insects; (2) by pollinating plant species; and (3) by dispersing seeds in the landscape.

Herbivorous insects can regulate the population dynamics of several plant species by direct or indirect effects on plant vigour, performance, competitiveness, recruitment, demography and fitness (Nowierski *et al*., 1999). Because the level of herbivory is not the same among plant species, or over time, herbivory may alter vegetation structure, energy flow and biogeochemical cycling, and often predispose some ecosystems to other disturbances (Schowalter, 2000). Thus, herbivorous insects can impact plant succession (Carson & Root, 1999).

Bird predation can play an important role in controlling populations of several groups of insects, including herbivorous ones (Welty & Baptista, 1988). Thus, birds can influence positively, although in an indirect way, the performance of plant species. Experimental studies supporting this relationship are still limited, but do demonstrate a clear effect of bird predation on populations of herbivorous insects with concomitant reduction of damage caused by these insects (e.g. Gunnarsson & Hake, 1999; Sipura, 1999; Murakami & Nakano, 2000). In a more detailed, experimental study, Strong *et al.* (2000) showed that bird predation decreased the abundance of herbivorous insects and levels of herbivory, but it did not increase plant biomass production during the following year. They concluded that the indirect effects of bird predation on plant performance (measured as biomass production) may depend on the plant species, abundance and composition of the herbivore community, and primary productivity of the ecosystem.

Birds are well known as pollinators of many species of plant, influencing both plant reproduction success and maintenance of population genetic variation (Jennersten, 1988). The number of the species involved in this kind of mutualistic interaction is impressive, as at least one-third of the plant families have at least one species pollinated by birds, and approximately 2000 bird species from 50 families feed on flowers and may act as pollinators in some way (Dias *et al*., 2000). However, we know very little about the consequences of this interaction for plant succession and ecological restoration (reviews in Gottsberger, 1989, 1993, 1996). Several studies indicate that there are important differences in the role played by birds in the pollination process of plant communities. For example, in tropical forests, bird pollinators are less important in dipterocarp forests in Sarawak than in Costa Rica forests because the irregular and ephemeral floral resources in dipterocarp forests do not adequately support many species of birds which require a continuous supply of energy rich resources (Momose *et al*., 1998).

Birds can affect vegetation succession by the dispersal of seeds. The lack of seeds is usually a critical factor constraining the restoration process in several terrestrial environments around the world, from temperate grassland and heathland communities (Bakker & Berendse, 1999) to tropical forests (Whittaker & Jones, 1994; Wunderle, 1997). Birds can play an important role in seed dispersal in these environments because the potential seed sources can usually be found in nearby remnant patches and birds are highly mobile (Debussche *et al*., 1982; McDonnell & Stiles, 1983; Guevara *et al*., 1986; Willson, 1986; Willson & Crome, 1989; McClanahan & Wolfe, 1993; Robinson & Handel, 1993; da Silva *et al*., 1996; Wunderle, 1997).

The fate of a seed dispersed by a bird is complex and is affected, among other factors, by an animal's behaviour, landscape configuration and seed source characteristics (Guevara *et al*., 1986; McClanahan & Wolfe, 1993; da Silva *et al*., 1996; Wunderle, 1997). Most studies on seed dispersal by birds are based on monitoring 'seed rains' in the study areas (e.g. Robinson & Handel, 1993), but other studies have focused directly on bird behaviour and movements. These latter studies are more important for developing effective restoration programmes. In landscapes composed of mixed forest and open lands, few fruit-eating bird species are able to cross the boundary between these two kinds of environments. For example, da Silva *et al.* (1996) found that of the 47 fruit-eating species found in second-growth forests of eastern Amazon, 18 species were recorded in abandoned pastures, but only three species occurred in active pastures. Fruit-eating species that move across open lands are usually small-bodied and have small gapes (bill width). As a result, they are only able to transport small seeds of shrubs and pioneer tree species (Moermond & Denslow, 1985; da Silva *et al*., 1996). Seed dispersal generated by birds in open lands is not random, but directed to isolated shrubs and trees (Duncan & Chapman, 1999). By providing artificial perches in the restoration site, it is possible to increase the number of visits by birds, thus increasing the local seed rain (McDonnell & Stiles, 1983; McClanahan & Wolfe, 1987; Robinson & Handel, 1993; Holl, volume 2). However, bird attraction can be greatly increased if fruiting trees are planted in the target area, because, in addition to providing perches and roosting sites, these trees or shrubs will provide food (da Silva *et al.*, 1996; but see Toh *et al.*, 1999 for a different viewpoint).

In general, forest birds do not move very far from the forest edge. Most movements are short and restricted to a narrow belt around the open land–forest boundary (da Silva *et al.*, 1996; Chapman & Chapman, 1999). Thus, one would be unlikely to attract fruit-eating birds by providing artificial perches if these perches are located very far from the forest edge (>200 m). Based on this finding, da Silva *et al.* (1986) predicted that seed dispersal from forest or woodland patches will decrease as the size of the open land increases. Thus, by understanding the spatial movements of fruit-eating birds, it is possible to predict the probable duration of a restoration project. When open land lacks trees or shrubs that will attract fruit-eating birds, forest recovery will proceed slowly from the forest edge into the open land. However, when shrubs and pioneer trees are available in the open land, forest recovery will be much more rapid because seed dispersal moves from the forest edge into open land as well as from existing shrubs and small trees towards the forest edge (but see Holl, volume 2). Because seed dispersal by birds is directed to specific vegetation components of open lands, these sites play an important role as centres of establishment and subsequent growth of bird-disseminated plant species (Guevara *et al.*, 1986; Vieira *et al.*, 1994).

Negative impacts

Negative impacts of birds on restoration have not been studied to a great degree. However, there are some studies that indicate that herbivorous and granivorous birds may negatively influence plant development, thus delaying vegetation succession.

In aquatic environments, grazing by herbivorous birds has been cited as an important factor in suppressing macrophyte development in shallow lakes undergoing restoration, thus delaying the attainment of the stable clear-water state (van Donk & Otte, 1996; Janse *et al.*, 1998). Perrow *et al.* (1997) examined this issue by evaluating the grazing pressure of coot (*Fulica atra*) on macrophytes in eastern England. They found that grazing pressure varied seasonally; pressure was low during spring, but 76 times greater in summer. Losses to grazing in both periods were negligible when compared to potential macrophyte growth rates documented in the literature. However, during autumnal senescence and in the winter months when some macrophyte species remained available, it was thought grazing by birds could impact the development and structure of macrophyte populations for subsequent growing seasons.

Granivorous birds are thought to have an important impact on early stages of tallgrass prairie succession by removing seeds found on the ground after burns or other disturbances. Howe & Brown (1999) tested this hypothesis through carefully designed experiments on recently ploughed ground. They found that granivorous birds (opportunistic finches and doves) reduced plant densities by 20% and grass biomass by 24% in high-density plantings, and reduced plant densities by 23% and grass biomass by 34% in low-density plantings. In high- and low-density plots, birds reduced species richness by 3% and 17% without influencing diversity. These results indicate that the effects of birds were more pronounced in high-density plantings and that they depressed tallgrass plant densities on open ground.

BIRDS AS INDICATORS OF RESTORATION SUCCESS

One major obstacle in the ecological restoration process is the definition of a reliable measure of success. Birds should be important in this aspect because, compared to other groups of organisms, they are generally well known and reasonably easy to identify. Birds are also well diversified and feed on several trophic levels, and are very sensitive to changes in habitat structure and composition, making them excellent indicators of ecological changes. In addition, several bird species can be used as symbols to increase public awareness and support to any restoration project. Because of these characteristics, birds have been used as indicators of restoration success in several kinds of situations, such as bogs after peat mining (Desrochers *et al.*, 1998), minimally restored tin strip-mines (Passell, 2000), wetlands (Brown & Smith, 1998) and a riparian meadow system (Dobkin *et al.*, 1998), and a we suggest that they are one of the best indicator groups to monitor the success of a restoration programme.

Where birds are used as the metric to measure success of ecological restoration, monitoring can include the entire bird community or it may be directed towards special families or species that will respond more precisely to the habitat changes effected by the restoration project. However, it is essential to know how birds respond to habitat changes brought about by restoration. For example, Brawley *et al.* (1998) monitored bird use at five tidal marshes in Connecticut, including both restored and reference marshes. They found that birds followed the same recovery pattern in that they detected changes in the vegetation, and macroinvertebrate and fish assemblages.

CONCLUDING REMARKS

When developing strategies to restore bird populations, we think the primary emphasis should be given to the processes that create a functional ecosystem rather than simply trying to re-establish bird populations to some predefined target. Especially in ecosystems in which natural disturbance is critical to maintaining ecosystem function, it is essential to recognise and incorporate these disturbance patterns and processes into management efforts.

The selection of appropriate techniques for ecosystem restoration will require an evaluation of site conditions. In general, the factors that limit, constrain or influence the natural recovery process are likely to vary from system to system. Thus, it is important to have a thorough understanding of the ecosystem before establishing restoration goals and objectives (MacMahon, 1997). Although this is preferable, it should also be possible to benefit from previous restoration efforts on similar types of ecosystems. By comparing restoration experiences within major global ecosystems, some patterns are likely to emerge which will be useful as general guidelines (e.g. Bakker & Berendse, 1999). Thus, we stress that the development of experimental, applied studies directed at developing predictive general recovery models and testing possible restoration scenarios for the world's major ecosystems (e.g. grasslands, wetlands, tropical forests) is a crucial step in the development of restoration ecology as a science.

Restoration of bird populations requires the use of a set of distinct methods. This suite of methods will vary between populations of a species or between distinct species, depending on the intrinsic and extrinsic features that caused the species' decline. To be successful, any restoration programme should take this variation into account.

We recognise that the role played by birds in the restoration process is only partially known; for example the importance of birds in some key processes such as pollination, predation on populations of herbivorous insects, and predation (herbivory or granivory) on some plant species has not been adequately addressed in relation to ecosystem restoration. However, studies so far show a pivotal role for birds in the process of dispersing seeds across relatively large distances, and thus, accelerating habitat restoration in terrestrial habitats. Therefore, studies on behaviour of fruit-eating birds in mosaics of natural and altered areas are helpful in refining models of natural vectors to enhance succession and ecosystem recovery.

Although the ways in which birds function in the restoration process are still poorly known, it is clear that bird community parameters, such as species diversity and composition, accurately reflect the major habitat changes due to succession. It is also clear that these changes in bird assemblages parallel those for other groups of organisms. Thus, birds are one of the best animal groups to monitor the success or failure of any restoration programme.

ACKNOWLEDGMENTS

J.M.C. da Silva received support from the Conselho Nacional de Desenvolvimento Científico e Tecnológico (CNPq), FACEPE, and Conservation International do Brasil. The opportunity to work on this manuscript was provided to Peter Vickery by the

Center for Biological Conservation, Massachusetts Audubon Society. Discussions and reviews by W.G. Shriver, A.L. Jones, D.W. Perkins, Ariadna Lopes and M. Tabarelli greatly improved this manuscript. We thank C. Jones, M.R. Perrow and C. Macias for comments that improved very much the first version of this chapter.

REFERENCES

Akçakaya, H.R. & Atwood, J.L. (1997). A habitat-based metapopulation model of the California Gnatcatcher. *Conservation Biology*, **11**, 422–434.

Anderson, J.G.T. & Devlin, C. M. (1999). Restoration of a multi-species seabird colony. *Biological Conservation*, **90**, 175–181.

Bakker, J.P. & Berendse, F. (1999). Constraints in the restoration of ecological diversity in grassland and heathland communities. *Trends in Ecology and Evolution*, **14**, 63–68.

Barclay, J.H. (1987). Augmenting wild populations. In *Raptor Management Techniques Manual*, eds. B.A. Pendleton, B.A. Millsap, K.W. Cline & D.M. Bird, pp. 239–247. Washington, DC: National Wildlife Federation.

Beissinger S.R. & Westphal, M.I. (1998). On the use of demographic models of population viability in endangered species management. *Journal of Wildlife Management*, **62**, 821–841.

BirdLife International (2000). *Threatened Birds of the World*. Barcelona, Spain and Cambridge: Lynx Editions and BirdLife International.

Brawley, A.H., Warren, R.S. & Askins, R A. (1998). Bird use of restoration and reference marshes within the Barn Island Wildlife Management Area, Stonington, Connecticut, USA. *Environmental Management*, **22**, 625–633.

Brown, J.H. (1995). *Macroecology*. Chicago, IL: University of Chicago Press.

Brown, S.C. & Smith, C.R. (1998). Breeding season bird use of recently restored versus natural wetlands in New York. *Journal of Wildlife Management*, **62**, 1480–1491.

Bruner, A.G., Gullison, R.E., Rice, R.E. & Fonseca, G.A.B. (2001). Effectiveness of parks in protecting tropical biodiversity. *Science*, **291**, 125–128.

Burnham, W. (1983). Artificial incubation of falcon eggs. *Journal of Wildlife Management*, **47**, 158–168.

Cade, T.J. (1986). Reintroduction as a method of conservation. *Raptor Research Report*, **5**, 72–84.

Cade, T.J. & Hardaswick, V.J. (1985). Summary of peregrine falcon production and reintroduction by the Peregrine Fund in the United States, 1973–1984. *Aviculture Magazine*, **91**, 79–92.

Cade, T.J. & Jones, C.G. (1993). Progress in restoration of the Mauritius kestrel. *Conservation Biology*, **7**, 169–175.

Cairns, J., Jr (ed.) (1995). *Rehabilitating Damaged Ecosystems*. London: Belhaven Press.

Carson, W.P. & Root, R.B. (1999). Top–down effects of insect herbivores during early succession: influence on biomass and plant dominance. *Oecologia*, **121**, 260–272.

Chapman, C.A. & Chapman, L.J. (1999). Forest restoration in abandoned agricultural land: a case study from East Africa. *Conservation Biology*, **13**, 1301–1311.

Collar, N.J., Gonzaga, L.P., Krabbe, N., Madroño Nieto, A., Naranjo, L.G., Parker, T.A. III & Wege, D.C. (1992). *Threatened Birds of the Americas: The ICBP/IUCN Red Data Book*. Cambridge: International Council for Bird Preservation.

Cooper, J.E. (1977). Veterinary problems of captive breeding and possible reintroduction of birds of prey. *International Zoo Yearbook*, **17**, 32–38.

Cunningham, A.A. (1996). Disease risks of wildlife translocations. *Conservation Biology*, **10**, 349–353.

da Silva, J.M.C., Uhl, C. & Murray, G. (1996). Plant succession, landscape management, and the ecology of frugivorous birds in abandoned Amazonian pastures. *Conservation Biology*, **10**, 491–503.

Debussche, M.J., Escarre, J. & Lepart, J. (1982). Ornithochory and plant succession in Mediterranean abandoned orchards. *Vegetatio*, **48**, 255–266.

Desrochers, A., Rochefort, L. & Savard, J.P.L. (1998). Avian recolonization of eastern Canadian bogs after peat mining. *Canadian Journal of Zoology*, **76**, 989–997.

Dias, B.S.F., Raw, A. & Imperatriz-Fonseca, V.L. (2000). International pollinators initiative: the São Paulo declaration on pollinators. http://www.bdt.org.br/polinizadores/declaration/

Dobkin, D.S., Rich, A.C. & Pyle, W.H. (1998). Habitat and avifaunal recovery from livestock grazing in a riparian meadow system of the northwestern Great Basin. *Conservation Biology*, **12**, 209–221.

Drewien, R.C. & Bizeau, E.G. (1977). Cross-fostering whooping cranes to sandhill crane foster parents. In *Endangered Birds: Management Techniques for Preserving Threatened Species*, ed. S.A. Temple, pp. 201–222. Madison, WI: University of Wisconsin Press.

Duncan, R.S. & Chapman, C.A. (1999). Seed dispersal and potential forest succession in abandoned agriculture in tropical Africa. *Ecological Applications*, **9**, 998–1008.

Engelhardt, K.A. M., Kadlec, J.A., Roy, V.L. & Powell, J.A. (2000). Evaluation of translocation criteria: case study with trumpeter swans (*Cygnus buccinator*). *Biological Conservation*, **94**, 173–181.

Fischer, J. & Lindenmayer, D.B. (2000). An assessment of the published results of animal relocations. *Biological Conservation*, **96**, 1–11.

Franzreb, K.E. (1997). Success of intensive management of a critically imperiled population of red-cockaded woodpeckers in South Carolina. *Journal of Field Ornithology*, **68**, 458–470.

Furness, R.W. & Greenwood, J.J.D. (eds.) (1993). *Birds as Monitors of Environmental Changes*. London: Chapman & Hall.

Fyfe, R.W. (1977). Reintroducing endangered birds to the wild: a review. In *Endangered Birds: Management Techniques for Preserving Threatened Species*, ed. S.A. Temple, pp. 323–329. Madison, WI: University of Wisconsin Press.

Gill, F.B. (1994). *Ornithology*, 2nd edn. New York: W.H. Freeman.

Gottsberger, G. (1989). Floral Ecology: report on the years 1985 (1984) to 1988. *Progress in Botany*, **50**, 352–379.

Gottsberger, G. (1993). Floral Ecology: report on the years 1988 (1987) to 1991 (1992). *Progress in Botany*, **54**, 461–504.

Gottsberger, G. (1996). Floral Ecology: report on the years 1992 (1991) to 1994 (1995). *Progress in Botany*, **57**, 368–415.

Grant, P.R. (1986). *Ecology and Evolution of Darwins's Finches*. Princeton, NJ: Princeton University Press.

Griffith, B., Scott, J.M., Carpenter, J.W. & Reed, C. (1989). Translocation as a species conservation tool: status and strategy. *Science*, **248**, 212–215.

Guevara, S., Purata, S.E. & van der Maarel, E. (1986). The role of remnant forest trees in tropical forest succession. *Vegetatio*, **66**, 77–84.

Gunnarsson, B. & Hake, M. (1999). Bird predation affects canopy-living arthropods in city parks. *Canadian Journal of Zoology*, **77**, 1419–1428.

Heyl, C.W., Bigalke, R.C. & Pepler, D. (1988). Captive rearing of the Cape francolin and prospects for stocking. *South African Journal of Wildlife Research*, **18**, 22–29.

Howe, H.F. & Brown, J.S. (1999). Effects of birds and rodents on synthetic tallgrass communities. *Ecology*, **80**, 1776–1771.

IUCN (1995). *IUCN/SSC Guidelines for Reintroductions*. http://iucn.org/themes/ssc/pubs/policy/reinte.htm.

Janse, J.H., van Donk, E. & Aldenberg, T. (1998). A model study on the stability of the macrophyte-dominated state as affected by biological factors. *Water Research*, **32**, 2696–2706.

Jennersten, O. (1988). Pollination in *Dianthus deltoides* (Caryophillaceae): effects of habitat fragmentation on visitation and seed set. *Conservation Biology*, **2**, 359–366.

Jones, C.G. & Swinnerton, K.J. (1997). A summary of the conservation status and research for the Mauritius kestrel *Falco punctatus*, pink pigeon *Columba mayeri* and echo parakeet *Psittacula eqques*. *Dodo*, **33**, 72–75.

King, W.B. (1981). *Endangered birds of the World: the ICBP Red Data Book*. Washington, DC: Smithsonian Institution Press.

Kleiman, D.G., Stanley-Price, M.R. & Beck, B.B. (1994). Criteria for reintroductions. In *Creative Conservation: Interactive Management of Wild and Captive Animals*, eds. P.J.S. Olney, G.M. Mace & A.T.C. Feistner, pp. 287–303. London: Chapman & Hall.

Kress, S.W. (1983). The use of decoys, sound recordings, and gull control for re-establishing a tern colony in Maine. *Colonial Waterbirds*, **6**, 185–196.

Kus, B.E. (1998). Use of restored riparian habitat by the endangered least Bell's vireo (*Vireo bellii pusillus*). *Restoration Ecology*, **6**, 75–82.

Kuyt, W.B. (1981). Nest fidelity, productivity and breeding habitat of whooping cranes. In *Crane Research around the World*, eds. J.C. Lewis & H. Masotomi, pp. 119–125. Baraboo, WI: International Crane Foundation.

Lack, D. (1966). *Population Studies of Birds*. Oxford: Clarendon Press.

Lawton, J. (1995) Population dynamic principles. In *Extinction Rates*, eds. J.H. Lawton & R.M. May, pp. 147–163. Oxford: Oxford University Press.

Luken, J.O. (1990). *Directing Ecological Succession*. London: Chapman & Hall.

MacMahon, J.A. (1997). Ecological restoration. In *Principles of Conservation Biology*, eds. G.K. Meffe & C.R. Carroll, pp. 479–511. Sunderland, MA: Sinauer Associates.

Margules, C.R. & Pressey, R.L. (2000). Systematic conservation planning. *Nature*, **405**, 243–253.

McClanahan, T.R. & Wolfe, R.W. (1993). Accelerating forest succession in a fragmented landscape: the role of birds and perches. *Conservation Biology*, **7**, 279–288.

McDonnell, M.J. & Stiles, E.W. (1983). The structural complexity of old-field vegetation and the recruitment of bird-dispersed plant species. *Oecologia*, **56**, 109–116.

McLennan, J.A., Potter, M.A., Robertson, H.A., Wake, G.C., Colbourne, R., Dew, L., Joyce, L., McCann, A.J., Miles, J., Miller, P.J. & Reid, J. (1996). Role of predation in the decline of kiwis, *Apteryx* spp., in New Zealand. *New Zealand Journal of Ecology*, **20**, 27–35.

Meine, C.D. & Archibald, G.W. (eds.) (1996). *The Cranes: Status Survey and Conservation Action Plan*. Gland, Switzerland: IUCN.

Moermond, T.C. & Denslow, J. (1985). Neotropical avian frugivores: patterns of behavior, morphology, and nutrition, with consequences for fruit selection. *Ornithological Monographs*, **36**, 865–897.

Momose, K., Yumoto, T., Nagamitsu, T., Kato, M., Nagamasu, H., Sakai, S., Harrison, R.D., Itioka, T., Hamid, A.A. & Inoue, T. (1998). Pollination biology in a lowland dipterocarp forest in Sarawak, Malaysia. 1: Characteristics of the plant–pollinator community in a lowland dipterocarp forest. *American Journal of Botany*, **85**, 1477–1501.

Murakami, M. & Nakano, S. (2000). Species-specific bird functions in a forest-canopy food web. *Proceedings of the Royal Society of London, Series N, Biological Sciences*, **267**, 1597–1601.

Myers, N., Mittermeier, R.A., Mittermeier, C.G., Fonseca, G.A.B. & Kent, J. (2000). Biodiversity hotspots for conservation priorities. *Nature*, **403**, 853–858.

Newton, I. (1980). The role of food limiting bird numbers. *Ardea*, **68**, 11–30.

Nowierski, R.M., Huffaker, C.B., Dahlsten, D.L., Letorneau, D.K., Janzen, D.H. & Kennedy, G.G. (1999). The influence of insects on plant populations and communities. In *Ecological Entomology*, eds. C.B. Huffaker & A.P. Gutierrez, pp. 585–642. New York: John Wiley.

Owens, I.P.F. & Bennett, P.M. (2000). Ecological basis of extinction risk in birds: habitat loss versus human persecution and introduced predators. *Proceedings of the National Academy of Sciences, USA*, **97**, 12144–12148.

Passell, H.D. (2000). Recovery of bird species in minimally restored Indonesian tin strip mines. *Restoration Ecology*, **8**, 112–118.

Pereira, S.L. & Wajntal, A. (1999). Reintroduction of guans of the genus *Penelope* (Cracidae: Aves) in reforested areas in Brazil: assessment by DNA fingerprint. *Biological Conservation*, **87**, 31–38.

Perrow, M.R., Schutten, J.H., Howes, J.R., Holzer, T., Madgwick, F.J. & Jowitt, A.J.D. (1997). Interactions between coot (*Fulica atra*) and submerged macrophytes: the role of birds in the restoration process. *Hydrobiologia*, **342**, 241–255.

Priddel, D., Carlile, N. & Wheeler, R. (2000). Eradication of European rabbits (*Oryctolagus cuniculus*) from Cabbage Tree Island, NSW, Australia, to protect the breeding habitat of Gould's petrel (*Pterodroma leucoptera leucoptera*). *Biological Conservation*, **94**, 115–125.

Reading, R.P., Clark, T.W. & Griffith, B. (1997). The influence of valuational and organizational considerations on the success of rare species translocations. *Biological Conservation*, **79**, 217–225.

Ricklefs, R.E. (1983). Comparative avian demography. *Current Ornithology*, **1**, 1–32.

Robbins, C.S., Dawson, D.K. & Dowell, B.A. (1989). Habitat area requirements of breeding forest birds of the middle Atlantic states. *Wildlife Monographs*, **103**, 1–34.

Robinson, G.R. & Handel, S.N. (1993). Forest restoration on a closed landfill: rapid addition of new species by bird dispersal. *Conservation Biology*, **7**, 271–278.

Safford, R.J. & Jones, C.G. (1998). Strategies for land-bird conservation on Mauritius. *Conservation Biology*, **12**, 169–176.

Sanz, V. & Grajal, A. (1998). Successful reintroduction of captive-raised yellow-shouldered Amazon parrots on Margarita Island, Venezuela. *Conservation Biology*, **12**, 430–441.

Schowalter, T.D. (2000). *Insect Ecology: An Ecosystem Approach*. San Diego, CA: Academic Press.

Seddon, P.J. (1999). Persistence without intervention: assessing success in wildlife reintroductions. *Trends in Ecology and Evolution*, **14**, 503.

Seixas, G.H.F. & Mourão, G.M. (2000). Assessment of restocking blue-fronted Amazon (*Amazona aestiva*) in the Pantanal of Brazil. *Ararajuba*, **8**, 73–78.

Sipura, M. (1999). Tritrophic interactions: willows, herbivorous insects and insectivorous birds. *Oecologia*, **121**, 537–545.

Snyder, N.R.F. (1978). Puerto Rico parrots and nest-site scarcity. In *Endangered Birds: Management Techniques for Preserving Threatened Species*, ed. S.A. Temple, pp. 47–54. Madison, WI: University of Wisconsin Press.

Snyder, N.F.R. & Snyder, H.A. (1989). Biology and conservation of the Californian condor. *Current Ornithology*, **6**, 175–267.

Soulé, M.E. & Terborgh, J. (1999). *Continental Conservation: Scientific Foundations of Regional Reserve Networks*. Washington, DC: Island Press.

Strong, A.M., Sherry, T.W. & Holmes, R.T. (2000). Bird predation on herbivorous insects: indirect effects on sugar maple saplings. *Oecologia*, **125**, 370–379.

Temple, S.A. (ed.) (1978). *Endangered Birds: Management Techniques for Preserving Threatened Species*. Madison, WI: University of Wisconsin Press.

Temple, S.A. (1986). The problem of avian extinctions. *Current Ornithology*, **3**, 453–485.

Toh, I., Gillespie, M. & Lamb, D. (1999). The role of isolated trees in facilitating tree seedling recruitment at a degraded sub-tropical rainforest site. *Restoration Ecology*, **7**, 288–297.

van Donk, E. & Otte, A. (1996). Effects of grazing by fish and waterfowl on the biomass and species composition of submerged macrophytes. *Hydrobiology*, **340**, 285–290.

Vickery, P.D., Hunter, M.L., Jr & Melvin, S.M. (1994). Effects of area requirements on the distribution of grassland birds in Maine. *Conservation Biology*, **8**, 1087–1097.

Vieira, I.C., Uhl, C. & Nepstad, D. (1994). The role of the shrub *Cordia multispicata* Cham. as a 'succession facilitator' in an abandoned pasture, Paragominas, Amazonia. *Vegetatio*, **115**, 91–99.

Warner, R.E. (1968). The role of introduced diseases in the extinction of the endemic Hawaiian avifauna. *Condor*, **70**, 101–120.

Welty, J.C. & Baptista, L. (1988). *The Life of Birds*. Orlando; FL: Saunders College Publishing and Harcourt Brace Jovanivich College Publishers.

Whittaker, R.J. & Jones, S.H. (1994). The role of frugivorous bats and birds in the rebuilding of a tropical forest ecosystem, Krakatau, Indonesia. *Journal of Biogeography*, **21**, 245–258.

Willson, M.F. (1986). Avian frugivory and seed dispersal in eastern North America. *Current Ornithology*, **3**, 223–279.

Willson, M.F. & Crome, F.H.J. (1989). Patterns of seed rain at the edge of a tropical Queensland rain forest. *Journal of Tropical Ecology*, **5**, 301–308.

Winkler, D.W. & Walters, J.R. (1983). The determination of clutch size in precocial birds. *Current Ornithology*, **1**, 33–68.

Wolf, C.M., Garland, T., Jr & Griffith, B. (1998). Predictors of avian and mammalian translocation success: reanalysis with phylogenetically independent contrasts. *Biological Conservation*, **86**, 243–255.

Wunderle, J.M., Jr (1997). The role of animal seed dispersal in accelerating native forest regeneration on degraded tropical lands. *Forestry Ecology and Management*, **99**, 223–235.

20 • Mammals

DAVID W. MACDONALD, TOM P. MOORHOUSE, JODY W. ENCK AND FRAN H. TATTERSALL

INTRODUCTION

Some 4600 mammal species are recognised world-wide, encompassing an extraordinarily diverse range of life history, morphology and ecological roles. For example, just over 900 species are bats (Chiroptera), making them the second largest order of mammals after the rodents, while around 85 species are in the entirely marine order, Cetacea. Mammals range from Savi's pigmy shrew (*Suncus etruscus*) which can weigh less than 3 g and live only a few months, to immense blue whales (*Balaenoptera musculus*) weighing around 100 tonnes, which some evidence suggests may live two centuries or more.

According to the estimates of the International Union for the Conservation of Nature (IUCN – The World Conservation Union), compiled in collaboration with the Committee on Recently Extinct Organisms, 83 species of mammals have become 'Extinct' in the last 500 years while at least another four are considered 'Extinct in the Wild', the most famous example being Przewalski's horse (*Equus przewalskii*). The 2000 IUCN *Red List of Threatened Species* lists some 1130 mammal species classed in the top three threat categories, including 180 species classed as 'Critically Endangered', 340 as 'Endangered' and 610 as 'Vulnerable'. In addition, 74 species are considered 'Lower Risk/Conservation Dependent', and another 602 species as 'Lower Risk/Near Threatened'. Altogether, more than a third of known mammal species are represented on the IUCN list, and at least a quarter face extinction in the near future, either directly or indirectly as a result of human activities. Restoration of mammal populations, species and communities, is therefore becoming an increasingly important element of mammal conservation.

The most universal threats to mammals are habitat loss and degradation, hunting and trade and, in some parts of the world, introduced exotics (Macdonald *et al.*, in press *a*). Restoring mammals, and the ecosystems within which they live, clearly requires us to address these problems, and as such becomes almost indistinguishable from conservation (see Macdonald *et al.*, chapter 4, this volume). Amongst the tools in the restoration ecologist's armoury are the legal and policy framework within which mammal conservation operates, including national and international legislation and agreements. Perhaps the most overarching piece of international legislation is the Convention on Biological Diversity (see Convention on Biological Diversity, 2001), now ratified by 176 countries. By requiring rehabilitation and restoration of degraded habitats, and promotion of recovery of protected species, the Convention specifically recognises the need for species and ecosystem restoration. By requiring, amongst other things, the development of protected areas, strategies for conservation and control of alien species, it provides some means of achieving this.

Strategies for conflict resolution are also an essential tool for mammal restoration (see Macdonald *et al.*, chapter 4, this volume). Wild mammals can damage a range of human interests such as livestock production, fisheries or food storage. In many cases, the extent and significance of this impact is hotly disputed, as the data are extremely difficult to obtain, and are confounded by current culling and other control measures, as well as the species' abundance (Macdonald *et al.*, 2000*a*). Monitoring and research are further vital elements of the restoration process. Monitoring is necessary both to detect and assess the extent of habitat degradation or species decline, and to evaluate the long-term success of restoration techniques. Research is frequently neglected, but is essential, first to understand the processes involved in degradation and decline, and second to devise means of reversing these processes.

Restoring mammal populations, species or communities involves a myriad of field techniques, including predator control, protective legislation, fencing, habitat management and restoration, translocation and reintroduction. In some cases, restoration of a natural mammal community is an essential component of wider habitat and species restoration, and vice versa. In this chapter, we focus on mammal restoration involving reintroduction techniques, using examples and case studies primarily from Britain and Europe. In most cases, restoration through reintroduction will also involve a range of other conservation measures.

Restoration through reintroduction

Reintroduction involves releasing a species into an area from which it has become extinct (IUCN, 1998). Logically, this might seem distinct from bolstering or supplementing the numbers of a species that is less abundant than desired, but the distinction is not always clear-cut, partly because of questions of scale, and partly because of common use of 'reintroduction' as an umbrella term that includes restocking and supplementing existing populations, and translocation (movement of animals from one area to another).

The restoration of British otter (*Lutra lutra*) populations is one example of a restocking and supplementation programme. Otter populations declined over much of Britain (and Western Europe) during the 1950s and 1960s, probably as a result of organochlorine pesticide poisoning (Strachan & Jefferies, 1996; Green & Green, 1997). Once widespread, by 1975, the otter was only found in northern Scotland and the west of England. The otter's recovery since then has been widely aided by restocking schemes (Strachan & Jefferies, 1996), and around 120 animals have been released since 1983. Objectives of the otter releases included repopulating areas where otters were extinct; strengthening existing, remnant, populations, which were believed otherwise to be doomed; testing whether otters were absent because of poor water quality or lack of colonists; and aiding natural colonisation by 'seeding' females in the path of dispersing males. The extent to which these individual aims were successful is debatable, particularly as there was rather poor monitoring at the local level. On a national level however, otters have re-established in many areas of the UK.

The first of our two detailed case studies, the reintroduction of the European beaver (*Castor fiber*) to Britain (Box 20.1), illustrates both a conventional

Box 20.1 Restoring beavers to Britain: exploring the planning process

The European beaver (*Castor fiber*) is Europes's largest and most distinctive rodent, with a unique aquatic lifestyle. The huge impact they can have on the landscape via their various building activities means they are also highly controversial. Beavers once occurred in wooded riparian or wetland areas throughout Europe, but were widely exterminated by over-hunting. In Britain, they were present in England, Wales and Scotland, but finally died out, in Scotland, in the sixteenth century. Over the last 20 years there has been regular discussion about the desirability and feasibility of reintroducing the European beaver to Britain (e.g. Yalden, 1986; Macdonald *et al.*, 1995).

Beavers are a 'keystone' species. Their building and feeding activities can greatly modify their wetland environments (e.g. Naiman & Melillo, 1984; Naiman *et al.*, 1986, 1988; Nummi, 1989; Cirmo & Driscoll, 1993). They can influence stream channel and hydrology, decrease current and increase retention of sediment and organic matter, in turn altering nutrient cycling and decomposition rates. This alters invertebrate community structure and biomass, with knock-on benefits for birds, fish and amphibians. By felling large trees, beavers can also substantial affect the structure and productivity of the riparian zone, and beavers' food preferences and differences in species' water tolerances lead to changes in riparian zone composition. Successional changes when beaver ponds are abandoned involve a complex pattern of wet or seasonally flooded meadow, open water, marsh, bog and flooded forest formation.

There are sound conservation and legal reasons for taking seriously a proposal to reintroduce beavers to the UK. The European beaver is still classified as 'Vulnerable' by the IUCN/SSC Rodent Specialist Group (Amori & Zima, 1993), and a reintroduction to Britain would further expand its range and help ensure the species' long-term survival. Reintroduction, especially of a keystone species such as the beaver (Macdonald *et al.*, 1995), also has ecological advantages on ecosystem functioning, recreating conditions and opportunities lost to other species.

As a member of the European Community, Britain has a responsibility to investigate the feasibility of reintroducing 'Annex IV' species such as the beaver (Article 22, EC Habitats and Species Directive, EC 92/43); these are native species extirpated through human activities in historical times, unable to naturally recolonise their former range. Under Article 11 (2) of the Convention of European Wildlife and Natural Habitats, the UK Government is bound to encourage reintroductions of native species where this would contribute to their conservation. Finally, The UK Biodiversity Action Plan (HMSO, 1994) written in response to the 1992 Rio Convention on Biodiversity, of which the UK is a signatory, aims to conserve and enhance biological diversity in the UK. Where practicable, it aims to enhance populations and natural ranges of native species.

Since the 1920s, European beavers have been successfully reintroduced in many countries in Europe and the former Soviet Union. Macdonald *et al.* (1995) estimated that 53% of 87 reintroductions for which population estimates were available after at least five years could be considered successful, in that numbers have increased through reproduction. Most failed schemes could clearly be attributed to release into unsuitable sites and inadequate planning. Indeed, many reintroductions, such as those in France and Switzerland, were notable for their lack of monitoring, strategic planning and standardised procedures (Stocker, 1985; Migot & Rouland, 1989). Hitherto, only the Netherlands reintroduction programme had had a well-structured approach, including a four-year public consultation and fact-finding period, a five-year experimental phase with an extensive associated research programme, and an assessment of population viability and overall project success (Nolet, 1994; Nolet & Baveco, 1996).

THE UK PROCESS

In Scotland

Following IUCN guidelines, Scottish Natural Heritage (the statutory body for conservation in Scotland) initially identified three stages in its consideration of beaver reintroduction: (1) to investigate the ecological feasibility of reintroducing the beaver; (2) to investigate its desirability; and (3) to establish a reintroduction protocol and programme (Cooper, 1996).

The pre-project stage of the potential reintroduction of the beaver to Britain began with Macdonald *et al.*'s (1995) review of its feasibility and desirability, and their conclusion that the idea merited further attention. This was followed by Scottish Natural Heritage's decision to investigate the possibility of reintroducing the beaver to Scotland, their commissioning of further research, investigating the history of the beaver in Scotland (Conroy & Kitchener, 1996), potential costs of a reintroduction (Tattersall & Macdonald, 1995), current habitat availability (Webb *et al.*, 1997), a method of assessing site suitability (Macdonald *et al.*, 1997, 2000*b*), interactions with fisheries (Collen, 1997) and impact on hydrology (Gurnell, 1997). The desirability of reintroducing beaver to Scotland was assessed by Scottish Natural Heritage through a broad-ranging public consultation, as required under Article 22(a) of the Habitats Directive (Scottish Natural Heritage, 1998). They concluded that it is feasible to reintroduce beaver, and that the majority of the public were in favour, although certain interest groups raised objections.

Following proposals from a Steering Committee, in 2000 the Scottish Natural Heritage board agreed to proceed with a trial release of twelve animals to see how beavers operate under current land management practices in Scotland (Scottish Natural Heritage, 2000). The foot-and-mouth crisis in Britain during 2001 prevented any further progress to date, but the project is planned to run over seven years, including two years of preparation and local consultation. One important hurdle remains: application to Scottish Executive Rural Affairs Department, for a licence to release beavers. During the five years the beavers are in the field, one full-time project officer plus some contract workers will carry out monitoring and research.

In England

In England, plans to restore beavers are also gathering momentum, but taking a different approach. Many of the beavers' impacts, such as raising the water table, increasing dead wood, reducing wooded cover and creating habitat mosaics, are attractive to conservationists managing wetland areas in Britain. Indeed, much practical conservation work, using human effort or livestock, is aimed to this effect. Kent Wildlife Trust has secured permission from English Nature and the Environment Agency to release beavers into the Ham Fen area in southeast England, to test the feasibility of using beaver as a wetland management tool. The hope is that as the beaver population grows, they will dam the principal waterway, causing the water level in the fen to rise, and allowing a range of semi-aquatic and aquatic plants to recolonise the surrounding grassland via both new beaver channels and existing, though choked, drainage ditches. The influence of beaver activity on both the large standing timber and coppice woodland in the core fen area will also be of importance to species diversity.

Legislation

Legally, any proposed reintroduction of beavers to Britain would have to be approved and licensed by the Department of the Environment, Food and Rural Affairs, because domestic legislation makes it illegal to release into the wild any animal which is not normally resident (Wildlife and Countryside Act 1981). Implicit in this would be approval by conservation agencies such as English Nature, Scottish Natural Heritage, Countryside Council for Wales and the Joint Nature Conservation Committee.

The lack of legal protection for beavers, as they are not normally resident in the UK, would need to be rectified under domestic legislation. On an international level, the European Union Habitats Directive requires 'proper consultation with the public concerned' prior to any reintroductions. Questions of acceptability and economic damage are of paramount importance.

THE HUMAN DIMENSION

Beavers reintroduced to Britain have the potential to cause damage to forestry, crops, agricultural land and river banks through their feeding, burrowing and damming. In Europe, beaver damage is usually localised, with minimal financial impact. Poplars and fruit trees are generally the most valuable trees felled by beavers, and softwoods are very rarely harmed. Where damage is unacceptable, dams can be destroyed or regulated with overflow pipes, and trees can be protected with heavy-gauge wire mesh. If damage spirals out of control, beavers are readily shot or trapped. Conversely, beaver may provide benefits such as flood control, water conservation and renewal, and stabilisation of groundwater levels. The human dimension of each of these had to be explored. For example in Scotland, anglers have been amongst the most vocal opponents of beaver reintroduction. Collen (1997) consulted all 62 Scottish District Salmon Fishery Boards and 60 angling clubs in Scotland on the proposal to reintroduce beaver. Of the 59 organisations which responded, 70% were against the reintroduction and 19% felt they did not have enough information to comment.

Many anglers believe that beaver dams will impede migrating salmonids, and that an increase in coarse woody debris resulting from beaver activity would be detrimental to fishing interests; some also expressed concern that beaver would damage riparian woodland planting schemes designed to enhance food resources for salmonids. Collen (1997) concluded that fisheries authorities would be unlikely to support the beaver's restoration unless the population was actively managed.

The likelihood of beavers damaging the Scottish fishing industry is probably remote. Reports of conflicts of interest come only from the Canadian beaver (*Castor canadensis*) in North America. In Europe, reintroduced beaver coexist with Atlantic salmon (*Salmo salar*), trout (*Salmo* spp.), pike (*Esox lucius*) and eels (*Anguilla anguilla*), and are generally thought to have a beneficial impact on fish stocks, by increasing habitat and invertebrate food (Collen, 1997). However, localised problems may occur when trout spawning areas are damaged by siltation, or, in hot weather, when the water in a beaver pond heats up sufficiently to deplete oxygen levels.

The consensus was that beaver are unlikely to cause serious damage in Scotland, and where they do cause damage they can probably be controlled and their impact minimised (Macdonald & Tattersall, 1999).

EVALUATING WHERE TO RELEASE BEAVERS

The British landscape of today is radically different from that of 400–1000 years ago, so an assessment of habitat availability is vital. Beavers are aquatic and require year-round access to water fringed with herbaceous and woody vegetation. The habitat size required to maintain a viable population will, clearly, depend on carrying capacity and estimated minimum viable population size (MVP). There is no consensus in the literature about either. Estimates of the total length of river needed to maintain a viable population vary from 6 km (Stocker, 1985) to 50 km (Heidecke, 1986), while in marshes the estimated minimum area required by one family is 10 ha (Ouderaa *et al.*, 1985).

Macdonald *et al.*'s (2000*b*) approach to identifying suitable release sites comprised four phases, combining geographic information system (GIS)-based analyses of habitat data with field survey and population modelling. Each phase identifies a list of potential sites which is progressively reduced in subsequent phases as more fine-scale assessments of suitability for release are considered. The four phases are as follows:

Phase 1: The national-scale identification of potential sites for releasing beavers and dispersal routes in Scotland, using GIS information

Phase 1 qualitative assessment suggested that at least three Scottish river systems might be suitable: the Spey river system north of Newtonmore and south of Grantown; Strath Dee from Braemar to Banchory; and the Tay river system including the main Tummel and Garry tributaries, north of Perth and especially north of Dunkeld. These three areas are heavily wooded, drain the Cairngorm mountains, and populations introduced to them would probably be in genetic contact with one another.

Studies by Webb *et al.* (1997) and Macdonald *et al.* (1997) provided more formal assessments. Macdonald *et al.* (1997) used a GIS to identify habitat suitable for beaver home ranges and habitat through which animals could disperse. They assumed that the minimum area capable of supporting beavers was either a contiguous block of woodland along more than 3.1 km of river bank, and/or blocks of marshland of greater than 10 ha. Dispersal habitats were assumed to be non-montane land through which a river corridor or stream system ran. Since the land surface of Scotland is extensively permeated by water courses the areas of non-dispersal, non-home-range habitat were confined to mountainous areas. Only 94 sites were identified as suitable and only three of these (Loch Insh, Glen Meinich and Glen Esk) had areas of contiguous habitat that could potentially support 10 or more families of beaver. Sites were widely separated and few large sites were within close proximity of each other.

Webb *et al.*'s (1997) analysis suggested that suitable areas were concentrated in a central block around the major river catchments of the Dee, Spey, Ness, Tay and Lomond. They estimated that, even without habitat restoration, the present landscape of Scotland could support 200–1000 beavers. These two desk studies indicated that the landscape in the region is well connected, offering considerable potential for beaver dispersal.

Phase 2: The design and field testing of a survey protocol and scoring system for site selection or rejection, using sites identified in Phase 1

In Phase 2 Macdonald *et al.* (2000*b*) selected a subsample of nine sites identified in Phase 1, and assessed habitat during a single, brief field survey. The physical parameters of the watercourse and its bank, availability of bank-side vegetation, human disturbance and adjacent land use were measured and graded into categories according to their impact on released beavers (Table B20.1). None of the sites emerged as ideal for beavers: of a potential score of 30, the highest score was 17.4.

Table B20.1. Criteria for scoring potential sites for reintroduction of beavers

(a)		Score X	Score 0	Score 1	Score 2
Water	Mean depth (m) across watercourse	<0.5 (still water only)	<1, >6	1–2, 4–6	2–4
	Width (m)	<0.075	<2, >300	2–10, 100–300	10–100
	Flow speed ($m\,s^{-1}$)	>1	0.6–1	0.3–0.6	0.3–0
	Fluctuation level (m)	>3 rapid	2–3	1.5–2	<1.5
	Pollution		Inorganic/high	Organic/low	None
Bank	Substrate	Rock/gravel	Sand/rock	Podsol	Peat/loam
	Angle (°)	>90	80–90	60–80	<60
	Height (m)		<0.5	0.5–1.5	>1.5
Vegetation	Emergent/in channel cover (%)		<40	40–60	>60
	Herbaceous cover within 100 m (%)	0–20	20–40	40–60	>60
	Shrub/hardwood cover within 20 m (%)	0–20	20–40	40–60	>60
	Shrub/hardwood cover 20–100 m (%)		<40	40–60	>60
	Average hardwood diameter (cm)		>20	10–20	<10
Human	Disturbance/mortality	Certain	High (>8)	Medium (4–8)	Low (<4)
	Potential for damage	Certain	High (4–8)	Medium (1–4)	Low (0)

(a) At each candidate release site 5–11 point sampling positions (selected at roughly 1 km intervals) were assessed for factors relating to the water, bank, vegetation and human impact. At each sampling point, each factor was given a score of X, 0, 1 or 2, and the scores were totalled. Where a criterion was ranked X, the score for that sampling point was 0, regardless of the quality of other factors. The mean total for all sampling points provided the overall score for each site.

Phase 3: Detailed site surveys and habitat mapping of the best sites identified in Phase 2 to provide a subjective assessment of the receiving potential of those sites for a release group of beavers

In Phase 3, for the three most promising sites carrying capacity was estimated assuming a minimum 3.1 km of suitable bank-side habitat per beaver colony of two adults and two to three subadults (Nolet & Baveco, 1996). To calculate the amount of suitable habitat available, the Environment Agency's River Corridor Survey methodology (a detailed habitat survey protocol), supplemented by existing data, was used to create an accurate vegetation map for land within 100 m of the water – the maximum distance beavers are thought to travel regularly from the bank in search of food (Stocker, 1985; Heidecke, 1989).

Phase 4: The development of a simulation model to explore the potential spread of an introduced beaver population from potential release sites identified in Phase 3

Next, we ran an individual-based population dynamics model which simulated life histories of individual beavers and their dispersal within the GIS-held landscape to predict how beavers released into three sites would survive and disperse. The model assumed a metapopulation structure, and was run under two dispersal and two carrying-capacity scenarios for each site, estimating the mean number of subpopulations, and the mean total metapopulation size in each year, following release of adults.

The mean number of subpopulations developing under each scenario at each site was in all cases

Table B20.1. Continues

(b)	Certain (Score X)	Score 0	Score 1	Score 2
Disturbance/ mortality	Road (B or larger) <100 m	Powered boats	Walkers/cyclists	Protected area
	Railway line <100 m	Commercial fishing	Anglers	No or restricted public access
	Intensive net/trap fishing <3 km	Hunting with hounds	Rowing boat	
		Shooting	Keepering	
		Built-up area <1 km	No formal protection	
		Development possible	Built-up area 1–5 km	
Potential for damage	Hardwood plantation <100 m and little alternative food	Farmed land <500 m	500 m–3 km	>3 km
	Road (B or larger) <100 m	Road/railway <500 m	500 m–3 km	>3 km
	Railway line <100 m	Plantations <500 m	500 m–3 km	>3 km
	Any intensively used land <50 m	Buildings <500 m	500 m–3 km	>3 km

(b) Human impact was assessed using a points system based on either the presence or absence of particular features, and/or their proximity. The presence of features likely to cause disturbance/mortality to beavers, or to be damaged by them merited a score of 0 points, while those unlikely to disturb or be damaged merited 2 points; intermediate factors merited 1 point. A low total score indicates high potential for damage or disturbance, and therefore scores 0; a large point total indicates low potential for damage or disturbance and therefore scores 2. The presence of one or more features certain to cause beaver mortality, or to be damaged by beavers, scored an X, and the entire sampling point returned a 0 score.

relatively low; at no time did the number of subpopulations exceed ten. The most obvious feature was the low level of colonisation away from all of the release sites. At one site no metapopulations went extinct under any of the dispersal or carrying-capacity scenarios. At the other two sites, metapopulations went extinct in 3–30% of all runs. In this example, carrying capacity and dispersal distance were varied, but one could equally vary, for example, numbers of animals, sex and age ratios, or compare sequential versus one-off releases at a particular site.

These results are interpreted with due caution as, for example, different population viability analysis (PVA) programmes may yield different results (e.g. Macdonald *et al.*, 1998). To emphasise this caution, we have explored alternative modelling approaches to test the sensitivity of these results (South *et al.*, 2000).

What do the results of the modelling tell us about the likely future of the beaver following reintroduction? Carrying capacities for the three most promising sites were 6–8 colonies, 2–4 colonies and 2–3 colonies. The results suggest that spread of the beaver following reintroduction at the selected sites is likely to be slow and relatively small scale.

DECIDE ON THE RELEASE STRATEGY

Using available information on life history characteristics and population ecology of beavers, Macdonald *et al.* (1995) ran VORTEX v8 (Lacy, 1993) – a stochastic, individual-based simulation of demographic events (e.g. birth, death, mating), environmental and stochastic variation in the frequency of these events – simulations to assess the time to extinction for a beaver population. The model assumed: (1) no inbreeding depression; (2) monogamous, density-dependent reproduction from three to 12 years; (3) an even sex ratio; (4) a maximum litter size of six; (5) 70% of males bred; and (6) maximum breeding age was 12 years.

Beavers were assumed to have a cyclical effect on their habitat: habitat quality diminished by 2% per annum for two years and then recovered at 2% per annum for the next ten years (these figures were arbitrarily chosen for demonstration purposes only). A loss of heterozygosity of 1% per generation (Soulé, 1986), and the risk that the population might become extinct within 200 years were considered tolerable. For each assessment 100 simulations were performed over 100 years, with a reporting interval of ten years.

Three sets of simulations were performed. The first assumed an area with a mean ±S.D. carrying capacity of 400±80 beavers (or one family of four per kilometre) in which there was a single release of 5, 10, 20, 30 or 50 pairs. Releases involving 20–50 pairs had a greater than 80% chance of survival over 100 years, while a release of five pairs had only a 40% survival rate over 100 years. The second and third simulations assumed a mean ±S.D. carrying capacity of 200±40 on 50 km of river with and without supplementary reintroductions of two pairs every second year for the first 20 years. Where the reintroductions were supplemented, there was a large improvement in the chances of survival for all populations: except for a release of 30 pairs, all supplemented reintroductions had a greater than 80% chance of survival after 100 years, while in unsupplemented reintroductions, only releases of 50 pairs had greater than 80% survival chance, and releases of five pairs had a less than 30% survival chance.

These preliminary results, particularly the comparison between supplemented and unsupplemented releases, demonstrate the usefulness of simulation modelling, which should play an important part in formulating a reintroduction strategy for beavers in Britain. In particular, the effects of creating a metapopulation by reintroducing beavers at several different sites should be explored.

CONCLUSIONS

In view of past European successes, and the large body of literature and practical expertise that has accumulated, reintroduction of the beaver to Britain is certainly feasible. Is it desirable, however? Canadian beavers can certainly cause extensive damage, but the effects of European beavers are less well documented. Ultimately it will be too complex to predict accurately the amount of damage that beavers are likely to cause, though we can be sure they will cause some. Sufficient political and public goodwill must be present either to pay compensation or absorb the costs of this damage. However, with careful selection of release sites, taking into account likely migration within and between watercourses, it should be possible to reduce the likelihood of damage occurring.

These results all suggest that to secure the long-term viability of reintroduced beavers in Scotland habitat creation is likely to be required. In the meantime, it is likely that reintroduced beavers would have to be the subject of metapopulation management. However, this conclusion does not nullify the conservation desirability of attempting to reintroduce beavers. Indeed, the conservation of many species, including for example, many bird species in the UK, is largely dependent on sustained intervention, and around the world acceptance of the need for active metapopulation management is increasingly widespread.

Many steps along the reintroduction route must now be consolidated. Habitat improvement and creation schemes require thought, as for successful reintroductions of beavers some sites may need to be restocked with fast-growing willow species. Monitoring of beavers and their impacts is essential, as summarised in Macdonald *et al.* (1998). Reintroduction projects need to quantify socio-economic indicators including measures of environmental impact, nuisance, ecotourism, educational penetration and public perception. Biological indicators also require quantification including measures of animal breeding success, of longevity, and of provisions for fail-safe mechanisms to mitigate any negative impacts of the reintroduction programme.

approach to restoring extinct species – that is, returning an extirpated species to the wild – and a more integrated approach to whole ecosystem restoration. Our second case study (Box 20.2), the restoration of the European mink (*Mustela lutreola*), illustrates an integrated approach to conservation involving captive breeding, experimental releases in safe havens, and, crucially, control of exotic American mink prior to wider reintroductions. We begin, however, with an exploration of some of the various stages involved in a reintroduction programme.

Box 20.2 Removing the cause of decline: the case of European mink

The European mink (*Mustela lutreola*) inhabits small, undisturbed watercourses with rapid currents and lush riparian vegetation, in forested areas. Its prey include fish, amphibians, small mammals and invertebrates (Sidorovich *et al.*, 1998). The species' range once extended from the Ural Mountains to eastern Spain and from central Finland to the Black Sea, but since the mid nineteenth century its range has dwindled (Maran & Henttonen, 1995; Maran *et al.*, 1998*a*). The European mink disappeared from Germany in the mid nineteenth century (Youngman, 1982), then from Switzerland (Gautschi, 1983) and, in the 1890s, from Austria (Novikov, 1939; Novikov *et al.*, 1970). Between the 1930s and 1950s European mink disappeared from Poland, Hungary, the Czech and Slovak Republics and probably also Bulgaria (Bárta, 1956; Szunyoghy, 1974; Schreiber *et al.*, 1989; Romanowski, 1990). Subsequently, in western Europe, an isolated enclave of European mink persisted between Brittany in France and Galicia in Spain, but recently mink have disappeared from the northern part of this enclave and appear in widespread decline in their remaining French range (Moutou, 1994). In the closing years of the twentieth century, the species teetered on the brink of extinction throughout most of what remained of its range.

In the European Union the European mink is listed in Annex II (species whose conservation requires the designation of special areas) and in Annex IV (species of community interest in need of strict protection) of the Directive on the conservation of natural habitats and of wild fauna and flora. The IUCN Action Plan for the Conservation of Mustelids and Viverrids (Schreiber *et al.*, 1989) nominates the European mink as a priority.

Since 1994, Macdonald *et al.* (in press *b*) have been collaborating with colleagues in Eastern Europe to develop a stepwise, integrated approach, first to discovering the cause of the European mink's decline, and then to restore its status (reviewed in Macdonald *et al.*, in press *b*). In brief, these steps are summarised below.

FORMULATE HYPOTHESES

Three families of hypotheses explain the European mink's decline. These are: (1) habitat loss or degradation (including pollution); (2) over-hunting; and (3) interspecific competition. These are elaborated in Maran *et al.* (1998*a*).

In the context of interspecific competition, between 1933 and 1963 American mink (*Mustela vison*) were deliberately released to establish a harvestable wild population in many localities in the former Soviet Union (Heptner *et al.*, 1967). Indeed, the releases of some 20 400 American mink were documented up to 1971 in nearly 250 sites (Pavlov & Korsakova, 1973). Male American mink are substantially heavier than their European counterpart, and even the female American mink is on a par with the male European mink.

TEST THE HYPOTHESES

Sidorovich established a study area in the Lovat River basin in northeastern Belarus where, from 1986 onwards, he had censused European and American mink. Otters and polecats were censused in small rivers, small streams and glacial lakes. Between 1995 and 1998, Sidorovich, Macdonald and colleauges radio-tracked nine European mink, 40 American mink and six polecats. This study revealed that before the arrival of the American mink (1986–9), the European mink population appeared to be relatively stable at a high density (Sidorovich *et al.*, 1995; Sidorovich, 1997). The first American mink was caught in this area at the end of 1988. Subsequently, numbers of European mink in the area declined between 1989 and 1992. In the winter of 1989/90, total mink density ranged from three to ten individuals per 10 km, of which 11.4% were American mink. By the spring of 1991, the proportion of the American mink had increased to 60% and the total mink density varied from six to 12 individuals per 10 km. By the spring of 1992, the total mink density had increased significantly to eight to 14 individuals per 10 km and the proportion of American mink was 80%.

We hypothesised that there was competition for food resources between European mink and the invasive American species. We explored this by faecal analysis of the species' diet (Sidorovich *et al.*, 1998). In American mink scats, small-mammals, amphibians and fish dominated and occurred about equally often; in European mink scats amphibians were by far the most important with fish also present in large numbers. For otters, fish dominated all other kinds of prey, closely followed by amphibians, whilst polecat scats were usually full of small-mammal remains, again followed in importance by amphibians. European mink were slightly more specialised than American mink, but between the two species there was a large overlap in diet, especially for rodents and frogs. Both these prey types were abundant, so competition for this resource appeared unlikely.

Inferences from further diet studies of both species as the American mink spread into Estonia and the European mink declined (Maran *et al.*, 1998*b*) suggested that the American species is aggressively dominant, ousting the European mink from the slow-flowing river areas first (the American species' preferred habitat), until it finally occupied all areas. This possibility gained support from the observation in captivity (Maran *et al.*, 1998*a*) of hostile interspecific relationships with American mink tending to be more aggressive than their European counterparts. During the two-year radio-tracking study, we recorded 12 instances of aggression from American mink towards European mink, and in five of these, the American mink chased the European mink for 200–500 m. In all 12 cases the European mink were driven away. This inferred aggression was initiated by male American mink towards European mink of both sexes (Sidorovich *et al.*, 1999).

In summary, while the introduction of American mink was not a sufficient explanation for the long-term decline of the European mink, our evidence strongly suggested that the introduced species was directly responsible for the Eueopean mink's recent plunge towards extinction (Macdonald *et al.*, in press *b*). The only circumstances in which European mink could survive would be in the absence of American mink. The next step was therefore to evaluate how this could be achieved.

PROPOSALS FOR INTERVENTION

Two options were identified. One, which might be applied locally, would be to hire wardens to live-trap both species of mink, selectively killing American specimens and releasing European ones. While the expense of such a programme would be prohibitive on a wide scale (and a permanent commitment), an inexpensive, rural labour force could make this feasible in certain special reserves. V. Sidorovich is undertaking pilot trials of this option in a Lovat River study area.

The second option was to create island sanctuaries for European mink. Many candidate islands are already populated by American mink, so this option is linked to the feasibility of first removing American mink. Some lessons could be learned from previous attempts to release European mink into sites free of American mink, but these have not been carefully planned and monitored, e.g. in the River Shingindira in Tadjikistan (Saudskj, 1989) and Valaam Island in Lake Ladoga, Leningrad (Tumanov & Rozhnov, 1993).

We undertook the experimental establishment of an island sanctuary to explore the practicalities of this exercise. This involved four steps: (1) the establishment of a captive-breeding programme to produce recruits for release; (2) selection of an experimental island; (3) the eradication of American mink; and (4) preparation, release and monitoring of European mink. Failure at any of these stages would result in terminating the sequence.

ESTABLISHMENT OF A CAPTIVE-BREEDING PROGRAMME

A captive-breeding programme was re-launched at Tallinn Zoo in 1992 (following an initial launch in 1984) under the auspices of the European Mink Conservation and Breeding Committee, thus putting the conservation efforts under international supervision and control (Maran, 1992, 1994). Amongst the tasks foreseen for the Breeding Committee were: (1) build up a long-term self-sustaining captive population; (2) establish a studbook guaranteed by IUCN/IUDZG; (3) integrate the European mink into the European Endangered Species Programme (EEP) breeding programme; and (4) negotiate funds from participating zoos and institutions.

One guiding principle was to connect the *in situ* and *ex situ* components of the programme by requiring all zoos participating in the captive breeding and receiving animals on loan to fund *in situ* actions. In 1999, 69 European mink kits were produced at Tallinn Zoo, and the number of co-operating institutions within the EEP had increased to 11.

SELECTING AN EXPERIMENTAL ISLAND

Hiiumaa island in the Baltic Sea, 22 km from the Estonian mainland, forms part of the West Estonian Archipelago Biosphere reserve, of UNESCO's Biosphere Reserve Network. The island (and islets) has an area of about 1000 km^2. Clark *et al.* (2001) emphasise the importance of clearly specified objectives. The objectives of the Hiiumaa experiment were: (1) successful introduction of the European mink to Hiiumaa, and prevention of the species extinction in Eastern Europe; (2) creation of an infrastructure in which to nurture the introduced island population; and (3) formulation of a workable plan for subsequent island refuges, possibly on a metapopulation management basis.

PLANNING AND IMPLEMENTING THE ERADICATION OF AMERICAN MINK

In mid 1997, a detailed survey of the island was conducted. This involved 75 sites stratified into three groups of 25 sites each: coastal sites, river sites and ditch sites, with additional surveys of lake shores. The proportion of positive sites in each habitat was used to indicate the extent of suitable habitat for mink, which was then multiplied by the population densities given in the literature. An estimated 105 to 203 American mink were present on the island. Two years later, in the winter of 1999, a snow-tracking survey suggested this was closer to 60 individuals.

To encourage public support, a publicity campaign was started in 1998, amongst the inhabitants of Hiiumaa, as well as in the national Estonian press and media. This involved considerable newspaper, TV and radio exposure. The plight of the European mink attracted considerable public sympathy in Estonia and a special postage stamp was issued. Nonetheless, the project team were sensitive to the value of American mink pelts to local communities, which might resent the loss of this resource. T. Maran, the leader of the captive-breeding programme, also addressed meetings of hunters and negotiated with representatives of the local hunting fraternities, and when the time came to remove the American mink in preparation for the introduction of the Europeans, there was a solid body of public opinion in favour of the project.

The effort to trap American mink began in November 1998. Diplomacy demanded that this was initiated with local hunters; 15 men volunteered for the task and each was provided with traps. By January 1999, they had enjoyed only meagre success, and therefore the local community agreed to hand over the task to a team of expert Belarussian trappers, headed by V. Sidorovich. By November 1999, 42 American mink had been killed, and snow-tracking revealed no further signs of this species on the island.

PREPARATION, RELEASE AND MONITORING OF EUROPEAN MINK

IUCN guidelines (1998) suggest that modelling should be used to determine the optimum release strategy for establishment of a viable population. The reintroduction of European mink was modelled using VORTEX v8 (Lacy, 1993). The data used to parameterise the models were derived from the Tallinn Zoo studbook. Ten different scenarios were tested, comprising five different supplementation regimes totalling 20 or 30 animals; 240 simulations were iterated 1000 times over 100 years. This exercise suggested that the optimum release number might be a little more than the 30 animals.

The individuals being prepared for release were allocated to two test groups: one was given regular periods in a large cage with 'natural' environment, including pond and live prey, and were deprived of all contacts with people; the others were raised in small cages, with artificial food and regular contact with the keeper. Both were familiarised with dummy radio-transmitters.

Two releases were planned. One in June 2000, using nine individuals, enabled the team to test all the protocols and logistics. The apparent success of this release cleared the way for the second release. Intensive monitoring will reveal the success of this experiment. If the lessons learnt from the Hiiumaa experiment continue to be promising, the next step will be a comparable exercise on Saaremaa, the neighbouring and much larger Baltic island.

THE STAGES OF A REINTRODUCTION PROGRAMME

World-wide there have been many attempts at reintroductions of mammals (Stanley-Price, 1991) and other taxa (Fyfe, 1977; Fischer & Lindenmayer, 2000), and the lessons gleaned from both successes and failures are crystallised in the IUCN's (1998) reintroduction guidelines, which are intended to 'help ensure that reintroductions achieve their intended conservation benefit and do not cause adverse side-effects of greater impact'. These guidelines identify three stages, summarised in Table 20.1. The first, pre-project stage includes both biological and socio-economic assessments of the feasibility and desirability of a reintroduction. The second stage covers planning, preparation and release, while the final stage involves post-release activities such as monitoring, research and evaluation. The viability of the project should be reassessed at each stage, and aborted if it appears to be unfeasible or undesirable. Below, we highlight, some of these stages. Many of the remainder are covered in our two case studies.

Resolving reasons for past declines

As our European mink case study (Box 20.2) makes clear, an important first stage in any reintroduction is to understand and remedy the original reasons for the species' decline (Sarrazin & Barbault, 1996). In the case of the otter in Britain, for example, the original cause of the decline was believed to be organochlorine poisoning, and the banning of dieldrin in various parts of the country was associated with otter population increases. Various voluntary and compulsory bans on the use of dieldrin on cereals, in sheep dips, in industry and finally on daffodil crops, were imposed from the 1962 through to 1989 (Simpson, 1999), and this, together with a programme of releases, contributed to a partial restoration of otter populations in Britain. However, despite considerable increases since the late 1970s, in 1994, 78% of riparian sites were still without otters, and populations in parts of England were very small and sparsely distributed (Strachan & Jefferies, 1996). During the late 1990s, habitat enhancement was used to further encourage the otter's natural expansion. Restoration measures aimed to increase cover by planting with shrubs and trees, such as hawthorn (*Crataegus* spp.), blackthorn (*Prunus spinosa*), ash (*Fraxinus excelsior*) and willow (*Salix* spp.), and to increase food by desilting old ponds or creating new ones (National Rivers Authority, 1993). Artificial holts were also built in areas of poor vegetation cover. Monitoring of these structures has been limited, but a recent assessment of a sample of 60 sites across England and Wales found that 72% had signs of otter use after five to seven years, and of these 37% had been used within the first year of construction (Strachan *et al.*, in press).

Habitat restoration is one common requirement, but others might include disease control and predator or competitor control. In Britain, the restoration of the red squirrel (*Sciurus vulgaris*) might require all three of these actions, making it a distant prospect. Red squirrels in Britain have been widely replaced by exotic grey squirrels (*S. carolinensis*) (Yalden, 1999), which are better able to exploit deciduous woodlands (Kenward & Holm, 1989, 1993), and transmit a deadly parapox virus (Sainsbury *et al.*, 2000). Furthermore, coniferous woodlands, the red squirrel's last refuge, are often unsuitable

Table 20.1. *Stages in a reintroduction, following IUCN (1998) guidelines*

Stage	Activities
1a. Pre-project activities: assessing biological feasibility	Study previous reintroductions Determine the species' critical needs Assess the species' impact on its environment Assess habitat availability and consider habitat restoration Assess likelihood of establishing a minimum viable population Establish criteria for selecting release sites Assess likely impact of previous causes of decline Assess availability of animals to be released
1b. Pre-project activities: assessing socio-economic and legal desirability	Assess likely long-term financial and political support Assess impacts, costs and benefits to local people Investigate means of reducing damage to property Assess attitudes of local communities through public consultation Check local, national and international legislation Acquire relevant permits
2. Planning, preparation and release	Obtain approval of relevant bodies Secure funding for all stages Construct a multidisciplinary team of experts Identify short- and long-term indicators of success Design a monitoring programme Establish policies on interventions Determine the release strategy Select final release sites Mount a public education/awareness campaign Arrange capture, travel and quarantine and screen animals for disease
3. Post-release activities	Implement the monitoring programme Carry out associated studies on released animals Collect and investigate mortalities Assess population viability Assess the need for further, extensive releases Evaluate project success and cost-effectiveness Evaluate socio-economic and environmental impact Continue habitat protection or restoration as necessary Intervene, revise or abandon project as necessary Disseminate information about the project

because they are not sufficiently diverse to provide squirrels with the tree seeds they eat year-round (Lurz *et al.*, 1995). Venning *et al.* (1997) have reviewed red squirrel translocations and recommended a set of guidelines for reintroductions following the advice given by the IUCN (1998). However, problems with disease in particular have hampered trial reintroductions.

Assessing availability of stock for reintroduction

IUCN guidelines stipulate that the individuals to be reintroduced should be as genetically similar as possible to the original, extinct, animals, or to any existing populations that are being supplemented. This is often difficult, particularly if there is a possibility that the extinct animals were a unique subspecies. The case of Barbary lions (*Panthera leo*) illustrates the role of modern molecular biology in resolving such restoration dilemmas. Barbary lions were first described in 1492, and appear to have been distinguished from the sub-Saharan lion by various differences, including a straighter line between the tip of the nose and the back of the head, a more rounded cheek and a narrower muzzle. The iris was pale yolky yellow, and both sexes had long body hair, the male with a dark mane extending behind the shoulder and along its belly with a well-developed elbow tuft. While classical taxonomy recognised Barbary lions as one of eight subspecies (Hemmer, 1974), it remains to be seen whether molecular techniques sustain this distinction. This distinction is relevant to restoration biology because of the extraordinary, and tragic, history of the Barbary lion (Yamaguchi, 2000).

History records that some of these lions were enormous (275 kg), and that they fed almost entirely on Arab livestock. The resulting conflict (paralleled in sub-Saharan Africa today) encouraged the nineteenth-century rulers of the Ottoman Empire to put a bounty on the lions and the policy of eradication was continued by the French authorities until the north African lion population became virtually extinct in the 1920s. The last lion was killed in Morocco in 1922. Barbary lions were widely accepted to be extinct. However, there had been a tradition that nomadic Berber high in the Atlas Mountains carried lions alive to the kings and sultans of Morocco, as a tribute of loyalty. Royal lions survived at the Sultan's palace in Marrakech, until King Hassan II realised the importance of this bloodline and ordered the construction of a new enclosure, which became the Rabat Zoo in 1973. Some 26 lions are held there. Furthermore, Moroccan rulers had given lions to other collections, and records reveal that some 60 of their descendents are now in 14 zoos around the world. Of course, two crucial questions arise with respect to restoring the lions: were the Barbary lions a distinct subspecies and, if so, has the genetic distinctiveness of their captive descendants been diluted through cross-breeding?

All of the existing 'Barbary' individuals have been graded into three categories showing the most to the least 'Barbary' lion traits. In the interim, only those in the first two categories will be used for captive breeding, but whether this distinction has any genetic justification is under investigation. A team from Oxford University, led by N. Yamaguchi, has collected hair and tissue samples from museum specimens known to be Barbary lions with the intent of isolating genetic markers that can be compared to the genetic traits of the extant, captive stock. This is a first step in the process of considering the reintroduction of lions to North Africa, a project developed in collaboration with the Moroccan government and which has identified a potential new park in a rocky mountainous area, dominated by oak woodland, between the Middle and High Atlas. Clearly, the human dimension is a major component of evaluations that lie ahead – restoring a major predator (and the prey base to support it) is a hugely ambitious proposition, but the Moroccan authorities are giving careful thought to the touristic potential, and associated employment and economic advantages to local communities.

Assessing social feasibility

The attitudes of local people, and their ability to cope with, or exploit, reintroduced mammals, is fundamental to the long-term success of a reintroduction project. These socio-economic aspects of reintroductions have generally not been explored as thoroughly as the biological aspects. Over the past decade, however, a range of tools has been developed with which to assess the 'human dimension' of reintroduction. These are discussed in detail in Macdonald *et al.* (chapter 4, this volume). Below, we illustrate one concept, 'community capacity' – an index of the degree to which a human community can anticipate and deal with impacts related to whatever changes it faces (Swanson, 1996) – with the proposed restoration of elk (*Cervus elaphus canadensis*) to New York State.

Attempts to restore elk to northern New York State during 1900–40 were unsuccessful, possibly due to disease and poaching (Severinghaus & Darrow, 1976) but by the 1990s there was renewed interest in elk restoration (Enck *et al.*, 1998). A combined biological and social feasibility study was initiated in 1996 (Enck *et al.*, 1998). That study applied a

Table 20.2. *Social infrastructure and planning characteristics of communities with low social feasibility index scores in three study areas of New York State in 1998*

	Reasons for having low potential social feasibility							
	Only had weak infrastructure[a]		Only lack planning		Both weak infrastructure and lack planning		All communities with low scores	
Area	*n*	%	*n*	%	*n*	%	*n*	%
Northern	22	29.3	43	57.4	10	13.3	75	100.0
Southwestern	21	30.4	29	42.0	19	27.6	69	100.0
Southeastern	3	21.4	10	71.5	1	7.1	14	100.0

[a]Weak infrastructure means the community had negative indicators associated with its current social and economic situation, and had negative indicators associated with trends in social and economic situation.

coarse filter to the entire land area of the state to eliminate areas proposed for elk reintroduction that managers thought would experience unacceptably high levels of elk emigration, elk mortality, or elk–human conflicts (Didier & Porter, 1999).

Three large areas of the state remained after application of the coarse elimination filter: a 16 800 km^2 northern area, a 4000 km^2 southwestern area, and 6300 km^2 in the southeast. A total of 279 communities (municipalities defined by state government as towns, villages or cities) occurred within these three areas. Enck *et al.* (1998) determined the relative capacity of each of these communities to respond positively to elk restoration (i.e. high, moderate improving, moderate worsening, or low) by applying their community capacity assessment methodology.

The 279 communities were pooled into a single group within each area. The southeastern area had both the highest number and percentage of communities having a high capacity while the southwestern area had the lowest number and percentage. Only 10–13% of communities in each of the three areas were designated as having a moderate and improving capacity. The southeastern area also had the largest aggregation of communities with high capacity. Communities with high capacity in the northern area were dispersed, a significant fact in light of the criteria developed by state wildlife managers to only consider areas greater than 500 km^2.

In all three areas, more than two-thirds of communities with low capacity were so designated because they lacked a formal planning mechanism (Table 20.2). This was especially evident in the southeastern area where 89% of low-capacity communities lacked planning mechanisms, which would affect their ability to determine how elk restoration might affect them. In the northern and southwestern areas, substantial percentages of low-capacity communities (44% and 58%, respectively) had a weak social infrastructure, which would affect their ability to turn ecotourism opportunities into reality.

Further interviews within the 26 high-capacity communities in the southeast generally supported the community capacity assessment but revealed great heterogeneity in their dominant economic sectors, demographic characteristics, linkages between communities, and dominant natural resource issues (Enck *et al.*, 1998). Nonetheless, they shared a high capacity to respond to change through their formal planning mechanisms, which identified community needs and opportunities to meet these. These communities were characterised by diffused leadership within several stakeholder groups, social cohesion that facilitated community action, some success at collaborating with neighbouring communities, a broad range of physical infrastructures to support residents, and visitors who might be interested in nature-based recreational opportunities.

Based on the community capacity assessment and identification of high biological feasibility in the same area (Didier & Porter, 1999), the local chapter of the non-governmental organisation Rocky Mountain Elk Foundation initiated discussions about the possibility of elk restoration with communities in the southeastern area. By summer 2000, seven communities in that area had passed resolutions supporting further consideration of elk restoration. A state environmental Quality Review Process (similar to an Environmental Impact Assessment) focused on elk restoration in the southeastern area was initiated in late 2000.

Experimenting with reintroduction

There have been surprisingly few attempts to use formal experiments to assess different reintroduction strategies, such as the use of captive-bred versus wild-caught animals. Experimental reintroductions also offer opportunities to test ecological hypotheses, but these have been largely unrealised (Bright & Morris, 2000). In Britain, work on dormice illustrates the potential in both these arenas.

At the end of the nineteenth century, hazel dormice (*Muscardinus avellanarius*) were considered common throughout southern England, and were present in Wales and northern England. By 1993, however, dormice had been lost from about half their range, especially in the north. Their decline has been linked with the loss and fragmentation of woodland habitats, and changes in management of those that remained (Bright & Morris, 1996). The loss and fragmentation of suitable habitat is particularly devastating to dormice because they are completely arboreal and sedentary, using an area of only about a hectare over the course of a year. They are reluctant to cross open ground or even gaps in hedgerows (Bright, 1998).

In a series of controlled field experiments, Bright & Morris (2000) found that dormice released in May or June had trouble finding enough food and quickly lost weight compared with those released in August or September. This not only provided invaluable guidance about the timing of reintroduction, but also provided evidence for the hypothesis that dormice are food-limited in summer. 'Hard-released' animals generally dispersed and ignored the food provided, while 'soft-released' animals remained at their release site (Bright & Morris, 2000). In hard releases, dormice were removed from their native site and released the same day; a nest box and food were provided. In the soft releases, animals were held in a pen at the release site to acclimatise, and thereafter could return to their nest boxes at will, where they were provisioned with food and water. Wild-caught animals were more likely to survive in the long-term than captive-bred animals (Bright & Morris, 2000). However, few wild dormice are available for translocation.

Setting priorities, objectives and measuring success

As the mood for manipulating mammals grows, the process of deciding which releases are desirable or legitimate or, more pragmatically, a funding priority, becomes more pressing (Griffith *et al.*, 1989; Hunter & Hutchinson, 1994).

Viable, self-sustaining, wild mammal populations are the ultimate goal for species reintroductions (Griffith *et al.*, 1989; Ebenhard, 1995; Wheeler, 1995; Sarrazin & Barbault, 1996; IUCN, 1998). To achieve this, a series of clear objectives should be set, and criteria with which to gauge the success of the project defined (Macdonald *et al.*, 1995; Wheeler, 1995). It should go without saying that measuring the success of reintroductions needs careful and long-term monitoring, and that an associated research programme is also vital to guide planning in a responsive way. Objectives should be quantitative and easily measured. For example, in an imaginary reintroduction project a short-term objective might be to release animals at two sites in the first year, while a long-term objective could be to establish a self-sustaining population at a network of five sites within five years. In this example, the success of the first objective might be measured as the number of animals actually released compared with the theoretical minimum viable population for the two sites, while criteria for success in the second might include predefined levels of breeding success, mortality and emigration. Sarrazin & Barbault (1996) prefer, as criteria for success, extinction probability estimates that combine population size, growth rate and growth rate variance. Other criteria for success could include levels of damage caused, and the attitudes of local people.

Restoration objectives and criteria for success might also range beyond the immediate species or habitat being restored. Charismatic, emblematic mammals can provide a persuasive flagship that facilitates the conservation of a community of plants and animal species and their habitat (Mallinson, 1995). A potent objective of restoration is to promote ecosystem functioning, recreating conditions and opportunities lost to other species when animals became extinct. For example, as part of the Dutch government's innovative and well-funded approach to large-scale nature conservation, the 5600-ha Oostvaarderplassen reserve (a polder created in 1968) is being managed with minimal human intervention (Kampf, 2000). Although still in the early stages, managers are using large herbivores to produce something approximating wilderness. Red deer (*Cervus elaphus*), tarpan (a semi-wild breed of pony from Poland) and heck (a type of cattle bred to recreate the extinct auroch) have been introduced, whereas roe deer (*Capreolus capreolus*) are naturally present. By the same token, in Britain, the restoration of large areas of natural wetland could be made very much easier with the help of reintroduced beaver, a keystone species whose activities can have a huge impact on hydrology and vegetation. This in turn could benefit other riparian mammals such as otters, water voles (*Arvicola terrestris*) and water shrews (*Neomys fodiens*). Variously radical approaches to the restoration of the British mammal fauna are discussed in Macdonald *et al.* (2000c), Macdonald & Tattersall (2001) and Harris *et al.* (2000).

ACKNOWLEDGMENTS

We gratefully acknowledge discussion with our colleagues in the WildCRU and, especially, helpful suggestions by Scott Henderson. Our work on European mink was sponsored by the Darwin Initiative, and on the beaver by Scottish Natural Heritage, the Peoples' Trust for Endangered Species and the International Trust for Nature Conservation, all of which we warmly acknowledge.

REFERENCES

Amori, G. & Zima, J. (1993). Threatened rodents in Europe: species status and some suggestions for conservation strategies. *Folia Zoologika*, **43**, 1–9.

Bárta, Z. (1956). The European mink (*Mustela lutreola* L.) in Slovenia. *Ziva*, **4**, 224–225.

Bright, P.W. (1998). Behaviour of specialist species in habitat corridors: arboreal dormice avoid corridor gaps. *Animal Behaviour*, **56**, 1485–1490.

Bright, P.W. & Morris, P.A. (1996). Why are dormice rare: a case study in conservation biology. *Mammal Review*, **26**, 189–195.

Bright, P.W. & Morris, P.A. (2000). Rare mammals, research and realpolitik: priorities for biodiversity and ecology. In *Priorities for the Conservation of Mammalian Diversity: Has the Panda Had its Day?*, eds. A. Entwistle & N. Dunstone, pp. 141–145. Cambridge: Cambridge University Press.

Cirmo, C.P. & Driscoll, C.T. (1993). Beaver pond geochemistry: acid neutralising capacity generation in a headwater wetland. *Wetlands*, **13**, 277–292.

Clark, T.W., Mattson, D., Reading, R.P. & B. Miller (2001). Interdisciplinary problem solving in carnivore conservation: an introduction. In *Carnivore Conservation*, eds. J.L. Gittleman, S.M. Funk, D.W. Macdonald & R.K. Wayne, pp. 223–240. Cambridge: Cambridge University Press.

Collen, P. (1997). *Review of the Potential Impacts of Reintroducing Eurasian Beaver* Castor fiber *L. on the Ecology and Movement of Native Fishes, and the Likely Implications for Current Angling Practices in Scotland*, Scottish Natural Heritage Review no. 85. Perth, UK: Scottish Natural Heritage.

Conroy, J.W.H. & Kitchener, A.C. (1996). *The European Beaver* (Castor fiber) *in Scotland: A Review of the Literature and Historical Evidence*, Scottish Natural Heritage Review no. 49. Perth, UK: Scottish Natural Heritage.

Convention on Biological Diversity (2001). http://www.biodiv.org/

Cooper, M. (1996). Reintroduction of the European beaver to Scotland. *Reintroduction News*, **12**, 11–12.

Didier, K.A. & Porter, W.F. (1999). Large-scale assessment of potential habitat to restore elk to New York State. *Wildlife Society Bulletin*, **27**, 409–418.

Ebenhard, T. (1995). Conservation breeding as a tool for saving animal species from extinction. *Trends in Ecology and Evolution*, **10**, 438–443.

Enck, J.W., Porter, W.F., Didier, K.A. & Decker, D.J. (1998). *The Feasibility of Restoring Elk to New York: A Final Report to the Rocky Mountain Elk Foundation*. Ithaca, NY: State University of New York, College of Agriculture and Life Sciences at Cornell University, and College of Environmental Science and Forestry in Syracuse, NY.

Fischer, J. & Lindenmayer, D.B. (2000). An assessment of the published results of animal relocations. *Biological Conservation*, **96**, 1–11.

Fyfe, R.W. (1977). Reintroducing endangered birds to the wild: a review. In *Endangered Birds: Management Techniques for Preserving Threatened Species*, ed. S.A. Temple, pp. 323–329. Madison, WI: University of Wisconsin Press.

Gautschi, A. (1983). Nachforschungen über den Iltis (*Mustela putorius* L.). *Schweiz Zeitschrift, Forstwes*, **134**, 49–60.

Green, J. & Green, R. (1997). *Otter Survey of Scotland 1991–1994*. London: Vincent Wildlife Trust.

Griffith, B., Johnston, C.A., Scott, J.M., Carpenter, J.W. & Reed, C. (1989). Translocation as a species conservation tool: status and strategy. *Science*, **245**, 477–480.

Gurnell, A.M. (1997). *Analysis of the Effects of Beaver Dam-Building Activities on Local Hydrology*, Scottish Natural Heritage Review no. 85. Perth, UK: Scottish Natural Heritage.

Harris, S., McLaren, G., Morris, M., Morris, P. & Yalden, D. (2000). Abundance/mass relationships as a quantified basis for establishing mammal conservation priorities. In *Priorities for the Conservation of Mammalian Diversity: Has the Panda Had its Day?*, eds. A. Entwistle & N. Dunstone, pp. 101–118. Cambridge: Cambridge University Press.

Heidecke, D. (1986). Bestandssituation und Schutz von *Castor fiber albicus*. *Zoologische Abhandlungen staatliches Museum für Tierkunde Dresden*, **41**, 111–119.

Heidecke, D. (1989). Ökologische bewertung von Biberhabitaten. *Saugertierkundliche Mitteilungen*, **3**, 13–28.

Hemmer, H. (1974). Untersuchungen zur Stammesgeschichte der Pantherkatzen. 3: Zur Artgeschichte des löwen *Panthera (Panthera) leo* (Linnacus 1758). *Veröffentlichungen der zoologischen Staatssammlung München*, **17**, 167–280.

Heptner, V.G., Naumov, N.P., Yurgenson, P.B., Sludsky, A.A. Chirkova, A.F. & Bannikov, A.G. (1967). *Mammals of the USSR*, Part 2, Vol. 1. Moscow: Nauk SSSR. (in Russian)

HMSO (1994). *Biodiversity, The UK Action Plan*. London: HMSO.

Hunter, M.L. & A. Hutchinson, A. (1994). The virtues and shortcomings of parochialism: conserving species that are locally rare but globally common. *Conservation Biology*, **8**, 1163–1165.

IUCN (1998). *IUCN/SSC Guidelines for Reintroductions*. Gland, Switzerland: IUCN/SSC Reintroduction Specialist Group, IUCN.

Kampf, H. (2000). The role of large grazing animals in nature conservation: a Dutch perspective. *British Wildlife*, **12**, 37–46.

Kenward, R.E. & Holm, J.L. (1989). What future for British red squirrels? *Biological Journal of the Linnean Society*, **38**, 83–89.

Kenward, R.E. & Holm, J.L. (1993). On the replacement of the red squirrel in Britain: a phytotoxic explanation. *Proceedings of the Royal Society of London, B*, **251**, 187–194.

Lacy, R.C. (1993). VORTEX: a computer simulation model for population viability analysis. *Wildlife Research*, **20**, 45–65.

Lurz, P.W.W., Garson, P.J. & Rushton, S.P. (1995). The ecology of squirrels in spruce dominated plantations: implications for forest management. *Forest Ecology and Management*, **79**, 79–90.

Macdonald, D.W. & Tattersall, F.H. (1999). Beavers in Britain: planning reintroduction. In: *Beaver Production, Management and Utilisation*, eds. P.E. Busher & R.M. Dzięciolowski, pp. 77–102. London: Kluwer.

Macdonald, D.W. & Tattersall, F.H. (2001). *Britain's Mammals: The Challenge for Conservation*. London: Mammals Trust.

Macdonald, D.W., Tattersall, F.H., Brown, E.D. & Balharry, D. (1995). Reintroducing the European beaver to Britain: nostalgic meddling or restoring biodiversity? *Mammal Review*, **25**, 161–200.

Macdonald, D., Maitland, P., Rao, S., Rushton, S., Strachan, R. & Tattersall, F. (1997). *Development of a Protocol for Identifying Beaver Release Sites*, Scottish Natural Heritage Research Survey and Monitoring Report no. 93. Perth, UK: Scottish National Heritage.

Macdonald, D.W., Mace, G. & Rushton, S. (1998). *Proposals for Future Monitoring of British Mammals*. London: Department of Environment Transport and Regions.

Macdonald, D.W., Tattersall, F.H., Johnson, P.J., Carbone, C., Reynolds, J.C., Langbein, J., Rushton, S.P. & Shirley, M.D.F. (2000*a*). *Managing British Mammals: Case Studies from the Hunting Debate*. Oxford: WildCRU.

Macdonald, D.W., Tattersall, F.H., Rushton, S., South, A.B., Rao, S., Maitland, P. & Strachan, R. (2000*b*). Reintroducing the beaver (*Castor fiber*) to Scotland: a protocol for identifying and assessing suitable release sites. *Animal Conservation*, **3**, 125–133.

Macdonald, D.W., Mace, G.M. & Rushton, S. (2000*c*). Conserving British mammals: is there a radical future? In *Priorities for the Conservation of Mammalian Diversity: Has the Panda Had its Day?*, eds. A. Entwistle & N. Dunstone, pp. 175–205. Cambridge: Cambridge University Press.

Macdonald, D.W., Bryce, J.M. & Thom, M. (in press *a*). Introduced mammals: do carnivores and herbivores usurp endemics by different mechanisms? In *Proceedings of the 2nd Vertebrate Pest Management Conference.*

Macdonald, D.W., Sidorovich, V.E., Maran, T. & Kruuk, H. (in press *b*). *The Darwin Initiative: European Mink,* Mustela lutreola: *Analyses for Conservation.* Oxford: WildCRU.

Mallinson, J.J.C. (1995). Conservation breeding programs: an important ingredient for species survival. *Biodiversity and Conservation*, **4**, 615–635.

Maran, T. (1992). The European mink, *Mustela lutreola*, Conservation and Breeding Committee (EMCC) founded. *Small Carnivore Conservation*, **7**, 20.

Maran, T. (1994). *Studbook for the European Mink*, Mustela lutreola *Linnaeus 1761*, Vol. 1. Tallinn, Estonia: European Mink Conservation and Breeding Committee, Tallinn Zoo.

Maran, T. & Henttonen, H. (1995). Why is the European mink, *Mustela lutreola*, disappearing? A review of the process and hypotheses. *Annales Zoologici Fennici*, **32**, 47–54.

Maran, T., Macdonald, D.W., Kruuk, H., Sidorovich, V. & Rozhnov, V.V. (1998*a*). The continuing decline of the European mink *Mustela lutreola*: evidence for the intraguild aggression hypothesis. In *Behaviour and Ecology of Riparian Mammals*, eds. N. Dunstone & M.L. Gorman, pp. 297–323. Cambridge: Cambridge University Press.

Maran, T., Kruuk, H., Macdonald, D.W. & Polma, M. (1998*b*). Diet of two species of mink in Estonia: displacement of *Mustela lutreola* by *M. vison. Journal of Zoology, London*, **245**, 218–222.

Migot, P. & Rouland, P. (1989). La réintroduction du castor en France: essai de synthese et reflexions. *Bulletin Mensuel del'Office National de la Chasse*, **132**, 35–43.

Moutou, F. (1994). Otter and mink in France. *Small Carnivore Conservation*, **10**, 18.

Naiman, R.J. & Melillo, J.M. (1984). Nitrogen budget of a subarctic stream altered by beaver (*Castor canadensis*). *Oecologia*, **62**, 150–155.

Naiman, R.J., Melillo, J.M. & Hobbie, J.E. (1986). Ecosystem alteration of boreal forest streams by beaver (*Castor canadensis*). *Ecology*, **67**, 1254–1269.

Naiman, R.J., Johnston, C.A. & Kelley, J.C. (1988). Alteration of North American streams by beaver. *BioScience*, **38**, 755–762.

National Rivers Authority (1993). *Otters and River Habitat Management*, Conservation Technical Handbook no. 3. Bristol, UK: National Rivers Authority.

Nolet, B.A. (1994). Return of the beaver to the Netherlands: viability and prospects of a reintroduced population. PhD thesis, University of Groningen, The Netherlands.

Nolet, B.A. & Baveco, J.M. (1996). Development and viability of a translocated beaver *Castor fiber* population in the Netherlands. *Biological Conservation*, **75**, 125–137.

Novikov, G.A. (1939). *The European Mink.* Leningrad. (in Russian)

Novikov, G.A., Airapetjantz, A.E., Pukinsky, Y.B., Strelkov, P.P. & Timofeeva, E.K. (1970). The European mink. In *Mammals of the Leningrad Region*, pp. 225–232. Leningrad, Russia: Leningrad University. (in Russian)

Nummi, P. (1989). Simulated effects of the beaver on vegetation, invertebrates and ducks. *Annales Zoologici Fennici*, **26**, 43–52.

Ouderaa, A. van der., Fey, D., Stelt, J. van der & Zadelhoff, F.J. van der (1985). *Bevertochten, Verslag van Twee Studiereizen in 1984*, Inspectie Natuurbehaoud Rapport no. 1985–2. Utrecht, Netherlands: Staatsbosbeheer. (In Dutch, with English summary).

Pavlov, M.A. & Korsakova, I.B. (1973). American mink (*Mustela lutreola* Brisson). In *Acclimatization of Game Animals in the Soviet Union*, ed. D. Kiris, pp. 118–177. Kirov, Russia: Volgo-Vjatsk.

Romanowski, J. (1990). Minks in Poland. *Mustelid and Viverrid Conservation*, **2**, 13.

Sainsbury, A.W., Nettleton, P., Gilray, J. & Gurnell, J. (2000). Grey squirrels have high seroprevalence to a parapoxvirus associated with deaths in red squirrels. *Animal Conservation*, **3**, 229–233.

Sarrazin, F. & Barbault, R. (1996). Reintroductions: challenges and lessons for basic ecology. *Trends in Ecology and Evolution* **11**, 474–478.

Saudskj, E.P. (1989). The European mink (*Mustela lutreola*) on the mountain rivers of Kuril and Tadzikistan. In *All-Union Conference Problems of Ecology in Mountain regions*, 9–13 October 1989, Dushanbe, Abstracts, pp. 48–52.

Schreiber, A.R., Wirth, R., Riffel, M. & van Rompaey, H. (1989). *Weasels, Civets, Mongooses, and Their Relatives: An Action Plan for the Conservation of Mustelids and Viverrids.* Gland, Switzerland: IUCN/SSC Mustelid and Viverrid Specialist Group.

Scottish Natural Heritage (1998). *Reintroduction of the European Beaver to Scotland: A Public Consultation.* Edinburgh: Scottish Natural Heritage.

Scottish Natural Heritage (2000). *Favoured Site for Beaver Trial Identified.* Press release, 21 September 2000. Edinburgh: Scottish Natural Heritage.

Severinghaus, C.W. & Darrow, R.W. (1976). Failure of elk to survive in the Adirondacks. *New York Fish and Game Journal*, **23**. 98–99.

Sidorovich, V.E. (1997). Mustelids in Belarus: evolutionary ecology, demography and interspecific relationships. Minsk, Belarus: Zolotoy Uley. (In Russian).

Sidorovich, V.E., Savchenko, V.V. & Budny, V.B. (1995). Some data about the European mink *Mustela lutreola* distribution in the Lovat river Basin in Russia and Belarus: current status and retrospective analysis. *Small Carnivore Conservation*, **12**. 14–18.

Sidorovich, V.E., Kruuk, H., Macdonald, D. W. & Maran, T. (1998). Diets of semi-aquatic carnivores in northern Belarus, with implications for population changes. *Symposia of the Zoological Society of London*, **71**. 177–190.

Sidorovich, V.E., Kruuk, H. & Macdonald, D.W. (1999). Body size and extinction: changes in European and American mink in Belarus. *Journal of Zoology, London*, **248**, 521–527.

Simpson V.R. (1999). *A Post Mortem Study of Otters* (Lutra lutra) *Found Dead in Southwest England*, Environment Agency RD Technical Report no. W148. Swindon, UK: Environment Agency.

Soulé, M. E. (1986). *Conservation Biology: The Science of Scarcity and Diversity.* Sunderland, MA: Sinauer Associates.

South, A., Rushton, S. & Macdonald, D.W. (2000). Simulating the proposed reintroduction of the European beaver (*Castor fiber*) to Scotland. *Biological Conservation*, **93**. 103–116

Stanley-Price, M.R. (1991). A review of mammal reintroductions and the role of the Reintroduction Specialist Group of IUCN/SCC. *Symposia of the Zoological Society of London*, **62**, 9–25.

Stocker, G. (1985). The beaver (*Castor fiber* L.) in Switzerland: biological and ecological problems of re-establishment. *Swiss Federal Institute of Forestry Research Reports*, **242**, 1–149. (In German, with English summary).

Strachan, R. & Jefferies, D.J. (1996). *Otter Survey of England 1991–1994: A Report on the Decline and Recovery of the Otter in England and on its Distribution, Status and Conservation in 1991–1994.* London: Vincent Wildlife Trust.

Strachan, R., Moorhouse, T. & Macdonald, D.W. (in press). Habitat enhancement for riparian mammals on agricultural land. In *Conservation and Conflict: Mammals and Farming in Britain*, eds. F.H. Tattersall & W.J. Manley. Otley, UK: Westbury Publishing.

Swanson, L.E. (1996). Social infrastructure and economic development. In *Rural Development Research*, eds. T.D. Rowley, D.W. Sears, G.L. Nelson, J.N. Reid & M.J. Yetley, pp. 103–119. Westport, CT: Greenwood Press.

Szunyoghy, J. (1974). Eine weitere Angabe zum Vorkommen des Nerzes in Ungarn, nebst einer Revision der Nerze des Karpatenbeckens. *Vertebratica Hungarica*, **15**, 75–82.

Tattersall, F. & Macdonald, D.W. (1995). A review of the direct and indirect costs of reintroducing the European beaver (*Castor fiber*) to Scotland, unpublished report to Scottish Natural Heritage, contract no. SNH/110/95/1BB, 28 April 1995.

Tumanov, I.L. & Rozhnov, V.V. (1993). Tentative results of release of the European mink into the Valaam Island. *Lutreola*, **2**, 25–27.

Venning, T., Sainsbury, A.W. & Gurnell, J. (1997). An experimental study on translocating red squirrels to Thetford Forest. In *The Conservation of Red Squirrels,* Sciurus vulgaris L., eds. J. Gurnell & P. W.W. Lurz, pp. 133–143. London: People's Trust for Endangered Species.

Webb, A., French, D.D. & Flitsch, A.C.C. (1997). *Identification and Assessment of Possible Beaver sites in Scotland*, Scottish Natural Heritage Research Survey and Monitoring Report no. 94, 1–16. Edinburgh: Scottish Natural Heritage.

Wheeler, B. D. (1995). Introduction: restoration and wetlands. In *Restoration of Temperate Wetlands*, eds. B. D. Wheeler, S. C. Shaw, W. Fojt, & R. A. Robertson, pp. 1–18. Chichester, UK: John Wiley.

Yalden, D. W. (1986). Opportunities for reintroducing British mammals. *Mammal Review*, **16**, 53–63.

Yalden, D. (1999). *The History of British Mammals.* London: T. & A. D. Poyser.

Yamaguchi, N. (2000). The Barbary lion and the Cape lion: their phylogenetic places and conservation. *African Lion Working Group News*, **1**, 9–11.

Youngman, P. M. (1982). Distribution and systematics of the European mink *Mustela lutreola* Linnaeus 1761. *Acta Zoologica Fennica*, **166**, 1–48.

Part 5 • Monitoring and appraisal

21 • Monitoring and appraisal

KAREN D. HOLL AND JOHN CAIRNS JR

INTRODUCTION

Discussing appropriate monitoring of restoration projects within a single chapter is a daunting task. The topic of ecological monitoring can and, indeed, has been the topic of entire books (e.g. Cairns *et al.*, 1982; Clarke, 1986; Spellerberg, 1991; Cairns & Niederlehner, 1995). Given the voluminous literature on the topic and that exact parameters to be monitored will be specific to the system of interest, most of this discussion focuses on questions of monitoring that are applicable to all projects, such as selecting reference systems, temporal and spatial scale, action thresholds, sample distribution, and desirable characteristics of monitoring parameters. Also discussed briefly are the parameters that have been suggested for monitoring specific ecosystem types and considerations of particular importance to those ecosystems; readers are referred to other chapters in this text, however, as well as other literature, to select monitoring criteria for specific ecosystems. Throughout, key points are illustrated with relevant case studies. We draw heavily on the literature in ecotoxicology, as both ecotoxicology and restoration ecology focus on stressed ecosystems and ecotoxicology has a longer history of rigorous monitoring requirements.

Terminology

Regrettably, monitoring is often used to describe three quite different activities: (1) sampling/surveying – gathering data at a particular point in time; (2) surveillance – a systematic and orderly gathering of specific data over a period of time; and (3) monitoring – surveillance undertaken to ensure that predetermined quality control conditions are being met. It is important to examine the definition of monitoring to take the steps necessary at all stages of the project planning and implementation (Fig. 21.1) to achieve a successful monitoring programme, rather than collecting endless data that are never used to evaluate project success.

First, goals must be clearly defined from the outset; otherwise, it is impossible to select appropriate quality control conditions (Pacific Estuarine Research Laboratory, 1990; National Research Council, 1992; Kondolf & Micheli, 1995; Sutter, 1996; Chapman, 1999; Ehrenfeld, 2000). This statement seems obvious, but various reviews of restoration projects (National Research Council, 1992; Kondolf & Micheli, 1995; Lockwood & Pimm, 1999) have found that lack of clear goals is surprisingly common. As is discussed elsewhere (this volume; National Research Council, 1992; Jackson *et al.*, 1995; Bradshaw, 1997) what constitutes restoration has been broadly debated and has to be agreed upon by all parties involved with respect to each restoration project. For example, the goal of a project may be: (1) to restore a single endangered species; (2) to restore the tallgrass prairie ecosystem; (3) to restore a buffer strip that will uptake agricultural nutrients; or (4) all of the above. Each of these goals is valid, but would require monitoring of different parameters for each goal to assess success. Goals should be sufficiently specific so that it is possible to evaluate whether they have been achieved. For example, the goal of 'restoring a biologically viable ecosystem' is impossible to test (Chapman, 1999), whereas achieving a certain percentage cover of native vegetation can be evaluated. The importance of clear goals for selecting conditions to monitor in evaluating success cannot be overemphasised.

Second, specific monitoring protocols must be outlined and criteria for success determined during the planning phase of the project (Fig. 21.1), and not after

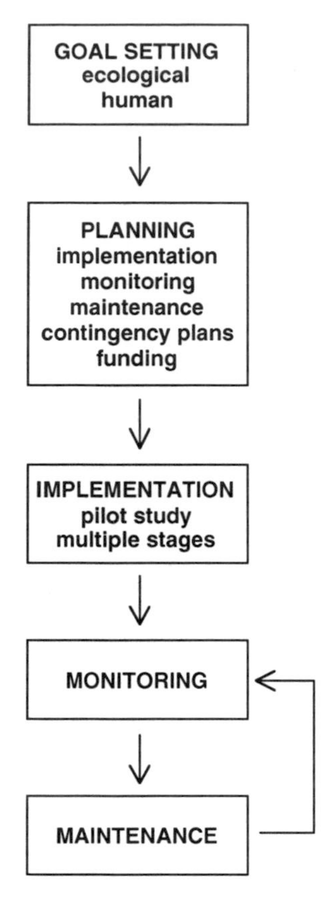

Fig. 21.1. Stages in restoration. Modified from Holl (1996).

implementation (Kondolf & Micheli, 1995). Without using 'milestones' or evaluation criteria to judge progress, it is impossible to generate early warnings that the restoration project is deviating from the expected trend lines. Monitoring is essentially a quality-control effort and is not productive without ecological specifications. Unfortunately, in many cases monitoring is an afterthought and is not discussed until after project implementation. Such timing is problematic because projects may not have been designed in an appropriate way for monitoring, funds may not have been allocated for this purpose, and appropriately trained personnel may not be available.

It is also essential at an early stage in project planning to develop a feedback loop of how monitoring will inform subsequent management decisions (Shabman, 1995) (Fig. 21.1). Restoration is often viewed as a product rather than an ongoing process. Monitoring is nearly useless if one views restoration as resulting in a final product. However, if restoration is viewed as an ongoing process resulting in self-maintaining dynamic systems, then monitoring is essential to success. If one acknowledges problems in the early phase and takes corrective actions, the process is likely to be much less costly than ignoring the problems.

Finally, it is important to note that monitoring will help to determine whether specific end points have been reached, but will not necessarily explain the underlying causes of the result. Certainly, monitoring will help to shed light on factors influencing success or failure of certain restoration strategies, but rigorously testing different restoration strategies or determining the reason for not reaching an end point requires carefully designed experiments (Sutter, 1996). For example, if plants show poor growth rates, it is difficult to assign the reason for this result (e.g. lack of nutrients, water stress, or competition) without experiments designed to control these variables. In this chapter, we focus on monitoring to determine whether ecosystem recovery is following an appropriate trajectory, but encourage land managers to consider consulting with scientists regarding appropriate experimental designs to test different management strategies and identify specific conditions that limit recovery. Such communication is critical to improving restoration efforts.

Funding

Most restoration projects are limited by available funding (Holl & Howarth, 2000). Given these constraints, agencies would often prefer to fund tangible projects, rather than monitoring efforts that might show that their projects are not achieving success (Kondolf, 1995). It is absolutely essential that monitoring be allocated a specific portion of a restoration budget, but this activity is often the first category cut from a project budget. Therefore, allocating a minimum percentage of funding to monitoring should be included in any legislation that governs restoration (Holl & Howarth, 2000).

Monitoring may actually save money in the long term. If problems are noted early, then corrective actions may be less costly. Also, monitoring of one project may serve to identify information that, if properly disseminated, may lead to greater overall success and less wasted cost in future restoration efforts (Kondolf, 1995). For example, between 1987 and 1991, 569 swift foxes (*Vulpes velox*) were released in Alberta and Saskatchewan, Canada, and their survival and movement patterns were monitored using radio collars (Carbyn *et al.*, 1994). Results of survival monitoring indicated that foxes released in the autumn had more than twice the survival rate of those released in spring, which was counter to the prediction of the investigators (Carbyn *et al.*, 1994). These data, although expensive to collect, are important in improving the success of future reintroduction programmes, particularly given the high costs of animal restoration projects.

Another example of the importance of expending money at the outset to test restoration strategies and monitor their effect before implementing them on a large scale is the high-profile Kissimmee River restoration project, which is documented in detail in a special issue of the journal *Restoration Ecology* (Vol. 5(3), 1995). The meandering 166-km river was channelised to a 90-km long, 100-m wide canal in the 1960s, draining over 12 000 ha of floodplain wetlands (Koebel, 1995). Due to public outcry over water-quality problems and loss of fish and wildlife, plans for restoration began almost immediately after channel construction (Koebel, 1995). Before attempting to restore the sinuosity of the river, both modelling and demonstration projects were implemented over a small portion of the river. Monitoring of hydrology, vegetation, invertebrates, fishes and birds demonstrated that the biotic communities were quite resilient when the appropriate hydrology was restored; this monitoring provided important information for implementing the full restoration project (Koebel, 1995). Although preliminary testing and monitoring efforts required substantial funding, few would argue their importance, particularly given the estimated US$8 billion that will be spent on the full restoration project (Enserink, 1999).

SAMPLING DESIGN

Selecting a reference system

One of the most problematic questions in designing monitoring plans is identifying a suitable reference site or criteria (White & Walker, 1997). For certain parameters, such as nutrient levels in water, ranges of acceptable levels may be well established. Unfortunately, this knowledge is not available for most parameters monitored. Therefore, reference systems must be selected to compare with restored areas. Usually, either historic reference data are used, baseline data are taken before a site is damaged, or restored sites are compared with other nearby sites that are judged to be more 'intact'; often, a combination of these comparisons is used. Each of these approaches presents a number of problems.

In many cases, historic data before the influence of humans are not available. Many ecosystems worldwide have been influenced by humans to some degree for thousands of years. Therefore, the selection of a reference historical state is subjective. For example, Holl & Cairns (1994) used the surrounding hardwood forest as a reference system with which to compare the success of mine reclamation efforts in southwestern Virginia. It is important to note that this forest, like nearly all hardwood forests in the eastern United States, was logged in the past. If historic data are available, there may be insufficient detail for the parameters of interest. In the case of trying to collect pre-disturbance data at the site of interest, often extensive damage has occurred before baseline data can be collected to provide a reference.

Selecting a nearby reference system with which to compare restoration of a single site is problematic. It is impossible to find an identical system, as other sites may be surrounded by different land uses; vary inherently along abiotic gradients, such as soil type or slope aspect; and may have been subjected to different disturbance regimes. This selection is particularly problematic for rivers (Norris *et al.*, 1996), as rivers reflect the geology and management practices of the entire watershed, which are impossible to replicate. Therefore, a good approach is to select a range of reference sites to assess the natural variation and define boundary conditions for parameters to be monitored.

Another problem with selecting a reference site or historical data for comparisons is deciding with which successional stage to compare a restored site. For example, if the goal of a project is to restore a hardwood forest, it is impossible to determine whether a recently restored site is on an appropriate trajectory towards a mature forest that may have taken 100–200 years to develop by only comparing the restored site to a mature forest. Therefore, it may be necessary to select sites that are in a range of successional stages, such as those that have recently burned or were disturbed at an earlier time period. Moreover, many ecosystems may naturally fluctuate through cycles related to disturbance regimes or climatic patterns. For example, in prairie restoration, is it more appropriate to select a recently burned site or a site that has not burned for five years as the reference system? Multiple years of data on reference systems are needed to encompass the temporal variability inherent to most ecological systems (Kondolf, 1995).

A final problem in selecting a reference system is that species and biotic communities are naturally patchily distributed spatially, as well as temporally. Therefore, reference systems must be sampled across a sufficient spatial scale to incorporate natural variability within systems. If variance is high, then more samples will be required to characterise the community. For composition data, one approach is to use a species–area curve to determine when additional sampling ceases to add many new species to the overall list.

Spatial scale

Ideally, all restoration projects would be monitored at large spatial scales, for long time periods, and high levels of detail (National Research Council, 1992). In reality, the opposite is usually true. Given typical personnel and budget constraints, a trade-off usually occurs between: (1) frequently sampling a number of parameters at a few points; (2) infrequently sampling many parameters at relatively few locations; and (3) infrequently sampling a few parameters at many locations (Michener & Houhoulis, 1997). As with most questions raised in this discussion, how best to allocate monitoring efforts over space and time will depend on the goals of the project and the ecosystem being monitored.

Ecosystem monitoring often must be conducted at large spatial scales that may include surrounding, non-restored areas, as human actions often affect ecosystems over large areas. For instance, surface mining causes increased nutrient levels in entire watersheds (Dick *et al.*, 1986). Therefore, monitoring success of mine reclamation will necessarily include water-quality monitoring in adjacent rivers. Likewise, recovery of reclaimed and restored areas is dependent on the composition of the surrounding landscape (e.g. Wolfe, 1990; Nepstad *et al.*, 1991; Anderson, 1993). For example, low nesting success of ground-nesting birds in restored riparian forests along the Sacramento River may be a result of predatory animals, such as skunks and raccoons, whose populations are high in the surrounding human-altered landscape (Small *et al.*, 1999).

The scale selected for monitoring must be appropriate for the population/community or process being monitored (Holl & Cairns, 1995). For example, the success of a reintroduction effort of a localised plant may not require monitoring outside the site. If the plant, however, is dependent on specialised pollinators, then monitoring the status of nearby populations may be necessary. Or if the survival of the plant is reduced by competition of invasive species, then monitoring the spread of nearby populations of the invasive species would be required. As another example, chemical contaminants of lakes may be transported over enormous distances in the atmosphere. Lake Siskiwit on Isle Royale is a national park and a wilderness area located more than 90 km from the nearest shoreline of western Lake Superior. Nevertheless, elevated polychlorinated dibenzodioxins (PCDDS) and PCBs have been found in fish from the lake (Czuczwa *et al.*, 1984). Identifying and monitoring the sources of these pollutants is essential to restoring this lake over the long term.

Fortunately, technologies that allow work at larger spatial scales are improving. Increasingly, geographic information systems (GIS) are being used to overlay land-use coverages with soil types, elevation, parcel ownership and other parameters to prioritise restoration efforts. Increasingly accurate mapping technology facilitates the mapping of monitoring points to locate them with reference to the surrounding landscape, which facilitates organisation and analysis of data. Remote sensing is being used to monitor plant cover at large spatial scales. For example, Phinn *et al.* (1996) used airborne, multispectral video data to assess quality of marsh restoration for the light-footed clapper rail (*Rallus longirostris levipes*) in southern California (see Zedler & Adam, volume 2), and Um & Wright (1998) used remote sensing to monitor revegetation along a pipeline in Scotland. While remote sensing may be useful in identifying different cover classes at large scales, field surveys will be needed to identify specific species or subtle changes in vegetation composition (Phinn *et al.*, 1996; Um & Wright, 1998). Remote sensing also helps focus on areas that need further field surveying.

Temporal scale

Another critical question in monitoring is scale of time, including length of monitoring and frequency. As stated previously, baseline monitoring should ideally be completed before disturbance or, at least, before initiation of restoration efforts. Such information is essential for comparing results. Baseline data before disturbance can be used as a reference for the end point of the restoration. Data from a site after disturbance can assist in ascertaining whether there has been significant change in the disturbed condition over time. Both can help to test appropriate sampling strategies and determine the number of samples required to detect a certain level of change (discussed below).

Time-frames for both monitoring and restoration are typically short (e.g. five to ten years). They are commonly limited by budget constraints, as well as the need to demonstrate compliance with regulatory standards, rather than being based on ecological principles. As a result, many restoration projects are aimed at achieving short-term goals, which may inhibit long-term ecosystem restoration. For example, aggressive, non-native, herbaceous species are often planted on reclaimed mined sites in the southeastern United States in an effort to minimise erosion and achieve five-year cover requirements (Holl & Cairns, 1994). These species, however, have been shown to inhibit the long-term development of the natural vegetational communities (Brenner *et al.*, 1984; Burger & Torbert, 1990).

Moreover, monitoring for only a few years is insufficient to determine whether ecosystems are self-maintaining. Rein (1999) monitored vegetation composition in areas removed from agricultural production and planted with native grasses to restore grassland and serve as buffer strips to uptake agricultural nutrients from nearby row-cropped fields. In the second year of the study, native grasses covered approximately 70% of the area and the project would have been deemed a success. By the fourth year after planting, however, native grasses covered $<5\%$ of the study area while non-native grasses and forbs covered the vast majority of the area, demonstrating how quickly restored areas can change. This study highlights the need for ongoing monitoring.

Ideally, monitoring should be continued until the ecosystem is self-regulating for some particular period of time (Cairns, 1991). Self-regulating means that the structural and functional attributes persist in the absence of whatever subsidies (introduction of new species, water or fertiliser, removal of exotics, etc.) may have been necessary during the initial restoration efforts. Unfortunately, the time period required for some ecosystems to achieve a self-regulating state may be well beyond what is politically feasible. For some grassland ecosystems, monitoring for 10–20 years may be sufficient, but many forested systems do not begin to resemble pre-disturbance condition until at least 50 years after restoration. As a minimum, monitoring should be continued until predetermined restoration goals are achieved. In fact, Lockwood & Pimm (1999) found that 48% of restoration projects reviewed ceased monitoring before goals were achieved, confirming that many restoration monitoring efforts are continued for too short a period of time.

Data must also be collected over a sufficient time period to incorporate natural cycles of variation (Sutter, 1996), such as fire intervals, periodic flooding or cyclical population fluctuations. Evaluating success can be problematic in ecosystems that are highly episodic in recruitment and growth, such as riparian or desert ecosystems. For example, most river restoration projects are designed to accommodate a certain level of flood (e.g. 10-year, 20-year intervals). It is impossible to evaluate the success of the restoration until such a flood occurs, which may be much longer than its average interval (Kondolf & Micheli, 1995). The natural variability of systems may not only affect rates of recovery, but also complicate comparisons with reference systems, as discussed earlier.

Another important question is how often to monitor parameters. The range may be wide, depending on ecosystem type and the parameter being monitored. For example, vegetation cover may only need to be measured annually, whereas multiple monitoring dates may be necessary to accurately identify all species. Monitoring nutrient levels in water may be necessary weekly because of high temporal fluctuation. In selecting frequency of monitoring, cost and labour availability certainly will be a factor, but it is important to consider the initial goals of the project and determine whether sufficient data will be collected to evaluate success of these goals. For example, if areas are being restored to serve as buffer strips, annual monitoring of nutrient levels is insufficient to evaluate that goal; whereas, if these same areas are being primarily restored for native grass cover, then measuring species composition once at the end of the growing season may be sufficient.

Early on in a restoration project, monitoring should be conducted more frequently to determine whether restoration efforts are proceeding along the predicted trajectory. With time, it may be possible to monitor less frequently (e.g. measure vegetation every other or every three years). There should be a mechanism in place to reactivate more regular monitoring if the parameters measured deviate beyond a critical threshold that requires some intervention.

Action thresholds

Monitoring data should have an acceptable level of precision to detect changes and initiate corrective actions over time, given natural variability (Sutter, 1996); precision specifically refers to the similarity of repeated measurements. In other words, it is essential to decide a priori how much variability for a predetermined value is acceptable or the degree of closeness to a reference system required for the project to be a success (National Research Council, 1992). This question can be challenging, given the natural variability of ecosystems. Replicate community samples are usually only 50–90% similar depending on community type (Gauch, 1982). Determining the acceptable range of values for a specific project depends on the goals of the project and weighing the relative importance of false positive and false negative results with respect to that project.

One purpose of a restoration monitoring system is to provide an early warning that recovery is not proceeding as predicted, either because original estimates or recovery rates were wrong or because ecosystem stress from some new source has occurred. Setting the corrective action threshold low enough to ensure early detection of deleterious change seems prudent, but can also produce false positive readings (Type I error). A false positive is an indication that some undesirable effect has occurred when, in fact, none has. Taking corrective action can be quite expensive. False positives are usually most numerous during the early developmental stages of monitoring and decrease substantially as experience is gained. Avoiding false positives by setting the action threshold well beyond the response threshold will probably result in false negatives.

A false negative is information that no deleterious effects have occurred when, in fact, some may have (Type II error). Thus, failure of ecological restoration occurs because no corrective action alert is produced when one is justified. This response is clearly a matter of prime importance in the design of all monitoring systems. The tolerable number of false negatives should be related to the resilience of the ecosystem being monitored. A highly resilient ecosystem needs less protection than one with low resiliency. Much more emphasis has been placed on false positives in environmental monitoring, whereas the consequences of false positives are usually less than those of false negatives (Fairweather, 1991). Not taking action when it is necessary can result in substantial costs at a later date.

It is well known that increasing sample size reduces variance and, therefore, increases the power of detecting an effect. The power of a test is the complement of false negatives (Type II error rate). The question is raised of how many samples are required to detect a particular magnitude of change. Taking too few samples will result in a large amount of uncertainty in evaluating success and may result in the need for costly corrective actions, but taking too many samples is a waste of limited time and resources. The number of samples needed depends on the amount of natural variance and the predetermined level of difference from reference conditions that has to be detected. A good strategy is to use some preliminary or baseline sampling to assess the amount of natural variability. A number of computer programs will calculate the power associated with different sample sizes regarding an estimate of variability. Norris *et al.* (1996) list various formulas for calculating power and give illustrative examples.

Distribution of samples

Since the number of samples is constrained by financial and time limitations, distribution of samples should be carefully considered. Most importantly, samples must be distributed over the entire area for which inferences need to be made (Fancy, 2000). For example, it is common to sample sites that are easily accessible by roads. But, if inferences are to be made for the success of an entire restored area, then samples must be selected from the full extent of the area being evaluated.

One basic assumption of most statistical analyses is that samples are randomly distributed over the entire area for which inferences are to be made. It is not uncommon to rely on experts to choose 'representative' areas, but such an approach violates the assumption of inferential statistics. In some cases, it may be appropriate to stratify sampling. In stratified sampling, areas are divided into units prior to randomising the sampling. The areas may be stratified (divided) according to soil type, aspect, moisture gradient or plant community type (Sutter, 1996), or riffles and pools in rivers (Norris *et al.*, 1996), and then a certain number of samples is randomly placed within each habitat type.

Data collection

Careful efforts should document monitoring protocols and monitoring site locations to facilitate repeating measurements in subsequent years. Sometimes, entire methodologies, and more often minor notes about modifications or assumptions, are not documented, which renders comparisons between years difficult if the same investigator is not coordinating the sampling. Also, vegetation changes often make it difficult to find the original survey sites in subsequent years. Increasingly, managers are using handheld global positioning systems (GPS) to locate sites using satellites. This makes relocating sampling plots in subsequent years much easier, even if markers are destroyed or removed.

Data management

A lengthy discussion of the statistics used in comparing restored areas with either historical data or a reference system is beyond the scope of this manuscript and is discussed in other publications (Pielou, 1986; Westman, 1991; Norris *et al.*, 1996; Michener & Houhoulis, 1997; Chapman, 1999). It is important to note, however, that data entry and analysis are time-consuming undertakings. They are often not budgeted in planning the costs of monitoring. As a consequence, many agencies collect reams of data that are never analysed. In planning a monitoring programme, the question of how the data will be analysed should be addressed. The US National Park Service recommends that 25–30% of a monitoring budget be allocated to data management and reporting (Fancy, 2000).

Selecting monitoring attributes

Ecological restoration is usually undertaken to re-establish attributes that were either impaired or destroyed. The National Research Council (1992) defines restoration as 'the return of an ecosystem to a close approximation of its condition prior to disturbance' (p. 18), although some projects may focus on single species or certain ecosystem functions. Criteria for monitoring will usually be those that validate or confirm the transitional stages of recovery, ideally ending with full ecosystem integrity and health. Selecting appropriate criteria for monitoring, however, is a challenging question. Since most ecosystems are quite complex, multiple lines of evidence at different levels of biological organisation are extremely useful. For example, at the population level, such criteria as genetic diversity, recruitment and age structure are important (Noss, 1990). At the community level, colonisation rates, species richness or ratio of native to exotic species may be measured. At the ecosystem level, illustrative examples of functional criteria are nutrient cycling, energy flow and habitat/resource partitioning. A complement of parameters should be selected with minimal intercorrelation (Westman, 1991).

Most complex systems have a certain amount of redundant information. Multiple confirming lines of evidence, which is another way of viewing information redundancy, are, however, good insurance against both false positives and false negatives that may occur if information is based on a single line of evidence. As stated earlier, incorrect decisions resulting from either false positives or false negatives are generally quite expensive. If redundant information results in even a fractional reduction in false positives and false negatives, it is virtually certain to have a value that exceeds its costs. Information redundancy has its highest value at the outset of a monitoring project when confidence in the reliability and accuracy of the measurements is justifiably in some doubt; over the life of a project, however, information redundancy should be regarded as backup to provide feedback on whether the system is functioning well.

In projects that aim to restore entire ecosystems, both functional and structural measurements should be measured. Structural measurements may include species composition, biomass or soil nutrients. Functional measurements range from nutrient flux to migration rates to hydrological flows. Structural and functional components may recover at different rates (Westman, 1991). Many projects monitor only structural measures of one or a few groups of species, such as plants, because of their ease of measurement, whereas other projects, such as mine reclamation, may focus primarily on functional parameters such as minimising soil loss and chemical leaching (Holl & Cairns, 1994). Not all structural and functional measurements are closely correlated, but they have complex interrelationships (Cairns & Pratt, 1986; Heckman, 1997). For example, research by Zedler and her colleagues that examined both structure and function of constructed saltmarshes in southern California (discussed further in Zedler & Adam, volume 2) demonstrated complex relationships between soil texture, nutrient cycling, cordgrass biomass, population size of a scale insect that eats cordgrass, and population size of a coccinellid beetle that feeds on the scale insects; these interrelated factors in turn influenced the suitability of the habitat for the endangered light-footed clapper rail, one of the focal species for creating the wetland. Likewise research by Holl and colleagues (Holl *et al.*, 2000), showed the interplay between soil nutrients, soil moisture, seed dispersal by birds, movement of mammals that eat seeds and seedlings, and competition with exotic pasture grasses on tropical forest restoration in Costa Rica. These examples illustrate the need to monitor a range of parameters in order to evaluate the recovery of an ecosystem.

For any given ecosystem type, numerous potential parameters could be monitored and all would provide some useful information. Time and financial constraints make it essential to select certain parameters to monitor. Cairns *et al.* (1993) reviewed the extensive literature of desirable characteristics of quality-control criteria, depending on whether the goal is to monitor compliance with restoration goals, provide insight into the reason for lack of compliance (diagnostic), or provide an early warning system that a restoration project is deviating from the desired trajectory (Table 21.1). Certain characteristics are important regardless of the goal of the monitoring, such as that they are cost-effective, readily measurable, cause minimum disruption to the system, and are at the appropriate scale. Some of these characteristics are inherently contradictory. A single parameter is unlikely to be both integrative and sensitive to specific stresses. For example, benthic (attached) communities are generally better indicators of local conditions in aquatic systems, whereas planktonic communities may integrate conditions across larger spatial scales (Cairns *et al.*, 1993). Parameters that are sensitive to stressors are important to serve as early-warning systems for problems. Ideally, changes in their values should precede those indicators that are most biologically and socially relevant (Cairns *et al.*, 1993). For example, monitoring nutrient levels in a wetland may not seem particularly important to the public, but may serve to indicate the potential for plant growth and,

Table 21.1. *Desirable characteristics for monitoring parameters*

Characteristic	Description
Biologically relevant	important in maintaining a balanced ecological community
Socially relevant	of obvious value to, and observable by, shareholders or predictive of a measure that is
Sensitive	sensitive to stressors without an all-or-none response to extreme or natural variability
Broadly applicable	usable at many sites
Diagnostic	helps explain the particular factor causing the problem
Measurable	capable of being operationally defined and measured, using a standard procedure with documented performance and low measurement error
Interpretable	capable of distinguishing acceptable from unacceptable conditions in a scientifically and legally defensible way
Cost-effective	inexpensive to measure, providing the maximum amount of information per unit
Integrative	summarises information from many unmeasured indicators
Historical or reference data available	data available to estimate variability, trends, and possibly acceptable and unacceptable conditions
Anticipatory	early warning, capable of providing an indication of deviation from a desired trajectory before serious harm has occurred
Non-destructive	causes minimal damage to ecosystem
Continuity	capable of being measured over time as the restored site matures
Appropriate scale	appropriate for the spatial scale of the restoration
Not redundant	provides unique information compared to other measures
Timely management	provides information quickly enough to initiate corrective action before extensive problems have occurred

Source: Modified from Cairns *et al.* (1993).

in turn, wildlife habitat. Parameters should be selected that provide a complement of information and are feasible to measure.

Standard methods

In any widespread activity, the monitoring procedures and protocols become standardised, and it is advisable to choose parameters from those for which measurement protocols are clearly outlined and widely used, particularly when measurements are made by people with less formal training than research investigators. Even research investigators should be obliged to collect some data using standard methods in order to facilitate comparisons between studies made in widely differing geographical areas or measurements made over considerable spans of time at the same place. Research investigators tend to be attracted by the latest methodologies and technologies which may provide substantial improvements in accuracy, precision and time efficiency over older techniques; at the same time, for studies covering large temporal and spatial spans, consistency in at least some measurement protocols over time is important.

It is often a shock to many ecologists when they find how important standard methods are in courts of law. Courts of law are particularly fond of standard methods for three reasons: (1) they usually represent a strong consensus by practitioners in a particular field as to how a particular measurement might be made; (2) each step of the analytic process is described in detail so that it is difficult to deviate inadvertently from the established methodology; and (3) because of their widespread use, especially in compliance to regulatory requirements, they are subjected to continual intense scrutiny, re-evaluation, and descriptions of situations that would produce spurious results. Standard methods give confidence to the general public, decision-makers and regulatory agencies, but should be carefully selected to answer the monitoring questions of interest.

Standard methods already exist for many of the chemical and physical parameters of interest to restoration ecologists. In fact, the mission of some organisations, such as the American Society for Testing and Materials, is entirely or substantially devoted to these activities. Biological standardised methods exist, but are not nearly as numerous as those in other categories.

The disadvantage of standard methods is their extreme rigidity, which has notable advantages, but is often frustrating when one tries to adjust them to the conditions of a specific site, many of which are unique. If one chooses to use non-standard methods for monitoring, it is usually advisable to run a few standard methods tests in parallel to be able to estimate what difference, if any, occurs in the numbers generated.

Indicator species

One overarching question applicable to monitoring restoration in all ecosystem types is whether or not indicator species should be used to evaluate restoration success. Obviously, the condition of each species and each ecosystem function cannot be monitored. Consequently, some restoration ecologists have tried to select species in the following categories: (1) the most sensitive species – i.e. one that is more sensitive to environmental stress than other species in the community; (2) indicator species – i.e. one that indicates specific conditions, either adverse or benign; and (3) representative species – i.e. one that responds to environmental conditions that correlate highly with other species in a community. Numerous species have been suggested as indicators of ecosystem health such as fish (Karr *et al.*, 1986), insects (Kremen *et al.*, 1993; Kerans & Karr, 1994), ants (Majer & Nichols, 1998) and birds (O'Connell *et al.*, 1998). Numerous criteria for selecting indicator species have been offered: (1) their life histories are well known; (2) they are easily surveyed and identified; (3) their populations fluctuate little in the absence of anthropogenic stress; and (4) they are sensitive to anthropogenic changes to the environment (Noss, 1990; McGeoch, 1998). In some cases, people have equated high species richness of indicator groups with high-quality habitat, but more often indices have been developed that use different functional groups to indicate habitat quality. For example, Karr *et al.* (1986) developed an index of biotic integrity that incorporates a number of metrics of fish communities: species richness, species in different habitat groups (e.g. benthic vs. pelagic species), trophic composition (omnivores, piscivores and insectivores), hybrids, and individuals with disease, tumours or other morphological anomalies.

Unfortunately, indicator species have proven to be problematic for a number of reasons. First, although each species indicates something, none indicates everything. Indicator species undoubtedly exist, but each indicates a different array of information. Otherwise, why preserve biodiversity? For example, abundant evidence in toxicological journals indicates that a species quite sensitive to compound A may be resistant to compound B. Further, Mayer & Ellersieck (1986) have shown that even extrapolations from one species to another are often unreliable. As a consequence, extrapolation from single species to higher levels of biological organisation, such as ecosystems, do not reveal a desirable level of statistical confidence (Cairns & Smith, 1996). Of course, if the objective is to restore a particular species, these deficiencies may be of no concern. If the goal is ecosystem restoration, however, these deficiencies deserve attention.

A third problem with indicator species is that few species or guilds of species act similarly across a range of ecosystems. For example, Erhardt (1985) and Murphy & Wilcox (1986) suggest the use of butterflies as sensitive indicators of plant species richness, whereas research by both Kremen (1992) and Holl (1995) disagrees with this assertion. Therefore, indicators will need to be tested for all ecosystems, which can be costly and time-consuming. Finally, different indicators respond to change at different scales. In the extreme case, using migratory songbirds as indicators of restoration success in temperate forests may mean little if their overwintering habitat has been impacted by deforestation in the tropics. On a smaller scale, Holl (1995) tested whether butterflies could serve as indicators of vegetation recovery on reclaimed coal surface mines. She found that adult butterflies were poor indicators of overall vegetation composition on the mines. Adults were found in disturbed areas with high abundance of primarily weedy, nectar-producing plants, even though their larvae often depend for food on later successional plants not present on reclaimed mines.

Indicator species may be appropriate to evaluate restoration success in some cases, but it is essential to specify which abiotic or biotic factors the species are being used to indicate (Landres *et al.*, 1988). Clearly, the idea of indicator species is attractive, because it would be much easier and less costly if a single, most sensitive species could be used. Given, however, that research on indicator species has provided highly mixed results on their utility, monitoring a carefully selected suite of parameters that answers specific management questions rather than just one group of species is recommended.

Monitoring invasive exotics

Invasive exotic species have a profound effect on both terrestrial and aquatic ecosystems worldwide by out-competing native species, altering disturbance regimes, increasing predation and disease and altering mutualistic relationships (Mackie *et al.*, 1989; Mills *et al.*, 1994; Hobbs & Humphries, 1995; Mack & D'Antonio, 1998). In some cases, invasive exotics are the primary focus of restoration efforts, such as removal of the bottle brush tree (*Melaleuca quinquenervia*), which has taken over an estimated 202 000–607 000 ha in the Everglades (Bodle *et al.*, 1994). In other cases, exotics may invade disturbed areas and alter the trajectory of succession or disturbance cycles (Mack & D'Antonio, 1998). Moreover, the appearance of invasive exotics, whether accidentally or deliberately introduced, may lead to further introduction of exotics in an effort to control them (e.g. Shireman & Maceina, 1981). Therefore, monitoring the spread of invasive exotics and the impacts of their removal is a consideration in designing restoration monitoring programmes in many ecosystems.

Monitoring the effects of invasive species and of the efforts to remove them is extremely difficult and is still in its developmental/exploratory stages. Ideally, removal and monitoring efforts lead to insights on reducing the problem, but they are unlikely to lead to total eradication. As a consequence, monitoring both the ecological effects of invasive exotics and the efficacy of control measures will almost certainly be an ongoing expense, justified by the enormous ecological and economic impacts caused by exotics. Given the extent and number of invasive exotic species, it will be necessary to prioritise those with the greatest impact and potential for spread for monitoring and removal efforts. Often, monitoring and removal of exotics focuses on the largest populations, but Moody & Mack (1988) have suggested that it is most effective to control 'satellite' populations to prevent the overall spread of invasive species. Therefore, monitoring beyond the focal populations of invasive species to detect new satellite populations is necessary (Hobbs & Humphries, 1995). Many of the monitoring techniques used for sampling the native biota will also serve to monitor for exotics, but, given their potential to spread, it may be necessary to monitor more widely for spreading populations. For aquatic systems, monitoring incoming sources of propagules, such as ballast water, is advisable (Kaiser, 1999). Monitoring for juvenile recruitment is also recommended, as high numbers of juveniles may indicate the potential for outbreaks.

Finally, efforts to remove exotics may have inadvertent effects on the native species. One well-publicised example concerns the exotic saltcedar (*Tamarix ramosissima*), which is lowering the water table and altering flooding regimes along many streams in the southwestern United States (Glausiusz, 1996). Its removal, however, has been controversial as it provides nesting habitat for the southwestern willow flycatcher (*Empidonax traillii extimus*). Therefore, the effects of removal efforts must be monitored, not only on the target species, but on the native ecosystem.

Social parameters

Although this discussion focuses on ecological monitoring of restoration, in many projects it is also important to measure social parameters, such as recreational opportunities, aesthetic qualities, or health and safety (Clarke, 1986; Westman, 1991). As much restoration is supported by public funds (Holl & Howarth, 2000), and restoration projects are often initiated by volunteers (Berger, 1985), how restoration projects affect people must be evaluated. These attitudes might be measured through monitoring of recreational use patterns, surveys of attitudes of nearby residents, or by community participation in management and decision-making (Kondolf & Micheli, 1995). For example, Barro & Bright (1998) surveyed residents of the Chicago area for their opinions regarding and knowledge of controversial efforts to restore pre-settlement conditions to nearby forests, and Schroeder (1998) analysed articles written for stewardship newsletters by volunteers working in north central Illinois to assess how participation in restoration efforts affected their lives. Quite often, volunteers are responsible for a large proportion of monitoring of restoration projects in public areas (Masters, 1997). Regardless of who collects monitoring data, it should and often is required by law to be made available to the public to increase its understanding of and confidence in restoration efforts.

THE BASICS OF MONITORING IN DIFFERENT ECOSYSTEM TYPES

Terrestrial ecosystems

A list of potential parameters for monitoring terrestrial restoration is given in Table 21.2. Of the numerous parameters that have been suggested, plant cover and composition are most commonly compared, probably because they are relatively easy to monitor. A number of monitoring systems for plants in both grasslands and forests have been developed. Grassland monitoring usually involves measuring the cover of individual species in quadrats or along transects. Forest monitoring usually includes measuring of diameter at breast height of trees and also sometimes includes measurements in the herb or shrub layers. Often, these species are evaluated according to origin (native or exotic) and range (localised species vs. widespread generalists). Numerous books and chapters outline plant sampling methods (e.g. Müller-Dombois & Ellenberg, 1974;

Table 21.2. *Potential parameters for monitoring terrestrial systems*

Genetic
Allelic diversity
Inbreeding depression
Gene flow
Species/Population
Abundance/density
Bioaccumulation
Birth, death and recruitment
Dispersal
Frequency
Growth rates
Population structure (age ratio; sex ratio)
Phenology
Survivorship
Size
Ecosystem
Carbon, nitrogen fixation
Decomposition rates
Disturbance intervals
Drainage patterns
Soil texture, nutrients and organic matter
Topography
Water quality of runoff
Community
Biomass
Occurrence of indicator species
Occurrence of rare species
Frequency of parasitism
Percentage cover
Amount of predation
Proportion of exotic species
Proportion of threatened and endangered species
Relative abundance of predators and prey
Species diversity
Species richness
Species composition
Vegetation structure
Landscape
Amount of edge
Connectivity
Fragmentation
Patch size
Proportion surrounded by different habitats

Sources: From Clarke (1986), Noss (1990), Spellerberg (1991).

Greig-Smith, 1983; Sutter, 1996; Masters, 1997; Sauer, 1998).

In some cases, other organisms such as birds or insects may be monitored. For example, quite detailed protocols have been established for monitoring birds through field surveys, mist-netting, and nest searches (Ralph *et al.*, 1993). Insects have been suggested as good indicators of environmental changes (Kremen *et al.*, 1993; Majer & Nichols, 1998), but standardised protocols for their sampling are not as well developed. With the exception of a few insect groups, such as butterflies and ants, difficulties of identification make their monitoring problematic.

Because much of terrestrial restoration is motivated by restoration of species that are legally protected, such as threatened and endangered species in the United States, a great deal has been written on requirements for monitoring of rare plant and animal species (Pavlik, 1994; Sutter, 1996; Fritts *et al.*, 1997; DeSante & Rosenberg, 1998). For endangered species, monitoring of restoration or reintroduction efforts should include the monitoring of survival at different life-history stages, as well as of other factors necessary to sustain long-term viability of the population, such as dispersal and genetic variation (Pavlik, 1996). For example, Pavlik & Espeland (1998) found that a reintroduced population of the rare serpentine annual plant (*Acanthomintha duttonii*) had similar plant size and seed set to a natural population, but that recruitment was quite low. They attribute low recruitment to seed predation and poor seed germination. With detailed demographic monitoring, they were able to isolate the cause for the low recruitment and focus future research and management efforts. At a minimum, recruitment and survival must be monitored over a number of years to determine whether the population is self-sustaining, as monitoring only adults requires much longer data sets to detect trends in population size, particularly for long-lived species (Pavlik, 1994; DeSante & Rosenberg, 1998). Obtaining data for animals can be quite expensive and time-consuming because of their mobility.

Soil monitoring is essential because of its tantamount importance in the establishment and survival of plant and animal species. Highly disturbed sites are often highly acidic and may be low in macronutrients, such as nitrogen, phosphorus and potassium, as well as micronutrients. In areas invaded by nitrogen-fixing exotics, such as bush lupine (*Lupinus arboreus*) in California dunes, nitrogen levels may be elevated and need to be reduced to restore native plant communities (Maron & Connors, 1996). If efforts are made to ameliorate soil conditions, ongoing monitoring of these parameters is necessary. Standardised methodologies are available for this monitoring (Smith & Mullins, 1991; Weaver, 1994; Sparks, 1996). Soil microbial communities are critical to restoring nutrient cycling processes and native plant communities (Allen *et al.*, 1999, this volume). Despite this fact, soil microbial communities are rarely monitored in restoration sites, except in highly contaminated sites, due to the difficulty of their identification. Parameters that are most often measured include total microbial biomass, certain enzyme levels, and microbial processes, such as respiration, nitrogen mineralisation and nitrogen fixation (Ross *et al.*, 1992; Brookes, 1995).

In some cases, functional measurements, such as nutrient mineralisation rates or productivity, are monitored, but such measurements are less common because of the time and equipment involved in such measurements. When ecosystem damage causes changes in fluxes of certain chemicals into the environment, then their monitoring may be required. For example, in the United States under the Surface Mine Control and Reclamation Act, mining companies are required to monitor water quality in the vicinity of active and recently reclaimed mines (Wali *et al.*, volume 2). Leachates from mining operations may contaminate both groundwater aquifers and surface waters for many years after the mining operations cease. This contamination may include acidification of the water when iron pyrite is present; but the more serious problem from the human health standpoint is undoubtedly the continual leaching of heavy metals, which requires ongoing monitoring. Often, restoration of highly disturbed sites is aimed at minimising erosion and sediment runoff into nearby water bodies. Therefore, sediment fluxes may be monitored by using sediment basins or other sediment capture structures.

Most monitoring protocols developed thus far for terrestrial ecosystems are for widespread ecosystems in the temperate zone, namely hardwood forests and grasslands. These measurement protocols may not be appropriate for other systems. For example, at least half of the tropical rainforests have been destroyed worldwide, and efforts are increasing to restore these ecosystems (various articles in *Restoration Ecology*, vol. 8(4), 2000; Holl, volume 2; Janzen, volume 2). Given the high diversity of tree species in these forests and the different strata of shrubs and trees, vegetation sampling regimes for the temperate zone are not likely to be appropriate. Likewise, efforts are increasing to restore montane and arid ecosystems (Bainbridge *et al.*, 1995; Chambers, 1997; Anderson *et al.*, volume 2; Tongway & Ludwig, volume 2). These systems often have low vegetation cover naturally (one common goal of restoration) and episodic establishment, which make judging project success complicated.

Wetlands

Establishment of suitable end points for wetland restoration is challenging because wetlands are transitional between terrestrial and aquatic systems and are highly variable temporally. Since a substantial number of wetland restoration projects are undertaken as mitigation, restoration ecologists have no choice but to become well acquainted with the requirements of legal compliance in their region. However, as many case studies have shown, evidence for legal compliance often falls short of what most ecologists would consider as restoration of ecological structure and function (Mitsch & Wilson, 1996; Simenstad & Thom, 1996; Zedler & Callaway, 1999). Wetlands have been variably defined over time in different countries (discussed in detail in Tiner, 1999). Generally, wetlands are defined by their periodically flooded conditions, and the soil characteristics and plant communities that result from these conditions (Tiner, 1999). Therefore, wetland monitoring generally focuses on hydrology, soils and vegetation. A thorough discussion of wetland monitoring is given in Pacific Estuarine Research Laboratory (1990) and Tiner (1999).

Hydrology is often measured as depth of standing water over an appropriate temporal period to assess fluctuations in water level. Many characteristics of flooding are critical, including depth of saturation, duration of saturation, and timing (Pacific Estuarine Research Laboratory, 1990; Tiner, 1999). For example, the periodicity of standing water may be of the order of hours for tidal wetlands, whereas water depth may change over an order of weeks for seasonally inundated wetlands such as vernal pools and prairie potholes.

Wetland hydrology influences the chemical and physical nature of wetland soils to a large extent. Wetland soils are typically low in oxygen and high in organic matter (Tiner, 1999). Basic measurements for monitoring wetland soils include texture, organic matter, toxic substances, pH and redox potential (Pacific Estuarine Research Laboratory, 1990). Redox potential is a measure of how oxidised the soils are, which plays an important role in the biogeochemical cycling of nitrogen and sulphur and the mobility of heavy metals. Past research (Langis *et al.*, 1991; Craft *et al.*, 1999) has demonstrated that it can be extremely challenging to restore nutrient cycling in wetlands. Therefore, measurements of certain nutrient cycling parameters, such as nitrogen fixation rates, litter decomposition, and nitrogen mineralisation, are important in determining whether wetland functions have been restored (Pacific Estuarine Research Laboratory, 1990).

Another parameter that is often measured in evaluating wetland success is plant community composition. Certain plant species have adaptations to flooded environments, such as hollow stems, transport of oxygen to roots from leaves, and roots near the surface. In local floras, wetland plant species are often categorised according to their requirement for inundated substrates. In the United States, a number of agencies jointly developed a manual that ranks plants on a scale based on the range of their dependence on wetland habitats (Reed, 1988, 1997). Atkinson *et al.* (1993) outlined a protocol for comparing whether restored or created wetlands are equally 'wet', in which they multiplied the cover value of each plant species by the indicator status of the plant (ranging from 1 for obligate wetland plants to 5 for obligate upland plants) and divided this by the total cover. As with all ecosystems, cover of invasive species and focal species must also be documented.

Wetlands play an important role in supporting food chains and also host a number of threatened and endangered species. Therefore, monitoring of various trophic levels may be important to evaluate success in achieving goals. Wetland restoration monitoring protocols may also include monitoring of species that play important functional roles, such as decomposers and shredders or pollinating and predatory insects (Pacific Estuarine Research Laboratory, 1990). Monitoring target species, such as rare birds or fishes, is important if increasing their population size is a goal of the project. Dahm *et al.* (1995) recommend, for the massive Kissimmee River restoration project (described previously) that hydrology, vegetation, aquatic invertebrates, birds and fish all be monitored to assess the success in restoring abiotic–biotic interactions and food-web interactions; but such an extensive monitoring plan would not be feasible for most wetland restoration projects.

Rivers

The need for a landscape-scale approach in restoration is especially critical in efforts to restore aquatic systems. Recovery of aquatic systems is dependent on restoration of natural levels of inputs, including water, sediments and chemicals, from terrestrial systems (Cairns & Heckman, 1996; Hobbs & Norton, 1996). Therefore, monitoring flows of materials into rivers and lakes is almost always necessary. If these inputs are corrected to a normal range of levels, riverine ecosystems, particularly, often recover quickly because of: (1) the adaptation of many species for rapid recolonisation and repopulation of disturbed areas; (2) the availability of unaffected upstream and downstream areas that serve as refugia for sources of colonisation; (3) high flushing rates of lotic systems that allow for the dilution of contaminated waters; and (4) the fact that riverine systems have been historically affected by many disturbances and are fairly resilient (Yount & Niemi, 1990; Cairns & Heckman, 1996).

Efforts to restore rivers focus on a range of goals, including restoring geomorphology, hydrology, water quality, biotic communities and/or single species, which will influence selection of parameters to be monitored. A number of parameters commonly monitored in river systems are listed in Table 21.3. For more detailed discussions of river restoration monitoring see National Research

Table 21.3. *Potential parameters for monitoring rivers*

Morphology
Meander geometry
Rates of bank erosion
Riffle-to-pool ratio
Width-to-depth ratios
Hydrology
Bank/stream storage
Bed material size
Groundwater flow and exchange processes
Quantity of discharge on annual, seasonal and episodic basis
Retention times
Sediment flux
Water velocity
Water depth
Water quality
Acidity/alkalinity
Dissolved toxics
Dissolved salts/conductivity
Dissolved oxygen
Nitrogen and phosphorus concentrations
Suspended sediments/turbidity
Water temperature
Biotic
Biotic interactions
Coarse woody debris
Habitat quality
Life-history stages of fish
Percentage vegetative cover
Presence of sensitive species
Production of algae, macrophytes, bacteria/fungi, invertebrates, fish
Standing stock of algae, macrophytes, bacteria/fungi, invertebrates, fish
Taxonomic richness and diversity
Trophic diversity

Sources: From National Research Council (1992), Kondolf & Micheli (1995), Armitage (1996), Norris *et al.* (1996), Federal Interagency Stream Restoration Working Group (1998).

Council (1992), Kondolf & Micheli (1995), Norris *et al.* (1996) and Federal Interagency Stream Restoration Working Group (1998). Rivers are dynamic systems and much of their channel movement occurs at peak flows. Therefore, it is necessary to monitor water flows over a sufficient period of time to include peak flow conditions in order to assess success of river restoration efforts (Kondolf & Micheli, 1995; Gore, 1996).

Restoring natural geomorphology is critical to restoring habitat for riparian and aquatic species (Kondolf & Micheli, 1995; Newson *et al.*, this volume; Jeppeson & Sammalkorpi, volume 2). Examples of efforts to restore natural meander patterns range from the Blanco River, in Colorado (where David Rosgen reduced the river's bank-full width from a 120-m-wide braided channel to a stable, 19.5-m-wide channel with a high pool-to-riffle ratio using soft engineering techniques [National Research Council, 1992, pp. 470–7] to the large-scale Kissimmee River restoration described earlier. Typically, stream channel surveys are used to measure width-to-depth ratios, meander geometry, riffle-to-pool ratio and bank erosion. Evaluating whether these parameters are within acceptable values depends on the natural system; in some cases, high bank erosion may be desirable, whereas in other cases it may signify a problem (Kondolf & Micheli, 1995).

Restoration of natural hydrological patterns is closely associated with geomorphology. Flow rates have been altered in many streams to control flooding, provide water storage and provide water for agricultural, domestic and industrial uses. Efforts to do controlled releases from dams, such as the Glen Canyon Dam on the Colorado River (Schmidt *et al.*, 1998), and restore more natural flooding regimes (Sparks *et al.*, 1998) are increasing. Commonly monitored parameters include seasonal quantity of discharge, with emphasis on high and low flow periods, and groundwater recharge. Peak flows mobilise sediment of various sizes, affecting both water quality and composition of bed substrate, which is important in determining habitat quality for a range of species.

In some cases, restoration of rivers focuses on reducing chemical inputs to restore water quality. For example, in the Willamette River in Oregon, efforts have been made to treat, reduce and dilute industrial effluent to improve habitat for anadromous fish species (National Research Council, 1992, pp. 170–1). Parameters for monitoring water quality have been extensively discussed in the ecotoxicology literature (Cairns *et al.*, 1978; American Public Health Association, 1995; Hoffman *et al.*, 1995; American Society for Testing and Materials, 2000); dissolved oxygen, suspended sediments, pH, nitrogen and phosphorus, and levels of other contaminants are examples. These values may be helpful in assessing extreme deviations from water-quality goals, but focusing on fragmented water-quality objectives, such as a concentration not to exceed x parts per million or the requirement that pH be above a certain level, are somewhat meaningless in isolation from other attributes because organisms respond to the aggregate conditions. Therefore, a number of different species such as macroinvertebrates, algae, protozoans and fish (Cairns, 1985; Cairns & Pratt, 1992; Armitage, 1996; Karr & Chu, 1998; Barbour *et al.*, 1999) have been used to integrate water-quality measurements. Cairns (1985) strongly argued for the need of multiple levels of measurement to assess aquatic community health.

As with terrestrial systems, many aquatic restoration projects focus on restoration of a single species – in particular, a species of fish. Typical monitoring parameters include those outlined for focal terrestrial species, including measuring a range of demographic parameters. In addition, certain habitat variables, such as appropriate bed substrate for spawning, riparian cover to reduce stream temperature, and large woody debris to provide safe areas from predators, may all require monitoring. If, however, appropriate geomorphology, hydrology and water quality are not restored, efforts to maintain single species will necessarily require ongoing subsidies to maintain populations.

Lakes

The fact that lakes occupy such a small fraction of the landscape belies their importance as environmental systems and resources for human use (National Research Council, 1992). Restoration of lakes is problematic because they are at the end of the line without the advantages of dilution and flushing mechanisms that rivers have; the restoration of lakes depends on proper land management to reduce inputs of nutrients, chemicals and sediments. Excessive nutrients can result in eutrophication and algal blooms. Suspended sediments impair light penetration, and, therefore, reduce plant productivity and have adverse effects on fisheries. Over time, sediment inputs result in reduced lake volume.

Most lake restoration has focused on controlling eutrophication and sedimentation and improving water quality; numerous methods have been used to restore the ecological condition of lakes, ranging from dredging to biological control to chemical inactivation of phosphorus (National Research Council, 1992; Søndergaard *et al.*, this volume; Jeppesen & Sammalkorpi, volume 2). Efforts to reduce inputs to lakes have resulted in surprisingly successful recovery in some cases. For example, almost immediately after nutrient loading to Lake Washington, USA, was reduced, lake transparency increased and phosphorus and chlorophyll levels decreased dramatically (National Research Council, 1992, pp. 118–19). Most commonly, lake monitoring programmes focus on water-quality parameters, paralleling those used in rivers (Table 21.3). High phosphorus levels are a particular concern in lakes since high levels often lead to eutrophication. Dissolved oxygen concentration is also regularly measured since low oxygen levels are common in highly eutrophied lakes. Other commonly measured water-quality parameters include water transparency and concentration of chlorophyll *a*, which is an indicator of algal biomass that signifies elevated nutrient levels. Lake trophic status can be altered by increased nutrients (bottom–up effects), as well as introduction of zooplanktivorous or piscivorous fish species (top–down effects: Perrow *et al.*, this volume; Jeppesen & Sammalkorpi, volume 2). Therefore, measures of trophic status are also used in monitoring the restoration of lakes (Hutchinson, 1975; Carlson, 1977; Kratzer & Brezonik, 1981).

Lakes trap both nutrients and toxics effectively, so an important step in the development of any monitoring programme is to determine both how these are partitioned in the lake and how episodic events might cause some nutrients and toxics, thought to be safely stored in the lake sediments, to be reintroduced into the water column. Both can be reintroduced into the water column by the activities of organisms, especially fish, and disturbance such as dredging and lake turnover, and can partition differentially through the food chain. As Hutchinson (1975) noted many years ago, lakes age in a variety of ways, even without anthropogenic effects. This ageing process must be taken into account in the monitoring programme because some changes are inevitable, even without the influence of human society.

Marine ecosystems

The world's oceans contain practically all of its water, yet only the most rudimentary knowledge is available of how the network of marine systems works or even what organisms inhabit them and their population and ecosystem dynamics (Hawkins *et al.*, volume 2). In particular relation to marine ecosystems, it is important to reiterate that robust goals for restoration cannot be developed unless one has good descriptions of both the structure and function of the system. Since the normal or stable state is imprecisely known, lack of evidence is often taken as no evidence of harm. Establishing the normal state as far as possible before activities begin is prudent. The degree to which such large systems vary is poorly understood, but the effects of changes in the oceans upon the climate of terrestrial systems, such as La Niña and El Niño, have been witnessed several times. Given this natural variability, setting end points for restoration is challenging, to say the least.

Most efforts to restore oceans have focused on coastal ecosystems such as coral reefs (Clark, volume 2), various types of wetlands (Zedler & Adam, volume 2) and seagrass beds (Fonseca *et al.*, volume 2), and clean-up of oil spills (Hawkins *et al.*, volume 2; Edwards, 1998). In such cases, many of the monitoring parameters discussed with respect to other ecosystems, such as vegetation composition (for wetlands), water quality and various faunal surveys, are relevant. Case histories in a range of systems have almost all shown that the cost of restoration exceeds the cost of prevention by at least an order of magnitude. Because of the rapid spread of contaminants and exotic species in the ocean, controlling their input becomes the only viable approach. For vast systems such as the oceans, ecological disequilibrium may inadvertently be caused by effects far from the focal area. Since these issues cross boundaries, international co-operation will be required to resolve them (National Research Council, 1997). For example, the Baltic Sea has been largely eutrophied. Resolving this problem has required co-ordinating control of nutrient inputs to the Baltic across a wide variety of political jurisdictions (Ferm, 1991).

CONCLUDING REMARKS

An enormous amount of material has been touched on in this discussion. Important points in planning monitoring for all ecosystems have been highlighted and questions of interest to specific ecosystems have been briefly discussed. In closing, a few points should be reiterated. First, restoration goals must be clearly stated at the outset. Otherwise, monitoring efforts are futile. Without clearly articulated goals, it is impossible to select parameters to monitor or criteria for comparison. Second, monitoring programmes will necessarily be specific to certain ecosystem types and project goals. Therefore, we have aimed to raise important questions in designing monitoring programmes, but have left them to be answered by those implementing restoration projects. Much of restoration implementation is handled by consulting firms that are not subject to a peer review process comparable to that of quality professional journals. We strongly recommend that managers of specific projects consult with a range of scientists in developing and interpreting the monitoring programme, in order for non-scientific audiences to have confidence in the monitoring and evaluation process (Zedler, 1988). Third, since the number of restoration projects is increasing exponentially and many have been initiated quite recently, monitoring protocols for restoration are necessarily somewhat experimental in nature. Therefore, practitioners and scientists must publish both their successes and failures and keep abreast of the literature. With time, sampling protocols are becoming increasingly standardised. As has been discussed, standardisation will help in comparing projects and developing confidence for restoration in decision makers and the general public.

ACKNOWLEDGMENTS

We are indebted to Eva Call for transcribing portions of this manuscript and to Darla Donald for editorial assistance. Financial support for preparation of parts of this manuscript was provided by the Cairns Foundation.

References

Allen, M.F., Allen, E.B., Zink, T.A., Harney, S., Yoshida, L.C., Sigüenza, C., Edwards, F., Hinkson, C., Rillig, M., Bainbridge, D., Doljanin, C. & MacAller, R. (1999). Soil microorganisms. In *Ecosystems of Disturbed Ground*, ed. L.R. Walker, pp. 521–544. Amsterdam: Elsevier.

American Public Health Association, American Water Works Association, and Water Environment Foundation (1995). *Standard Methods for the Examination of Water and Wastewater*, 19th edn, eds. A.D. Eaton, L.S. Clesceri & A.E. Greenberg. Washington, DC.

American Society for Testing and Materials (2000). *Annual Book of ASTM Standards*, vol. 11.01, *Water*. Philadelphia, PA.

Anderson, A.N. (1993). Ants as indicators of restoration success at a uranium mine in tropical Australia. *Restoration Ecology*, **1**, 156–167.

Armitage, P.D. (1996). Prediction of biological responses. In *River Biota*, eds. G. Petts & P. Calow, pp. 231–252. Oxford: Blackwell.

Atkinson, R.B., Perry, J.E., Smith, E. & Cairns, J., Jr (1993). Use of created wetland delineation and weighted averages as a component of assessment. *Wetlands*, **13**, 185–193.

Bainbridge, D.A., Fidelbus, M. & MacAller, R. (1995). Techniques for plant establishment in arid ecosystems. *Restoration and Management Notes*, **13**, 190–197.

Barbour, M.T., Gerritsen, J., Snyder, B.D. & Stribling, J.B. (1999). *Rapid Bioassessment Protocols for Use in Streams and Wadeable Rivers: Periphyton, Benthic Macroinvertebrates, and Fish*, 2nd edn, EPA 841-B-99-002. Washington, DC: US Environmental Protection Agency, Office of Water.

Barro, S.C. & Bright, A.D. (1998). Public views on ecological restoration: a snapshot from the Chicago area. *Restoration and Management Notes*, **16**, 59–65.

Berger, J.J. (1985). *Restoring the Earth*. New York: Knopf.

Bodle, M.J., Ferriter, A.P. & Thayer, D.D. (1994). The biology, distribution, and ecological consequences of *Melaleuca quinquenervia* in the Everglades. In *Everglades: The Ecosystem and Its Restoration*, eds. S.M. Davis & J.C. Ogden, pp. 341–355. Delray Beach, FL: St Lucie Press.

Bradshaw, A. (1997). What do we mean by restoration? In *Restoration Ecology and Sustainable Development*, eds. K.M. Urbanska, N.R. Webb & P.J. Edwards, pp. 8–16. Cambridge: Cambridge University Press.

Brenner, F.J., Werner, M. & Pike, J. (1984). Ecosystem development and natural succession in surface coal mine reclamation. *Minerals and Environment*, **6**, 10–22.

Brookes, P.C. (1995). The use of microbial parameters in monitoring soil pollution by heavy metals. *Biology and Fertility of Soils*, **19**, 269–279.

Burger, J.A. & Torbert, J.L. (1990). Mined land reclamation for wood production in the Appalachian region. In *Proceedings of the 1990 Mining and Reclamation Conference and Exhibition*, vol. 1, eds. J. Skousen, J. Sencindiver & D. Samuel, pp. 159–163. Morgantown, WV: West Virginia University.

Cairns, J., Jr (1985). *Multispecies Toxicity Testing*. New York: Pergamon Press.

Cairns, J., Jr (1991). Developing a strategy for protecting and repairing self-maintaining ecosystems. *Journal of Clean Technology and Environmental Science*, **1**, 1–11.

Cairns, J., Jr & Heckman, J. R. (1996). Restoration ecology: the state of an emerging field. In *Annual Review of Energy and the Environment*, vol. 21, ed. R.H. Socolow, pp. 167–187. Palo Alto, CA: Annual Reviews, Inc.

Cairns, J., Jr & Niederlehner, B.R. (1995). *Ecological Toxicity Testing: Scale, Complexity, and Relevance*. Boca Raton, FL: Lewis Publishers.

Cairns, J., Jr & Pratt, J.R. (1986). Ecological consequence assessment: effects of bioengineered organisms. *Water Resources Bulletin*, **22**, 171–182.

Cairns, J., Jr & Pratt, J.R. (1992). A history of biological monitoring using benthic macroinvertebrates. In *Freshwater Biomonitoring and Benthic Macroinvertebrates*, eds. D. M. Rosenberg & V. H. Resh, pp. 10–27. New York: Chapman & Hall.

Cairns, J., Jr & Smith, E.P. (1996). Uncertainties associated with extrapolating from toxicologic responses in laboratory systems to the responses of natural systems. In *Environmental Problem Solving*, ed. J. Lemons, pp. 188–205. Cambridge, MA: Blackwell.

Cairns, J., Jr, Dickson, K.L. & Maki, A.W. (eds.) (1978). *Estimating the Hazard of Chemical Substances to Aquatic Life*, STP657. Philadelphia, PA: American Society for Testing Materials.

Cairns, J., Jr, Buikema, A.L., Jr, Cherry, D.S., Herricks, E.E., Matthews, R.A., Niederlehner, B. R., Rodgers, J.H., Jr & van der Schalie, W.H. (1982). *Biological Monitoring in Water Pollution*. Oxford: Pergamon Press.

Cairns, J., Jr, McCormick, P.V. & Niederlehner, B.R. (1993). A proposed framework for developing indicators of ecosystem health. *Hydrobiologia*, **263**, 1–44.

Carbyn, L.N., Armbruster, H.J. & Mamo, C. (1994). The swift fox reintroduction program in Canada from 1983 to 1992. In *Restoration of Endangered Species: Conceptual Issues, Planning, and Implementation*, eds. M.L. Bowles & C.J. Whelan, pp. 247–271. Cambridge: Cambridge University Press.

Carlson, R.E. (1977). A trophic state index for lakes. *Limnology and Oceanography*, **22**, 361–369.

Chambers, J.C. (1997). Restoring alpine ecosystems in the western United States: environmental constraints, disturbance characteristics, and restoration success. In *Restoration Ecology and Sustainable Development*, eds. K.M. Urbanska, N.R. Webb & P.J. Edwards, pp. 161–187. Cambridge: Cambridge University Press.

Chapman, M.G. (1999). Improving sampling designs for measuring restoration in aquatic habitats. *Journal of Aquatic Ecosystem Stress and Recovery*, **6**, 235–251.

Clarke, R. (1986). *The Handbook of Ecological Monitoring*. Oxford: Oxford University Press.

Craft, C., Reader, J., Sacco, J.N. & Broome, S.W. (1999). Twenty-five years of ecosystem development of constructed *Spartina alterniflora* (Loisel marshes). *Ecological Applications*, **9**, 1405–1419.

Czuczwa, J.M., McVeety, B.D. & Hites, R.A. (1984). Polychlorinated dibenzo-p-dioxins and dibenzofurans in sediments from Siskiwit Lake, Isle Royale. *Science*, **226**, 568–569.

Dahm, C.N., Cummins, K.W., Valett, H.M. & Coleman, R.L. (1995). An ecosystem view of the restoration of the Kissimmee River. *Restoration Ecology*, **3**, 225–238.

DeSante, D.F. & Rosenberg, D.K. (1998). What do we need to monitor in order to manage landbirds? In *Avian Conservation: Research and Management*, eds. J.M. Marzluff & R. Sallabanks, pp. 93–106. Washington, DC: Island Press.

Dick, W.A., Bonta, J.V. & Haghiri, F. (1986). Chemical quality of suspended sediment from watersheds subjected to surface coal mining. *Journal of Environmental Quality*, **15**, 289–293.

Edwards, A. (1998). Rehabilitation of coastal ecosystems. *Marine Pollution Bulletin*, **37**, 371–372.

Ehrenfeld, J.G. (2000). Defining the limits of restoration: the need for realistic goals. *Restoration Ecology*, **8**, 2–9.

Enserink, M. (1999). Plan to quench the Everglades' thirst. *Science*, **285**, 180.

Erhardt, A. (1985). Diurnal Lepidoptera: sensitive indicators of cultivated and abandoned grassland. *Journal of Applied Ecology*, **22**, 849–861.

Fairweather, P.G. (1991). Statistical power and design requirements for environmental monitoring. *Australian Journal of Marine and Freshwater Research*, **42**, 555–568.

Fancy, S.G. (2000). Monitoring natural resources in our national parks. http:/www.nature.nps.gov/sfancy/

Federal Interagency Stream Restoration Working Group (15 federal agencies of the US government) (1998). *Stream Corridor Restoration: Principles, Processes, and Practices*, GPO Item no. 0120-A; SuDocs no. A 57.6/2:EN 3/PT.653. Springfield, VA: National Technical Information Service.

Ferm, R. (1991). Integrated management of the Baltic Sea. *Marine Pollution Bulletin*, **23**, 533–540.

Fritts, S.H., Bangs, E.E., Fontaine, J.A., Johnson, M.R., Phillips, M.K., Koch, E.D.& Gunson, J.R. (1997). Planning and implementing a reintroduction of wolves in Yellowstone National Park and central Idaho. *Restoration Ecology*, **5**, 7–27.

Gauch, H.G., Jr (1982). *Multivariate Analysis in Community Ecology*. Cambridge: Cambridge University Press.

Glausiusz, J. (1996). Trees of salt: tamarisk trees are drying up American west. *Discover*, **17**, 30–31.

Gore, J.A. (1996). Responses of aquatic biota to hydrological changes. In *River Biota*, eds. G. Petts & P. Calow, pp. 209–230. Oxford: Blackwell.

Greig-Smith, P. (1983). *Quantitative Plant Ecology*, 3rd edn. Berkeley, CA: University of California Press.

Heckman, J.R. (1997). Restoration of degraded land: a comparison of structural and functional measurements of recovery. PhD thesis, Virginia Polytechnic Institute and State University, Blacksburg, VA.

Hobbs, R.J. & Humphries, S.E. (1995). An integrated approach to the ecology and management of plant invasions. *Conservation Biology*, **9**, 761–770.

Hobbs, R.J. & Norton, D.A. (1996). Towards a conceptual framework for restoration ecology. *Restoration Ecology*, **4**, 93–110.

Hoffman, D.J., Rattner, B.A., Burton, G.A. & Cairns, J., Jr (eds.) (1995). *Handbook of Ecotoxicology*. Boca Raton, FL: Lewis Publishers.

Holl, K.D. (1995) Nectar resources and their effect on butterfly population dynamics on reclaimed coal surface mines. *Restoration Ecology*, **3**, 76–85.

Holl, K.D. (1996) Restoration ecology: some new perspectives. In *Preservation of Natural Diversity in Transboundary Protected Areas: Research Needs/Management Options*, eds. A. Breymeyer & R. Noble, pp. 25–35. Washington, DC: National Academy Press.

Holl, K.D. & Cairns, J., Jr (1994). Vegetational community development on reclaimed coal surface mines in Virginia. *Bulletin of the Torrey Botanical Club*, **121**, 327–337.

Holl, K. D. & Cairns, J., Jr (1995). Landscape indicators in ecotoxicology. In *Handbook of Ecotoxicology*, eds. D. J. Hoffman, B. A. Rattner, G. A. Burton & J. Cairns, Jr, pp. 185–197. Boca Raton, FL: Lewis Publishers.

Holl, K.D. & Howarth, R.B. (2000). Paying for restoration. *Restoration Ecology*, **8**, 261–267.

Holl, K.D., Loik, M.E., Lin, E.H.V. & Samuels, I.A. (2000) Restoration of tropical rain forest in abandoned pastures in Costa Rica: overcoming barriers to dispersal and establishment. *Restoration Ecology*, **8**, 339–349.

Hutchinson, G.E. (1975). *A Treatise on Limnology*, vol. 3. New York: John Wiley.

Jackson, L.L., Lopoukhine, N. & Hillyard, D. (1995). Ecological restoration: a definition and options. *Restoration Ecology*, **3**, 71–75.

Kaiser, J. (1999). Stemming the tide of invading species. *Science*, **285**, 1836, 1838–1839, 1841.

Karr, J.R. & Chu, E. (1998). *Restoring Life in Running Waters*. Washington, DC: Island Press.

Karr, J.R., Fausch, K. D., Angermeier, P.L., Yant, P.R. & Schlosser, I.J. (1986). *Assessing Biological Integrity in Running Waters: A Method and its Rationale*, Illinois Natural History Survey Special Publication no. 5. Champaign, IL: Illinois Natural History Survey.

Kerans, B.L. & Karr, J.R. (1994). A benthic index of biotic integrity (B-IBI) for rivers of the Tennessee Valley. *Ecological Applications*, **4**, 768–785.

Koebel, J.W., Jr (1995). An historical perspective on the Kissimmee River Project. *Restoration Ecology*, **3**, 149–159.

Kondolf, G.M. (1995). Five elements for effective evaluation of stream restoration. *Restoration Ecology*, **3**, 133–136.

Kondolf, G.M. & Micheli, E.R. (1995). Evaluating stream restoration projects. *Environmental Management*, **19**, 1–15.

Kratzer, C.R. & Brezonik, P.L. (1981) A Carlson-type trophic state index for nitrogen in Florida lakes. *Water Resources Bulletin*, **17**, 713–715.

Kremen, C. (1992). Assessing the indicator properties of species assemblages for natural areas monitoring. *Ecological Applications*, **2**, 203–217.

Kremen, C., Colwell, R.K., Erwin, T.L., Murphy, D.D., Noss, R.F. & Sanjayan, M.A. (1993). Terrestrial arthropod assemblages: their use in conservation planning. *Conservation Biology*, **7**, 796–808.

Landres, P.B., Verner, J. & Thomas, J.W. (1988). Ecological uses of vertebrate indicators: a critique. *Conservation Biology*, **2**, 316–328.

Langis, R., Zalejko, M. & Zedler, J.B. (1991). Nitrogen assessments in a constructed and a natural salt marsh of San Diego Bay. *Ecological Applications*, **1**, 40–51.

Lockwood, J.L. & Pimm, S.L. (1999). When does restoration succeed? In *Ecological Assembly Rules*, eds. E. Weiher & P. Keddy, pp. 363–392. Cambridge: Cambridge University Press.

Mack, M.C. & D'Antonio, C.M. (1998). Impacts of biological invasions on disturbance regimes. *Trends in Ecology and Evolution*, **13**, 195–198.

Mackie, G.L., Gibbons, W.N., Muncaster, B.W. & Gray, I.M. (1989). *The Zebra Mussel,* Dreissena polymorpha*: A Synthesis of European Experiences and a Preview for North America*. Toronto, Canada: Environment Toronto.

Majer, J.D. & Nichols, O.G. (1998). Long-term recolonization patterns of ants in Western Australian rehabilitated bauxite mines with reference to their use as indicators of restoration success. *Journal of Applied Ecology*, **35**, 161–182.

Maron, J.L. & Connors, P.G. (1996). A native nitrogen-fixing shrub facilitates weed invasion. *Oecologia*, **105**, 302–312.

Masters, L.A. (1997). Monitoring vegetation. In *The Tallgrass Restoration Handbook*, eds. S. Packard & C.F. Mutel, pp. 279–301. Washington, DC: Island Press.

Mayer, F.L., Jr. & Ellersieck, M.R. (1986). *Manual of Acute Toxicity: Interpretation and Data Base for 210 Chemical and 66 Species of Freshwater Animals*, Resource Publication no. 160. Washington, DC: US Department of the Interior.

McGeoch, M.A. (1998). The selection, testing and application of terrestrial insects as bioindicators. *Biological Reviews of the Cambridge Philosophical Society*, **73**, 181–201.

Michener, W.K. & Houhoulis, P.F. (1997). Detection of vegetation changes associated with extensive flooding in a forested ecosystem. *Photogrammetric Engineering and Remote Sensing*, **63**, 1363–1374.

Mills, E.L., Leach, J.H., Carlton, J.T. & Secor, C.L. (1994). Exotic species and integrity of the Great Lakes. *BioScience*, **44**, 666–676.

Mitsch, W.J. & Wilson, R.F. (1996). Improving the success of wetland creation and restoration with know-how, time, and self-design. *Ecological Applications*, **6**, 77–83.

Moody, M.E. & Mack, R.N. (1988). Controlling the spread of plant invasions: the importance of nascent foci. *Journal of Applied Ecology*, **25**, 1009–1021.

Müller-Dombois, D. & Ellenberg, H. (1974). *Aims and Methods of Vegetation Ecology*. New York: John Wiley.

Murphy, D.D. & Wilcox, B.A. (1986) Butterfly diversity in natural habitat fragments: a test of the validity of vertebrate-based management. In *Wildlife 2000, Modeling Habitat Relationships of Terrestrial Vertebrates*, eds. J. Verner, M.L. Morrison & C.J. Ralph, pp. 287–292. Madison, WI: University of Wisconsin Press.

National Research Council (1992). *Restoration of Aquatic Ecosystems: Science, Technology, and Public Policy*. Washington, DC: National Academy Press.

National Research Council (1997). *Striking a Balance: Improving Stewardship of Marine Areas*. Washington, DC: National Academy Press.

Nepstad, D.C., Uhl, C. & Serrao, E.A.S. (1991). Recuperation of a degraded Amazonian landscape: forest recovery and agricultural restoration. *Ambio*, **20**, 248–255.

Norris, R.H., McElravy, E.P. & Resh, V.H. (1996) The sampling problem. In *River Biota*, eds. G. Petts & P. Calow, pp. 184–208. Oxford: Blackwell.

Noss, R.F. (1990). Indicators for monitoring biodiversity: a hierarchical approach. *Conservation Biology*, **4**, 355–374.

O'Connell, T.J., Jackson, L.E. & Brooks, R.P. (1998). A bird community index of biotic integrity for the mid-Atlantic highlands. *Environmental Monitoring and Assessment*, **51**, 145–156.

Pacific Estuarine Research Laboratory (1990). *A Manual for Assessing Restored and Natural Coastal Wetlands with Examples from Southern California*, Report no. T-CSVCP-021. La Jolla, CA: California Sea Grant.

Pavlik, B.M. (1994) Demographic monitoring and the recovery of endangered plants. In *Restoration of Endangered Species*, eds. M.L. Bowles & C.J. Whelan, pp. 322–349. Cambridge: Cambridge University Press.

Pavlik, B.M. (1996) Defining and measuring success. In *Restoring Diversity: Strategies for Reintroduction of Endangered Plants*, eds. D.A. Falk, C.I. Millan & M. Olwell, pp. 127–155. Washington, DC: Island Press.

Pavlik, B.M. & Espeland, E.K. (1998). Demography of natural and reintroduced populations of *Acanthomintha duttonii*, an endangered serpentinite annual in northern California. *Madroño*, **45**, 31–39.

Phinn, S.R., Stow, D.A. & Zedler, J.B. (1996). Monitoring wetland habitat restoration in southern California using airborne multispectral video data. *Restoration Ecology*, **4**, 412–422.

Pielou, E.C. (1986). Assessing the diversity and composition of restored vegetation. *Canadian Journal of Botany*, **64**, 1344–1348.

Ralph, C.J., Geupel, G.R., Pyle, P., Martin, T.E. & DeSante, D.F. (1993). *Handbook of Field Methods for Monitoring Landbirds*, General Technical Report PSW-GTR-144. Albany, CA: Pacific Southwest Research Station, Forest Service, US Department of Agriculture.

Reed, P.B., Jr (1988). *National List of Plant Species that Occur in Wetlands: 1988 National Summary*, Biological Report: no. 88 (24). Washington, DC: US Fish and Wildlife Service.

Reed, P.B., Jr (compiler) (1997). *Revision of the National List of Plant Species that Occur in Wetlands*. Washington, DC: US Fish and Wildlife Service.

Rein, F.A. (1999). Vegetative buffer strips in a Mediterranean climate: potential for protecting soil and water resources. PhD thesis, University of California, Santa Cruz.

Ross, D.J., Speir, T.W., Cowling, J.C. & Feltham, C.W. (1992). Soil restoration under pasture after lignite mining: management effects on soil biochemical properties and their relationships with herbage yields. *Plant and Soil*, **140**, 85–97.

Sauer, L.J. (1998). *The Once and Future Forest*. Washington, DC: Island Press.

Schmidt, J.C., Webb, R.H., Valdez, R.A., Marzolf, G.R. & Stevens, L.E. (1998). Science and values in river restoration in the Grand Canyon. *BioScience*, **48**, 735–747.

Schroeder, H.W. (1998). Why people volunteer: an analysis of writings. *Restoration and Management Notes*, **16**, 66–67.

Shabman, L.A. (1995). Making watershed restoration happen: what does economics offer? In *Rehabilitating Damaged Ecosystems*, 2nd edn, ed. J. Cairns, Jr, pp. 35–47. Boca Raton, FL: CRC Press.

Shireman, J.V. & Maceina, M.J. (1981). The utilization of grass carp, *Ctenophyaryngodo idella* Val. for hydrilla control in Lake Baldwin, Florida. *Journal of Fish Biology*, **19**, 629–636.

Simenstad, C.A. & Thom, R.M. (1996). Functional equivalency trajectories of the restored Gog-le-hi-te estuarine wetland. *Ecological Applications*, **6**, 38–56.

Small, S., DeStaebler, J., Geupel, G.R. & King, A. (1999). *Songbird Responses to Riparian Restoration on the Sacramento River System: Report of the 1997 and 1998 Field Season to The Nature Conservancy and US Fish and Wildlife Service*. Stinson Beach, CA: Point Reyes Bird Observatory.

Smith, K.A. & Mullins, C.E. (eds.) (1991) *Soil Analysis: Physical Methods*. New York: Marcel Dekker.

Sparks, D.L. (ed.) (1996). *Methods of Soil Analysis*, Part 3, *Chemical Methods*. Madison, WI: Soil Science Society of America.

Sparks, R.E., Nelson, J.C. & Yin, Y. (1998). Naturalization of the flood regime in regulated rivers. *BioScience*, **48**, 706–720.

Spellerberg, I.F. (1991). *Monitoring Ecological Change*. Cambridge: Cambridge University Press.

Sutter, R.D. (1996). Monitoring. In *Restoring Diversity: Strategies for Reintroduction of Endangered Plants*, eds. D.A. Falk, C.I. Millan & M. Olwell, pp. 235–264. Washington, DC: Island Press.

Tiner, R.W. (1999). *Wetland Indicators: A Guide to Wetland Identification, Delineation, Classification, and Mapping*. Boca Raton, FL: Lewis Publishers.

Um, J.S. & Wright, R. (1998). A comparative evaluation of video remote sensing and field survey for revegetation monitoring of a pipeline route. *Science of the Total Environment*, **215**, 189–207.

Weaver, R.W. (ed.) (1994). *Methods of Soil Analysis*, Part 2, *Microbiological and Biochemical Properties*. Madison, WI: Soil Science Society of America.

Westman, W.E. (1991). Ecological restoration projects: measuring their performance. *Environmental Professional*, **13**, 207–215.

White, P.S. & Walker, J.L. (1997). Approximating nature's variation: selecting and using reference information in restoration ecology. *Restoration Ecology*, **5**, 338–349.

Wolfe, R.W. (1990). Seed dispersal and wetland restoration. In *Accelerating Natural Processes for Wetland Restoration after Phosphate Mining*, pp. 51–95. Bartow, FL: Florida Institute of Phosphate Research.

Yount, J.B. & Niemi, G.J. (eds.) (1990). Recovery of lotic communities and ecosystems following disturbance: theory and application. *Environmental Management*, **14**, 515–762.

Zedler, J.B. (1988). Salt marsh restoration: lessons from California. In *Rehabilitating Damaged Ecosystems*, ed. J. Cairns Jr, pp. 123–138. Boca Raton, FL: CRC Press.

Zedler, J.B. & Callaway, J.C. (1999). Tracking wetland restoration: do mitigation sites follow desired trajectories? *Restoration Ecology*, **7**, 69–73.

Index

Page numbers in italics refer to information in a box, figure or table